T0188259

ENVIRONMENTAL DEGRADATION *of* Advanced *and* Traditional Engineering Materials

ENVIRONMENTAL DEGRADATION *of* Advanced *and* Traditional Engineering Materials

Edited by
Lloyd H. Hihara
Ralph P.I. Adler
Ronald M. Latanision

CRC Press
Taylor & Francis Group
Boca Raton London New York

CRC Press is an imprint of the
Taylor & Francis Group, an **informa** business

CRC Press
Taylor & Francis Group
6000 Broken Sound Parkway NW, Suite 300
Boca Raton, FL 33487-2742

© 2014 by Taylor & Francis Group, LLC
CRC Press is an imprint of Taylor & Francis Group, an Informa business

First issued in paperback 2019

No claim to original U.S. Government works

ISBN 13: 978-0-367-44585-0 (pbk)
ISBN 13: 978-1-4398-1926-5 (hbk)

This book contains information obtained from authentic and highly regarded sources. Reasonable efforts have been made to publish reliable data and information, but the author and publisher cannot assume responsibility for the validity of all materials or the consequences of their use. The authors and publishers have attempted to trace the copyright holders of all material reproduced in this publication and apologize to copyright holders if permission to publish in this form has not been obtained. If any copyright material has not been acknowledged please write and let us know so we may rectify in any future reprint.

Except as permitted under U.S. Copyright Law, no part of this book may be reprinted, reproduced, transmitted, or utilized in any form by any electronic, mechanical, or other means, now known or hereafter invented, including photocopying, microfilming, and recording, or in any information storage or retrieval system, without written permission from the publishers.

For permission to photocopy or use material electronically from this work, please access www.copyright.com (http://www.copyright.com/) or contact the Copyright Clearance Center, Inc. (CCC), 222 Rosewood Drive, Danvers, MA 01923, 978-750-8400. CCC is a not-for-profit organization that provides licenses and registration for a variety of users. For organizations that have been granted a photocopy license by the CCC, a separate system of payment has been arranged.

Trademark Notice: Product or corporate names may be trademarks or registered trademarks, and are used only for identification and explanation without intent to infringe.

Visit the Taylor & Francis Web site at
http://www.taylorandfrancis.com

and the CRC Press Web site at
http://www.crcpress.com

During the preparation of this manuscript, we greatly appreciated the understanding and sacrifices made by our families:

Lori, and daughters Eliesse and Cassidy Hihara;
Susan Adler; and Carolyn Latanision.

Contents

SECTION I   Metals

Preface

Corrosion is ubiquitous: all engineering systems are subject to environmental degradation in service environments, whether these systems are used for national defense or to save and improve the quality of life of individuals (medical devices of all kinds); to meet our energy needs on this planet; to provide clean air; to transport water, energy products, and other objects of our commercial world (pipelines, oil tankers, automobiles, aircraft, etc.); and many others including the vast spatial presence of infrastructure systems. From heart stents to nuclear electric generating stations, corrosion is part of our world. What remains a persistent, resource-consuming reality in the engineering enterprise is that engineering systems are built of materials that are subject to environmental degradation that ultimately must be repaired or replaced. Whether an airframe, integrated circuit, bridge, prosthetic device, or implantable drug-delivery system, the chemical stability of the materials of construction of such systems continues to be a key element in determining their useful life. To put the detrimental effects of corrosion into perspective, the overall annual cost of metallic corrosion on a global basis was estimated to be 3.8% of gross world output or $1.9 trillion (based on the year 2004). The losses for the United States were estimated to be approximately 30% of the global losses (Bhaskaran et al. 2005).

This book provides a comprehensive treatment of the environmental degradation of traditional and advanced engineering materials, covering metals, polymers, ceramics, composites, and natural materials. This coverage of environmental degradation goes beyond the classical definition of the corrosive degradation of metals that was defined as the "destructive conversion of a metal into oxides and metallic salts under the influence of certain chemical reactions between the metal and the corroding agent" (D Van Nostrand Co. 1944). Ironically, the broadening of environmental degradation to encompass all classes of materials beyond this previously narrow definition of corrosion that only applied to metals has been considerably expanded by legislative rather than conventional scientific actions, most notably by the U.S. Congress in 2002 when they "redefined" corrosion to include more than only metallic materials: "The term 'corrosion' means the deterioration of a material or its properties due to a reaction of that material with its chemical environment." The specific motivation was that Congress was made aware of the high cost of corrosion and its negative impact on the readiness, availability, and long-term durability on military material and infrastructure. This concern led to the passage of "The Bob Stumpf National Defense Authorization Act for Fiscal Year 2003" (107th U.S. Congress 2002).

Most reference volumes that are available today focus on one specific type or class of materials and do not provide detailed, practical information regarding the fundamental degradation processes at a technical level appropriate for nonspecialists. This book also offers a general reference source for anyone interested in the environmental degradation of the major types or classes of materials. The goal is to enable the reader to easily find essential information related to the degradation mechanisms of a particular material group. It will be relevant not only to those interested in metallic corrosion, but also to researchers and practitioners in the areas of ceramics, polymers, composites, and natural materials. The redefinition of corrosion to include all materials was an important stimulus, leading to the publication of *Environmental Degradation of Advanced and Traditional Engineering Materials*. Thus, the organization of this book differs from the classical coverage of corrosion that is limited to metals. It covers four generic classes of materials: (1) metals, (2) polymers, (3) ceramics and glassy materials, and (4) other natural materials. We have included information on fundamental degradation processes with sufficient basics for an engineer/scientist to understand the major degradation mechanisms for classes of materials outside of their specialty. In short, this book will fill the need for a comprehensive and broad-based reference source covering the full spectrum of engineering materials. For each particular class of material, general information will be provided to give the reader (1) practical background, basic properties, and fundamental principles on the environmental degradation process; (2) degradation characteristics for some specific alloys or compositions; (3) guidelines on how to protect against degradation; (4) testing procedures; and (5) a list of further reading on the topic.

It is expected that this book will serve as a trusted resource to generations of engineers. The need for such a book is greater now than ever before. It appears that much of the expertise in the area of materials selection and performance is being lost as current staff retires and are not being replaced with similarly experienced personnel, as many of the metal producers have responded to the global economy of the past decade and more.

In addition, the interest of young people in engineering education, including corrosion engineering, is also in decline (Committee on Assessing Corrosion Education 2009). When the global economy recovers from impending uncertainty and the meltdown of the recent past, nations with a strong manufacturing base that creates products of value to the market are likely to respond most quickly. But, this will require an educated and informed engineering workforce. Industrialized nations all over the world are on the brink of losing this technological infrastructure through retirement, the decline of traditional manufacturing industries, and declining student interest. Without a means of capturing this expertise in a useful form, the next generation of engineers is going to find a gap in their knowledge base. This book will be of value in that context. Every industrialized nation must have the capacity and intellectual strength necessary to design, manufacture, and maintain either contemporary engineering systems or emerging engineering systems that may find their way into the marketplace of the future.

References

107th U.S. Congress (2002). Public Law 107-314. *The Bob Stumpf National Defense Authorization Act for Fiscal Year 2003*, 2658.

Bhaskaran, R., N. Palaniswamy et al. (2005). Global cost of corrosion—A historical review. In *ASM Handbook*, Vol. 13B *Corrosion: Materials*, S. D. Cramer and B. S. Covino Jr. (eds.), Materials Park, OH, ASM International, pp. 619–628.

Committee on Assessing Corrosion Education (2009). Assessment of corrosion education. National Materials Advisory Board; Division of Engineering and Physical Sciences; National Research Council of the National Academies; National Academy of Sciences.

D. Van Nostrand Co. (1944). *Van Nostrand's Scientific Encyclopedia*, p. 317.

Acknowledgments

We are grateful for the invaluable assistance provided by Dr. Raghu Srinivasan, who helped in preparing many figures and made editorial changes by checking and modifying chapter formats during the compilation of this work.

Editors

Dr. Lloyd H. Hihara is a professor of mechanical engineering at the University of Hawaii at Manoa (UHM). He is the director of the Hawaii Corrosion Laboratory that he established to conduct corrosion research in the varied Hawaiian microclimates, which are some of the most spatially diverse on Earth. One of his main interests is the correlation between accelerated corrosion testing in the laboratory and corrosion behavior in natural environments. He also investigates the corrosion of metal–matrix composites; the compatibility between ceramics, composites, and metals; corrosion in biofuels; hybrid ceramic–polymer coatings; and corrosivity sensors. He conducts corrosion research on projects sponsored by the U.S. Air Force, Army, Marines, and Navy; the Department of Energy; as well as industry. He is a recipient of the Guy Bengough Award from the Institute of Materials, Minerals and Mining and a Presidential Young Investigator Award from the National Science Foundation. He is also a Boeing Company A.D. Welliver Faculty Fellow.

Dr. Hihara received his BS in mechanical engineering from UHM and an SM in mechanical engineering and a PhD in metallurgy from the Massachusetts Institute of Technology. Dr. Hihara and his wife, Lori, live in Honolulu, HI.

Dr. Ralph P.I. Adler is a retired research metallurgist from the Army Research Laboratory (ARL), where his primary focus was on the conduct and administration of multiorganizational basic and applied research programs in corrosion science and engineering. At ARL, his hands-on projects covered the evaluation and characterization of the corrosion of materials used in structural components for defense applications involving monolithic and eutectic metal alloys as well as metal–matrix composites.

Throughout his tenure at ARL, Dr. Adler represented the Army on high-level Department of Defense (DoD) panels providing technical knowledge of Army materials technology programs. Since the mid-1990s, he has served on and led several corrosion-related planning committees for the annual Army Corrosion Summit and the biennial DoD Corrosion Conference, with additional responsibilities as the editor of several of those DoD corrosion conference proceedings.

Professional and academic engagement has been a lifetime priority for Dr. Adler. Previously, he was the corrosion subject matter expert for several Army agencies and was a frequent advisor for industrial, professional society, and academic organizations, where some collaborations led to coauthored joint publications. He also served on the National Materials Advisory Board's "Assessing Corrosion

Education" panel for the National Academy of Sciences that resulted in the 2009 National Academies Press publication *Assessment of Corrosion Education*.

Dr. Adler earned a DEng in metallurgy from Yale University and received his BS and MS in metallurgical engineering from Stanford University. He is the author of numerous publications and holds patents in a variety of technical areas with commercial and military applications. Dr. Adler and his wife, Susan, live in Wellesley Hills, MA.

Dr. Ronald M. Latanision is a corporate vice president of Exponent, Inc., an engineering consulting company. Prior to joining Exponent, Dr. Latanision was the director of The H.H. Uhlig Corrosion Laboratory in the Department of Materials Science and Engineering at MIT and held joint faculty appointments in the Department of Materials Science and Engineering and in the Department of Nuclear Engineering. He is now an emeritus professor at MIT. In addition, he is a member of the National Academy of Engineering and a fellow of ASM International, NACE International, and the American Academy of Arts and Sciences. From 1983 to 1988, Dr. Latanision was the first holder of the Shell Distinguished Chair in Materials Science. He was a founder of Altran Materials Engineering Corporation, established in 1992, and led the Materials Processing Center at MIT as its director from 1985 to 1991.

Dr. Latanision's research interests are focused largely on the areas of materials processing and in the corrosion of metals and other materials in aqueous (ambient as well as high temperature and pressure) environments. He specializes in corrosion science and engineering, with particular emphasis on materials selection for contemporary and advanced engineering systems and in failure analysis. His expertise extends to electrochemical systems and processing technologies, ranging from fuel cells and batteries to supercritical water power generation and waste destruction. His research interests include stress corrosion cracking and hydrogen embrittlement of metals and alloys, water and ionic permeation through thin polymer films, photoelectrochemistry, and the study of aging phenomena/life prediction in engineering materials and systems. Dr. Latanision is a member of the International Corrosion Council and serves as coeditor in chief of *Corrosion Reviews*, with Professor Noam Eliaz of Tel-Aviv University. He also serves as the editor in chief of the National Academy of Engineering quarterly, *The Bridge*.

Dr. Latanision has served as a science advisor to the U.S. House of Representatives Committee on Science and Technology in Washington, District of Columbia. He has also served as a member of the Advisory Committee to the Massachusetts Office of Science and Technology, an executive branch office created to strengthen the Commonwealth's science and technology infrastructure with emphasis directed toward future economic growth. Dr. Latanision has served as a member of the National Materials Advisory Board of the National Research Council (NRC) and currently serves as a member of the NRC's Committee on Undergraduate Science Education. He hosts the annual Siemens Westinghouse Science and Technology Competition on the MIT campus. In June of 2002, Dr. Latanision was appointed by President George W. Bush to membership on the U.S. Nuclear Waste Technical Review Board, a position in which he continues to serve in the administration of President Barack Obama. Dr. Latanision and his wife, Carolyn, live in Winchester, MA.

Contributors

Hitoshi Asahi
Nippon Steel & Sumitomo Metal Corporation
Tokyo, Japan

Neal S. Berke
Tourney Consulting Group LLC
Kalamazoo, Michigan

Anil K. Bhowmick
Rubber Technology Centre
Indian Institute of Technology, Kharagpur
Kharagpur, India

Nick Birbilis
Department of Materials Engineering
Monash University
Melbourne, Victoria, Australia

Anusuya Choudhury
Rubber Technology Centre
Indian Institute of Technology, Kharagpur
Kharagpur, India

Lawrence Coulter
U.S. Air Force Research Laboratory
Hill Air Force Base, Utah

Guy D. Davis
Consultant in Materials Science
Baltimore, Maryland

Suresh Divi
Titanium Metals Corporation (TIMET)
TIMET-Henderson Technical Laboratory
Henderson, Nevada

Marc A. Edwards
Department of Civil and Environmental
    Engineering
Virginia Polytechnic Institute and State
    University
Blacksburg, Virginia

Victoria J. Gelling
Department of Coatings and Polymeric
    Materials
North Dakota State University
Fargo, North Dakota

Larry Gintert
Beechcraft Corporation
Wichita, Kansas

James Grauman
Titanium Metals Corporation (TIMET)
TIMET-Henderson Technical Laboratory
Henderson, Nevada

Carolyn M. Hansson
Department of Mechanical and Mechatronics
    Engineering
University of Waterloo
Waterloo, Ontario, Canada

Koji Hashimoto
Institute for Materials Research
Tohoku University
Sendai, Japan

Larry L. Hench
Department of Materials Science and
    Engineering
University of Florida
Gainesville, Florida

and

University of Central Florida
Orlando, Florida

and

Imperial College London
London, United Kingdom

Lloyd H. Hihara
Department of Mechanical Engineering
University of Hawaii at Manoa
Honolulu, Hawaii

Bruce R.W. Hinton
Department of Materials Engineering
Monash University
Melbourne, Victoria, Australia

Shahzma J. Jaffer
AECL–Chalk River Laboratories
Chalk River, Ontario, Canada

Haruhiko Kajimura
Research and Development Center
Nippon Steel & Sumikin Stainless Steel
    Corporation
Hikari, Yamaguchi, Japan

Didier Lesueur
Lhoist Recherche et Développement
Nivelles, Belgium

Laura Mammoliti
Lafarge Canada Inc.
Mississauga, Ontario, Canada

Tracy D. Marcotte
CVM Engineers
King of Prussia, Pennsylvania

James H. Michel
Copper Development Association Inc.
New York, New York

Kent R. Miller
Department of Polymer Engineering
The University of Akron
Akron, Ohio

Jeffrey J. Morrell
Department of Wood Science & Engineering
Oregon State University
Corvallis, Oregon

Drew Pavlacky
Department of Coatings and Polymeric Materials
North Dakota State University
Fargo, North Dakota

Dennis W. Readey
Department of Metallurgical and Materials
    Engineering
Colorado School of Mines
Golden, Colorado

and

Department of Materials Science and
    Engineering
University of Illinois at Urbana-Champaign
Urbana, Illinois

Raul B. Rebak
GE Global Research
Schenectady, New York

David Roylance
Department of Materials Science and
    Engineering
Massachusetts Institute of Technology
Cambridge, Massachusetts

Margaret Roylance
U.S. Army Natick Soldier Research
    Development and Engineering Center
Natick, Massachusetts

Barbara A. Shaw
Department of Engineering Science and
    Mechanics
Pennsylvania State University
University Park, Pennsylvania

Richard G. Sibbick
Grace Construction Products
W. R. Grace & Co
Cambridge, Massachusetts

Elizabeth Sikora
Department of Engineering Science and
    Mechanics
Pennsylvania State University
University Park, Pennsylvania

R.K. Singh Raman
Department of Mechanical and Aerospace
    Engineering
and
Department of Chemical Engineering
Monash University
Melbourne, Victoria, Australia

Mark D. Soucek
Department of Polymer Engineering
The University of Akron
Akron, Ohio

Raghu Srinivasan
Department of Mechanical Engineering
University of Hawaii at Manoa
Honolulu, Hawaii

Chad Ulven
Department of Mechanical Engineering and
    Applied Mechanics
North Dakota State University
Fargo, North Dakota

Chris Vetter
Department of Coatings and Polymeric
    Materials
North Dakota State University
Fargo, North Dakota

Xiaojiang Wang
Department of Polymer Engineering
The University of Akron
Akron, Ohio

Jack Youtcheff
United States Department of Transportation-
    Federal Highway Administration
Turner-Fairbank Highway Research Center
McLean, Virginia

Xiaoge Gregory Zhang
Zinnovate Inc.
Toronto, Canada

I

Metals

1

Forms of Metallic Corrosion: An Overview

Raghu Srinivasan
University of Hawaii at Manoa

Lloyd H. Hihara
University of Hawaii at Manoa

1.1 Introduction

The definition of metallic corrosion is the "destructive attack of a metal by chemical or electrochemical reaction with its environment" (Uhlig and Revie 1985). Whether or not a metal will corrode depends on the thermodynamic properties of the reaction where corrosion may occur whenever the Gibbs free energy of the reaction is negative. The rate of the corrosion reaction, however, is not directly proportional to the negative magnitude of the Gibbs free energy but is more dependent on the kinetics of that reaction that can be controlled by either the rate of electron transfer, or the protectiveness of the naturally formed oxide film, or the rate of oxygen diffusion in the electrolyte, or other similar rate-limiting mechanisms. Once corrosion happens, it can take on various forms, such as being either spatially uniform or localized resulting in pitting or cracking, or it could be coupled with mechanical or physical mechanisms leading to degradation such as erosion–corrosion. The taxonomic structure of this chapter will first cover the basic thermodynamic and kinetic aspects of metallic corrosion to be followed by a brief description of various different forms of metallic corrosion.

1.2 Thermodynamics

For the case of corrosion occurring in an aqueous medium, the resultant electrochemical process involves both anodic (or oxidation) and cathodic (or reduction) reactions. This also applies to atmospheric corrosion where the electrochemical reactions take place under thin electrolyte films. The dissolution of metal M is represented by the anodic reaction:

$$M \rightarrow M^{n+} + ne^-. \tag{1.1}$$

The electrons produced by the anodic reaction must then be consumed by a cathodic reaction for corrosion to proceed. Two predominant cathodic reactions in aqueous corrosion are either oxygen reduction (Equations 1.2 and 1.3) or hydrogen evolution (Equations 1.4 and 1.5), respectively. Hydrogen evolution is also referred to as proton reduction (Equation 1.4) and water reduction (Equation 1.5).

$$O_2 + 4e^- + 4H^+ \rightarrow 2H_2O \, (\text{acidic solution}) \tag{1.2}$$

$$O_2 + 4e^- + 2H_2O \rightarrow 4OH^- \, (\text{neutral or basic solution}) \tag{1.3}$$

$$2H^+ + 2e^- \rightarrow H_2 \, (\text{acidic solution}) \tag{1.4}$$

$$2H_2O + 2e^- \rightarrow H_2 + 2OH^- \, (\text{neutral or basic solution}) \tag{1.5}$$

Oxygen reduction can only occur in aerated solutions, whereas proton reduction can occur in both deaerated and aerated solutions. A metal need not be immersed in an aqueous solution for such corrosion to occur. Water may condense on surfaces at values of relative humidity much lower than 100% due to temperature fluctuations (e.g., condensation on a glass of cold water on a warm day) or due to the presence of hygroscopic impurities. The atmosphere can contain airborne impurities such as NaCl and other salts in concentrations that may vary by location (Uhlig and Revie 1985). Corrosion often initiates under these hygroscopic impurities that settle onto a material surface.

For corrosion to occur, the cell potential E_{cell} of the corrosion reaction must be positive, which corresponds to a decrease in Gibbs free energy. E_{cell} is the difference in the equilibrium potentials of the half-cell reduction ($E_{Reduction}$) and oxidation ($E_{Oxidation}$) reactions:

$$E_{cell} = E_{Reduction} - E_{Oxidation}. \tag{1.6}$$

The equilibrium potential of a half-cell reaction

$$v_o O + ne \rightleftharpoons v_R R \tag{1.7}$$

is given by the Nernst equation

$$E = E^\circ - \frac{RT}{nF} \ln \frac{(a_o)^{v_o}}{(a_R)^{v_R}}, \tag{1.8}$$

where
    O is the oxidized species
    R is the reduced species
    v_o is the coefficient of O
    v_R is the coefficient of R
    E° is the standard potential of the half-cell reaction
    R is the universal gas constant (8.314 J/mol°K)
    T is the temperature in °K
    n is number of electrons transferred in the half-cell electrochemical reaction
    F is the Faraday (96,487 C/mol of electrons)
    a_O is the activity of species O
    a_R is the activity of species R

For example, for the corrosion of a metal (Equation 1.1) with hydrogen evolution (Equation 1.4) as the cathodic reaction, E_{cell} would be calculated as follows:

$$E_{cell} - E_{H^+/H_2} - E_{M^+/M} \tag{1.9}$$

As the value of E_{cell} only indicates if the reaction is thermodynamically possible or impossible, knowledge of cell kinetics is required to obtain the rates of that corrosion reaction.

1.3 Kinetics

In the study of aqueous corrosion, polarization diagrams (Gellings 1976) are used to determine the rates of metal dissolution and oxygen reduction or proton reduction. The thermodynamic driving force for that electrochemical reaction is measured as potential on the vertical axis of the polarization diagram. The kinetics of the electrochemical reaction is measured as current on the horizontal axis of the polarization diagram. Notice that the anodic and cathodic reactions (Equations 1.1 through 1.5) involve the transfer of electrons, and therefore their rates are proportional to that current.

Depending on the metal and environment, metals may have active, passive, or active–passive electrochemical behavior. Active metals and alloys do not form protective passive films and show increasing anodic currents (Figure 1.1) as the potential is increased above the open-circuit or corrosion potential (E_{Corr}). Passive metals form protective passive films and have very low dissolution currents (Figure 1.2) at potentials more positive than E_{Corr}. Active–passive metals (e.g., Cr, Ni) (Fontana and Greene 1978)

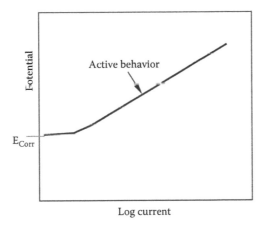

FIGURE 1.1   Anodic polarization diagrams showing active behavior.

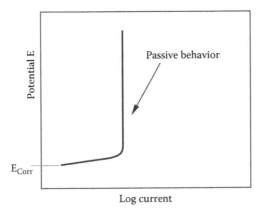

FIGURE 1.2   Anodic polarization diagrams showing passive behavior.

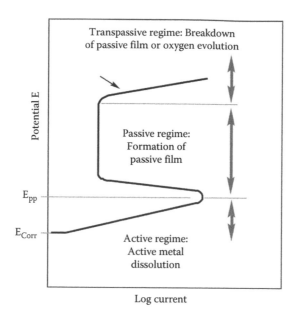

FIGURE 1.3   Anodic polarization diagrams showing active–passive behavior.

generally show active, passive, and transpassive regimes on the anodic polarization diagram (Figure 1.3). In the active regime from E_{Corr} to the primary passive potential (E_{pp}), the metal dissolution rate increases as the potential increases. In the passive regime, a protective oxide passive film forms on the metal at potentials greater than E_{pp} causing the current to dramatically decrease. In the transpassive regime at more noble potentials, the anodic current again increases either due to passive-film breakdown leading to metal dissolution or the evolution of oxygen (Uhlig and Revie 1985).

Aggressive anions, commonly from the halide group (e.g., Br^-, Cl^-), can prevent passivation, resulting in localized corrosion or pitting of the metal surface. Pitting commences when the electrode potential exceeds the critical value known as the pitting potential (E_{Pit}) (Uhlig and Revie 1985), where this also decreases with increasing aggressive-anion concentrations (Figure 1.4) (Galvele 1978).

In corrosion, the predominant cathodic reactions are hydrogen evolution and oxygen reduction. In deaerated environments, only the hydrogen evolution can occur; but in aerated environments, both hydrogen evolution and oxygen reduction may occur. Hence, on a polarization diagram, only hydrogen evolution will be seen in deaerated environments, while both hydrogen evolution and oxygen reduction may be seen in aerated environments (Figure 1.5). The current of the oxygen reduction curve will saturate when the rate at which oxygen molecules reach the cathodic sites is limited by diffusion. The diffusion-limited current can increase (Figure 1.5) with higher oxygen concentrations and thinning of the diffusion layer (e.g., by agitation of the electrolyte in bulk solutions or by evaporation for thin electrolyte films).

When corrosion occurs, the amount of electrons generated by metal dissolution is equal to that consumed by the cathodic reactions to conserve charge. Hence, the corroding metal assumes the corrosion potential (E_{Corr}) where the anodic and cathodic reactions occur at the same rate, that is, the corrosion current (I_{Corr}) (Figure 1.6).

The value of I_{Corr} can be converted into mass loss DM using Faraday's law:

$$\Delta M = \frac{I_{Corr}t}{nF} A_r,\qquad (1.10)$$

where

    t is the duration of time that the corrosion current is flowing
    A_r is the atomic mass of the metal

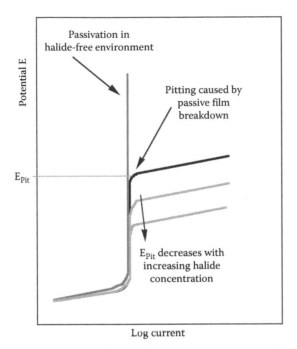

FIGURE 1.4 Anodic polarization diagram showing decreasing pitting potentials as the aggressive-ion (e.g., halide) concentration increases.

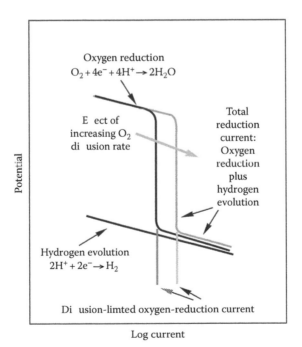

FIGURE 1.5 Cathodic polarization diagrams in deaerated environment showing hydrogen evolution and in aerated environment showing both hydrogen evolution and oxygen reduction.

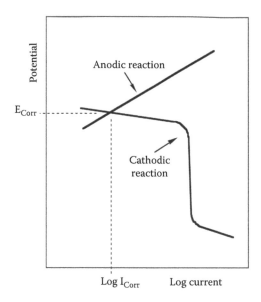

FIGURE 1.6 The intersection of the anodic and cathodic polarization diagrams shows the corrosion potential and the corrosion current.

The corrosion rate $\dot{M}$ can be calculated if the mass loss is divided by the time t and the surface area of the metal A_{Surf}:

$$\dot{M} = \frac{I_{Corr}}{nF} A_r \frac{1}{A_{Surf}} \tag{1.11}$$

Some commonly used corrosion rate units in terms of mass loss are grams per square meter per day (gmd) and milligrams per square decimeter per day (mdd).

If corrosion occurs uniformly over the metal surface, the corrosion rate $\dot{M}$ can be converted into a depth of penetration rate:

$$\dot{d} = \frac{I_{Corr}}{nF} A_r \frac{1}{\rho A_{Surf}}, \tag{1.12}$$

where ρ is the density of the metal. Commonly used units for the rate of penetration are millimeters per year (mm/y), inches per year (ipy), and mils (1/1000th of an inch) per year (mpy).

1.4 Forms of Metallic Corrosion

Corrosion can take on various metallic morphologies and can be affected by various environmental influences. As a result, corrosion generally has been classified based on its form and environmental driving forces. The major forms of corrosion have often been categorized as uniform corrosion, galvanic corrosion, crevice corrosion, pitting corrosion, intergranular corrosion, stress corrosion cracking (SCC), corrosion fatigue, hydrogen embrittlement, erosion corrosion, and dealloying. These corrosion forms in addition to microbiological influenced corrosion (MIC) will be briefly described. MIC was included because under certain circumstances, biological agents can also severely enhance corrosion. Except for uniform corrosion, most of the other forms of corrosion are insidious and can be relatively difficult to detect (Fontana and Greene 1978). It is also possible for two or more forms of corrosion to occur simultaneously on the same structural component. The ability to recognize where the different forms of corrosion can occur and their driving forces can help prevent or mitigate corrosion through improved/proactive maintenance and material selection/design practices.

1.4.1 Uniform Corrosion

As the name suggests, uniform corrosion or general corrosion occurs over the entire exposed surface and proceeds at a relatively uniform rate. Uniform corrosion is the most common form of corrosion and the resulting mass loss of the surface is easy to visually identify. Uniform corrosion can be prevented or mitigated by proper material selections, or the application of protective coatings or cathodic protection, or the use of inhibitors for closed systems (Fontana and Greene 1978). Atmospheric corrosion of plain carbon steel (Figures 1.7 and 1.8) is a good example of uniform corrosion.

FIGURE 1.7   Uniform corrosion of plain carbon steel corrosion test coupons.

FIGURE 1.8   Uniform corrosion of plain carbon steel I-beam and bolts. (Photo courtesy of Ryan Sugamoto.)

1.4.2 Galvanic Corrosion

Galvanic corrosion is defined as accelerated corrosion of a metal due to electrical contact with a more noble conductive material in a corrosive environment. Examples could be steel–copper or aluminum–graphite couples. The three conditions that are necessary for galvanic corrosion to take place are as follows: (1) at least two different conductive materials must be involved; (2) these materials must be in electrical contact with each other so that electric current can flow between them; and (3) the coupled materials must be in contact with an electrolyte so that an ionic current can flow between them. If any one of these three conditions is removed, galvanic corrosion will not occur (Oldfield 1988).

In the galvanic couple, the metal that originally has the lower corrosion potential is more active and acts as the anode where oxidation (or corrosion) takes place at an accelerated rate (Equation 1.1). The metal that originally has the higher corrosion potential is more noble and acts as the cathode where oxygen reduction (Equations 1.2 and 1.3) and/or hydrogen evolution takes place (Equations 1.4 and 1.5). The relative position of the corrosion potential of many different alloys exposed to aerated seawater is given in the galvanic series (Table 1.1) (Fontana and Greene 1978). This table can be used as a guide in selecting alloys for a specific application if the actual corrosion potentials are not readily available. The farther apart that two alloys are positioned in the galvanic series, the higher the driving force will be

TABLE 1.1 Galvanic Series of Some Metals and Alloys in Seawater Selected from Fontana and Greene

Platinum
Gold
Graphite
Titanium   ↑ Noble
Silver
HASTELLOY C (62 Ni, 17 Cr, 15 Mo)
18-8 Austenitic stainless steels (passive condition)
Stainless steel (Fe, 11%–30% Cr) (passive condition)
Inconel (80 Ni, 13 Cr, 7 Fe) (passive condition)
Nickel (passive condition)
Monel (70 Ni, 30 Cu)
Cupronickel alloys (60–90 Cu, 40–10 Ni)
Bronzes (Cu–Sn)
Copper
Brasses (Cu–Zn)
Inconel (80 Ni, 13 Cr, 7 Fe) (active condition)
Nickel (active condition)
Tin
Lead
Lead–tin solders
18-8 Austenitic stainless steels (active condition)
Stainless steel (Fe, 13% Cr) (active condition)
Cast iron   Active
Mild steel and iron
Cadmium
Aluminum alloys
Zinc   ↓
Magnesium and magnesium alloys

Source: Fontana, M.G. and Greene, N.D., *Corrosion Engineering*, McGraw-Hill, Inc., New York, 1978.

for galvanic corrosion to occur. The galvanic corrosion rates, however, cannot be determined from their relative positions in this galvanic series. If the two chosen metals are very close to each other in the galvanic series, there is a possibility that they could alternate from being an anode to being a cathode, and vice versa, which may result in unpredictable galvanic corrosion behavior. Some active–passive metals occupy two positions in the galvanic series—a more negative position where active corrosion is occurring and a more noble position where the metal is passivated. The galvanic series is also specific to aerated seawater, and therefore the relative positions of the corrosion potentials may differ in other environments, for example, in freshwater.

Nowadays, new alloys and electrically conductive materials (composites, polymers, and ceramics) are often used together in material systems to reduce weight and increase performance; hence, it is always a good practice to experimentally determine the corrosion potentials of the materials in the environment of intended use if they are not available in the galvanic series.

The galvanic series also gives no information on the corrosion rates of any anodic material. The galvanic corrosion rate of the anode material can be determined by the magnitude of the galvanic current flowing between the cathode and the anode (Figure 1.9). According to Faraday's law (Equation 1.10), the amount of current flowing between the anode and the cathode is directly proportional to the amount of anodic material that is dissolved (Revie 2011).

The galvanic corrosion rate of the anode material can be exacerbated as the relative area of the cathode is increased, generating a larger galvanic current.

The galvanic effect of two dissimilar materials can be predicted using polarization curves. Cathodic and anodic polarization curves for materials A (cathode) and B (anode) in the same electrolyte/solution are plotted together in Figure 1.10. The intersection of the extrapolated anodic and cathodic curves identifies the normal corrosion current (I_{corr}) of the cathode ($I_{Corr,A}$) and anode ($I_{Corr,B}$) materials in the uncoupled state. When materials A and B are electrically connected, the potential difference between the materials causes a current to flow between them, which could be measured using an ammeter. This measured current is equal to the galvanic current (I_{Galv}) that is identified at the intersection of the cathodic curve of material A (the cathode) and the anodic curve of material B (the anode). The galvanic current between the electrodes is higher than the normal corrosion current $I_{Corr,B}$ of material B, which is the anode in the galvanic couple. Hence, the corrosion current of material B will be higher when it is coupled to material A compared to when it is in the uncoupled state. Regarding material A, its dissolution current will decrease from $I_{Corr,A}$ in the uncoupled state to a lower value $I_{Diss,ACoup}$ that is dependent on the potential of material A when in the coupled state. Another important factor in the galvanic couple is the ohmic or IR loss. Current flowing between the two electrically connected materials experiences a voltage loss that is given by the Ohm's law: $E = IR$. The electrolyte resistivity, the resistivity of materials A and B, and the distance between the materials A and B all contribute to the resistance in the IR term. The galvanic corrosion current will decrease from the maximum $I_{Galv,Max}$

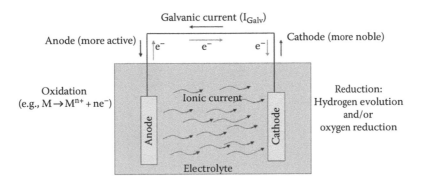

FIGURE 1.9 Electrolytic cell showing flow of ionic and electric current.

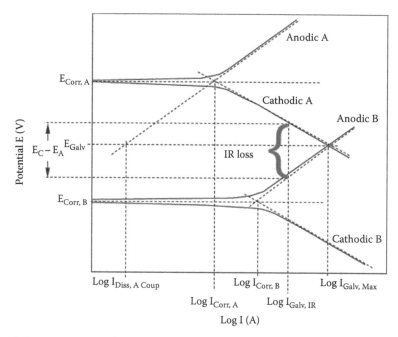

FIGURE 1.10 Polarization curves showing the galvanic corrosion mechanism. Metal A is the cathode, and metal B is the anode in the galvanic couple.

value to a lower value $I_{Galv,IR}$ if the IR loss is greater than zero (Figure 1.10). It should also be noted that the galvanic current does not account for local corrosion on the anode that is caused by cathodic reactions simultaneously occurring on the anode. Depending on the materials in the galvanic couple and the environment, the local corrosion on the anode may or may not be negligible. The total corrosion of the anode in the galvanic couple is equal to the galvanic component plus the local component:

$$I_{Total} = I_{Galv} + I_{Local} \tag{1.13}$$

A severe case of galvanic corrosion occurred on a galvanized steel nipple (the anode) coupled to a copper tube (the cathode) on a water heater (Figure 1.11). Galvanic corrosion initially occurred on the inside of

FIGURE 1.11 Galvanized steel nipple coupled to a copper tube on a water heater.

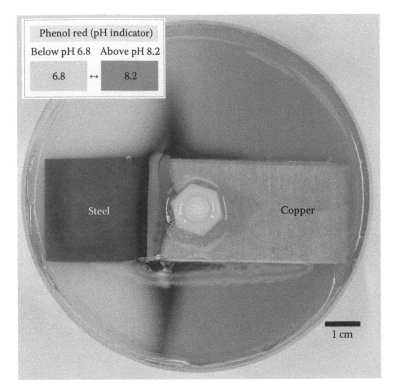

FIGURE 1.12 Plain carbon steel–copper galvanic couple placed in an artificial-seawater gel containing phenol red pH indicator.

the galvanized nipple (where both materials were exposed to the electrolyte) causing corrosion around the pipe threads, leading up to a leak and eventual corrosion on the external pipe surface (Figure 1.11).

During galvanic corrosion, the electrolyte medium around the cathode will become alkaline due to the generation of hydroxyl (OH^-) ions (Equations 1.3 and 1.5) or the consumption of protons (Equations 1.2 and 1.4) that will lead to the generation of OH^- due to the ionization of water; the medium around the anode will become acidic when the dissolved metal cations undergo hydrolysis and produce excess protons. This pH distribution is demonstrated when a plain carbon steel–copper galvanic couple was placed in an artificial-seawater gel (made with agar) containing a phenol red pH indicator that is yellow below pH 6.8 and red above pH 8.2. The electrolytic medium around the copper (cathode) was pink due to the generation of hydroxyl (OH^-) ions, and the electrolyte medium around the steel (anode) was yellow (Figure 1.12).

Galvanic corrosion can be prevented by eliminating the direct electrical contact between the dissimilar metals by using insulating gaskets, washers, etc. If electrical contact cannot be avoided, galvanic corrosion can be minimized by (1) selecting similar metals or those that are close in position in the galvanic series, (2) making additional thickness allowances for the more active metal in the couple, (3) keeping the cathode-to-anode area ratio low to reduce the galvanic current, or (4) coating the cathode to reduce the cathode-to-anode area ratio and suppress the galvanic current.

1.4.3 Crevice Corrosion

Localized corrosion that takes place within a crevice such as at gasket surfaces and at overlapping zones from riveting, bolting, or welding for various applications is termed as crevice corrosion (Fontana and Greene 1978). An opening as small as a few thousands of an inch or less is sufficient for some liquids to enter and initiate crevice corrosion (Szklarska-Smialowska 1986). At the onset of crevice corrosion,

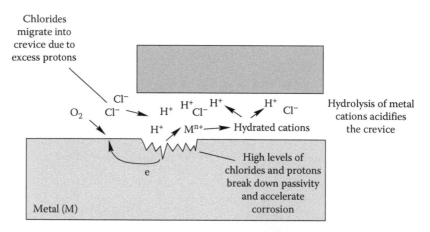

FIGURE 1.13 Anodic dissolution within the crevice produces metal cations. The hydrolysis of the metal cations produces solvated protons that acidify the solution within the crevice. The buildup of excess negative charge in the crevice attracts anions such as chlorides that can make the crevice environment very corrosive.

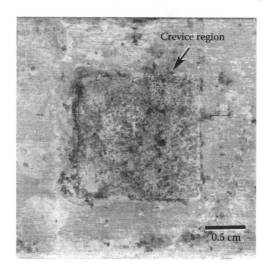

FIGURE 1.14 Crevice corrosion at the interface of a 6061-T6 Al alloy that was coupled to an insulative silicon nitride ceramic in humid marine environment.

normal anodic dissolution of the metal and oxygen reduction can take place both inside and outside of the crevice (Fontana and Greene 1978). Over a period of time, as oxygen is depleted in the crevice, only anodic dissolution of the metal takes place within the crevice, forming metal cations. The hydrolysis of the metal cations generates solvated protons and makes the crevice more acidic. In addition, to counteract the buildup of excess positive charge within the crevice, anions such as chlorides (if present) will migrate into the crevice (Figure 1.13). The combination of high levels of chloride ions and the acidic solution can break down passivity and accelerate corrosion in the crevice, as shown for a 6061-T6 Al coupon that was coupled to a silicon nitride ceramic piece (Figure 1.14).

1.4.4 Pitting Corrosion

Pitting is another form of localized corrosion where small pits and holes form on the surface and can penetrate deep into the bulk. Pitting can be a dangerous form of corrosion, as much of the corrosion damage is hidden beneath the surface within pits of varying depths.

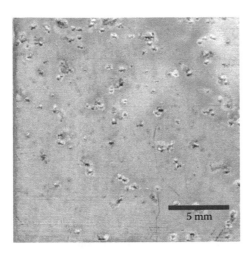

FIGURE 1.15 Pitting of a 6061-T6 Al alloy exposed in an aerated 3.15 wt.% NaCl solution at 30°C.

Passive metals such as aluminum (Figure 1.15), stainless steel, and nickel alloys are susceptible to pitting corrosion. The protective oxide films that form on the surface of passive metals can breakdown locally in the presence of aggressive ions (e.g., halogens, perchlorates, nitrates) so the exposed metal surface corrodes locally forming a small cavity. Once the pit has been initiated, the same mechanism as described in crevice corrosion takes place. Thus, localized anodic dissolution takes place inside of the pit, and the cathodic reaction (e.g., oxygen reduction) primarily takes place on the adjacent outer surface. The anodic dissolution of the metal and hydrolysis of the metal cations in the pit cause the electrolyte within the pit to become more acidic by the generation of solvated protons. To preserve charge neutrality within the pit, anions such as chlorides migrate into the pit causing the solution within the pit to become more corrosive.

The resistance of a passive metal to pitting is characterized by E_{Pit} above which the dissolution rate dramatically increases (Figure 1.4). The value of E_{Pit} generally decreases as the concentration of aggressive anions in the solution increases, as shown for 6061-T6 Al in deaerated chloride solutions (Figure 1.16). Whether or not the metal will pit in the open-circuit condition depends on whether the operative

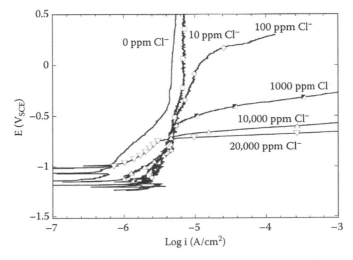

FIGURE 1.16 Anodic polarization diagrams of 6061-T6 Al in deaerated sodium chloride solutions of various Cl⁻ ion concentrations. Notice that the pitting potential decreases with increasing Cl⁻ ion concentration. Scan rate = 1 mV/s, 30°C.

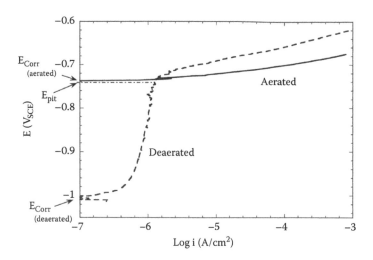

FIGURE 1.17 Anodic polarization of 6061-T6 Al in aerated and deaerated 3.15 wt.% NaCl solutions. In the deaerated solution, E_{Corr} is lower than E_{Pit} and, therefore, pitting does not occur spontaneously in the open-circuit condition, whereas, in the aerated solution, E_{Corr} and E_{Pit} are coincident and pitting occurs spontaneously in the open-circuit condition. Scan rate = 1 mV/s, 30°C.

cathodic reaction can polarize the metal to its pitting potential. For example, 6061-T6 Al will not pit in the open-circuit condition in deaerated chloride solutions because proton reduction alone cannot polarize 6061-T6 Al to its pitting potential (Figure 1.17). In aerated solutions, however, the oxygen reduction reaction polarizes 6061-T6 Al to its pitting potential in the open-circuit condition (Figure 1.17).

1.4.5 Intergranular Corrosion

Preferential corrosion at grain boundaries in a metal is called intergranular corrosion. This type of corrosion is metallurgically influenced. The grain boundaries of certain alloys can be less resistant to corrosion in comparison to the bulk due to the presence of impurities, the enrichment of alloying elements, or even the depletion of alloying elements in the grain-boundary region. Some heat treatments that alter the microstructure can have a significant effect on intergranular corrosion. For example, iron impurities in aluminum alloys can cause intergranular corrosion by segregating in the grain boundaries (Fontana and Greene 1978). Some stainless steel alloys become susceptible to intergranular corrosion when heat treated in certain temperature ranges that cause chromium carbide formation at the grain boundaries. The formation of chromium carbides depletes the local matrix in the grain boundaries of sufficient chromium to impart corrosion resistance (Fontana and Greene 1978).

1.4.6 Stress Corrosion Cracking

SCC is a form of corrosion that occurs when susceptible alloys are exposed to certain environments where they are also subjected to a tensile stress that could be either applied or residual. This type of corrosion leads to the formation of transgranular or intergranular cracks that generally grow perpendicular to the tensile stress axis (Anderson 2005). By its nature, SCC can lead to catastrophic failures if the stress corrosion cracks reach the critical length for fast fracture to occur. In SCC, the crack velocity ranges from $\approx 10^{-9}$ to 10^{-6} m/s (Jones 1992). Pressure vessels, pipe work, and stressed components are susceptible to SCC. The mechanism of the SCC process is usually divided into three stages (Figure 1.18): The first stage is crack initiation, the second stage is subcritical crack growth, and the last or third stage is crack propagation to final fast fracture (Uhlig and Revie 1985).

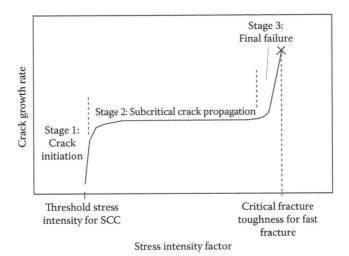

FIGURE 1.18 Three stages of SCC.

Cracks often initiate from pits, but a pit is not necessarily a requirement for crack initiation. Environmental parameters that influence the rate of crack growth in aqueous solutions are temperature, pressure, solute species, solute concentration and activity, pH, electrochemical potential, solution viscosity, and convection (Jones 1992). SCC is also influenced by the magnitude of stress, stress state (plane stress or plane strain), metallurgical condition, alloy composition, and crack geometry (Jones 1992).

1.4.7 Corrosion Fatigue

Cracking due to the combination of cyclic stress and corrosion is termed corrosion fatigue (Phull 2003). The fatigue crack growth rate will increase in a corrosive environment when compared to the normal fatigue crack growth rate in an inert environment. In addition, the number of cycles to failure (N) in corrosion fatigue is dependent on the stress frequency, whereas, for regular fatigue, the number of cycles to failure is generally independent of the stress frequency.

Corrosion fatigue can be cycle dependent, time dependent, or a combination of the two (McEvily and Wei 1973). The crack growth rate in cycle-dependent corrosion fatigue is proportional to an acceleration factor (ϕ) given by Equation 1.14 (Anderson 2005):

$$\frac{da}{dN}_{\text{Corrosion Fatigue}} = \phi \cdot \frac{da}{dN}_{\text{inert}} \tag{1.14}$$

The crack growth rate in time-dependent corrosion fatigue depends on the loading frequency (f) and it is given by Equation 1.15 (Anderson 2005):

$$\frac{da}{dN}_{\text{Corrosion Fatigue}} = \frac{da}{dN}_{\text{inert}} + f \cdot \frac{da}{dt} \tag{1.15}$$

where da/dt is the average environment crack growth rate over a loading cycle.

Corrosion fatigue of most metals, however, exhibits both cycle-dependent and time-dependent behavior. Equation 1.16 gives the general expression by combining both the equations (Anderson 2005).

$$\frac{da}{dN}_{\text{Corrosion Fatigue}} = \phi \cdot \frac{da}{dN}_{\text{inert}} + f \cdot \frac{da}{dt} \tag{1.16}$$

1.4.8 Hydrogen Embrittlement

Hydrogen damage or hydrogen embrittlement is not a form of corrosion damage, but it is included in this section because it can be caused indirectly by corrosion. When atomic hydrogen diffuses into the metal structure and collects at microscopic defects, the metal may become brittle leading to mechanical fracture. This process is termed as hydrogen embrittlement.

Atomic hydrogen is readily available in high-temperature moist atmospheres. Atomic hydrogen is also available as a result of the hydrogen evolution reaction induced during corrosion, cathodic protection, or electroplating. During the reduction of protons, molecular and atomic hydrogen may form. If the formation of the molecular hydrogen is slow when compared to the formation of atomic hydrogen, any excess atomic hydrogen on the surface can quickly diffuse into the metal and collect at microscopic defects, causing the metal to lose its ductility (Piron 1991). A metal that has been embrittled by atomic hydrogen can be restored to some extent to original properties by baking the atomic hydrogen out of the microstructure at elevated temperatures.

1.4.9 Erosion Corrosion

Erosion corrosion or the accelerated corrosion attack in a metal can occur if a fluid of high velocity impinges on the metal surface and continuously damages its protective oxide layer (Fontana and Greene 1978). The flow of air bubbles, abrasive particles suspended in the liquid, or turbulent flow of liquids can mechanically damage the protective oxide layer. Cavitation corrosion is a form of erosion–corrosion. When fluids flow over a discontinuity in the geometric surface of a substrate (e.g., the internal wall surface of a pipe), fluid separation can occur leading to a rapid decrease in local pressure causing vapor bubbles to nucleate. As the bubbles travel downstream and as the pressure increases, the bubbles implode generating high local temperatures and surface stresses, which can damage the protective oxide film on the metal surface. This leads to high localized corrosion rates. Another form of erosion–corrosion is fretting corrosion that occurs at the interface of two adjacent components subjected to vibration. The vibration can cause an oscillating sliding motion at the interface resulting in the mechanical breakdown of the protective oxide layer and accelerated corrosion at that interface. Fretting corrosion is common at surfaces of clamps, press fits, keyways, etc.

1.4.10 Dealloying

Dealloying is the selective dissolution of one or more electrochemically active species in an alloy. In copper–zinc alloys, the phenomenon is called dezincification where zinc is preferentially leached, leaving behind a weakened porous copper structure that often retains its original shape (Uhlig and Revie 1985). The dezincified structure may appear to be only tarnished, but corrosion damage can penetrate deep within the microstructure (Figure 1.19). A similar process occurs in copper–aluminum alloys with aluminum being selectively corroded (Uhlig and Revie 1985). In gray cast iron (a commonly used material for water distribution piping in the past that remains in place today after decades of service), a process known as graphitic corrosion occurs, which is associated with the selective removal of the more active iron leaving behind a network of graphite flakes (Uhlig and Revie 1985). In noble alloys such as gold–silver or gold–copper, where silver or copper is selectively leached, the phenomenon is referred to as parting (Uhlig and Revie 1985).

1.4.11 Microbiologically Influenced Corrosion

Microbiologically influenced corrosion (MIC) occurs when microorganisms such as bacteria, fungi, and microalgae accelerate a corrosion reaction or alter the mechanism for corrosion (Little and Lee 2007). For example, anaerobic sulfur-reducing bacteria (SRB) can accelerate the corrosion rate of buried steel structures in the absence of oxygen. SRB are known to produce hydrogen sulfide and accelerate

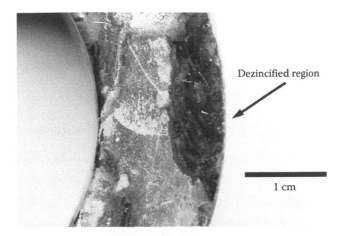

FIGURE 1.19 Brass component with dealloying corrosion. The darker regions are dezincified.

the corrosion rate (Little and Lee 2007). Aerobic bacteria living in the presence of oxygen such as *Acidithiobacillus thiooxidans* oxidizes sulfur and sulfides into sulfuric acid that accelerates the corrosion process (Little and Lee 2007). Organic acids such as oxalic, acetic, and lactic acids are produced by fungi, which can increase the corrosion rate (Little and Lee 2007). Algae can also produce organic acids, which can also be utilized by other microorganisms.

1.5 Summary

The study of thermodynamics and kinetics are important in determining if and how fast metallic corrosion can occur. Corrosion also develops into various forms (e.g., uniform, pitting, cracking) that depend on the metal species, the environment, and the presence or absence of physical or mechanical influences (e.g., tensile stresses, vibrational loads). Proper material selection and design of components to eliminate potential sources of physical and chemical corrosion degradation mechanisms or the additional use of protective treatments and coatings or cathodic protection systems as well as proactive maintenance cycles can significantly increase the useful service life of components subject to corrosive environments. For greater detailed explanations about metallic corrosion, the reader can consult the excellent texts given in the Suggested Readings.

References

Anderson, T. L. 2005. *Fracture Mechanics*, Taylor & Francis Group, Boca Raton, FL.

Fontana, M. G. and N. D. Greene. 1978. *Corrosion Engineering*, McGraw-Hill, Inc., New York.

Galvele, J. R. 1978. Present state of understanding of the breakdown of passivity and repassivation, in R. P. Frankenthal and J. Kruger, Eds., *Passivity of Metals*, The Electrochemical Society, Inc., Princeton, NJ, pp. 285–327.

Gellings, P. J. 1976. *Introduction to Corrosion Prevention and Control for Engineers*, Delft University Press, Delft, the Netherlands.

Jones, R. H. 1992. *Stress-Corrosion Cracking: Materials Performance and Evaluation*, ASM International, Novelty, OH.

Little, B. J. and J. S. Lee. 2007. *Microbiologically Influenced Corrosion*, John Wiley & Sons, New York.

McEvily, A. J. and R. P. Wei. 1973. Fracture mechanics and corrosion fatigue, in O. Devereux, A. J. McEvily, and R. W. Staehle, Eds., *Corrosion Fatigue: Chemistry, Mechanics, and Microstructures*, NACE International, Houston, TX.

Oldfield, J. W. 1988. Electrochemical theory of galvanic corrosion, in H. P. Hack, Ed., *Galvanic Corrosion, STP 978*, American Society for Testing and Materials, West Conshohocken, PA.

Phull, B. 2003. Evalulating corrosion fatigue, in S.D. Cramer and B.S. Covino Jr., Eds., *ASM Handbook Corrosion: Fundamentals, testing, and Protection, Volume 13A*, ASM International, Materials Park, OH, pp. 625–638.

Piron, D. L. 1991. *The Electrochemistry of Corrosion*, NACE International, Houston, TX.

Revie, R. W. 2011. *Uhlig's Corrosion Handbook*, Wiley, New York.

Szklarska-Smialowska, Z. 1986. *Pitting Corrosion of Metals*, National Association of Corrosion Engineers, Houston, TX.

Uhlig, H. H. and R. W. Revie. 1985. *Corrosion and Corrosion Control*, John Wiley & Sons, Inc., New York.

Suggested Readings

Gellings, P. J. 1976. *Introduction to Corrosion Prevention and Control for Engineers*, Delft University Press, Delft, the Netherlands.

Fontana, M. G. 1986. *Corrosion Engineering*, 3rd edn., McGraw-Hill, Inc., New York.

Jones, D. A. 1996. *Principles and Prevention of Corrosion*, 2nd edn., Prentice Hall, Upper Saddle River, NJ.

Revie, R. W. and H. H. Uhlig. 2008. *Corrosion and Corrosion Control*, 4th edn., John Wiley & Sons, Inc., New York.

2

Crystalline Alloys: Magnesium

Barbara A. Shaw
Pennsylvania State
University

Elizabeth Sikora
Pennsylvania State
University

2.1 Introduction to Magnesium and Magnesium Alloy Corrosion

Magnesium is the third most commonly used structural metal (steel is first and aluminum is second). While the amount of steel and even aluminum used is far greater than that of Mg, Mg alloys are very attractive because of their low density and high strength-to-weight ratios. Mg is 75% lighter than steel and 33% lighter than Al. With today's emphasis on energy costs, fuel economy, and our environment, Mg is becoming even more attractive because it is readily recycled and its increased use could lead to better fuel efficiency for cars/aircraft and lighter, more energy-efficient equipment in factories. Other desirable properties of Mg alloys include good castability, high stiffness-to-weight ratios, high thermal conductivity, high electrical conductivity, good weldability under controlled atmospheres, and high damping capacity. On the other hand, magnesium has a relatively low melting temperature and possesses a strong thermodynamic driving force for corrosion (you may recall that it is the most active structural metal in the EMF series and galvanic series) where a typical surface film does not present a very protective kinetic barrier to corrosion in many environments. However, despite these rather significant shortcomings, the interest in magnesium and magnesium alloys is quite high because of their low density and the abundant availability of Mg (2.7% of the earth's crust is comprised of Mg-rich minerals, and Mg ions are the third most abundant ion in seawater).

Magnesium has been used commercially since the mid-1800s, and until the last quarter of the twentieth century, the market for Mg has fluctuated significantly (rising during wars and falling off afterward). The market for Mg still fluctuates but more steady growth has been noted in the past 30 years since alloys with improved corrosion resistance and higher-temperature characteristics have become available and since the automotive industry has been searching for lighter-weight materials that translate into better fuel efficiency. When used in suitable applications and when adequately insulated from more noble metals, the corrosion rate of newer magnesium alloys can be similar to that of mild steel (Arlhac and Chaize 1997). Current approaches for improving the corrosion resistance of Mg alloys include careful control of impurities, enhancement of protective properties of surface films by choice of alloying additions, barrier protection provided by sophisticated coatings/surface layers, and using novel processing methods for the production of alloys.

2.2 Background on Magnesium and Magnesium Alloys

Commercial production of Mg started in the mid-nineteenth century in Paris where at the time its primary use was in photography (as a brilliant white light source for flash photography). Production then spreads to England and Germany. By 1900 worldwide production of Mg was 10 ton/year. The United States and several other countries began producing Mg, and production rose to 32,000 ton in 1939 and increased dramatically to 228,000 in 1944, but slipped back to 10,000 ton after WWII (Mordike and Ebert 2001). While the amount of Mg produced over the years has had its ups and downs, worldwide production of primary (not recycled) Mg climbed to 360,000 ton in 1998 (at a price of U.S. $3.6/kg), and as Table 2.1 illustrates, it reached its highest point ever with a production of 792,000 ton in 2007. As a result of the recession in 2008 with the automotive sector being hit extremely hard, production

TABLE 2.1 Production of Primary Magnesium from 2002 to 2008 (Thousands of Metric Tons)

	2002	2003	2004	2005	2006	2007	2008
United States	35	43	43	43	43	43	52
Brazil	7	6	11	6	6	18	18
Canada	86	50	55	54	50	16	0
China	232	354	450	470	490	627	559
France	0	0	0	0	0	0	0
Israel	34	30	33	28	28	25	30
Kazakhstan	10	14	14	20	20	21	20
Norway	10	0	0	0	0	0	0
Russia	52	45	45	45	50	37	35
Ukraine	0	0	0	2	2	3	3
Serbia	2	2	4	2	1	2	2
Total	468	544	655	670	693	792	719
Annual change	+4%	+16%	+20%	+2%	+3%	+14%	−9%

Source: Patzer, G., The magnesium industry today ... the global perspective, *Proceedings of the Magnesium Technology 2010*, TMS, Warrendale, PA, pp. 85–90, 2010.

TABLE 2.2   Primary Magnesium
Production

Years 2009 (Thousands of Tons)	
United States	45
Brazil	15[a]
Canada	0[a]
China	470[b]
France	0[a]
Israel	30[a]
Kazakhstan	20[a]
Norway	0[a]
Russia	30[a]
Ukraine	3[a]
Serbia	2[a]
Total	615
Annual change	−15%

Source: International Magnesium Association.
2010, http://www.intlmag.org/files/yend2009.pdf,
Accessed August 30, 2010.
   Production source documents: [a]U.S. Geological
Survey, 2010; [b]China Magnesium Association.

diminished in 2009 to 615,000 ton (International Magnesium Association 2010). At the present time, China is the largest producer of primary Mg as revealed in Table 2.2. Over the past two decades, primary production of Mg has been switching from countries with high power costs (like the United States, Canada, Europe, and Australia) to countries with lower power costs (like China). Production of primary Mg ceased recently in France and Norway, and as a result, Mg is no longer being produced in the European Union. Magnesium is readily recycled and because of this, it is considered a "green" metal by the automotive industry.

The structural uses for unalloyed Mg are limited; as a result, Mg is alloyed with other elements (like Al, Zn, Mn, rare earths [Res], Zr, and a few other elements) for most engineering applications. Naming of Mg alloys usually follows the ASTM standard 275 naming method using a three-part letter–number–letter system. The first code of letters indicates the two principal alloying elements (with the highest concentration first). Alloys with the designation AM would be Mg alloys containing Al and Mn additions. AZ-designated alloys contain Al and Zn. WE-designated alloys containing Y (the W in the designation) and REs (the E in the designation). A table showing the code letters used to designate commercial Mg alloys is presented in Table 2.3. The second part of the designation is a group of numbers giving the weight percentages of the two alloying additions (rounded off to the nearest whole number and in the same order as the code letters). For example, AZ91 contains 9 wt.% Al and 1 wt.% Zn. The third part of the alloy designation is again a letter and distinguishes between alloys having the same nominal designation. An X designates experimental alloys. As an example, AZ91C refers to the third specific composition registered having the nominal composition of 9 wt.% Al and 1 wt.% Zn. Temper designations also follow the same ASTM method and consist of a letter plus one or more digits. As an example, AZ91C-F denotes the same alloy as described earlier that is in the as-fabricated or as-cast condition. A number of common Mg alloys and their engineering characteristics can be found in handbooks (Polmear 1999). Because Mg alloys are easily cast and because castings are less expensive than wrought products, cast Mg alloys are much more commonly used than wrought ones.

The most common alloying addition to Mg is Al and the second most common alloying addition is Zn. The addition of both Al and Zn to Mg results in the popular AZ alloys with AZ91 being the most commonly used Mg alloy. Small amounts of Mn are almost always present in Mg alloys containing Al to

TABLE 2.3 Code Letters for the Designation System of Magnesium Alloys, Part A; Temper Designation for Magnesium Alloys, Part B

Letter	Alloying Elements
Part A	
A	Aluminum
C	Copper
E	RE metals
H	Thorium
K	Zirconium
L	Lithium
M	Manganese
Q	Silver
S	Silicon
W	Yttrium
Z	Zinc
Part B	
F	As fabricated
O	Annealed, recrystallized (wrought products only)
H	Strain hardened
T	Thermally treated to produce stable tempers other than F, O, or H
W	Solution heat treated (unstable tempers)

Source: Cramer, S.D. and Covino, B.S., Jr., Eds., *ASM Metals Handbook*, Vol. 13B, ASM International, Novelty, OH, p. 18, 2005.

enhance grain structure. For alloys that do not contain Al and Mn, Zr is added instead. A table showing some of the most common cast and wrought alloys is presented in Table 2.4.

Mg alloys with RE additions that have been developed over the past decade or two have improved high-temperature strength and creep resistance and are among the most corrosion-resistant Mg alloys. These alloys can be more costly and thus find applications in the aerospace field as well as other areas where better corrosion resistance or higher-temperature capabilities are needed. A few of these alloys (most notably WE43) are also being considered for possible biomedical applications. These more corrosion-resistant alloys include WE43 (Mg–Y–RE–Nd) and EV31 (formerly Elektron 21, Mg–Nd–Gd–Zr).

Overcoming some of the engineering challenges for Mg alloys (like further reductions in corrosion rate for biomedical applications or expanding the temperature range for automotive alloys) will likely

TABLE 2.4 Some Common Mg Alloys

Cast alloys	AE42	EV31	HK31	QE22	WE43	ZK51
	AM20		HZ32	QH21	WE54	ZK61
	AM50					ZE41
	AM60					ZC63
	AS21					
	AS41					
	AZ63					
	AZ81					
	AZ91					
Wrought alloys	AZ31	Electron 675	HK31			ZK60
	AZ61		HK21			
	AZ80					

require the use of novel processing methods and perhaps the use of Mg metal matrix composites (MMCs). As an example, vapor deposition allows one to significantly expand the range of solubility for alloying additions making alloys with much improved passive films and corrosion resistance possible. By adding reinforcements (like SiC) to magnesium or a Mg alloy, MMCs with improved high-temperature creep resistance will become available.

2.3 Applications of Magnesium and Magnesium Alloys

2.3.1 Uses

Magnesium and magnesium alloys are used in a number of areas including housings and components for consumer products (like laptop computers, cameras, and cell phones) and alloying additions for other metals, electrochemical processes (like sacrificial anodes for the protection of pipelines and hot-water heaters), processing of metals (like nodular iron) and chemical processes (like the Grinard process for production of organic and organometallic compounds), structural applications (like automotive and aircraft components), and pyrotechnics (like fireworks and flares). Magnesium oxide (magnesia) is commonly used in pharmaceuticals like milk of magnesia and Epsom salts. In fact, Mg is a vital element needed in the body. Mg is an essential element for the human body; the daily recommended intake of Mg for adults is about 300–400 mg. Mg^{+2} is the fourth most abundant cation in the human body. The unusual mechanical properties of Mg alloys, like low density, high strength-to-weight ratio, and an elastic modulus similar to that of human bone (allowing one to avoid stress shielding common with other metal implants), make them an ideal choice for bioimplants. While in the human body, Mg alloys will undergo gradual beneficial dissolution releasing Mg ions, which play an essential role in metabolism processes, are stabilizers for DNA and RNA structures (Hartwig 2001), and facilitate the healing of bone tissue.

2.3.1.1 As an Alloying Addition and Assorted Other Processing Uses

The largest use of magnesium is as an alloying addition in the production of Al alloys as shown in Figure 2.1 (Dunlop 2002). Mg is also used as an alloying addition in Zn die castings and other Zn products. Other alloys, like Ni, Ni–Cu, and Cu–Ni–Zn, utilize Mg as an alloying addition. Mg acts as a scavenger in the production of metals like steel (where it is used for desulfurization), Cu-based alloys (where it is a deoxidizer), and Pb (where it is used to remove bismuth). Mg is used in the Kroll process for the production of Ti and Zr. Magnesium has long established uses in a variety of

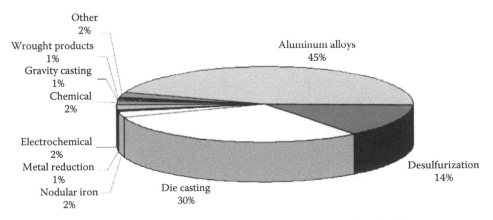

FIGURE 2.1 Worldwide consumption of primary Mg. (From Dunlop, G., Australian R&D on the production and application of magnesium alloys, H.I. Kaplan, Ed., *Proceedings of the Magnesium Technology 2002*, TMS, Warrendale, PA, 2002.)

chemical and electrochemical processes including the purification of gases (like Ar and H), deoxygenation and dechlorination of boiler water, sacrificial anodes in hot-water heaters and underground structures, batteries, and photographic engraving plates. Likewise, there is the long-standing use of Mg powder in pyrotechnics like fireworks, high-energy fuels, incendiary devices, and night aerial photography.

2.3.1.2 Automotive Applications

The automotive industry has been steadily increasing the use of magnesium alloy components to lower the weight of its vehicles for improving fuel efficiency. One of the primary automotive uses of magnesium is in transmission casings; because these components are necessarily large, the weight savings realized with magnesium are substantial. The current average use of 11–13 lb of magnesium alloys per vehicle is expected to rise to 99–353 lb per vehicle. This reduction in weight has been estimated to result in a 1% boost in fuel efficiency translating into 100,000 barrels of oil per day (Critical Strategic Metals 2010). Some current and potential applications for Mg alloys in the automotive sector are presented in Table 2.5.

2.3.1.3 Military Applications

Earlier it was mentioned that Mg production in the past has spiked during war times and receded after wars, so the military has a long history of interest in and use of Mg and Mg alloys. Some military applications of Mg alloys include radar equipment, portable ground equipment, decoy flare ordnance, helicopter transmission and rotor housings, and Stingray torpedoes.

2.3.1.4 Medical Applications

Mg is used in medical equipment like wheelchair attachments (where the weight savings improve mobility for the handicapped) and housings for hospital electronics (easier mobility of these devices).

TABLE 2.5 Automotive Uses for Mg (Current and Potential)

Body structures	*Powertrain*
Instrument panel beams	4WD transfer cases
Radiator supports	Transmission cases/housings
Seat frames	Clutch housings
Center consoles	Axle carriers
Steering column	*Engine components*
Brackets	Head covers
Steering wheel	Intake manifolds
Armatures	Oil pans
Front of dash panels	Bed plates
Doors	Engine block
Lift gates	Engine mount brackets
Head lamp bezels	Alternator brackets
Roof bows	A/C compressors
Mirror frames	*Chassis*
Spare tire carriers	Front subframe
Shifter housings	Engine cradle
	Rear subframe

Source: Patzer, G., The magnesium industry today ... the global perspective, *Proceedings of Magnesium Technology 2010*, TMS, Warrendale, PA, pp. 85–90, 2010.

A potential medical application that has spurred substantial activity in the global Mg research community in the past 5–8 years (Witte et al. 2005, 2007a,b, Staiger et al. 2006, Rettig and Virtanen 2008, 2009) is the use of Mg alloys in resorbable or bioabsorbable implant materials for humans. In the last 5 years alone, there have been hundreds of papers published worldwide concerning the use of Mg alloys in the human body. This is an application for Mg alloys where the alloy should corrode (albeit in a slow and controlled manner). Two of the biggest areas of interest for resorbable Mg alloy implants are in stents and orthopedic fixation devices. Much of this research is being done in Europe, China, and Australia. A clinical trial of resorbable Mg alloys stents was conducted in Europe (in 2004) on 63 patients with promising results. Thus interest in Mg alloys with relatively low but controllable corrosion rates has been growing since this trial was conducted (Di Mario et al. 2004). Since the number of cardiac stents and orthopedic fixation devices used in the world each year is in the hundreds of thousands to millions of range, the inclusion of Mg alloys in such devices could account for an entirely new market for high-end Mg alloys. If newer processing methods like vapor deposition are used to produce the resorbable Mg-based biomaterials, it is possible to use the inherent advantages of this processing method (like low-impurity concentrations, very small grain sizes, tailoring of the alloy structure and composition, net-shaping capabilities, and greatly extended solid solubilities to improve corrosion resistance) to produce single-phase alloys (like Mg–Ti and Mg–Y–Ti) that are impossible to produce with conventional processing. Recent research has shown that these vapor-deposited alloys have lower and more controllable corrosion rates than the conventional Mg alloys (like WE43) for bioapplications (Shaw et al. 2008, Petrilli 2009).

2.3.1.5 Potential Use as a Hydrogen Storage Material

Magnesium is also being considered for use as a material to store hydrogen. Magnesium stores hydrogen through the formation of MgH_2 (in a reversible reaction) that is capable of storing large amounts of hydrogen (7.6 wt %). However, when pure Mg is used for hydrogen storage, the rate of hydrogen absorption and desorption is low because the diffusion rate of hydrogen atoms is slow. Therefore, finding ways of improving the hydration kinetics of Mg has been one of the main research challenges for this application. Improvements can be attained by structural refinement and/or by changes in the chemical composition of these alloys. The ability to form nanostructured Mg films by physical vapor deposition (PVD) is especially exciting since it has been established that nanometric structures enhance hydriding rates (Murray et al. 1995, Aizawa et al. 2001, Au 2005) by increasing surface areas and rates of diffusion. While nanostructuring improves the hydrogen storage kinetics, the thermodynamics of this process do not change. Recently, a novel approach involving processing Mg by equal-channel angular pressing (ECAP) resulted in obtaining favorable parameters of hydrogen storage kinetics and thermodynamics (Skripnuk et al. 2010). This process seems to be very appealing since it produces ultrafine grain Mg alloys in amounts much higher than those obtained by high-energy ball milling. Another way of improving hydration kinetics is by changes in chemical composition. It is believed that the presence of Ni has a catalytic effect on hydrogen dissociation and lowers the desorption temperature by 60°C (David 2005). The Mg_2Ni combines with hydrogen to form Mg_2NiH_4 hydride, which, while having much faster hydration kinetics, has a lower hydration capacity (3.6 wt.% H) than pure Mg (Kuji et al. 2002, Nakano et al. 2005, Rojas et al. 2005). Since the addition of heavy transition metals reduces storage capacity and adds overall weight to the storage material, it seems that a better solution would be to add a light metal that has low affinity to hydrogen, like aluminum. Recently synthesized, nanocrystalline magnesium alanates, $Mg_{2-x}Al_xNi$ and $Mg(AlH_4)_2$ containing 9.3 wt.% hydrogen, have shown promise as a hydrogen storage material (Fichtner et al. 2004, Wang et al. 2004, Pranevicius and Milcius 2005). PVD could also be used to create new, nanocrystalline thin-film alloys containing Mg, Ni, and Al or Li for hydrogen storage applications. This is especially attractive since vapor deposition allows for the creation of nonequilibrium alloys with a variety of compositions. Also, vapor-deposited films could be hydrogenated by ion beam-assisted deposition. In this approach, hydrogen is incorporated into the structure of the alloy during deposition.

2.3.2 Challenges

As with any strategic metal, the future use of Mg is controlled by its price and availability. Government legislation mandating better gas mileage for cars as well as the increasing cost of oil has renewed interest in and for the use of magnesium in the automotive sector. In a 2005 publication (Winzer et al. 2005), it was reported that Mg usage in the automotive sector had undergone 20% annual growth in the previous 10 years. In most high-volume applications, like the automotive market, magnesium's price as compared to that of Al determines its usage. At times during the past decade, the price per pound for Mg and Al has been close.

As noted earlier, the producers of primary Mg have been changing in the past decade or two. Many countries that were once leaders in the production of Mg (like the United States and Canada) now rely on others (like China) for primary Mg. As energy cost rises, the desire for lighter vehicles and machinery increases, and the use of low density of Mg and Mg alloys becomes increasingly attractive. In the past, the need for and the production of Mg during times of war has soared. With production of Mg now largely in the hands of developing countries, one has to wonder what implications this might have in the future concerning the availability of Mg.

Based on the world's increasing need for more energy-efficient transportation and manufacturing, one would think that this should result steadily increased utilization of Mg alloys in cars, aircraft, and machinery. In addition the use of Mg alloys in consumer goods like cell phones, laptops, and small electronics also appears to be a rapidly expanding market. With increased use of Mg, its efficient recycling also needs to be taken into account. Magnesium in a number of forms (like scrap magnesium from its original processing) is already being recycled and reused in the production of secondary (recycled) Mg alloys. Thus, Mg is considered a "green" alloy in the automotive sector. Recycling of Mg from consumer goods and scraped vehicles is not as well established, but it is a growing industry. In 2015 European guidelines prescribe a recycling rate of 95%, meaning that Mg automotive components will need to be removed from scrapped cars and recycled (Antrekowitsch et al. 2002). Production of Mg, like that of any other metal, has its environmental concerns (like reducing greenhouse gas emissions of SF_6 and the proper disposal of the by-products of Mg processing) that need to be addressed, and these topics remain active areas of research.

2.4 General Properties of Magnesium and Magnesium Alloys

2.4.1 Physical

Pure Mg has a hexagonal close-packed (HCP) crystal structure (with the axial ratio of c/a = 1.624), an atomic weight of 24.305 g, an atomic diameter of 0.320 nm, a density of 1.738 g/cm^3, a melting point of 650°C (1202°F), and a boiling point of 1090°C (1994°F). Magnesium's atomic diameter is such that it favors solid solubility with a number of elements. Al, Zn, and Mn were the earliest alloying additions to Mg. Today, the alloying additions most commonly added to Mg (in alphabetical order) are Al, Ag, Be, Ca, Cu, Li, Mn, Ni, RE, Si, Sn, Th, Zn, Zr, and Y.

2.4.2 Mechanical

Trends in Mg alloy development for improving mechanical properties are illustrated in Figure 2.2 (Mordike and Ebert 2001). These trends include increasing specific strength, increasing ductility, improving creep resistance, and the development of composite materials with improved elastic modulus, thermal expansion properties, and wear and creep resistance. Nominal compositions and mechanical properties for some Mg alloys are presented in Table 2.6.

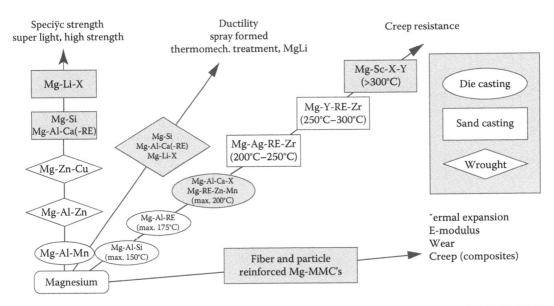

FIGURE 2.2 Alloy development directions. (From Mordike, B.L. and Ebert, T., *Mater. Sci. Eng. A*, 302, 37, 2001.)

2.4.3 Thermal

Magnesium is a low-melting-temperature metal (T_m = 650°C), and as a result, elevated temperatures adversely affect magnesium alloy mechanical properties. In addition to corrosion, elevated temperature mechanical properties (especially creep resistance) have been mentioned as a characteristic of concern related to the more widespread use of Mg alloys, especially in the automotive sector. Newer Mg alloys have adequate properties up to temperatures in the range of 300°C. Elevated temperature data for Mg and Mg alloys can be found in handbooks such as the *ASM Specialty Handbook on Mg and Mg Alloys*. Table 2.7 shows yield strength, tensile strength, and elongation data as a function of temperature for several of the more popular Mg alloys.

With a melting temperature of 650°C, it is also not surprisingly that Mg and its alloys are susceptible to creep at elevated temperatures. Figure 2.2 showed the evolution of creep resistance in Mg alloys and how the upper limit temperature has been increasing with the development of new classes of Mg alloys. Creep data for some Mg alloys are given in Figure 2.3.

As Figure 2.2 also reveals, Mg-based MMCs are a way to improve the high-temperature creep resistance of Mg, and research is underway in this area (Tian et al. 2010).

2.5 Structure of Magnesium and Magnesium Alloys

By virtue of the active nature of Mg, second phases significantly impact the corrosion of Mg. Al is the most common alloying addition to Mg, and while it has an equilibrium solubility limit of 12.7 wt.%, most commercial Al-containing alloys contain a β-$Mg_{17}Al_{12}$ phase as well as other phases arising from impurities and other alloying additions. The Al-rich solid solution, α-phase, is anodic to these other phases and results in corrosion of the matrix of the grain in preference to these other phases. As a result, Mg and Mg alloys do not undergo intergranular attack; they undergo the opposite of this where the attack of the matrix leaves the cathodic precipitates at the grain boundaries unattacked.

TABLE 2.6 Mechanical Properties of Some Common Mg Alloys

Alloy	Al	Mn[a]	Th	Zn	Zr	Other[b]	TS MPa	TS ksi	YS Tensile MPa	YS Tensile ksi	YS Compressive MPa	YS Compressive ksi	YS Bearing MPa	YS Bearing ksi	Elongation in 50 mm (2 in.) (%)	Shear MPa	Shear ksi	Hardness HR[c]
Sand and permanent mold castings																		
AM100A-T61	10.0	0.10	...	...	...	...	275	40	150	22	150	22	...	...	1	...	...	69
AZ63A-T6	6.0	0.15	...	3.0	...	...	275	40	130	19	130	19	360	52	5	145	21	73
AZ81A-T4	7.6	0.13	...	0.7	...	...	275	40	83	12	83	12	305	44	15	125	18	55
AZ91C and E-T6[d]	8.7	0.13	...	0.7	...	...	275	40	145	21	145	21	360	52	6	145	21	66
AZ92A-T6	9.0	0.10	...	2.0	...	...	275	40	150	22	150	22	450	65	3	150	22	84
EQ21A-T6	...	...	...	...	0.7	1.5 Ag, 2.1 Di	235	34	195	28	195	28	...	...	2	...	...	65–85
EZ33A-T5	...	...	...	2.7	0.6	3.3 RE	160	23	110	16	110	16	275	40	2	145	21	50
HK31A-T6	...	...	3.3	...	0.7	...	220	32	105	15	105	15	275	40	8	145	21	55
HZ32A-T5	...	...	3.3	2.1	0.7	...	185	27	90	13	90	13	255	37	4	140	20	57
K1A-F	...	...	...	...	0.7	...	180	26	55	8	...	...	125	18	1	55	8	...
QE22A-T6	...	...	...	...	0.7	2.5 Ag, 2.1 Di	260	38	195	28	195	28	...	...	3	...	...	80
QH21A-T6	...	...	1.0	...	0.7	2.5 Ag, 1.0 Di	275	40	205	30	...	...	...	...	4	...	...	...
WE43A-T6	...	...	...	...	0.7	4.0 Y, 3.4 RE	250	36	165	24	...	...	...	...	2	...	...	75–95
WE54A-T6	...	...	...	...	0.7	5.2 Y, 3.0 RE	250	36	172	25	172	25	...	...	2	...	...	75–95
ZC63A-T6	...	0.25	...	6.0	...	2.7 Cu	210	30	125	18	...	...	...	...	4	...	...	55–65
ZE41A-T5	...	...	...	4.2	0.7	1.2 RE	205	30	140	20	140	20	350	51	3.5	160	23	62
ZE63A-T6	...	...	...	5.8	0.7	2.6 RE	300	44	190	28	195	28	...	...	10	...	...	60–85
ZH62A-T5	...	...	1.8	5.7	0.7	...	240	35	170	25	170	25	340	49	4	165	24	70
ZK51A-T5	...	...	...	4.6	0.7	...	205	30	165	24	165	24	325	47	3.5	160	23	65
ZK61A-T5	...	...	...	6.0	0.7	...	310	45	185	27	185	27	...	...	...	170	25	68
ZK61A-T6	...	...	...	6.0	0.7	...	310	45	195	28	195	28	...	...	10	180	26	70

	Al	Mn	Zn	Zr	Other												
Die castings																	
AE42-F	4.0	0.1	...	...	2.5 RE	230	34	145	21	145	21	...	...	11	...	...	60
AM20-F	2.1	0.1	...	...	...	210	31	90	13	90	13	...	...	20	...	...	45
AM50A-F	4.9	0.26	...	...	...	230	33	125	18	125	18	...	...	15	...	...	60
AM60A and B-F[e]	6.0	0.13	...	...	...	240	35	130	19	130	19	...	...	13	...	...	65
AS21-F	2.2	0.1	...	...	1.0 Si	220	32	120	17	120	17	...	...	13	...	...	55
AS41A-F[f]	4.2	0.20	...	...	1.0 Si	240	35	140	20	140	20	...	...	15	...	...	60
AZ91A, B, and D-F[g]	9.0	0.13	0.7	...	...	250	36	160	23	160	23	...	...	7	140	20	70
Forgings																	
AZ318-F	3.0	0.20	1.0	...	...	260	38	170	25	...	...	...	...	15	130	19	50
AZ61A-F	6.6	0.15	1.0	...	...	295	43	180	26	125	18	...	...	12	145	21	55
AZ80A-T5	8.5	0.12	0.5	...	...	345	50	250	36	195	28	...	...	6	160	23	72
AZ80A-T6	8.5	0.12	0.5	...	...	345	50	250	36	170	25	...	...	11	172	25	75
MIA-F	...	1.2	...	...	...	250	36	160	23	...	...	...	...	7	110	16	47
ZK31-T5	...	...	3.0	0.6	...	290	42	210	30	...	...	...	...	7	...	...	...
ZK60A-T5	...	...	5.5	0.45[a]	...	305	44	215	31	160	23	285	41	16	165	24	65
ZK61-T5	...	...	6.0	0.8	...	275	40	160	32	...	...	...	...	7	...	...	...
ZM21-F	...	0.5	2.0	...	...	200	29	125	18	...	...	...	...	9	...	...	...
Extruded bars and shapes																	
AZ10A-F	1.2	0.2	0.4	...	...	240	35	145	21	69	10	...	...	10	...	...	...
AZ318 and C-F[h]	3.0	0.20	1.0	...	...	255	37	200	29	97	14	230	33	12	130	19	49
AZ01A-F	6.5	0.15	1.0	...	...	305	44	205	30	130	19	285	41	16	140	20	60
AZ80A-T5	8.5	0.12	0.5	...	...	380	55	275	40	240	35	...	...	7	165	24	80
MIA-F	...	1.2	...	...	...	255	37	180	26	83	12	195	28	12	125	18	44
ZC71-T6	...	0.5	6.5	...	1.25 Cu	295	43	324	47	...	...	...	...	3	...	...	70–80
ZK21A-F	...	...	2.3	0.45[a]	...	260	38	195	28	135	20	...	...	4	...	...	...
ZK31-T5	...	...	3.0	0.6	...	295	43	210	30	...	...	...	...	7	...	...	...

(continued)

TABLE 2.6 (continued)   Mechanical Properties of Some Common Mg Alloys

Alloy	Composition (%)						Tensile Strength		Yield Strength						Elongation in 50 mm (2 in.) (%)	Shear Strength		Hardness HR[c]
	Al	Mn[a]	Th	Zn	Zr	Other[b]	MPa	ksi	Tensile		Compressive		Bearing			MPa	ksi	
									MPa	ksi	MPa	ksi	MPa	ksi				
ZK40A-T5	...	...	...	4.0	0.45[a]	...	275	40	255	37	140	20	...	...	4	...	...	...
ZK60A-T5	...	...	...	5.5	0.45[a]	...	350	51	285	41	250	36	405	59	11	180	26	82
ZM21-F	...	0.5	...	2.0	...	...	235	34	155	22	...	...	...	...	8	...	...	...
Sheet and plate																		
AZ31B-H24	3.0	0.20	...	1.0	...	...	290	42	220	32	180	26	325	47	15	160	23	73
ZM21-0	...	0.5	...	2.0	...	...	240	35	120	17	...	...	...	...	11	...	...	...
ZM21-H24	...	0.5	...	2.0	...	...	250	36	165	24	...	...	...	...	6	...	...	...

Source: Avedesian, M.M. and Baker, H., Eds., *ASM Specialty Handbook: Mg and Mg Alloys*, ASM International, Novelty, OH, 1999.

[a] Minimum.

[b] RE, rare earth; Di, didymium (a mixture of RE elements made up chiefly of neodymium and praseodymium).

[c] 500 kg load, 10 mm ball.

[d] Properties of C and E are identical, but AZ91E castings have 0.17% minimum Mn and maximum contaminant levels of 0.005% Fe, 0.0010% Ni, and 0.015% Cu.

[e] Properties of A and B are identical, but AM60B castings have maximum contaminant levels of 0.005% Fe, 0.002% Ni, and 0.010% Cu.

[f] Properties of A and XB are identical, but AS41B castings have maximum contaminant levels of 0.0035% Fe, 0.002% Ni, and 0.002% Cu.

[g] Properties of A, B, and D are identical, except that 0.30% maximum residual Cu is allowable in AZ91B, and AZ91D castings have maximum contaminant levels of 0.005% Fe, 0.002% Ni, and 0.030% Cu.

[h] Properties of B and C are identical, but AZ31C has 0.15% minimum Mn, 0.1% maximum Cu, and 0.03% maximum Ni.

TABLE 2.7 Elevated Temperature Tensile Properties for a Variety of Sand-Cast Mg Alloys

Alloy	Temper	Temperature (°C)	Tensile Yield Strength		Tensile Strength		Elongation (%)
			MPa	ksi	MPs	ksi	
AZ81A	T4	20	85	12	275	40	15
		95	90	13	255	37	22
		150	85	12	195	28	32
		200	75	11	140	20	30
		260	70	10	95	14	25
AZ91C	T4	20	85	12	275	40	14
		95	95	14	235	34	26
		150	95	14	195	28	30
		200	90	13	140	20	30
	T6	20	130	19	275	40	5
		95	130	19	255	37	24
		150	115	17	185	27	31
		200	95	14	140	20	33
AZ92A	T5	20	110	16	180	26	2
		95	110	16	165	24	2
		150	95	14	160	23	4
		200	74	11	140	20	15
		260	55	8	110	16	32
		315	30	4	60	9	61
	T6	20	145	21	275	40	2
		95	145	21	255	37	25
		150	115	17	195	28	35
		200	85	12	115	17	36
		260	55	8	75	11	33
		315	35	5	55	8	49
EQ21A	T6	20	195	28	261	38	4
		100	189	27	230	33	10
		150	180	26	211	31	16
		200	170	25	191	28	16
		250	152	22	169	25	15
		300	117	17	132	19	10
		325	92	13	105	15	9
		350	60	9	78	11	14
EZ33A	T5	20	105	15	160	23	3
		95	105	15	160	23	5
		150	95	14	150	22	10
		200	85	12	145	21	20
		260	70	10	125	18	31
		315	55	8	85	12	50
QE22A	T6	20	205	30	275	40	4
		95	195	28	235	34	...
		150	185	27	205	30	...

(continued)

TABLE 2.7 (continued)   Elevated Temperature Tensile Properties for a Variety of Sand-Cast Mg Alloys

| Alloy | Temper | Temperature (°C) | Tensile Yield Strength | | Tensile Strength | | Elongation (%) |
			MPa	ksi	MPs	ksi	
		200	162	24	185	27	…
		260	110	16	140	20	…
		315	60	9	85	12	…
WE54A	T6	20	200	29	275	40	5
		250	170	25	225	33	7.5
ZE41A	T5	20	140	20	205	30	5
		95	130	19	185	27	8
		150	115	17	165	24	15
		200	95	14	130	19	29
		260	70	10	95	14	40
		315	55	8	75	11	43
ZE63A	T6	20	175	25	290	42	…
		100	130	19	235	34	…
		150	110	16	185	27	…
		200	95	14	130	19	…
ZK51A	T5	20	165	24	275	40	8
		95	145	21	205	30	12
		150	115	17	160	23	14
		200	90	13	115	17	17
		260	60	9	85	12	16
		315	40	6	55	8	16

Source: Avedesian, M.M. and Baker, H., Eds., *ASM Specialty Handbook: Mg and Mg Alloys,* ASM International, Novelty, OH, p. 178, 1999.

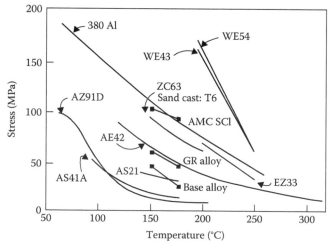

FIGURE 2.3 Creep data for some Mg alloys. Stress for 0.1% creep strain in 100 h. (From Dunlop, G., 2002, *Proceedings of the Magnesium Technology 2002,* TMS, Warendale, PA.)

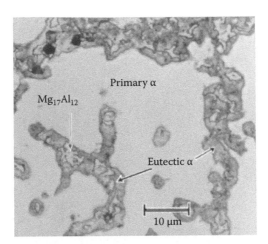

FIGURE 2.4 Microstructure of die-cast AZ91D. (From Song, G. and Atrens, A., *Adv. Eng. Mater.*, 5(12), 837, 2003.)

The microstructure of die-cast AZ91D is shown in Figure 2.4 (Song and Atrens 2003). Gesing et al. show that AZ91D typically contains the distribution of phases identified in Figure 2.5 (Gesing et al. 2010). Figure 2.6 shows an image of AZ91E after 4 h in 5% NaCl that reveals attack of the α-matrix leaving behind the β-precipitates and a bit of α-phase close to the β-phase (Song 2003). This α-phase close to the β-phase was not attacked because it had a much higher Al concentration than the α-phase in the interior of the grain.

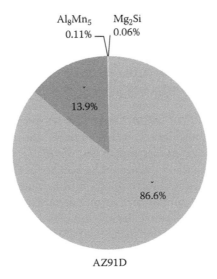

FIGURE 2.5 Distribution of phases in AZ91D by weight percent calculated from mass balance of alloy and individual phase average compositions. (From Gesing, A.J. et al., Development of recyclable Mg-based alloys: AZ91D and AZC1231 phase information derived from heating/cooling curve analysis, *Proceedings of the Magnesium Technology 2010*, TMS, Warrendale, PA, p. 98, 2010.)

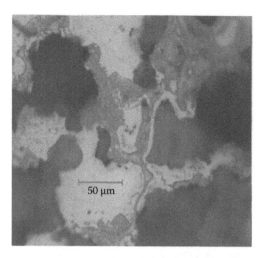

FIGURE 2.6   Morphology of corrosion attack on AZ91E after 4 h immersion in 5% NaCl. (From Song, G., *JCSE*, 6, paper C104, 2003.)

2.6 Degradation of Magnesium and Magnesium Alloys (Chemical and Electrochemical)

2.6.1 Thermodynamics of Corrosion of Mg Alloys

The electrochemical thermodynamics of magnesium is usually presented in the form of a Pourbaix diagram. Figure 2.7 is the potential–pH diagram of the magnesium–water system (Pourbaix 1974, p. 141). Magnesium dissolves at low and neutral pH values at potentials of interest for most, if not all, applications. At high pH values a magnesium hydroxide (brucite) film forms that is only semi-protective. This semi-protective or quasi-passive film is of low density and undergoes compressive rupture resulting in constant exposure of fresh metal (Froes et al. 1989).

The surface film that forms on Mg is dull gray in appearance. A number of investigations into the structure of the magnesium metal/oxide interface using transmission electron microscopy (TEM) have been conducted (Nordlien et al. 1995, 1996, 1997). Native oxide films were found to be 20–50 nm thick, while films grown at 65% RH and 30°C were 100–150 nm thick. Films grown in distilled water for 48 h consisted of three layers: an inner cellular layer (0.4–0.6 μm thick), a dense intermediate layer (20–40 nm thick), and a platelike outer layer (1.8–2.2 μm thick). The intermediate layer is believed to be the original air-formed oxide, while the inner layer is assumed to be responsible for corrosion resistance. Improvements in corrosion resistance (as measured by stability of the inner layer, assumed to correspond to resistance to dehydration, as measured upon electron irradiation) hit a plateau at 4 wt.% Al. At this concentration and above, the aluminum concentration in the inner layer (in the form of aluminum oxide) reaches 35 wt.%. RE elements decrease the amount of hydration in the inner layer, as measured by a decrease in the speed phase of separation using the TEM observations. However, the inner cellular layer may be an artifact of electron irradiation, as the cellular layer only appears upon electron radiation. (The cellular layer is originally indistinguishable from the middle, allegedly dense layer.)

One can get an idea of the influence of alloying additions on improving the corrosion resistance of Mg by simply overlaying Pourbaix diagrams for the alloying additions on the Mg diagram. Figure 2.8 shows the Pourbaix diagrams for Al, Y, Ti, and Zr. By looking at these elements as alloying additions and superimposing them on the diagram for Mg, one could get an idea of how the passive region for Mg could be extended by alloying. Yttrium is added to Mg (in alloys like WE43) to improve the protective nature of its "quasi-passive" film. Indeed, x-ray photoelectron spectroscopy (XPS) of these

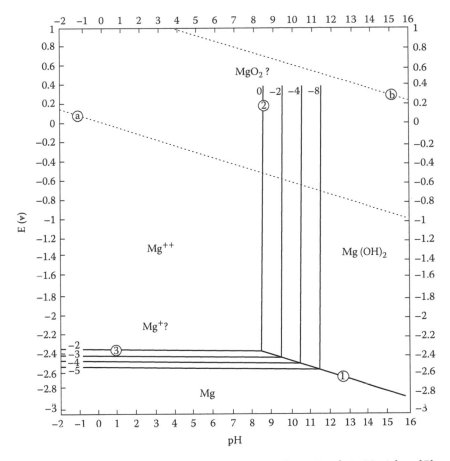

FIGURE 2.7 Pourbaix diagram for the Mg–water system at 25°C. (From Pourbaix, M., *Atlas of Electrochemical Equilibria in Aqueous Solutions*, NACE CEBELCOR, Pergamon Press, New York, p. 141, 1974.)

more protective films on Mg–Y-containing alloys has revealed the presence of Y_2O_3 in the surface film (Heidersbach 1998, Wolfe 2005).

In the design of new alloy systems, computational thermodynamics (typically referred to as calculation of phase diagrams, CALPHAD) is now being used to aid in the development of Mg alloys with improved high-temperature properties. Such modeling has been used to aid in assessing the creep resistance of Mg–Al–Ca alloys (Ozturk et al. 2003) and Mg–Ca–Sn (Arroyave and Liu 2006) and in the development of Mg–Al–X alloys with improved high-temperature properties.

2.6.2 Kinetics of Corrosion of Mg Alloys

2.6.2.1 Corrosion Rate Measurements

While thermodynamic data provide the basis for understanding corrosion tendencies and aid in the design of new alloys, kinetics provides the actual corrosion rate data needed to assess performance of a material in a given environment. Corrosion rate measurements on Mg and Mg alloys are accomplished via a number of methods including weight-loss measurements in actual or simulated environments, hydrogen capture experiments that convert the amount of hydrogen evolved to a corrosion rate, and electrochemical methods (including polarization resistance and electrochemical impedance spectroscopy [EIS]). When the rates of corrosion for Mg and Mg alloys are expected to be high, one should make these

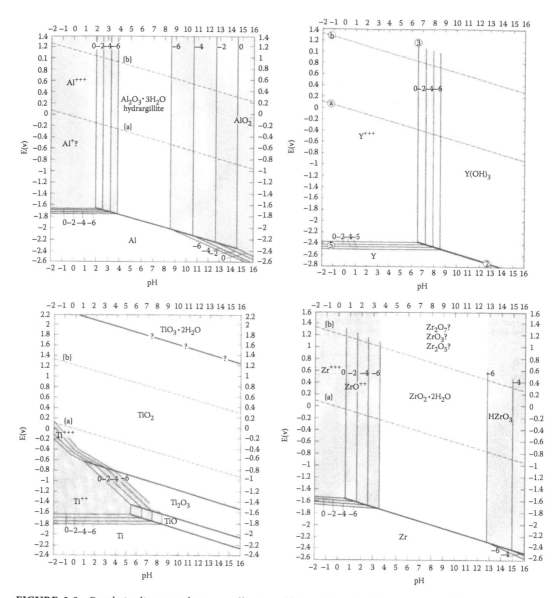

FIGURE 2.8 Pourbaix diagrams for some alloying additional that should improve passivity in Mg. (From Pourbaix, M., *Atlas of Electrochemical Equilibria in Aqueous Solutions*, NACE CEBELCOR, Pergamon Press, New York, 1974.)

measurements via the hydrogen capture or weight-loss methods since corrosion rates obtained via electrochemical methods can be confounded by the negative difference effect as described in detail in several references (Song and Atrens 2007). An example of the underestimation of the corrosion rate for pure Mg via electrochemical methods because of the negative difference effect (NDE) is shown in Figure 2.9 (Makar and Kruger 1990).

2.6.2.2 Corrosion in Atmospheric Environments

Corrosion rates for Mg and Mg alloys vary significantly depending on the alloy and the environment. Table 2.8 shows a difference of more than seven orders of magnitude in the corrosion for pure Mg depending on the environment (Hawke et al. 1999). Typically, Mg and Mg alloys are not used in

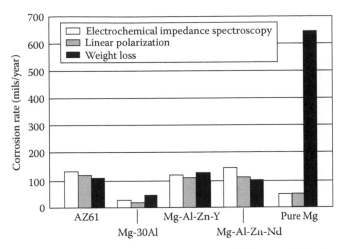

FIGURE 2.9 Corrosion rates for several Mg alloys measured via different methods showing a large discrepancy between values obtained for pure Mg in electrochemical and weight-loss measurements. (From *ASM Metals Handbook*, Cramer, S.D. and Covino, B.S., Jr., Eds., Vol. 13B, ASM International, Novelty, OH, p. 207, 2005.)

TABLE 2.8 Commercially Pure Magnesium Corrosion Rate in a Variety of Media

Medium	Corrosion Rate	
	mm/year	mils/year
Humid air	1.0×10^{-5}	0.0004
Humid air with condensation	1.5×10^{-2}	0.6
Distilled water	1.5×10^{-2}	0.6
Distilled water exposed to acid gases	0.03–0.3	1.2–12
Hot deionized water (100°C) (14 days stagnant immersion)	16	640
Hot deionized water inhibited with 0.25 NaF	5.5×10^{-2}	2.2
Seawater	0.25	10
3 M $MgCl_2$ solution	300	12,000
3 M NaCl (99.99% high-purity magnesium with <10 ppm Fe)	0.3	12
Grades 9980, 9990, 9991, 9995, 9998 except for NaCl solution.		

Source: Avedesian, M.M. and Baker, H., Eds., *ASM Specialty Handbook: Mg and Mg Alloys*, ASM International, Novelty, OH, p. 195, 1999.

extremely corrosive environments like 3 M $MgCl_2$, and much more reasonable corrosion rates like those shown in Tables 2.9 and 2.10 are noted under atmospheric corrosion conditions. In air free of chlorides, magnesium is covered by a partially protective gray surface film that gives limited protection. Corrosion of Mg increases with increasing relative humidity. Atmospheric attack in damp situations is usually superficial. When chlorides, sulfates, and other species that promote moisture retention on the surface are present, this can promote corrosion and pitting. The corrosion rates under marine atmospheric conditions are much lower than that observed in a salt spray cabinet as Figure 2.10 reveals for three AZ alloys. Corrosion of Mg in all types of environments is significantly influenced by the concentration of impurities in the alloy, as will be discussed. Corrosion rates for two die-cast alloys (AZ91D and AM60A) in salt fog and 5% NaCl immersion are shown in Figure 2.11 (*ASM Metals Handbook* 2005, p. 209).

TABLE 2.9   Corrosion Rates for Some Mg Alloys Exposed to the Atmosphere

UNS No.[a]	Alloy and Temper	Rural mm/year	Rural mils/year	Industrial mm/year	Industrial mils/year	Marine Rural mm/year	Marine Rural mils/year
M11311	AZ31B-H24 (10 ppm Fe)	0.013	0.52	0.025	1.0	0.017	0.69
M11312	AZ31CO (70 ppm Fe)[b]	0.012	0.46	0.025	1.0	0.038	1.5
M13310	HK31A.H24	0.018	0.73	0.030	1.2	0.016	0.64
M13210	HM21A.T8	0.020	0.80	0.032	1.3	0.022	0.88
M16100	2E10A.O	0.022	0.88	0.030	1.2	0.028	1.1
Extrusions							
M11311	AZ31B.F	0.013	0.53	0.025	1.0	0.019	0.77
M13312	HM31A.F	0.018	0.70	0.035	1.4	0.020	0.80
M16600	2K60A-T5	0.017	0.66	0.032	1.3	0.025	1.0
Castings							
M11630	AZ63A-T4	0.0086	0.34	0.022	0.88	0.019	0.76
M11914	AZ91C-T6 (350 ppm Fe)[b]	0.0043	0.17	0.015	0.62	0.022	0.88
M11914	AZ91C-T6 (10 ppm Fe)	0.0027	0.11	0.014	0.57	0.0064	0.25
M11920	AZ92A-T6	0.0094	0.37	0.020	0.80	0.025	1.0
M12330	EZ33A-T5	0.020	0.79	0.040	1.6	0.028	1.1
M13310	HK31A-T6	0.017	0.67	0.035	1.4	0.028	1.1
M13320	HZ32A-T5	0.015	0.61	0.038	1.5	0.028	1.1
M16620	ZH62A-T5	0.015	0.58	0.040	1.6	0.041	1.6
M16510	ZK51A-T5	0.014	0.57	0.035	1.4	0.025	1.0
Site average	0.014	0.014	0.030	1.2	0.024	1.0	1.0

Source: Cramer, S.D. and Covino, B.S., Jr., Eds., *ASM Metals Handbook*, Vol. 13B, ASM International, Novelty, OH, p. 211, 2005.

[a] Possible UNS composition equivalent of the tested alloy is listed. In some cases, more than one UNS number is registered to alloy designation.

[b] Iron content would exceed specification.

TABLE 2.10   Corrosion Rates for Some AZ Alloys in Atmospheric Environment and Salt Spray Cabinet

Alloy[a]	Iron (ppm)	Rural mm/year	Rural mils/year	Industrial mm/year	Industrial mils/year	Marine mm/year	Marine mils/year	20% NaCl Spray mm/year	20% NaCl Spray mils/year
(1) AZ31	10	0.012	0.5	0.025	1.0	0.038	1.5	...	...
(2) AZ31	10	0.013	0.51	0.025	1.0	0.018	0.7	...	...
Ratio of (1) to (2)	7	0.9	0.9	1.0	1.0	2.2	2.2	...	...
(3) AZ91	350	0.0043	0.17	0.016	0.6	0.022	0.87	95	3740
(4) AZ91	10	0.0028	0.11	0.014	0.55	0.0064	0.25	0.71	27.9
Ratio of (3) to (4)	35	1.5	1.5	1.1	1.1	3.5	3.5	134	134

Source: Cramer, S.D. and Covino, B.S., Jr., Eds., *ASM Metals Handbook*, Vol. 13B, ASM International, Novelty, OH, p. 211, 2005.

[a] (1) AZ31C sheet, (2) AZ31B-H24 sheet, (3) AZ91C-T6 cast plate (Fe/Mn – 0.15), (4) AZ91C.T6 cast plate (FeMn – 0007).

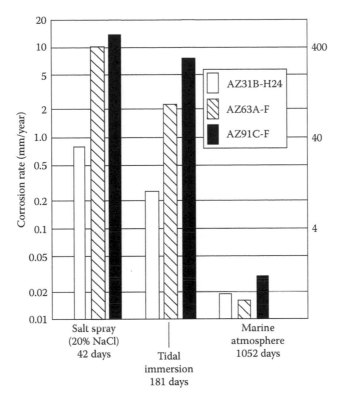

FIGURE 2.10 Corrosion rates for some AZ alloys in various chloride environments. Note the high rates for salt spray cabinet testing as opposed to marine atmosphere. (From Cramer, S.D. and Covino, B.S., Jr., Eds., *ASM Metals Handbook*, Vol. 13B, ASM International, Novelty, OH, p. 209, 2005.)

2.6.2.3 Corrosion in Solutions

Magnesium undergoes corrosion in aqueous solutions with the amount that varies depending on the solution, as well as its volume, temperature, and movement. In water a sparingly soluble $Mg(OH)_2$ film forms on the Mg surface. Because of its low open-circuit potential, dissolved oxygen typically plays a minimal role in the corrosion of Mg and Mg alloys. Corrosion increases when the flow velocity of the aqueous solution in contact with Mg is high enough to influence the surface hydroxide film. Mg and its alloys are subject to severe corrosion in chloride solutions. A look at the Pourbaix diagram for Mg reveals that its surface film is most stable at high pH, and as a result, Mg is rather resistant to corrosion by alkalis if the pH is above 10.5 (Song and Atrens 2003). Fluoride ions are inhibitors for Mg corrosion and decreased corrosion is a result of the formation of a MgF_2 surface film. All mineral acids, with the exception of chromic and hydrofluoric acid, attack Mg. Organic acids, especially those containing polar reactive groups, can attack Mg. At room temperature, many organic liquids, like ethyl alcohol, methylated spirits, oils, and degreasing agents, do not attack Mg (Song and Atrens 2003).

2.6.2.4 Protection of Mg from Corrosion during Shipment

In the transportation of Mg and Mg alloys, care must be taken to prevent significant surface corrosion while in transit. Thus, it is recommended that Mg and Mg alloys be transported in sealed containers with a desiccant (Song et al. 2006)

2.6.2.5 Factors That Exert a Significant Influence on the Corrosion of Mg

As mentioned earlier in this chapter, corrosion resistance is a primary factor limiting the more widespread use of Mg alloys. While much has been done to improve the corrosion resistance of Mg alloys

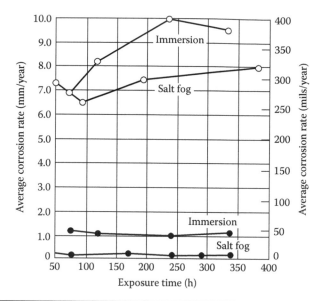

FIGURE 2.11 Corrosion rates for AZ91D and AM60A in salt fog cabinet test and immersion in 5% NaCl as a function of exposure time. (From Cramer, S.D. and Covino, B.S., Jr., Eds., *ASM Metals Handbook*, Vol. 13B, ASM International, Novelty, OH, p. 209, 2005.)

	Analysis of die-cast plates (%)	
	AM60A (○)	AZ91D (●)
Aluminun	6.2	9.7
Zinc	0.09	0.74
Manganese	0.22	0.19
Nickel	0.003	0.0018
Iron	0.005	0.006
Copper	0.03	0.0067

since their first commercial usage, gaining an understanding of the issues associated with the corrosion of Mg and Mg alloys is of extreme importance in their successful use. The issues of most importance in the corrosion of Mg and Mg alloys include impurities, galvanic coupling corrosion, pitting, and stress corrosion cracking (SCC).

2.6.2.5.1 Impurities

The factor with the strongest influence on corrosion of Mg and Mg alloys is the concentration of its impurities. In the 1940s Hanawalt published a seminal paper on the effect of impurities on the corrosion behavior of Mg (Hanawalt et al. 1942). Hanawalt's paper is likely the most widely cited publication on Mg corrosion. Specific tolerance limits for impurities such as iron, nickel, and copper in Mg were quantified. In addition, the importance of Mn in mitigating these effects was identified in this chapter. Corrosion of Mg and Mg alloys increases by several orders of magnitude when the tolerance limits for Ni, Fe, and Cu are exceeded. The level of impurities that is tolerable is a function of each particular Mg alloy and its processing. The influence of impurities like Ni, Cu, and Fe is schematically illustrated in Figure 2.12. Consequently, Mg corrosion can be reduced dramatically by reducing/eliminating impurities. Improvements in the corrosion resistance of newer Mg (alloys developed in the past 20–30 years) are based on careful control of impurities. A summary of these tolerance limits for several Mg alloy limits can be found in the literature and are shown in Table 2.11 (Song and Atrens 1999). The presence

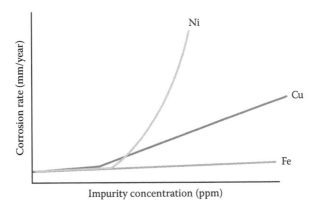

FIGURE 2.12 Schematic illustration of corrosion rate as a function of impurity-level concentration.

TABLE 2.11 Tolerance Limits for Ni, Cu, and Fe in Some Mg Alloys

Specimen	Condition	Tolerance Limits		
		Fe	Ni (ppm)	Cu (ppm)
Pure Mg		170 ppm	5	1000
Pure Mg		170 ppm	5	1300
AZ91		20 ppm	12	900
AZ91		0.032 Mn	50	400
AZ91	High pressure (F)	0.032 Mn	50	400
AZ91	Low pressure (F)	0.032 Mn	10	400
AZ91	Low pressure (T4)	0.035 Mn	10	100
AZ91	Low pressure (T6)	0.046 Mn	10	400
AZ91B			<100	<2500
AZ91	Die casting	0.032	50	400
AZ91	Die casting	50 ppm	50	700
AZ91	Die casting	0.032 Mn	50	700
AZ91	Gravity casting	0.032 Mn	10	400
AM60	Die casting	0.021 Mn	30	10
AS41		0.01 Mn	40	200

Source: Song, G. and Atrens, A., *Adv. Eng. Mater.*, 1, 11, 1999.

of nonmetallic inclusions such as oxides and chlorides also degrades corrosion resistance. The critical concentration of chloride for pit initiation for a number of alloys in both the cast and wrought conditions has been reported to be in the range of 2×10^{-3} to 2×10^{-2} M NaCl (Mitovic-Scepanovic and Brigham 1992).

2.6.2.5.2 Galvanic Coupling Corrosion

Galvanic corrosion is and should be a source of serious concern in the use of Mg and Mg alloys since Mg is the most active structural metal. (Recall its position in the EMF and galvanic series.) The origin of the problem is the large difference in potential between magnesium (the active component) and those metals and alloys to which it is coupled (the more noble component). Since it is the most active structural metal, it is effectively always the anode in a galvanic couple. In order to mitigate galvanic corrosion, one needs to eliminate one of the following: (1) the anode (or anodic sites), (2) the cathode

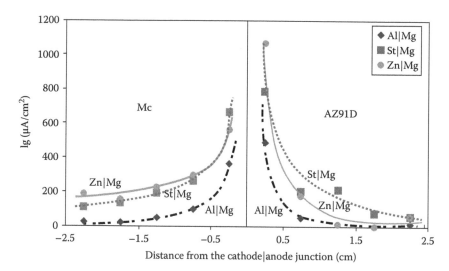

FIGURE 2.13 Galvanic current density as a function of distance from the couple interface for AZ91D coupled to Zn, steel, or aluminum. (From Song, G., *JCSE*, 6, paper C104, 2003.)

(or cathodic sites), (3) the electrolyte, or (4) the electrical connection between the anodic member and the cathodic member in the couple. Steel, galvanized steel, and aluminum are all popular structural materials, and galvanic corrosion of Mg when coupled to these metals is of interest. Figure 2.13 shows galvanic currents for Mg coupled (individually) to each of these metals. In all cases, it is clear to see that the influence of the galvanic couple decreases as a function of distance from the bimetallic interface and that, as expected, the highest current densities (and therefore corrosion rates) would be associated with the coupling of steel and Mg. Some metals that are more galvanically compatible with Mg alloys are 5052 Al, 5056 Al, 6061 Al, and Sn-, Zn-, or Cd-plated ferrous alloys (Eliezer and Alves 2002, p. 275). The relative galvanic compatibility with some common metals and alloys with Mg is presented in Figure 2.14. A number of examples of good design details for mitigating Mg galvanic corrosion

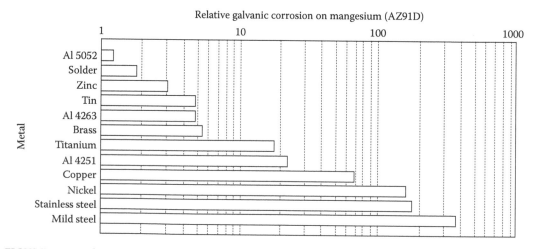

FIGURE 2.14 Relative galvanic compatibility of some metals with Mg. Galvanic assemblies were immersed in a 5% NaCl solution with an anode to cathode area ratio of approximately 1.8. (From Hydro Magnesium, http://www.nanomag.us/Hydro_Mg_Brochure_Corrosion_and_Finishing_of_Mg.pdf, Accessed August 30, 2010.)

can be found in the *Metals Handbook*, Volume 13 B in the chapter on Corrosion of Mg. Designs that specifically avoid entrapment of electrolyte can also be useful in mitigating this type attack.

Methods used to prevent galvanic corrosion of Mg and Mg alloys include careful attention to design and assembly details, especially for fasteners and connectors, coupling only with materials compatible with Mg, and the use of coatings and electrical isolation. Designs also need to allow for good drainage of water and/or condensation and involve the use of nonconductive isolation materials at fastener and connector sites and other couplings with dissimilar metals.

2.6.2.5.3 Breakdown of the Surface Film

D. Eliezer and H. Alves' review chapter discusses the formation of surface films on Mg and Mg alloys under different conditions (Eliezer and Alves 2002, p. 275). In neutral or alkaline salt solutions, corrosion of Mg and Mg alloys involves pitting. Pitting in Mg and *single*-phase Mg alloys (α-alloys) is not quite the same as in stainless steels (Song and Atrens 2007). Pits in single-phase Mg alloys are typically shallower than those on stainless steel and grow laterally to cover the surface. As opposed to the autocatalytic nature of pits in stainless steels, there is evidence that pits in single-phase Mg may be self-limiting due to the increase in pH from the cathodic hydrogen reduction reaction and the increased stability of the Mg surface film at high pH. Pitting in Mg–Al alloys is quite different and typically tied to selective attack that follows the $Mg_{17}Al_{12}$ network, leading to undermining and grain fallout.

Since magnesium corrosion is relatively insensitive to oxygen concentration differences, crevice corrosion has not been reported to be much of an issue (Makar and Kruger 1993). Filiform corrosion is observed on Mg and Mg alloys. In this "worm track" form of corrosion, the head of the track is the anode and the tail of the track is the cathode. This form of corrosion is typically associated with coated Mg alloys, but has been noted on vapor-deposited pure Mg (Wolfe 2005, p. 85) and on uncoated AZ91 (Nisancioglu et al. 1990).

2.6.2.5.4 Concurrent Presence of Stress and a Corrosive Environment

It has been reported that all Mg alloys have some degree of susceptibility to SCC (Busk 1986, p. 256). While there is not much documentation of failures in service, failures in laboratory tests occur at tensile loads less than 50% of yield strength in environments causing very little observable corrosion (ASM Handbook 2005). This is likely because the stresses that have been applied to Mg alloy components in the past were relatively low. With increased use of Mg alloys in automotive applications (where the load-bearing capabilities of the alloys are increasing), some believe that it is likely that more instances of SCC for Mg alloys will rise. An extensive review of SCC in Mg and Mg alloys is available in the paper by Weinzer et al. (2005). This reference contains a lengthy table of SCC tendencies for common Mg alloys. A much shorter table showing some of the alloy + environment susceptibilities that have been documented (and where information regarding SCC resistance is published) is shown in Table 2.12. SCC susceptibility is high for Mg alloys containing more than a threshold value of 0.15%–2.5% Al (Hawke et al. 1999). This tendency for cracking increases with increasing Al, as illustrated in Figure 2.15. Zinc additions also increase SCC susceptibility and not surprisingly AZ alloys (the most commonly used Mg alloys) have a high susceptibility for SCC. Higher-Al-content alloys like AZ61, AZ80, and AZ91 can be quite susceptible to SCC in atmospheric and more severe environments. Figure 2.16 shows tensile stresses causing failure versus exposure time data for AZ91C in a rural atmosphere and highlights the susceptibility of this common alloy in a rather benign environment. Mg–Al–Zn alloys with lower Al concentrations, like AZ31, are generally more resistant to SCC. Mg–Zn alloys containing either Zr or REs and no Al (like nZK60 or ZE10) show an intermediate susceptibility to SCC. For Mg alloys without Al or Zn, the susceptibility to SCC appears to be the lowest.

TABLE 2.12   Environment + Alloy Combination Leading to SCC for Mg Alloys Is Presented in Part A

Alloy	Metallurgical Condition, Environment	SCC Resistance
Part A		
Mg2%Mn	Industrial atmosphere	"For practical purposes not susceptible to SCC"
Mg2%Mn0.24%Ce	Industrial atmosphere	"For practical purposes not susceptible to SCC"
Mg alloys with Al > 1.5%		Magnesium alloy containing more than 1.5% Al is susceptible to SCC. Wrought alloys appear more susceptible than cast alloys. While there is little documentation of service SCC of castings, laboratory tests can cause SCC at tensile loads less than 50% of yield stress in environments causing negligible to low actual stresses in service in the past.
AZ91C AZ31	Industrial atmosphere	"Very slightly susceptible to SCC"
AZ61, AZ91	Industrial atmosphere	"Highly susceptible to SCC"
AZ61	Humid atmosphere	No SCC if RH < 95%
AZ61 sandwiched between sheets of Mg2%Mn	Marine atmosphere	No SCC for an applied stress less than 210 MPa
AZ81 sandwiched between sheets of Mg2%Mn	Atmosphere	No SCC in 200 days for an applied stress less than 150 MPa, while unprotected sheets cracked in a few days
AZ61	Mg alloy in service in atmospheric exposure	All service SCC examined by Dow Chemical Co has been attributed to residual stresses
AZ31, AZ61	Aircraft wing skins in naval aircraft	"Good service experience," few SCC cases, service SCC attributed to residual stress
Part B		
Alloy, Environment	σ_{SCC}	
HP Mg, 0.5%KHF$_2$	60%YS	
Mg2Mn, 0.5%KHF$_2$	50%YS	
MgMnCe, air, 0.001 N NaCl, 0.01 Na$_2$SO$_4$	85%YS	
ZK60A-T5, rural atmosphere	50%YS	
QE22, rural atmosphere	70–80%YS	
HK31, rural atmosphere	70–80%YS	
HM21, rural atmosphere	70–80%YS	
HP AM60, distilled water	40–50%YS	
HP AS41, distilled water	40–50%YS	
AZ31, rural atmosphere	40%YS	
AZ61, costal atmosphere	50%YS	
AZ63-T6, rural atmosphere	60%YS	
HP AZ91, distilled water	40–50%YS	

Source: Winzer, N. et al., *Adv. Eng. Mater.*, 7(8), 659, 2005.
Note: Threshold stress values for some of the combinations listed in part A are presented in part B.

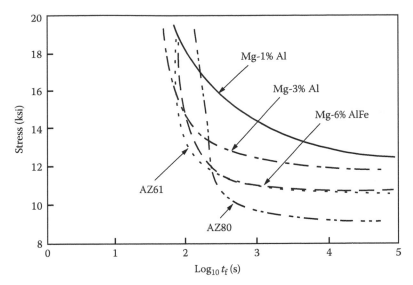

FIGURE 2.15 Stress versus time to failure for magnesium–aluminum alloys in aqueous 40 g/L NaCl + 40g/L Na$_2$CrO$_4$. (From Cramer, S.D. and Covino, B.S., Jr., Eds., *ASM Metals Handbook*, Vol. 13B, ASM International, Novelty, OH, p. 212, 2005.)

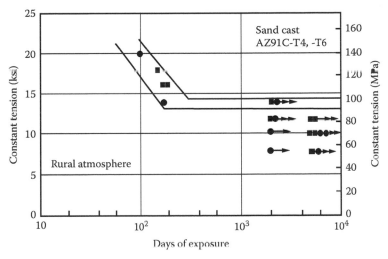

FIGURE 2.16 Stress corrosion of sand-cast AZ91C (T4 and T6) in rural atmosphere. (From Cramer, S.D. and Covino, B.S., Jr., Eds., *ASM Metals Handbook*, Vol. 13B, ASM International, Novelty, OH, p. 212, 2005.)

In general, it has been found that cast alloys appear to be less susceptible than wrought alloys. With increasing use of Mg in load-bearing automotive applications, it has been suggested that more SCC research in Mg alloys is critical (Song and Atrens 2007).

2.7 Degradation of Specific Magnesium-Based Alloys

At this point, it should be obvious that in all Mg alloys, the presence of trace amounts of contaminants plays a key role in the corrosion resistance of that alloy. This includes trace amounts of impurities that are incorporated during the production of the alloy and any metal particles subsequently picked up on the surface of the Mg or Mg alloy.

In general, one can classify commercial Mg alloys into two categories: (1) those containing Al and (2) those without Al (but containing small quantities of Zr to refine grain structures). A large number of Mg alloys are available, and more comprehensive reviews of degradation of specific alloy systems can be found elsewhere. The following is a brief overview of corrosion for a few alloy systems. In the past decade, a large amount of development work has begun on alloy systems that are not yet commercially available to support the increased interest in Mg alloys from the automotive section and new niche markets like biomaterials. The expansion of the Mg market into the biomaterials area will likely require alloys with even better corrosion resistance that of the commercial RE alloys. Corrosion data for some of these novel new Mg alloys are also presented as follows.

2.7.1 Mg–Al Alloys

In general, it is believed that Al improves the corrosion resistance of Mg–Al alloys, but its effect is highly influenced by the morphology and distribution of β-phase (Feliu et al. 2009). For the commercial Al-containing alloys where the microstructure is typically Mg rich, the α-solid solution matrix contains an Al-rich secondary β-phase ($Mg_{17}Al_{12}$). Within the case of single-phase (α only) alloys, increasing the Al content in these alloys increases the corrosion resistance presumably by forming a bit more protective surface film. In two-phase Mg–Al alloys where the β-phase is a very efficient cathode, the corrosion rates of these alloys increase with the content of the β-phase precipitate. Figure 2.17 shows the influence of β-phase on the corrosion of binary, two-phase, Mg–Al alloys. The corrosion rate dramatically increased going from a Mg–1% Al (solid solution alloy with no β) to a Mg–5% alloy (an alloy with some β-precipitates along the grain boundaries), but stayed about the same even when more precipitates were present as in the Mg–10% Al alloy. At high volume fractions, it has been suggested that the β-precipitates provide some barrier protection, thus slowing corrosion of the more anodic α-matrix (Song and Atrens 2003). Other phases also precipitate preferentially at the grain boundaries resulting in accelerating corrosion through microgalvanic coupling with the matrix. This is especially true for Fe-rich phases, like Fe–Al (Song and Atrens 2003). Pardo et al. investigated the corrosion behavior of Mg–Al alloys in 3.5% NaCl at room temperature and concluded that corrosion attack was initiated through the formation of microgalvanic couples at the magnesium matrix/Al–Mn and $Mg_{17}Al_{12}$ intermetallic compound interfaces, followed by formation of a $Mg(OH)_2$, a non-protective layer (Pardo et al. 2008).

2.7.2 Mg–Al–Zn Alloys

Mg–Al–Zn alloys are among the most commonly used Mg alloys, and of these AZ alloys, AZ91 ranks the highest in terms of use. Hillis expanded on the earlier foundational work of Hanawalt by identifying tolerance levels for impurities to include commercial alloys and processing, and this resulted in

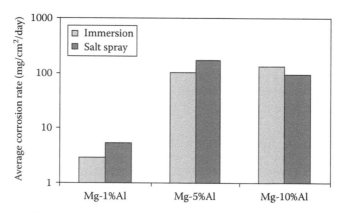

FIGURE 2.17   Influence of β-phase in Mg–Al alloys on corrosion rate. (From Song, G., *JCSE*, 6, paper C104, 2003.)

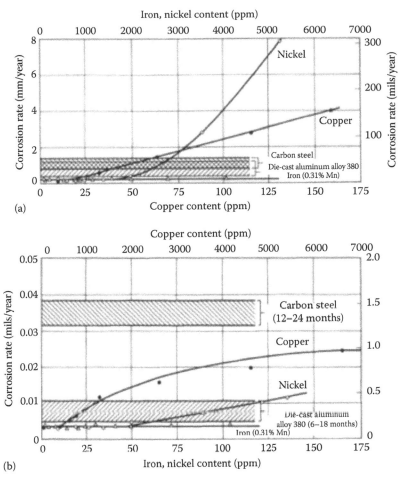

FIGURE 2.18 Corrosion rate of AZ91 as a function of impurity concentration (in ppm). (From *ASM Metals Handbook*, Cramer, S.D. and Covino, B.S., Jr., Eds., Vol. 13B, ASM International, Novelty, OH, p. 208, 2005.)

establishing tolerance limits for impurities in AZ91 (Hillis 1983). Figure 2.18 shows the dependence of corrosion rate on Cu, Fe, and Ni content as a function of impurity concentration in AZ91.

When Zn is added to Mg–Al alloys, the roles that the second phase constituents, especially the β-phase, play in corrosion of the alloy change. The improved corrosion resistance of alloys AZ80 and AZ91 has been attributed to the formation of semi-protective Al-rich oxide layer on the alloy's surface and to the presence of a fine β-phase network that acts as a barrier to progression of the corrosion damage. Ballerini et al. studied the corrosion of AZ91D in salt spray ASTM B117 tests and also confirmed that higher corrosion resistance of finer microstructure was due to the higher amount of aluminum-rich β-phase that was observed on the alloy surface after the test (Ballerini et al. 2005).

Microstructure also plays an important role in the biodegradation behavior of magnesium–aluminum alloys used under physiological conditions. Wen et al. investigated the corrosion behavior of AZ91D, AZ91, AZ61, and pure Mg through immersion tests in modified simulated body fluids (SBFs). The results showed that the microstructure and the Al content in the α-Mg (Al) matrix significantly affected their corrosion properties. Alloy AZ91D exhibited the highest corrosion resistance that was attributed to the uniform and continuous distribution of β-precipitates along the grain boundaries in this alloy (Wen et al. 2009). On the other hand, Kannan (2010) reported that the microstructural difference between the fine-grained die-cast and coarse-grained sand-cast AZ91 alloys had no significant effect on their degradation.

However, he suggested that the sand-cast AZ91 alloy is better suited for bioapplications because the high volume fraction of β-precipitates in die-cast alloys is more stable in SBF, and the fast dissolution of fine grains in that alloy leaves behind those precipitates that compromise the mechanical stability of the alloy.

The corrosion resistance of AZ alloys in chloride-containing solutions depends on pH and Cl concentration. In general, Mg and its alloys corrode at low rates in alkaline or poorly buffered sodium chloride solutions; but the corrosion rate increases with higher chloride ion concentrations (Altun and Sen 2004). Wang et al. studied the influence of chloride, sulfate, and bicarbonate anions on the corrosion behavior of AZ31 alloy. The results of this study are presented in the form of "corrosion maps." It was shown that AZ31 underwent passivation and corrosion in various solutions; the passive behavior is strongly influenced by the type of anions present. The broadest passivation zone was obtained in Na_2SO_4 solution; however the lowest corrosion rates were recorded in $NaHCO_3$ (Wang et al. 2010).

2.7.3 Yttrium and Rare Earth-Containing Alloys

With the addition of Y and RE elements to Mg and Mg alloys, corrosion resistance is further improved. In single-phase Mg–Y alloys, Y enhances the protective nature of the surface film. In commercial alloys with much more complicated microstructures, the Y still enhances the protective nature of the film, but the $Mg_{24}Y_5$ intermetallic degrades corrosion resistance due to microgalvanic cells acting between the intermetallic particles and the matrix phase. Which of the two processes wins out in this competition depends on the electrolyte (Liu et al. 2010). The RE additions have a beneficial influence on creep resistance, age-hardening response, and castability and are also believed to play a role in enhancing the protective capabilities of the surface film. In addition, the electrochemical potentials of the RE intermetallics are close to that of Mg and, as a result, are not likely to initiate the action of microgalvanic cells within the alloy (Eliezer and Alves 2002). RE additions also assist in removing Fe during the processing of molten Mg. In a study of binary Mg–RE additions, Birbilis et al. found that the volume fraction of the of intermetallic phases increased with increasing amount of either Ce, La, or Nd in the alloy (Birbilis et al. 2009). These RE additions increased the strength of the alloys, but as the volume fractions of intermetallic phases increased, so did corrosion rate.

2.7.4 WE43

WE43 is often reported to be one of the most corrosion-resistant commercial Mg alloys. With alloying addition of Y, RE, and Zr, its producer reports a corrosion rate of 0.1–0.2 mg/cm²/day (10 mpy) as measured in ASTM B117 (salt fog cabinet testing) (Elektron WE43). For the wrought alloy similar corrosion rates during ASTM B117 exposures were reported, and rates of 0.1 and 0.023 mg/cm²/day were cited for seawater immersion and intermittent salt spray, respectively. This alloy maintains good long-term mechanical properties at temperatures up to 250°C and thus finds applications in the aerospace industry, on high-performance cars, and possibly as biomedical implants. Others have reported similar corrosion rates for WE43 after immersion in 3.5% NaCl for several days (Rzychon et al. 2007). In many applications, this corrosion rate for WE43 is high enough to require that surface treatments and/or coatings be applied prior to use. Environmentally assisted cracking in cast WE43-T6 has been reported to initiate in atmospheric air at brittle intermetallic phases and propagate transgranularly, in a manner similar to that observed in other high-strength Mg alloys.

2.7.5 EV31

EV31 (formerly Elektron 21) is a newer Mg alloy with Gd, Nd, RE, Zr, and Zn alloying additions. Like WE43, this alloy has good creep properties up to 250°C and has applications similar to that of WE43, but its production costs are lower. As a result, it is gaining widespread acceptance for aerospace applications. The corrosion rates for EV31 are a little higher than those for WE43 with its producer reporting

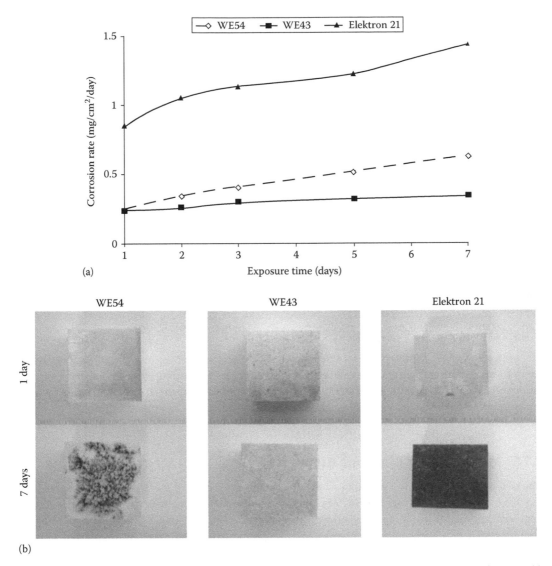

FIGURE 2.19 Corrosion rates for EV31, WE42, and WE54 after immersion testing in 3.5% NaCl are shown in (a), and images of representative specimens after testing are shown in (b). (From Rzychon, T. et al., *JAMME*, 21(1), 51, 2007.)

0.13–0.37 mg/cm²/day (10–30 mpy) for ASTM B117 exposures. A comparison of the corrosion resistance of EV31, WE43, and WE54 after immersion in 3.5% NaCl for 7 days is presented in Figure 2.19. An extensive discussion of the influence of alloying additions on the properties of EV31 has been prepared by Lyon et al. (2005). The structural evolution of EV31 as a function of aging times is also available in the literature (Riontiono et al. 2008). In many applications, especially aerospace applications, EV31 is typically coated prior to use.

2.7.6 Novel Mg Alloys and Composites

2.7.6.1 Nonequilibrium Vapor-Deposited Mg Alloys

In order to produce Mg alloys with further improvements in corrosion resistance, newer production approaches are need. One such approach is vapor deposition. In this approach the metals are condensed from vapors, rather than being solidified from liquids; so dramatically different types of Mg alloys can be produced. Alloy combinations like Mg–Ti, where several weight percent Ti are in solid solution with Mg, that is not possible with conventional processing can be made. This allows one to make alloys with better enhancements to passivity. Recall the Pourbaix diagram for passivity presented earlier, and envision how passivity could perhaps be enhanced over a broader range of pH values by superimposing this diagram on that of Mg. Vapor deposition is now used in large-scale production of metalized food packages and in the microelectronic industry and could be tailored to produce thicker alloys with highly tailorable structures and chemistries for other industries. Dodd and Gardiner (1996) explored the use of vapor deposition in the 1990s for the production of vapor-deposited Al and Mg alloys. Today others are exploring the use of vapor deposition to make nonequilibrium Mg–Y and Mg–Ti, Mg–Y–Ti, and other Mg-based alloys with significantly enhanced corrosion resistance (Shaw et al. 2008). Through the incorporation of passivity-enhancing elements like Ti and Y into the alloy and into its surface film, more passive alloys are produced. Figure 2.20 illustrates the galvanic corrosion benefits of producing a Mg alloy with a higher open-circuit potential and more passive behavior (Wolfe 2005). In this figure, the value of the crossover point for the anodic and cathodic curves is proportional to the rate of galvanic corrosion. In these vapor-deposited alloys, that crossover point is significantly lower than it is in the commercial WE43 alloy. While these vapor-deposited materials are still in the development stages, they hold great promise for future niche applications like biomedical implants.

2.7.6.2 Metal Matrix Composites

The evaluation of Mg MMCs has been underway for the past couple of decades and is being explored as a way to reduce the coefficient of thermal expansion and improve the strength, modulus, and wear resistance at both room temperature and elevated temperatures. Reinforcements like SiC, boron, and graphite are of increasing interest. Special care needs to be used in the design of and manufacturing of composites because of the galvanic issues between Mg and some of the reinforcing materials (Chan et al. 1998).

2.7.7 Mg Alloys for Bioapplications

The task of designing or even selecting a Mg alloy suitable for a desired application is even more daunting when considering the case of using Mg alloys for biodegradable implants. The purpose of using biodegradable implants is to support healing and regeneration of tissue with a material that subsequently degrades away or, in other words, corrodes in a complex environment. One has to take into account a relationship between that material and the response of the human body: If the presence of that implant induces local inflammation, the products of that inflammation can also affect the degradation process.

While considering corrosion of Mg and its alloys in the human body, the influence of aggressive ions and large biomolecules must be taken into account. The 3.5% or 5% NaCl solutions used in conventional accelerated assessments of the corrosion resistance of Mg alloys is too high to mimic the concentration of chlorides in the body fluids; however, even lower NaCl concentrations can still influence the stability of Mg (Mueller et al. 2009). The presence of sulfates, phosphates, and carbonates, usually considered as beneficial to lowering corrosion of Mg, may have an opposite effect in the human body (Xin et al. 2008). In addition, there are conflicting results regarding the role of amino acids and proteins on stability of Mg implants.

Mueller et al. reviewed results from different corrosion studies of Mg and Mg alloys used for bioapplications and concluded that the enormous variability of data found in the literature is due to the different experimental conditions; thus he points out that only the "in vivo" data should be relied upon for bioapplications (Mueller et al. 2010).

Last but not least, the question arises about what selection rules should be used to identify the right magnesium alloy for bioapplications. The alloys used so far (AZ61, AZ91, WE43) were designed for other unrelated technical applications. There are many reasons one could argue whether those alloys

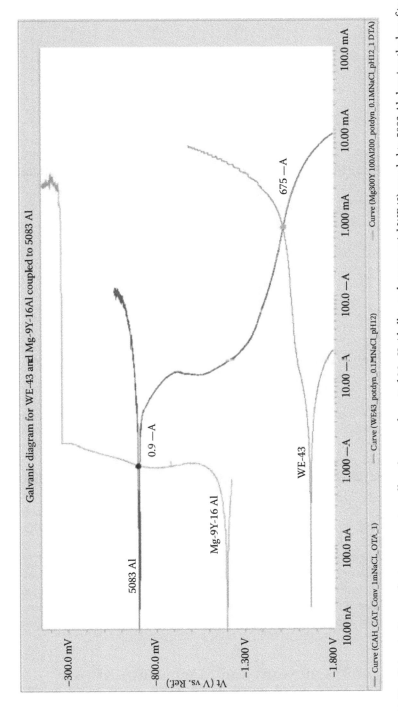

FIGURE 2.20 Galvanic diagram for two magnesium alloys (a vapor-deposited Mg–Y–Al alloy and commercial WE43) coupled to 5083 Al showing the benefits of passive Mg on galvanic corrosion. (From Wolfe, R.C., Imparting passivity to vapor-deposited magnesium alloys, PhD dissertation, The Pennsylvania State University, University Park, PA, 2005.)

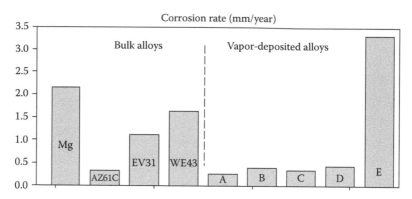

FIGURE 2.21 The corrosion rates calculated from linear polarization measurements for bulk and thin film Mg-based alloys in HBSS at 37C. For the vapor-deposited alloys: (A) Mg–Ti, (B–D) Mg–Y–Ti, and (E) Mg–Y.

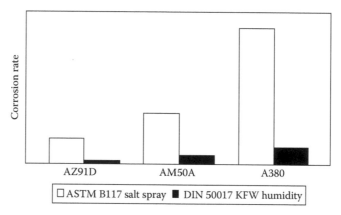

FIGURE 2.22 Comparison of ASTM B117 and DIN humidity cabinet test results. (From Hydro Magnesium, http://www.nanomag.us/Hydro_Mg_Brochure_Corrosion_and_Finishing_of_Mg.pdf, Accessed August 30, 2010.)

are the right choice (toxicity of alloying elements, mismatched mechanical properties between these commercial alloys, and the biomaterials with which they will interface). The use of these previously developed alloys for bioapplications has already identified that the corrosion rates of these alloys are still too high, so that further refinements in alloys (or coatings) or even totally new alloys may be needed. To date, some of the most promising Mg alloys from a bioapplication perspective are vapor-deposited Mg–Ti and Mg–Y–Ti alloys. Their electrochemical behavior characteristics in Hanks' solution (HBSS) at 37°C have been compared to some commercially available Mg alloys as shown in Figure 2.21. WE43 and EV31 are not currently used for bone plates and screws, but there is interest in Europe for using WE43 for stents and orthopedic applications. By varying not only the composition but to some extent, the structure and morphology of the alloy, one can obtain alloys with a significant range in their dissolution rates. Among commercial alloys shown in Figure 2.22, the lowest corrosion rate is shown by AZ61C; however, the presence of Al is usually not desired from a biological perspective.

2.8 Protection of Magnesium and Magnesium Alloys

In order to protect surfaces of Mg and its alloys from the detrimental effects of aggressive environments, several coating techniques have been developed. The most frequently used industrial surface treatments for corrosion protection include anodizing, conversion coatings (CCs), and electroplating. Other methods, designed to improve, along with corrosion resistance, surface properties of Mg and its alloys (albeit well developed and studied), are not widely used commercially. Among these are thermal spray, PVD, and polymer coatings.

Depending on final application, the requirements for coatings properties may vary, but ideally, the coatings on Mg should be strong and thick enough to be resistant to scratching and flexible to allow for deformation by the substrate without cracking of that coating. It is essential for the coating to have sufficient adhesion to the substrate, have a high hardness, be environmentally friendly, and possess good wear, corrosion, and fatigue resistance properties. To make matters even more complicated, most processes involving deposition of coatings on Mg are carried out at relatively high temperatures. This in turn may lead to annealing of the not thermally stable Mg and its alloys which could result in degradation of their mechanical properties. Another set of problems arise for coatings specifically designed for biomedical applications, since these coatings must be compatible with and not toxic or detrimental to the human body; thus the use of any biomedical coating technology containing Cr, Al, or some RE metals must be prohibited.

2.8.1 Anodizing

Anodizing is a well-established technique (an exhausting and excellent review of the whole spectrum of anodizing treatments was written by Blawert et al. 2006) used to protect surfaces of light metals. This process electrolytically forms a thick and stable oxide (anodized) layer on the substrate surface by applying either voltage or current (DC or AC). Anodizing produces an adherent, hard, thermally stable layer that is also heat, wear, and corrosion resistant. Furthermore, this oxide layer is also electrically insulating, an especially important fact for protecting Mg, a very electrochemically active metal. It is important to point out that the resulting layer is very porous and must have further posttreatment to seal these pores.

The oxide film formed on a Mg surface depends on the anodizing potential. As the voltage increases, a porous oxide film is formed. With intermediate potentials the electric field strength reaches the critical value beyond which the film is broken due to tunneling ionization. Small, luminescent sparks are observed, moving rapidly across the oxide surface; large-arc discharges arise as the voltage is further increased; the thermal ionization is partially blocked by the buildup of negative charge in the thickening oxide film. This constitutes the so-called microarcing effect where the film is gradually fused and alloyed with elements present in the electrolyte. At still higher voltages these microarcs transform into powerful arcs that can destroy the substrate. Currently, most of the anodizing processes take place in the intermediate-voltage spark discharge region.

Plasma electrolytic oxidation (PEO) is another form of anodizing. It differs from the classical one by the application of very high voltages (above 100 V); as a result, a plasma is created locally. On Mg, the high temperatures reached within the plasma enable the formation of a dry ceramic surface layer of Mg oxides. One of the major advantages of this process is its high throwing power and use of environmentally friendly electrolytes, especially for current commercial PEO processes such as Keronite, Anomag, Magoxid, or Algan. The thickness and the morphology of anodic layers formed by PEO depend on electrolyte chemistry (composition, concentration) and on applied current density (Cao et al. 2007, Arrabal et al. 2008). Barchiche et al. studied the corrosion resistance of plasma-anodized AZ91D in 3 M KOH with 0.5 M KF and 0.25 M $Na_3PO_4 \cdot 12H_2O$; by applying low current densities (5 mA/cm^2) and longer anodizing times (30 min), a doubly beneficial electrochemical effect was observed: a large inhibition of the cathodic processes and the formation of a very wide, stable passive region. The authors claim that those favorable features were due to decreased porosity in the film formed under those conditions (Barchiche et al. 2007). Arrabal et al. investigated the PEO of various Mg alloys in a silicate/phosphate electrolyte using AC current. This very detailed research revealed the mechanism and parameters for the growth rate of the anodic layer; it was also found that regardless of the substrate, the coatings consisted of MgO and Mg_2SiO_4 with incorporation of alloying elements. Electrolyte species were also present, mainly in a porous layer at the coating surface. The corrosion rate of PEO-treated alloys was reduced by 2–4 orders of magnitude (Arrabal et al. 2009).

Mg and its alloys can be anodized in almost any solution that does not cause Mg and the coating to dissolve faster than it is formed. However, the use of an environmentally friendly electrolyte is strongly encouraged, and that influences restrictions on the application of some industrial anodizing processes. In the following text we present the most prominent, commercially available anodizing treatments (Blawert et al. 2006).

2.8.1.1 Dow-17

Introduced by Dow Chemical Company almost 50 years ago, it uses an electrolyte composed of sodium chromate, fluoride, and phosphoric acids (pH5) at or above 160°F where the voltage is not supposed to exceed 100 V. The thickness of the coating can vary between 5 and 75 μm; the color, depending on coating thickness, ranges from light to dark green. The composition of the film is mainly MgF_2, $NaMgF_3$, $Mg_{x+y/2}O_x(OH)_y$ with small amounts of Cr_2O_3. However, the presence of chromates in the electrolyte greatly impacts the safety of the whole process.

2.8.1.2 HAE

Invented by Harry A. Evangelides is an anodizing process that takes place in an electrolyte containing potassium permanganate, potassium fluoride, trisodium phosphate, potassium hydroxide, and aluminum hydroxide (pH 14) at 20°C–30°C. The thickness varies from 5 to 75 μm and the color, depending on thickness, ranges from light tan to dark brown.

2.8.1.3 Anomag

This process was developed by Magnesium Technology in New Zealand; the process produces smooth layers and possesses an exceptional throwing power. The environmentally friendly electrolyte contains an aqueous solution of ammonia and sodium ammonium phosphate; the coating's thickness can vary between 5 and 25 μm.

2.8.1.4 Magoxid-Coat

This process was conceived in Russia and further developed in Germany. In this PEO process, the discharge and formation of an oxygen plasma in the electrolyte results in short-lived melting of some parts at the substrate surface that in turn results in the formation of an oxide–ceramic layer. The low-alkali aqueous electrolyte contains borate or sulfate ions, phosphate, and fluoride or chloride ions at a preferred pH of 8–9. Anodizing is carried out with direct currents of 1–2 A/dm² and voltages up to 400 V. The coating reaches thickness of 15–20 μm; its color varies from white to light gray but can be black if the electrolyte contains appropriate additives.

2.8.1.5 Tagnite

This process was developed in the Unites States in the 1990s by Technology Applications Group. The alkaline electrolyte (without chromium or heavy metals) operates at ambient temperature and consists of an aqueous solution with hydroxide and fluorides and silicates. The coating is applied under high voltages above 300 V; its thickness varies between 2 and 23 μm.

2.8.1.6 Keronite

This technology started in Russia and was further developed in the United Kingdom by Isle Coat. The novelty of this PEO process is in the application of electrically bipolar (positive and negative) pulsed electrical currents of a specific waveform in a nonhazardous, low-concentration alkaline solution. The coating formed is mainly composed of $MgAl_2O_4$ together with SiO_2 and SiP; its thickness can vary between 10 and 80 μm. The Keronite coating can be further impregnated with polymers, metals, and ceramics to create surfaces with other desired properties (better corrosion and wear resistance).

2.8.2 Conversion Coatings

Another very popular technology for coating Mg and its alloys is the chemical or electrochemical treatment of their surfaces to form a chemically bonded layer of substrate metal oxides, chromates, phosphates, or other compounds. The most popular among them are chromate CCs, developed by the DOW Corporation, which has received extensive attention because of its excellent corrosion-resistant films and the simplicity of the coating process. The mechanism of protection offered by these chromate CCs is due to the dissolution of the metal surface and resulting increase in pH close to metal/electrolyte interface. The increase in pH results in precipitation on the substrate surface of a thin complex chromium metal–gel containing hexavalent and trivalent chromium compounds (Grey and Luan 2002). However, due to the toxicity of the treatment solution (chromium compounds being considered carcinogenic), the use of these chromate CCs is being restricted. The search for Cr^{+6}-free surface treatments for Mg and its alloys has led to the development of more environmentally friendly CC treatments like permanganate, phosphate–permanganate, stannate, and RE metal baths (Zheng and Liang 2007, Zhang et al. 2009). The properties of CCs are closely related to their microstructures and compositions that, in turn, strongly depend on the composition of the Mg-based alloys, the pretreatment surface cleaning process, the type and composition of the solution, and the corresponding operating parameters such as solution temperature and type and degree of agitation. When treated in stannate baths, the surface of the Mg alloys is covered with an almost continuous protective layer of crystalline $MgSnO_3 \cdot 3H_2O$ or $MgSnO_3 \cdot H_2O$ and/or magnesium–tin oxides or mixed $MgSn(OH)_6$ hydroxides (Lin et al. 2006). On the other hand, the permanganate–phosphate coatings were found to be characterized by a network of cracks, associated with hydrogen release or CC dehydration effects (Umahara et al. 2003). The composition of the permanganate–phosphate coating naturally depended on the composition of the Mg alloy and the treatment bath, consisting of either phosphates of Mg with oxides or hydroxyoxides of Mn and Al.

Because of the similarity of permanganate to chromate, there is a possibility for replacing the hazardous chromate treatments with environmentally friendly permanganate. Zheng reported that the CC formed on AZ91 in permanganate–rare earth metal salt (REMS) bath showed better corrosion resistance than that obtained in chromate-based treatment (Zheng and Liang 2007). It was also found that permanganate–phosphate treatments may lead to a coating at least equivalent or more protective than the chromate treatment (Umahara et al. 2003).

Recently, the effect of REMSs on corrosion of Mg has been investigated (Takenaka et al. 2007, 2008). The CCs that formed on Mg alloys AZ31, AZ61, AZ91, and AM60 by immersion in a solution with $Mg(NO_3)_2$ and $La(NO_3)_3$ had corrosion resistances that were significantly improved. Also Rudd et al. studied the corrosion protection of pure Mg and alloy WE43 after treatment in RE metal solutions and especially cerium salts in an 8.5 pH buffered solution (Rudd et al. 2000). It is believed that cerium salts inhibited the dissolution of Mg and its alloys by formation and precipitation of cerium oxide or hydroxide over cathodic sites on the metal surface and that these in turn led to a decrease in the rates of the reduction reaction. Following this interest in the inhibiting properties of cerium salts, a new, chemical CC treatment based on an aqueous solution of cerium, zirconium, and niobium compounds at pH 4 was designed (Ardelean et al. 2008). The new coating significantly improves the corrosion resistance of AZ91 and AM50 magnesium alloys.

In an increased effort to find more environmentally friendly (or so-called natural) compounds, Frignani et al. studied CCs on AZ31 alloys formed by immersion in aqueous solutions of sodium salts of monocarboxylic acids (Frignani et al. 2010). It is believed that the anions of the monocarboxylic acids should bind to the oxide-covered metal surfaces and form highly protective, hydrophobic layers (Liu et al. 2006). Indeed, the monocarboxylic acids inhibited corrosion of Mg alloys in saline solutions and hindered corrosion of AZ31 alloys in synthetic cooling water and also offered excellent protection of that alloy for very long times (over 800 h) in diluted sulfate solutions. Another new, pollution-free protective coating for Mg is the CC formed in phytic acid (Cui et al. 2008). Phytic acid ($C_6H_{18}O_{24}P_6$), extracted from grains, has capabilities of chelating with many metal ions. Although it is mainly used as corrosion inhibitor in paints, it was found that when it was employed as the main reactant to form CCs on AZ91 alloys, it significantly reduced the rate of corrosion in 3.5% NaCl aqueous solutions.

2.8.3 Electroplating

Electroplating is a relatively economical and convenient method for altering surface properties of metals. In this process a metal salt in solution is reduced to its metallic form on the substrate surface. There are two types of plating process: electroplating, where electrons for the reduction reaction are supplied from an external source, and electroless (chemical) plating where electrons are supplied by a reducing agent in the solution or by the substrate itself (Grey and Luan 2002). Some applications for Mg alloys plated with Cu–Ni–Cr are limited to use in interior and some mild exterior environments since these coatings do not have a very high corrosion resistance; magnesium plated with Ni has found applications in the computer and electronic industries, and Ni–Au coatings on Mg have been used in space applications due to their good optical reflectance properties.

Because Mg alloys are very active electrochemically, it is very difficult to obtain a good protective coating by electroplating in aqueous plating solutions. Typically the surface of Mg and its alloys is usually covered with oxides or hydroxides that form in the air or upon immersion in water. The low electrical conductivity and bonding ability of this oxide/hydroxide layer further impedes efficient electroplating of Mg surfaces. Another problem arises from the inhomogeneous microstructure of Mg alloys that can produce uneven coatings. For example, in nickel plating it is believed that nickel initially nucleates on the β-phase ($Mg_{17}Al_{12}$) and then spreads over the eutectic and primary α-(magnesium solid solution) phases. The metal coating on Mg must be pore-free due to the significant potential difference between the Mg substrate and its plated metal coating (Huang et al. 2008). And of course the whole electroplating process has to be environmentally friendly. This is especially challenging for electroless deposition where plating solutions have a limited bath life and low regeneration rate.

It is clear then that one of the most difficult parts in plating Mg is to select a suitable pretreatment process that properly prepares the surface of Mg for coating. Many pretreatments for preparing the surface of the Mg alloy have been proposed, but only two processes (Zhang et al. 2009) have been successfully used to produce desirable property metal coatings on Mg alloys. These pretreatment routes are either of the following:

1. Degrease→ pickling → activation → zinc immersion → cyanide–copper plating → common plating
2. Degrease → pickling → activation → electroless nickel plating → common plating

Both these use either cyanide–copper plating or electroless nickel plating as the pre-plating step, but both have negative implications since cyanides are toxic to humans and can pollute the environment, whereas the nickel-plating bath process has a very high production cost.

CCs can provide a stable base for further treatments and in fact have been used as a pretreatment for electroplating processes. For example, a stannate CC can effectively replace chromate coating, is environmentally friendly, and after proper activation can be employed as a pretreatment process for electrodeposition of zinc. The experimental results showed that this procedure resulted in a greatly improved corrosion-resistant coating on AZ91D magnesium alloy (Zhang et al. 2010).

To address some of the previously mentioned problems, a new trend in electroless plating of Mg alloys has emerged: composite plating. It is accomplished by the addition of solid particles (SiO_2, CeO_2, ZrO_2, Al2O3, diamond) into an electroless plating bath. These solid particles deposit concurrently with the metal coating and improve its properties. It was found that when Mg alloy AZ91D was plated with a Ni–P–ZrO_2 composite coating (Song et al. 2007, 2008), it had much better corrosion resistance than the monolithic Ni–P coating. Another way to improve the corrosion resistance of the electroless Ni–P layer is to process it as a multilayer coating. It can be especially effective if each layer has a different phosphorous content and therefore has different electrochemical properties since E_{corr} of a Ni–P alloy coating becomes more positive with increasing P content. When this type of coating was deposited on AZ91D (Gu et al. 2005), the composition of the final coating varied from the substrate to surface. Where Ni–P deposited by direct electroless, Ni–P plating formed the first layer, followed by a Ni–9.5% P layer and then Ni–5.4% P as the outer layer. The electrochemical evaluation results revealed that for this hybrid coating, the surface layer would be corroded preferentially and so protects the substrate.

2.8.4 Organic/Polymer Coatings

Organic coatings may include a number of different processes designed to enhance corrosion resistance and wear properties or simply improve the cosmetic appearance. Organic coating application systems can use painting, powder coating, polymer plating, or the sol–gel process. One effective surface finish was obtained by *polymer plating* where triazine disulfide polymers were deposited by electrochemical polymerization on the surface of Mg alloys (Kang et al. 2005). This polymer-plated AZ91 Mg alloy exhibited excellent corrosion resistance. The *sol–gel process* consists of the simultaneous hydrolysis and condensation polymerization of metal alkoxides; it seems promising for preparing adherent, chemically inert metal oxide coatings at low temperatures (below 200°C). Sol–gel coatings consisting of CeO_2 components for corrosion protection and ZrO_2 for increased wear resistance were deposited on AZ91 and AZ31 with different surface finishes. The resulting coatings exhibited superior corrosion resistance when applied over machined surfaces, but severe corrosion was observed when applied on rolled and sand-blasted surfaces (Phani et al. 2005). Recently, highly effective composite CeO_2/TiO_2 thin films were successfully deposited on the surface of AZ91D magnesium alloy with a sol–gel process. Since Mg alloys cannot withstand the relatively low-pH TiO_2 sol–gel, CeO_2 thin films were first deposited as the inner layer (Fan et al. 2009). This composite coating exhibited excellent corrosion resistance and hydrophilicity.

2.8.5 Thermal Spray

Protective coatings can be produced from the gas/solid phase (as opposed to those discussed earlier, which were deposited from wet phase). In thermal spray methods, the coating material (metal, polymer, or ceramic particulates) is preheated to almost its melting point and then sprayed using a nozzle with an auxiliary gas phase directed onto the substrate without significantly heating the substrate. Although these coatings are relatively easy to deposit, the cost of the equipment is rather high, and the coating usually requires an additional sealing step after deposition due to the inherent porosity of the thermal spray coating. Another way to improve the functionality of thermal spray coatings is with a thermal (Pokhmurska et al. 2008) or mechanical posttreatment. Cold-pressing treatments applied to Al–11Si thermally sprayed coatings on Mg–Al alloys improved their homogeneity and adhesion, significantly reduced porosity, and increased surface hardness, but the coating still exhibited significant galvanic corrosion (Pardo et al. 2009).

2.8.6 Physical Vapor Deposition

The PVD process involves deposition of atoms or molecules from the vapor phase. In general, PVD techniques are well suited to produce hard and smooth surfaces for wear and corrosion protection, but this deposition process on Mg and its alloys still faces some challenges. Specifically, during deposition the Mg substrate temperature has to be kept below 180°C where the existing Mg microstructure remains thermally stable and there is also some concern about maintaining appropriate substrate/coating adhesion. Notably good results were obtained while vapor depositing pure Mg onto the surfaces of Mg alloys (Yamamoto et al. 2001).

2.9 Specific Standardized Testing of Magnesium and Magnesium Alloys

2.9.1 Cabinet Testing

ASTM B117 (salt fog testing) is an old but still commonly used cabinet test for assessing resistance to corrosion in chloride environments. In the automotive industry, this test is now being replaced by cabinet tests with environments of more representatives of those encountered by automobiles, such as the GM9540 or the SAE J2334 protocols. Large differences in the measured accelerated corrosion rates of Mg between ASTM B117 and other cabinet tests have been noted; see Figure 2.22 (Hydro Magnesium 2010).

2.9.2 Other Common ASTM Tests

ASTM G 44 (SCC resistance) is a commonly used method for testing the SCC susceptibility of metallic alloys including Mg alloys.

Electrochemical corrosion test methods like G-5 (potentiostatic and potentiodynamic anodic polarization), G-59 (potentiodynamic polarization resistance measurements), and G106 (EIS) and G102 (calculating corrosion rates from electrochemical measurements) are commonly used to evaluate corrosion rates for Mg alloys.

A few ASTM methods are commonly used for testing bioimplant alloys (ASTM F2129 and F746). However, these methods are meant for testing corrosion-resistant materials and not necessarily Mg or Mg alloys. As a result, these methods are not usually applicable for testing of Mg alloys designed to corrode in the human body where body fluids have compositions significantly different from the conventionally used testing solutions.

2.10 Summary

Mg and its alloys are seeing increased consumption since there is increased interest in expanding their use in the transportation sector, for use as containers in portable electronics, and in the growing niche market of bioimplants. This chapter has identified that the corrosion of Mg is significantly influenced by impurities, galvanic coupling, and design details. The effective use of Mg alloys often requires the use of coatings and surface treatments. Novel surface coatings and alloys using nonconventional processing show promise for producing alloys with enhanced corrosion resistance.

Acknowledgments

Ryan Wolfe, John Petrilli, Sean Pursel, and Zi-Kui Liu.

References

Aizawa, T., T. Kuji, and H. Nakano. 2001. Non-equilibration of nanostructured Mg_2Ni by bulk mechanical alloying, *Materials Transactions*, 42, 1284–1292.

Altun, H. and S. Sen. 2004. Studies on the influence of chloride ion concentration and pH on the corrosion and electrochemical behavior of AZ63 magnesium alloy, *Materials & Design*, 25, 637–643.

Antrekowitsch, H., G. Hanko, and P. Ebner. 2002. Recycling of different types of magnesium scrap, in *Proceedings of the Magnesium Technology 2002*, TMS, Warrendale, PA, 2002.

Ardelean, H., I. Frateur, and P. Marcus. 2008. Corrosion protection of magnesium alloys by cerium, zirconium and niobium-based conversion coatings, *Corrosion Science*, 50, 1907–1918.

Arlhac, J.M. and J.C. Chaize. 1997. New magnesium alloys and protections in new helicopters, in G.W. Lorimer, Ed., *Proceedings of the 3rd International Magnesium Conference*, Manchester, U.K., Institute of Materials, London, U.K., pp. 213–230.

Arrabal, R., E. Matykina, T. Hashimoto et al. 2002. Characterization of AC PEO coatings on magnesium alloys, *Surface and Coatings Technology*, 203, 2207–2220.

Arrabal, R., E. Matykina, F. Viejo et al. 2009. Corrosion resistance of WE43 and AZ91D magnesium alloys with phosphate PEO coatings, *Corrosion Science*, 50, 1744–1752.

Arroyave R. and Z.-K. Liu. 2006. Intermetallics in the Mg-Ca-Sn ternary system: Structural, vibrational and thermodynamic properties from first principle, *Physical Review B*, 74, 174118.

ASM Metals Handbook, Vol. 13B. 2005. Cramer, S.D. and B.S. Covino, Jr., Eds., ASM International, Novelty, OH.

ASM Specialty Handbook: Mg and Mg Alloys. 1999. Avedesian, M.M. and H. Baker, Eds., ASM International, Novelty, OH.

Au, M. 2005. Hydrogen storage properties of magnesium based nanostructured composite materials, *Materials Science and Engineering B*, 117, 37–44.

Ballerini, G., U. Bardi, R. Bignucolo et al. 2005. About some corrosion mechanics of AZ91D magnesium alloy, *Corrosion Science*, 47, 2173–2184.

Barchiche, C.-E., E. Rocca, C. Juers et al. 2007. Corrosion resistance of plasma anodized AZ91D magnesium alloy by electrochemical methods, *Electrochimica Acta*, 53, 417–425.

Birbilis, N., M.A. Easton, A.D. Sudholz et al. 2009. On the corrosion of binary magnesium-rare earth alloys, *Corrosion Science*, 51(3), 683–689.

Blawert, C., W. Dietzel, E. Ghali, and G. Song. 2006. Anodizing treatments for magnesium alloys and their effect on corrosion resistance in various environments, *Advanced Engineering Materials*, 8, 511–533.

Busk, R.S. 1986. *Magnesium Products Design*, Marcel Dekker, New York, p. 256.

Cao, F.H., J.L. Cao, Z. Zhang et al. 2007. Plasma electrolytic oxidation of AZ91D magnesium alloy with different additives and its corrosion behavior, *Materials und Corrosion*, 58, 696–703.

Chan, W.M., F.T. Cheng, Leung, L.K. et al. 1998. Corrosion behavior of magnesium alloy AZ91 and its MMC in NaCl solution, *Corrosion Reviews*, 16(1–2), 43–52.

Critical Strategic Metals. 2010. http://www.criticalstrategicmetals.com/wp-content/uploads/2010/06/CSM-June_2010.pdf, Accessed August 30, 2010.

Cui, X., Q. Li, Y. Li et al. 2008. Microstructure and corrosion resistance of phytic acid conversion coatings for magnesium alloy, *Applied Surface Science*, 255, 2098–2103.

David, E. 2005. An overview of advanced materials for hydrogen storage, *Journal of Materials Processing Technology*, 162–163, 69–177.

Di Mario, C., H. Griffiths, O. Goktekin et al. 2004. Drug-eluting bioabsorbable magnesium stent, *Journal of Interventional Cardiology*, 17, 391–395.

Dodd, S.B. and R.W. Gardiner. 1996. Vapour condensation applied to the development of corrosion resistant magnesium, in G.W. Lorimer, Ed., in *Proceedings of the 3rd International Magnesium Conference*, Manchester, U.K., April 10–12, 1996, The Institute of Materials, London, U.K., 1997, pp. 271–284.

Dunlop, G. 2002. Australian R&D on the production and application of magnesium alloys, in H.I. Kaplan, Ed., *Proceedings of the Magnesium Technology 2002*, TMS, Warrendale, PA.

Elektron WE43B, Datasheet: 467A, Magnesium Elektron, Manchester, U.K., www.magnesium-elektron.com/data/downloads/467B%20for%20PC%20111202.pdf (accessed August 16, 2010).

Eliezer, D. and H. Alves. 2002. Corrosion and oxidation of magnesium alloys' in M. Kuts, Ed., *Handbook of Materials Selection*, pp. 267–291. Wiley & Sons, Inc., New York.

Fan, J.M., Q. Li, W. Kang et al. 2009. Composite cerium oxide/titanium oxide thin films for corrosion protection of AZ91D magnesium alloy via sol-gel process, *Materials and Corrosion*, 60, 438–443.

Feliu Jr., S., A. Pardo, M.C. Merino et al. 2009. Correlation between the surface chemistry and the atmospheric corrosion of AZ31, AZ80 and AZ91D magnesium alloys, *Applied Surface Science*, 255, 4102–4108.

Fichtner, M., J. Engel, O. Fuhr et al. 2004. Nanocrystalline aluminum hydrides for hydrogen storage, *Materials Science and Engineering B*, 108, 42–47.

Frignani, A., V. Grassi, F. Zucchi, and F. Zanotto. 2010. Mono-carboxylate conversion coatings for AZ31 Mg alloy protection, *Materials and Corrosion*, 62, 995–1002, DOI: 10.1002/maco.200905615, Accessed August 30, 2010.

Froes, F.H., Y.-W. Kim, and S. Krishnamurthy. 1989. Rapid solidification of lightweight metal alloys, *Materials Science and Engineering A*, 117(9), 19–32.

Gesing, A.J., N.D. Reade, J.H. Sokolowski et al. 2010. Development of recyclable Mg-based alloys: AZ91D and AZC1231 phase information derived from heating/cooling curve analysis, in *Proceedings of the Magnesium Technology 2010*, TMS, Warrendale, PA, p. 98.

Grey, J.E. and B. Luan. 2002. Protective coatings on magnesium and its alloys—A critical review, *Journal of Alloys and Compounds*, 336, 88–113.

Gu, C., J. Lian, Z. Jiang et al. 2005. Multilayer Ni-P coating for improving the corrosion resistance of AZ91D magnesium alloy, *Advanced Engineering Materials*, 7, 1032–1036.

Hanawalt, J.D., C.E. Nelson, and J.A. Peloubet. 1942. Corrosion studies of magnesium and its alloys, *Light Metals*, 5, 30–36.

Hartwig, A. 2001. Role of magnesium in genomic stability, *Mutation Research: Fundamental and Molecular Mechanisms of Mutagenesis*, 475, 113–121.

Hawke, D.L., J.E. Hillis, M. Pekguleryuz et al. 1999. Corrosion behavior, in M.M. Avedesian and H. Baker, Eds., *Magnesium and Magnesium Alloys*, ASM International, Materials Park, OH, pp. 194–210.

Heidersbach, K.L. 1998. An evaluation of the corrosion performance of magnesium-yttrium and yttrium-magnesium nonequilibrium alloys, PhD dissertation, The Pennsylvania State University, University Park, PA.

Hillis, J.E. 1983. The effects of heavy metal contamination on magnesium corrosion performance, *Light Metal Age*, June, pp. 25–29.

Huang, C.A., T.H. Wang, T. Weirich, and V. Neubert. 2008. Electrodeposition of a protective copper/nickel deposit on the magnesium alloy AZ31, *Corrosion Science*, 50, 1385–1890.

Hydro Magnesium. 2010. http://www.nanomag.us/Hydro_Mg_Brochure_Corrosion_and_Finishing_of_Mg.pdf, Accessed August 30, 2010.

International Magnesium Association. 2010. http://www.intlmag.org/files/yend2009.pdf, Accessed August 30, 2010.

Kang, Z., K. Mori, and Y. Oishi. 2005. Surface modification of magnesium alloys using triazine dithiols, *Surface and Coatings Technology*, 195, 162–167.

Kannan, M.B. 2010. Influence of microstructure on the in-vitro degradation behavior of magnesium alloys, *Materials Letters*, 64, 739–742.

Kuji, T., H. Nakano, and T. Aizawa. 2002. Hydrogen absorption and electrochemical properties of $Mg_{2-x}Ni$ (x = 0–0.5) alloys prepared by bulk mechanical alloying, *Journal of Alloys and Compounds*, 330–336, 590–596.

Lin, C.S., H.C. Lin, K.M. Lin, and W.C. Lai. 2006. Formation and properties of stannate conversion coatings on AZ61 magnesium alloy, *Corrosion Science*, 48, 93–109.

Liu, M., P. Schmutz, P.J. Uggowitzer et al. 2010. The influence of Y on the corrosion behavior of binary M-Y alloys, *Corrosion Science*, 52(11), 3687–3701.

Liu, Y., Z. Wu, S. Zhou, and L. Wu. 2006. Self-assembled monolayers on magnesium alloy surfaces from carboxylate ions, *Applied Surface Science*, 252, 3818–3827.

Lyon, P., T. Wilkis, and I. Syed. 2005. The influence of alloying elements and heat treatment upon the properties on Electron 21 (EV31A) alloy, in *Proceedings of the Magnesium Technology 2005*, TMS, Warrendale, PA, pp. 303–308.

Makar, G.L. and. J. Kruger. 1990. Corrosion studies of rapidly solidified magnesium alloys, *Journal of the Electrochemical Society*, 137, 414–421.

Makar, G.L. and J. Kruger. 1993. Corrosion of magnesium, *International Materials Reviews*, 38(3), 138–153.

Mitovic-Scepanovic, V. and R.J. Brigham. 1992. Localized corrosion initiation on magnesium alloys, *Corrosion*, 48(9), 780–784.

Mordike, B.L. and T. Ebert. 2001. Magnesium properties-applications-potential, *Materials Science and Engineering A*, 302, 37–45.

Mueller, W.D., Lorenzo Fernendez, M. de Mele, M.L. Nascimento, and M. Zeddies. 2009. Degradation of magnesium and its alloys: Dependence on the composition of the synthetic biological media, *Journal of Biomedical Materials Research*, 90A, 487–495.

Mueller, W.D., M.L. Nascimento, L. Fernendez, and M. de Mele. 2010. Critical discussion of the results from different corrosion studies of Mg and Mg alloys for biomedical applications, *Acta Biomaterials*, 6, 1749–1755.

Murray, J., H. Miller, P. Bird et al. 1995. The effect of particle size and surface composition on the reaction rates of some hydrogen storage alloys, *Journal of Alloys and Compounds*, 231, 841–845.

Nakano, S., Sh. Yamaura, S. Uchinashi et al. 2005. Effect of hydrogen on the electrical resistance of melt-spun $Mg_{90}Pd_{10}$ amorphous alloy, *Sensors and Actuators B* 104, 75–79.

Nisancioglu, K., O. Lunder, and T.K. Aune. 1990. Corrosion mechanism of AZ91 Mg alloy, in *Proceedings of the 47th World Magnesium Association*, McLean, VA, 1990, pp. 43–50.

Nordlien, J.H., K. Nisanciogu, S. Ono et al. 1996. Morphology and structure of oxide films formed on MgAl alloys by exposure to air and water, *Journal of the Electrochemical Society*, 143(8), 2564–2572.

Nordlien, J.H., K. Nisanciogu, S. Ono et al. 1997. Morphology and structure of water-formed oxides on ternary MgAl alloys, *Journal of the Electrochemical Society*, 144(2), 461–466.

Nordlien, J.H., S. Ono, M. Noburo, and K. Nisancioglu. 1995. Morphology and structure of oxide films formed on Magnesium by exposure to air and water, *Journal of the Electrochemical Society*, 142(10), 3320–3322.

Ozturk K., Y. Zhong, A.A. Lua et al. 2003. Creep resistant Mg-Al-Ca alloys: Computational thermodynamics and experimental investigation, *JOM*, 55(11), 40–44.

Pardo, P., M.C. Merino, P. Casajus et al. 2009. Corrosion behavior of Mg-Al alloys with Al-11Si thermal spray coatings, *Materials and Corrosion*, 60, 939–948.

Pardo, A., M.C. Merino, A.E. Coy et al. 2008. Corrosion behavior of magnesium/aluminum alloys in 3.5wt.% NaCl, *Corrosion Science*, 50, 823–834.

Patzer, G. 2010. The magnesium industry today…the global perspective, in *Proceedings of the Magnesium Technology 2010*, TMS, Warrendale, PA, pp. 85–90.

Petrilli, J.D. 2009. Evaluating the in vitro corrosion behavior and cytotoxicity of vapor deposited magnesium alloys, MS thesis, The Pennsylvania State University, University Park, PA.

Phani, A.R., F.J. Gammel, T. Hack et al. 2005. Enhanced corrosion resistance by sol-gel-based ZrO_2-CeO_2 coatings on magnesium alloys, *Materials and Corrosion*, 56, 77–82.

Pokhmurska, H., B. Wielage, T. Lampke et al. 2008. Post-treatment of thermal spray coatings on magnesium, *Surface and Coatings Technology*, 202, 4515–4524.

Polmear, P. 1999. *ASM Specialty Handbook on Mg and Mg Alloys*, ASM International, Novelty, OH, p. 16.

Pourbaix, M. 1974. *Atlas of Electrochemical Equilibria in Aqueous Solutions*, NACE CEBELCOR, Pergamon Press, New York.

Pranevicius, L.L. and D. Milcius. 2005. Synthesis of $Mg(AlH_4)_2$ in bilayer Mg/Al thin films under plasma immersion hydrogen ion implantation and thermal desorption processes, *Thin Solid Films* 485, 135–140.

Rettig, R. and S. Virtanen. 2008. Time-dependent electrochemical characterization of the corrosion of a magnesium rare-earth alloy in simulated body fluids, *Journal of Biomedical Materials Research Part A*, 85, 167–175.

Rettig, R. and S. Virtanen. 2009. Composition of corrosion layers on a magnesium rare-earth alloy in simulated body fluids, *Journal of Biomedical Materials Research Part A*, 88, 359–369.

Riontiono, G., D. Lussana, M. Massazza et al. 2008. Structure evolution of EV31 Mg alloy, *Journal of Alloys and Compounds*, 463, 200–206.

Rojas, P., S. Ordonez, D. Serafini et al. 2005. Microstructural evolution during mechanical alloying of Mg and Ni, *Journal of Alloys and Compounds*, 391, 267–276.

Rudd, A.L., C.B. Breslin, and F. Mansfeld. 2000. The corrosion protection afforded by rare earth conversion coatings applied to magnesium, *Corrosion Science*, 42, 275–288.

Rzychon, T., J. Michalska, and K. Kiebus. 2007. Corrosion resistance of Mg-RE-Zr alloys, *JAMME*, 21(1), 51–54.

Shaw, B.A., E. Sikora, and S. Virtanen. 2008. Fix, heal and disappear: A new approach to using metals in the human body, *Interface*, 17(2), 45–49.

Skripnuk, V.M., E. Rabkin, Y. Estrin et al. 2010. The possible route to making magnesium fit for hydrogen storage in automotive applications, in *Proceedings of the Magnesium Technology 2010*, TMS, Warrendale, PA.

Song, G. 2003. Investigation on corrosion of Mg and its alloys, *JCSE*, 6, paper C104.

Song, G. and A. Atrens. 1999. Corrosion mechanisms of magnesium alloys, *Advanced Engineering Materials*, 1, 11–33.

Song, G. and A. Atrens. 2003. Understanding magnesium corrosion—A framework for improved alloy performance, *Advanced Engineering Materials*, 5(12), 837–858.

Song, G. and A. Atrens. 2007. Recent insights into the mechanism of Magnesium corrosion and research suggestions, *Advanced Engineering Materials*, 9(3), 177–183.

Song, G., S. Hapugpda, D. St. John, and C. Bettles. 2006. Simulation of atmospheric environments for storage and transport of Mg and its alloys, in *Proceedings of the Magnesium Technology 2006*, TMS, Warrendale, PA, 2006, pp. 3–6.

Song, Y.W., D.Y. Shan, and E.H. Han. 2007. Comparative study on corrosion protection properties of electroless Ni-P-ZrO$_2$ and Ni-P coatings on AZ91D magnesium alloy, *Materials and Corrosion*, 58, 506–510.

Song, Y.W., D.Y. Shan, and E.H. Han. 2008. High corrosion resistance of electroless composite plating coatings on AZ91D magnesium alloys, *Electrochimica Acta*, 53, 2135–2143.

Staiger, M.P., A.M. Pietak, J. Huadmai et al. 2006. Magnesium and its alloys as orthopedic biomaterials: A review, *Biomaterials*, 27, 1728–1734.

Takenaka, T., Y. Narazaki, N. Uesaka, and M. Kawakami. 2008. Improvement of corrosion resistance of magnesium alloys by surface film with rare earth element, *Materials Transactions*, 49, 1071–1076.

Takenaka, T., T. Ono, Y. Narazaki et al. 2007. Improvement of corrosion resistance of magnesium metal by rare earth elements, *Electrochimica Acta*, 53, 117–121.

Tian, J., W. Li, L. Han, and J. Peng. 2010. Creep behavior of an AZ91 Mg alloy reinforced with Al silicate short fibers, *Advanced Materials Research*, 97–101, 492–495.

Umahara, H., M. Takaya, and S. Terauchi. 2003. Chromate-free surface treatments for magnesium alloys, *Surface and Coatings Technology*, 169–170, 666–669.

Wang, L., T. Shinohara, and B.-P. Zhang. 2010. Influence of chloride, sulfate and bicarbonate anions on the corrosion behavior of AZ31 magnesium alloy, *Journal of Alloys and Compounds*, 496, 500–507.

Wang, L.B., J.B. Wang, H.T. Yuan et al. 2004. An electrochemical investigation of Mg$_{1-x}$Al$_x$Ni (x = 0–0.6) hydrogen storage alloys, *Journal of Alloys and Compounds*, 385, 304–308.

Weinzer, N., A. Atrens, G. Song et al. 2005. A critical review of the stress corrosion cracking (SCC) of Mg alloys, *Advanced Engineering Materials*, 7(8), 659–693.

Wen, Z., C. Wu, Ch Dai, and F. Yang. 2009. Corrosion behaviors of Mg and its alloys with different al contents in a modified simulated body fluid, *Journal of Alloys and Compounds*, 488, 392–399.

Winzer, N., A. Atrens, G. Song et al. 2005. Critical review of the stress corrosion cracking (SCC) of magnesium alloys, *Advanced Engineering Materials*, 7(8), 659–693.

Witte, F., V. Kaese, H. Haferkamp et al. 2005. In vivo corrosion of four magnesium alloys and the associated bone response, *Biomaterials*, 26, 3557–3562.

Witte, F., H. Ulrich, C. Palm et al. 2007a. Biodegradable magnesium scaffolds: Part II: Peri-implant bone remodeling, *Journal of Biomedical Materials Research*, 81A, 757–765.

Witte, F., H. Ulrich, M. Rudert et al. 2007b. Biodegradable magnesium scaffolds: Part I: Appropriate inflammatory response, *Journal of Biomedical Materials Research*, 81A, 748–756.

Wolfe, R.C. 2005. Imparting passivity to vapor deposited magnesium alloys, PhD dissertation, The Pennsylvania State University, University Park, PA.

Xin, Y., K. Huo, H. Tao et al. 2008. Influence of aggressive ions on the degradation behavior of biomedical magnesium alloy in physiological environment, *Acta Biomaterials*, 4, 2008–2015.

Yamamoto, A., A. Watanabe, K. Sugahara et al. 2001. Improvement of corrosion resistance of magnesium alloys by vapor deposition, *Scripta Materialia*, 44, 1039–1042.

Zhang, S.Y., Q. Li, B. Chen et al. 2010. Electrodeposition of zinc on AZ91D magnesium alloy pre-treated by stannate conversion coatings, *Materials and Corrosion*, 61(10), 860–865.

Zhang, Z., G. Yu, Y. Quyang et al. 2009. Studies on influence of Zn immersion and fluoride on nickel electroplating on magnesium alloy AZ91D, *Applied Surface Science*, 255, 7773–7779.

Zheng, R.F. and C.H. Liang. 2007. Conversion coating treatment for AZ91 magnesium alloys by a permanganate-REMS bath, *Materials and Corrosion*, 58, 193–197.

Further Readings

ASM Metals Handbook, Vol. 13B. 2005. S.D. Cramer and B.S. Covino, Jr., Eds., ASM International, Novelty, OH.

ASM Specialty Handbook: Mg and Mg Alloys. 1999. M.M. Avedesian and H. Baker, Eds., ASM International, Novelty, OH.

Blawert, C., W. Dietzel, E. Ghali, and G. Song. 2006. Anodizing treatments for magnesium alloys and their effect on corrosion resistance in various environments, *Advanced Engineering Materials*, 8, 511–533.

Grey, J.E. and B. Luan. 2002. Protective coatings on magnesium and its alloys—A critical review, *Journal of Alloys and Compounds*, 336, 88–113.

Mueller, W.D., M.L. Nascimento, L. Fernandez, and M. de Mele. 2010. Critical discussion of the results from different corrosion studies of Mg and Mg alloys for biomedical applications, *Acta Biomaterials*, 6, 1749–1755.

Song, G. and A. Atrens. 1999. Corrosion mechanisms of magnesium alloys, *Advanced Engineering Materials*, 1, 11–33.

Song, G. and A. Atrens. 2003. Understanding magnesium corrosion—A framework for improved alloy performance, *Advanced Engineering Materials*, 5(12), 837–858.

Song, G. and A. Atrens. 2007. Recent insights into the mechanism of Magnesium corrosion and research suggestions, *Advanced Engineering Materials*, 9(3), 177–183.

Staiger, M.P., A.M. Pietak, J. Huadmai et al. 2006. Magnesium and its alloys as orthopedic biomaterials: A review, *Biomaterials*, 27, 1728–1734.

Weinzer, N., A. Atrens, G. Song et al. 2005. A critical review of the stress corrosion cracking (SCC) of Mg alloys, *Advanced Engineering Materials*, 7(8), 659–693.

3

Crystalline Alloys: Aluminum

Nick Birbilis
Monash University

Bruce R.W. Hinton
Monash University

3.1 Introduction

Aluminum (Al) as an engineering material ranks only behind ferrous alloys (in tonnage used) while growth in production continues to increase. The global tonnage in 2007 was 60 million tons, of which ~65% was provided by primary production and ~35% million by recycled scrap (aluminium.org).

The key to today's extensive use of Al lies in its strength-to-density ratio, toughness, and in part, its corrosion resistance. Such versatility makes it suitable for a wide range of applications from low-end commodity uses to essential construction material for aircraft and space vehicles. Transportation, largely aerospace applications, has provided the greatest stimulus for alloy development and corrosion research that continues today (Polmear 1995). Al and its alloys offer a diverse range of desirable properties that can be matched precisely to the demands of each application by the appropriate choice of composition, temper, fabrication, and processing mode (Davis 1999). In an ever-evolving era of "green" engineering, Al is increasingly the metal chosen for reducing the weight (and hence emissions as well as fuel efficiency) from the world's rapidly expanding fleet of vehicles. The generally stable and reasonable price of Al has made it the commodity "light metal."

From a corrosion perspective, Al has been a successful metal. A number of Al alloys can be satisfactorily deployed in environmental/atmospheric conditions, and the corrosion protection industry for Al continues to respond to market needs. In the past half century, major corrosion issues addressed include the localized corrosion of Al alloys containing magnesium, stress corrosion cracking (SCC) of alloys used in aerospace applications, galvanic corrosion of Al in architectural and automotive applications, and most recently, the filiform corrosion of painted Al sheet in both architectural and automotive applications.

This chapter will cover the pertinent aspects related to the corrosion, and the corrosion protection, of Al and its alloys. While an attempt is made to give the reader a holistic overview of the key technical aspects for a comprehensive insight into the topic as it presently stands, the size of this chapter is not a replacement to dedicated monographs on the specific topics at hand nor the ever-evolving journal literature that represents the state of the art. The structure of this chapter presents the basic aspects of Al metallurgy (which are ultracritical in understanding Al alloy corrosion), followed by the manifestations of corrosion upon Al alloys, and ending with corrosion prevention strategies that are specific to Al alloys.

3.2 Metallurgical Aspects of Aluminum Alloys

3.2.1 Aluminum Production

In regard to Al production from ore, readers are directed to dedicated texts (Grjotheim and Welch 1988, Grjotheim and Kvande 1993). Al is produced using the Hall–Héroult process, involving electrolysis of Al oxide dissolved in cryolite that was independently discovered by Charles Martin Hall in the United States and Paul Héroult in France in 1886.

The raw material for Al production is bauxite, which is refined to alumina. Global bauxite production has increased from 144 million tons worldwide in 2002 to over 200 million tons in 2008. Recent growth in demand for Al has led to a predicted increase in annual demand by ~7 p.a. to 2020, suggesting that global production will be nearly doubled by the completion of the next decade owing to growth markets such as in developing nations.

This latter point brings with it an important notion in that industry average emissions associated with primary Al production are 9.73 kg CO_2e/kg with 55% of this from electricity generation (aluminium.org). Recycling of Al requires 95% less energy than that required for primary Al production and recycling of used Al products generates only 0.5 kg of CO_2e/kg of Al produced (Schlesinger 2006). As a result, the use of recycled Al will be a critical issue in coming years; these have future ramifications for corrosion and will need addressing. Presently, in order to meet the mechanical and corrosion performance requirements of many alloy and product specifications, much of the recycled metal must be blended or diluted with primary metal to reduce impurity levels. The result is that, in many cases, recycled metal tends to be used primarily for lower-grade casting alloys and products (Das 2006).

3.2.2 Properties of Aluminum Alloys

Al has several unique properties, its lightness being principally important. A number of other functional properties such as conductivity (electrical and thermal) and corrosion resistance also make it a popular material. It was mentioned by Vargel (2004) that as early as the 1930s, naval architects had realized that Al was light enough and corrosion resistant enough to be useable for shipbuilding. In fact, since the 1960s, all high-speed ferries have been built using Al–Mg alloys. The basic physical properties for Al alloys are given in Table 3.1.

The diverse and exacting technical demands made on Al alloys in different applications are met by the considerable range of alloys available; each of which has been designed to provide various combinations of properties. These include strength/weight ratio, corrosion resistance, workability, castability, high-temperature properties, and even weldability. The basic functional properties for a selection of Al alloys are given in Table 3.2.

TABLE 3.1 The Physical Properties of Aluminum

Atomic number	13
Atomic weight	10.0
Atomic volume	26.97
Valency	3
Crystal structure	Face-centered cubic
Interatomic distance	2.863 Å
Electrochemical equivalent	0.3354 g/Ah
Density at 293 K	2700 kg/m^3
Melting point	931 K
Sp. heat at 293 K	896 J/kg K
Latent heat of fusion	387 kJ/kg
Coeff. of linear exp. (293–393 K)	0.61×10^{-6} m/K
Thermal conductivity at 273 K	214 W/m K
Elec. vol. resistivity at 293 K	2.7–3.0 $\mu\Omega$ cm
Elec. vol. conductivity at 293 K	63%–57% IACS

TABLE 3.2 Properties of Selected Aluminum Alloys

Alloy	Temper	Wrought/Cast	Density (g/cm^3)	Electrical Conductivity (%IACS)	Yield Strength (MPa)	Ultimate Tensile Strength (MPa)	Elongation (%)
1199	O	W	2.71	60	10	45	50
3003	H14	W	2.73	50	145	152	8
5005	H38	W	2.70	52	200	186	5
5052	H38	W	2.68	35	290	255	7
2024	T861	W	2.77	30	490	517	6
6061	T6	W	2.7	43	276	310	12
7075	T6	W	2.80	22	503	572	11
201.0	T4	C (sand cast)	2.80	30	215	365	20
356.0	T51	C (sand cast)	2.69	41	140	175	2
413.0	F	C (die cast)	2.66	39	140	300	2.5

Source: Adapted from Polmear, I.J., *Light Alloys: Metallurgy of the Light Metals*, 3rd Edn., Arnold, London, U.K., 1995; Davis, J.R., Ed., *Corrosion of Aluminum and Aluminum Alloys*, ASM International, Materials Park, OH, 1999; *Corrosion of Aluminum*, Vargel, C., Copyright 2004, with permission from Elsevier; Hatch, J.E., *Aluminum: Properties and Physical Metallurgy*, ASM International, Materials Park, OH, 424 p., 1984.

3.2.3 Physical Metallurgy: Alloys and Tempers

Properties of Al alloys (mechanical, physical, and chemical) depend on alloy composition and microstructure as determined by casting conditions and thermomechanical processing. As a result, an understanding of the microstructures in Al alloys is critical to an understanding of the subsequent corrosion performance. While certain metals alloy with Al rather readily (Van Horn 1967), comparatively few have sufficient solubility to serve as major alloying elements. Of the commonly used alloying elements, magnesium, zinc, copper, and silicon have significant solubility (and form a function in nearly all the Al alloys used commercially), while a number of additional elements (with less that 1% total solubility) are also used to confer important improvements to alloy properties. Such elements include manganese, chromium, zirconium, titanium, and scandium (Hatch 1984, Polmear 1995)—these are used for inoculation and grain refinement purposes.

The very low yield strength of pure Al (~10 MPa) necessitates alloying for subsequent engineering applications. The simplest strengthening technique is solution hardening, whereby alloying additions have appreciable solid solubility over a wide range of temperatures and remain in solution after any thermal cycles (this is the basis for Mg, and to a lesser extent, Si additions). Solid solution strengthening can lead to strength increases of about a factor of three (and this can be further improved by cold working).

Discovered only a century ago (Polmear 2004), the most significant increase in strength for Al alloys is derived from age (precipitation) hardening—which can result in strengths beyond 800 MPa. In fact, more recently, ultrafine-grained, nanocrystalline, and specialty alloys with novel solute architecture are proving that strengths in excess of 1 GPa are indeed possible for Al alloys (Liddicoat et al. 2009, Zhang et al. 2010). In the classical sense, age hardening requires a decrease in solid solubility of alloying elements with decreasing temperature. The age-hardening process can be summarized by the following stages: solution treatment at a temperature within a single-phase region to nominally dissolve the alloying element(s), quenching or rapid cooling of the alloy to obtain what is termed a supersaturated solid solution, and decomposition of the supersaturated solid solution at ambient or moderately elevated temperature to form finely dispersed precipitates.

The fundamental aspects of decomposition of a supersaturated solid solution are complex, particularly at the early stages when formation of atomic clusters occur (Raviprasad et al. 2003, Kovarik et al. 2006, Winkelman et al. 2007). Typically, however, Guinier–Preston (GP) zones and intermediate phases are formed as precursors to the equilibrium precipitate phase (Hatch 1984; Figure 3.1 reveals a typical micrograph showing precipitate particles).

Properties can be enhanced further by careful thermomechanical processing that may include heat treatments like duplex aging and retrogression and re-aging (Polmear 1995, Li et al. 2009). Maximum hardening in commercial alloys is often achieved when the alloy is cold worked by stretching after quenching and before aging, increasing dislocation density and providing more heterogeneous nucleation sites for precipitation.

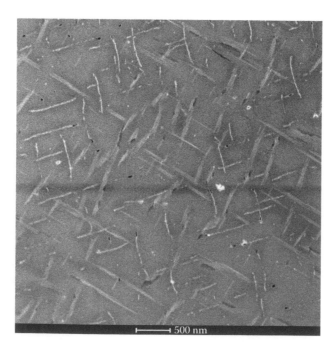

FIGURE 3.1   Dark-field STEM of coarse Al_2CuMg precipitate particles in an Al–Cu–Mg alloy (2xxx series)—imaged down <100> zone axis. (Photo courtesy of Nick Birbilis.)

Yield strength increases may also be achieved by grain refinement, exploiting the Hall–Petch relationship (Dieter 1998). In the case of Al however, the Hall–Petch coefficient is not as great as in other metal systems (Dieter 1998), and the process of grain refinement is used not principally for strength development but for a range of other property enhancements. Grain refinement in Al, alloys is achieved by additions of small amounts of low solubility additions such as Ti and B to provide grain nuclei and by recrystallization control using dispersoids. Trace alloying additions of Cr, Zr, or Mn promote submicron-sized insoluble dispersoid particles that subsequently can restrict or pin grain growth.

The microstructures developed in Al alloys are complex and incorporate a combination of equilibrium and nonequilibrium phases. Typically, commercial alloys can have a chemical composition incorporating as many as ten alloying additions. It is prudent, from a corrosion point of view, to understand the role that impurity elements have on microstructure. While not of major significance to alloy designers, impurity elements such as Fe, Mn, and Si can form insoluble compounds. These constituent particles are comparatively large and irregularly shaped with characteristic dimensions ranging from 1 to ~10 μm. These particles are formed during alloy solidification and are not appreciably dissolved during subsequent thermomechanical processing. Rolling and extrusion tend to break up and align constituent particles within the alloy. Often, constituents are found in colonies made up of several different intermetallic compound types. Because these particles are rich in alloying elements, their electrochemical behavior is often significantly different than the surrounding matrix phase. In most alloys, pitting is associated with specific constituent particles present in the alloy (Buchheit 1995, Szklarska-Smialowska 1999, Cavanaugh et al. 2009b). A range of alloying elements are found in constituent particles (typical example seen in Figure 3.2); examples include Al_3Fe, Al_6Mn, and Al_7Cu_2Fe.

To give the reader some context that the families of precipitates and constituent particles can exist on different length scales (often difficult to image simultaneously), a low-magnification scanning transmission electron microscope (STEM) micrograph is included in Figure 3.3. The different particle types

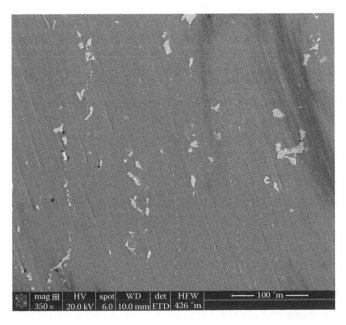

mag ⊞	HV	spot	WD	det	HFW	⟶ 100 ˝m ⟶
350 ×	20.0 kV	6.0	10.0 mm	ETD	426 ˝m	

FIGURE 3.2 SEM image of constituent particles in AA2124-T3 imaged in backscattered electron mode. (Photo courtesy of Nazatul Sukiman.)

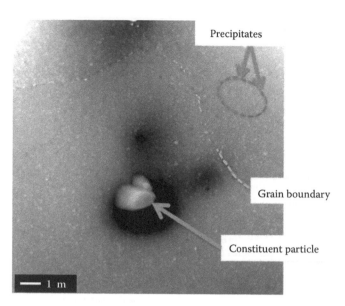

FIGURE 3.3   STEM micrograph of constituent particles in a grain populated with precipitates in AA7075-T6. (Photo courtesy of Nick Birbilis.)

have a different influence on the initiation (constituents) and propagation (precipitates) of corrosion within Al alloys.

The International Alloy Designation System (IADS) gives each wrought alloy a four-digit number where the first digit is assigned on the basis of the major alloying element(s), as is summarized in Figure 3.4, along with the associated temper description.

Cast Al alloy designations adopt the Aluminum Association notation system summarized in Table 3.3. Casting compositions are described by a four-digit system that incorporates three digits followed by a decimal. The 0.0 decimal indicates the chemistry limits applied to an alloy, the 0.1 decimal indicates the chemistry limits for ingot used to make the alloy casting, and the 0.2 decimal indicates ingot composition but with somewhat different chemical limits. Some alloy designations include a letter. Such letters, which precede an alloy number, distinguish between alloys that differ only slightly in percentages of impurities or minor alloying elements (e.g., 356.0, A356.0, and F356.0). Temper designations adopted by the Aluminum Association are similar for both wrought and cast Al alloys.

A brief overview of the Al alloy classes is given here, to familiarize the reader with the key role of the principal alloying elements.

3.2.3.1 Pure Aluminum

Corrosion resistance of Al increases with increasing metal purity. Iron and silicon are the two main impurities in unalloyed Al (i.e., the 1xxx series alloys). The Fe to Si ratio is usually found to be close to 2. The use of the 99.8% and 99.99% grades is usually confined to those applications where very high corrosion resistance or extraordinary ductility is required. General-purpose alloys for lightly stressed applications are approximately 99% pure Al.

3.2.3.2 Copper- and Copper–Magnesium-Containing Alloys

Copper is one of the most common of the alloying additions to Al, since it has both good solubility and a significant strengthening effect by its promotion of age-hardening response. Copper is added as a major alloying element in the 2xxx series alloys. These alloys were the foundation of the modern aerospace

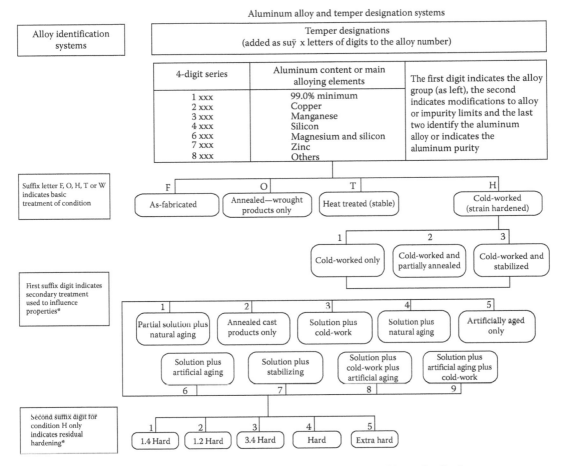

FIGURE 3.4 Wrought Al alloy and temper designations.* denotes that these additional suffix designations are not always quoted. (Adapted from Polmear, I.J., *Light Alloys: Metallurgy of the Light Metals*, 3rd Edn., Arnold, London, U.K., 1995.)

TABLE 3.3 Cast Aluminum Alloy Designations

Alloy Class	Designation
Aluminum of ≥99.0%	1xx.x
Al + Copper	2xx.x
+ Silicon (with copper and/or magnesium)	3xx.x
+ Silicon	4xx.x
+ Magnesium	5xx.x
(Series is unused)	6xx.x
+ Zinc	7xx.x
+ Tin	8xx.x
+ Other elements	9xx.x

Source: Adapted from Polmear, I.J., *Light Alloys: Metallurgy of the Light Metals*, 3rd Edn., Arnold, London, U.K., 1995.

construction industry and, for example, AA2024 (Al–4.4Cu–1.5Mg–0.8Mn), can achieve strengths of up to 520 MPa depending on temper (Hatch 1984). Copper, while ennobling E_{corr}, is actually detrimental to corrosion resistance.

3.2.3.3 Manganese-Containing Alloys

Manganese has a relatively low solubility in Al but improves its corrosion resistance when in solid solution. Additions of manganese of up to 1% form the basis for an important series of non-heat-treatable (3xxx series) wrought alloys, which have good corrosion resistance, moderate strength (i.e., AA3003 tensile strength ~110 MPa), and high formability and amenability to deep drawing.

3.2.3.4 Magnesium-Containing Alloys

Magnesium has significant solubility in Al and imparts substantial solid solution strengthening and improved work-hardening characteristics. The 5xxx series alloys (containing <7% Mg) do not age harden, but have excellent strain-hardening capability. Nominally, the corrosion resistance of these weldable alloys is good, and their mechanical properties make them ideally suited for structural use in aggressive conditions. These alloys are used both for boat and shipbuilding. Fully work-hardened AA5456 (Al–4.7Mg–0.7Mn–0.12Cr) has a tensile strength of 385 MPa. Such alloys make up the low to medium strength class of Al alloys. Figure 3.5 shows rather markedly the increase in strength that can be achieved with the addition of Mg and the attendant change in elongation (Figure 3.5 represents only fully annealed alloys; hence, even greater strengths are possible via cold work). This range of physical property variation comes with no loss to corrosion properties.

3.2.3.5 Silicon- and Magnesium–Silicon-Containing Alloys

Silicon additions alone can lower the melting point of Al while simultaneously increasing fluidity and thus is largely the basis of Al casting alloys and the associated shape-casting industry. These alloys are increasing in importance in automotive applications for engine and drive train components.

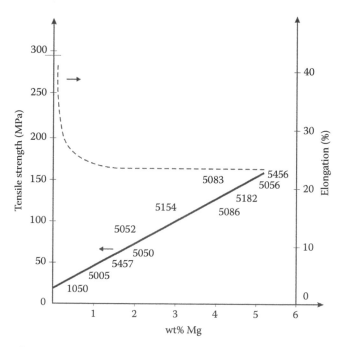

FIGURE 3.5 Effect of Mg on the properties of 5xxx series alloys. (Adapted from *Corrosion of Aluminum*, Vargel, C., 2004, with permission from Elsevier.)

Corrosion issues with Al casting alloys are rare. Adding Mg + Si in combination can also allow alloys to be age hardened, with the heat-treatable Al–Mg–Si being predominantly structural materials, where they have a high resistance to corrosion, immunity to SCC, and weldability. 6xxx series alloys are used in extruded form for architectural applications, although increasing amounts of rolled automotive sheet are being produced. Magnesium and silicon additions are made in balanced amounts to form quasi-binary Al–Mg$_2$Si alloys, or excess silicon additions are made beyond the level required to form Mg$_2$Si. Alloys containing magnesium and silicon in excess of 1.4% develop higher strength upon aging. Such alloys have very good suitability for surface treatments.

3.2.3.6 Zinc- and Zinc–Magnesium-Containing Alloys

The Al–Zn–Mg alloy system provides a range of commercial compositions, primarily where strength is the key requirement. Al–Zn–Mg–Cu alloys have traditionally offered the greatest potential for age hardening and as early as 1917, a tensile strength of 580 MPa was achieved; however, such alloys were not suitable for commercial use until their high susceptibility to SCC could be moderated. In spite of this, such alloys do still have a low resistance to corrosion. Al–Zn–Mg alloys can be welded in instances where no Cu is added. Aerospace needs in the postwar era led to the introduction of a range of high-strength alloys of which AA7075 (Al–5.6Zn–2.5Mg–1.6Cu–0.4Si–0.5Fe–0.3Mn–0.2Cr–0.2Ti) is the most well known and then was superseded by AA7050 and now AA7150 in recent years. The high-strength 7xxx series alloys derive their strength from the precipitation of η-phase (MgZn$_2$) and its precursor forms. The heat treatment of the 7xxx series alloys is complex, involving a range of heat treatments that have been developed to balance strength and SCC performance (Sprowls 1978).

3.2.3.7 Lithium-Containing Alloys

Lithium is soluble in Al to about 4 wt% (corresponding to a value of 16 at%). Li-containing alloys have very high specific strength and stiffness and readily respond to heat treatment. The ultralightweight aspect of such alloys has instigated present research, and development has intensified due to their potential for widespread usage in aerospace applications.

Recent studies have focused on Al–Cu–Li, Al–Li–Mg, and Al–Li–Cu–Mg alloys (including the use of minor Ag additions). These alloys derive their strength principally from age hardening, involving intermetallics such as Al$_2$CuLi and S' phase resulting in strengths in excess of 700 MPa (Rao and Ritchie 1992). The Airbus A380 and the Boeing Dreamliner are comprised of some portion of Al–Li-based alloys (produced by Alcoa in the United States). These alloys, if not prescribed in the appropriate temper and composition, can also have the lowest corrosion resistance of all Al alloys, so that susceptibility to intergranular corrosion (IGC) is the most challenging present issue for this family of alloys.

3.2.4 Processing of Aluminum Alloys

3.2.4.1 Castings

Cast products are usually produced in foundries from prealloyed metal supplied from secondary smelters (although certain high-performance castings are made from primary metal). The three most commonly used processes are sand casting, permanent mold casting, and die casting (including high-pressure die casting). The latter can also contain high levels of porosity due to entrapped gas and are not easily welded or heat treated.

Despite their wide use, little work has been carried out to date to understand the corrosion behavior of cast Al alloys. Due to their high content of alloying elements such as silicon, iron, and magnesium, casting alloys may have a higher density of coarse intermetallic and constituent particles compared to wrought alloys. Processing parameters like cooling rate and pouring temperature, and even minor alloying element content variation, leads to significant changes in the microstructure of these alloys, with further impurity pickup and solute segregation requiring attention.

3.2.4.2 Direct Chill Casting

This is a semicontinuous process used for the production of rectangular ingots or slab for rolling to plate, sheet, and foil and cylindrical ingots or billet for extruded rods, bars, shapes, hollow sections, tube, wire, and rod. Most of production is from primary Al and process scrap or selected postconsumer scrap. Direct chill (DC) casting is the first step in the production of Al alloys prior to the value adding thermomechanical treatments, and while it may appear to be a topic not requiring discussion in such a chapter, it is important for the reader to realize that corrosion performance of an alloy is dictated by each step in the entire processing cycle, from the initial casting operation on up.

Thermodynamic considerations often fail to correctly predict the phase content and solid solution content of an as-cast microstructure because of the nonequilibrium nature of solidification during DC casting. This is important as alloy corrosion properties are controlled by solid solution levels and intermetallic phase amount, crystallography, and morphology; all of the aforementioned depend on complex kinetic competitions during nucleation and growth. While such factors presently remain production issues, it is important for the corrosion engineer to appreciate that a detrimental microstructure has its genesis from the initial casting step, and subsequent production steps cannot be arbitrarily assumed to produce a product with a uniform compositional spread or a satisfactory constituent population. Many cases of dramatic and insidious alloy corrosion can be traced back to non-optimized or poor quality control production operations.

3.2.4.3 Hot and Cold Rolling

Rolling blocks or slabs that are up to 30 ton in weight are heated to temperatures up to 400°C–500°C and passed through a reversing breakdown mill using large reductions per pass to reduce the slab gauge. The slab surface undergoes intense shear deformation during this process and a grain refined surface layer (GRSL) is developed. Hot rolling deforms the original cast structure and the as-cast grains are elongated in the rolling direction. The elongated microstructure developed during hot rolling can have a profound effect on corrosion properties like SCC and exfoliation corrosion. Such propagation processes are well known to be dependent on the orientation with respect to hot rolling, with the short transverse direction being the most susceptible (Zhao and Frankel 2007a).

The GRSL results from mechanical processing and is enhanced where rolling requires multiple passes. It contains a mixture of recrystallized fine metallic grains that can incorporate oxide fragments and porosity (Leth-Olsen et al. 1998, Afseth et al. 2001). Grain sizes as fine as 40–400 nm have been reported. The presence of GRSLs has been linked to the susceptibility of several Al alloys to filiform corrosion susceptibility. As a result, metal finishing and surface treatments are required to consider removal of surface processed layers (Leth-Olsen et al. 1997, Mol et al. 2002). There are various reports on the depth of the modified surface region ranging from 1 to 8 μm. The GRSL is thought to be metastable as aging has been reported to lead to changes in the oxide composition and environment (Viswanadham et al. 1980). Specifically for commodity alloys, the subsequent heat treatment of the 3xxx alloys have been demonstrated to precipitate manganese-containing particles that renders the surface layer extremely surface active (Scamans et al. 2003).

3.2.4.4 Extrusion

During extrusion, a hydraulic ram forces a preheated billet against a die, forcing the metal through the die opening. Control of temperature and speed are important to avoid problems associated with overheating. However, most of these problems (i.e., streaking, surface reflectivity, color) relate to surface defects seen during finishing operations such as anodizing rather than direct corrosion issues.

3.3 Corrosion of Aluminum Alloys

3.3.1 Forms and Causes of Corrosion

Al is a very reactive metal with a high affinity for oxygen. Nevertheless, Al is highly resistant to most atmospheres and chemicals. This resistance is due to the protective character of the passive (Al oxide) film that forms on the metal surface—which rapidly reforms if damaged. The oxide inhibits corrosion because it is both insoluble in a wide range of environments and a good insulator. The oxide film on Al attains a thickness of about 10 Å on freshly exposed metal in seconds. Oxide growth can be modified by impurities and alloying additions and is accelerated by increasing temperature and humidity (or immersion in water).

Corrosion of Al is an electrochemical process that involves the dissolution of Al metal, taking place once the oxide film has been dissolved or damaged. Al is amphoteric in nature, its oxide film being stable in neutral conditions but soluble in acidic and alkaline environments. The thermodynamic stability of Al oxide films is expressed by the potential versus pH (Pourbaix) diagram seen in succeeding text (Pourbaix 1966).

This Pourbaix diagram indicates the theoretical circumstances where Al should dissolve (forming Al^{3+} at low pH values and AlO_2^- at high pH values), passivity due to hydrargillite, that is, $Al_2O_3 \cdot 3H_2O$ (at near-neutral pH values) and supposed immunity (at highly negative potentials).

Some caveats, however, exist with respect to interpretation of the practical performance of Al that is not reflected in Figure 3.6.

Firstly, the nature of the oxide varies according to temperature, and more than ~75°C, boehmite ($Al_2O_3 \cdot H_2O$) is the stable form.

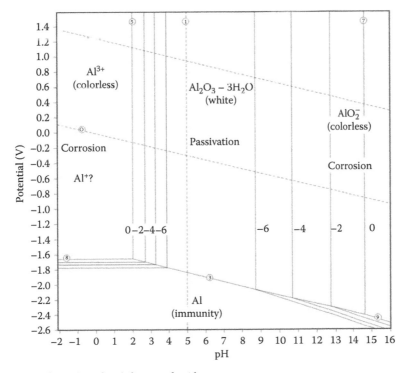

FIGURE 3.6 Potential–pH (Pourbaix) diagram for Al.

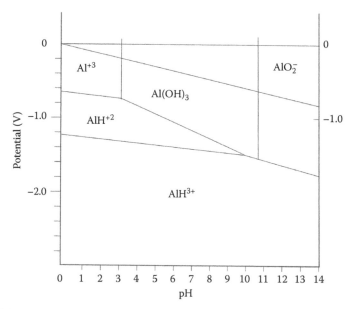

FIGURE 3.7 Modified potential–pH diagram for Al accounting for hydride formation. (Adapted from Perrault, G.G., *J. Electrochem. Soc.*, 126(2), 199, 1979.)

Secondly, it also noted that the potential–pH diagram does not indicate an important property of Al, that is, its ability to become passive in strongly acid solutions of high-redox potential such as concentrated nitric acid, which can produce a pseudo-passive Al nitrate film that can protect the Al from further attack (Singh et al. 1992).

Finally, it must also be emphasized that in a practical sense, Al does not, in practice, display a region of immunity per se. This was discussed by Perrault (1979), who calculated and presented a modified Pourbaix diagram accounting for the possibility of hydride formation shown in Figure 3.7. Such hydride formation will compromise the notion of immunity, and furthermore, the cathodic reaction upon the surface of Al at such negative potentials is also likely to increase pH such that a regime of Al dissolution evolves. As a result, in a practical context, the immunity region for Al is unlikely to be realized.

3.3.2 Effects of Microstructure on Corrosion

Al alloys are susceptible to all forms of corrosion, and hence, a brief mention of these manifestations is given in the context of Al in this section. Corrosion is an electrochemical process, and hence, the corrosion potentials achieved at the surface of different Al alloys are of considerable importance. Also, the difference between the potential of Al alloys and other metals is important, as is the relationship between the potential of microstructural constituents within a single alloy (Figure 3.8).

The major forms of corrosion of Al are as follows.

3.3.2.1 General Corrosion

As a general rule, dissolution occurs in acid or alkaline solutions (as per the Pourbaix diagram), but there are specific exceptions. For instance, in concentrated nitric acid, the metal is passive and the kinetics of the process are controlled by ionic transport through the oxide film. Also, inhibitors (such as silicates) permit the deployment of Al in certain alkaline solutions (to pH < 11.5). In instances where some corrosion may occur, Al may be preferred to other metals because its corrosion products are colorless (i.e., architectural applications).

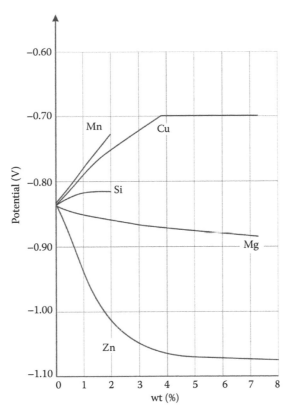

FIGURE 3.8 Variations realized in the corrosion potential of Al as a function of binary alloying additions of Mn, Cu, Si, Mg, and Zn. (Adapted from *Corrosion of Aluminum*, Vargel, C., Copyright 2004, with permission from Elsevier.)

3.3.2.2 Pitting and Localized Corrosion

Pitting is the most commonly encountered form of corrosion for Al alloys. In certain near-neutral solutions, a pit once initiated will continue to propagate into the metal due to acidification of the solution within the pit. The acidic conditions limit the formation of a protective oxide film that usually prevents pit growth (Frankel 1998). As such, pitting arises when localized/aggressive environments break down the nominally passive film. Pits may form at scratches, mechanical defects, second-phase particles (principally the case for Al alloys), or stochastic local discontinuities in the oxide film (for pure or solid solution Al alloys). Since by definition pitting refers to local loss passivity, pitting can only occur in the near-neutral pH range for Al.

Solutions containing halide ions (the most common of which are chlorides) are detrimental. Chlorides facilitate the breakdown of the film by forming $AlCl_3$, which is also usually present in the solution with the pits. When Al ions migrate away from pits, Al_2O_3 precipitates as a membrane, further isolating and intensifying local acidity, and autocatalytic metal loss results. While the shape of pits can vary rather significantly depending on alloy type and environment, pit cavities are nominally hemispherical and have ratio of diameter to depth that is nominally >1. This distinguishes pits from other forms of corrosion such as intergranular or exfoliation corrosion. Pitting is strongly influenced by alloy type and microstructure as discussed later in this chapter.

Al alloys display the classical aspects of pitting: (1) The Al oxide is soluble in the presence of halide ions. (2) When pitting occurs, the acidic conditions that evolve make the process autocatalytic, since Al readily dissolves at low pH (unlike other alloys, like say, the corrosion-resistant classes of stainless steel or nickel). (3) Regions of high cathodic activity will cause local alkalinity increase, rendering the Al susceptible to local dissolution where the pH is alkaline (this is obviously not the case for pure forms or alloys Fe or Zn where passivation occurs at high pH). As a result of this, Al is highly prone to all forms of localized attack with respect to other metal systems.

A review of all aspects of Al pitting is beyond the scope of this chapter and was given by Smialowska (Szklarska-Smialowska 1999, 2005). It is worthy to note, however, that as stated in the review, from a detailed mechanistic point of view, the processes on the atomic level that lead to the breakdown of the film due to halide interaction are presently not understood in sufficient detail. In near-neutral pH environments, pitting potentials for Al and its alloys have been reproducibly measured by many investigators (Kaesche 1963, Wall and Martinez 2003, Birbilis and Buchheit 2005). However, it is also important to mention that the potentio-dynamically determined pitting potential yields no information regarding the number and size of pits that may form upon a given alloy. Practical evidence for this can be ascertained by the fact that while Cu solutes has the ability to ennoble the pitting potential of Al, it is also capable of increasing the number of pit sites that ultimately occur—ultimately rendering Cu detrimental to corrosion performance.

In order to study Al pitting in more detail and to discriminate pitting statistics between alloys, certain researchers have recently studied current oscillations at a constant anodic potential below the pitting potential (Pride et al. 1994, Trueman 2005, Kim and Buchheit 2007, Cavanaugh et al. 2009a, Ralston et al. 2009). The occurrence of these oscillations is explained by the formation and repassivation of nano-/micropits, termed "metastable" pits. These investigations have been largely carried out to understand the processes that lead to the formation of stable pits. It is posited in such works that there is a direct correlation between the number of metastable pits that can be detected and the number of stable pits that are likely to form. Such methods are an emerging area for pitting studies in Al alloys and represent a growing body of fundamental works (Ralston et al. 2010, Soltis et al. 2010).

Al alloy microstructures develop as a result of alloy composition and thermomechanical treatment. From a corrosion perspective, the dominant features of alloy microstructure are the grain structure and the distribution of second-phase intermetallic particles, including the constituent, dispersoid, and precipitate families. Such particles have electrochemical characteristics that differ from the behavior of the Al matrix rendering alloys susceptible to localized forms of corrosion attack in a manner that the community has termed micro-galvanic corrosion.

Several studies over the years have been carried out in order to assess the effect of intermetallic particles on the corrosion susceptibility of specific Al alloys (Zamin 1981, Mazurkiewicz and Piotrowski 1983, Pryor and Fister 1984, Nişancioğlu 1990, Scully et al. 1993, Birbilis and Buchheit 2005). In the 1990s, Buchheit reported the corrosion potential values for intermetallic phases common to Al alloys mainly in chloride-containing solutions (Buchheit 1995). More recently, various groups have focused on the electrochemical properties of Fe-containing intermetallics and Cu-containing intermetallics (Ilevbare et al. 2004, Schneider et al. 2004), and this has been expanded into a more comprehensive treatise covering a variety of common intermetallics present in commercial Al alloys (both wrought and cast) (Birbilis and Buchheit 2005, 2008). An abridged summary of the results of such a study is shown later in Table 3.4. Such potentials are time dependent and very much environment dependent (Birbilis et al. 2006), however, serve as very useful classification tool.

The identification and structural characterization of intermetallic particles present in Al alloys has been quantified by particle extraction techniques combined with electron probe microanalysis and by scanning and transmission electron microscopy combined with x-ray microanalysis, micro-diffraction, and electron backscatter diffraction (EBSD) (Birbilis and Buchheit 2008).

Intermetallic particles in Al alloys may be either anodic or cathodic relative to the matrix. As a result, two main types of pit morphologies are typically observed. Circumferential pits appear as a ring around a more or less intact particle or particle colony and the corrosion attack is mainly in the matrix phase. This type of morphology arises from localized galvanic attack of the more active matrix promoted by the more noble (cathodic) particle as is shown in Figure 3.7, which also shows the phenomenon with an image collected via optical profilometry.

TABLE 3.4 Corrosion Potentials for Intermetallic Particles Commonly Present in Al Alloys

Phase	Corrosion Potential (mV$_{SCE}$) in Quiescent 0.1 M NaCl
Al$_3$Fe	−539
Al$_2$Cu	−665
Al$_6$Mn	−779
Al$_3$Ti	−603
Al$_{32}$Zn$_{49}$	−1004
Mg$_2$Al$_3$	−1013
MgZn$_2$	−1029
Mg$_2$Si	−1538
Al$_7$Cu$_2$Fe	−551
Al$_2$CuMg	−883
Al$_{20}$Cu$_2$Mn$_3$	−565
Al$_{12}$Mn$_3$Si	−810
Al–4%Cu	−602

The second type of pit morphology is due to the selective dissolution of the constituent particle. Pits of this type are often deep and may have the remaining remnants of the particle in them (Figure 3.7 also shows this phenomenon). This morphology has been interpreted as particle fallout, selective particle dissolution in the case of electrochemically active particles, or in the case of some Cu-bearing particles, particle dealloying and non-faradaic liberation of the Cu component.

Localized corrosion activity is however a complex phenomenon that is still under active research. In the context of Al alloys, localized corrosion leads to local pH gradients as recently studied in detail by Ilevbare and Schneider (Ilevbare et al. 2004, Schneider et al. 2004). Cathodic sites of enhanced oxygen reduction generate hydroxyl ions promoting local pH increase, which can then modify the subsequent rate and morphology of corrosion propagation. The precise morphology of particle-induced pitting is important for emerging damage accumulation models (Cavanaugh et al. 2009b). For these models to be predictive, it is necessary to develop a comprehensive, self-consistent accounting of this type of pitting. In cases where the electrochemical characteristics of constituent particles have been rigorously characterized, they have been found to have much more complicated behavior than categorized by simple characterizations like "noble" or "active"—presenting forms of attack such as incongruent dissolution and dealloying.

Recent studies have shown that the surface of rolled, ground, or machined Al alloys have deformed surface layers (GRSLs) that range in thickness from 100 to 200 nm up to several microns (Scamans et al. 2002, Zhou et al. 2003) and that the presence of these layers has a strong effect on the initiation of corrosion (a video of such damage can be found at Zhao and Frankel, http://hdl.handle.net/1811/24655). These deformed layers are characterized by ultrafine grains formed due to high levels of shear strain at the Al alloy surface. The surface layer grains are purported to be ~50 to 100 nm in diameter and possibly stabilized by oxide particles on their boundaries and are more susceptible to corrosion than the underlying bulk alloy.

Deformed layers on AA3005 and similar architectural alloys are activated by the preferential precipitation of manganese-containing dispersoids, whereas the nucleation of precipitates during aging gives rise to similar layers on AA6016 and AA6111 automotive and AA7075 aerospace alloys. Such layers were also observed to occur following mechanical polishing in other instances (Anawati et al. 2010). While the observation of such GRSLs is reported following surface preparation for laboratory studies, it remains unclear as to the role of such layers in dictating the corrosion events that subsequently lead to the various forms of corrosion propagation. For example, the scanning electron microscope (SEM) image in Figure 3.9 (representing many such images in the open literature) shows that in spite of a mechanical

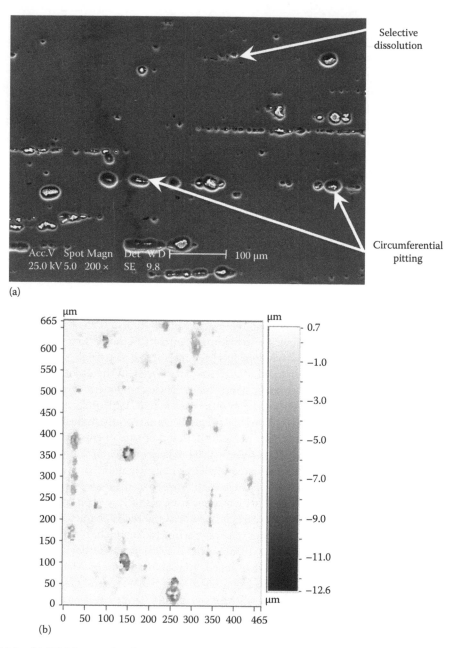

(a)

(b)

FIGURE 3.9 (a) SEM image of early-stage corrosion development upon AA7075-T651 immersed in quiescent 0.01 M NaCl at 40°C for 1 day. (b) Corresponding optical profilometry image, surface height variation denoted by scale on the contour bar. (Photo courtesy of Mary K. Cavanaugh.)

polish prior to and during subsequent SEM examination exposure, the corrosion morphology remains as may be expected with respect to the features present in the microstructure. Furthermore, the ability of researchers to obtain high-quality electron backscatter diffraction (EBSD) information from ground surfaces also raises the question as to whether or not GRSLs from polishing can exist in all alloys (such refinement would require significant alloying to be present) and whether the GRSL will contribute to corrosion other than superficially. Such questions presently remain to be answered.

3.3.2.3 Bimetallic or Galvanic Corrosion

Al is anodic to most other metals when coupled within an electrolyte, the exceptions being only Mg for the structural metals and in some cases, Zn (where Zn coatings are used). While galvanic corrosion is largely an engineering design issue and can be solved by design, some mention will be made here owing to the Al alloys being significantly less noble compared to other systems. Table 3.5 shows the corrosion potentials for a range of non-heat-treatable and heat-treatable (wrought) Al alloys based on measurements made according to ASTM G69.

Contact with ferrous metals will accelerate attack on Al, but in some natural waters and other special cases, Al can be protected at the expense of ferrous materials—particularly when the Al is passive. Titanium appears to behave in a similar manner to steels. Stainless steel in contact with Al will nominally increase attack on Al, notably in seawater or marine environments, but the high electrical resistance of the two surface oxide films can minimize bimetallic effects in less aggressive or high-resistivity environments. Where Al is coupled to copper, or exposed to metallic copper contamination (such as in water systems), corrosion of the Al is very rapid. This is because Cu is particularly efficient at supporting cathodic reactions (e.g., oxygen and water reduction), more so than Fe. Limiting cathodic currents measured for pure copper are reported to be in the vicinity of 1.5 mA/cm^2, whereas limiting currents on pure Al are up to three orders of magnitude lower (approx. <10 µA/cm^2) (Seegmiller et al. 2004).

3.3.2.4 Dealloying

Dealloying is somewhat analogous to galvanic corrosion but occurs where there is an accelerated/preferential loss of one or more elements from an intermetallic phase or from the wider alloy surface. Under electrochemical control, intermetallic phases have been demonstrated to dealloy as a result

TABLE 3.5 Corrosion Potentials for Al Alloys in 1 M NaCl Containing 9 ± 1 g/L H_2O_2 (ASTM G69)

Alloy	Corrosion Potential (V_{SCE})
Al (99.999)	−0.75
1100	−0.74
1199	−0.75
2014-T6	−0.69
2024-T3	−0.60
2024-T8	−0.71
3003	−0.74
3004	−0.75
5052	−0.76
5083	−0.78
5154	−0.77
6060	−0.71
6061-T4	−0.71
6061-T6	−0.74
6063	−0.74
7039-T6	−0.84
7055-T77	−0.75
7075-T6	−0.74
7075-T7	−0.75
7475-T7	−0.75
7079-T6	−0.78
8090-T3	−0.70
8090-T7	−0.75

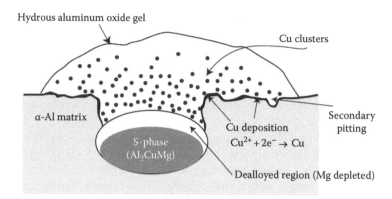

FIGURE 3.10 Dealloying of Al$_2$CuMg phase contained within an Al alloy matrix. Anodic polarization of particle (or conversely open-circuit exposure in conductive electrolytes) results in preferential loss of Al and Mg. A Cu-rich network, which coarsens with age and is susceptible to breakup during hydrodynamic flow, releases small Cu-rich particles (diameter approximately 10–100 nm). The electrically isolated particles are dissolved, making Cu ions available for replating as elemental Cu onto cathodic sites. The replated cathodic sites serve as efficient local cathodes that stimulate secondary pitting. (After Buchheit, R.G. et al., *J. Electrochem. Soc.*, 147(1), 119, 2000.)

of both anodic and cathodic mechanisms. Anodic polarization can result in the generation of localized acidity that is able to preferentially attack certain elements, removing them from a bulk phase. Conversely, cathodic polarization can remove elements susceptible to dissolution under alkaline conditions and as noted in the previous section may also replate or redeposit and concentrate noble metals on cathodic sites.

For Al alloys, Cu enrichment at the interface has been of both scientific and industrial interest. One of the most studied examples of dealloying concerns S-phase (Al$_2$CuMg), which is commonly found in 2xxx series alloys and has been shown to initially exhibit a less noble corrosion potentials to the alloy matrix. Under neutral pH conditions, Mg and Al are preferentially dissolved from the Al$_2$CuMg phase, resulting in a Cu-enriched high surface area remnant, which in turn exhibits potentials significantly noble to the matrix. Figure 3.10 illustrates the enrichment of an Al$_2$CuMg particle in copper, its release into solution and redeposition onto adjacent surface areas. Ultimately, the form that copper takes on the surface is thought to be important in determining the corrosion performance of alloys such as AA2024. The redistribution of copper has been demonstrated to enhance the cathodic kinetics of a surface and rather efficiently increase the rate of corrosion.

In the case of dealloying from a bulk phase (such as the matrix), the physical structure of a surface has been predicted by percolation theory to be dependent upon the dissolution rate and concentration of the noble elements in the phase (Sieradzki 1993, Newman and Sieradzki 1994). Rapid dissolution rates lead to more porous network structures, where there is a possibility that unoxidized fragments enriched in the more noble metal will be released into solution, whereas slow dissolution allows surface diffusion and relaxation processes to maintain a stable surface structure (Buchheit et al. 2000, Vukmirovic et al. 2002, Muster et al. 2008).

The accumulation of nobler metals such as iron and copper at the surface of Al alloys is problematic even in the absence of specific intermetallic phases such as Al$_2$CuMg. It has been shown that Cu accumulation on the surface of AA2024 also arises from the Cu in the Al solid solution (Vukmirovic et al. 2002). In a range of corrosive environments, Al has been shown to preferentially oxidize, resulting in the buildup of copper within a layer approximately 2–5 nm thick at the alloy surface. In environments where Al alloys continually experience anodic dissolution, it has been suggested that alloys with a wide range of copper concentrations (0.06–26 at%) can display copper enrichment at the metal–oxide interface (Habazaki et al. 1997, Koroleva et al. 1999, Garcia-Vergara et al. 2004).

3.3.2.5 Crevice Corrosion

If a crevice is formed between two Al surfaces or between the surfaces of Al and a nonmetallic material (i.e., a polymer), localized corrosion can, and most often will, occur within the crevice in the presence of electrolyte. For Al alloys, mention is made here—since crevice corrosion can be a very problematic form of damage in an engineering sense. Real constructions involving Al include riveting, welded lap joints, valve seats, or even deposits that arise in service (Roberge 1999)—all very common since Al is used extensively in the aerospace and automotive sectors. The general rules for the severity of crevice corrosion are presently under active research for several metal alloy systems, Al included (Chen et al. 2008). Typically, in Al, tighter crevices lead to more rapid initiation of attack (owing to less electrolyte and a steeper oxygen concentration profile being achieved more rapidly).

3.3.2.6 Filiform Corrosion

Filiform corrosion may be considered as a specific type of differential-aeration cell corrosion that occurs from defects where bare metal is exposed on painted or coated Al surfaces. It is worth mentioning here, since as recently as the 1970s and 1980s, filiform corrosion was not considered as a damage mode upon Al. Much like crevice corrosion, a differential-aeration cell drives filiform attack with an anodic head growing under a coating and a cathodic tail where oxygen is reduced. The filiform filaments are filled with corrosion products that can include alumina gel and partially hydrated corrosion products. Filiform corrosion is now a recognized and routinely encountered corrosion issue for Al alloys in aging aircraft. This is because aircraft are routinely painted for corrosion protection with polyurethanes and more complex coating systems (Coleman et al. 2007, McMurray et al. 2007, Schneider et al. 2007). It is important to note that filiform corrosion is not observed in cases where the Al has been anodized or conversion coated. Recent work has shed some light on the mechanistic aspects of filiform corrosion as for Al specifically, revealing that susceptibility to filiform corrosion in architectural and automotive applications is due in most if not all cases to the presence of a deformed layer on the surface of the sheet (McMurray et al. 2010).

3.3.3 Intergranular Forms of Corrosion

IGC is a phenomenon where the precise mechanisms have been under debate for almost half a century (Hunter et al. 1963). While in a simple view, we can consider IGC a special form of microstructurally influenced corrosion; IGC can be summarized as a process whereby the grain boundary "region" of the alloy is anodic to the bulk or adjacent alloy microstructure.

Corrosion activity may develop because of some heterogeneity in the grain boundary structure. In Al–Cu alloys, precipitation of Al_2Cu particles at the grain boundaries leaves the adjacent solid solution anodic and more prone to corrosion. With Al–Mg alloys, the opposite situation occurs, since the precipitated phase Mg_2Al_3 is less noble than the solid solution (Searles et al. 2001). Hence, the behavior of the grain boundary region may be influenced by either the grain boundary particles themselves or the chemistry of the precipitate-free zone. Serious intergranular attack in Al alloys (whether the grain boundary phase is either anodic or cathodic in nature) may however be avoided, provided that correct processing and heat-treatment conditions are observed.

In the case of the Al–Mg system, almost all commercial alloys are supersaturated as the magnesium solubility at room temperature is less than 1 wt% (but significantly greater at elevated temperatures). This effectively means that for alloys with more than a few wt% magnesium, even mildly elevated service temperatures can lead to grain boundary precipitation (viz., sensitization of grain boundaries) and an IGC susceptibility. The extent of this sensitization may be approximately deduced from the apparent continuity or extent of Mg_2Al_3 precipitation at the boundaries. Apparently, continuous boundaries correspond to high levels of sensitization to IGC.

The level of sensitization to IGC of fabricated sheets, plates, or extrusions can be easily determined by measurement of the NAMLT value (ASTM G67). This is a 24 h exposure to nitric acid and a measurement

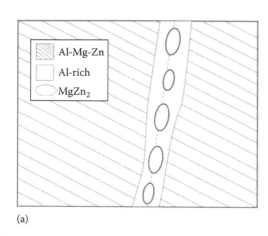

(a)   (b)

FIGURE 3.11 (a) Hypothetical grain boundary in an Al–Zn–Mg alloy. This schematic indicates the different chemistry that exists in the grain interior, solute-depleted zone (precipitate-free zone), and grain boundary precipitates—giving rise to electrochemical heterogeneity localized at the grain boundary region. (b) STEM (HAADF) image of high-angle grain boundary in AA7075-T651, revealing grain boundary precipitates ($MgZn_2$). (Photo courtesy of Nick Birbilis.)

of weight loss due to IGC and loss of grains. A NAMLT value of more than 30 g/cm² of surface is necessary before an alloy should be considered to become sensitized to IGC or cracking. The specific test method used to evaluate susceptibility to IGC depends on alloy type. For 5xxx series alloys, the NAMLT method (ASTM G67) is adopted, while for 2xxx and 7xxx series alloys, ASTM G110 is most common, employing testing in sodium chloride solutions containing hydrogen peroxide.

In the case of Al–Zn–Mg alloys, where the precipitated phase is the highly anodic ($MgZn_2$) η-phase, IGC occurs readily. Again, however, susceptibility to IGC is strongly dependent on the heat-treatment condition and its effect on grain boundary solute segregation and the morphology and composition of the grain boundary precipitate and the surrounding alloy matrix (Knight 2003). The most resistant heat treatments are based on the use of over-aging to the T7 treatment or more complex heat treatments that involve retrogression and re-aging to minimize the trade-off between alloy strength and IGC resistance (Sprowls 1978, Polmear 1995, Lynch et al. 2008).

The images in Figure 3.11 help rationalize the origins of IGC, using the Al–Zn–Mg system as a model.

IGC differs from pitting corrosion. While IGC may initiate from a pit, the propagation of IGC is more rapid than pitting corrosion, and while both may have a deleterious effect on corrosion fatigue (CF), the sharper tips produced by intergranular attack are drastic stress concentrators that may reduce the number of cycles to failure.

Exfoliation corrosion (Davis 1999, Zhao and Frankel 2007) of Al alloys is also frequently due to IGC. It generally occurs where the alloy microstructure has been heavily deformed (i.e., by rolling) and the grain structure has been flattened and extended in the direction of working. IGC attack from transverse edges and pits then runs along grain boundaries parallel to the major working direction of that alloy. Exfoliation is characterized by the leafing off of layers of relatively uncorroded metal grains caused by the swelling of corrosion product in the intergranular layers. Exfoliation is nominally only observed on aircraft components such as around riveted or bolted structural members. Testing for exfoliation corrosion is carried out by a number of ASTM tests, including the acidified salt spray test (ASTM G85), the ASSET immersion test (ASTM G66), and most commonly, the EXCO immersion test (ASTM G34).

3.3.4 Environmentally Assisted Cracking

Environmentally assisted cracking (EAC) is a term that incorporates SCC, liquid metal embrittlement (LME), CF, and hydrogen embrittlement (HE).

3.3.4.1 Stress Corrosion Cracking

SCC is a time-dependent intergranular fracture mode that requires the combined presence of a susceptible alloy, a sustained tensile stress and a corrosive environment.

The minimum tensile stress required to cause SCC in susceptible alloys is usually small and significantly less than the macroscopic yield stress (Scamans and Holroyd 1986, Scamans et al. 1987). The susceptibility to SCC initially placed restrictions on the use of high-strength Al alloys and until relatively recently had restricted the use of 7xxx series alloys to specific applications. There are several theories postulated for the mechanism of SCC. The main theories are either corrosion dominated where cracking is due to preferential corrosion along the grain boundaries by anodic dissolution (i.e., analogous to IGC) or hydrogen dominated where cracking along grain boundaries is enhanced by absorbed atomic hydrogen. The origin of this hydrogen is the result of IGC itself (so generated by corrosion in acidified cracks) and it is considered that the presence of absorbed hydrogen weakens grain boundaries. Again, this is an area under present fundamental research, and while several mechanisms may be simultaneously in action, the dominant mechanism is leading to failure (if there is one dominant mechanism) and is under debate (Lynch et al. 2008).

SCC development depends on both the duration and magnitude of applied tensile stress. Fracture mechanics tools for the determination of crack growth rates are commonly used in the evaluation of SCC resistance for Al alloys (Summerson and Sprowls 1974). Such tests suggest a minimum (threshold) stress intensity that is required for cracking to develop. Residual stresses in Al products that may arise as a result of quenching and cold working may also play an important role in SCC should the level of residual stress be significant.

SCC is an important phenomenon for Al alloys in instances when the high-strength class of Al alloys is deployed (since inevitably the use of high-strength materials will correlate with an application where high stresses will be applied). SCC of 7xxx alloys occurs in water and water vapor in addition to chloride-containing electrolytes. Most other susceptible alloys fail only due to exposure to environments containing chloride ions. An elegant micrograph showing an example of intergranular SCC in AA7079-T651 is seen in Figure 3.12.

Low-strength and relatively pure Al alloys are not susceptible to SCC. The alloys most prone to SCC are the 7xxx, 2xxx, and higher-strength 5xxx series alloys (i.e., those that have grain boundaries populated by precipitates). Most service failures involving SCC have occurred in the short transverse

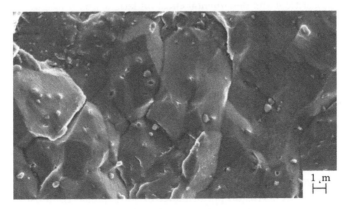

FIGURE 3.12 HR-SEM micrograph of a SCC fracture surface taken from an AA7079-T651 DCB specimen exposed to 75% relative humidity (specimen was from the T/6 position from 3 in. thick rolled plate). (Photo courtesy of Steven P. Knight.)

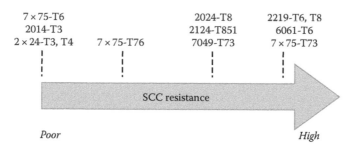

FIGURE 3.13 Qualitative SCC resistance of Al alloys (in the rolled plate form). (Adapted from Polmear, I.J., *Light Alloys: Metallurgy of the Light Metals*, 3rd Edn., Arnold, London, U.K., 1995; original source: U.S. Air Force Research Laboratory, Wright Patterson Air Force Base, OH.)

direction (Summerson and Sprowls 1974). The relative resistance of certain Al alloys to SCC is summarized later in Figure 3.13 (adapted from Davis [1999]). In order to do the topic justice, the reader is referred to specific monographs on the topic (Sedriks 1990, Jones 1992).

3.3.4.2 Liquid Metal Embrittlement

LME is not commonly observed in the general usage of Al alloys. It can be defined as a mode of attack that results in a complete loss of alloy ductility of a solid metal, well below the normal yield stress, as a result of the surface being wet by a liquid metal. LME-induced fractures are generally intergranular in Al alloys.

It is well known that gallium (Ga) in contact with Al can result in disintegration of the Al (or Al alloy) into individual grains, placing serious restrictions on Ga usage (viz., it cannot be carried onto an aircraft). In addition, mercury can also embrittle Al and its alloys. Al has also been noted as being embrittled appreciably by indium, sodium, tin–zinc, and lead–tin alloys.

3.3.4.3 Corrosion Fatigue Cracking

CF can be defined as the combination of cyclic plastic deformation with localized corrosion activity. The interaction of each of these mechanisms, and the transition from initiation to propagation, is a matter under present research and of considerable technological importance. Corrosion pits have been routinely observed to nucleate crack growth in structures subject to fatigue loading (Van der Walde et al. 2005). In fact, a recent paper reveals that in a fatigue study of Al alloy 7075-T6, all specimens in that study fractured from cracks associated with pitting (Jones and Hoeppner 2005). The relevance of this to the health of aircraft structures is critical. Indeed, the numerous studies referenced within Davis (1999) support the notion that pitting has a critical and detrimental effect on fatigue life. A fatigue crack can initiate from a corrosion pit or surface flaw when the flaw reaches a critical size at which the stress intensity factor reaches a threshold for fatigue cracking or conversely when the rate of fatigue crack growth exceeds that of pit growth. Reports have suggested that the critical flaw size to initiate a fatigue crack can be as small as an alarming 40 μm. Since CF is a phenomenon that involves a mechanical component (and not corrosion alone), significant work exists in the mechanical engineering field, including several technical publications, dedicated monographs, and handbooks (Hoeppner 1979, Kondo 1989, Hertzberg 1995).

3.3.4.4 Hydrogen Embrittlement

Hydrogen can readily dissolve in Al and all Al alloys, both in the molten state and during elevated temperatures close to the alloy melting temperature (where the atmosphere may include water vapor or hydrocarbons). The hydrogen dissolved within molten Al results in porosity in cast products but does not have an influence on corrosion performance or subsequent HE. Most importantly, however, an accumulation of literature is beginning to show that there is experimental evidence for hydrogen generated during corrosion penetrating into grain boundaries (the most rapid diffusion path and region of free volume in a crystalline material) and leading to embrittlement (Kondo 1989, Osaki et al. 2006).

The mechanism by which hydrogen causes the embrittlement is difficult to study (or image) and is postulated to act by either facilitating enhanced dislocation emission ahead of an advancing crack or by bond weakening and enhanced plasticity ahead of an advancing crack.

Understanding HE is important in unraveling the mechanism and modes of many failure processes for Al alloys; however, thus far, it has not been a key factor that has restricted the usage of Al. No specific tests for assessing HE susceptibility of Al alloys exist.

3.3.4.5 Environmental Influences upon Corrosion

The discussion of the performance of Al and its alloys in a range of environments is beyond the scope of this chapter. In essence, the presentation of information thus far allows the reader to appreciate (1) the pH-dependent behavior of Al and (2) the role of microstructure on the types and manifestations of corrosion that occur. Specific chemical combinations present within various electrolytes can influence what has been covered thus far by alteration of local chemistry. As a result, the reader is directed toward alternate sources for the role of environment, with an appropriate starting point being the many dedicated and comprehensive chapters in Vargel (2004).

3.4 Corrosion Prevention Strategies

The durability of Al and Al alloy products, assemblies, and components in a service environment depends on many factors such as the type of alloy used, the design and fabrication considerations, any protective coatings used, the environmental conditions, the maintenance programs, and the inspection requirements. In relation to design and fabrication, it is important that possible moisture entrapment areas be avoided and then, if still present, provisions for adequate drainage be considered, especially in areas such as where welds and galvanic couples and stresses (SCC) may be present. Designing to avoid corrosion is an extensive topic that has been dealt with comprehensively elsewhere (Roberge 1999, ASM Handbook 2002).

The choice of alloy for a particular component may depend on factors other than corrosion resistance such as load-bearing requirements, fatigue resistance and ease of fabrication, and even the price of that metal alloy. Over the years, Al alloys with good mechanical properties and adequate corrosion resistance have been developed for various applications in many industries. However, the nonhomogeneous nature of the microstructure of these alloys ensures that they will never be immune from some corrosion. In order to avoid corrosion, protective coatings are employed and where possible, modifications made to local operating environments. These preventative measures coupled with a well-planned maintenance program and adequately trained inspection teams provide the basis for a comprehensive corrosion prevention strategy. In the following section, methods and strategies are outlined that may be used to prevent and limit the amount of corrosion damage Al alloys may sustain in service. These methods are considered in the context of a simple model of the corrosion process.

Figure 3.14 is a simplified schematic of a localized corrosion process on an Al alloy. The process may be either corrosion pitting, crevice, intergranular, galvanic, or filiform corrosion, or SCC.

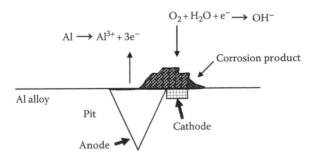

FIGURE 3.14 Schematic of a corrosion pit on an Al alloy surface. The cathode represents an intermetallic particle.

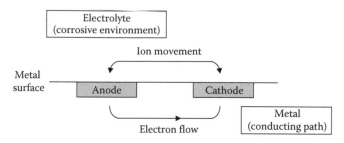

FIGURE 3.15 Schematic drawing of the electrochemical corrosion circuit.

Figure 3.14 shows the local anode and cathode separated, with Al going into solution at the anode as Al^{3+} ions. This anodic reaction is accompanied by a cathodic reaction, for example, reduction of oxygen, at the nearby cathode, which may be an adjacent intermetallic particle, or the mouth of a pit, crevice, or crack. As a result of the development of these anodic and cathodic sites, electron flow occurs in the metal and ionic transport of charge occurs in the moisture layer. Thus, an electrochemical circuit is established as shown schematically in Figure 3.15.

The rate at which an Al alloy corrodes is controlled by the slowest of these processes. The Al will not corrode to produce ions faster than the cathode can consume the released electrons or faster than the rate at which anions can be transported through the electrolyte. These basic principles help to understand how corrosion can be controlled or prevented.

One preemptive way to prevent corrosion is in the design phase where the designers can choose an Al alloy with a microstructure more resistant to corrosive attack. That is one containing fewer electrochemically active intermetallic particles such as in the 5000 series alloys compared with the 2000 or 7000 series alloys. Because of other material property considerations and requirements, it is not always possible to design a component using more corrosion-resistant alloys. The nature of the electrolyte in contact with the metal part can be changed to reduce its aggressiveness by removing or diluting dissolved ions such as chlorides, changing the conductivity, and altering the pH. Also, the additional application of inhibitors can be used to produce protective films on the surface. Reducing the concentration of chloride ions in the environment, for example, during washing or rinsing, maintenance operations will reduce the efficiency of ionic transport that affects both anodic and cathodic processes. Thus, modification of the environment and the electrolyte resulting from that environment forms a significant part of any corrosion prevention strategy.

A major part of any corrosion prevention strategy is the exclusion from or the restriction of the access of moisture and conducting ions to the metal surface, through the use of paint coatings. A coating not only acts as a barrier to moisture, it also can provide a high-resistance pathway to the movement of ions when that barrier has been permeated by water, thus restricting the flow of ions in the electrochemical circuit as depicted in Figure 3.15. In the event that a coating is damaged or deteriorates over time and fails mechanically, such breaches allow moisture to penetrate to the exposed metal substrate. Inhibitor pigments contained in the pretreatment layer or paint coating or subsequently applied during some preventative maintenance action such as detergent washing and application of corrosion-inhibiting compounds (CICs) (to be discussed later) can provide additional corrosion protection. Some inhibitors can act to prevent corrosion by shutting down potential electrochemical reactions on plain surfaces and inside crevices, pits, and cracks. Other inhibitors may produce protective layers on the surface either by combination with the metal to form an oxide film, for example, or by adsorption of molecules onto the surface.

3.4.1 Corrosion Inhibitors

Corrosion inhibitors are chemical substances that have some solubility in water and that prevent or slow corrosion in aggressive aqueous environments. The presence of an inhibitor may decrease the rate of the anodic process (anodic inhibitor), the cathodic process (cathodic inhibitor), or both processes

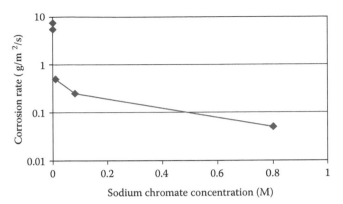

FIGURE 3.16 The effects of chromate concentration on the corrosion rate of AA7075-T651 in 0.1 M NaCl solution. (After Hinton, B.R.W., *Corrosion*, 66(8), 1, 2010.)

(mixed inhibitor). Anodic inhibitors function by reacting with Al to produce a thin passive film that leads to a decrease in the corrosion rate. Hexavalent chromates, for example, zinc strontium and barium chromates, are extremely effective inhibitors for Al alloys (Frankel and McCreery 2001). As Figure 3.16 shows, a very small amount of chromate in a sodium chloride solution can reduce the corrosion rate of a susceptible alloy AA7075-T651 to almost zero (Hinton 2010). Chromate does this by producing a passive corrosion-resistant oxide film on the Al surface. This is the result of the chromate ion adsorbing at corrosion sites across the alloy surface where it is reduced to form a mixture of trivalent chromium oxide and Al oxide. This film reduces the activity of both anodic sites and cathodic sites (Frankel 2001). At the anodic sites, the passive film prevents dissolution of the Al, and at the cathodic sites, the film prevents oxygen adsorption and restricts access of electrons to oxygen for oxygen reduction to occur.

Cathodic inhibitors typically affect the rate of the oxygen reduction reaction and/or the hydrogen evolution reaction. Common cathodic inhibitors for Al alloys are those that react with hydroxyl ions produced at cathodic sites to form an insoluble complex hydroxide layer (corrosion product) at those sites (see Figure 3.14). This film provides a barrier between the cathodic site and the electrolyte restricting access of oxygen or hydrogen ions to the site. Thus, ionic flow in the electrochemical circuit is restricted and the rate of corrosion reduced. Zinc salts and the salts of the rare earth metals (Hinton et al. 1984a, Hinton 1989) may act as effective cathodic inhibitors. The latter by allowing complex rare earth metal oxides/hydroxides to precipitate at cathodic sites.

Many organic compounds such as sulfonates, phosphonates, amines, and quaternary ammonium salts are also inhibitors of corrosion for Al alloys (Hinton et al. 1984b,c). They function by chemisorbing onto the metal surface either at anodic, cathodic sites, or both. The chemisorption process involves an electronic interaction between the inhibitor and the metal via a donation of electrons through atoms such as nitrogen, oxygen, and sulfur in functional groups on the inhibitor molecule. The adsorbed layer effectively acts as a barrier to the environment by preventing access of water, chloride ions, and oxygen.

Beneficial synergistic effects through mixed inhibition are possible using mixtures of various inhibitors or through the design of a particular compound. For many years, zinc chromate has been used effectively for the inhibition of corrosion on Al. The zinc forms a precipitate of hydroxide at the cathodic sites, while the chromate produces a passive film at the anodic sites. Forsyth and coworkers have used the known inhibiting effects of the carboxylate ions (Mercer 1980) (organic inhibitors) and the proven effectiveness of the rare earth metal salts (cathodic inhibitors) (Hinton et al. 1984a) to produce compounds having a synergistic inhibiting effect (Markley et al. 2007, Forsyth et al. 2008).

While many compounds are known to be effective inhibitors of Al alloys, they are not as effective as chromates. Chromates are used across all industries as Al corrosion inhibitors, but their use is becoming increasingly restricted over concerns for occupational health and safety and environmental pollution. Hexavalent chromates are toxic and known human carcinogens.

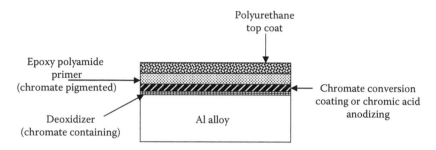

FIGURE 3.17 The various layers of a typical corrosion protection system used for Al alloy aircraft components.

The various actions of inhibitors as described earlier are utilized in a range of protection systems that form the basis of corrosion prevention strategies. They are (1) present in conversion coatings and anodized films, (2) incorporated into paint coatings, (3) present in temporary corrosion protection coatings, and (4) included in detergent washing formulations. These strategies are all discussed in the following sections.

3.4.2 Conversion Coatings

Al alloys are widely used in numerous applications for building products, in automobiles, ships and aircraft, and for food and beverage cans. In all of these applications, the alloys are exposed to a variety of outdoor and indoor environments. The natural passive oxide on Al alloys is insufficient to protect from corrosion in these environments. Therefore, a range of protection schemes and technologies are required to prevent degradation under these conditions. For many engineering applications, and certainly the most demanding ones (e.g., aircraft components), a corrosion protection system consisting of multiple coating layers is used (see Figure 3.17). Each layer contributes to the functionality of the overall coating system. Functions include promotion of coating adhesion (to the underlying alloy), flexibility, low permeability to moisture and oxygen, chemical resistance, abrasion resistance, electromagnetic signature control, and corrosion protection. This section describes the various layers that make up a corrosion protection system that forms part of an overall prevention strategy.

3.4.2.1 Pretreatments

Paint coatings do not readily adhere to the natural oxide film present on Al alloys. In order to achieve good adhesion, a complex pretreatment process is required that ultimately produces a modified oxide film.

The first stage common to most pretreatment processes is surface preparation. It is possibly the most important step in applying a protective system. If not carried out correctly, any coating system will soon fail. Generally, surface preparation consists of several processes that could include solvent or vapor degreasing, an alkaline clean to remove greases and oils, and several intermediate water rinses. The alkaline clean may have an etching effect on the surface providing keying points for subsequent paint coatings, but it can also leave a residue or smut on the surface. These processes are followed by deoxidizing, which involves treatment with an inhibited acid solution (e.g., phosphoric acid containing a chromate inhibitor). The purpose of deoxidizing is to (1) remove any smut left from the cleaning process, (2) remove the existing natural oxide film, and (3) produce a uniform oxide film ready for the next stage that may be either a conversion coating or an anodizing process. References (Biestek and Weber 1976, Wernick et al. 2007) provide detailed descriptions of these pretreatment processes and the characteristics of the coatings they produce.

3.4.2.2 Chromate Conversion Coating

The nature and use of chromate conversion coatings (CCCs) were first reported by Bauer and Vogel in 1915 (Wernick et al. 2007). They are used, for example, with both cast and wrought Al alloy architectural fittings and aircraft components. The coatings provide a layer of corrosion protection and improve the adhesion of any paint coatings. They also have a low resistivity that allows them to be used in the electronics industry, for example, for waveguides.

CCCs are produced by immersion, spraying, or swabbing with an acid solution containing the salts of hexavalent chromium ions (chromates). Common trade names for the process are Alodine™ and Iridite™. It is generally accepted (Treverton and Davies 1977, Katzman et al. 1979) that the chromate ions in the acid solution react with the Al surface resulting in a deposition of a complex film of hydrated chromium oxides. Two types of CCC solutions are used.

The chromate–phosphate solution contains chromium trioxide, phosphoric acid, and hydrofluoric acid. It is used for coating architectural trim, such as screen door mesh and window frames, and for caravan and mobile home exterior panels. The coating has a dark-green appearance and consists of an outer layer of hydrated chromium phosphate with a layer of Al oxide adjacent to the metal substrate (Newhard 1979).

The most commonly used CCC is produced by the chromium–chromate solution. This solution contains chromium trioxide, hydrofluoric acid, and potassium ferricyanide. The role of hydrofluoric acid and ferricyanide is to assist with the deposition process (Treverton and Davies 1977). This chromating solution is used on many Al alloy aircraft components. When applied to AA2024, a common aerospace alloy, the coating formed consists of an outer layer of hydrated chromium oxides and hydroxides with an inner layer of chromium oxide (Hughes et al. 1997). Hughes et al. (1997) have found that the film also retains some soluble hexavalent chromate ions.

The oxide film produced by the chromium–chromate solution is typically around 0.5–1 μm thick (Biestek and Weber 1976). This film is very adherent and can provide protection for alloys such a 2000 and 7000 series in excess of 168 h in the very severe industrial standard neutral salt spray (NSS) test (ASTM Standard B 117 1987). The presence of the film on the surface acts as a barrier to any environment and inhibits the dissolution of Al and retards the rate of any reaction at cathodic sites. One of the advantages of the CCC process is that the availability of retained hexavalent chromate ions to be leached from any break in the coating provides a self-healing property. The chromate ions act to inhibit corrosion at the breakdown site.

The application times for treatment with the solution to produce the CCC may vary depending on the product and the alloy but can be up to several minutes. Figures 3.18 and 3.19 show that for two high-strength alloys, an optimum CCC time was observed. With too long a contact time, both the corrosion protection performance, as measured by the time for pits to be detected on a specimen surface exposed to the NSS test, and the adhesion of a paint coating were reduced. In relation to Figure 3.19, adhesive failure is when the locus of fracture area in the pull off adhesion test is at the paint layer/chromated substrate interface. A desirable low-percentage adhesive failure measurement means that in the adhesion

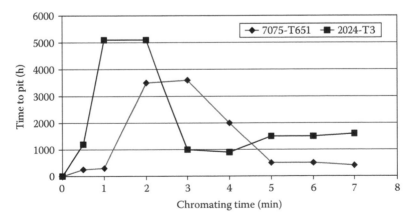

FIGURE 3.18 The effects of immersion time on the corrosion protection performance in the NSS of CCCs on AA7075-T651 and AA2024-T3. (After Trathen, P.N. et al., The variability of the chromate conversion coating process and the effect on corrosion protection performance, in *Proceedings of the Conference of Australasian Corrosion Association*, Newcastle, Australia, paper 33-08, 1993.)

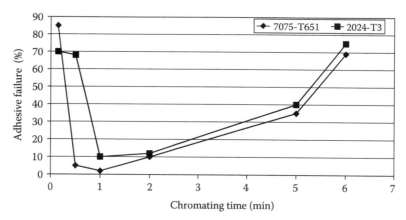

FIGURE 3.19 Percentage of adhesive failure as a function of chromating time for AA7075-T651 and AA 2024-T3 coated with an epoxy–polyamide paint. (After Trathen, P.N. et al., The variability of the chromate conversion coating process and the effect on corrosion protection performance, in *Proceedings of the Conference of Australasian Corrosion Association*, Newcastle, Australia, paper 33-08, 1993.)

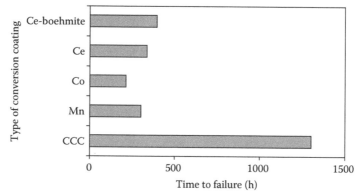

FIGURE 3.20 The times to failure in the NSS test for AA7075-T651 with various non-CCCs. (After Hughes, A.E. et al., *ATB Metallurgie*, 43(1–2), 265, 2003.)

test, the locus of "cohesive" fracture occurs within the paint layer (Trathen et al. 1993). Any decrease in corrosion protection performance and any increase in the adhesive failure value with longer immersion times are thought to be due to the formation of a thicker coating and the subsequent spalling and cracking of the coating. It is worth commenting on the differences in the times to failure for AA7075-T651 in Figure 3.18 and the result in a different study shown in Figure 3.20. These studies used different pretreatment processes and chromating times. The differences demonstrate the importance of these two parameters and also the known variability in the NSS test.

3.4.2.3 Non-Chromate Conversion Coatings

Due to environmental, health, and safety concerns, there is increasing legislative pressure in various parts of the world to limit and prevent the use of chromates due to their inherent toxicity and carcinogenicity. The fact that chromium is a heavy metal also causes disposal problems when discarding CCC process solutions. Many alternatives to CCCs for use on Al alloys have been discussed (Hinton 1991a,b, Buchheit and Hughes 2003, Hughes et al. 2003).

Comprehensive studies of the corrosion protection performance of various commercially available alternatives in the NSS test (Chalmer 1995, Buchheit and Hughes 2003, Hughes et al. 2003) and with paint coatings in place (Placzankis et al. 2003) have been carried out in recent years. Hughes et al. (2003) have investigated four alternative processes all of which produce a metal–oxide or metal–Al oxide conversion coating. They compared the performance of the alternative coatings with that of a commercial CCC process. The alternative processes produced oxide films of (1) cobalt (Roland and Hart 1996), (2) manganese (Bibber 1996), (3) cerium (Hughes et al. 2000), and (4) a mixture of Ce and Al oxide (boehmite) (Trueman et al. 2000). Figure 3.20 shows the results from NSS corrosion tests on AA7075-T651 specimens with the various coatings. The time to failure was taken as the first appearance of a specified number of optically detectable pits, as specified in U.S. Military Specifications MIL-DTL-5541F (2006) and MIL-DTL-81706 (2006). These results show that the time to failure for each of the alternative coatings was much less than that for the CCC. However, the results were sufficient to pass the criteria set down in those specifications. The chromate-free coatings included a step to seal the oxide layer. While the sealed coatings obviously improved the corrosion protection performance, it was found that sealing had a detrimental effect on the adhesion of subsequent paint coatings (Hughes et al. 2003).

Buchheit and coworkers (Buchheit and Martinez 1998, Buchheit et al. 2002) have developed a conversion coating based on compounds known as hydrotalcites. These are naturally occurring claylike minerals, for example, hydromagnesite, $Mg_6Al_2(OH)_{16} \cdot CO_3 \cdot 4H_2O$. These compounds were formed as a coating on Al alloy surfaces by immersion in an alkaline lithium salt solution. The resulting film was sealed in a cerium nitrate solution. Buchheit et al. (2002) have shown that these coatings protected AA2024-T3 in the NSS test for 168 h as required by the U.S. Specifications (MIL-DTL-5541F 2006, MIL-DTL-81706 2006). However, under constant immersion conditions in a NaCl solution, the corrosion resistance as measured by electrochemical impedance was an order of magnitude less than that observed for a CCC.

It has been concluded by Hughes et al. (2003) and others (Chalmer 1995, Eichinger et al. 1997) that while many of these alternative conversion coatings are promising, none offer a ready-made alternative to CCCs taking into account both corrosion protection performance and adhesion of paint coatings. Until an alternative is available with acceptable performance for the high-strength Al alloys in particular, or when specific legislation is enacted that completely prohibits their use, CCCs will continue to be used.

3.4.3 Anodizing

Commercial anodizing processes that produce anodic coatings on Al alloys were developed in the 1920s. They are used for corrosion protection and decorative purposes, as well as to provide hard wear-resistant surfaces, electrical insulating layers, and to improve paint adhesion. Typical applications include architectural fittings, automotive trim, furniture and home appliances, and interior and exterior aircraft components.

The coatings are produced by anodic polarization of the Al alloy in electrolytes of chromic acid or sulfuric acid with proprietary additives. Chromic acid anodizing is widely used in the aerospace industry. Other electrolytes based on oxalic or phosphoric acid may be used (Gabe 1972), especially as a pretreatment prior to adhesive bonding. For most applications, anodizing produces an oxide film thickness of between 5 and 30 µm depending on the type of acid electrolyte (Wernick et al. 2007). (Natural oxide films on Al are typically around 0.05 µm.) The film consists of a thin nonporous barrier layer (<1 µm) at the metal–oxide interface and a much thicker outer porous layer. Because the oxide film is an integral part of the metal substrate, the adhesion of anodized films is extremely good. Organic and inorganic dyes may be added to the anodizing process solution to provide a colored finish. For wear resistance, a special class of coating known as "hard anodic coatings" with thicknesses of 50–100 µm may be produced using a sulfuric acid electrolyte at low temperatures. These "hard films" can be brittle and prone to cracking.

Even though anodized coatings provide better corrosion protection than CCCs, their use on products subjected to cyclic loading may be restricted due to cracks that may develop in the coating causing adverse effects on fatigue strength. Such cracks can lower the fatigue strength of anodized Al alloys by

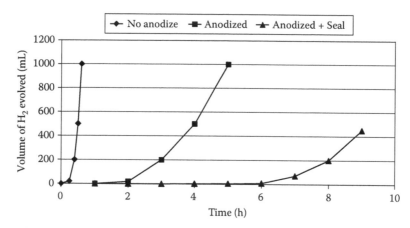

FIGURE 3.21 The corrosion rate in an HCl solution for an Al–Cu alloy with (i) no anodized coating, (ii) an anodized coating, and (iii) an anodized and sealed coating. (After Biestek, T. and Weber, J., *Electrolytic and Chemical Conversion Coatings: A Concise Survey of Their Production, Properties and Testing*, Portcullis Press Ltd., Surrey, U.K., 1976.)

30%–50% (Cochran and Sprowls 1979, Stevenson 2002). Because of this effect, CCC and anodized coatings may be separately used on different components in an aircraft structure depending on structural and loading considerations.

The effectiveness of anodic films in providing corrosion protection was demonstrated by Champion and Spillett (1956) over 50 years ago. They studied the corrosion protection performance of a thin anodic film (6.5 μm thick) on 99.5% Al in both an industrial and marine atmospheric environment. It was found that compared with the case of no anodizing present, the anodic coating increased the time to failure in both environments by 50%. With the anodized film present, time to failure was increased to 2.5 years for the industrial environment and to 1 year for marine exposure. Failure was measured as the time to a fixed arbitrary level of surface degradation on the specimens.

The corrosion protection performance of anodized coatings may be improved by sealing in boiling water or in solutions of nickel or cobalt salts (Biestek and Weber 1976). Figure 3.21 shows the corrosion rate of an Al–Cu alloy in a hydrochloric acid solution as indicated by the change in volume of hydrogen gas liberated from the alloy surface as a function of time. The data clearly show the increased corrosion protection provided by the presence of the anodized coating and the improvement made by sealing in boiling water. Sealing has the effect of growing the walls of any pores in the oxide film, thus reducing any porosity (Biestek and Weber 1976, Wernick et al. 2007).

The thickness of an anodic film also has a significant effect on corrosion protection performance. Results from Mader 1972 in Figure 3.22 show the number of pits counted on specimens of AA1100 exposed in an industrial atmospheric environment for 8 years, as a function of coating thickness. The specimens were sulfuric acid anodized and sealed in boiling water. These data indicate that the number of pits decreased by several orders of magnitude with increasing film thickness. Similar behavior was observed for in less aggressive environments (Cochran and Sprowls 1979). It is more difficult to achieve high levels of protection with anodized coatings in marine atmospheres (Cochran and Sprowls 1979) and under constant immersion conditions in salt solutions (Danford 1994) on alloys such as AA2024 and 7075 than it is on the less corrosion-prone alloys such as 1000 series. This is probably related to the absence of intermetallic particles in the microstructure and their effects on the structure of the anodized film.

As with CCCs, the presence of the anodized layer on the surface acts as a barrier to any environmental agents, inhibits the dissolution of Al, and retards the rate of any reaction at cathodic sites.

Concerns about chromates have also led to the development of boric–sulfuric acid anodizing as a replacement for chromic acid anodizing, and this process is being used on some aircraft components (Boeing 2010).

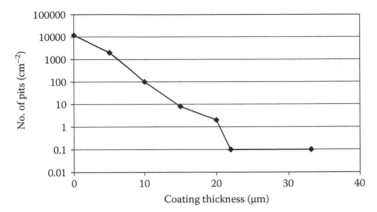

FIGURE 3.22 The effect of the thickness of the anodized coating on its corrosion protection performance as measured by the number of pits that developed on the surface of AA1100 after exposure for 8 years in an outdoor industrial environment. (After Mader, O.M., *Metals Mater.*, 7, 303, 1972.)

However, recent studies with AA7075 have shown that in terms of corrosion protection performance, boric–sulfuric acid anodizing is not as effective as chromic acid anodizing (Thompson et al. 1999).

3.4.4 Organic (Paint) Coatings

Painting (organic coatings) is the most common method used to protect Al alloys when the application requires prolonged exposure to outdoor environments. They are variously referred to as varnishes, lacquers, enamels, primers, finishing, and topcoats. Paints consist of volatile and nonvolatile components. The volatile components are solvents, while the nonvolatiles are the film-forming ingredients in the paint. The latter include polymeric resins (the back bone of the paint), pigments, oils, plasticizers, fillers, and extenders. All of these ingredients produce the properties of a good paint coating, which include uniform coverage, low porosity, low permeability to water and oxygen, good adhesion, and resistance to mechanical damage and chemical attack. Pigments and additives help reduce moisture permeation, provide a source of corrosion inhibitor, provide resistance to infrared (IR) and ultraviolet (UV) radiation degradation, and add color. Plasticizers help prevent brittleness during application and during service, while extenders assist with the coverage of the painted surface. Lambourne, Hare, Schweitser, and Wicks (Lambourne 1987, Hare 1994, Schweitzer 2007, Wicks et al. 2007) all provide detailed descriptions of paint coating chemistry, paint properties, application methods, and mechanisms of protection.

A long service life for a coating depends on (1) the type chosen for the application; (2) the surface preparation (pretreatment); (3) its ability to resist environmental factors such as UV and IR radiation, moisture, and oxygen; and (4) the coating remaining adherent to the substrate during its service life.

Most paint coatings for Al alloys consist of several layers of different paint types or several of the same type. For the optimum performance of the coating, it is common to put a primer coat directly onto the base alloy pretreated with an anodizing or a CCC. A finishing or topcoat may be applied over the primer. The role of the primer is to provide good adhesion for the topcoat and to act as a source of corrosion-inhibiting pigments. The topcoats are used to provide the decorative properties, such as color and gloss, and in the case of military vehicles and aircraft, camouflage and low-IR radiation signatures. It also acts to protect the primer from mechanical damage and atmospheric radiation. In some cases where a rough surface is to be painted (e.g., a casting), another intermediate coating between the primer and the topcoat (undercoat) may be used. These various layers including the pretreatment make up the complete corrosion protection system.

In recent years, environmental concerns about air quality have generated interest in the use of paint systems containing lower levels of volatile organic compounds (VOCs). This has led to water replacing volatile organics as the solvent (i.e., waterborne technology). Another means through which low VOC levels may be achieved is with high solids technology. This involves the use of low-molecular-weight solvents, which do not evaporate as the coating cures, but react with the polymer resins and form part of the polymeric backbone of the paint. A review by Spadafora et al. (1997) provides a summary of developments in the areas of low VOC and high solids technology in relation to aerospace protection systems.

The selection of a paint system takes into account many factors such as the size and shape of the Al article or structure to be coated, the appearance and protection levels required, and economic factors. Therefore, the coating system chosen for indoor decorative products will be different to that for architectural fittings and will be different to that for automotive and aircraft components. It is not always possible to use the absolute best paint system for Al products in terms of performance and properties, as constraints such as cost of coating, safe working conditions, and environmental regulations must also be considered.

Corrosion prevention through the use of paint coatings is due to both a barrier effect provided by the paint layer and active corrosion protection. The barrier prevents contact of the underlying Al substrate with the environment, and this function is responsible for most of the protection provided by the coating. As simple as the application of coatings may seem, correct application is difficult to implement in practice. Coatings can contain defects that are formed during the application process or may develop during service. UV radiation can significantly degrade the integrity of paint coatings. Thus, the rate degradation of a paint coating system determines the useful life of any product, component or structure, or the length of time before maintenance and repainting is required.

No paint coating barrier is totally effective in preventing access of water, oxygen, and soluble ions to the underlying metal. Data and calculations by Mayne (1995) and Hare (Hare, 1994) have shown that a wide range of polymeric coatings on steel are unable to prevent enough of these reactants from reaching the steel substrate to prevent the cathodic reaction of oxygen reduction from occurring. Therefore, some of the protection effect of a coating must also be due to the coating having a high resistance to ionic flow that is a necessary part of the corrosion process (refer Figure 3.15). This resistance reduces the effective current between any anodes and cathodes that may develop beneath the coating. The development of cracks and flaws in the coating will obviously accelerate the transport of the necessary reactants for corrosion to the underlying metal.

A large degree of the protection provided by a paint coating is due to pigments included in the paint formulation acting as corrosion inhibitors. The solubilization of corrosion-inhibiting pigments and their transport with moisture to the metal coating interface allows the inhibitor to prevent or arrest either the anodic or cathodic reaction on the metal surface that forms the corrosion process. As discussed earlier, strontium, zinc, and barium chromates have been the most common of these inhibitors used in paint formulation for over 80 years. Typical concentrations of these chromates are around 15% by weight, and their solubilities are very low (Hare 1994). When the rate of transport of moisture is matched to the solubility of the inhibitor ions, enough of these ions can migrate to the metal surface to inhibit corrosion. A clever paint formulation has the rate of release of the inhibitors at a sufficient level such that the reservoir is not rapidly exhausted. A high concentration of inhibitor insures that the inhibitor lasts for many years. However, the effects of inhibitor concentration on mechanical properties of the paint coating are also important considerations in paint formulation.

3.4.4.1 Paint Coating Systems: Buildings

A large-volume market for Al alloys is the building industry. Al alloy architectural components include window frames, facades, louvers, curtain walls, storm doors, and balustrades. These applications see a range of environments with various levels of severity. Anodized coatings with no overlaying paint coating are widely used for interior use. They can be produced in various colors with matt or gloss finishes. Where higher levels of corrosion protection are required, three types of paint coatings are typically used in the building industry:

1. For high-volume outdoor structures and architectural applications, it is common to use solvent or water-based paints using epoxy, acrylic, polyester, or polyurethane resins. These may be used with or without primers to provide increased adhesion. However, a primer coat adds to the cost. When the use of chromate is not permissible, zinc phosphate is a commonly used inhibitor pigment.

2. When long-term outdoor exposure performance is required (e.g., for high-rise buildings), solvent-based coatings using the polymer polyvinylidene difluoride (PVDF) as the resin are a frequent choice. These coatings are very durable, resistant to UV radiation, but are expensive as several coats are generally required for optimum protection.

3. Powder coating is a technology that applies dry paint powder to an Al alloy product when it is lowered into a fluidized bed of powder that may or may not be electrostatically charged. The product is then placed in an oven; the powder particles melt with associated chemical cross-linking, coalesce, and form a high-density continuous film. The powders are based on polyester, polyurethane, acrylic or epoxy resins, or mixtures. The coatings may be applied over CCCs that provide added adhesion and a source of inhibiting chromate ions. They contain almost no VOCs and are therefore environmentally friendly. They have a high density, are tough and durable, and provide excellent corrosion protection for architectural Al alloys. Because of these properties, powder coatings are ideal for buildings in areas where access may be difficult. They are also used widely for marine and automotive applications.

3.4.4.2 Paint Coating Systems: Aircraft

Many internal and external aircraft structural components are made from 2000, 6000, or 7000 series Al alloys. Aircraft operate in a very diverse range of environments arising from the atmosphere in which they fly and the airports from which they operate. They may be exposed to salt contamination near seacoasts, to high-humidity tropical conditions, and dusty dry locations. The environment inside an aircraft structure may be different to that outside, when flying or on the ground where the aircraft is parked. In these operating environments, the alloys mentioned earlier are all susceptible to corrosion.

The corrosion protection systems used on aircraft generally consist of an anodized or CCC pretreatment, a primer, and a finishing topcoat. In some areas of internal structure, a primer only is applied, but in areas where moisture and aggressive contaminants collect such as under floors, around toilets, and galleys, several layers of topcoat may be applied. The main purpose of the pretreatment process is to provide a coating with some level of corrosion protection capability and improved adhesion for the following paint coatings. The pretreatment process is most important because it removes contaminants, which may affect the adhesion of the paint coatings.

The primer is the primary source of protection in the system and provides improved adhesion of subsequent topcoats. The one most commonly used in aircraft structure is based on a two-part epoxy–polyamide polymer. It consists of an epoxy resin binder and a polyamide resin hardener. Titanium dioxide is added for durability and opacity, extenders such as silica are added to fill out the coating, and pigments for color. The most common corrosion inhibition pigment used in this primer is strontium chromate. This pigment provides a source of the powerful hexavalent chromate inhibitor ions. These ions may leach from the coating particularly at points where the coating is damaged or cracking occurs and there provide protection.

For external aerodynamic surfaces on aircraft, such as fuselages, tails and wings, and in landing wheel bays, and for internal areas where aggressive environments collect as described earlier, a polyurethane topcoat is applied over the primer. This topcoat is a two-part system consisting of an isocyanate resin and a hydrolated polyester hardener. The function of the topcoat is to protect the primer against UV radiation, to provide chemical resistance, and to be durable and flexible. Polyurethane topcoats are generally applied to a dry film thickness of around 50 µm.

The U.S. Navy (USN) has developed a self-priming polyurethane topcoat with low VOCs and with no chromate inhibitors to replace the strontium chromate epoxy–polyamide primer and polyurethane topcoat. The self-priming topcoat possesses the adhesion- and corrosion-inhibiting properties of the primer–topcoat system it replaces, and it has the same durability. This technology has been applied to many USN aircraft. However, the long-term effectiveness of the topcoat in all operating environments is yet to be established.

3.4.4.3 Paint Coating Systems: Army Ground Vehicles

An analogous multifunctional coating system to the USN coating has been developed by the U.S. Army Research Laboratory (Escarsega et al. 1999, 2004). This coating is universally applied to U.S. Army ground vehicles. It is a two-part water-based polyurethane chemical agent resistant coating (CARC) that is chromate-free and low in VOC content and does not contain hazardous air pollutants. In addition to providing corrosion protection, it has other functionalities including low observable properties (visible and IR), high abrasion resistance and durability, and the ability to withstand chemical agent decontamination treatments. The CARC coating standards are covered by two U.S. Department of Defense (DoD) Specifications "Chemical Agent Resistant Coating (CARC) System Application Procedures and Quality Control Inspection" MIL-DTL-53072C U.S. DoD, Washington, DC, June 2003 and "Coatings Aliphatic Polyurethane, Single Component, Chemical Agent Resistant" MIL-DTL-53039B, June 2005.

3.4.4.4 Paint Coating Failure

As indicated earlier, corrosion will not occur until the protective coating system fails. The development of corrosion on a typical aircraft component provides a good illustration of how failure of a paint coating may occur. All paints on aircraft structure will admit moisture over time, but they will also degrade due to weather exposure, temperature excursions from high-altitude flight, soil accumulation, and stressing. Evidence of degradation is discoloration, chalking, and cracking; the latter occur particularly in areas of structure where flexing can occur.

On a component where there is very little movement or stress, the coating may fail at sites where defects were present from the time of application (e.g., insufficient paint coverage at sharp corners) or where flaws develop due to chalking and cracking associated with prolonged exposure to UV radiation, rain, and airborne contaminants. With a component under operating stress, many factors contribute to the mechanical failure of paint coatings. All paints age, harden, and crack with time due to constant exposure to extremes of temperature, IR, and UV radiation (Wicks et al. 2007). These environmental factors break down the molecular structure of the coating and under stress they crack. Cracks will admit moisture through the coating and at a faster rate than by normal diffusion. The stresses present that cause the coatings to crack arise from operating or working loads on the component such as the constant aerodynamic buffeting during flight and cyclic stressing during pressurization cycles. Over fastener heads, at holes, and along joint lines, thin films of paint of the order of 50–70 µm will be subjected to significant stress levels. Within the joints (e.g., inside fastener holes and on mating surfaces), the constant working of the joint due to the operational loads will produce a rubbing and fretting action. The absorption of moisture into the coatings and the freeze/thaw cycle due to the aircraft alternating between time on the ground and at high altitude will also produce an internal source of cyclic stress. All of these stresses, in combination with the degradation of mechanical properties of the thin polymeric coating layer, will eventually produce many cracks and flaws. It is the combination of years of operation and environmental exposure that will eventually cause the protective coatings to fail mechanically. The network of cracks formed will allow all constituents of the atmospheric environment described earlier to penetrate through to the substrate metal.

Even though moisture and airborne contaminants such as salt may accumulate within cracks that have developed in coatings, corrosion will not occur immediately. As mentioned in the previous section, the epoxy–polyamide primer not only provides a barrier coating as part of the total protective scheme but also acts as a source of chromate inhibiting ions. Small quantities of chromate ions are also available from the CCC and from any chromate sealed anodized layer. Chromate ions leached from the primer and conversion coating at the point of failure migrate to any exposed metal and form a protective film of chromium oxide (Frankel and McCreery 2001). However, this film, while protective, will also break down over time, due to the movement of the joints and fastener through rubbing and fretting. While additional chromate ions from the coating may be available to repair the damaged oxide film, due to the continual ingress of moisture or the continued presence of existing moisture through the processes describe earlier, gradual depletion of the chromate ions from the primer by leaching occurs. At a point in time when all available chromate ions of sufficient concentration to prevent corrosion are consumed, the protective capability of the coating system will be lost and corrosion will occur.

3.4.4.5 Non-Chromate-Inhibiting Pigments for Paints

The projected limited use of chromates into the future due to their inherent toxicity, and carcinogenicity has been discussed earlier. Many nontoxic alternatives to chromates as pigments in paints such as zinc phosphate, borates, molybdates, and silicates have been investigated (Twite and Bierwagen 1998). Although these pigments have been shown to provide some level of inhibition (Mayne 1995), none have been shown to be as effective as chromates (Eichinger et al. 1997, Twite and Bierwagen 1998). However, some combinations of inhibitors (Spadafora et al. 1997) have provided levels of inhibition similar to that provided by chromates when included in coatings.

The salts of the rare earth metals such as cerium and lanthanum are known to be very effective inhibitors of corrosion of Al (Hinton 1989, Forsyth et al. 2008). Over recent years, considerable research has been conducted on the protective performance of coatings such as epoxy–polyamides containing rare earth metal salts (e.g., cerium sulfate and formate) (Smith et al. 1996), cerium nitrate (Morris et al. 1999), and cerium diphenyl phosphate (Hughes et al. 2005, Markley 2008). These research programs typically used Al alloys such as AA2024 or 7075 coated with an epoxy–polyamide or a polyurethane coating containing a scribe through to the substrate metal. The coated specimens were exposed to aggressive test environments such as the NSS test and the extent of underfilm corrosion away from the scribe measured with time. While all of these formulations showed significant levels of protection, none were found to be as effective as chromate pigmented coatings.

Hydrotalcites (HT) described earlier in the section on chromate-free conversion coatings have been modified and assessed as potential non-chromate pigments for coatings (Williams and McMurray 2003). Inhibiting anions such as carbonates, nitrates, and chromates were absorbed onto the hydrotalcite compound, hydromagnesite, and their effectiveness as pigments for inhibiting filiform corrosion on Al alloy AA2024 in a high-humidity environment was assessed. The delaminated area due to filiform corrosion that developed adjacent to a scribe in a polyvinyl butyral coating was used as a measure of effectiveness. As Figure 3.23 shows, the HT pigments provided significant resistance to the development of filiform corrosion after 7 days exposure, but the performance did not match that of a chromate-pigmented coating.

Eichinger et al. (1997) have concluded in a recent review that there are no nontoxic alternatives that match the protection performance of chromates as pigments in coatings.

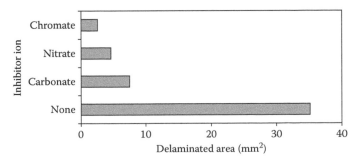

FIGURE 3.23 Area of delamination due to filiform corrosion on AA 2024-T3 as a function of inhibitor ion incorporated into a hydrotalcite pigment. (After Williams, G. and McMurray, H., *Electrochem Solid-State Lett.*, 6(3), B9, 2003.)

3.4.5 Alclad Coatings

Another layer that may form part of a protection scheme for Al alloys is metallic rather than nonmetallic. As discussed in an earlier section, the presence of various microstructural phases in the surfaces of Al alloys can affect corrosion resistance. A simple but very effective way of overcoming this problem for components made from sheet stock is through the use of cladding, known as Alclad. This term was originally a trademark of the Alcoa Company but is now used as a generic term to describe sheet produced by metallurgically bonding high-purity Al surface layers to high-strength Al alloy core material by hot rolling. The cladding thickness is typically between 2% and 10% of the sheet thickness. The clad alloy for use on 2000 series alloys is generally around 99.5% Al, while that for 7000 series alloys is 99% Al with 1% zinc. Because of the absence of microstructural phases such as intermetallic particles in the clad alloy, it has a higher corrosion resistance than the core alloy. Alclad sheet is commonly used by the aircraft industry for external surfaces such as fuselage skins.

3.4.6 Temporary Corrosion Protection

Even with a comprehensive protective scheme of the types described earlier, coating systems will eventually fail given time. It is now common practice to apply an additional layer of protection using CICs as a form of preventative maintenance for articles and components prone to corrosion. This is common practice with aircraft structural components particularly in areas where moisture gathers such as beneath floors and around galleys and toilet areas. CICs are organic formulations consisting of a carrier solvent containing dissolved oil, wax or resin, and some organic-based corrosion inhibitors such as phosphonates and sulfonates. Similar oil-based products for domestic use are WD40™ and RP7™ that are multipurpose lubricating and penetrating sprays. CICs are typically sprayed onto the structure after the full protective system is in place. When the solvent evaporates, a film of oil, wax, or resin is left together with the inhibitors. In reference to Figure 3.15, they provide a barrier layer for protection and a source of organic corrosion inhibitors when moisture is present. Some CICs also have water-displacing and repellent properties, which enable them to be used to prevent moisture accessing exposed metal at cracks in paint coatings and sealants within joints and holes or even to remove moisture. Many CICs remain mobile on the surface for some time and can provide protection when cracks form in the underlying paint coating. These products do have excellent corrosion prevention properties (Figure 3.24; Hinton et al. 1996, 2000). Figure 3.24 shows that a sprayed on layer of various commercially available CICs was able to provide corrosion protection for AA2024-T3 specimens exposed for 10 months on a navy ship.

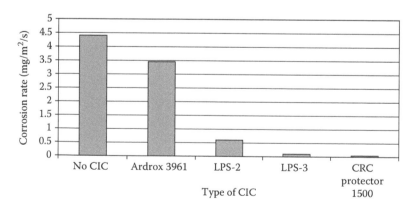

FIGURE 3.24 The effect of CIC on the corrosion rate of AA 2024-T3 exposed on a navy ship for 10 months. (After Hinton, B.R.W. et al., The protection performance of corrosion prevention compounds on aluminum alloys in laboratory and outdoor environments, in *Proceedings of the 4th Joint DoD/FAA/NASA Conference on Conference Aging Aircraft*, DoD/FAA/NASA, St. Louis, MO, 2000.)

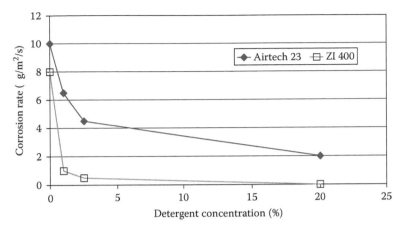

FIGURE 3.25 The effects of additions of various detergents to the 0.1 M NaCl test solution on the corrosion rate of AA7075-T651. (After Hinton, B.R.W. et al., Washing detergents and corrosion inhibition, in *Proceedings of the Corrosion and Prevention 1995*, Australasian Corrosion Association, Melbourne, Australia, 1995.)

The best protection was observed for the thicker resin-based coatings. It has been shown that CICs are also able to arrest the growth of existing corrosion process including SCC (Salagaras et al. 2001). The rate of SCC in an AA7079-T651 alloy exposed to a high-humidity environment with NaCl contaminant present was reduced by almost two orders of magnitude when the crack was treated with an oily based CIC LPS-2. While CICs can provide acceptable levels of corrosion protection, they do have a limited effective life and therefore are only temporary coatings and need frequent replacement.

Another strategy for temporary corrosion protection and arrest is the use of the common maintenance action of periodic washing. Detergent washing formulations contain surfactants some of which are very surface active and penetrating and have corrosion-inhibiting properties (Hinton et al. 1995, Salagaras et al. 2000). Figure 3.25 shows constant immersion corrosion rate data obtained for AA7075-T651 in NaCl solution and the effects of various concentrations of commercially available detergents when added to the test solution. The detergent ZI400™ reduced the corrosion rate to almost zero at a concentration of 15%. Figure 3.26 shows data obtained from a galvanic corrosion test with a copper electrode and an electrode of AA7075-T651 (equal areas) in 0.1 M NaCl solution. The current flowing in the cell is related to the corrosion rate of the Al alloy. It can be seen that the addition of a detergent (ZI 400) to the solution (at a 10% concentration), where galvanic corrosion was occurring, reduced the current in one case by almost an order of magnitude. The spike in the

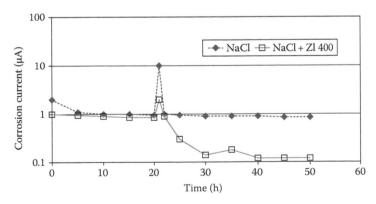

FIGURE 3.26 The effect of an addition of detergent ZI 400 on the corrosion current between a galvanic couple of A7075 and Cu in 0.1 M NaCl. (After Hinton, B.R.W. et al., Washing detergents and corrosion inhibition, in *Proceedings of the Corrosion and Prevention 1995*, Australasian Corrosion Association, Melbourne, Australia, 1995.)

data was due to a perturbation of the system following the addition of the detergent. These data demonstrate not only the corrosion-inhibiting ability of detergents but also their ability to retard or slow the rate of an existing corrosion process. With such detergents, cleaning and a corrosion preventative maintenance would be combined. Experience in the Australian Defence Force suggests that both CICs and some detergents are capable of arresting existing corrosion processes on aircraft components.

3.4.7 Maintenance and Training

One of the essentials for an effective corrosion prevention strategy is a well-trained maintenance workforce. Traditionally, corrosion prevention courses for these workers cover the various types of corrosion and examples of their appearance; the ways in which corrosion can be prevented through the use of design, alloy selection, and coatings; and how corrosion can be repaired. With this type of instruction, maintainers are trained to look for corrosion. That is, maintenance personnel are trained to look for the "failed state" where corrosion damage has occurred. Unfortunately, finding the "failed state" is too late, because expensive repairs and component replacement are then required. Better training should involve instruction in identifying (1) the precursors of corrosion, (2) indicators of a deleterious environmental presence, (3) the signs of potential coating degradation and failure, and (4) the actions to be taken to prevent further corrosion damage before the next inspection period (Miah and Hinton 2007). Typical precursor examples are chipped, cracked, and degraded paint coatings, water stains, pooled water, and surface contaminants. Recognition of these features should form a part of any training that also includes procedures to be put in place to delay the start/progress of corrosion. This proactive approach to corrosion prevention maintenance may mean in the short-term extra maintenance hours, because the existing culture does not include looking for these precursors to corrosion. However, in the long term, less time would be spent repairing corrosion damage and associated repainting operations. This would help reduce system downtime and the labor hours used during periodic vehicle maintenance cycles as well as increase the safety of all structures, equipment, and materiel used in civilian and military applications.

3.5 Future Trends and Research Avenues

Research into the corrosion and protection of Al alloys continues rather vigorously owing to the relevance of Al in the aircraft and military sector of most nations, along with the growth in demand for Al as developing markets emerge.

Just some of the present Al corrosion challenges are

1. The ramifications from the elimination of chromates

   The imminent and total replacement of chromate-based protection systems continues to stimulate coating research. Present research into coating technologies and corrosion inhibition methodologies is shedding new light on the way that traditional protective systems work. This is allowing coatings to be developed to optimize functions such as galvanic protection, passivation, and to introduce self-healing mechanisms into coatings. Issues that inevitably will arise and need to be dealt with include requests by a manufacturer to their customers to accept a product with a performance less than that shown by a chromate-containing system suggesting a lesser quality product. However, it should be kept in mind that current acceptance tests such as the salt spray test and the underfilm corrosion test were developed for the chromate-containing systems. It is not logical to expect that a non-chromate system should comply with the requirement of those tests.

2. The tolerance of increased impurity levels due to the increased use of recycled metal

   As previously suggested, the remelting of Al requires only 5% of the energy needed to extract Al from its ore. As a result of rising energy costs, increased environmental awareness, and the increased demand for Al, the usage of recycled Al will also dominate the future Al market. This is already reflected in the fact that Al scrap has maintained a higher value than ferrous scrap. Impurity elements such as iron, tin, lead, and trace amounts of other metals will influence the mechanical and corrosion properties of end product. Understanding the role and threshold of impurities will be critical in the expanded use of recycled Al for engineered products.

3. The integration of Li into alloys while retaining corrosion resistance

   Al enjoys widespread use because of its favorable strength-to-density ratio. Al can be made both lighter and stronger, via alloying with Li. As is well known in terms of electrochemistry, Li is highly reactive and can compromise corrosion performance if not engineered correctly into the alloy. Research understanding the role of Li in terms of corrosion performance, particularly EAC, will be critical to the expanded use of what are known as the ultralight- and ultrahigh-strength class of Al alloys.

4. The sensitization of non-heat-treatable Al alloys

   The continued use of 5xxx series alloys in marine- and land-based vehicles as an alternative to ferrous products is based upon such alloys providing good levels of corrosion resistance. Natural aging has revealed that atmospheric exposure (particularly in warm climates) can lead to sensitization "in service." As a result, research understanding low-temperature aging/sensitization processes and the precise impact upon corrosion will remain a hot topic.

5. Al-based metallic glasses

   Metallic glasses are purported to have a high resistance to corrosion. The fundamental limits of this resistance are presently under research in many alloy systems; however, in the context of Al— which is capable of forming low-density metallic glasses—there is significant research needed on the glass-forming ability itself. A range of processing friendly Al-based metallic glasses is yet to be developed.

6. Advanced coatings deposited using HVOF, cold spray, pulsed thermal spray, and laser cladding

   Recent works on sonic and supersonic cold spray coatings and laser cladding present an exciting opportunity to be able to surface coat Al with effectively any metal, metal alloy, or metal matrix composite. Furthermore, such coatings have the possibility to be used in the context of repair of damaged sections in factories and depots and even extended to field repair applications. It is expected that such technology may find use in the repair of military assets in the near future; however, effective deployment of the technology will require significant and targeted fundamental research.

7. Development of analytic corrosion modeling software for both evaluation of phenomenological corrosion behavior as well as prediction of corrosion failure of bare and coated metallic substrates

Corrosion will not occur generally until a protective coating fails. Currently, models that will reliably predict coating failure under the combined action of environmental and mechanical variables do not exist. Although several research programs worldwide are beginning to address this requirement, producing these models presents challenges for corrosion scientists and chemists. Once coating failure occurs, and corrosion-inhibiting pigments have been depleted to a level where protection is no longer possible (available from laboratory studies), it is reasonable to assume that corrosion will quickly occur. Also, corrosion models are required that will predict the depth and distribution of corrosion damage with time, whether it is pitting or IGC of metallic surfaces. The major challenge for any coating failure and corrosion prediction models is obtaining validation in a wide range of environments, with various alloy types for various sections and geometries under laboratory and service conditions.

Acknowledgments

The authors gratefully acknowledge the assistance of Ms. Julie Fraser (Monash University) for the careful production of diagrams, figures, and tables. Additionally, all those who provided material for the figures within the chapter are also very gratefully acknowledged.

References

Afseth, A., J.H. Nordlien, G.M. Scamans, and K. Nisancioglu. 2001. *Corrosion Science*, 43, 2093–2109.

Aluminium.org, http://www.world-aluminium.org, Accessed December 2010.

Anawati, B., N.H. Graver, Z. Zhao, G.S. Frankel, J.C. Walmsley, and K. Nisancioglu. 2010. *Journal of the Electrochemical Society*, 157(10), C313–C320.

ASM Handbook, Vol. 13. 2002. ASM International, Materials Park, OH.

ASTM. 1987. *Standard Method for Salt Spray (Fog) Testing B117-90*, ASTM, Philadelphia, PA.

Bibber, J.W. 1996. Conversion coating with nitrate for corrosion resistant coating on Al alloys, U.S. Patent No. 5554231.

Biestek, T. and J. Weber. 1976. *Electrolytic and Chemical Conversion Coatings: A Concise Survey of Their Production, Properties and Testing*, Portcullis Press Ltd., Surrey, U.K.

Birbilis, N. and R.G. Buchheit. 2005. *Journal of the Electrochemical Society*, 152(4), B140–B151.

Birbilis, N. and R.G. Buchheit. 2008. *Journal of the Electrochemical Society*, 155(3), C117.

Birbilis, N., M.K. Cavanaugh, and R.G. Buchheit. 2006. *Corrosion Science*, 48(12), 4202.

Boeing. 2010. Preferred materials and processes, http://www.boeingsuppliers.com/environmental/chromic.html.

Buchheit, R.G. 1995. *Journal of the Electrochemical Society*, 142, 3994.

Buchheit, R.G. and A.E. Hughes. 2003. Chromate and chromate free conversion coatings, in *ASM Handbook*, Vol. 13A, ASM International, Materials Park, OH, pp. 720–735.

Buchheit, R.G. and M.A. Martinez. 1998. Corrosion protective coatings for metallic materials, U.S. Patent No. 5,756,218.

Buchheit, R.G., M.A. Martinez, and L.P. Montes. 2000. *Journal of the Electrochemical Society*, 147(1), 119–124.

Buchheit, R.G. et al. 2002. *Corrosion*, 58(1), 3.

Cavanaugh, M.K., N. Birbilis, and R.G Buchheit. 2009a. *ECS Transactions*, 16(52), 1.

Cavanaugh, M.K., R.G. Buchheit, and N. Birbilis. 2009b. *Engineering Fracture Mechanics*, 76, 641–650.

Chalmer, P.D. 1995. Alternatives to chromium for metal finishing, in Final report, National Center for Manufacturing Sciences, Ann Arbor, MI.

Champion, F.A. and E.E. Spillett. 1956. *Sheet Metal Industry*, 33, 25.

Chen, Z.Y., F. Cui, and R.G. Kelly. 2008. *Journal of the Electrochemical Society*, 155(7), C360–C368.

Cochran, W.C. and D.O. Sprowls. 1979. Anodic coatings for aluminum, in H. Leidheiser, Ed., *Corrosion Control by Coatings*, Princeton Science Press, Princeton, NJ, p. 170.

Coleman AJ, H.N. McMurray, and G. Williams. 2007. *Electrochemical and Solid State Letters*, 10(5), C35–C38.

Danford, M.D. 1994. *The Corrosion Protection of Several Al Alloys by Chromic Acid and Sulfuric Acid Anodizing*, NASA, Marshall Space Flight Center, Huntsville, AL.

Das, S.K. 2006. *Light Metal Age*, 6, 26–33.

Davis, J.R., Ed. 1999. *Corrosion of Aluminum and Aluminum Alloys*, ASM International, Materials Park, OH.

Dieter, G.E. 1998. *Mechanical Metallurgy*, McGraw-Hill, London, U.K.

Eichinger, E., J. Osborne, and T.V. Cleave. 1997. *Metal Finishing*, 95, 36–41.

Escarsega, J.A., Chesonis, K., Crawford, D.M. 2004. Chemical agent resistant coatings (CARC) reach higher levels of performance, *AMPTIAC Quarterly* (U.S. Army Research Laboratory, Aberdeen Proving Ground, MD/AMPTIAC, Rome, NY), 8, 96–100.

Forsyth, M. et al. 2008. *Corrosion*, 64(3), 191–197.

Frankel, G.S. 1998. *Journal of the Electrochemical Society*, 145(6), 2186–2198.

Frankel, G.S. 2001. Mechanism of Al alloy corrosion and the role of chromate inhibitors, AFOSR Multidisciplinary University Research Initiative, Ohio State University, Columbus, OH.

Frankel, G.S. and R.L. McCreery. 2001. *Inhibition of Al Alloy Corrosion by Chromates*, The Electrochemical Society Interface Winter Edition, Philadelphia, PA, pp. 34–38.

Gabe, D.R. 1972. *Principles of Metal Surface Treatment and Protection*, Pergammon Press, Oxford, U.K.

Garcia-Vergara, S.G., F. Colin, P. Skeldon, G.E. Thompson, P. Bailey, T.C.Q. Noakes, H. Habazaki, and K. Shimizu. 2004. *Journal of the Electrochemical Society*, 151(1), B16–B21.

Grjotheim, K. and H. Kvande, Eds. 1993. *Introduction to Aluminum Electrolysis*, 2nd Edn., Aluminum-Verlag, Dusseldorf, Germany.

Grjotheim, K. and B.J. Welch. 1988. *Aluminum Smelting Technology*, 2nd Edn., Aluminum-Verlag, Dusseldorf, Germany.

Habazaki, H., K. Shimizu, P. Skeldon, G.E. Thompson, G.C. Wood, and X. Zhou. 1997. *Corrosion Science*, 39(4), 731–737.

Hare, C.H. 1994. *Protective Coatings—Fundamentals of Chemistry and Composition*, Technology Publishing Co., Pittsburgh, PA, SSPC 94-17.

Hatch, J.E. 1984. *Aluminum: Properties and Physical Metallurgy*, ASM International, Materials Park, OH, 424 p.

Hertzberg, R.W. 1995. *Deformation and Fracture Mechanics of Engineering Material*, 4th Edn., Wiley, New York.

Hinton, B.R.W. 1989. New approaches to corrosion inhibition with rare earth metal salts, in *Corrosion 89*, NACE, New Orleans, LA, Paper No. 170.

Hinton, B.R.W. 1991a. Corrosion prevention and chromates, the end of an era? Part 2, *Metal Finishing*, 89(10), 15.

Hinton, B.R.W. 1991b. Corrosion prevention and chromates, the end of an era? Part 1, *Metal Finishing*, 89(9), 55.

Hinton, B.R.W. 2010. The prevention and control of corrosion in aircraft components—Changes over four decades, *Corrosion*, 66(8), 1–15.

Hinton, B.R.W., D.A. Arnott, and N.E. Ryan. 1984a. The inhibition of aluminum alloy corrosion by cerous cations, *Metals Forum*, 7(4), 211–212.

Hinton, B.R.W., N.E. Ryan, and P.N. Trathen. 1984b. Inhibition of corrosion and stress corrosion cracking in high strength aluminum alloys by surface active agents, in *Proceedings of the 9th International Congress on Metallic Corrosion*, Toronto, Ontario, Canada.

Hinton, B.R.W., P.N. Trathen, and H. Chin Quan. 1984c. Effect of inhibitors on the stress corrosion cracking of an Al-Zn-Mg-Cu alloy, *Corrosion Australasia*, 9(1), 4–10.

Hinton, B.R.W. et al. 1995. Washing detergents and corrosion inhibition, in *Proceedings of the Corrosion and Prevention 1995*, Australasian Corrosion Association, Melbourne, Australia.

Hinton, B.R.W. et al. 1996. Prevention and control of corrosion on aircraft structure with corrosion prevention compounds, in *Proceedings of the 4th International Aerospace Corrosion Control Symposium*, Jakarta, Indonesia.

Hinton, B.R.W. et al. 2000. The protection performance of corrosion prevention compounds on aluminum alloys in laboratory and outdoor environments, in *Proceedings of the 4th Joint DoD/FAA/NASA Conference on Aging Aircraft*, DoD/FAA/NASA, St. Louis, MO.

Hoeppner, D.W. 1979. Model for prediction of fatigue lives based upon a pitting corrosion fatigue process, in J.T. Fong, Ed., *Fatigue Mechanisms*, in *Proceedings of the ASTM-NBS-NSF Symposium*, Kansas City, MO, ASTM STP675, p. 841.

Hughes, A.E., R.J. Taylor, and B.R.W. Hinton. 1997. Chromate conversion coatings on 2024 Al alloy, *Surface and Interface Analysis*, 25(4), 223–234.

Hughes, A.E. et al. 2000. Accelerated cerium based conversion coatings, in *Proceedings of the Corrosion 2000 Research Topics Symposium: Surface Conversions of Aluminum and Ferrous Alloys for Corrosion Resistance*, NACE International, Denver, CO, pp. 47–66.

Hughes, A.E. et al. 2003. Characterisation of various conversion coatings on 2024-T3, *ATB Metallurgie*, 43(1–2), 265–272.

Hughes, A.E. et al. 2005. Towards replacements of chromate inhibitors by rare earth systems, in *Proceedings of the First World Congress on Corrosion in the Military*, Sorrento, Italy.

Hunter, M.S., G.R. Frank, and D.L. Robinson. 1963. in *Proceedings of the 2nd International Congress on Metallic Corrosion*, New York, p. 66.

Ilevbare, G.O., O. Schneider, R.G. Kelly, and J.R. Scully. 2004. *Journal of the Electrochemical Society*, 151, 453.

Jones, R.H., Ed. 1992. *Stress-Corrosion Cracking*, ASM International, Materials Park, OH.

Jones, K. and D.W. Hoeppner. 2005. Pit-to-crack transition in pre-corroded 7075-T6 aluminum alloy under cyclic loading, *Corrosion Science*, 47, 2185–2198.

Kaesche, H. 1963. Investigation of uniform dissolution and pitting of aluminum electrodes, *Werkstoffe und Korrosion*, 14, 557.

Katzman, H.A. et al. 1979. *Applied Surface Science*, 2, 416.

Kim, Y. and R.G. Buchheit. 2007. *Electrochimica Acta*, 52, 2437–2446.

Knight, S.P. 2003. A review of heat treatments, in *Proceedings of the Australasian Corrosion Association 2003*, Melbourne, Victoria, Australia,

Kondo, Y. 1989. Prediction of fatigue crack initiation life based on pit growth, *Corrosion*, 45, 7.

Koroleva, E.V., G.E. Thompson, G. Hollrigl, and M Bloeck. 1999. *Corrosion Science*, 41, 1475–1495.

Kovarik, L. et al. 2006. Origin of the modified orientation relationship for S(S″)-phase in Al-Mg-Cu alloys, *Acta Materialia*, 54(7), 1731–1740.

Lambourne, R. 1987. *Paint and Surface Coatings: Theory and Practice*, Ellis Horwood Ltd. and John Wiley and Sons, Chichester, U.K.

Leth-Olsen, H., J.H. Nordlein, and K. Nisancioglu. 1997. *Journal of the Electrochemical Society*, 144, L196–L197.

Leth-Olsen, H., J.H. Nordlein, and K. Nisancioglu. 1998. *Corrosion Science*, 40, 2051–2063.

Li, J.F., N. Birbilis, C.X. Li, Z.Q. Jia, B. Cai, and Z.Q. Zheng. 2009. *Materials Characterization*, 60(11), 1334–1341.

Liddicoat, P.V., X.-Z. Liao, and S.P. Ringer. 2009. *Materials Science Forum*, 618–619, 543–546.

Lynch, S.P., S.P. Knight, N. Birbilis, and B.C. Muddle. 2008. Stress corrosion cracking of Al-Zn-Mg-Cu alloys: Effects of composition and heat treatment, in *Proceedings of the 2008 International Hydrogen Conference: Effects of Hydrogen on Materials*, Jackson Hole, WY.

Mader, O.M. 1972. Critical review of aluminum in the building industry, *Metals and Materials*, 1972(7), 303–307.

Markley, T. 2008. PhD thesis in materials engineering, Monash University, Melbourne, Australia.

Markley, T. et al. 2007. Synergistic corrosion inhibition in mixed rare earth diphenyl phosphate systems, *Electrochemical and Solid-State Letters*, 10(12), C72.

Mayne, J.E.O. 1995. The mechanism of protective action of paints, in L.L. Sheir, R.A. Jarman, and G.T. Burstein, Eds., *Corrosion*, Vol. 2, Butterworth-Heinemann, London, U.K., p. 14.22.

Mazurkiewicz, B. and A. Piotrowski. 1983. *Corrosion Science*, 23, 697.

McMurray, H.N. et al. 2007. Scanning Kelvin probe studies of filiform corrosion on automotive aluminum alloy AA6016, *Journal of the Electrochemical Society*, 154(7), C339–C348.

McMurray, H.N., A. Holder, G. Williams, G.M. Scamans, and A.J. Coleman. 2010. *Electrochimica Acta*, 55(27), 7843–7852.

Mercer, A.D. 1980. in *Proceedings of the 5th European Symposium on Corrosion Inhibitors*, Ferrara, Italy.

Miah, S. and B.R.W. Hinton. 2007. A corrosion related reliability centered maintenance pilot program for the Royal Australian Air Force C-130 J-30 Hercules Aircraft, in *Proceedings of the Tri-Services Corrosion Conference*, NACE International, Denver, CO.

MIL-DTL-5541F. 2006. *Military Specification Chemical Conversion Coatings on Aluminum and Aluminum Alloys*, U.S. DoD, Washington, DC.

MIL-DTL-81706. 2006. *Military Specification Chemical Conversion Materials for Coating Aluminum and Aluminum Alloys*, U.S. DoD, Washington, DC.

Mol, J.M.C., J.H. de Wit, and S. Van der Zwaag. 2002. *Journal of Materials Science*, 37, 2755–2758.

Morris, E. et al. 1999. Evaluation of non-chrome inhibitors for corrosion protection of high-strength aluminum alloys, *Polymeric Materials Science and Engineering*, 81, 167–168.

Muster, T., A.E. Hughes, and G.E. Thompson. 2008. Copper distributions in aluminium alloys. In I.S. Wang, Ed., *Corrosion Research Trends*, Nova Science Publishers, New York, pp. 35–106.

Newhard, N.J. 1979. Conversion coatings—Chromate and non-chromate, in H. Leidheiser, Ed., *Corrosion Control by Coatings*, Science Press, Princeton, NJ.

Newman, R.C. and K. Sieradzki. 1994. *Science*, 263, 1708–1709.

Nişancioğlu, K. 1990. *Journal of the Electrochemical Society*, 137, 69.

Osaki, S., H. Kondo, and K. Kinoshita. 2006. Contribution of hydrogen embrittlement to SCC process in excess Si type Al-Mg-Si alloys, *Materials Transactions*, 47(4), 1127–1134.

Perrault, G.G. 1979. *Journal of Electrochemical Society*, 126(2), 199–204.

Placzankis, B.E., C.E. Miller, and C.A. Matzdorf. 2003. *ASTM B117 Screening of Non-Chromate Conversion Coatings on Al Alloys 2024, 2219, 5083 and 7075 Using DoD Paint Systems*, US Army Research Labs, Aberdeen, MD.

Polmear, I.J. 2004. Aluminum alloys—A century of age hardening, *Materials Forum*, 28, 1–14.

Polmear, I.J. 1995. *Light Alloys: Metallurgy of the Light Metals*, 3rd Edn., Arnold, London, U.K., 1995.

Pourbaix, M. 1966. *Atlas of Electrochemical Equilibria in Aqueous Solutions*, NACE International, Denver, CO.

Pride, S.T., J.R. Scully, and J.L. Hudson. 1994. *Journal of the Electrochemical Society*, 141(11), 3028–3040.

Pryor, M.J. and J.C. Fister. 1984. *Journal of the Electrochemical Society*, 131, 1230.

Ralston, K.D., N. Birbilis, M.K. Cavanaugh, M. Weyland, B.C. Muddle, and R.K.W. Marceau. 2010. *Electrochimica Acta*, 55, 7834–7842, http://dx.doi.org/10.1016/j.electacta.2010.02.001

Ralston, K.D., N. Birbilis, and C.R. Hutchinson. 2009. in *Proceedings of the Conference on Corrosion and Prevention '09*, Coffs Harbour, Australia.

Rao, K.T.V. and R.O. Ritchie. 1992. Fatigue of aluminum lithium alloys, *International Materials Reviews*, 37(4), 153–185.

Raviprasad, K. et al. 2003. Precipitation processes in an Al-2.5Cu-1.5Mg (wt.%) alloy microalloyed with Ag and Si. *Acta Materialia*, 51(17), 5037–5050.

Roberge, P.R. 1999. *Handbook of Corrosion Engineering*, McGraw-Hill, New York.

Roland, W.A. and R. Hart. 1996. Chromium free metal pre-treatment processes, in *Proceedings of the Interfinish 96 World Congress*, Institute of Metal Finishing, Coventry, U.K., pp. 265–277.

Salagaras, M., P. Bushell, and B.R.W. Hinton. 2000. The use of corrosion preventative compounds for controlling the rate of stress corrosion cracking on aluminum alloy 7075, Corrosion Control Report No. 5/2000, Defence Science and Technology Organisation, Australia.

Salagaras, M. et al. 2001. The arrest of corrosion with CPCS, in *Proceedings of the 5th Joint DoD/FAA/NASA Conference on Aging Aircraft*, DoD/FAA/NASA, St. Louis, MO.

Scamans, G.M., A. Afseth, G.E. Thompson, and X. Zhou. 2002. Ultra-fine grain sized mechanically alloyed surface layers on aluminum alloys, in *Proceedings of the 8th International Conference on Aluminum Alloys, Their Physical and Mechanical Properties*, Cambridge, U.K., pp. 1461–1466.

Scamans, G.M., A. Afseth, G.E. Thompson, and X. Zhou. 2003. *ATB Metallurgie*, 43, 90–94.

Scamans, G.M. and N.J.H. Holroyd. 1986. Stress-corrosion of aluminum aerospace alloys, *Journal of the Electrochemical Society*, 133(8), C308–C308.

Scamans, G.M., N.J.H. Holroyd, and C.D.S. Tuck. 1987. *Corrosion Science*, 27(4), 329–347.

Schlesinger, M. 2006. *Aluminum Recycling*, CRC Press, Boca Raton, FL.

Schneider, O., G.O. Ilevbare, R.G. Kelly, and J.R. Scully. 2004. *Journal of the Electrochemical Society*, 151, 465.

Schneider, O. et al. 2007. In situ confocal laser scanning microscopy of AA2024-T3 corrosion metrology—III. Underfilm corrosion of epoxy-coated AA2024-T3, *Journal of the Electrochemical Society*, 154(8), C397–C410.

Schweitzer, P.A. 2007. *Corrosion Engineering Handbook. Corrosion of Linings and Coatings, Cathodic and Inhibitor Protection and Corrosion Monitoring*, 2 Edn., CRC Press, Boca Raton, FL.

Scully, J.R., T.O. Knight, R.G. Buchheit, and D.E. Peebles. 1993. *Corrosion Science*, 35, 185.

Searles, J.L., P.I. Gouma, and R.G. Buchheit. 2001. *Materials Transactions A*, 32A, 2859.

Sedriks, A.J. 1990. *Stress Corrosion Cracking: Test Methods*, National Association of Corrosion Engineers, Houston, TX.

Seegmiller, J.C., R.C. Bazito, and D.A. Buttry. 2004. *Journal of the Electrochemical Society*, 7(1), B1–B4.

Sieradzki, K. 1993. *Journal of the Electrochemical Society*, 140(10), 2868–2872.

Singh, D.D.N., R.S. Chaudhary, and C.V. Agarwal. 1992. *Journal of the Electrochemical Society*, 129(9), 1869–1874.

Smith, C.J.E. et al. 1996. Research into chromate-free treatments for the protection of aluminum alloys, in *Proceedings of the AGARD SMP "Environmentally Compliant Surface Treatments of Materials for Aerospace Applications,"* Florence, Italy, paper 8-1.

Soltis, J., D.P. Krouse, N.J. Laycock, and K.R. Zavadil. 2010. *Corrosion Science*, 52(3), 838–847.

Spadafora, S. et al. 1997. Aerospace finishing systems for naval aviation, in *Proceedings of the 42nd International SAMPE Symposium on Naval Air Systems Team Advanced Materials*, SAMPE, Anaheim, CA.

Sprowls, D.O. 1978. High strength aluminum alloys with improved resistance to corrosion and stress corrosion cracking, *Aluminum*, 54(3), 214–217.

Stevenson, M.F. 2002. Anodizing, in *ASM Handbook*, Vol. 5, *Surface Engineering*, ASM International, Materials Park, OH.

Summerson, T.J. and D.O. Sprowls. 1974. Corrosion behavior of aluminum alloys, in *International Conference of the Hall-Heroult Process Vol. III of Conference Proceedings*, Engineering Materials Advisory Services Ltd. (University of Virginia), Charlottesville, VA, 1576–1662.

Szklarska-Smialowska, Z. 1999. *Corrosion Science*, 41, 1743.

Szklarska-Smialowska, Z. 2005. *Pitting and Crevice Corrosion*, NACE International, Houston, TX, 650 pp.

Thompson, G.E. et al. 1999. *Corrosion*, 55(11), 1052–1061.

Trathen, P.N. et al. 1993. The variability of the chromate conversion coating process and the effect on corrosion protection performance, in *Proceedings of the Conference of Australasian Corrosion Association*, Newcastle, Australia, Paper No. 33-08.

Treverton, J.A. and N.C. Davies. 1977. *Metals Technology*, 4, 480.

Trueman, A.R. 2005. Determining the probability of stable pit initiation on aluminium alloys using potentiostatic electrochemical measurements, *Corrosion Science*, 47(9), 2240–2256.

Trueman, A.R. et al. 2000. Corrosion protection performance and EIS characterisation of a thickened oxide conversion coating containing cerium ions formed on AA2024, in H. Terryn, Ed., in *Proceedings of the International Symposium on Aluminum Surface Science Technology*, UMIST, Manchester, U.K., pp. 270–276.

Twite, R.L. and G.P. Bierwagen. 1998. Review of alternatives to chromate for corrosion protection of aluminum aerospace alloys, *Progress in Organic Coatings*, 33(2), 91–100.

Van der Walde, K. et al. 2005. Multiple fatigue crack growth in pre-corroded 2024-T3 aluminum, *International Journal of Fatigue*, 27, 1509–1518.

Van Horn, K.R., Ed. 1967. *Aluminum*, Vol. 1, ASM International, Materials Park, OH.

Vargel, C. 2004. *Corrosion of Aluminum*, Elsevier, London, U.K.

Viswanadham, R.K., T.S. Sun, and J.A.S. Green. 1980. *Corrosion*, 36, 275–278.

Vukmirovic, M.B., N. Dimitrov, and K. Sieradzki. 2002. *Journal of the Electrochemical Society*, 149(9), B428–B439.

Wall, F.D. and M.A. Martinez. 2003. A statistics-based approach to studying aluminum pit initiation— Intrinsic and defect-driven pit initiation phenomena, *Journal of the Electrochemical Society*, 150(4), B146–B157.

Wernick, S., R. Pinner, and P.G. Sheasby. 2007. *The Surface Treatment and Finishing of Aluminum and Its Alloys*, 6th Edn., Finishing Publications Ltd and ASM International, Materials Park, OH.

Wicks, Z.W., F.N. Jones, and S.P. Pappas. 1999. *Organic Coatings: Science and Technology*, 2nd Edn., Wiley, New York.

Wicks, Z.W., F.N. Jones, and S.P. Pappas. 2007. *Organic Coatings: Science and Technology*, 3rd Edn., Wiley, New York.

Williams, G. and H. McMurray. 2003. Anion-exchange inhibition of filiform corrosion on organic coated AA2024-T3 aluminum alloy by hydrotalcite-like pigments, *Electrochemical and Solid-State Letters*, 6(3), B9–B11.

Winkelman, G.B., K. Raviprasad, and B.C. Muddle. 2007. Orientation relationships and lattice matching for the S phase in Al-Cu-Mg alloys, *Acta Materialia*, 55(9), 3213–3228.

Zamin, M. 1981. *Corrosion*, 37, 627.

Zhang, J., N. Gao, and M.J. Starink. 2010. *Materials Science and Engineering A* 527(15), 3472–3479.

Zhao, Z. and G.S. Frankel. 2007a. *Corrosion Science*, 49(7), 3064–3088.

Zhao, X.Y. and G.S. Frankel. 2007b. Quantitative study of exfoliation corrosion: Exfoliation of slices in humidity technique, *Corrosion Science*, 49(2), 920–938.

Zhao, Z. and G.S. Frankel, http://hdl.handle.net/1811/24655 (accessed June 11, 2013).

Zhou, X., G.E. Thompson, and G.M. Scamans. 2003. *Corrosion Science*, 45(8), 1767–1777.

4

Crystalline Alloys: Titanium

Suresh Divi
*Titanium Metals
Corporation (TIMET)*

James Grauman
*Titanium Metals
Corporation (TIMET)*

4.1 Introduction

Titanium is a special and unique metal because it provides a higher specific strength with excellent corrosion resistance as compared to other structural metals. It is a commonly used metal in the aerospace industry based on its good strength-to-weight ratio and durability. In industrial markets such as the chemical process industry (CPI) where corrosion resistance is the key property, titanium has proved to be far superior to stainless steels and other engineering metals.

This chapter on titanium is intended to provide sufficient and significant corrosion information whereas more specific details can be obtained from the provided literature references. At the same time, this chapter is not intended to provide an all-inclusive source of corrosion data for every existing environment, rather for selected environmental conditions, such as oxidizing and reducing, where titanium provides real application benefits. A good source for comprehensive titanium corrosion data is the *Metals Handbook*, Volume 13 (Schutz 2003), and *Handbook of Corrosion Data* (Craig 1995), both published by the ASM International. The purpose of this chapter is to provide basic information related to degradation of titanium and its alloys in its primary use environments. The principal focus is on aqueous corrosion, highlighting the metal's behavior in specific media as well as the corrosion protection mechanisms of titanium. In addition to the degradation behavior of titanium, its history, alloy classes, and physical and mechanical properties are also provided.

4.2 Background

Titanium is a shiny, dark-gray metal that is classified as a nonferrous and light metal with the symbol Ti and atomic number 22 and atomic weight 47.9 in the transitional metal group IV on the periodic table. It is the ninth most abundant element in the earth's crust and the fourth most abundant structural metal.

TABLE 4.1   Comparison of Structural Metal Properties

	Element				
Item	Al	Mg	Ti	Fe	Ni
% of Earth crust	8.13	2.09	0.44	5	0.008
Density (gm/cm^3)	2.77	1.66	4.42	7.75	8.86
Melting point (°C)	660	650	1668	1535	1454

Source: International Titanium Association (ITA), *The Fundamentals of Titanium*, ITA, Denver, CO, 2008.

The relatively high melting point (1668°C) makes it useful as a refractory metal also. The titanium metal density is 4.42 gm/cm^3 lying between aluminum and steel. Table 4.1 shows the earlier mentioned characteristics of titanium metal compared with other important metals (Donachie 2000).

Titanium was discovered by William Gregor in 1791 (Leyens and Peters 2003) while he was examining the black magnetic sand that contains ilmenite (FeTiO$_3$). Due to the strong tendency of titanium metal to react with oxygen and nitrogen and the lack of technology, production of pure titanium was almost impossible in the eighteenth and nineteenth centuries. More than 100 years after its discovery, Matthew Hunter first extracted pure titanium metal from its ore in 1910 using sodium to reduce the intermediate product titanium tetrachloride (TiCl$_4$) to produce pure titanium. However, in 1932 Wilhelm Kroll was the first person to demonstrate the commercial extraction of titanium by reducing TiCl$_4$ using magnesium instead of sodium in an inert gas atmosphere. This process was named "Kroll's process." Due to the rising demand for titanium in various sectors, mainly aerospace and industrial, alternate methods of titanium production have been studied (Fox and Yu 2011), but Kroll's process remains the most widely used method today and will be for the foreseeable future.

The most common natural titanium ores are highly stable (Ti^{+4}) oxides: namely, ilmenite (FeTiO$_3$) and rutile (TiO$_2$), and these ores are mostly found in mineral sands and beach sands, respectively. Rutile is the preferred ore for titanium metal extraction due to its high content (>97%) of TiO$_2$. Titanium extraction from its ores involves reaction with chlorine to produce titanium tetrachloride (TiCl$_4$), which is subsequently distilled to increase its purity; then the purified TiCl$_4$ is reduced with molten magnesium to produce metallic titanium. Metallic titanium obtained at this stage has a sponge-like porous appearance being labeled titanium "sponge." This sponge is further purified under vacuum to remove the by-products. Finally, it is crushed and sized for subsequent melting/alloying operations. Figure 4.1 shows a typical sponge particle obtained from the vacuum distillation process (VDP) at Titanium Metals Corporation (TIMET).

Once the granulated sponge titanium is blended with alloying elements and compacted, two- or three-stage vacuum arc remelting (VAR) and/or a single cold-hearth (electron-beam or plasma-arc) melting process is typically utilized to produce commercial-sized ingots. Figure 4.2 shows a 16 metric ton commercially pure (CP) titanium slab produced by electron beam (EB) melting at TIMET. Classic open-air, high temperature primary and secondary thermomechanical processing (e.g., forging, rolling, and extrusion) produces traditional mill product forms/shapes. Depending on the alloy type and condition, final cold processing such as cold rolling, cold pilgering, cold drawing, and/or cold-forming is used where possible. Some examples of mill products are billet, bar, sheet, strip, tube, and plate. Figure 4.3 shows some of these typical titanium mill products.

Soon after Kroll's demonstration, in the early 1950s titanium metal was industrially produced in the United States, United Kingdom, Japan, and the former USSR. Due to the constant demand of titanium metal from aerospace and other markets, the production rate has continued to rise over the years. As of today, the world titanium metal production is more than tens of thousands of tons a year (ITA 2008).

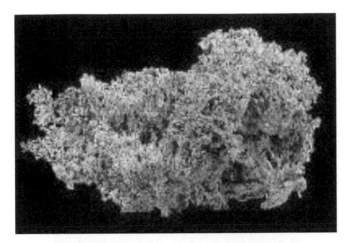

FIGURE 4.1 Pure titanium sponge, mass reduced to <13 mm particle size, obtained during VDP production at TIMET. (Photo Courtesy of TIMET VDP Department.)

FIGURE 4.2 Approximately 16 metric ton EB slab produced at TIMET Morgantown. (Photo Courtesy of Dave Tripp, TIMET Morgantown.)

Yearly production quantities of titanium along with other important engineering metals and alloys are given in Table 4.2. One of the reasons for the lower production volumes of titanium as compared to other engineering metals is due to the higher costs associated with this complex titanium metal extraction process. However, continuous efforts to lower the cost of extracting and fabricating titanium are being made through new methods and improvements, realizing that cost reductions will be an important key necessary for broader titanium usage (Boyer et al. 1994).

(a)

(b)

(c)

FIGURE 4.3 Titanium mill products produced at TIMET Toronto mill: (a) coil, (b) plate, and (c) bar. (Photo courtesy of TIMET Sheet and Plate Departments.)

TABLE 4.2 World Annual Production of Engineering Metals/Alloys

Engineering Metal/Alloy	World Annual Production (Million Tons)
Steel	800
Aluminum	20
Stainless steel	16
Copper	12
Titanium	0.1

4.3 Titanium Benefits, Applications, and Limitations

4.3.1 Benefits

Titanium and its family of alloys can exhibit a very unique, synergistic combination of properties and features of design benefit for a wide range of end users. Primary design benefits are listed in Table 4.3.

Other, secondary features of titanium that can be synergistic with the primary features listed in Table 4.3 for certain specific applications are listed in Table 4.4.

4.3.2 Applications

The first and foremost user of titanium and its alloys is the aerospace industry. Titanium was first used in flight on the Douglas X-3 Stiletto in 1952 (ITA 2008). The aerospace market for titanium includes

TABLE 4.3 Design Benefits of Titanium Alloys Based on Their Synergistic Combination of Properties and Features

Property or Feature	Comments
Elevated strength-to-density ratio	High structural efficiency
Lower density	Roughly half that of steel, nickel, and copper alloys
Exceptional corrosion resistance	Superior resistance to chlorides, seawater, sour and oxidizing acidic media, and MIC resistant
Elevated erosion/erosion–corrosion/cavitation resistance	In seawater, brines, most aqueous media
Resistant to corrosion fatigue	Maintains high fatigue life in seawater, sour brines
Galvanic, thermal, and physical compatibility with carbon/carbon and graphite-reinforced composites	For seawater/marine/aircraft applications
Lower elastic modulus	Approximately half that of steel/Ni alloys; good for deflection controlled components, e.g., springs and offshore risers
High operational thermal conductivity	Maintains high heat transfer in heat exchangers when designed properly

TABLE 4.4 Secondary Features of Titanium That Can Be Synergistic with the Primary Table 4.3 Features in Certain Applications

Property or Feature	Comments
Low thermal expansion	Less than steels, Ni, Cu, and Al alloys
Essentially nonmagnetic	
Elevated shock and ballistic impact resistance	Better survivability of structures and vehicles
Excellent cryogenic properties	Depends on alloy type
High melting point	Relative to nonferrous and nonrefractory structural alloys
Very short radioactive half-life	After irradiation
Elevated anodic pitting potentials	In aqueous chlorides and other salt solutions
Nontoxic, nonallergenic, and fully biocompatible	Scrap recyclable; has no harmful by-products when used in the living body or the environment
Most commercial alloys are readily fabricable	Machinable and weldable using standard methods/equipment with minor adjustments

commercial aircraft, military aircraft, and spacecraft where it is used in both airframes and engines where its high strength-to-weight ratio and mechanical properties are highly desirable. In contrast, titanium applications in nonaerospace industries are primarily driven by its exceptional corrosion resistance making it suitable for a wide array of applications (Froes et al. 1998; Leyens and Peters 2003) such as in the chemical process, petrochemical, and geothermal industries. In industrial markets, titanium is used extensively as heat exchanger materials. As an example, a power plant in Arizona used approximately 2132 km (7 million feet) of titanium tubing for condensing heat exchangers in the early 1990s without reported corrosion failures (Poulsen 1994). Over the years, the demand for titanium in industrial markets has been on the rise due to its longer service life. Good dimensional stability of titanium electrodes is attracting the mining industry for extraction operations such as electrowinning and electrorefining of metals such as copper and nickel. Titanium also proved to be the right material for outlet ductwork and stacks of flue gas desulfurization (FGD) plants by providing excellent resistance to these highly aggressive conditions (Peacock and Grauman 1995).

Due to good biocompatibility, titanium alloys have been used in a variety of biomedical devices for many years (Brown and Lemons 1996). Titanium prostheses have been used successfully in humans from the 1950s. Titanium and its alloys have, to a large extent, replaced the conventional biomedical implant materials such as cobalt–chrome, 316 stainless steel, and chromium–cobalt–molybdenum

TABLE 4.5 Traditional Titanium Alloy Applications

Legend Code	Application Area
Aerospace	
AD	Aircraft ducting, hydraulic tubing, misc.
AF	Airframe components
LG	Landing gear components
PT	Pressure tanks for space transportation systems
SS	Space vehicles/structures, missile components
Nonaerospace	
AC	Anode/cathode/cell components
AP	Air pollution control equipment
AR	Architectural, roofing
AU	Automotive components
BA	Ballistic armor
CG	Consumer products (watches, eye glass frames, etc.)
CP	Chemical processing equipment
DS	Desalination, brine concentration/evaporation
FG	Flue gas desulfurization
FP	Food processing/pharmaceutical
GB	Geothermal brine energy extraction
HE	Hydrometallurgical extraction/electrowinning
HR	Hydrocarbon refining/processing
JF	Jewelry and fashion
MI	Medical implants/devices, surgical instruments
NS	Navy ship/submersible components
OP	Offshore hydrocarbon production/drilling
PB	Pulp/paper bleaching/washing equipment
PD	Hydrocarbon production/drilling
PP	Power plant cooling system components
SR	Sports/recreational equipment

due to their unique combination of properties such as superior biocompatibility, higher strength, lower density and modulus, and enhanced corrosion resistance. In situ tests in body fluids have shown that titanium alloys not only provide high corrosion resistance but also allow for good bonding with body tissues (Brown and Lemons 1996). In the 1980s, approximately 120,000 hip joints and 72,000 knee joints made from titanium alloys were surgically implanted in the United States alone (Poulsen 1994).

Titanium is also used in the food and pharmaceutical industries where avoiding metal contamination is a major concern (Krvuchek and Volynskii 1976). Titanium corrosion resistance does not allow any metal salts to form during food processing that would otherwise destroy the flavor, color, and quality of those processed products. Even if small traces of titanium salt were to form on titanium containers, they would be innocuous upon ingestion. Other applications based on titanium's low modulus of elasticity along with high yield stress range from automotive coil springs (Diem 2001) to recreational equipment like golf clubs (Froes 1999). Some of the areas of aerospace and commercial industrial usage of titanium and its alloys are listed in Table 4.5. Figure 4.4 illustrates some of the titanium products used in these various industries.

4.3.3 Limitations/Challenges

Despite a myriad of attractive, useful design features, designers and end users should account for and address the following potential limitations associated with titanium as seen in Table 4.6.

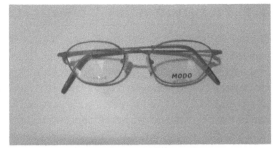
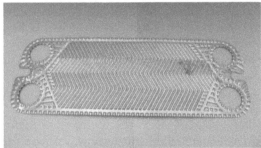

FIGURE 4.4 Components made of titanium and its alloys, in clockwise direction: automotive coil springs, eye glass frames, turbine, plate heat exchanger panel. (Photo Courtesy of TIMET, Henderson Technical Lab [HTL].)

TABLE 4.6 Limitations/Challenges Associated with Some Titanium Alloy Applications

Limitations/Challenges	Comments
Strong tendency to gall/cold-weld on metal-to-metal sliding/fraying contact surfaces	Can be avoided with anti-galling surface treatments on fasteners, connections, bearings, etc.
Weldable only to itself or other titanium alloys (and other reactive metal alloys), but not to iron-, nickel-, copper-, or aluminum-based alloys	When dissimilar metal joining is required must use mechanical connections
Fusion welding requirements	Ensure proper inert gas shielding and joint cleanliness
Limit exposure in air to below 594°C (1100°F) due to brittle alpha-case formation/growth and excessive oxidation	Can be avoided by surface reconditioning of components by removal of diffused-in, interstitial-rich surface layers after high temp. thermal treatments in air
Compensate for the significant reduction in strength and/or creep resistance with increasing service temperature	Can be avoided by lowering the allowable stress factor during the component design phase
Avoid the few specific environmental degradation conditions that promote ignition/burning (enriched-oxygen gas or liquid oxygen, water-lean N_2O_2 or red fuming HNO_3, dry halogens [Cl_2, Br_2])	Do not use titanium alloy components in specific degradable conditions by avoiding enriched-oxygen gas or liquid oxygen, water-lean N_2O_2 or red fuming HNO_3, dry halogens (Cl_2, Br_2) service environments
Avoid exposure of titanium components to dry or near-anhydrous methanol	These are conditions where SCC susceptibility is very severe
Avoid exposing titanium components to concentrated solutions of strong reducing mineral acids	These are service conditions where severe general corrosion in concentrated HCl, HBr, H_2SO_4, H_3PO_4 solutions can be expected
Avoid exposure of titanium components to acid fluoride solutions	This will prevent severe etching/hydriding by HF solutions

4.4 Titanium Alloy Classifications and Microstructures

Similar to iron and nickel, titanium is also an allotropic element existing in more than one crystallographic form. Titanium has a hexagonal close packed (hcp) crystal structure at room temperature and transforms to a body-centered cubic (bcc) structure above the transition temperature (β_t) of 888°C. Hexagonal titanium is referred to as alpha (α) phase and body-centered titanium is referred to as beta (β) phase. Addition of specific alloying elements either increases or decreases the beta transus temperature, where the beta transus is a very important factor that affects processing and heat treatment operations. Based on the relative amount of these phases present in a particular alloy, titanium alloys are classified as α, near α, $\alpha + \beta$, and β alloys. Some of the α stabilizers in titanium are aluminum, oxygen, and nitrogen. Some of the β stabilizers are vanadium, molybdenum, niobium, silicon, and chromium. A complete list of α and β stabilizers are given in Table 4.7. It is worthwhile to note that there are two classes of beta phase stabilizing elements: one is isomorphous, containing elements that are miscible in beta phase, while the other solute group forms eutectoid phase systems with titanium. Increasing the quantity of α stabilizers promotes the volume fraction of alpha phase, and similarly, increasing the quantity of β stabilizers promotes the beta phase. Figure 4.5 shows a pseudo phase diagram illustrating the variation of phases as the beta stabilizer is added to the titanium system (Lütjering and Williams 2007).

Chemical composition and microstructure are the most important factors that contribute to attaining certain properties of titanium alloys. The chemical composition of each titanium alloy determines its properties and the volume fraction of the α and β phases present in the microstructure. For example, increasing amounts of α stabilizers increases the creep strength and provides good weldability.

TABLE 4.7   Titanium α and β Phase Stabilizers

α Stabilizers	β Stabilizers	
	Isomorphous	Eutectoid
Al, Ga, Ge, La, Ce, C, N, O	V, Nb, Ta, Mo, Rh, Zr, Ha, Sn	Cu, Ag, Au, In, Pb, Bi, Cr, W, Mn, Fe, Co, Ni, U, H, Si

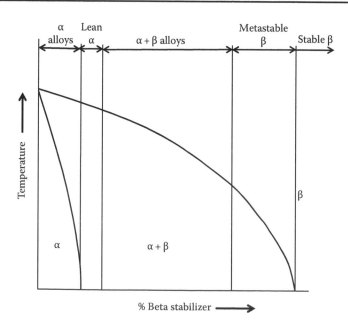

FIGURE 4.5   Pseudo phase diagram of titanium showing different phases. (From Lütjering, G. and Williams, J.C., *Titanium*, Springer, Berlin, Germany, 2007.)

Similarly, increases in β stabilizers provide higher strength, good strain rate sensitivity, and better formability (Davis 2000).

CP grades are unalloyed titanium available in the market where just oxygen and iron are the distinguishing impurities; mainly they are classified into four different types based on the tensile strength obtained from these impurities. CP and other CP-modified grades are considered to be the alpha alloys. Alpha and near-alpha alloys contain larger amount of alpha stabilizer solutes and very low concentrations of beta stabilizers. These alpha and near-alpha alloys are weldable and provide good corrosion resistance. The most widely used alpha alloy is CP Grade 2 (UNS No. R50400) where the mill-annealed microstructure consists of equiaxed alpha grains. The alpha–beta alloys, as the name suggests, contain both phases and provide for good combinations of strength, toughness, and formability. The most famous and widely used alpha–beta alloy is Ti-6Al-4V (UNS No. R56400); the aerospace industry itself consumes more than 80% of this alloy. The microstructure of the Ti-6Al-4V in mill-annealed condition consists of equiaxed alpha phase (bright phase) in a transformed beta matrix (dark phase). Since alpha–beta alloys respond to heat treatment, Widmanstätten, lamellar, or other microstructural elements can be produced depending on strength or other property requirements. The beta alloys contain a substantial amount of beta stabilizers and are used where very high tensile strength is required. A good example of a beta alloy is TIMETAL'21S, commonly referred to as Beta-21S (Parris and Bania 1990). This alloy offers high strength, good cold formability, room and high temperature corrosion resistance, oxidation resistance, and creep resistance (Grauman 1990a, 1992; Parris and Bania 1992). The mill-annealed microstructure of Beta-21S consists of recrystallized beta grains. Figure 4.6 shows the typical mill-annealed microstructures of the alpha (CP Grade 2), alpha + beta (Ti-6Al-4V), and beta (Beta-21S) titanium alloys. At the same magnification the beta grain size (in Beta-21S) is larger than the alpha grain

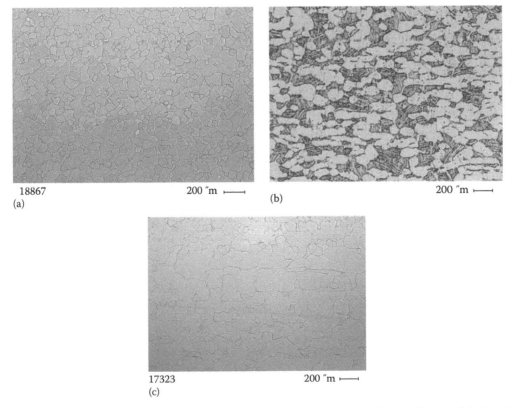

18867   200 ″m ⊢——⊣   (b)   200 ″m ⊢——⊣
(a)

17323   200 ″m ⊢——⊣
(c)

FIGURE 4.6 Mill-annealed microstructure of titanium alloys: (a) CP Grade 2, (b) Ti-6Al-4V, and (c) Beta-21S. (Photo Courtesy of TIMET, HTL.)

TABLE 4.8 Designations and Typical Applications of Common Commercial Titanium Alloys

Common Name	ASTM Grade (UNS No.)	Nominal Composition (wt%)	Typical Applications[a]
Alpha alloys			
CP 1	1 (R50250)	Unalloyed titanium	AC, CG, CP, DS, HE, HR, FP, MI, PP, NS
CP 2	2 (R50400)	Unalloyed titanium	AC, AD, AP, AR, CG, CP, DS, FP, HE, HR, MI, NS, PB, PP, OP, SR
CP 3	3 (R50550)	Unalloyed titanium	CP, NS, PP
CP 4	4 (R50700)	Unalloyed titanium	AC, AD, CP, SR
Ti–Pd	7/11 (R52400/R52250)	Ti-0.15Pd	AC, AP, CP, PS, HE, HR, PB
Ti–lean Pd	16/17 (R52402/R52252)	Ti-0.06Pd	AC, AP, CP, DS, HE, HR, PB
Ti–Ru	26 (R52404)	Ti-0.1Ru	AC, AP, CP, DS, HE, HR, PB
Grade 12	12 (R53400)	Ti-0.3Mo-0.8Ni	CP, DS, GB, HE, HR, OP
Near-alpha alloys			
Ti-3-2.5	9 (R56320)	Ti-3Al-2.5V	AD, CG, NS, SR
Ti-3-2.5-Ru	28 (R56323)	Ti-3Al-2.5V-0.1Ru	CP, GB, HE, OP, PD
Ti-5-2.5	6 (R54250)	Ti-5Al-2.5Sn	GT, SS
Ti-5-1-1-1	32 (R55111)	Ti-5Al-1Sn-1Zr-1V-0.8Mo	NS
Ti-8-1-1	(R54810)	Ti-8V-1V-1Mo	GT
Ti-6-2-4-2-S	(R54620)	Ti-6Al-2Sn-4Zr-2Mo-0.1Si	AF, AU, GT, SS
Alpha–beta alloys			
Ti-6-4	5 (R56400)	Ti-6Al-4V	AD, AF, AU, BA, CG, GT, HE, LG, NS, PD, SR, SS
Ti-6-4 ELI[b]	23 (R56407)	Ti-6Al-4V (0.13 max O)	AF, MI, BA, NS, OP, SS
Ti-6-4-Ru	29 (R56404)	Ti-6Al-4V-0.1Ru (0.13 max O)	CP, DS, GB, OP, PD
Ti-6-6-2	(R56620)	Ti-6Al-6V-2Sn-0.6Fe-0.6Cu	AF
Ti-6-2-4-6	(R56260)	Ti-6Al-2Sn-4Zr-6Mo	GT, PD
Ti-6-22-22	(R56222)	Ti-6Al-2Sn-2Zr-2Mo-2Cr-0.15Si	AF, SS
Ti-17	(R58650)	Ti-5Al-2Zr-2Sn-4Mo-4Cr	GT
Near-beta alloys			
Ti-10-2-3	(R56410)	Ti-10V-2Fe-3Al	AF, LG
Ti-5-5-5-3	(–)	Ti-5Al-5Mo-5V-3Cr	AF, LG
Beta alloys			
Ti-15-3-3-3	(R58153)	Ti-15V-3Sn-3Cr-3Al	AD, AF, SR
Ti Beta-C™	19 (R58640)	Ti-3Al-8V-6Cr-4Zr-4Mo	GB, LG, NS, PD, SS
Beta-21S	21 (R58210)	Ti-15Mo-2.7Nb-3Al-0.25Si	AD, GT, MI

[a] See legend in Table 4.5 for application code.
[b] ELI, extra-low interstitial grade.

size (in CP Grade 2); beta alloys can also be aged (producing alpha precipitates) for improved strength (Boyer and Hall 1992; Upadhyaya et al. 1992). Designations, nominal chemical composition, and typical applications of the commercial titanium alloys are given in Table 4.8.

4.5 Physical and Mechanical Properties

Titanium and its alloys exhibit very unique combinations of physical and mechanical properties compared to other common engineering metals/alloys. Titanium alloys are essentially nonmagnetic for practical purposes and exhibit relatively high melting points (1590°C–1660°C). Table 4.9 shows the physical properties of titanium. The density and elastic moduli of titanium are roughly half those of

TABLE 4.9 Physical Properties of Titanium

Coefficient of friction—0.8 at 40 m/min	Heat of Vaporization—9.83 MJ/kg
Coefficient of thermal expansion—8.64×10^{-6}/°C	Melting point—1668°C
Color—dark gray	Modulus of elasticity—14.9×10^6 psi
Covalent radius—1.32 Å	Poisson's ratio—0.41
Density—4.51 g/cm³	Specific gravity—4.5
Electrical conductivity—3% IACS (copper 100%)	Specific heat (at 25°C)—0.518 J/kg °C
Hardness—R_B 70–74	Specific resistance—554 μΩ cm
Heat of fusion—440 kJ/kg	Thermal conductivity—20.2 W/m·°C

TABLE 4.10 Mechanical and Physical Properties of Common Industrial Titanium Alloys

Ti Alloy ASTM Grade No.	Min. 0.2% Yield Strength, MPa (ksi)	Elastic Modulus (GPa)	Density (g/cm³)	Thermal Conductivity (W/m·°C)	Specific Heat (J/kg·°C)	Coeff. Thermal Exp. (10⁻⁶/°C)
Gr. 1, 11	172 (25)	103	4.51	20.8	520	9.2
Gr. 2, 7, 16, 26	276 (40)	104	4.51	20.8	520	9.2
Gr. 12	345 (50)	103	4.51	19.0	544	9.6
Gr. 3	379 (55)	105	4.51	19.7	520	9.2
Gr. 9, 18, 28	483 (70)	107	4.48	8.3	544	9.9
Gr. 5	827 (120)	115	4.43	6.6	565	9.5
Gr. 23, 29	759 (110)	114	4.43	7.3	565	9.5
Gr. 19	1103 (160)[a]	102	4.82	6.2	515	9.5

[a] Typical ST + aged condition.

iron and nickel-based alloys and the coefficient of thermal expansion is also lower than practically all other metals. Strength and, therefore, strength-to-density depend on the specific Ti alloy and its metallurgical condition, but are typically much higher on a unit volume basis than most industrial structural metals/alloys. Since corrosion is the primary emphasis of this chapter, some typical properties for common, corrosion-resistant industrial titanium alloys are presented in Table 4.10. The "work horse" CP Grade 2 is the most widely used alloy in industrial markets due to its good combination of mechanical and corrosion properties (Grauman 1998a).

4.6 Corrosion

4.6.1 Corrosion Resistance: Passivation

Like any other passive metal, titanium's corrosion resistance stems from the passive oxide layer present on the metal surface. The passive layer that forms on the surface is very stable, highly adherent, chemically resistant, and continuous. This naturally formed oxide layer is very thin (less than ten nanometers) and transparent (thus macroscopically invisible), but highly resistant to many aggressive environments. The high equilibrium electrode potential of titanium (-1.63 V$_{HE}$ for Ti/Ti^{2+}) itself signifies that it is a very reactive metal. The standard electrode potential and corrosion potential of titanium in NaCl solutions is compared with other metals in Table 4.11 (Lacombe 1976).

This beneficial titanium dioxide surface film forms spontaneously due to the high affinity of titanium metal for oxygen (Mote et al. 1958; Andreeva 1964; Tomashov 1974; Schutz 2003; Been and Grauman 2011). The naturally formed stable titanium oxide present in an aqueous environment is rutile (TiO_2), but it may also contain a combination of suboxides such as Ti_2O_3 and TiO. The TiO_2 content is very high at the oxide/electrolyte interface while the suboxides (Ti_2O_3 and TiO) are more prevalent nearer the metal/oxide interface. The stable outer oxide can be either a crystalline or amorphous form of TiO_2

TABLE 4.11 Comparison of Standard Electrode Potentials and Corrosion Potentials of Common Metals in 0.5 M NaCl Solution

Metal	V (SHE)		
	Standard Electrode Potential for M^{n+}/M	Standard Potential for MxOy/M at pH 6	Corrosion Potential in Aerated 0.5 M NaCl Solution pH ~6
Magnesium	−2.37	−1.27	−1.40
Aluminum	−1.63	−1.84	−0.57
Titanium	−1.66	−1.62	+0.18
Zinc	−0.76	−0.77	−0.78
Chromium	−0.74	−0.93	−0.11
Iron	−0.44	−0.45	−0.44
Copper	+0.34	+0.12	+0.06

Source: Lacombe, P., Corrosion and oxidation of Ti and Ti alloys, in *Proceedings of the 3rd International Conference on Titanium*, Plenum Publications Corporation, New York, 1976, pp. 847–880.

SHE, standard hydrogen electrode; V, volts.

(rutile/anatase) or a combination of both. Due to its reactive nature, titanium readily oxidizes during exposure to air or moisture and also in aqueous and nonaqueous electrolytes (Bomberger 1969).

All passive metals adsorb and react with oxygen present in the atmosphere almost instantaneously when bare metal surfaces are freshly exposed (Jones 1996). However, passivation kinetics are very important and these rates can be represented by the heat of oxide formation. The heat of titanium oxide formation lies between tantalum and iron where Table 4.12 provides a list of these heats of metal oxide formation. Also when damaged either through mechanical abrasion or by any other means, this titanium oxide film can re-heal itself instantaneously; in contrast, this may not be the case for iron. In most environments, with the presence of a few parts per million (ppm) of oxygen or water, an exposed titanium surface can repassivate itself quite easily. This mechanism works analogously for aqueous environments, but has some limitations for anhydrous conditions where there is an absence of oxygen.

This tenacious titanium oxide film is stable over a wide range of pH and potentials; this phenomena can be noted in the potential (Eh)–pH system diagram for titanium and water at room temperature, as shown in Figure 4.7 (Pourbaix 1974). There is a wide regime of stability for the nontoxic titanium oxide (TiO_2) in aqueous solutions, thus explaining its wide usage in biomedical applications; the specific passive TiO_2 range important for biomedical applications is highlighted (shaded zone) in the aforementioned Pourbaix diagram. Titanium oxide is also very resistant to high anodic potential/currents in most aqueous solutions. Figure 4.8 shows a hypothetical anodic polarization plot of titanium and steel in chloride solutions. Under similar conditions in chloride solutions, titanium passivity is far more stable and covers a much wider range of potentials than steel. However, at negative potentials (less than −1 V) and at lower pH levels, that is, in strong reducing acids, the titanium oxide film is not stable and forms passive

TABLE 4.12 Heat of Formation of Some Metal Oxides

Metal	Heat of Formation of Lowest Oxide (kcal/g mol)
Tantalum	500.0
Titanium	217.4
Iron	64

Source: Wood, J.D., The characterization of particulate debris obtained from failed orthopedic implants (research), College of Material Engineering, San Jose State University, San Jose, CA, 1993.

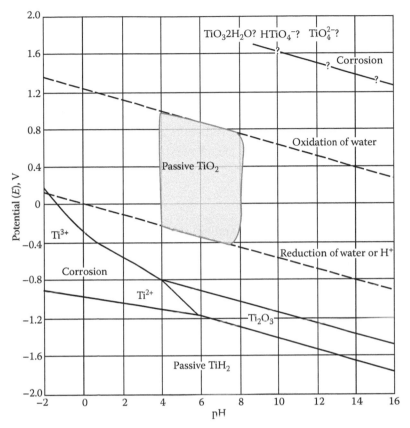

FIGURE 4.7 Eh–pH diagram (Pourbaix) of titanium–water system at 25°C, with preferable region (shaded area) for biomedical applications. (From Pourbaix, M., *Atlas of Electrochemical Equilibria in Aqueous Solutions*, National Association of Corrosion Engineers, Houston, TX, 1974.)

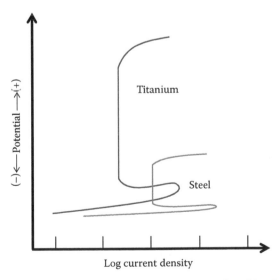

FIGURE 4.8 Hypothetical anodic polarization plots of titanium and steel in chloride solutions; note wide range of passive oxide stability for titanium.

titanium hydride (TiH_2). This hydride film initially formed on the surface can lead to hydrogen diffusion inward with elevated temperature exposures that may result in hydrogen embrittlement. Further theory behind the corrosion resistance of titanium oxide, its semiconductor nature, and its use as a valve metal is detailed in the literature (Chappell and Leach 1978; Yu 2003; Been and Grauman 2011).

4.6.2 Corrosion Resistance: Various Media

The stability of the oxide film on titanium depends on in-service environmental conditions. This section presents titanium behavior, based on the passive or active nature of the oxide film in various media (Tomashov 1963).

4.6.2.1 Seawater

Studies have shown that titanium is fully resistant to natural seawater (Schutz 1988a; Schutz and Grauman 1988; Baker and Grauman 1991; Mountford 1996; Mountford and Grauman 1997; Mountford 2001; Mountford 2002). In seawater, titanium alloys are immune to all forms of localized corrosion, and withstand seawater impingement and flow velocities in excess of 30 m/s (100 ft/s) due to titanium's excellent erosion–corrosion resistance. Most impurities such as sulfides have no detrimental effect on titanium performance in seawater; additionally, polluted waters at ports or harbors, which can have detrimental effects on some other corrosion-resistant metals, have no detrimental effect on titanium alloys (Mountford 2002). Corrosion rates measured on titanium exposed to marine atmospheres showed extremely low values, 0.0003 mm/year. Figure 4.9 compares the qualitative seawater corrosion resistance of titanium and two other metals commonly used in marine applications (Schutz and Scaturro 1991). In addition, the fatigue strength and toughness of alpha titanium alloys are unaffected by seawater and these alloys are also immune to seawater stress corrosion at ambient conditions. Unalloyed titanium (CP grades) and other alpha alloys have demonstrated total reliability in many marine and naval applications not only due to their excellent corrosion resistance but also due to good abrasion and cavitation resistance. Titanium tubing has been used with great success for more than 40 years in seawater-cooled heat exchangers in the chemical, oil refinery, and desalination industries (Gray 1964; Beavers 1986).

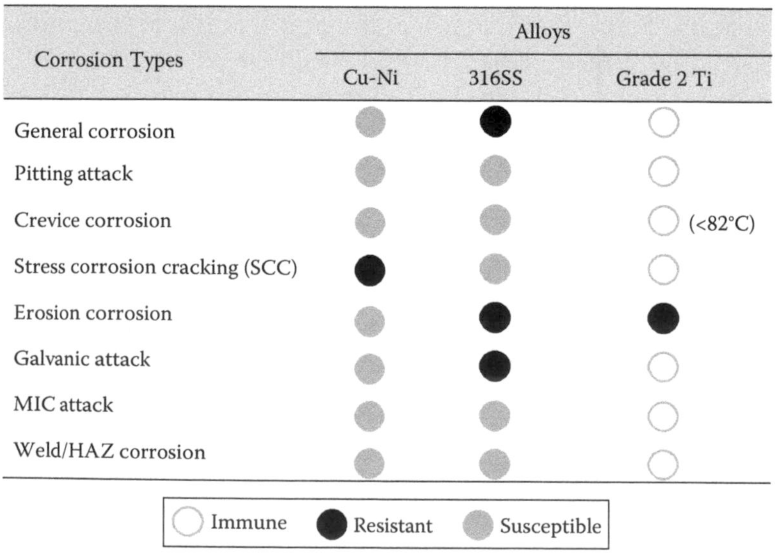

Corrosion Types	Alloys		
	Cu-Ni	316SS	Grade 2 Ti
General corrosion	Susceptible	Resistant	Immune
Pitting attack	Susceptible	Susceptible	Immune
Crevice corrosion	Susceptible	Susceptible	Immune (<82°C)
Stress corrosion cracking (SCC)	Resistant	Susceptible	Immune
Erosion corrosion	Susceptible	Resistant	Resistant
Galvanic attack	Susceptible	Resistant	Immune
MIC attack	Susceptible	Susceptible	Immune
Weld/HAZ corrosion	Susceptible	Susceptible	Immune

○ Immune   ● Resistant   ◐ Susceptible

FIGURE 4.9   Qualitative seawater corrosion resistance of titanium compared with two other metals used in similar seawater application areas. (From Schutz, R.W. and Scaturro, M.R., *J. Naval Eng.*, 175, 1991.)

4.6.2.2 Fresh Water/Steam

Titanium alloys are highly resistant to natural waters and steam to temperatures in excess of 316°C (Hughes and Lamborn 1960). Water streams containing oxides of iron or manganese or traces of sulfides, sulfates, or carbonates have no influence on the performance of titanium. Titanium remains totally unaffected by chlorination treatments used to control biofouling, which in contrast typically results in severe localized corrosion of stainless steel. Microbes present in water systems, especially stagnant water, can be harmful to many metals, and over time, these microbes deposit on metal surfaces leading to general or localized corrosion damage to the metal. Titanium metal is immune to microbiologically influenced corrosion (MIC) in all water systems (Little et al. 1991) and is often endorsed as the most resistant metal to MIC.

4.6.2.3 Salt Solutions

Titanium and titanium alloys exhibit excellent resistance to salt solutions due to their stable passive oxide layer over a wide range of pH and temperatures. Performance is outstanding when both oxidizing anionic and cationic salts are present. Oxidizing anionic salts, such as nitrates, molybdates, chromates, permanganates, and vanadates, or oxidizing cationic salts, such as ferric, cupric, and nickel compounds, cause no corrosion damage to titanium. Also, the resistance of titanium to metal chlorides, bromides, and other halides, except fluorides, is excellent. However, crevice corrosion at elevated temperatures in acidic salt solutions is a concern (Covington 1976; Grauman 1990b). CP Grade 2 titanium in 42 wt% $MgCl_2$ or 20 wt% $CaCl_2$ above 200°C is susceptible to crevice corrosion (Schutz 1986). Depending on concentrations and temperatures, either Grade 12 or Grade 7 titanium is the preferred grade under these conditions.

4.6.2.4 Organic Compounds and Acids

Performance of titanium and its alloys is excellent in organic compounds. They are highly resistant to hydrocarbons, chlorohydrocarbons, fluorocarbons, ketones, aldehydes, ethers, esters, amines, alcohols, and most other organic compounds. However, titanium performance can be compromised when anhydrous organic compounds are present due to instability of the oxide film under those service conditions. Specifically, in the case of absolute methanol, cases of stress corrosion cracking (SCC) of titanium equipment were reported (Sedriks and Green 1971). However, whenever trace quantities of oxygen or moisture are present in organic streams, such conditions are typically adequate to maintain the passivity of titanium. Similarly, with at least 1 wt% of water in methanol, conditions sufficient to prevent the depassivation and SCC in titanium alloys were reported (Haney and Wearmouth 1969; Chem et al. 1971). Acetic, tartaric, stearic, lactic, and tannic acids are considered to be weak organic acids and have little to no effect on titanium corrosion. For example, the corrosion rate of titanium in 99.5 wt% acetic acid at 100°C is nil (TIMET 1997). Due to its high corrosion resistance, titanium equipment has traditionally been used for production of terephthalic acid, acetaldehyde, and adipic acid. However, under nonaerated conditions, higher concentration of some organic acids can destabilize the passive oxide layer resulting in higher corrosion rates of titanium, especially in the case of formic acid. In addition to formic acid, strong organic acids such as oxalic, sulfamic, and trichloroacetic can impair the corrosion resistance of titanium. Titanium performance in these strong organic acids depends on concentration, temperature, degree of aeration, and if possible inhibitors are present. In the case of nonaerated formic or sulfamic acid environments, only Grade 7 titanium alloy provided sufficient corrosion resistance.

4.6.2.5 Chlorine Chemicals and Chlorine Solutions

Titanium is fully resistant to chlorine chemicals such as chlorites, hypochlorites, chlorates, perchlorates, and chlorine dioxide. Titanium alloys, especially the CP grades, have been used to handle these chemicals in the pulp and paper industry for many years without evidence of corrosion (McMaster and Kane 1970). Titanium components have provided excellent service for many years without any noticeable degradation effects in sodium and calcium hypochlorite solutions (Bomberger 1969). Similarly excellent performance with chloride salt solutions and other brines over the full concentration range, especially as temperatures increase, has been reported where near nil corrosion rates can be expected in brine media over the pH range of 3–11. However, localized corrosion becomes a limiting factor for using titanium components operating in the lower pH and higher temperature ranges with these solutions.

4.6.2.6 Halogen Compounds

Titanium alloys are highly resistant to wet (aqueous) chlorine, bromine, iodine, and other chlorine chemicals because of their strongly oxidizing natures. In many chloride- and bromide-containing environments, titanium has cost-effectively replaced stainless steels, copper alloys, and other metals where severe localized corrosion and SCC were experienced (Bomberger 1969).

4.6.2.7 Alkaline Media

Titanium and its alloys are highly resistant to alkaline media. At sub-boiling temperatures, titanium performance is excellent in alkaline solutions of sodium hydroxide (NaOH), potassium hydroxide (KOH), calcium hydroxide ($Ca(OH)_2$), magnesium hydroxide ($Mg(OH)_2$), and ammonium hydroxide (NH_4OH). Near nil corrosion rates are expected in boiling hydroxides of calcium, magnesium, and ammonium. Table 4.13 enumerates the corrosion rate of unalloyed titanium in various alkaline solutions. At higher concentrations of potassium hydroxide solutions, the corrosion rate can become severe (Tomashov 1963; Covington 1982; Schutz 2003). In addition, there is a chance of excessive hydrogen uptake and eventual embrittlement of titanium alloys in hot, strongly alkaline media, especially when pH is over 12 and the temperature is above 80°C. However, even at higher temperatures, titanium resists pitting, stress corrosion, or the conventional caustic embrittlement than many stainless steel alloys exhibit in these solutions.

4.6.2.8 Oxidizing Acids

In general, titanium has excellent resistance to oxidizing acids over a wide range of temperatures and concentrations. Most notable of the acids in this category are nitric (HNO_3), chromic (H_2CrO_4), perchloric ($HClO_4$), and hypochlorous ($HClO$, wet Cl_2) (TIMET 1997).

4.6.2.8.1 Nitric Acid

For over 50 years, titanium has been employed to a large extent for handling nitric acid due to its corrosion resistance over a wide range of concentrations, temperatures, and conditions (Thomas 1986). Because nitric acid is such a strong oxidizing agent, the trivalent titanium ion in solution is quickly oxidized to the tetravalent state, providing a potent inhibitor in the acid solution. Over the years, titanium has replaced stainless steel that was subject to a significant amount of uniform and intergranular corrosion (Millaway 1965). Titanium and its alloys offer excellent corrosion resistance to a

TABLE 4.13 Corrosion Rates of Unalloyed Titanium in Highly Alkaline Solutions

Medium	Concentration (wt%)	Temperature (°C)	Corrosion Rate (mm/year)
Sodium carbonate	20	Boiling	Nil
Ammonium hydroxide	70	Boiling	Nil
Sodium hydroxide	10	Boiling	0.02
Sodium hydroxide	73	Boiling	0.13
Potassium hydroxide	10	Boiling	0.13
Potassium hydroxide	50	Boiling	2.7

Source: Tomashov, D., *Corrosion and Protection of Titanium (Government Scientific)*, Technical Publication of Machine-Building Literature, Moscow, Russia, 1963.

TABLE 4.14 Corrosion Rates of CP Titanium in Nitric Acid at Low (Sub-Boiling) Temperatures

HNO₃ Concentration (wt%)	Temperature (°C)	Average Corrosion Rate (mm/year)
10	27	0.005
30	27	0.004
40	27	0.002
50	27	0.001
60	27	0.005
70	27	0.003
10	40	0.005
20	40	0.015
30	50	0.016
40	50	0.037
50	60	0.040
60	66	0.040
70	70	0.005

Source: Tomashov, D., *Corrosion and Protection of Titanium (Government Scientific)*, Technical Publication of Machine-Building Literature, Moscow, Russia, 1963.

wide range of nitric acid concentrations at sub-boiling temperatures. Table 4.14 shows the corrosion rates of CP titanium in nitric acid.

At boiling temperatures and above, titanium's corrosion resistance is very sensitive to nitric acid purity where significant general corrosion may occur in hot pure solutions or vapor condensates of nitric acid. Certain metallic cations added in trace amounts to these hot concentrated nitric acid solutions appear to inhibit corrosion of titanium. The presence of some transition metals (Cr, Fe, Cu, etc.) and precious metal (Pt, Pd, etc.) ions can completely inhibit the corrosion at higher concentrations of boiling nitric acid (Takamura et al. 1970). The effect of certain cations on the corrosion of titanium in 65 wt% boiling nitric acid is given in Table 4.15 (Millaway et al. 1964).

Even titanium's own corrosion product (Ti^{+4}) is highly inhibitive; an 80 mg/L addition of titanium ions (Ti^{4+}) to 68 wt% boiling nitric acid can produce a 100-fold reduction in corrosion rate. Titanium often exhibits outstanding performance in recycled nitric acid streams such as reboiler loops due to

TABLE 4.15 Corrosion Rate of CP Titanium in 65 wt% of Boiling Nitric with Various Cation Addition

Metallic Cation[a]	Corrosion Rate (mm/year)
Sb^{5+}	Wt. gain
As^{5+}	0.11
Ba^{2+}	0.02
Cu^{2+}	0.18
Cr^{3+}	0.58
Fe^{3+}	0.11

Source: Millaway, E.E. et al., Titanium in nitric acid (internal report), TIMET, Titanium Metallurgical Laboratory, Dallas, TX, 1964.

[a] Concentration of cation in solution is 1.25 g/L.

metallic ion contaminations (Degnan 1975). Titanium reactors, reboilers, condensers, heaters, and thermowells have been used in solutions containing 10%–70% HNO_3 at temperatures from boiling to 315°C (600°F).

4.6.2.8.2 Fuming Nitric Acid

The corrosion resistance of titanium in white fuming nitric acid is excellent. However, it can corrode severely in red fuming nitric acid. It can also create a pyrophoric reaction product resulting in ignition of the titanium and potentially serious accidents (Ambrose 1955). A high content (over 6%) of nitrogen dioxide (NO_2) and very low water content (less than 1%) in red fuming nitric acid develops a rapid oxidation reaction generating very high amounts of exothermic heat. When red fuming nitric acid attacks a titanium surface, it generates fine titanium particles that ignite spontaneously. Increases in the nitric acid concentration will make this situation even worse. Accordingly titanium and its alloys should never be used when it can be in contact with red fuming nitric acid solutions (TIMET 1997).

4.6.2.8.3 Chromic Acid

Similar to nitric acid, the effect of chromic acid on titanium or titanium alloys is also negligible; the corrosion rate of titanium in boiling 10 wt% chromic acid is less than 0.003 mm/year. Even at much higher concentrations (50 wt%) of boiling chromic acid, titanium maintains its passive nature (Covington 1982).

4.6.2.9 Reducing Acids

Unlike oxidizing media, the corrosion resistance of titanium and its alloys in reducing media is highly sensitive to acid concentration, temperature, alloy chemistry, and other factors (Schutz 2003). Titanium alloys are generally very resistant to mildly reducing acids but can display severe limitations in strongly reducing acids. Examples of reducing acids include acetic ($C_2H_4O_2$), adipic ($C_6H_{10}O_4$), lactic ($C_3H_6O_3$), hydrofluoric (HF), hydrochloric (HCl), hydrobromic (HBr), phosphoric (H_3PO_4), and sulfuric (H_2SO_4). In mild conditions, at low concentrations and low temperatures, titanium provides excellent corrosion protection. Relatively pure, strong reducing acids, such as hydrochloric, hydrobromic, sulfuric, and phosphoric, can accelerate general corrosion of titanium near room temperature; additionally, deaerated solutions can further increase corrosion rates. Corrosion guidelines given in the literature for alpha titanium alloys over a wide range of temperatures and concentrations (iso-corrosion diagram) in pure, naturally aerated acid solutions are very helpful in selecting the right titanium alloy for specific applications (Covington 1982). Among all the titanium alloys, Grade 7 and other Pd-containing alloys offer dramatically improved corrosion resistance under these severe conditions. In fact, Ti–Pd alloys often compare quite favorably to super nickel alloys in dilute reducing acids.

4.6.2.9.1 Hydrochloric Acid

Hydrogen chloride (HCl) is a highly corrosive, strong mineral acid with many industrial uses, especially in the chemical industry as a reagent. Titanium alloys may be considered for use with HCl after other refractory metals and some nickel super alloys. Grade 2 titanium provides good corrosion resistance up to 7 wt% HCl at room temperature, Grade 12 extends that up to 9 wt%, and Grade 7 enhances the range to 27 wt% (TIMET 1997). However, the corrosion resistance of these alloys reduces significantly at higher temperatures, especially at boiling. Table 4.16 compares the corrosion rates of these three titanium alloys at room and boiling temperatures. Only Grade 7 titanium provides good corrosion resistance in boiling solutions and only up to certain limited concentrations.

4.6.2.9.2 Sulfuric Acid

Titanium displays resistance to corrosion in very dilute solutions of sulfuric acids at low temperatures but is not corrosion resistant to sulfuric acid at higher concentrations and temperatures. Grade 2 is completely immune to corrosion in 20 wt% sulfuric acid at 0°C and 5 wt% at room temperature. However,

TABLE 4.16 Corrosion Rates of Titanium Alloys in Naturally Aerated HCl Solutions

HCl (wt%)	Temperature (°C)	Corrosion Rate (mm/year)		
		Grade 2	Grade 12	Grade 7
1	Room	Nil	0.005	0.003
2	Room	Nil	0.003	0.006
3	Room	0.013	0.013	0.01
1	Boiling	2.16	0.036	0.02
2	Boiling	7.11	0.254	0.046
3	Boiling	14	10.2	0.069

TABLE 4.17 Corrosion Rates of Titanium Alloys in Naturally Aerated H_2SO_4 Solutions

H_2SO_4 (wt%)	Temperature (°C)	Corrosion Rate (mm/year)		
		Grade 2	Grade 12	Grade 7
0.54	Boiling	6.401	0.015	0.003
2.16	Boiling	24.0	19.3	0.097
5.4	Boiling	52	61.21	0.757

it severely corrodes in as little as 0.5 wt% sulfuric acid at boiling temperatures. At room temperature, higher concentrations of sulfuric acid (greater than 10 wt%) will cause the protective oxide film to become unstable and dissolve, thereby allowing active corrosion of the titanium. Corrosion rates of titanium alloys in various concentrations of boiling sulfuric acid solutions are given in Table 4.17. The presence of ferric or certain other metal ions in sulfuric acid will inhibit the corrosion of titanium and its alloys. Addition of 16 g/L of ferric ion to boiling 20 wt% sulfuric acid can reduce the corrosion rate dramatically from 62 to 0.13 mm/year (Tomashov 1963).

4.6.2.9.3 Hydrofluoric Acid

Hydrofluoric acid is the most aggressive reducing media for titanium and titanium alloys. Fluoride ions in HF form highly stable, soluble complexes with titanium resulting in very high corrosion rates. These soluble complexes form over a wide range of concentrations and temperatures. The dissolution rate of the titanium increases with either an increase in concentration of HF or temperature or a combination of both. Any other low pH (<7) solution containing fluoride ions is also detrimental to the titanium corrosion resistance. For example, the presence of ammonium fluoride (NH_4F) in reducing acids such as HCl or H_2SO_4 accelerates the corrosion rate of titanium due to formation of free fluoride ions and eventual oxide film dissolution. In addition to the corrosion, the formation of a blue-gray titanium hydride film occurs. Therefore, titanium is not recommended for use with hydrofluoric acid solutions or in fluoride-containing solutions below pH 7. Titanium cannot be recommended for plants where conditions permit active, uncomplexed fluorides to persist (Thomas and Bomberger 1983).

One useful application of HF is for the acid pickling of titanium. A mixture of HF and HNO_3 is typically used to pickle or etch titanium alloys.

4.6.2.9.4 Other Reducing Acids

Titanium is resistant to naturally aerated solutions of phosphoric acids up to 30 wt% concentrations at room temperatures. Grade 12 and the palladium-containing titanium alloys offer superior resistance over CP grade titanium alloys. Grade 7 displays excellent performance up to 80 wt% phosphoric acid at room temperature. All these alloys will corrode in boiling solutions, with Grade 7 being the most resistant, offering good titanium corrosion resistance up to about 6 wt% boiling phosphoric acid solutions.

Titanium offers better corrosion resistance in mixed acid solutions. The corrosion rate can be reduced significantly with a mixture of acids; whenever nitric acid is added to hydrofluoric or hydrochloric acid, it brings down the corrosion rates. Also, titanium is immune to aqua regia ($3HCl: 1HNO_3$) at ambient conditions. However, increases in the concentration of reducing acids in that mixture can accelerate the corrosion rates. High concentrations of inorganic acids such as 50 wt% HI or HBr acid have no influence on titanium corrosion at room temperature.

4.6.2.10 Liquid Metals and Fused Salts

Titanium has good corrosion resistance in contact with molten aluminum, sodium, potassium, magnesium, and other liquid metals at moderate temperatures, but it can dissolve rapidly at higher temperatures. Titanium exposed to gallium at 400°C has around a 0.1 mm/year corrosion rate and the corrosion rate is 1 mm/year when it is exposed to magnesium at 800°C (Bomberger 1969). Some liquid metals such as silver and mercury may also cause SCC of titanium and its alloys at higher temperatures. Titanium demonstrates good corrosion resistance in contact with molten chloride salts in the presence of oxygen, but high rates of titanium corrosion have been reported when in contact with fused sodium and potassium hydroxides, peroxides, and other salts (TIMET 1997). Proper guidelines should be followed when titanium is used for handling fused salts (Schutz 2003).

4.6.2.11 Gases

The tenacious oxide film on the surface of the titanium provides corrosion resistance not only in aqueous environments but also in gaseous atmospheres (Covington and Schutz 1981). Titanium alloys are totally resistant to all forms of atmospheric corrosion regardless of pollutants present in marine, rural, or industrial locations. These include nitrogen, carbon dioxide, carbon monoxide, sulfur dioxide, and hydrogen sulfide. Titanium is widely used to handle moist or wet chlorine gas and has earned a reputation for outstanding performance in this service. The strongly oxidizing nature of moist chlorine passivates titanium resulting in very low corrosion rates. However, dry chlorine gas rapidly attacks titanium, and ignition is possible if the moisture content is very low. Therefore, use of titanium in chlorine has to follow guidelines considering the moisture content and temperature (Schutz 2003). Moist or dry ammonia gas or ammonia water (NH_4OH) solutions will not corrode titanium up to their boiling points and above (Jones and Wilde 1977). Sulfur dioxide and hydrogen sulfide, either wet or dry, have no effect on titanium (Hanson 1984). Dry sulfur dioxide at room temperature has no effect on titanium Grade 2 performance. Titanium alloys, especially Grade 7, used in FGD scrubbers of coal-fired power plants containing the earlier mentioned gases displayed outstanding performance (Schutz and Grauman 1986b; Peacock and Grauman 1995).

Titanium has excellent resistance to gaseous oxygen and air at temperatures up to 370°C (700°F). Above this temperature, titanium begins to form colored surface oxide films, which thicken slowly with time. Above 650°C (1200°F) or so, titanium alloys suffer from lack of long-term oxidation resistance and will become brittle due to the increased diffusion of oxygen into the metal. Figure 4.10 gives an oxidation profile for a new alpha–beta alloy (TIMETAL® 54M) in pure oxygen at 760°C (Gudipati and Divi 2012). Rapid oxidation occurred within 24 h of exposure forming around a 15 μm thick oxide layer. In oxygen, titanium is subject to ignition, which can occur with oxygen concentrations above 35% or at pressures over 25 bar (350 psig) when a fresh surface is created (Schwartzberg et al. 1957; Littman and Church 1959). Oxygen ignition must be considered as a severe safety hazard, and proper guidelines must be followed whenever titanium is utilized in an oxygen-rich environment. Nitrogen reacts much more slowly with titanium than oxygen; however, above 800°C (1400°F), excessive diffusion of nitrogen into the metallic phase may cause metal embrittlement (TIMET 1997).

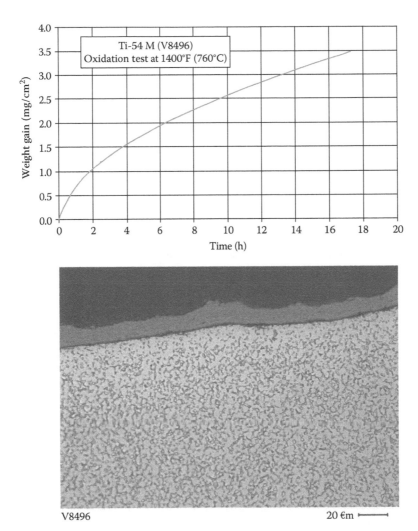

FIGURE 4.10   High temperature oxidation (weight gain) profile for a titanium alloy (TIMETAL 54M) at 760°C in pure oxygen; microstructure with 15 μm oxide layer. (From Gudipati, P. and Divi, S., High temperature oxidation study of TIMETAL® 54 M alloys, TIMET Internal Data, 2012.)

4.6.3 Types of Titanium Corrosion

As mentioned in the previous sections, titanium and its alloys provide excellent corrosion resistance against a wide variety of chemicals. However, like any other metal, titanium is also subject to some specific types of corrosion in certain environments. The damage that one usually encounters in titanium alloys can be classified into six types. Each of these corrosion types are explained in the following sections (Revie and Uhlig 2011).

4.6.3.1 General Corrosion

Titanium and its alloys can suffer from general corrosion in specific environments due to oxide film breakdown leading to accelerated mass loss rates. The rate of damage depends on conditions such as solution pH, temperature, and the nature of the environment. Such damage to titanium can be from a chemical or an electrochemical reaction in which the metal transitions from the elemental to the ionic form (Yu 2003). This basic anodic reaction results in generating either the trivalent (Ti^{3+}) or tetravalent

TABLE 4.18 Typical Environments in which Titanium Corrodes
Severely/Ignites

Environment	Corrosion Mode
Strong reducing acids	General corrosion
Warm alkaline peroxide solutions	General corrosion
Dry chlorine gas	General corrosion/ignition/burning

(Ti^{4+}) ions depending on whether the reaction occurs under reducing (trivalent) or neutral to oxidizing (tetravalent) conditions. The trivalent ion forms soluble titanium compounds, which dissolve into solution and typically produce a violet/purple color. The tetravalent ion forms nonsoluble hydrated oxides, which usually re-precipitate onto the titanium surface, thus inhibiting or reducing further corrosion. The corresponding cathode reaction involves the reduction of ionic species. In the case of reducing acid solutions, the main cathodic reaction is the reduction of hydrogen ions to molecular hydrogen. In oxidizing or other environments, the main cathodic reaction involves the reduction of the oxidizing agent, usually oxygen or water. The electrochemical nature of these reactions can play a significant role in controlling the titanium corrosion kinetics.

Titanium's general corrosion rates are determined by using weight loss data following ASTM standard G31 (ASTM Standards 2012). Electrochemical polarization tests are also used as supplemental tests to calculate the weight loss using ASTM standards G3, G5, G59, and G102 (ASTM Standards 2012). Test methods for the evaluation of titanium corrosion damage are given in the literature (Schutz and Covington 1981b; Liening 1986; Schutz 1986; Baboian 2005). Some important environments where titanium general corrosion susceptibility exists are listed in Table 4.18.

4.6.3.2 Crevice Corrosion

Titanium alloys are often selected for industrial applications that involve exposure to chloride environments due to their superior corrosion resistance to that environment. However, with higher temperature exposures in these solutions, crevice corrosion of titanium can become a limiting factor. This is not only a problem associated with chlorides but it also applies to other halide environments (Schutz 1995; Schutz and Grauman 1986a; Schutz and Grauman 1986c).

The mechanism for crevice corrosion of titanium is due to development of an oxygen-depleted reducing acid solution within tight crevices (Griess 1968; Schutz 1988b). Since oxygen and other oxidizing species are consumed more rapidly in crevices than in bulk solutions, the metal potentials within crevices become very active with respect to free metal surfaces and thus an electrochemical cell (battery) is created where metal within a crevice acts as the anode (and corrodes) and the free metal surface acts as the cathode (Liening 1983). To maintain ionic neutrality, chlorides in the bulk solution diffuse into the crevice and form titanium chlorides that are not stable under these conditions so they readily hydrolyze to form HCl. This in turn causes the pH level within the crevice to decrease, producing an accelerated localized attack on the titanium surface within the crevice. This synergistic mechanism accounts for the often rapid failures seen for titanium components subject to crevice corrosion attack.

The factors that influence titanium crevice corrosion include the temperature, physical nature of the crevice, metal surface conditions, metal potential, pH and solution chemistry, and alloy composition (Covington 1982; TIMET 1997). However, crevice attack has not been observed on titanium alloys below 70°C.

For crevice corrosion testing, samples with controlled tight crevice gaps that promote crevice corrosion initiation due to oxygen depletion are required. Titanium metal-to-metal contact surfaces often produce larger crevice gaps that are usually less susceptible. However, extremely tight ($<5 \times 10^{-5}$ cm) crevices created between titanium metal and nonmetallic gaskets (usually PTFE), readily initiate crevice corrosion under the proper conditions. A sheet sandwich-type crevice assembly is utilized during testing to give more consistent results. The classic multi-crevice washer assembly used for testing stainless steel and other passive metals in chloride solutions is not recommended for titanium as they cannot

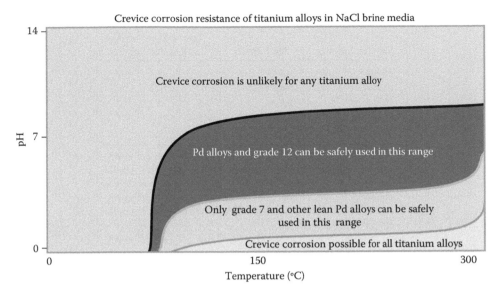

FIGURE 4.11 Crevice corrosion resistance range of certain industrial titanium alloys. (From Schutz, R., Titanium alloy crevice corrosion: Influencing factors and methods of prevention, *Proceedings of the Sixth World Conference on Titanium*, Les Éditions de Physique, Cannes, France, 1988b, pp. 1917–1922.)

achieve the appropriately sized crevice gap geometries. Titanium crevice testing procedures including certain types of gaskets, test temperatures, solutions, minimum crevice gaps, and other important parameters are well documented elsewhere (Diegle 1981; Bergman and Grauman 1992; Schutz 1992a,b; Babulan 2005). The crevice corrosion resistance of titanium alloys varies with pH, where a decreasing pH strongly increases the susceptibility to crevice attack.

Certain solute elements can enhance the crevice corrosion resistance of titanium alloys. Precious metals (Pt, Pd, Ru, etc.) and other alloying elements (Ni, Mo, etc.) can significantly broaden the pH and temperature limits for providing improved crevice corrosion resistance. Figure 4.11 shows the general guidelines for enhancing the crevice corrosion resistance of titanium alloys in sodium chloride brines (Schutz 1988b).

In addition to proper alloy selection, surface treatments using precious metal coatings and proper cleaning to remove surface iron contamination are some of the preventative methods most useful in reducing crevice corrosion failures (Schutz et al. 1984). Also, the specification of elastic rather than plastic or hard gasket materials could be an effective measure to prevent crevice corrosion. Another proactive strategy is to incorporate some nickel, copper, molybdenum, or palladium into the gasket by either coating or doping.

4.6.3.3 Pitting Corrosion

Pitting corrosion is a form of localized corrosion in passive metals. Fortunately, titanium exhibits excellent resistance to pitting corrosion in most environments so it is rarely the limiting factor for titanium applications. Due to this attribute, titanium is extensively used as dimensionally stable anodes (DSA) for chlor-alkali cells and recovery of metal oxides, etc. (Tomashov 1963).

Pitting occurs whenever the potential of the metal exceeds the breakdown potential of the protective oxide film covering the titanium surface. The pitting potential of titanium and titanium alloys in phosphate and sulfate media is over +80 V (SCE). This anodic pitting potential (extrinsic) depends on several factors such as chemistry of media, temperature, and titanium alloy type. For example, the pitting potential for CP titanium in chloride solutions at room temperature is around +10 V, but it can dropdown to +5 V at boiling temperatures (Schutz 2003). Titanium is one of the premier metals when considering its pitting corrosion resistance in sea water, providing a far greater pitting resistance than stainless steels. Thus, spontaneous pitting of titanium is almost never observed in sea water service.

TABLE 4.19   Repassivation Potential of the Titanium Alloys

Titanium Alloy	Alloy Type	Repassivation Potential (V vs. Ag/AgCl)[a]
Grade 2	Alpha	6.2
Ti-6-2-4-6	Near beta	2.5
Beta-C	Beta	2.7
Beta-21S	Beta	2.8
Ti-64	Alpha + beta	1.8

Source: Grauman, J.S., BETA-21S: A high-strength corrosion-resistant titanium alloy, *Proceedings of the TDA International Conference*, Orlando, FL, 1990a; Grauman, J.S., Influence of surface condition on the resistance of titanium to chloride pitting, in: *Advances in Localized Corrosion*, NACE International, Houston, TX, 1990b, pp. 331–334.

[a] Test conditions: 5 wt% boiling NaCl (pH 3.5), applied current density 200 mA/cm[2].

The anodic pitting potential is measured by using electrochemical techniques specified in the ASTM standards. ASTM G3, G5, and G59 standards are usually used for these tests (ASTM Standards 2012). Another property, repassivation potential, can be used as more conservative measure of pitting resistance. Instead of using the traditional potentiodynamic anodic polarization or cyclic polarization method, this method uses the galvanostatic method to measure the minimum potential at which pitting can occur. Details of the test method can be found in the literature (Baboian 2005).

Repassivation potentials measured for various titanium alloys are given in Table 4.19 (Grauman 1990a). Overall, the repassivation potential of titanium alloys in chloride solutions is much better (>1 V) than stainless steels.

Despite Titanium's inherent pitting resistance, the inadvertent presence of stray voltages due to electrical grounding in aggressive environments can lead to localized pitting failures of titanium equipment. An example of pitting corrosion failure of titanium tubing due to an unintentionally applied potential is shown in Figure 4.12.

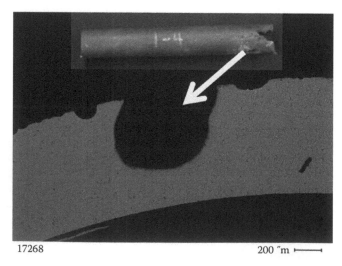

17268   200 ˝m ├───┤

FIGURE 4.12   Photomicrograph of the failed tube (inset) containing large pits (>50 μm), due to unintentional applied potentials. (From Divi, S., *Examination of Titanium Condenser Tubes* (failure analysis no. 3351), TIMET, Henderson Technical Lab (HTL), Dallas, TX, 2010.)

4.6.3.4 Stress Corrosion Cracking

SCC is a form of environmentally assisted cracking (EAC) where cracking failure is caused by the combination of tensile stress and a corrosive environment. Stresses can be from externally applied loads or could be an internal stress from cold work deformation during processing or residual stresses developed during fabrication/welding. Titanium generally has excellent resistance to SCC in natural waters and most chemical environments but the most notably resistant are the CP and modified CP grades, the most widely used industrial titanium alloys. Aerospace titanium alloys such as Ti-6Al-4V are moderately susceptible to SCC in various specific chemical environments. In chloride solutions, SCC failures of alloys such as Ti-6Al-4V have been observed in laboratory testing but very few cases have been documented from field applications. Figure 4.13 shows a scanning electron microscope (SEM) fractograph of a Ti-6Al-4V specimen subjected to slow strain rate testing (SSRT) in seawater at room temperature (Divi and Grauman 2011). The fractograph shows a lower energy cleavage fracture for the specimen exposed to simulated seawater at room temperature, whereas a specimen tested in air will exhibit a typical ductile failure.

In addition to SCC failures due to chlorides, Ti-6Al-4V and other aerospace alloys under stress are also sensitive to halogenated hydrocarbons. Halide ion concentrations (Cl^-, Br^-, and I^-) in the environment also play a role in initiating and propagating SCC cracking, where the SCC susceptibility increases with halide concentration. Alloying elements also influence the SCC susceptibility; with higher aluminum contents (>5 wt%) and the increased presence of hard Ti–Al intermetallic phases causes a decrease in the SCC resistance (TIMET 1997). Similarly, higher concentrations of interstitial solute elements also degrade the SCC resistance. The oxygen content in Ti-6Al-4V can be a concern relating to the SCC susceptibility; for that reason in certain service applications where high chloride concentrations can be expected, the Ti-6Al-4V extra-low interstitial (ELI) alloy should be used instead of the normal Ti-6Al-4V alloy. Similarly, CP grades with higher oxygen contents (>0.25%) that are also susceptible to SCC failures should not be used for service in seawater environments. Table 4.20 gives the environments in which titanium alloys are susceptible to SCC (Schutz 2003).

Hot salt SCC, where residual chloride salts on hot titanium metal can pit the surface that eventually cracks under high tensile loads, has also been extensively studied on high temperature titanium aerospace components (Jackson and Boyd 1968; Rideout et al. 1968; Liening 1983). Stress relieving and

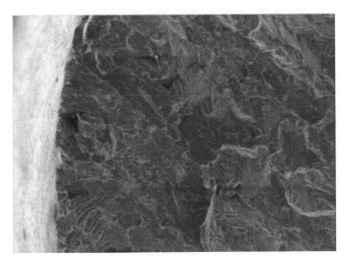

FIGURE 4.13 SEM fractograph showing cleavage fracture characteristics for Ti-6Al-4V alloy SSRT test sample in seawater at 200× magnification. (From Divi, S. and Grauman, J.S., Environmentally assisted cracking study of a newly developed titanium alloy, *Proceedings of the CORROSION'11*, NACE International, Houston, TX, 2011, Paper 11288.)

TABLE 4.20 Environments Known to Promote SCC of Titanium Alloys under Ambient Conditions

Environment	Titanium Alloys
Highly susceptible	
Methyl alcohol (anhydrous)	All commercial grades
Ethyl alcohol (anhydrous)	Ti-8Al-1Mo-1V, Ti-5Al-2.5Sn
Nitric acid (red fuming)	Ti, Ti-8Mn, Ti-6Al-4V, Ti-5Al-2.5Sn
Moderately susceptible	
Ethylene glycol	Ti-8Al-1Mo-1V
Freons (fluorinated hydrocarbons)	Ti-8Al-1Mo-1V, Ti-5Al-2.5Sn, Ti-6Al-4V
Methylene chloride, trichloroethylene, carbon tetrachloride	Ti-8Al-1Mo-1V, Ti-5Al-2.5Sn
Chlorinated diphenyl	Ti-5Al-2.5Sn
Silver (solid) and AgCl	Ti-7Al-4Mo, Ti-5Al-2.5Sn, Ti-6Al-6V-2Sn
Monomethylhydrazine	Ti-8Al-1Mo-1V, Ti-6Al-4V
Reasonably susceptible	
Seawater/NaCl solution	Unalloyed Ti (with >0.25% O)
	Ti-3Al-11Cr-13V, Ti-5Al-2.5Sn, Ti-8Mn,
	Ti-6Al-4V, Ti-6Al-6V-2Sn,
	Ti-7Al-2Nb-1Ta, Ti-4Al-3Mo-1V,
	Ti-8Al-1Mo-1V, Ti-6Al-Sn-4Zr-6Mo
Distilled water	Ti-8Al-1Mo-1V, Ti-5Al-2.5Sn,
	Ti-11.5Mo-6Zr-4.5Sn

Source: Schutz, R., Corrosion of titanium and titanium alloys, in *CORROSION*, ASM International, Metals Park, OH, 2003.

proper cleaning to eliminate any chlorides/halides from metal surfaces are very important SCC mitigation procedures that are especially important to fabricators finishing up titanium components (Lane et al. 1966). CP grades and other modified grades (Grade 12, 16, 26, etc.) are immune to this type of corrosion. Ti-6Al-4V, CP Grade 4, and other alloys are moderately resistant while alloys with higher aluminum contents like Ti8Al-1Mo-1V are least resistant to this form of attack (Bomberger 1969).

SCC testing methods for titanium and its alloys as well as relevant specimen configurations are well documented in the literature (Schutz 2003). However, comparison of results from experiments using different specimen configurations can often lead to wrong conclusions; for relevant validation, the test specimen configuration (smooth or notched) and testing method (statically or dynamically loaded) should be appropriately chosen based on the expected in-service application conditions.

Besides SCC and hydrogen embrittlement, the third type of EAC is corrosion fatigue. Corrosion fatigue is a special case of stress corrosion, and it is due to the result of the combined action of alternating or cyclic mechanical forces/stresses and a corrosive environment. Vast amounts of laboratory studies, especially for aerospace alloys, are available in the literature (Speidel 1983; Lampman 1996).

4.6.3.5 Hydrogen Absorption/Embrittlement

Titanium alloys provide excellent corrosion resistance in most aqueous environments over a wide range of pH and temperatures, but as mentioned earlier, at very low pH and negative potentials, formation of titanium hydride (TiH_2) takes place on the metal surface. Titanium hydrides appear as dark, acicular needle-like structures in the metal microstructure. At high temperatures the titanium hydride diffuses into the metal, and as it precipitates, the crystal lattice becomes locally strained causing the metal to become brittle. Hydrogen absorption and eventual embrittlement is also possible in high pressure and high temperature anhydrous gas streams (Paton and Williams 1974). In this case, a small quantity (~2%) of moisture or oxygen is sufficient to stabilize the oxide film, which then becomes an excellent barrier to hydrogen gas absorption. Still, titanium is not recommended for use in pure hydrogen gas due to the potential of severe hydriding and embrittlement whenever the protective oxide film is disrupted.

Solubility limit of titanium hydride varies by alloy type. In alpha and near-alpha titanium, which contains an hcp structure, the maximum solubility is around 150 ppm. In beta alloys, with a bcc structure, the maximum solubility increases dramatically to around 2000–3000 ppm. The solubility limit in alpha–beta alloys lies between these two with the limit varying somewhat depending on the alloying content. As a beta-stabilizing element, hydrogen solubility increases in the amount of beta phase. The high solubility of hydrogen in beta alloys makes them much less susceptible to hydrogen embrittlement. On the other hand, alpha alloys are much more sensitive to hydrogen embrittlement. In alpha–beta alloys, excessive hydrogen absorption causes titanium hydride precipitates in the alpha phase. Thus, once the bulk concentration of hydrogen is above the solubility limit for that particular alloy, hydride precipitates will form. Small amounts of hydride precipitation are not harmful to the titanium structure, but increasing levels will begin to degrade the mechanical properties such as ductility and toughness.

Laboratory studies and field experiences have revealed the factors/conditions that influence hydrogen uptake and possible embrittlement in titanium. There are three factors that must exist concomitantly for titanium hydriding to occur. Formation of nascent, or atomic, hydrogen on a titanium surface is the first factor. This hydrogen can be formed from cathodic protection at excessively high impressed currents, or corrosion of titanium, or at a galvanic couple (titanium as cathode), or due to dynamic abrasion of the metal surface. As mentioned earlier, the second factor is oxide film disruption whenever the solution pH is less than 3 or greater than 12, or when an impressed potential more negative than −0.7 V (SCE: saturated calomel electrode) is present; in seawater applications it can be reduced to −1 V depending on the conditions, or whenever a mechanical film disruption occurs (Covington 1979; Schutz and Grauman 1989). Metal temperature is the third factor; it has to be above 80°C where accelerated kinetics permits the surface formed hydrides to diffuse into the metal causing embrittlement. The combined effects of these three factors are put together in a Venn diagram as shown in Figure 4.14. Hydriding and hydrogen degradation can be avoided by eliminating one or more of these factors.

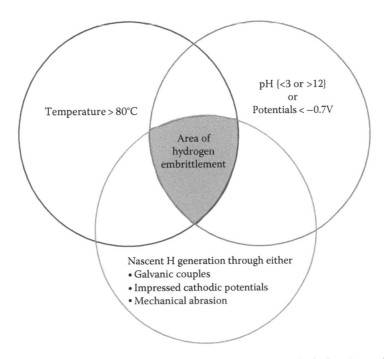

FIGURE 4.14   Combined effects of three most important factors for titanium hydriding (Venn diagram).

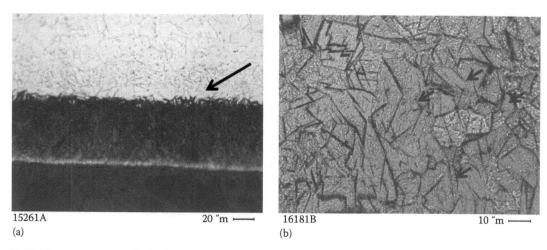

15261A   20 ″m ⊢——⊣   16181B   10 ″m ⊢——⊣
(a)   (b)

FIGURE 4.15 Titanium hydride (black) layer formed on the surface of titanium exposed to acidic solution (a), completely hydrated titanium at higher temperature (b). (From Divi, S., (Internal) TIMET, Henderson Technical Lab (HTL), Dallas, TX, 2009.)

Despite all this, titanium alloys, especially alpha alloys, are being widely used with great success in the chemical processing industry and for other applications where hydrogen-containing and/or hydrogen-generating conditions exist such as either cathodic protection systems or when galvanic couples are present (Covington and Parris 1976). However, due to the combined effect of the earlier mentioned factors/conditions, some hydriding failures of titanium have occurred in service (Schutz and Grauman 1989). Hydriding of surface condensers in power plants due to cathodic protection systems at overzealous negative potentials and hydriding of titanium heat exchanger tubes due to galvanic coupling between titanium tubes and carbon steel tubes, sheets, shells, and/or baffles are some examples of those failures. Figure 4.15 shows a Grade 7 microstructure (500×) with a 40 μm surface hydride film (left photo), formed from an impressed cathodic potential (lab study) under highly acidic conditions at 90°C and a completely hydrided (embrittled) Grade 7 titanium at 100 × magnification (right photo) (Divi 2009).

Titanium susceptibility to hydrogen uptake and embrittlement can be determined by simple galvanic coupling tests or cathodic charging tests. Cathodic charging and other tests specifically developed for titanium are provided in the literature (Schutz and Covington 1981b).

4.6.3.6 Galvanic Corrosion

Galvanic corrosion of titanium itself as a result of its contact with other metals is rarely observed. In salt solutions, be it either a NaCl solution or seawater, titanium is passive and provides excellent galvanic corrosion resistance. In seawater, titanium is located at the passive/noble end of the galvanic series next to platinum and graphite as shown in Figure 4.16. In almost all environments, titanium and its alloys will act as cathodes in a galvanic couple. This can, however, have the effect of accelerating the corrosion of its anodic metal couple if that metal (anode) is active in that environment. When titanium is galvanically coupled in seawater with other metals such as copper alloys and carbon steels, these anodes very actively corrode; similarly, aluminum and magnesium alloys moderately corrode under certain conditions (Bomberger 1969). The active metal's corrosion rate depends on factors like the relative surface areas of each metal forming the galvanic couple, the temperature, the medium's chemistry and solution flow rate, as well as the concentration of any dissolved cathodic depolarizers like oxygen. If the area of the cathode (titanium) is greater than area of the anode (active metal), then corrosion of the active metal will be even more severe. As the media's pH drops, higher corrosion rates of the active metal will also ensue. Higher flow rates and exposure temperatures will also accelerate the attack of the

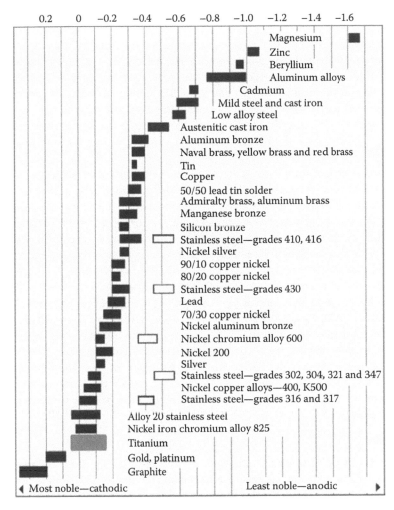

FIGURE 4.16 Galvanic series of various metals in seawater at room temperature. Unshaded boxes represents metals in active state. Black shaded boxes represents metals in passive state. Light shaded box just highlighting titanium. (From Fontana, M.G. and Greene, N.D., *Corrosion Engineering*, McGraw-Hill, New York, 1978.)

anodic member. Further, when titanium is galvanically coupled with active metals, the deleterious possibility of hydrogen absorption by the titanium is significant. The severity of the hydride formation and possible embrittlement of titanium is proportional to the temperature of the galvanic couple, the acidity of the media, and the rate of anode corrosion (which is directly proportional to the amount of hydrogen generated on titanium). However, whenever titanium is coupled to passive metals, such as stainless steels or nickel alloys, there generally will not be any corrosion of the anodic member of the galvanic couple or hydrogen embrittlement of titanium.

Titanium coupled with noble metals provides for anodic protection of the titanium in mildly reducing to neutral or oxidizing solutions (Grauman 1998b).

The most successful strategies to eliminate galvanic degradation is to use more resistant (compatible) passive metals coupled with titanium, or an all-titanium construction, or by using a dielectric material in the joint to insulate/isolate the adjoining dissimilar metals. Other approaches for mitigating galvanic corrosion that have also been effective include using coatings and linings and employing cathodic protection (Mountford 2002).

4.6.4 Degradation Protection

Simply put, titanium corrosion protection is mainly based on the stability of its oxide film; thus, with increased passivation the degradation protection can be expanded into more aggressive aqueous reducing acid media. There are various methods to enhance the oxide stability by effectively shifting the alloy's potential toward the positive (noble) direction. These methods include titanium alloying with certain solute elements, using anodizing and thermal oxidation pretreatments, or by applying platinum group metal appliqués (PGMA), and/or adding corrosion inhibitors to the exposure environments.

4.6.4.1 Alloying

Alloying titanium with certain solute elements is the best method of achieving improved corrosion resistance. The general principle of alloying for better corrosion resistance is to achieve further/improved chemical stability of the passive layer. Minor amounts of Pd (0.05%–0.15%) solute additions to the CP grades of titanium produces the best enhancement of corrosion resistance in these CP titanium alloys. The primary mechanism in providing enhanced corrosion resistance by these additions of precious metal solutes is that it facilitates cathodic depolarization by lowering the hydrogen overvoltage that in turn shifts that alloy's corrosion potential toward the noble direction thus increasing the oxide film passivity. This sort of benefit is also observed, although to a lesser extent, by the addition of other elements such as Ni, Mo, and Nb, alone or in combination, to commercial titanium alloys (Tomashov 1963; Andreeva et al. 1972). Figure 4.17 shows the corrosion rates of various titanium alloys in boiling hydrochloric acid solutions. Recent studies have shown that the interstitial element carbon can be a useful alloying element in enhancing the corrosion resistance of titanium when added in limited (≤2%) amounts, without impairing ductility (Grauman 2005; Grauman and Fox 2007; Grauman et al. 2010). This new Ti–C alloy system is very promising in terms of providing a series of inexpensive, highly corrosion-resistant titanium alloys. These new alloys should soon become commercially available in standard mill product forms.

4.6.4.2 Inhibitors

Industrial acid streams often contain contaminants that are oxidizing in nature and, thus, can passivate titanium alloys in normally aggressive acid media (Covington 1972). Metal ion concentration levels as low as 20–100 ppm can effectively inhibit corrosion. Potent inhibitors for titanium in reducing

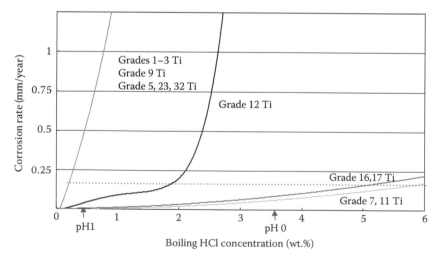

FIGURE 4.17 Corrosion rates of titanium alloys in boiling acid solutions, showing the effect of alloying on corrosion resistance. (From Grauman, J.S., PGMA—A corrosion protection method well suited for use of Titanium within the CPI, *Proceedings of the Corrosion Solutions Conference*, ATI Wah Chang, Coeur d'Alene, ID, 2003a.)

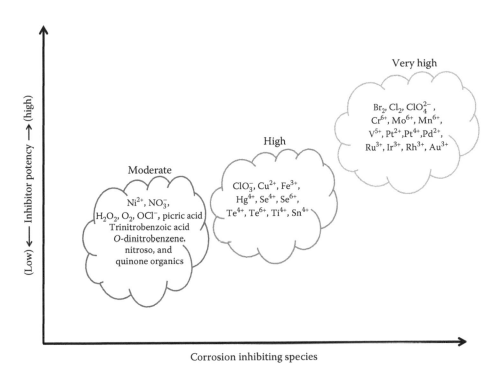

FIGURE 4.18 Corrosion inhibition species.

acid media are dissolved oxygen, chlorine, bromine, nitrate, chromate, permanganate, molybdate, and multivalent transition metal ions, such as ferric (Fe^{+3}), cupric (Cu^{+2}), and nickel (Ni^{+2}), and the precious metal ions (Gupta 1982; Schutz 2003). It is this potent metal ion inhibition property that permits titanium to be successfully utilized for equipment handling hot HCl and H_2SO_4 acid solutions such as in metallic ore leaching processes. Corrosion inhibiting species, based on their level of inhibition, are shown in Figure 4.18.

4.6.4.3 Thermal Oxidation

Among the various protective methods, thermal oxidation was found to be a very cost-effective and simple method to generate a thicker, more protective oxide layer (>10 µm) on pretreated titanium surfaces. This additional thermally formed thicker oxide layer not only provides excellent corrosion resistance but also increases surface hardness and improves wear resistance (Schutz and Covington 1981a). The thermal oxide is formed within the pretreatment temperature range of 600°C–800°C where the exposure time is generally limited to a few minutes (2–10 min) in air. This window of time and temperature avoids any property degradation due to oxygen intrusion. Any thermal treatment done above 800°C for prolonged times will result in oxygen diffusion into the metal, whereupon subsequent applied stresses could generate cracks leading to failure of the metal (Leyens and Peters 2003). An optimized thermal treatment produces an adherent, homogeneous, and sufficiently thick protective surface oxide layer.

4.6.4.4 Anodic Protection

The anodic protection of titanium is achieved with the external application of anodic potentials/currents that shifts the corrosion potential from the active region into the passive region. This type of protection can function very effectively for titanium due to its wide range of passivity in the Eh–pH diagram. The shift in potential provides for a significantly reduced (>100-fold) corrosion rate of titanium. This technique has been very useful for titanium alloys used in reducing acid environments (Schutz and Covington 1981a). In strong reducing acids, the impressed potentials can extend the range by about

+4 V (vs. V_{SHE}). However, impressed potentials beyond the protection zone can result in breakdown of the passive layer resulting in localized (pitting) corrosion in reducing environments such as sulfuric, phosphoric, and hydrochloric acids. Stray currents and complex titanium components can present challenges to this protection method.

4.6.4.5 Appliqués

PGMA is a very effective, yet simple degradation protection method for protecting titanium in hot reducing acids. This method uses a platinum group metal (Pt, Pd, Ru, etc.) as an appliqué, instead of an alloying agent to stabilize the titanium oxide film. A minor surface area (with a 1:125 ratio) appliqué of a platinum group metal can serve to protect a large surface area of Grade 2 titanium. The protection appears to be of a galvanic nature, where the platinum group metal ennoblizes the entire surface of the Grade 2 over great distances (throwing power), making it behave as if it were a Grade 7 titanium material. Due to this advantage, this method proves to be useful whenever the cost of the Grade 7 titanium alloy becomes too expensive for certain applications. Further details related to this method are given elsewhere (Grauman 2003a,b; Grauman et al. 2003; Divi 2009; Divi and Grauman 2010). One drawback of this system is that it will not work in wet/dry cycling or vapor phase environmental areas. Figure 4.19 shows the corrosion rate suppression by using PGMA in boiling HCl solutions.

4.6.4.6 Surface Coatings

Surface coatings are used on titanium to improve the hardness, wear resistance, and corrosion performance. Wear resistance and surface integrity of the titanium is critical in biomedical and some military applications (Leyens and Peters 2003). Surface hardening of Ti and its alloys, by diffusion of interstitial elements such as nitrogen, carbon, and boron, is employed to improve the wear resistance. These interstitial elements react with titanium to form hard compound layers at the surface and thereby enhance the wear/abrasion properties. In addition to wear resistance, these surface films can provide excellent corrosion resistance. Ion implantation, ion plating, sputter deposition, and other alternative vapor phase methods are used to inject atoms into the near surface or to deposit coatings on the surface of titanium (Cotell et al. 1994). Precious metal coatings on titanium surfaces applied through electroplating or ion implantation also provide excellent corrosion resistance. In particular, these coatings are very effective in inhibiting crevice corrosion (Moroishi and Miyuki 1980).

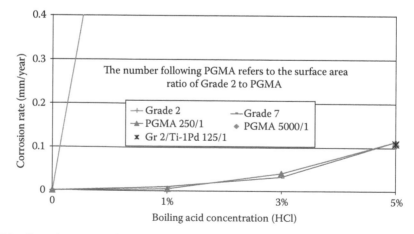

FIGURE 4.19 Corrosion rates of titanium alloys and PGMA in boiling acid hydrochloric acid. (From Grauman, J.S., *Titanium for the Chemical Process Industry: An overview for TIMET employees*, 2009.)

4.7 Summary

Titanium is a unique and distinguished metal among engineering metals in terms of providing exceptional corrosion resistance across a wide spectrum of environments, thanks to the stability of its passive surface oxide film. The combination of specific strength and excellent corrosion resistance makes titanium and its alloys reliable materials for many diverse applications where the use of other structural metals like stainless steel is not feasible. Titanium is utilized as a high performance metal in market sectors such as aerospace, chemical industry, and medical engineering where its highly reliable performance and extended service life metrics outweigh the initially higher costs of these alloys when compared to conventional structural metals such as steel and aluminum alloys, which are commonly used in commodity products, structures, and systems. In particular, it is the correct material for any seawater service. By recognizing and respecting its strengths and weaknesses, as highlighted throughout this chapter, material and design engineers will successfully facilitate its expanded use into new and advanced applications where products, equipment, and systems made from titanium alloys will successfully provide unexcelled and reliable performance.

References

Ambrose, P.M. 1955. Investigation of accident involving titanium and red fuming nitric acid, U.S. Department of the Interior, Bureau of Mines, Washington, DC.

Andreeva, V. 1964. Behavior and nature of thin oxide films on some metals in gaseous media and in electrolyte solutions, *Corrosion*, 20, 35–46.

Andreeva, V., A.I. Glukhova, S.N. Dontsov, I.S. Moiseeva, and L.V. Melnikova. 1972. Corrosion resistance and electrochemical and mechanical properties of Nb-Ti-Ta and Nb-Ti-Cr ternary alloys, *Protection of Metals*, 8, 369–372.

ASTM Standards. 2012. *Annual Book of ASTM Standards 2012, Wear and Erosion; Metal Corrosion*, ASTM International.

Baboian, R. 2005. *Corrosion Tests and Standards: Application and Interpretation*, ASTM International, West Conshohocken, PA.

Baker, G. and J.S. Grauman. 1991. Solving some seawater corrosion problems with titanium, *Sea Technology*, 32(4), 50.

Beavers, J. 1986. Corrosion of metals in marine environments, Battelle-Columbus Division, Metals and Ceramics Information Center, Columbus, OH.

Been, J. and J. Grauman. 2011. Titanium and titanium alloys, in *Uhlig's Corrosion Handbook*, R.W. Revie (Ed.), 3rd edn., John Wiley & Sons, Inc., New York, pp. 861–878.

Bergman, D. and J. Grauman. 1992. The detection of crevice corrosion in titanium and its alloys through the use of potential monitoring, in *Proceedings of the TITANIUM-92 Science and Technology*, TMS, San Diego, CA.

Bomberger, H.B. 1969. Titanium, in *Home Study and Extension Courses*, Harold D. Kessler (Ed.), ASM International, Niles, OH.

Boyer, R. and J.A. Hall. 1992. Microstructure-property relationships in titanium alloys (critical review), in *Proceedings of the TITANIUM'92: Science and Technology*, TMS, San Diego, CA, pp. 77–88.

Boyer, R., G. Welsch, and E.W. Collings. 1994. *Materials Properties Handbook Titanium Alloys*, ASM International, Materials Park, OH.

Brown, S.A. and J.E. Lemons. 1996. *Medical Applications of Titanium and Its Alloys*, ASTM International, West Conshohocken, PA.

Chappell, M.J. and J.S.L. Leach. 1978. Passivity and breakdown of passivity of valve metals, in *Passivity of Metals*, R. Frankenthal and J. Kruger (Eds.), Electrochemical Society, Princeton, NJ.

Chem, C.M., H.B. Kirkpatrick, and H.L. Gegel. 1971. Cracking of titanium alloys in methanolic and other media, in *Proceedings of the International Symposium on Stress Corrosion Mechanisms in Titanium Alloys*, Georgia Institute of Technology, Atlanta, GA.

Cotell, C.M., J.A. Sprague, F.A. Smidt, and ASM International. Handbook Committee. 1994. *ASM Handbook: Surface Engineering*, ASM International, Materials Park, OH.

Covington, L.C. 1972. The role of multi-valent metal ions, in *Proceedings of the Titanium Science and Technology*, Plenum Publications Corporation, New York, Cambridge, MA, pp. 2395–2403.

Covington, L.C. 1976. Pitting corrosion of titanium tubes in hot concentrated brine solutions, ASTM STP 576, ASTM International, West Conshohocken, PA, pp. 147–157.

Covington, L.C. 1979. The influence of surface conditions and environment on the hydriding of titanium, *Corrosion*, 35, 378–382.

Covington, L.C. 1982. Corrosion resistance of titanium, TIMET Corporation, Dallas, TX.

Covington, L.C. and W.M. Parris. 1976. The resistance of titanium tubes to hydrogen embrittlement in surface condensers, in *Proceedings of the NACE Corrosion*, NACE, Houston, TX.

Covington, L.C. and R.W. Schutz. 1981. Resistance of titanium to atmospheric corrosion, in *Proceedings of the CORROSION*, NACE International, Toronto, Canada, Paper 113.

Craig, B.D. 1995. *Handbook of Corrosion Data*, ASM International, Materials Park, OH.

Davis, J.R. 2000. *Corrosion Understanding the Basics*, ASM International, Materials Park, OH.

Degnan, T.F. 1975. Materials for handling hydrofluoric, nitric and sulfuric acids, in *Process Industries Corrosion*, NACE International, Houston, TX, pp. 228–230.

Diegle, R.B. 1981. Electrochemical cell for monitoring crevice corrosion in chemical plants, in *Proceedings of the CORROSION'81*, NACE International, Toronto, Canada, Paper 154.

Diem, W. 2001. Titanium in the auto industry, *Auto Technology*, 5, 36–37.

Divi, S. 2009 (Internal). PGMA corrosion protection system for use of Ti in CPI. TIMET, Henderson Technical Lab (HTL), Dallas, TX.

Divi, S. 2010. Examination of titanium condenser tubes (failure analysis no. 3351), TIMET, Henderson Technical Lab (HTL), Dallas, TX.

Divi, S. and J.S. Grauman. 2009. PGMA corrosion protection system for use of Ti in CPI, in *Proceedings of TEG 120X Meeting, CORROSON'09*, Atlanta, GA.

Divi, S. and J.S. Grauman. 2010. PGMA—A corrosion protection method for titanium and its recent experiences within the CPI, in *Proceedings of the Material Science and Technology-10*, TMS, Houston, TX, pp. 373–382.

Divi, S. and J.S. Grauman. 2011. Environmentally assisted cracking study of a newly developed titanium alloy, in *Proceedings of the CORROSION'11*, NACE International, Houston, TX, Paper 11288.

Donachie, M.J. 2000. *Titanium: A Technical Guide*, ASM International, Materials Park, OH.

Fontana, M.G. and N.D. Greene. 1978. *Corrosion Engineering*, McGraw-Hill, New York.

Fox, S. and K.O. Yu. 2011. Recent changes and developments in titanium extraction, in *Proceedings of the 12th World Conference on Titanium*, Beijing, China.

Froes, F.H. 1999. Will the titanium golf club be tomorrow's mashy niblick? *Journal of Metals*, 51, 18–20.

Froes, F.H., P.G. Allen, and M. Ninomi. 1998. *Non-Aerospace Applications of Titanium*, TMS, San Diego, CA.

Grauman, J.S. 1990a. BETA-21S: A high-strength corrosion-resistant titanium alloy, in *Proceedings of the TDA International Conference*, Orlando, FL.

Grauman, J.S. 1990b. Influence of surface condition on the resistance of titanium to chloride pitting, in *Advances in Localized Corrosion*, H.S. Isaacs (Ed.), NACE International, Anaheim, CA, pp. 331–334.

Grauman, J.S. 1992. Corrosion behavior of TIMETAL-21S for non-aerospace applications, in *Proceedings of the TITANIUM'92: Science and Technology*, TMS, San Diego, CA, pp. 2737–2742.

Grauman, J.S. 1998a. Shedding new light on titanium in CPI construction, *Chemical Engineering*, 105, 106–111.

Grauman, J.S. 1998b. Titanium-properties and applications for the chemical process industries, *Encyclopedia of Chemical and Design*, J.J. McKetta (Ed.).

Grauman, J.S. 2003a. PGMA—A corrosion protection method well suited for use of Titanium within the CPI, in *Proceedings of the Corrosion Solutions Conference*, ATI Wah Chang, Coeur d'Alene, ID.

Grauman, J.S. 2003b. PGMA—A novel corrosion protection method for titanium, in *Proceedings of the Ti-2003 Science and Technology*, Wiley-VCH, Weinheim, Germany, pp. 2107–2114.

Grauman, J.S. 2005. A new titanium alloy for CPI, in *Proceedings of the Corrosion Solutions Conference*, ATI Wah Chang, Sun River, OR.

Grauman, J.S. 2009. Titanium for the chemical process industry: An overview for TIMET employees, Dallas, TX.

Grauman, J.S., R. Adams, and J. Miller. 2003. Titanium article having improved corrosion resistance, United States Patent 6,607,846.

Grauman, J.S. and S. Fox. 2007. Development of a new corrosion resistant titanium alloy for the chemical process industry, in *Proceedings of the Ti-2007 Science and Technology*, The Japan Institute of Metals, Kyoto, Japan, pp. 1213–1216.

Grauman, J.S., S. Fox, and S. Nyakana. 2010. Titanium alloy having improved corrosion resistance and strength, US Patent 776257 B2.

Gray, K.O. 1964. Material protection, *MAPRA*, 3, 46.

Griess, J.C. 1968. Crevice corrosion of titanium in aqueous salt solutions, *Corrosion*, 24, 96–109.

Gudipati, P. and S. Divi. 2012. High temperature oxidation study of TIMETAL® 54 M Alloys, TIMET internal data, Dallas, TX.

Gupta, V.P. 1982. Process for decreasing the rate of titanium corrosion, US Patent 4321231.

Haney, E.G. and W.R. Wearmouth. 1969. Effect of pure methanol on the cracking of titanium, *Corrosion*, 25, 87.

Hanson, B.J. 1984. Titanium in hydrogen sulfide atmospheres at elevated temperatures, in *Industrial Applications of Titanium and Zirconium: Third Conference, STP 830*, ASM International, Metals Park, OH, pp. 19–28.

Hughes, P.C. and I.R. Lamborn. 1960. Contamination of titanium by water vapor, *Journal of the Institute of Metals*, 89, 165–168.

International Titanium Association (ITA). 2008. *The Fundamentals of Titanium*, ITA, Denver, CO.

Jackson, J.D. and W.K. Boyd. 1968. Stress corrosion cracking in titanium alloys, in *Proceedings of the Science, Technology and Application of Titanium*, Pergamon Press, London, U.K., pp. 267–282.

Jones, D.A. 1996. *Principles and Prevention of Corrosion*, Prentice-Hall, Upper Saddle River, NJ.

Jones, D.A. and B.E. Wilde. 1977. Corrosion performance of some metals and alloys in liquid ammonia, *Corrosion* 33, 46.

Krvuchek, V.G. and V.V. Volynskii. 1976. Corrosion and electrochemical behavior of Ti alloys in media of the food industry, *Zashchita Metallov*, 12, 683–684.

Lacombe, P. 1976. Corrosion and oxidation of Ti and Ti alloys, in *Proceedings of the 3rd International Conference on Titanium*, Plenum Publications Corporation, New York, Moscow, Russia, pp. 847–880.

Lampman, S.R. 1996. *ASM Handbook 19. Fatigue and Fracture*, ASM, Metals Park, OH.

Lane, I.R., J.L. Cavallaro, and A.G. Morton. 1966. Stress corrosion cracking of titanium, in *STP 397*, ASTM International, West Conshohocken, PA.

Leyens, C. and M. Peters. 2003. *Titanium and Titanium Alloys: Fundamentals and Applications*, Wiley-VCH [John Wiley Distributor], Weinheim, Germany.

Liening, E.L. 1983. Unusual corrosion failure of titanium chemical process equipment, in *Proceedings of the CORROSION'83*, NACE International, Anaheim, CA, Paper 15.

Liening, E.L. 1986. Electrochemical corrosion testing techniques, in *Process Industries Corrosion*, NACE International, Houston, TX, pp. 85–122.

Little, B., P. Wagner, and F. Mansfeld. 1991. Microbiologically influenced corrosion of metals and alloys. *International Materials Review*, 36, 253–272.

Littman, F.E. and F.M. Church. 1959. Reactions of metals with oxygen and steam, Stanford Research Institute to Union Carbide Nuclear Co., Final Report AECU-4092.

Lütjering, G. and J.C. Williams. 2007. *Titanium*, Springer, Berlin, Germany.

McMaster, J.A. and R.L. Kane. 1970. The use of titanium in the pulp and paper industry, in *Proceedings of the Fall Meeting of the Technical Association of the Pulp and Paper Industry*, Denver, Co.

Millaway, E.E. 1965. Titanium: Its corrosion behavior and passivation, *Material Protection and Performance*, 4, 16–21.

Millaway, E.E., R.L. Powell, and S.M. Weiman. 1964. Titanium in nitric acid (internal report), TIMET, Titanium Metallurgical Laboratory, Dallas, TX.

Moroishi, T. and H. Miyuki. 1980. Effect of several ions on the crevice corrosion of titanium, in *Proceedings of the Titanium'80-Science and Technology*, The Metallurgical Society (TMS), Kyoto, Japan.

Mote, M., R. Hooper, and P. Frost. 1958. The engineering properties of commercial titanium alloys (TML report no. 92), Battelle Memorial Institute, Titanium Metallurgical Laboratory, Columbus, OH.

Mountford, J.A. 1996. Basics and benefits of titanium for sea service—A review, in *Proceedings of the Annual Surface Ships Corrosion Control Conference*, Louisville, KY.

Mountford, J.A., 2001. Titanium meeting the challenge of the new millennium. Presented at the *CORROSION'2011*, NACE International, Houston, TX, p. Paper 01329.

Mountford, J.A. 2002. Titanium-properties, advantages and applications solving the corrosion problems in marine service, in *Proceedings of the CORROSION*, NACE International, San Diego, CA, Paper 02170.

Mountford, J.A. and J.S. Grauman. 1997. Titanium for marine applications, in *Proceedings of the International Workshop on Advanced Materials for Marine*, American Bureau of Shipping, New York, New Orleans, LA, pp. 107–128.

Paton, N.E. and J.C. Williams. 1974. Effect of hydrogen on titanium and its alloys, in *Hydrogen in Metals*, ASM International, Metals Park, OH, pp. 409–431.

Parris, W.M. and P.J. Bania. 1990. Beta-21S: A high temperature metastable beta titanium alloy, in *Proceedings of the TDA International Titanium Conference*, TDA, Orlando, FL.

Parris, W.M. and P.J. Bania. 1992. Oxygen effect on the mechanical properties of TIMETAL 21S, in *Proceedings of the 7th International Titanium Conference*, TMS, San Diego, CA, pp. 153–160.

Peacock, D.K. and J.S. Grauman. 1995. Titanium-installation and operation of titanium linings in flue gas desulfurization ductwork and stacks, in *Proceedings of the CORROSION'95*, NACE International, Orlando, FL, Paper 252.

Poulsen, E.R. 1994. (Technical). TIMET, Henderson Technical Lab (HTL), Dallas, TX.

Pourbaix, M. 1974. *Atlas of Electrochemical Equilibria in Aqueous Solutions*, National Association of Corrosion Engineers, Houston, TX.

Revie, R.W. and H.H. Uhlig. 2011. *Uhlig's Corrosion Handbook*, Wiley InterScience (Online service), Wiley, Hoboken, NJ.

Rideout, S.P., R.S. Ondrejcin, and M.R. Louthan. 1968. Hot-salt stress corrosion cracking of titanium alloys, in *Proceedings of the Science, Technology and Application of Titanium*, Pergamon Press, London, U.K., pp. 307–320.

Schutz, R.W. 1986. Titanium, in *Process Industries Corrosion-The Theory and Practice*, NACE International, Houston, TX, p. 503.

Schutz, R.W. 1988a. Titanium: An improved material for shipboard and offshore platform seawater systems, in *Proceedings of the UK CORROSION*, NACE International, Brighton, U.K., pp. 285–300.

Schutz, R. 1988b. Titanium alloy crevice corrosion: Influencing factors and methods of prevention, in *Proceedings of the 6th World Conference on Titanium*, Les Éditions de Physique, Cannes, France, pp. 1917–1922.

Schutz, R.W. 1988c. Titanium alloy crevice corrosion: Influencing factors and methods of prevention, in *Proceedings of the 6th World Conference on Titanium*, Les Éditions de Physique, Cannes, France, pp. 1917–1922.

Schutz, R.W. 1992a. Understanding and preventing crevice corrosion of titanium alloys: Part I, *Material Performance*, 31, 57–62.

Schutz, R.W. 1992b. Understanding and preventing crevice corrosion of titanium alloys: Part II, *Material Performance*, 31, 54–56.

Schutz, R. 1995. Developments in titanium alloy environmental behavior, in *Proceedings of the TITANIUM'95: Science and Technology*, The University Press, Cambridge, U.K., pp. 1860–1870.

Schutz, R. 2002. Utilizing titanium to successfully handle chloride process environments, *CIM Bulletin*, 84–88.

Schutz, R. 2003. Corrosion of titanium and titanium alloys, in *CORROSION*, S.D. Cramer and B.S. Covino, Jr. (Eds.), ASM International, Metals Park, OH.

Schutz, R.W. and L.C. Covington. 1981a. Effect of oxide films on the corrosion resistance of titanium, in *Proceedings of the CORROSION'81*, NACE International, Ontario, Canada, Paper 61.

Schutz, R.W. and L.C. Covington. 1981b. Guidelines for corrosion testing of titanium, in *Industrial Applications of Titanium and Zirconium*, E.W. Kleefisch (Ed.), ASTM-STP, ASTM International, West Conshohocken, PA, pp. 59–70.

Schutz, R. and J.S. Grauman. 1986a. Fundamental corrosion characterization of high-strength titanium alloys, in *Industrial Applications of Titanium and Zirconium*, C.S. Young and J.C. Durham (Eds.), ASM International, West Conshohocken, PA, pp. 130–143.

Schutz, R.W. and J.S. Grauman. 1986b. Performance of titanium in aggressive zones of a closed-loop FGD scrubber, in *Proceedings of the CORROSION'86*, NACE International, Houston, TX, Paper 357.

Schutz, R.W. and J.S. Grauman. 1986c. Selection of titanium alloys for concentrated seawater, NaCl and MgCl$_2$ Brines, in *Proceedings of the Titanium-Titanium Products and Applications*, International Titanium Association, Denver, CO, San Francisco, CA.

Schutz, R. and J.S. Grauman. 1988. Localized corrosion behavior of titanium alloys in high temperature seawater service, in *Proceedings of the CORROSION*, NACE, Houston, TX, Paper 162.

Schutz, R.W. and J.S. Grauman. 1989. Determination of cathodic potentials limits for prevention of titanium tube hydride embrittlement in salt water, in *Proceedings of the CORROSION'89*, NACE, Houston, TX, Paper 110.

Schutz, R.W., J.S. Grauman, and J.A. Hall. 1984. Effect of solid solution iron on the corrosion behavior of titanium, in *Proceedings of the TITANIUM Science and Technology*, Deutshe Gesellscharft Für Metallkunde E.V., Munich, Germany, pp. 2617–2624.

Schutz, R.W. and M.R. Scaturro. 1991. An overview of current and candidate titanium alloy applications on U.S. navy surface ships, *Journal of Naval Engineers*, 175–191.

Schwartzberg, F., F. Holden, H. Ogden, and R. Jaffee. 1957. The properties of titanium alloys at elevated temperatures (TML No. 82), Battelle Memorial Institute, Titanium Metallurgical Laboratory, Columbus, OH.

Sedriks, A.J. and J.S.A. Green. 1971. Stress corrosion of titanium in organic liquids, *Journal of Metals*, 23, 48.

Speidel, M.O. 1983. Stress corrosion cracking and corrosion fatigue fracture mechanics, in *Corrosion in Power Generating Equipment*, M.O. Speidel and A. Atrens (Eds.), Plenum Publications Corporation, New York, pp. 85–132.

Takamura, A., K. Arakawa, and Y. Moriguchi. 1970. Corrosion resistance of titanium and titanium-5% tantalum alloys in hot concentrated nitric acid, in *The Science, Technology and Applications of Titanium*, R.A. Jafee and H.M. Burte (Eds.), Pergamon Press, New York, pp. 207–210.

Thomas, D.E. 1986. Titanium alloy corrosion resistance in nitric acid solutions, Technical Program from the 1986 International Conference, in *Proceedings of the Titanium 1986—Products and Applications*, Titanium Development Association, San Francisco, CA.

Thomas, D.E. and E.B. Bomberger. 1983. The effect of chlorides and fluorides on titanium alloys in simulated scrubber environments, *Material Performance*, 22, 229–236.

TIMET. 1997. *Corrosion Resistance of Titanium*, TIMET Corporation, Dallas, TX.

Tomashov, D. 1963. *Corrosion and Protection of Titanium (Government Scientific)*, Technical Publication of Machine-Building Literature, Moscow, Russia.

Tomashov, D. 1974. The passivation of alloys on titanium bases, *Electrochimica Acta*, 19, 159–172.

Upadhyaya, D., D.M. Blackketter, C. Suryanarayana, and Froes, F.H. 1992. Microstructure and mechanical properties of beta-21S titanium alloy, in *Proceedings of the Titanium'92: Science and Technology*, TMS, San Diego, CA, pp. 447–454.

Wood, J.D. 1993. The characterization of particulate debris obtained from failed orthopedic implants (research), San Jose State University, College of Material Engineering, San Jose, CA.

Yu, S. 2003. Corrosion resistance of titanium alloys, in *Corrosion: Fundamentals, Testing, and Protection, ASM Handbook*, S.D. Cramer and B.S. Covino, Jr. (Eds.), ASTM International, Materials Park, OH, pp. 703–711.

5

Crystalline Alloys: Plain-Carbon and Low-Alloy Steels

Hitoshi Asahi
*Nippon Steel & Sumitomo
Metal Corporation*

5.1 Introduction

Most of the metallic materials utilized for construction parts in terms of weight consist of plain-carbon or low-alloy steel. Steel is cheap, has a relatively high strength-to-weight ratio, and is easily cut, formed, and welded. Furthermore, the properties of steel can be widely changed using small amounts of alloying additions and heat treatments. Thus, steel is very suitable for construction of structures. Applications, however, can often be limited for various reasons, such as low-temperature embrittlement, fatigue, creep at elevated temperatures, and environmental degradation. This chapter describes the environmental degradation of steel, from general corrosion to environmental cracking. If engineers who use steel understand the mechanisms of corrosion in steel and the specific environmental and material factors affecting this corrosion, they can better design that system by selecting the proper material. This leads to avoiding accidents and economic losses caused by environmental degradation. From the viewpoint of preventing environmental degradation, low-alloy, corrosion-resistant steel, such as "weathering steel," and metal-plating technologies for general corrosion prevention as well as the selection of special steels resistant to environmental cracking are explained. This chapter highlights the use of modern steel as an example.

5.2 Background

General corrosion has been the biggest problem for steel because steel is easily corroded in wet ambient environments. To prevent general corrosion, stainless steel that contains Cr of 13% or more as explained in Chapter 6 has been developed but generally is too expensive to be used extensively. Depending on environmental conditions, plain-carbon and low-alloy steels can be safely used for a known duration. Therefore, knowing the relationship between environmental conditions and corrosion behavior is very important. Moreover, steel that contains small amounts of alloying elements has resistance to general corrosion in moderate environments. One such steel that is resistant to atmospheric corrosion is known as "weathering steel" and can contribute to an extreme reduction in the maintenance cost of bridges, buildings, and so on. Other than general corrosion that is slight in many cases, brittle failure has also been experienced and is often related to corrosion. Different from general corrosion, such cracking occurs suddenly and unexpectedly, especially in the case of hydrogen embrittlement (HE) cracking, as later explained. Environmental cracking occurs under the combined conditions of environment, steel type, and stress level. Therefore, if at least one of these conditions can be avoided, environmental cracking will not occur. Thus, it is very important to understand both corrosion behavior and environmental cracking in order to use steel safely and economically.

5.3 Applications

5.3.1 Uses

Plain-carbon and low-alloy steels are widely used as construction materials. Often, steel is classified primarily by their semifinished shape and processing method such as hot-rolled sheets (thin plates), cold-rolled sheets, heavy plates (thick plates), pipes and tubes, shaped steel, bars, and rods. Examples for uses are (a) sheets for automobile and household appliances; (b) heavy plates for ships, bridges, and buildings; (c) pipes for pipelines, oil/gas wells, and chemical plants; and (d) shaped steel for buildings. Steel is also classified in terms of strength, for example, by (ultimate) tensile strengths (TSs) such as 490 and 980 MPa.

5.3.2 Challenges

Major challenges exist for using low-alloy steels to more corrosive environments and the application of higher-strength steels to more hostile environments, where environmental cracking can occur. From the viewpoint of steel production, the challenge is to provide inexpensive corrosion-resistant steel by adding the smallest amounts of alloying elements. Often, some combination additions of different alloying elements give unexpected effects. For environmental cracking, adding small amounts of certain alloying elements provides a higher-strength steel that is less susceptible to cracking. Meanwhile, from the viewpoint of utilization, avoiding environmental degradation through more suitable selection of material and better structural design, even for inexpensive steel, is important.

5.4 General Properties

Steel is basically an alloy of Fe and C, where generically it is called iron whenever C < 0.02 mass% and steel above this carbon content. Conventionally manufactured steels also contain Si, Mn, N, P, S, and O. Among these elements, N, P, S, and O are generally considered as impurities, and their concentration are kept as low as possible within economically allowable restrictions. Generally, C, Si, Mn, P, and S are the five major elements in steel and are specified in terms of these contents. Alloying elements such as Ni, Cr, Mo, Ti, Nb, V, and B are often selectively added to improve properties such as strength, toughness, and sometimes corrosion resistance. Thus, the classification and chemistry of steels vary widely providing a large selection of steels enabling consumers to match their specific combination of performance and cost requirement.

5.4.1 Mechanical

The strength of steel for general service ranges from 300 to 600 MPs, but sometimes higher values of 1000 MPa and rarely 2000 MPa can be provided. Some steel is specified by its yield strength (YS), while others are specified by (ultimate) TS values. Because steel is generally used as a structural component, ductility and toughness are also essential. Both these latter properties tend to decrease with an increase in strength but can be improved through metallurgical control. In order to maintain these properties for high-strength steel, optimum chemistry and better control of hot-working conditions and heat-treatment conditions are necessary.

5.4.2 Physical

Iron shows very typical properties peculiar to metals. The intrinsic surface of steel is shiny. It possesses good thermal conductivity and low electric resistance: a typical value for the thermal conductivity at 273 K is 83.5 Wm^{-1}/K, and that for electric resistivity is 8.9×10^{-8} $\Omega \cdot$m at 273 K. Iron is ferromagnetic at room temperature and becomes paramagnetic above 1043 K.

5.4.3 Thermal

The strength of steel generally decreases with an increase in temperature. Figure 5.1 illustrates changes in strength from ambient temperature to 900 K for C–Mn steel that is often used for machine structures.

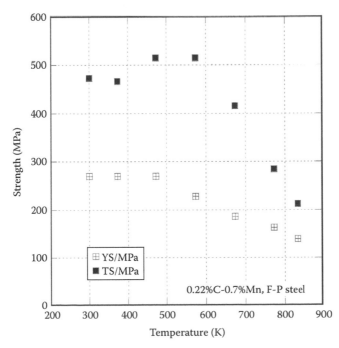

FIGURE 5.1 Systematic changes of the strength of a low-alloy steel as a function of testing temperatures (0.22%C–0.7%Mn, ferrite–pearlite steel).

Note that this trend for decreasing the strength with increasing temperature is affected by both the chemical composition and the microstructure. The addition of Mo typically suppresses such decreases in strength.

5.5 Structure

5.5.1 Atomic

At room temperature, the major stable phase of steel is ferrite that has a body-centered cubic (bcc) lattice usually containing a small amount of precipitates, such as carbides. When steel is heated above 1000 K, it transforms into the austenitic phase that has a face-centered cubic (fcc) lattice. Figure 5.2 shows the low-carbon section of the Fe–C equilibrium phase diagram. If steel is cooled from a high temperature at a very slow cooling rate, the transformation proceeds according to the diagram. Under practical conditions (when the cooling rate is faster), the transformation is suppressed below the equilibrium transformation temperature, and another phase that is not denoted in the phase diagram appears. Thus, whenever steel is rapidly cooled from the austenite region, which is called quenching, the steel transforms instead into its martensite or bainite phase, neither of which exist in the Fe–C equilibrium phase diagram. Martensite and bainite have bcc or tetragonal lattice structure, and both are very hard.

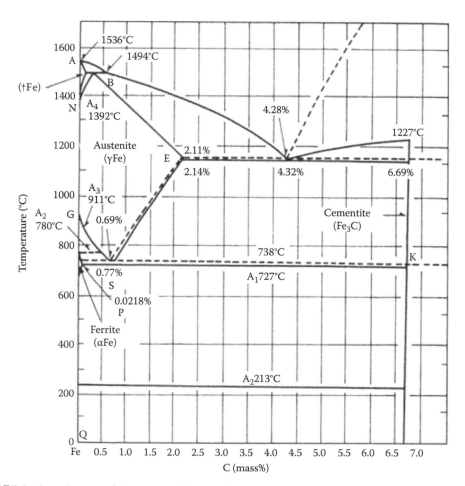

FIGURE 5.2   Low-C region of the Fe–C equilibrium phase diagram.

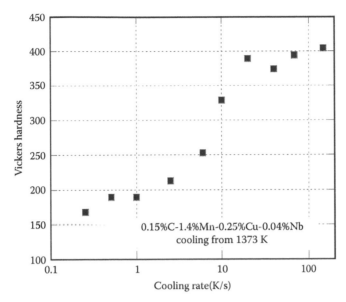

FIGURE 5.3 Effects of cooling rate on the hardness of a low-alloy steel (0.15%C, 1.4%Mn, 0.25%Cu, 0.04%Nb steel, cooled from 1373 K [austenite region]).

5.5.2 Microstructure

Steel microstructures vary widely and are influenced by chemical composition, hot-working conditions, as well as the thermal processing history, even if the basic crystal structure remains bcc. The typical steel cooled from austenite at a relatively slow rate has ferrite–pearlite microstructure. Whenever steel is rapidly cooled from its austenite phase to room temperature, that steel transforms into martensite, a very hard phase. Depending on the cooling rate, steel develops various microstructures even when the chemistry and austenitizing condition are constant. Figure 5.3 illustrates an example of the effect of variations of the cooling rate on the hardness for steel cooled from 1373 K (austenite). The hardness soars at cooling rates exceeding 10 K/s. Thus, variation in microstructure and/or hardness can be achieved by making specific changes in production conditions.

The thermomechanical control process (TMCP) uses carefully controlled hot-working conditions and accelerated cooling conditions to achieve required properties and is currently widely applied to steel production, especially for steel plates and sheets. Figure 5.4 schematically illustrates the TMCP from slab reheating through accelerated cooling. The controlled rolling can change grain size, and accelerated cooling can change microstructure, as shown in the figure. The TMCP improves mechanical properties, such as strength and toughness for a large class of steels with lower alloying content, thus producing a higher-quality steel at lower production costs.

5.6 Degradation Principles for Materials

5.6.1 Thermodynamics

When steel is immersed into an aqueous solution, localized inhomogeneities on the steel surface cause anodic and cathodic regions to form, thus resulting in a corrosion reaction between adjacent anode and cathode cells. This process is schematically illustrated in Figure 5.5 for the case in which hydrogen reduction serves as the cathodic partial process. Fe dissolves into the solution, as ferrous ions (Fe^{2+}) at the anode sites, and a hydrogen reduction reaction (and/or oxygen reduction reaction if dissolved oxygen [DO] is present) occurs at cathode sites where electrons generated from the anode reaction are

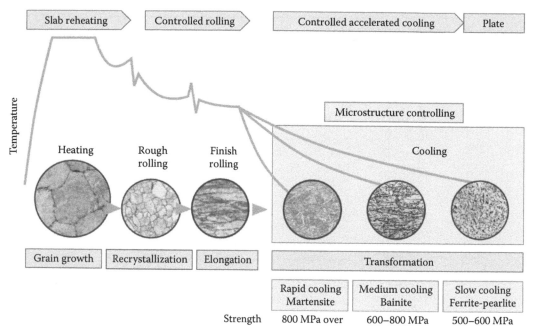

FIGURE 5.4 Schematic illustration of TMCP. (Photo Courtesy of Nippon Steel & Sumitomo Metal Corporation, Tokyo, Japan.)

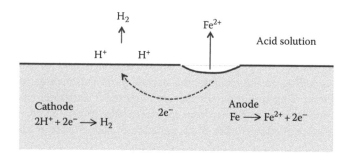

FIGURE 5.5 Schematic of the corrosion process for Fe in a deaerated acid solution.

consumed. The anodic reaction is known as an "oxidation reaction," while the cathodic reaction is known as a "reduction reaction." Because the standard oxidation reduction (redox) potential of Fe (Fe = Fe^{2+} + $2e^-$) is −0.440 V versus the standard hydrogen electrode [SHE], Fe can be oxidized (corroded) if the reduction reaction exists at a potential higher than the redox potential. When Fe is immersed in an acid solution, hydrogen evolution occurs:

$$Fe = Fe^{2+} + 2e^- \text{ oxidation reaction}$$

$$2H^+ + 2e^- = H_2 \text{ reduction reaction}$$

If DO exists in the solution, the cathode reaction is accelerated by the following reaction, which is also known as depolarization:

$$2H^+ + 1/2\ O_2 + 2e^- \rightarrow H_2O$$

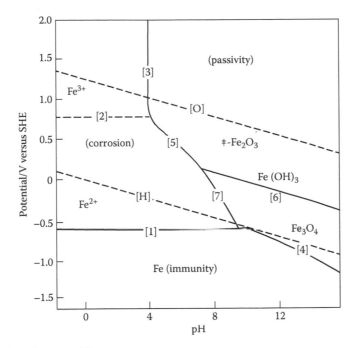

FIGURE 5.6 Pourbaix diagram of the Fe system.

The thermodynamics of this corrosion phenomena have been generalized by means of a potential-pH diagram, often called a Pourbaix diagram. Figure 5.6 shows a Pourbaix diagram for the Fe–H_2O system. Important reactions and their potentials, which were used to draw the Pourbaix diagram, are tabulated in Table 5.1. The diagram is divided into three regions; immunity, corrosion, and passivity. Here, the iron ion (Fe^{2+}) concentration in the solution is assumed to be 10^{-6}. In the region between (H) and (1) boundaries, hydrogen-evolution-type corrosion occurs because both Fe^{2+} and H_2 exist. Corrosion through oxygen reduction occurs in the region surrounded by (O), (1), (5), and (7). However, corrosion is suppressed in the region on the right side of the (3), (5), (7), and (4) boundaries because the buildup of stable solid corrosion products suppresses the corrosion reaction. This diagram only indicates whether or not a reaction is thermodynamically possible with a specific combination of potential and pH, but it does not provide any information about the corrosion rate. It should be noted that the pH value is that at the steel surface itself. Thus, whenever a corrosion reaction proceeds allowing the corrosion product to cover the steel, the actual pH value on the steel surface can sometimes be very different from that far from the pH of the bulk solution far from that steel surface.

TABLE 5.1 Representative Reactions and Their Potentials in the Fe–H_2O System

	Reaction	Potential/V
[O]	$2H_2O = O_2 + 4H^+ + 4e^-$	$1.228 - 0.0591\,pH + 0.0148 \log pO_2$
[H]	$2H^+ + 2e^- = H_2$	$-0.0591\,pH - 0.0295 \log pH_2$
[1]	$Fe^{2+} + 2e^- = Fe$	$-0.441 + 0.0295 \log[Fe^{2+}]$
[2]	$Fe^{3+} + e^- = Fe^{2+}$	$0.771 + 0.0591 \log[Fe^{3+}]/[Fe^{2+}]$
[4]	$Fe_3O_4 + 8H^+ + e^- = 3Fe + 4H_2O$	$-0.085 - 0.0591\,pH$
[5]	$Fe(OH)_3 + 3H^+ + e^- = Fe^{2+} + 3H_2O$	$1.057 - 0.177\,pH - 0.0591 \log[Fe^{2+}]$
[6]	$3Fe(OH)_3 + H^+ + e^- = Fe_3O_4 + 5H_2O$	$0.276 - 0.0591\,pH$
[7]	$Fe_3O_4 + 8H^+ + 2e^- = 3Fe^{3+} + 4H_2O$	$0.983 - 0.236\,pH - 0.0886 \log[Fe^{2+}]$

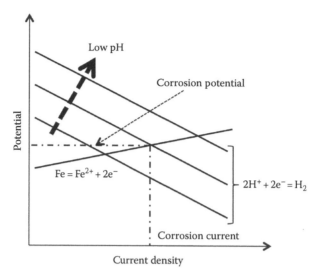

FIGURE 5.7 Effects of pH on corrosion current density in hydrogen reduction systems.

5.6.2 Kinetics

It is important in industrial application to know how fast corrosion proceeds. The corrosion rate is explained by polarization curves. When steel is polarized in a deaerated acid solution, an anodic polarization curve and a cathodic polarization curve are obtained as schematically illustrated in Figure 5.7. The intersection of the two curves corresponds to the corrosion potential and the corrosion current density. Then, the corrosion rate can be calculated from that corrosion current density value using Faraday's law. With a decrease in the pH of the solution, the cathodic polarization curve shifts upward as the result of an increase in equilibrium hydrogen electrode potential, with the dependence of −0.059 pH. Thus, as shown in Figure 5.7, the corrosion current density value becomes larger. Consequently, the corrosion rate increases with any decrease in pH in a deaerated acid solution. If DO is present in the solution, the cathodic reaction is accelerated by the following reaction (depolarization):

$$2H^+ + 1/2\,O_2 + 2e^- \rightarrow H_2O$$

This reaction is controlled by the supply of DO to the steel surface. Figure 5.8 shows the effect of the increasing supply of oxygen on polarization curves where corrosion proceeds through depolarization. Since the vertical position of the cathode polarization curve is controlled by the magnitude of the oxygen supplied, the corrosion current density increases with an increase in oxygen supply, either by a higher DO content or with a higher solution flow rate.

In the following sections, some examples of how practical environments influence corrosion rates are shown.

5.6.2.1 Dissolved Oxygen

The corrosion rate of plain C steel in air-saturated water with a neutral pH is initially as much as 0.5 mm/year, and then it gradually decreases as corrosion product builds up on the steel surface and finally saturates at 0.1 mm/year or so. At steady state, the volume of DO that reaches the steel surface is proportional to the DO concentration. In deep water, where the DO concentration is very low, the corrosion rate can be very small.

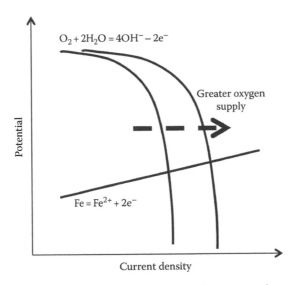

FIGURE 5.8 Effects of oxygen supply on corrosion current density in oxygen reduction system.

5.6.2.2 pH

As is apparent from the Pourbaix diagram, the corrosion rate is presumed to change with the pH of the solution, but practically, it varies in slightly different ways. An example of the pH dependence on the corrosion rate in water exposed to air is semiquantitatively illustrated in Figure 5.9 (Whitman et al. 1924). The corrosion rate between a pH of 4 and 10 is almost constant. This is because the actual pH on the steel surface is maintained at approximately 9.5, irrespective of the pH in the bulk solution away from the steel surface, because the steel surface is covered by corrosion product (mainly consisting of $Fe(OH)_3$). At a pH of 4 or lower, the corrosion product is dissolved, and, thus, the pH on the steel surface can decrease leading to a higher corrosion rate. At a pH of 10 or higher, the corrosion rate decreases due to the presence of protective films.

5.6.2.3 Temperature

Whether the reduction process is the hydrogen-evolution or oxygen-reduction reaction, the corrosion rate increases with a rise in temperature. However, whenever corrosion proceeds through oxygen

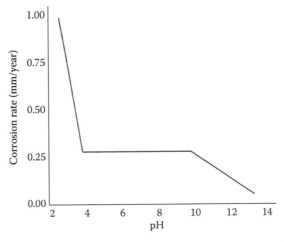

FIGURE 5.9 Effects of pH on corrosion rate in water exposed to air. (Semi-schematically illustrated based on the data in Whitman, W. et al., *Ind. Eng. Chem.*, 16, 665, 1924.)

reduction in an open system, the corrosion rate drops just below the boiling temperature (higher than approximately 90°C) because the amount of DO rapidly decreases due to the systematic reduction in oxygen solubility as the water temperature increases.

5.6.2.4 Chloride Ions

NaCl, or the Cl⁻ ion, is known to accelerate corrosion, and this is probably because the presence of increasing concentration of Cl⁻ makes the corrosion product more unstable or less dense. However, since very high Cl⁻ concentrations lower the oxygen solubility, it will actually reduce the corrosion rate.

5.6.3 Forms of Corrosion

The occurrence of general corrosion, uneven corrosion, and pitting on steel depends on the combination of the steel type and environment. General corrosion refers to circumstances when corrosion proceeds homogeneously, that is, when corrosion rates are almost constant throughout the exposed surface. Uneven corrosion refers to cases in which some areas become more corroded than others and includes groove corrosion. In the case of pitting corrosion, small localized pits are formed separately from each other, while the majority of the surface is almost not corroded at all. Typical pitting, however, is expected only under limited conditions for carbon and low-alloy steels.

Figure 5.10 shows schematic drawings of crevice and galvanic corrosion. Crevice corrosion can occur in solutions containing Cl⁻ and in geometries such as shown in Figure 5.10a where the gap between metal surfaces is narrow but not so narrow for the solution to penetrate into it. In the interior of the crevice, although oxygen is consumed quickly, Fe oxidation continues through the electrons supplied from the corrosion reaction on the steel surface. This is presumed to result in a lower pH in the gap, leading to crevice corrosion. Crevice corrosion should be avoided by carefully designing structures so as to eliminate crevices. Galvanic corrosion occurs when two adjacent pieces of steel with different corrosion potentials are electrically connected. One of the two pieces becomes much more corroded than if it was uncoupled. Figure 5.10b shows an example of galvanic corrosion near a weld metal. Steel close to the weld metal zone is selectively corroded because the weld metal is more noble than the base metal. This is generally true because the weld metal overall contains more alloying elements than the base steel does, and care should be taken when selecting the filler metal alloy to be used for the welding operation. Regarding galvanic corrosion, the combination of small less noble area and large noble area should be avoided because the less noble area is rapidly corroded.

Figure 5.11 shows the cross sections of electric resistance-welded (ERW) pipes used for water transportation. In Figure 5.11a, uneven corrosion localized along the weld line has occurred causing (weld) groove corrosion. This problem in ERW pipes occurs due to different corrosion potentials between the ERW zone and the rest of the pipe. Possible reasons include a difference in microstructure and the amount of MnS and S in solution in the weld zone and the pipe base metal. Groove corrosion has been

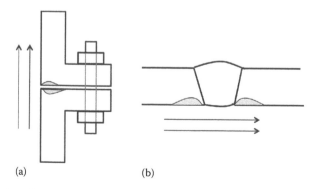

(a) (b)

FIGURE 5.10 Illustrations showing crevice and galvanic corrosion: (a) Crevice corrosion and (b) galvanic corrosion.

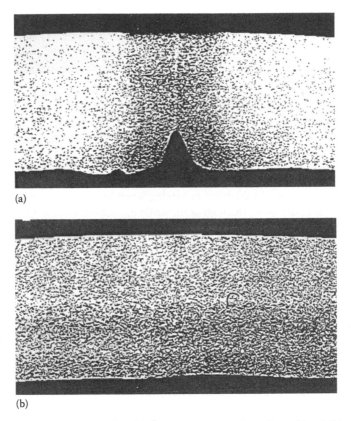

(a)

(b)

FIGURE 5.11 Examples of groove corrosion: (a) Groove corrosion along the weld and (b) no groove corrosion along the weld (antigroove corrosion ERW pipe).

solved today through specifying pipe steels with a lower S content that also contain additional special alloying elements, such as Cu, and with the application of postwelding seam heat treatment (Kato et al. 1978). When antigroove corrosion protection measures are used for ERW pipe, groove corrosion is eliminated, as indicated in Figure 5.11b. Both of the pipes used were exposed to the same environment for the same duration.

Atmospheric corrosion is another type of corrosion that occurs primarily due to the presence of oxygen and airborne moisture. Intervals of alternating wetting/drying can affect the corrosion rate. Atmospheric corrosion of steel is the single generic corrosion process that practically accounts for the biggest proportion of all economic losses caused by corrosion due to the vast amount of steels used. Atmospheric corrosion is greatly affected by specific environmental conditions, and these are usually classified into three general categories: (a) marine (or coastal), (b) industrial, and (c) rural. In coastal areas, steel is heavily corroded because of the presence of airborne sodium chloride. Industrial conditions are typically more corrosive than rural conditions because SO_2 that is generated and emitted into the atmosphere forms sulfuric acid. Steels where the corrosion rate is suppressed due to the formation of stable protective corrosion products have been developed by adding a small amount of certain alloying elements. This class of steel is called "weathering steel" and will be explained further in Section 5.7.2.3. Of course, no atmospheric corrosion occurs when there is no water source.

5.6.4 Environmental Cracking

Steel breaks when external stress exceeds the TS. Steel, however, may suffer from cracking at a stress lower than the YS when certain environmental and stress limits are exceeded. In general, higher-strength steel is more susceptible to environmental cracking, and higher stresses accelerate this form of premature cracking. Environmental cracking is divided into two categories: "stress corrosion cracking" (SCC) and "hydrogen embrittlement" (HE). SCC is defined narrowly to occur due to localized anodic dissolution. Therefore, it is often called "active path corrosion" or APC. Cracking morphology, namely, either intergranular cracking or transgranular cracking, differs depending on the steel and the service environment. SCC is a more serious problem in stainless steel but occurs sometimes in low-alloy steel under specific combinations of the environment and type of steel used. SCC is affected by corrosion, but, generally, the extent of corrosion itself is very slight. HE cracking is featured by the fact that prior to crack initiation, there is a time delay (incubation time) after hydrogen is incorporated into steel. HE is a very serious problem, especially in high-strength, low-alloy steel, because it occurs suddenly and often without any warning, causing catastrophic failures. SCC and HE can also be suppressed using specially developed steels.

HE is manifest by a reduction in ductility occurring when hydrogen is contained in steel. Atomic hydrogen (H) is capable of diffusing through steel, while molecular hydrogen (H_2) cannot. Thus, HE is caused by atomic hydrogen. HE requires only a very small amount of hydrogen to occur, often less than 1 ppm and at even lower concentrations for higher strength of steels. There are two kinds of HE, as classified by the hydrogen uptake mechanism. In the first case, hydrogen is included during steel production and fabrication, such as melting, welding, heat treatment, and plating, and no additional hydrogen uptake is expected. In the second case, hydrogen penetrates into steel during service more or less continuously. This latter type is a form of environmental degradation. Unlike most embrittlement phenomena, HE is enhanced by slow strain rates. Steel is most susceptible to HE at approximately room temperature and is much less susceptible at low or high temperatures.

In the vicinity of room temperature, hydrogen atoms can rapidly diffuse in steel through interstitial sites present in the steel crystallographic lattices. This type of mobile atomic hydrogen is considered to contribute to HE, while trapped hydrogen does not. Some mechanisms have been proposed, but none of them have been proven. The representative ones are as follows:

1. Dissolved hydrogen degrades interatomic bonds (the lattice decohesion theory) (Oriani and Josephic 1974).
2. Dissolved hydrogen makes it easier for dislocation multiplication and motion (Beachem 1972).

Intergranular and transgranular cracking can also occur during HE. Intergranular cracking tends to appear mostly in higher-strength steels, and transgranular cracking is more common for lower-strength steels.

Furthermore, SCC and HE respond differently to changes in potential, as shown in Figure 5.12. Cathodic polarization prolongs the time to failure of SCC and shortens that for HE, theoretically.

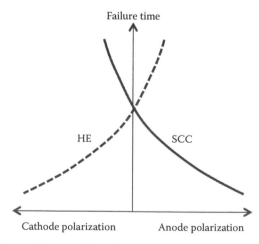

FIGURE 5.12 Schematic drawing showing effects of polarization on failure times for HE and SCC.

Therefore, cathodic protection is an effective method for preventing SCC, whereas it may accelerate the HE failure process. Although the stress that causes SCC or HE can be externally applied or residual, it should be noted that only tensile stresses can cause SCC and HE failures, while compressive stress does not. The effect of metallurgical factors on susceptibility to SCC and HE is explained from the viewpoints of chemistry and microstructure in Section 5.7.2. The following specific examples in Section 5.7 closely relate to the methods used to prevent SCC or HE failures.

5.7 Degradation of Specific Systems

5.7.1 Plain-Carbon Steel in Various Environments

Plain-carbon steel compositions basically consist of small percentages of C, Si, Mn, and unavoidable impurities and are not otherwise intentionally alloyed. Therefore, strength is generally low, at a nominal level of 400 MPa for TS. Steel corrodes when it contacts water. Corrosion rates are influenced by DO concentrations, pH levels, flow rates, temperature, and other solutes dissolved in the water itself, previously explained in Section 5.6.2. General corrosion most commonly occurs, but uneven corrosion sometimes occurs under specific conditions. More attention should be paid to uneven corrosion because penetration, such as through the wall thickness of a pipe or sheet, can often take place causing structural failures within an unexpectedly short time. In the following, an example of uneven corrosion that is specific to CO_2 dissolved in water is shown. Corrosion features looking like worm-eaten spots are known as mesa corrosion (Tomoe 1998), and it occurs in steel immersed in water with dissolved CO_2. Because the solubility of CO_2 in water is high, but the dissociation ratio from $CO_2 + H_2O$ to $H^+ + HCO^{3-}$ is very low, once H^+ is consumed by the corrosion reaction, H^+ must be supplied by proceeding with a dissociation reaction. Thus, the corrosion rate for this CO_2 corrosion mechanism does not depend on pH but on CO_2 concentration. On the other hand, if iron carbonate ($FeCO_3$) starts to precipitate due to an increase in the HCO_3^- concentration, the corrosion can almost be suppressed. Whenever the environmental water conditions are near the boundary between whether or not $FeCO_3$ precipitates, coexisting corroded and uncorroded areas will form on the steel surface, resulting in uneven corrosion.

Natural water often contains various kinds of minerals that can sometimes suppress corrosion. This primarily occurs when $CaCO_3$ precipitates onto the steel surface, thus becoming a barrier to further diffusion of DO toward the steel surface. Therefore, corrosion is often less severe for a water transportation and storage systems that convey the so-called hard water. Sometimes, instead, blockages due to thick layers of $CaCO_3$ occur, thus becoming another problem causing pipes to become clogged.

Obviously, corrosion does not occur when water is absent. Therefore, the corrosion rate of steel in dry areas such as in deserts is very low. As a matter of course, corrosion does occur whenever steel gets wetted by rain/snow or when dew forms on the steel surface at temperatures lower than the dew point. In addition, if airborne sulfur oxide or salt dissolves into that water, corrosion rates will be accelerated.

5.7.2 Low-Alloy Steel

5.7.2.1 General

Low-alloy steel contains small amounts of alloying elements such as Cr, Mo, Ni, Cu, Nb, and Ti in order to improve various properties such as strength, toughness, ductility, and, sometimes, corrosion. These alloying elements affect environmental degradation in two ways. If used to increase the mechanical strength properties, they will increase that steel's susceptibility to HE and SCC. The other motive is to improve/change the corrosion resistance primarily through change in corrosion potential and change in the stability of the corrosion product. The general corrosion of low-alloy steel is very similar to that of plain-carbon steel, except for the special cases of some low-alloy corrosion-resistant steel including weathering steel, in which some chemical elements stabilize the corrosion product and reduce the corrosion rate in the atmospheric environment. In the case of a hydrogen reduction system, some precipitates or nonmetallic inclusions lower the hydrogen overvoltage and increase the corrosion rate, due to the acceleration of the cathode reaction. For example, the size and volume of carbides that act as a cathodic site change the corrosion rate. With an increase in carbide content, for example, by tempering from the as-quenched state or through an increase in carbon content, the corrosion rate would increase. Extremely high-purity steel with few cathodic sites is known to have a very low corrosion rate in an acid solution. Further, TMCP steel is expected to be less corroded than traditional steel because TMCP steel generally contains less carbon, and its microstructure is homogeneous. TMCP steels are important in automobiles, ships, pressure vessels, line pipes, and other applications. On the other hand, such phenomena are not expected for oxygen reduction corrosion because the corrosion rate is controlled by the supply of oxygen.

Seawater contains about 3.5% salt (mainly NaCl), has pH of about 8, and is corrosive to steel. Especially, note that, the most severe corrosion occurs in the splash zone because such parts are alternately wetted and dried and because the seawater is more aerated (contains more DO).

Corrosion tests on pipes with varied Cr contents for seawater injection were performed using a flow tester, under environmental conditions where DO was kept at 1 ppb or lower but soars to 1400 ppb every 24 h. This simulates seawater for injection being steadily deaerated; however, an upset of the deoxidizing plant sometimes occurs giving rise to 1400 ppb exposures. As Figure 5.13 illustrates, the corrosion rate rapidly drops through the addition of 1% Cr, and further Cr additions gradually decrease the corrosion rate. Hence, 5% Cr steel is very resistant to general corrosion, but localized corrosion does occur (Nose et al. 2001). In this manner, very good general corrosion resistance often leads to localized corrosion susceptibility.

Sulfuric acid dew corrosion occurs in flu gas-treating equipment at heavy oil-fired power plants, coal-fired power plants, and so on. Generally, if exhaust gas contains as little as 1 ppm of SO_3, sulfuric acid dew points exceed 100°C, and thus heavy corrosion can occur where temperatures are lower than the dew point. Steel containing a small amount of Cu (0.3%) and Sb (0.01%) shows good corrosion resistance to sulfuric acid because both Cu and Sb reduce anodic reaction kinetics, and the Cu_2Sb formed on the

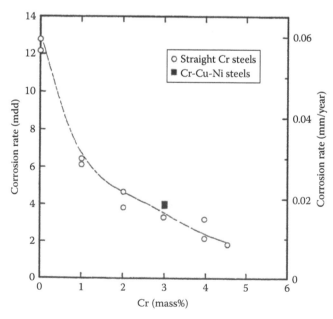

FIGURE 5.13 Effects of Cr content on corrosion rate in deaerated seawater with intermittent increase in DO. (From Nose, K. et al., Corrosion properties of 3%Cr steels in oil and gas environments, in *Proceedings of the Corrosion 2001*, NACE International, Houston, TX, 2001, Paper No. 01082.)

steel surface suppresses the cathodic reaction. Recently, low-alloy steel resistant to not only sulfuric acid dew corrosion but also hydrochloric acid dew corrosion has been developed (Usami et al. 2004).

5.7.2.2 HSLA (High-Strength Low-Alloy) Steel

Environmental cracking is more important for HSLA steels because high-strength steels are generally more susceptible to environmental cracking than low-strength steels. Some examples of SCC caused by caustic (alkaline) cracking and carbonate/bicarbonate (CO_3^{2-}/HCO_3^-) cracking are explained in detail in the following.

In the past, water in a boiler was often pH controlled at around 10 or 11 through the use of caustic soda (NaOH). NaOH is not particularly corrosive, but the boilers often suffered from cracking that was labeled alkaline cracking or caustic cracking. This form of SCC is an intergranular type that occurs at a potential of −1000 to −450 mV versus SCE. The presumed mechanism is as follows: the magnetite (Fe_3O_4) that forms on the steel surface is repeatedly broken up by stress (or strain), while self-repairing causes localized anodic dissolution, resulting in SCC. The fact that NaOH stabilizes the Fe_3O_4 results in this form of SCC.

CO_3^{2-}/HCO_3^- cracking is often observed on the external surface of asphalt-coated pipelines that are buried and cathodically protected. This type of cracking is called "high-pH-type external SCC." Figure 5.14 shows the cross-sectional photo of a line pipe where high-pH-type SCC occurred. This type of SCC is of an intergranular type, with cracks proceeding along the ferrite grain boundaries, as is apparent in Figure 5.14 (Asahi et al. 1999). This SCC tends to occur in line pipes near pump stations where the temperature is relatively high, between 70°C and 80°C, and also where the pressure fluctuation is relatively large. The potential where this SCC occurs is between −600 and −700 mV versus SCE. Steel with a multiphase microstructure consisting of a soft phase (ferrite) and a hard phase (pearlite), such as ferrite–pearlite steel, is more susceptible to this high-pH-type SCC. However, steel with a homogeneous microstructure such as those produced by TMCP is less susceptible to SCC (Asahi et al. 1999). This SCC often occurs in old line pipes that are made of ferrite–pearlite steel and are external-coated with asphalt or coal tar, where a specific environment to this high-pH-type SCC tends to be formed.

The other types of SCC occur when these steels are in contact with nitrate solutions, cyanide solutions, and liquid ammonium.

Steel containing hydrogen can fail by HE in a brittle-like manner at a stress level lower than the yield stress. HE susceptibility increases for those steels with higher strength. Thus, HE becomes a more important consideration when using higher-strength steels from an industrial viewpoint. HE phenomena are variously named depending on the specific mechanism of hydrogen penetration into the steel in question. Hydrogen reduction reactions generate hydrogen atoms (H). Most of them, however, recombine, becoming hydrogen molecules (H_2) that escape where only a very low fraction of H atoms

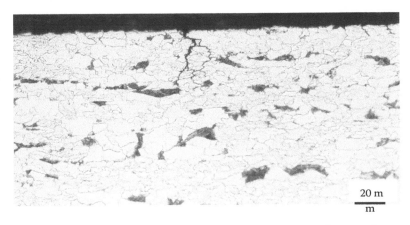

20 m
m

FIGURE 5.14 Intergranular cracking by high-pH-type external SCC. (From Asahi, H. et al., *Corrosion*, 55, 644, 1999.)

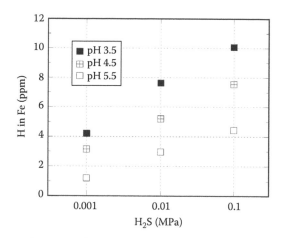

FIGURE 5.15 Effects of pH and H$_2$S partial pressure on hydrogen entry into steel.

actually penetrate into the steel. Whenever the so-called poisons, such as H$_2$S or arsenious acid, are present in the solution, the fraction of atomic hydrogen penetrating into the steel remarkably increases. Figure 5.15 illustrates the effects of pH and H$_2$S partial pressure on the atomic hydrogen content in steel immersed in each solution. Experimentally, the solution was saturated with H$_2$S by bubbling H$_2$S containing gas into the solution. Since the partial pressure of H$_2$S is linearly related to H$_2$S concentration in the solution, the hydrogen content in steel increases with either the increase of H$_2$S partial pressure or the decrease of pH (an increase in H$^+$ concentration). H$_2$S often exists in oil and gas reservoir and is known as a "sour environment." When steel is exposed to this sour environment, a lot of hydrogen can penetrate into the steel, resulting in the possibility of HE. This phenomenon is known as sulfide stress cracking (SSC). SSC is most relevant to oil country tubular good (OCTG) used in sour wells. Figure 5.16 contains some tensile-type SSC test data results using the NACE standard TM0177-2005 method A protocol, where each specimen was tensile stressed and the time to failure was recorded. From this graph, the threshold stress, σth, can be determined as the highest stress at which no failure occurs in 720 h of testing. Figure 5.17 shows the relation between the YS and σth for four generic steels. The values of σth start to drop whenever the strength level exceeds a certain value about 700 MPa in that local environment. Also, it is here that the fracture mode changes from a transgranular type to a mixed form consisting of both transgranular and intergranular cracking and then changes to a fully intergranular mode

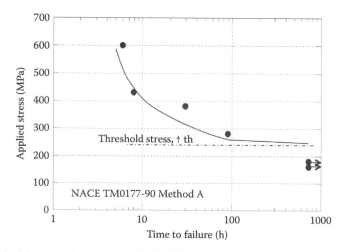

FIGURE 5.16 Methodology for obtaining SSC threshold stress.

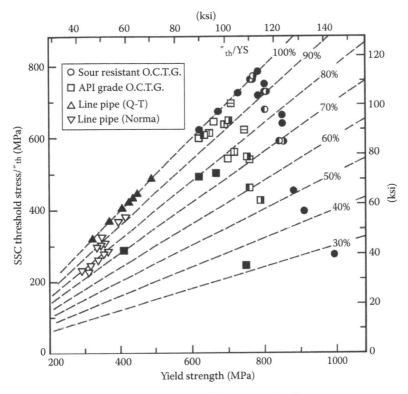

FIGURE 5.17 SSC threshold stress against YS in NACE TM0177 solution A.

of failure at very high YSs. Figure 5.18 is a scanning electron micrograph image depicting the mixed fracture-type mode. In Figure 5.17, the scatter of σth values at constant YSs is shown. The shape of each data plot symbol corresponds to one of several different types of steel with a unique chemistry and microstructure. Special steels with high resistance to SSC have been developed by optimizing chemistry and microstructure. In order to prevent the occurrence of intergranular cracking, Mn and P contents

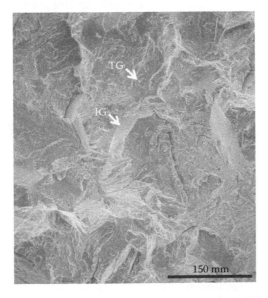

FIGURE 5.18 Mixture of intergranular (IG) and transgranular (TG) cracks in SSC.

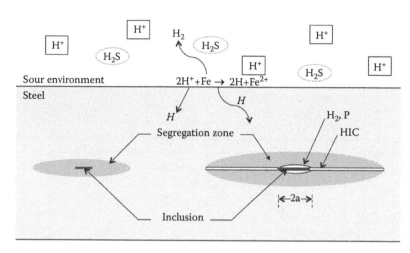

FIGURE 5.19 Schematic illustration showing how HIC occurs.

are reduced and alloying additions of Mo are made for chemical compositions. Also, a microstructure consisting of tempered martensite has been found to be resistant to SSC (Asahi et al. 1989).

There is another type of HE occurring in sour environments known as "hydrogen-induced cracking" or HIC. In HIC, a crack occurs parallel to the rolling surface of steel exposed to a sour environment without any external stress being applied. HIC often occurs in pipelines transporting sour gas consisting of natural gas and H_2S. The mechanism of HIC is schematically shown in Figure 5.19. Hydrogen atoms that are generated by corrosion reaction penetrate into the steel and diffuse to steel/nonmetallic inclusion interfaces. There, two hydrogen atoms recombine into molecular H_2, where its accumulation generates pressure along that the interface. This internal pressure causes a stress that makes the resulting crack progress. As is obvious, HIC occurs more easily at larger inclusions where MnS is a typical nonmetallic inclusion that can initiate HIC. Figure 5.20 shows a scanning electron micrograph image that depicts HIC initiating from an elongated MnS inclusion. Because MnS easily deforms and elongates during hot rolling, this reduction in elongated MnS inclusions is an effective method of preventing HIC occurrence. For that purpose, the S content (an impurity element) should be deliberately reduced during steel-making processing, and furthermore, the remaining S should be fixed as CaS, which is hardly elongated during hot rolling, by the addition of Ca. Low-C steel with a homogeneous microstructure produced through TMCP is also essential to superior HIC-resistant properties. Thus, HIC-resistant steels that are produced by modern steel production processes such as TMCP feature high purity (low S content, less than 10 mass ppm), low C (0.04–0.06 mass%), and the homogeneous microstructure that are necessary for steels with HIC-resistant properties.

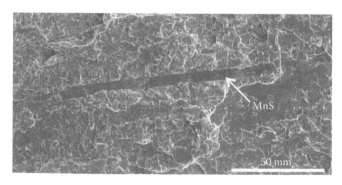

FIGURE 5.20 HIC fracture surface extending around an elongated MnS inclusion. (Photo courtesy of Mituru Sawamura.)

Besides a sour environment, hydrogen can penetrate into steel in many other ways. Even under normal atmospheric corrosion conditions, once any SO_x molecules in air can dissolve into water, sulfuric acid forms lowering the pH in this, leading to a higher entry rate of hydrogen into steel. Under such conditions, high-strength bolts often suffer from cracking. For such cracking to occur, σth begins to decrease for steels with ultimate tensile stress around 1300 MPa and above as shown in a graph similar to Figure 5.18 even where the hydrogen content is low, compared with cases seen in sour environments.

5.7.2.3 Weathering Steel

Small amounts of Cu (less than 1%) increase resistance to atmospheric corrosion because the Cu helps to form a more dense and protective rust layer. Small alloying additions of Ni, Cr, and P also have similar effects. These alloying effects have been quantitatively formulated as a weathering index or WI (ASTM standard 2010), where

$$WI = 26.01Cu + 3.88Ni + 1.20Cr + 1.49Si + 17.29P - 7.29Cu \times Ni - 9.10Ni \times P - 33.39Cu^2$$

(all concentrations are in mass% units) (ASTM G101 2010)

Higher WI values mean better weathering properties of steels.

Low-alloy steels containing these elements that enhance resistance to atmospheric corrosion are classified as weathering steels. The use of weathering steels for open-air steel structures, such as buildings and bridges, results in greatly reduced maintenance costs. However, weathering steels can corrode anomalously faster in coastal areas where the amount of airborne salt is higher, allowing accumulation of chloride ions at the rust and steel interface; this causes a loose layer of nonadherent/nonprotective rust to form.

Figure 5.21 illustrates the effects of typical alloying elements on corrosion rates for long-exposure tests in coastal areas where deposition rate of airborne salt was as high as 1.3 mg/dm²/day (Kihira et al. 2000). As is obvious from this figure, Ni additions remarkably suppress these corrosion rates; however, a small amount of Cr, up to 2%, will increase the corrosion rate. Ni additions suppress the active dissolution of steel in seawater. Based on such consideration on desirable/undesirable additions, certain weathering steels have been developed that can be used even in coastal areas. The compositions of the major desirable chemical elements are 0.1% C, 3% Ni, and 0.4% Cu. Also, Cr is eliminated, and P is

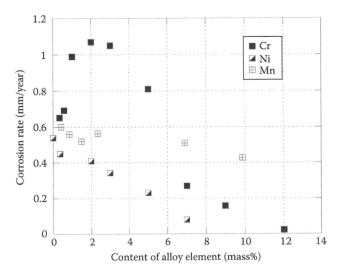

FIGURE 5.21 Effects of Cr, Ni, and Mn on corrosion rates of iron-based binary alloys exposed to coastal environments, for 1 year. (Composed from graphs in reference material, Kihira, H. et al., *Electrochem. Soc. Proc.*, 99(26), 127, 2000.)

TABLE 5.2 Chemical Compositions of One Weathering Steel Suitable for Coastal Atmosphere Exposure

Code	C	Si	Mn	P	S	Cu	Ni	Cr	Pcm
SMA490W-mod	0.10	0.20	0.61	0.006	0.002	0.40	3.07	0.02	0.21

Pcm = C + Si/30 + Mn/20 + Cu/20 + Ni/60 + Cr/20 + Mo/15 + V/15 + 15B

TABLE 5.3 Mechanical Properties of Weathering Steels Suitable for Coastal Atmosphere Exposure

Code	Thickness (mm)	Tensile Tests			Charpy Impact Test	
		YS (MPa)	TS (MPa)	Elongation (%)	Temp. (°C)	Energy (J)
SMA490W-mod	9	439	531	28	0	163
	25	405	512	26	0	230
	80	418	543	34	0	263

reduced to improve low-temperature toughness and weldability. Typical chemical compositions and mechanical properties are tabulated in Tables 5.2 and 5.3. Pcm in Table 5.2 is a carbon equivalent, and the value of 0.21 means that this steel possesses good weldability (Ito and Bessho 1968). The rust structure of this steel after exposure to coastal environments for nine years consisted of bilayers and is similar to the typical protective rust observed in conventional weathering steels in milder environments. The compositions of Ni, Cu, and Na are enriched in the inner adhesive layer, of ultrafine Fe_3O_4, with Cl present in the outer layer. After nine years of exposure in coastal areas, the average penetration for this anti-airborne salinity weathering steel is about 0.2 mm, while that for conventional weathering steel is about 3 mm (Kihira et al. 2000).

5.8 Degradation Protection

5.8.1 Surface Treatment

Surface roughness may affect the corrosion rate itself. Since smooth surfaces effectively have less actual surface area than rougher surfaces, they present significantly less reaction interface area than rougher surfaces that result in a proportionately smaller amount and rate for Fe dissolution. The formation of a more stable and dense rust layer, such as Fe_3O_4, also retards corrosion in mild environments. Whenever environmental cracking could be a problem, surface treatment to introduce compressive residual stress by shot blasting, for example, is sometimes effective for mitigating it.

5.8.2 Coatings

Coatings used on steel surfaces can provide a satisfactory barrier between steel and the environment and are widely used to prevent corrosion. Among them, the application of metallic coatings is described in this chapter, while the technologies of polymer coatings are included in Chapter 16. Typical metals used for coating are Zn, Sn, Cr, Cu, Ni, and Al. These metal-coated steels are widely applied to automobiles, buildings, household appliances, and cans used for foods. One method used in metal coating consists of an electrodeposition process or electroplating, where the steel part to be coated is immersed in a solution containing the metal ion to be plated and current is passed between the steel part and another electrode. Another method known as hot dip coating coats the steel part with molten metal. Galvanized steel (Zn plating) is a popular example. The thickness of the coating produced by the hot dip process is greater than that from electroplating.

There are two types of metal coatings: barrier types and sacrificial types. Zn coating is a sacrificial type, while many other metal coatings are barrier types. In the following, Sn and Zn coatings are described.

Sn is a good corrosion-resistant metal and provides an effective barrier. However, once a Sn layer is either removed by corrosion or breached by mechanical damage or the presence of some other defects and the steel substrate is directly exposed to the environment, any remaining adjacent Sn layer will cause accelerated localized corrosion of Fe, due to a galvanic effect, because Sn is nobler than Fe in water. Sn-plated steel can become a sacrificial type, in case of some environments where food cans are handled.

However, Zn is less noble than Fe in water. If Zn is removed by corrosion or breached so that the steel substrate is directly exposed to the environment, the Zn layer surrounding that exposed region begins to sacrificially dissolved and suppresses dissolution (ionization) of Fe through a galvanic effect. Furthermore, the corrosion product of Zn, known as the white rust that sometimes forms on steel, also suppresses further steel corrosion. When a coated steel is heat treated, it forms an alloy zone at the interface between the coated metal and the steel. In order to further improve corrosion resistance, some combination of phosphate treatment, chromate treatment, and paints is sometimes applied to Zn-plated steel. The corrosion prevention of automobile bodies (longer life) is attributed to the widespread use of Zn-coated steel.

5.8.3 Metallurgical Methods

Metallurgical corrosion prevention methods have been explained in Sections 5.6 and 5.7. Some low-alloy steels, such as weathering steel, provide long service in specific environments. Some carefully produced steels have been designed to be less susceptible to SCC or HE.

5.9 Specific Standard Tests

There is no one special test method that will be explained in detail. For measuring corrosion rates, a very primitive but basic method is the measurement of the weight loss of a steel block exposed to the environment, and then normalize the data by just dividing the weight loss by the surface area and the test duration time. As for the test specimen preparation and dimensions and the stress application methods for measuring SCC and HE phenomena, typical ASTM testing procedures (ASTM Standard 2010) should be used. For example, these include descriptions of constant load-type tensile tests, bent beam tests, and fracture mechanics type tests. NACE standards also provide many specific test methods (NACE standard TM0177 2005), for example.

5.10 Summary

Iron and steel constitute a large fraction of the metallic materials that are used as construction parts and structures; thus, the economic loss due to corrosion and environmental cracking is enormous. The corrosion mechanism of steel is explained briefly in this chapter, and the effects of environmental conditions on the corrosion rate are described. Even in low-alloy steels, corrosion can be suppressed by the addition of small amounts of alloying elements for applications in a specific moderate environments, such as weathering steels. On the other hand, care during design and manufacturing to prevent crevice corrosion and galvanic corrosion is also very important. Regarding environmental cracking, two types were considered: SCC type and HE type. Metallurgical control can suppress environmental cracking. Additionally, using knowledge for preventing the combination of a specific steel type and a known aggressive environment can be used to prevent environmental cracking.

Nomenclature and Units

[Conversion factor for steel corrosion rates] 0.1 mm/year = 19.2 mdd (mg/dm^2/day) = 1.92 gmd (g/m^2/day)

DO   dissolved oxygen
HE   hydrogen embrittlement
HIC   hydrogen-induced cracking

SCC   stress corrosion cracking
SCE   saturated calomel electrode
SHE   standard hydrogen electrode
SSC   sulfide stress cracking
TMCP   thermomechanical control process

Acknowledgments

I acknowledge Professor Ronald Latanision and Dr. Misao Hashimoto of Nippon Steel & Sumitomo Metal Corporation for providing me with the opportunity of contributing to this book.

References

Asahi, H., Y. Sogo, M. Ueno, and H. Higashiyama. 1989. Metallurgical factors controlling SSC resistance of high-strength, low-alloy steels, *Corrosion*, 45, 519–527.

Asahi, H., T. Kushida, M. Kimura, H. Fukai, and S. Okano. 1999. Role of microstructures on stress corrosion cracking of pipeline steels in carbonate-bicarbonate solution, *Corrosion*, 55, 644–652.

ASTM Standard. 2010. Section 3, Vol. 03.02, *Corrosion of Metals; Wear and Erosion*, ASTM, West Conshohocken, PA.

Beachem, C. 1972. A new model for hydrogen-assisted cracking, *Metallurgical Transactions*, 3, 437.

Ito, Y. and K. Bessho. 1968. Cracking parameter of high strength steels related to heat affected zone cracking, *Journal of the Japan Welding Society*, 37, 983.

Kato, C., Y. Otoguro, S. Kado et al. 1978. Grooving corrosion in electric resistance welded steel pipe in sea water, *Corrosion Science*, 18, 61.

Kihira, H., A. Usami, K. Tanabe et al. 2000. Corrosion and corrosion control in saltwater environment, *Electrochemical Society Proceedings*, 99(26), 127–136.

NACE standard test method TM0177-2005, NACE, Houston, TX.

Nose, K., H. Asahi, N. Perry et al. 2001. Corrosion properties of 3%Cr steels in oil and gas environments, in *Proceedings of the Corrosion 2001*, NACE International, Houston, TX, 2001, Paper No. 01082.

Oriani, R. and P. Josephic. 1974. Equilibrium aspects of hydrogen-induced cracking of steels, *Acta Metallurgica*, 22, 1065.

Tomoe, Y. 1998. Corrosion in petroleum development and related case studies, in K. Kaneda, Ed., *Corrosion and Corrosion Resistant Materials in the Oil and Gas Industry*, JNOC, Chiba, Tokyo, Japan, pp. 3–21.

Usami, A., M. Okushima, S. Sakamoto et al. 2004. Nippon Steel technical report no. 90, Tokyo, Japan, p. 25.

Whitman, W., R. Rusell, and V. Altieri. 1924. Effect of hydrogen-ion concentration on the submerged corrosion of steel, *Industrial and Engineering Chemistry*, 16, 665.

Further Readings

About Steel

Honeycombe, R.W.K. and H.K.D.H. Bhadeshia. 1995. *Steels, Microstructure and Properties*, 2nd Edn., Edward Arnold, London, U.K.

Leslie, W.C. 1981. *The Physical Metallurgy of Steels*, McGraw-Hill, Inc., New York.

Pickering, F.B. 1978. *Physical Metallurgy and the Design of Steels*, Applied Science Publishers Ltd., London, U.K.

About Corrosion

Fontana, M.G. 1987. *Corrosion Engineering*, 3rd Edn., McGraw-Hill, Inc., New York.

Revie, W. and H.H. Uhlig. 2008. *Corrosion and Corrosion Control*, Wiley-Interscience, New York.

6

Crystalline Alloys:
Stainless Steels

Haruhiko Kajimura
Nippon Steel & Sumikin
Stainless Steel Corporation

6.1 Introduction

Stainless steels are used for a wide variety of applications owing to their good corrosion resistance. Stainless steel is the name given to a group of alloys of the ferrous (Fe) system where the primary alloying element that makes stainless steel resistant to corrosion is chromium (Cr). Wide varieties of stainless steel alloys have been developed through the addition of various other elements. This chapter outlines the types of stainless steel, their properties, corrosion characteristics, and protective measures.

6.2 Background

The most important alloying element for stainless steel, Cr, was discovered in the eighteenth century. Then, M. Faraday and many other researchers studied alloying with Cr for corrosion protection of iron and steel, but it was as recently as in the twentieth century that what is now called stainless steel was more fully developed. A patent (Duetsches REICH No. 304159) for an alloy that would later develop into type 304 stainless steels was applied for in Germany in 1912; this was the most important industrial patent in the history of stainless steel. Many other alloys were developed thereafter as steel refining technology advanced to form the present lineup of stainless steels for different applications.

6.3 Applications

Stainless steel with different chemical compositions and shapes is used for widely varied applications. In terms of shape, stainless steel products are divided into flat products and bar and wire products where flat products are further divided into coil and sheet products. Coil products are less than about 6 mm in thickness, and plate products are usually thicker than 6 mm. Surface properties are important for coil and sheet products; different kinds of surface finishes are employed for different applications. An outline of stainless steel products with different properties and shapes for different applications as well as their latest, newly developed uses is given in the succeeding text.

6.3.1 Typical Applications

Because of its excellent corrosion resistance, high-temperature strength, and design features, stainless steel is used as a functional or structural material for kitchen facilities, home electric appliances, railway carriages, automotive parts, equipment for chemical plants, and other widely varied applications. Table 6.1 shows typical applications of different product types of stainless steel.

Coil and sheet products are found in our daily lives. Indoors, we use stainless steel cutlery, pots and pans, and kitchen knives as well as sinks and other equipment for household and commercial kitchens. Besides, thanks to its good appearance and cleanliness, the applications for stainless steel have expanded to electric appliances such as for refrigerator doors and washing machine tubs. Outdoors, it is used for railway carriages (where because of its higher strength, one can consequently use lighter-weight components). Stainless steel is also used for mufflers and other parts for automobile exhaust systems because of its heat resistance. In consideration of its excellent corrosion resistance, stainless steel plates are used to fabricate reactor parts for nuclear power plants, towers and vessels for chemical plants, tanks for marine vessels, desalination plants, etc. Bar and wire products are used for fasteners such as bolts and screws, components for electronic devices such as hard disks, and many types of springs.

Coil and sheet products of stainless steel are often used for making the most of excellent design features in addition to high corrosion resistance, and therefore, the surface properties are of great importance for these products. A wide variety of surface treatment methods are employed in consideration of the design features required for different applications. Table 6.2 shows the principal surface treatment methods for different final applications. These products are mostly used after annealing and pickling (2D finish) or after additional light cold rolling at a reduction of several percent (2B finish). Treatments for mirror, polishing, dulling (low-gloss) finishes, or embossing are also utilized according to the final application. Figure 6.1 shows typical surface conditions obtainable through these different production processes.

TABLE 6.1 Microstructures and Applications for Stainless Steel Products by Type of Primary Form

Primary Form	Microstructure	Appliances
Coil and sheet	γ: Austenitic stainless steel	Train carriages, kitchen facilities, building materials
	α: Ferritic stainless steel	Home electronic appliances, hot water systems, automotive parts (mufflers and other exhaust system parts)
	M: Martensitic stainless steel	Cutting tools, brake disks
Plate	γ: Austenitic stainless steel	Chemical plant equipment, seawater facilities
	D: Duplex stainless steel	Chemical tankers, salt works, desalination plants
Bar and wire	γ: Austenitic stainless steel	Bolts, nuts, screws
	M: Martensitic stainless steel	Valves, bearings

TABLE 6.2 Surface Finishes of Cold Rolled Products and Applications

AISI, JIS	EN 10088-2	Description	Notes	Appliances
2D	2D	Cold rolled, annealed, and descaled	Uniform matt finish	General appliances
2B	2B	Cold rolled, annealed, descaled, and lightly rolled with smooth	Smooth finish, brighter than finish No. 2D	General appliances, building material
BA	2R	Cold rolled and bright annealed	Cold rolled surface with metallic luster retained through annealing	Home electronic appliances, kitchen facilities, decorations
No. 4	1J/2J	Polished with fine grits	Surface finish of a not-too-reflective, unidirectional texture given by polishing with successively finer abrasives of #180 to #240	Building material, train carriages

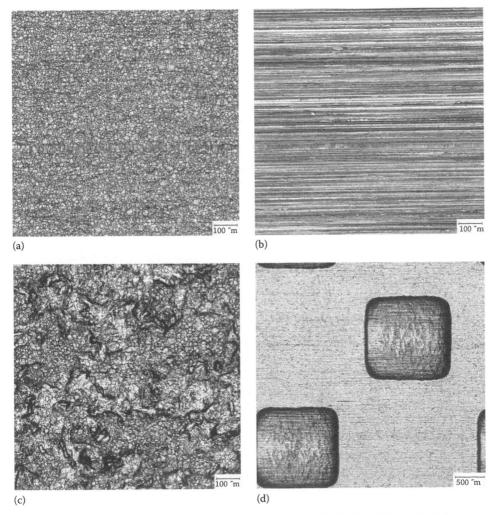

(a)   (b)   (c)   (d)

FIGURE 6.1 Examples of popular types of surface finishes (type 304): (a) No. 2B, (b) No. 4, (c) dulling treatment, and (d) embossing.

FIGURE 6.2   Application of high-purity ferritic stainless steel (type 430J1L) to a special-design kitchen sink.

6.3.2 Latest Applications

With increasing concerns about global warming, the use of stainless steel has been expanded to cover the fields of energy conservation and the development of new energy sources, typically involving systems such as fuel cells and photovoltaic power devices. In fuel cells, stainless steel is used for the reformers to generate and for tanks to store hydrogen gas. For photovoltaic units, it is used for the baseplates of solar panels and their mountings. In addition, in response to the demand to increase the water supply, a larger amount of stainless steel is now used for desalination plants.

On the other hand, there is a new trend from the viewpoint of resource saving, which focuses on using lesser amounts of materials required for each application: high-purity ferritic and lean duplex stainless steels containing lower amounts of Ni and Mo alloying solutes—to help decrease the depletion of these elements from the world supply—have been developed in line with this philosophy, and their use is being promoted. Figure 6.2 shows a kitchen sink made of such high-purity ferritic stainless steel.

6.4 Material Properties

6.4.1 Mechanical Properties

As other variations of Fe–Cr alloys, many grades of stainless steel have been developed using additional alloying elements such as Ni, Mo, Cu, Al, and Si to enhance various combinations of corrosion resistance, mechanical properties, strength, formability, and other characteristics. In terms of metallographic and crystallographic structure, stainless steel is classified into the following groups: (a) steels of the Fe–Cr system, which are further divided into type 430 steels having a ferritic structure and type 410L steels having a martensitic structure, and (b) steels of the Fe–Cr–Ni system, which are divided further into type 304 steels having an austenitic structure and type 2205 steels having a duplex structure of ferritic and austenitic phases. Tables 6.3 and 6.4 list typical chemical compositions and tensile properties for these groups of steel. Since formability mainly relates to press forming of thin coil and sheet products of types 304 and 430 steels, the items of Table 6.4 related to formability have entries only of these steel grades. Type 304 steels exhibit high strength and ductility with excellent formability, especially in stretch formability indicated by the Erichsen value (ISO 8490/ASTM E 643-84).

TABLE 6.3 Chemical Compositions of the Four Stainless Steel Classes/Types

		C	Si	Mn	Ni	Cr
Austenitic stainless steel	Type 304	0.05	0.5	0.8	8	18
Ferritic stainless steel	Type 430	0.06	0.3	0.8	—	17
Martensitic stainless steel	Type 410	0.02	0.5	0.5	—	12
Duplex stainless steel	Type 2205	0.02	0.5	1.8	5	22

TABLE 6.4 Tensile Properties and Formabilities

	0.2% Poof Strength (MPa)	Tensile Strength (MPa)	Elongation (%)	r-Value	Limit Drawing Ratio	Erichsen Value
Type 304(1.0)[a]	314	618	59	1.0	2.05	13.4
Type 430(1.0)[a]	343	500	30	1.1	2.10	9.7
Type 410(4.5)[a]	410	510	28	—	—	—
Type 2205(6.0)[a]	590	780	35	—	—	—

[a](): Thickness of tested specimen (mm).

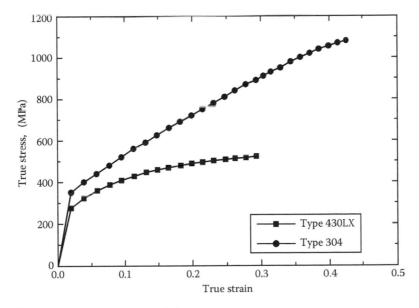

FIGURE 6.3 Comparison of work hardening behavior.

Figure 6.3 compares type 304 with 430LX (a grade of ferritic stainless steel having high ductility), in terms of work hardening behavior. Although type 304 steels harden significantly during mechanical working, they have high ductility thanks to transformation-induced plasticity (TRIP) (Zackay et al. 1967; Gerberich et al. 1968) resulting from the transformation of the austenitic phase into a martensitic phase under strain.

6.4.2 Physical and Thermal Properties

The previous subsection explained that stainless steel was classified in terms of chemical composition and metallographic/crystallographic structure into characteristic groups such as ferritic and austenitic

TABLE 6.5 Physical Properties

	Magnetism	Density (10^3 kg/m³)	Elastic Modulus (MPa)	Electric Conductivity (10^{-9} Ωm)	Thermal Conductivity (W/m·K, 0°C–100°C)	Thermal Expansion Ratio (10^{-6} K^{-1}, 0°C–100°C)
Type 304	No	7.93	193,000	720	16.3	17.2
Type 430	Yes	7.70	200,000	600	23.9	10.4
Type 410	Yes	7.75	200,000	570	24.9	9.9
Type 2205	Yes	7.82	189,000	850	19.0	13.7

Source: AISI, *Steel Product Manual Stainless and Heat Resisting Steels,* Iron and Steel Society, Inc., 1990; Japanese Standard Association, *JIS Handbook,* JIS G 4310, 1992, p. 1375; Japan Stainless Steel Association, *Databook of Stainless Steels,* 2000, p. 4.

stainless steels. The physical properties of these groups are determined substantially by their crystallographic structure and alloy contents as shown in Table 6.5. While the physical properties of ferritic and martensitic stainless steels are similar to each other, those of austenitic steels are significantly different, especially in their magnetic and thermal properties. Austenitic stainless steels (type 304) are not magnetic and have the lowest thermal conductivity values and the largest thermal expansion coefficients of the steels listed in this table. Duplex stainless steels (type 2205) consisting of ferrite and austenite phases exhibit physical properties midway between those of the ferritic (type 430) and austenitic (type 304) steels, except for their resistivity.

6.5 Metallographic Structure

6.5.1 Atomic Bonding

Cr stabilizes the ferritic phase, and, therefore, steels of the Fe–Cr system maintain their body-centered cubic (BCC) crystal structure (see Figure 6.4) throughout the course of cooling. Ni is a strong austenite-forming element, and, therefore, steels of the Fe–Cr–Ni system maintain their austenitic phase (the face-centered cubic [FCC] crystal structure) even at room temperature. It has to be noted, however, that, depending on the chemical composition, the austenitic phase of Fe–Cr–Ni steels may become unstable with superimposed stress causing the austenite to transform into ferrite.

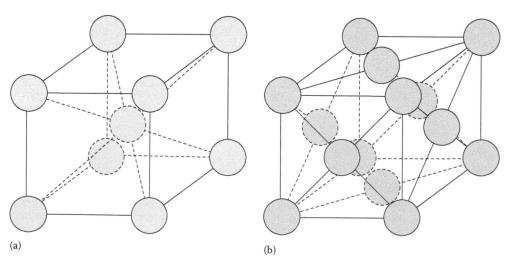

(a)   (b)

FIGURE 6.4 Crystal structures of stainless steel: (a) BCC (ferritic stainless steel) and (b) FCC (austenitic stainless steel).

6.5.2 Microstructure

Figure 6.5 shows photomicrographs of different types of stainless steels taken at sections in the rolling direction. The structure of type 304 after solution heat treatment is that of a typical *austenitic stainless steel*, where the pairs of parallel lines within the crystal grains represent annealing twin interfaces. Type 430 is a typical *ferritic stainless steel*; it exhibits an annealed structure with black carbide particles scattered in a ferritic matrix. Gray and white layers are in clear contrast in the dual-phase structure of the *duplex stainless steel*; the gray zones are the ferritic, and the white ones are the austenitic grains that were elongated in the rolling direction.

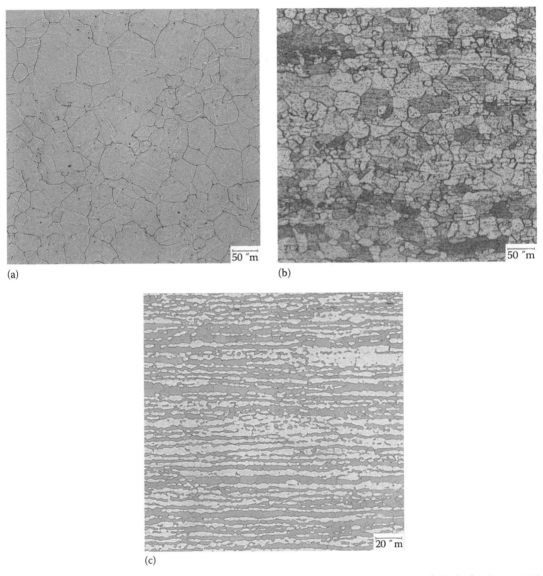

(a)

(b)

(c)

FIGURE 6.5 Typical microstructures of (a) ferritic (type 304), (b) austenitic (type 430), and (c) duplex (type 2205) stainless steels.

6.6 Principles of Material Degradation

6.6.1 Thermodynamics

Corrosion of metal materials is classified, in terms of environmental conditions, into either wet or dry corrosion. This subsection focuses on wet corrosion. As stated earlier, stainless steels are iron-system alloys containing Cr as the principal alloying element with additional solute elements such as Ni. To understand the corrosion behavior of a metal material, it is important to understand the dissolution behavior of each of the alloying elements. Wet corrosion occurs as a result of an electrochemical reaction. Generally, the reaction by which a metal (M) dissolves in an aqueous solution as an ion (M^{n+}) is expressed as follows:

$$M = M^{n+} + ne^{-} \tag{6.1}$$

The Gibbs free-energy change ΔG of this equation is given as follows:

$$\Delta G = \Delta G^0 + RT \ln K \tag{6.2}$$

where
  ΔG^0 is the standard Gibbs free-energy change
  R is the gas constant
  T is the absolute temperature
  K is the equilibrium constant for Equation 6.1

Then, where F is Faraday's constant and E is the electrode potential, $-\Delta G$ is equal to maximum work nFE of the reaction.
  Consequently, the following equation is derived from Equation 6.2:

$$E = \left(\frac{-\Delta G^0}{nF}\right) - \left(\frac{RT}{nF}\right) \ln K \tag{6.3}$$

Here, if normal electrode potential E^0 is given as

$$E^0 = \left(\frac{-\Delta G^0}{nF}\right) \tag{6.4}$$

then Equation 6.3 can be expressed in the following manner:

$$E = E^0 - \left(\frac{RT}{nF}\right) \ln K \tag{6.5}$$

Equation 6.5 (the Nernst equation [Nernst 1899]) expresses the equilibrium potential of an electrochemical reaction when a normal hydrogen electrode is used. A potential-pH diagram is obtained by plotting the equilibrium potential as a function of pH in a coordinate system. Figure 6.6 shows an example of a potential-pH diagram of the $Cr-H_2O$ system (Pourbaix 1974) when the soluble ion concentration is 10^{-6} mol/kg. In this diagram, (a) the zone where a metal element is stable in a metallic state is called an immunity zone; (b) the zone where it is stable in an ionic state, a corrosion zone; and (c) the zone where it is stable in the form of oxides or hydroxides, a passivation zone (Pourbaix 1974).

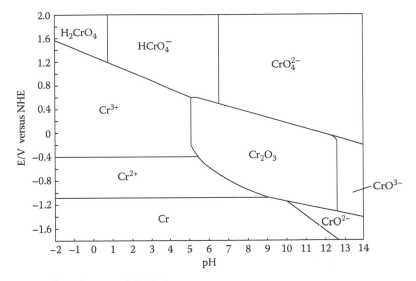

FIGURE 6.6 Potential-pH diagram of Cr–H$_2$O.

6.6.2 Kinetics

Figure 6.7 schematically illustrates local cells. Equation 6.1 expresses the reaction for dissolution of a metal material, which is an oxidation reaction. This type of reaction is called an anodic reaction. On the same metal surface where an anodic reaction is taking place, a reduction reaction expressed by Equation 6.6 takes place when the solution is acidic, and another reduction reaction expressed by Equation 6.7 takes place when the solution is neutral; this reduction reaction is called a cathodic reaction:

$$M \rightarrow M^{n+} + ne^- \tag{6.1}$$

$$nH^+ + ne^- \rightarrow \left(\frac{n}{2}\right)H_2 \tag{6.6}$$

$$\left(\frac{n}{4}\right)O_2 + \left(\frac{n}{2}\right)H_2O + ne^- \rightarrow nOH^- \tag{6.7}$$

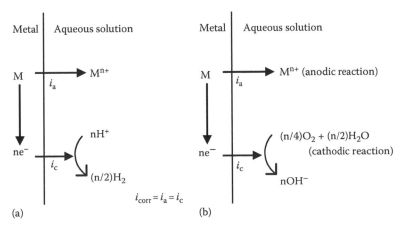

FIGURE 6.7 Models of local cells: (a) hydrogen evolution type and (b) oxygen consumption type.

In this relation, the site where an anodic reaction takes place is called an anode, and that where the cathodic reaction takes place the cathode; the electric current flowing at an anode is called the anodic current (i_a), and that flowing at a cathode, the cathodic current (i_c). On a metal surface in contact with a solution, locally formed anodes and cathodes are directly connected through the metal, forming innumerable small cells. These small cells are local cells, and the currents flowing between them are corrosion currents (i_{corr}):

$$i_{corr} = i_a = i_c \tag{6.8}$$

Evans (1946) and Wagner and Traud (1938) propounded the concept that corrosion of a metal material proceeded through this mechanism of local cells. The case of general corrosion where anodes and cathodes are distributed on the surface densely and homogeneously is called the Wagner model, while for the case of localized corrosion where anodes and cathodes are relatively distant from each other, the Evans model.

To understand the electrochemical reactivity of a metal material in an aqueous solution, studying its polarization curve obtainable by measuring the current while controlling the potential can be effective. Figure 6.8 shows an example polarization curve. The potential of a metal material in a naturally corrosive condition without application of external potential is called its corrosion potential. In the case where corrosion of a metal material as expressed by Equation 6.1 is in progress, when the potential is raised above its corrosion potential, the current increases exponentially with the increase in potential; in this case, the metal is in an active state. With a further increase in potential, in some cases, the rate of dissolution of the metal into the solution may decrease when the potential rises above a certain level. This region where the metal dissolution rate decreases is called passivation, and the condition where the rate of dissolution is very low, the passive state.

Metals that can have a passive state are Fe, Cr, Ni, Al, and Ti, and when in a passive state, the surface of the material is covered with a passive film composed of oxides as expressed in the following (Müller and Schwabe 1933):

$$M + nH_2O \rightarrow MO_n + 2nH^+ + 2ne^- \tag{6.9}$$

If a chemically stable passive film is obtained for an alloy made by adding another metal that exhibits passivation, that alloy can become an excellent corrosion-resistant material. A typical example is stainless steel, a group of alloys of Fe and Cr. Here, an amorphous (McBee and Kruger 1972) passive film, a few nanometers in thickness (Sugimoto and Matsuda 1980), containing condensed Cr (Hashimoto et al. 1979) and bound water (Okamoto and Shibata 1965), forms on the surface of that stainless steel

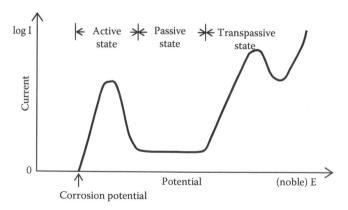

FIGURE 6.8 Schematic anodic polarization curve of stainless steel.

(Okamoto and Shibata 1978). This passive Cr oxide film forms when the content of Cr in that steel is about 12% or more (Asami et al. 1978).

When there are anions capable of destroying the passive film in a corrosive environment, the film may be locally destroyed even if the potential is in the passive state. Anions that can destroy the passive film of stainless steel include Cl⁻, Br⁻, and SCN⁻ (Galvele 1978). On the other hand, an important characteristic of the passive film is its self-regenerating ability. If the portion destroyed by the anions is repaired immediately through this self-regenerating ability, thus the integrity of the material can be maintained, but otherwise some other types of localized corrosion such as pitting corrosion will result. It must be noted here that the protectiveness of the passive film is also influenced by the chemical composition of the alloy and any nonuniformities such as the distribution of nonmetallic inclusions, grain boundaries, and solute segregation. Also note that, in a potential range below the passive state, the passive film dissolves through reduction and, in a range above the passive state, transpassive dissolution occurs, and as a result, the corrosion resistance of the stainless steel can be lost.

6.6.3 Form of Corrosion

Figure 6.9 shows the types for corrosion of stainless steels. Modes of corrosion are roughly classified into general and localized corrosion; the latter is responsible for the major amount of the damage to stainless steel. Localized corrosion occurs in forms such as pitting corrosion, crevice corrosion, intergranular corrosion, and stress corrosion cracking (SCC). Whereas *pitting corrosion* is a type of corrosion where only a limited area of a surface is dissolved to form a small pit, *crevice corrosion* is another form, but here only the material portion forming a crevice is dissolved; note both of these types of corrosion occur in closed environments.

Figure 6.10 schematically shows a corrosion pit. When the concentration of metal ions in the solution inside a pit or crevice increases, chloride ions (Cl⁻) enter from outside to maintain the electric neutrality of the solution, then form metal chlorides (refer to Equation 6.10) that then change into hydroxide through a hydrolysis reaction, lowering the local pH level (refer to Equation 6.11). Whenever a high concentration of chloride ions (Cl⁻) and low pH are maintained through the earlier processes, pitting or crevice corrosion continues to develop:

$$M + nCl^- \rightarrow MCl_z + ne^- \tag{6.10}$$

$$MCl_n + nH_2O \rightarrow M(OH)_n + nH^+ + nCl^- \tag{6.11}$$

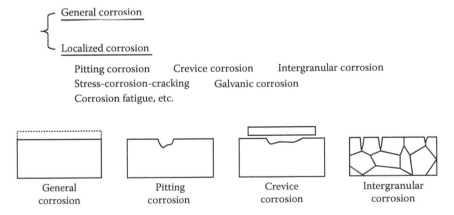

FIGURE 6.9 Forms of corrosion for stainless steels.

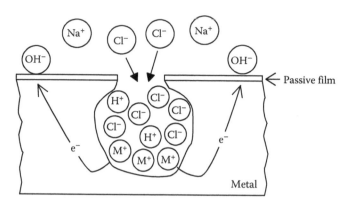

FIGURE 6.10   Schematic diagram of a corrosion pit.

Intergranular corrosion is a mode of corrosion wherein a metal material dissolves selectively at crystal grain boundaries. With stainless steel, this often occurs at weld joints, especially at those of austenitic stainless steel. For the most part, intergranular corrosion of stainless steel is caused by selective dissolution of Cr-depleted zones generally due to the localized formation of Cr carbides at grain boundaries. Any processing condition that leads to Cr depletion at grain boundaries is called sensitization.

SCC is a phenomenon where a crack initiated at a localized corrosion site propagates under tensile stress; it often occurs with austenitic stainless steel. When the crack propagation results from material dissolution due to an anodic reaction, it is called *active path corrosion-type stress corrosion cracking* (APC-SCC), and when it results from material dissolution due to a cathodic reaction, it is called *hydrogen embrittlement-type stress corrosion cracking* (HE-SCC); the latter is also called hydrogen embrittlement. In a variation of APC-SCC, a crack propagates owing to repeated mechanical destruction of the film and its self-repair ability; this type of SCC is called *tarnish rupture-type stress corrosion cracking* (TR-SCC). From another crack propagation point of view, when a crack propagates across crystal grains, it is called *transgranular stress corrosion cracking*, and when it propagates along grain boundaries, it is called *intergranular stress corrosion cracking*.

6.7   Degradation of Specific Systems

6.7.1   Introduction

As stated earlier, the corrosion resistance of stainless steel is due to the formation of a passive film composed mainly of Cr oxide on the surface, but the corrosion resistance also depends on the metallic microstructure and the presence of certain alloying elements. Localized corrosion often occurs, and the localized corrosion mode between the austenitic, ferritic, martensitic, and duplex stainless steels is different. With respect to pitting and crevice corrosion, Cr, Mo, and N are known to be effective in suppressing the occurrence of these types of corrosion. Thus, the corrosion resistance of stainless steel can be expressed in terms of the presence and amounts of these elements. The pitting resistance equivalent number (PREN = %Cr + 3.3 × %Mo + 16 × %N) is a widely used indicator for this metric.

The corrosion behavior for each class of stainless steels with respect to localized corrosion is outlined in the succeeding text.

6.7.2   Austenitic Stainless Steel

6.7.2.1   Pitting and Crevice Corrosion

The resistance of austenitic stainless steel to pitting and crevice corrosion in the presence of chloride ions is lowered as the temperature rises. As seen in Figure 6.11, the critical corrosion temperature,

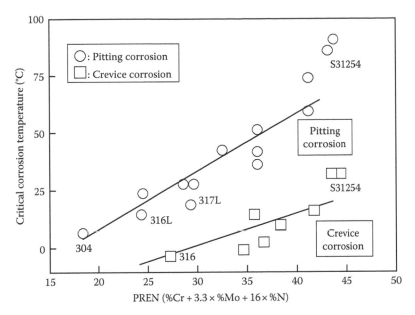

FIGURE 6.11 Relationship between the critical temperature for pitting and crevice corrosion and PREN.

the lowest temperature at which this kind of corrosion occurs, tends to be higher with increasing PREN (Lorenz and Medawar 1969; JSSA 1995). The corrosion rate of austenitic stainless steel containing Ni is often lower than that of comparable ferritic stainless steel. The resistance to pitting and crevice corrosion has recently been remarkably improved: some of the latest developments called super stainless steels have PREN values equal or greater than 40.

6.7.2.2 Intergranular Corrosion

Intergranular corrosion sometimes constitutes serious problems with austenitic stainless steel. It often occurs after welding work, mainly at the heat-affected zone (HAZ), those portions adjacent to the weld that are heated between 400°C and 800°C. This is because, as shown in Figure 6.12, C in steel forms carbides ($M_{23}C_6$) with Cr near grain boundaries, and as a result, the local Cr content near these boundaries falls to below 12 mass% (e.g., Cr-depleted zones), thus lowering the corrosion resistance. This is called sensitization to intergranular corrosion. In extreme cases, crystal grains may fall out, leading to a dangerous drop in strength. A time–temperature-sensitization (TTS) curve visualizes the combined conditions of heating temperature and time with respect to sensitization. As seen in Figure 6.13, it is possible to identify from these TTS curves under which conditions intergranular corrosion is likely to occur (Rocha 1966). The zone of sensitization shifts to the lower-temperature and longer-time side with decreasing C content. The most effective practical measure for preventing intergranular corrosion from occurring is to use steels containing as little C as possible (such as 304L and 316L) or those containing Nb, Ti, or other elements that form carbides more readily than Cr does. The reason for this selection is that, when steel contains Nb, Ti, or the like, carbides such as NbC or TiC preferentially form, effectively decreasing the amount of C in solid solution, thus making the formation of Cr carbides more difficult.

6.7.2.3 Stress Corrosion Cracking

SCC is a mode of corrosion that occurs under concurrent specific conditions involving the steel alloying content, the specific corrosive environment, and any acting or residual tensile stress. With austenitic stainless steel, chloride ions locally destroy the passive film on the surface, which, together with consequent activated dissolution of the steel material, then leads to cracking of the material, where the crack propagates in the direction normal to the direction of stress and the activated dissolution continues at

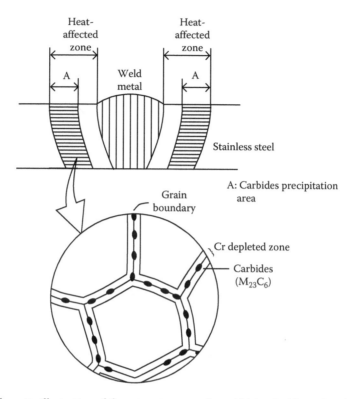

FIGURE 6.12 Schematic illustration of the microstructure of a weld joint (weld metal and HAZ) of austenitic stainless steel.

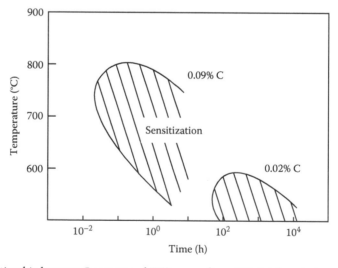

FIGURE 6.13 Relationship between C content and TTS curve of type 304 stainless steel.

the leading edge of the crack. In most cases, the crack propagates by branching through a repetition of the earlier processes as shown in Figure 6.14.

SCC can develop in either the transgranular or the intergranular mode; for sensitized steel, it proceeds in the intergranular mode or along grain boundaries. SCC occurs more frequently in an aqueous environment with higher concentrations of chloride ions and at higher temperatures. Another factor

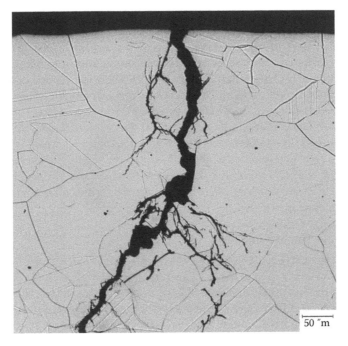

50 ″m

FIGURE 6.14 Typical stress corrosion cracking of austenitic stainless steel in a high-temperature chloride ion solution.

triggering SCC is the existence of dissolved oxygen or other oxidizing agents. As for the effects of steel chemistry, one must note that different alloying elements may prove effective in preventing SCC, but these also depend on the specific composition of the test liquid used for experiments that simulate the expected corrosive environment. For instance, in an environment of a boiling 42% solution of $MgCl_2$, SCC sensitivity is determined principally by Ni content: here, a steel containing 8% Ni is most sensitive to SCC, but it can be suppressed with steels having a Ni content is 45% or more. SCC often initiates from pitting or crevice corrosion sites, and, thus, addition of Mo or other similar elements effective in preventing these types of localized corrosion modes can also be effective in preventing this kind of SCC. To prevent SCC, it is effective to take preventative measures to avoid any of its causal factors, namely, selecting certain steel alloys not sensitive to the expected service/corrosive environments, as well as reducing existing tensile stresses. For instance, stress relief annealing is often effective after forming or welding processes to reduce the level of residual tensile stresses resulting from forming or welding operations; sometimes, however, it is not always practicable, as when this heat treatment operation leads to the deformation of object pieces. In certain cases, shot peening of the surface to produce residual compressive stress can sometimes be effective. Control of environmental factors (to reduce the amounts of chloride and dissolved oxygen, temperature, etc., in the service environment) and the application of a corrosion inhibitor can also be effective depending on the environmental conditions.

6.7.3 Ferritic Stainless Steel

6.7.3.1 Pitting and Crevice Corrosion

As similarly explained in Section 6.7.1, the resistance of ferritic stainless steel to pitting corrosion is also determined by the alloying content of Cr, Mo, and N solutes as the equation for the PREN value shows. However, the solute level for N has little effect on resistance to pitting corrosion. For this reason, the pitting corrosion resistance of ferritic stainless steel is nearly the same as that of austenitic stainless steel containing the equivalent amounts of Cr and Mo. It has to be noted, however, that because the solubility limit of N in ferritic stainless steel is far lower than that in austenitic steel, one cannot expect that any deliberate addition of N to ferritic steel will improve its pitting corrosion resistance.

The resistance to crevice corrosion of ferritic stainless steel is known to improve with increases in the contents of Cr, Mo, and N (Suutala and Kurkela 1984). It follows, therefore, that high-purity ferritic stainless steel containing Cr and Mo exhibits excellent crevice corrosion resistance even in a dilute aqueous chloride solution. However, once crevice corrosion does occur in ferritic stainless steel, its rate of corrosion is markedly higher than that of austenitic stainless steel containing the same amounts of Cr and Mo. This is because ferritic stainless steel does not contain much Ni, which is effective in lowering the active dissolution rate of stainless steel. Since the occurrence of crevice corrosion is considered to decrease when the gap of a crevice is wider than 40 μm (Shiobara and Kawaguchi 1983), when using ferritic stainless steel in an environment where crevice corrosion is likely, it is important to either eliminate crevices or make those gaps wider.

6.7.3.2 Intergranular Corrosion

In general, when welded, parts made from type 430 ferritic stainless steel containing roughly 0.05% C are prone to intergranular corrosion at the HAZ and the weld metal zone. This is due to the low solubility limit for C and N in this type of stainless steel. During processing at steel works, this type of steel often undergoes a long period of heat treatment to cause C and N to precipitate as a Cr carbonitride uniformly distributed in the matrix phase. During welding, the carbonitride precipitates melt in the substrate metal in which temperature is above 900°C. After rapid cooling, the C and N eluted from the metal form Cr carbonitride precipitates preferentially near grain boundaries. As a result, the concentration of Cr near grain boundaries is reduced, and the welded steel becomes sensitized to corrosion at HAZ. To prevent the intergranular corrosion of ferritic stainless steel resulting from the earlier processing, using a more purified steel and adding stabilizing elements such as Ti and Nb can be effective (Abo et al. 1977). Regarding steel purification concepts, decreasing the total content of C and N to less than 0.014% is very effective in suppressing sensitization. Because Nb has a stronger affinity for C and N than does Cr, its addition is very effective in suppressing the sensitization of weld joints. Specifically, when the C + N content is 0.015% or more, the addition of Nb at 12 times the amount of C + N or more is recommended; however, when C + N is less than 0.015%, smaller Nb additions are sufficient. Addition of Ti has nearly the same effects as addition of Nb (Streicher 1977).

6.7.3.3 Stress Corrosion Cracking

Unlike austenitic stainless steel, ferritic stainless steel is resistant to SCC. This is presumably because the rate of localized corrosion of ferritic stainless steel is so large (larger than that of austenitic stainless steel) that cracks cannot easily develop. Although some researchers report that ferritic stainless steel containing small amounts of Cu and Ni will undergo cracking in a boiling solution of $MgCl_2$ (Steigelwald et al. 1977), this type of steel is sufficiently resistant to SCC in a normal atmosphere or in an environment of city water or other common aqueous solutions of neutral chlorides.

6.7.4 Duplex Stainless Steel

Duplex stainless steel containing 20%–30% Cr, 1%–8% Ni, up to 4% Mo, and roughly 0.1%–0.3% N demonstrates good corrosion resistance and mechanical properties. However, because this type of steel is prone to age hardening in the temperature range of 475°C and aging-induced precipitation of the sigma phase, nitrides, and the like at high temperatures, these phenomena cause some limitations to their use at high temperatures. In addition, care must be taken concerning deterioration of the mechanical properties in the HAZ due to the said precipitation. As shown in Table 6.6, the temperature range where precipitates form in duplex stainless steel is from 600°C to 1000°C.

TABLE 6.6 Details of Some Precipitates in Duplex Stainless Steel

Phase	Chemical Formula	Structure	Temperature Range of Formation (°C)
Sigma phase (σ)	Fe–Cr–Mo	Tetragonal	750–1000
Chi phase (χ)	$Fe_{36}Cr_{12}Mo_{10}$	BCC	700–950
Cr nitride (M_2N)	Cr_2N	Hexagonal	650–900
Cr carbide ($M_{23}C_6$)	$Cr_{23}C_6$	FCC	600–850

6.7.4.1 Thermal Aging Embrittlement

Owing to its high Cr concentration, the ferritic phase of duplex stainless steel is prone to 475°C embrittlement and sigma (σ)-phase embrittlement during thermal aging. The said 475°C embrittlement results in the deterioration of impact toughness due to development of fine modulated structures where either Cr or Fe is separately and locally enriched during aging in the temperature range from 400°C to 550°C as a result of mutual repulsion between these two atom species. The σ-phase embrittlement, on the other hand, is the kind of age hardening that leads to a rapid fall in impact toughness caused by precipitation of a hard σ phase and/or χ phase especially when duplex stainless steel containing great amounts of Cr and Mo is exposed to temperatures in the 700°C–1000°C range.

6.7.4.2 Lowered Corrosion Resistance of the Weld Metal and the Heat-Affected Zones

Precipitation readily occurs at the weld joints of duplex stainless steel during the high-temperature welding operation where the resultant precipitation leads to deterioration of corrosion resistance. In those portions that undergo rapid cooling from high temperatures, the nitride of Cr tends to form precipitates inside ferrite grains and at grain boundaries creating localized Cr-depleted zones around them that lowers the corrosion resistance. With high-alloy duplex stainless steel, the σ-phase precipitate forms during slow cooling or reheating of the weld metal and HAZ; thus, Cr- and Mo-depleted zones exist around the σ-phase precipitates, causing the local corrosion resistance to deteriorate.

6.7.5 Martensitic Stainless Steel

Martensitic stainless steel including type 410 steel is characterized by high strength after quenching where a martensitic structure is formed during the quenching operation. Thus, it is necessary to control the amount of Cr, essential for corrosion resistance, to a certain limit. Therefore, the Cr content of this kind of steel is approximately 12%, lower than that of austenitic stainless steel (type 304, etc.).

Applications for martensitic stainless steels are generally limited to shafts, bolts, cutting tools, and other high-strength machine parts for use in mild environments. The main problems for martensitic stainless steel are due to corrosion such as rusting and delayed fracture after long exposure times of bolts and other high-strength structural members.

6.7.5.1 Rusting

The rate and amount of rusting for martensitic stainless steel depend on the service environment, alloying elements, metallographic structure, and surface conditions; basically, its resistance to rusting improves, as in the case of austenitic stainless steel, as the PREN value (= %Cr + 3.3 × %Mo + 16 × %N) increases. Care must be taken, however, since the metallographic structure is also significant in influencing the rusting phenomena. Figure 6.15 shows the effects of the PREN number, the amount of δ-ferrite, and cooling rate on the pitting potential of martensitic stainless steel. In the case of natural air-cooled steels, when the steel does not have a δ-ferrite skin, the pitting potential becomes higher as the PREN number increases; however, when the material has a δ-ferrite skin, the pitting potential is drastically influenced by the metallographic structure, regardless of the PREN value. It must be noted,

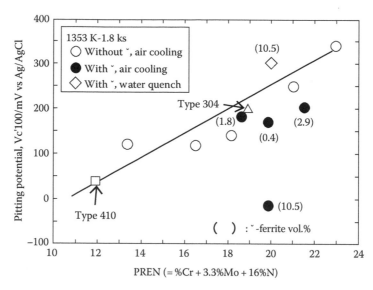

FIGURE 6.15   Effect of PREN, δ-ferrite, and cooling rate on the pitting potential of martensitic stainless steel.

however, that when martensitic stainless steel containing δ-ferrite is water cooled, the pitting potential remains as high as that of steel having no δ-ferrite. This occurs because Cr-depleted zones form at the boundaries between austenite and δ-ferrite; thus, for lower cooling rates during quenching, great quantities of Cr carbide precipitates form at these boundaries. Rusting corrosion for this type of stainless steel also greatly depends on surface conditions, and for this reason, the final production operation before shipment from steel works is a passivation treatment in a bath of nitric acid or the like to provide good rusting resistance.

6.7.5.2 Delayed Fracture

Delayed fracture in martensitic stainless steels is dependent on corrosion in the service environment and the stress imposed during use. It is often a type of failure caused by hydrogen produced

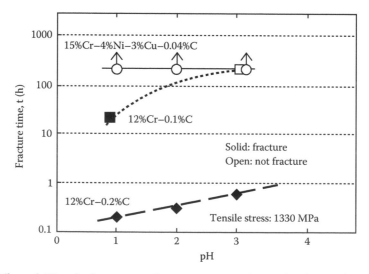

FIGURE 6.16   Effects of pH on the fracture time for martensitic stainless steel under tensile stress correction.

during localized or general corrosion that diffuses into the steel. Brittle fracture is seen along pre-austenite grain boundaries at the surface of a delayed fracture. Figure 6.16 shows the relationship between pH and time to fracture of typical grades of martensitic stainless steel under tensile stress of 1330 MPa. Although the fracture time varies with different steel grades, the lower the pH value, or the more corrosive the environment, the more rapidly the failure occurs owing to hydrogen originating from corrosion. This indicates that care must be taken when using martensitic stainless steel in high-tensile-stress, corrosive environments where hydrogen can be generated, because then the possibility of delayed fracture is high.

6.8 Protection against Degradation

Because of its excellent corrosion resistance due to the presence of a protective passive film on the surface, stainless steel is often used without any coating. The service/use conditions for stainless steels, however, have diversified over the years, yet various kinds of surface treatments combined with better design features have been developed to provide better corrosion resistance as well. Thus, various surface treatment measures such as reformation of the passive film, coating with metal or paint, and the use of cathodic protection (which is widely applied to marine structures) that have been utilized are explained in the succeeding text.

6.8.1 Surface Treatment

Processes for reforming the passive film on stainless steel surfaces that improve corrosion resistance and develop coloring are explained here.

The excellent corrosion resistance of stainless steel in natural environments is due to the normal formation of a passive film on the surface. There are cases when an additional passivation treatment is applied to this film to further improve its protective performance. A well-known method is to immerse stainless steel products in a bath of nitric acid, a strong oxidizing agent (ASM Metals Handbook 1987) whereby Cr is further concentrated in the film, while MnS and other nonmetallic inclusions are dissolved and removed, thus enhancing the corrosion protection function of the film. In fact, some papers reported that after an emery grinding and passivation treatment of type 304 steel, the Cr concentration at the surface increased from 20% to approximately 60% (Omata and Yukawa 1978). However, since this improvement in corrosion resistance is due to the existence of a very thin protective film on the surface, the effects of the passivation treatment may be lost when grinding or other machining is applied subsequently.

The passive film on the stainless steel surface is usually a few nanometers thick and does not exhibit any color; whenever the film is thicker, color appears as a result of interference patterns with visible light (with wavelength in the 340–780 nm range). The most common industrial treatment method is the wet oxidation process; the INCO method (Evans et al. 1972) is practiced worldwide. As Table 6.7 shows, this method consists of a coloring stage using a bath of a CrO_3–H_2SO_4 solution followed by a hardening stage using cathodic electrolysis in a bath of a CrO_3–H_3PO_4 solution; possible colors are blue, gold, red, and green.

TABLE 6.7 Coloring Conditions of Stainless Steels (INCO Method)

Treatment Process	Coloring	Hardening
Solutions	H_2SO_4—500 g/L	CrO_3—250 g/L
	CrO_3—250 g/L	H_3PO_4—2.5 g/L
Temperature	80°C–85°C	RT
Conditions	Immersing: 8–20 min	Electrolyzing: 0.2–0.4 A/dm², 5–10 min

6.8.2 Coating

Coatings are applied to stainless steel surfaces for two purposes: first, as sacrificial protection to improve corrosion resistance and, second, to improve its appearance for good design features. For sacrificial protection purposes, one example is aluminum coating applied to type 409 steel for automotive uses. Even though type 409 steel contains as little as 11% Cr, a corrosion resistance comparable to that of type 304 steel is obtainable due to the sacrificial protection provided by the aluminum coating. This material is used not only for automobiles but also for roofs on buildings (Maki et al. 1996). Zinc coating has also been used for the same purposes (Kawaguchi et al. 1996).

Coating is often applied to materials for the roofs and external panels of buildings and home electric appliances to enhance design features in addition to corrosion resistance. In the case of building applications, silicon polyester and fluorine resins are often used as coating materials for color-coated stainless steels; of these, the latter resin with excellent weather resistance is preferred especially for use on large buildings.

The latest surface appearance trend in designing home electric appliances is to make the most of the metallic color of stainless steel. An example is clear-coated stainless steel sheets with pre-coated layers of various types of transparent coating materials that often are used for refrigerator doors and other similar applications. The variety of clear coatings for stainless steel sheets includes those with the metallic stainless steel color; these provide enhanced resistance to corrosion, smudging, and fingerprint markings for good design features, as well as other clear-coated sheets with additional colorings using pearlescent or chromatic pigments. These coating layers on these products are usually 2–15 μm thick, which are resins of the acrylic, epoxy, and polyester systems being selected for the coating for the desired properties required for the final application. Besides these, other coating products having special functions such as the addition of a lubricant into the coating layer to improve sheet workability could also eliminate the use of other surface protective films; in other cases, special clear-coated sheets containing an antibacterial agent in the coating layer have been produced (Yano et al. 1998).

6.8.3 Cathodic Protection

Electric protection is a practical method for stopping or slowing the progress of corrosion of metal materials usually in an aqueous environment by changing their potential by applying an external electric current. Cathodic protection is widely employed and capable of shifting the potential of objects in the baser direction to a state of immunity. With this method, the protective potential or current is applied to the object to be protected by having it in contact with a baser metal or by applying a direct electric current from the outside. The former method is called voltaic anode protection (or sacrificial anode protection); Zn or Mg is mainly used for these anodes, and in an environment containing chlorides, aluminum alloy anodes are used. For the latter method, impressed current protection, high-silicon steel, graphite, platinum, or the like is used as the inert anode. Maintenance of the protective system is essential if satisfactory effects from cathodic protection are to be obtained; thus, it is necessary to periodically confirm that the protective potential and current are being continuously applied and check the sacrificial anodes and replace them when they have been totally consumed.

6.9 Specific Standardized Tests

Principal standards on corrosion-testing methods for stainless steels set forth by the International Organization for Standardization (ISO) and the American Society for Testing and Materials (ASTM) are listed in Table 6.8.

TABLE 6.8 Main Standards of Corrosion-Testing Method for Stainless Steels

No.	Title
ISO	
3651	Determination of Resistance to Intergranular Corrosion of Stainless Steels
7539	Stress Corrosion Testing
11463	Evaluation of Pitting Corrosion
11782	Corrosion Fatigue Testing
12732	Electrochemical Potentiokinetic Reactivation Measurement Using the Double Loop Method
15324	Evaluation of Stress Corrosion Cracking by the Drop Evaporation Test
17475	Electrochemical Test Methods—Guidelines for Conducting Potentiostatic and Potentiodynamic Polarization Measurements
17864	Determination of the Critical Pitting Temperature under Potentiostatic Control
ASTM	
A262	Practice for Detecting Susceptibility to Intergranular Attack in Austenitic Stainless Steels
G28	Test Methods of Detecting Susceptibility to Intergranular Corrosion in Wrought, Nickel-Rich, Chromium-Bearing Alloys
G36	Practice for Evaluating Stress-Corrosion-Cracking Resistance of Metals and Alloys in a Boiling Magnesium Chloride Solution
G46	Guide for Examination and Evaluation of Pitting Corrosion
G48	Test Methods for Pitting and Crevice Corrosion Resistance of Stainless Steels and Related Alloys by Use of Ferric Chloride Solution
G61	Test Method for Conducting Cyclic Potentiodynamic Polarization Measurements for Localized Corrosion Susceptibility of Iron-, Nickel-, or Cobalt-Based Alloys
G78	Guide for Crevice Corrosion Testing of Iron-Base and Nickel-Base Stainless Alloys in Seawater and Other Chloride-Containing Aqueous Environments
G108	Test Method for Electrochemical Reactivation (EPR) for Detecting Sensitization of AISI Type 304 and 304 L Stainless Steels
G129	Practice for Slow Strain Rate Testing to Evaluate the Susceptibility of Metallic Materials to Environmentally Assisted Cracking
G150	Test Method for Electrochemical Critical Pitting Temperature Testing of Stainless Steels

6.10 Summary

The mechanical properties and corrosion resistance of typical kinds of stainless steel, used for many industrial products in our daily lives, have been described earlier. Different types of stainless steel used for a multitude of applications have widely varied corrosion resistances; hence, it is impossible to give an exhaustive description of them all within the pages allotted to this chapter. Readers are encouraged to study further referring to specialized technical literature listed in the following.

Nomenclature and Units

ΔG Gibbs free-energy change
ΔG^0 standard Gibbs free-energy change
E electrode potential
F Faraday constant
i_a anodic current
i_c cathodic current

i_{corr} corrosion current (= i_a = i_c)
K equilibrium constant
M metal
M^{n+} metal ion
R gas constant
T absolute temperature

Abbreviations

HAZ   heat-affected zone(s)
IGC   intergranular corrosion
pHd   depassivation pH
PREN   pitting resistance equivalent number
SCC   stress corrosion cracking
TRIP   transformation-induced plasticity
TTS   time–temperature sensitization

References

Abo, H., M. Nakata, S. Takekuma, M. Onoyama, H. Ogawa, and H. Okada. 1977. *Stainless Steel*, 77, 35–47.

Asami, K., K. Hashimoto, and S. Shimodaira. 1978. *Corrosion Science*, 18, 151.

ASM Metals Handbook. 1987. *Corrosion*, Vol. 13, ASM International, Metals Park, OH, p. 552.

Evans, U.R. 1946. *Metallic Corrosion, Passivation and Protection*, Edward Arnold & Co., Tokyo, Japan.

Evans, T.E., A.C. Hart, H. James, and V.A. Smith. 1972. *Transactions of the Institute of Metal Finishing*, 50, 77.

Galvele, J.R. 1978. III. Breakdown and repassivation. In R.P. Frankenthal and J. Kruger., Eds., *Passivity of Metals, The Corrosion Monograph Series*, The Electrochemical Society, Inc., Princeton, NJ, p. 285.

Gerberich, W.W., P.L. Hemmings, M.D. Merz, and V.F. Zackey. 1968. *Transactions of the American Society for Metals*, 61, 843.

Hashimoto, K., K. Asami, and K. Teramoto. 1979. *Corrosion Science*, 19, 3.

JSSA. 1995. *Handbook of Stainless Steel*, Nikkan Kogyo Shimbun Ltd., Tokyo, Japan, p. 622.

Kawaguchi, H., K. Watanabe, and W. Harada. 1996. *Nisshin Technical Report*, 74, 45–52.

Lorenz, K. and G. Medawar. 1969. *Thyssen Forschung*, 1, 97.

Maki, J., T. Omori, K. Asakawa, S. Higuchi, and N. Okada. 1996. *Shinnittetsu Giho*, 361, 52–56.

McBee, C.L. and J. Kruger. 1972. *Electrochemical Acta*, 17, 1337.

Müller, E. and K. Schwabe. 1933. *Zeitschrift fur Elektrochemie*, 39, 414.

Müller, E. and K. Schwabe. 1933. *Zeitschrift fur Elektrochemie*, 39, 815.

Nernst, W. 1899. *Zeitschrift für Physikalische Chemie*, 4, 129.

Okamoto, G. and T. Shibata. 1965. *Nature*, 206, 1350.

Okamoto, G. and T. Shibata. 1978. VI. Metal and alloys. In R.P. Frankental and J. Kruger, Eds., *Passivity of Metals*, The Electrochemical Society, Inc., Princeton, NJ, p. 646.

Omata, H. and K. Yukawa. 1978. In *Proceedings of the 19th Japan Conference Corrosion Symposium*, JSCE, Tokyo, Japan, pp. 6–9.

Pourbaix, M. 1974. *Atlas of Electrochemical Equilibria in Aqueous Solutions*, National Associations of Corrosion Engineers, Houston, TX.

Rocha, J.H. 1966. *DEW Technical Report*, 2, 16.

Shiobara, K. and K. Kawaguchi. 1983. In *Proceedings of the 48th Japan Conference Corrosion Symposium*, JSCE, Tokyo, Japan, pp. 44–47.

Steigelwald, R.F., A.P. Bond, H.J. Dundas, and E.A. Lizlov. 1977. *Corrosion*, 33, 279.

Streicher, M.A. 1977. *Stainless Steel*, 77, 1.

Sugimoto, K. and S. Matsuda. 1980. *Materials Science and Engineering*, 42, 181.
Suutala, N. and M. Kurkela. 1984. *Stainless Steel*, 84, 240–247.
Wagner, C. and W. Traud. 1938. *Zeitschrift fur Elektrochemie*, 44, 391.
Yano, H., Y. Udagawa, T. Sakai, and H. Entani. 1998. *Nisshin Technical Report*, 78, 90–96.
Zackay, V.F, E.F. Parker, D. Fahr, and R. Busch. 1967. *Transactions of the American Society for Metals*, 60, 252.

Further Readings

General Properties

Angel, T. 1954. *Journal of Iron and Steel Institute*, 177, 165.
Chao, H.C. 1967. *Transactions of the American Society for Metals*, 60, 37.
Peckering, F.B. 1978. *Physical Metallurgy and Design of Steels*, Applied Science Publishers, London, U.K., p. 228.
Peckering, F.B. 1978. *Physical Metallurgy and Design of Steels*, Applied Science Publishers, London, U.K., p. 231.
Peckner, D. and I.M. Bernstein. 1977. *Handbook of Stainless Steel*, McGraw-Hill Book Company, New York.
Schaeffler, A.L.1949. *Metal Progress*, 56, 620.

Degradation Principles for Materials

Davis, J.R. 2000. *Corrosion*, ASM International, Metals Park, OH.
Evans, U.R. 1963. *An Introduction to Metallic Corrosion*, Edward Arnold & Co, London, U.K.
Marcus, P. 2002. *Corrosion Mechanisms in Theory and Practice*, 2nd Edn., Marcel Dekker, Inc., New York.
Perez, N. 2004. *Electrochemistry and Corrosion Science*, Springer, New York.
Pourbaix, M. 1973. *Lecture on Electrochemical Corrosion*, Plenum Press, New York.
Uhlig, H.H. 1963. *Corrosion and Corrosion Control*, John Wiley & Sons, Inc., New York.

Degradation of Specific Systems

Schweitzer, P.A. 2004. *Encyclopedia of Corrosion Technology*, 2nd Edn., Marcel Dekker, Inc., New York.
Sedriks, A.J. 1996. *Corrosion of Stainless Steels*, 2nd Edn., National Associations of Corrosion Engineers, Houston, TX.

7

Crystalline Alloys: Nickel

Raul B. Rebak
GE Global Research

7.1 Introduction

Nickel (Ni) is a versatile metallic element since it is used as a component in hundreds of alloys; it is used as a corrosion-resistant plating product and also as a catalyst. Approximately 61% of the Ni produced worldwide is used in the fabrication of stainless steels, which contain by weight approximately 10% Ni (Nickel Institute 2010). Only about 12% of the world production of Ni is used in the fabrication of Ni-based alloys also called high-Ni alloys. Over 90% of Ni-containing products are recycled at the end or their useful life, and there are no limits on how many times the Ni metal can be recycled (Nickel Institute 2010).

Ni alloys are solid solutions of the element Ni and other alloying elements. In general, the minimum amount of Ni in Ni alloys is in the order of 50% by mass; however, some alloys, such as alloy 800 (N08800) and alloy 28 (N08028), are classified in the family of nickel alloys even though they may contain less than 35% Ni. Large percentages of alloying elements can be added to Ni to produce a vast variety of alloys, some of which may be specially tailored for specific applications. The resulting

Ni alloys still maintain the face-centered cubic (fcc), gamma, or austenitic microstructure of pure Ni. In contrast, iron alloys (stainless steels) cannot dissolve as much alloying elements as Ni without precipitating secondary phases. The first commercial Ni alloy was the Ni–copper (Cu) Monel, which was commercially introduced in the early 1900s. Since then, Ni has been alloyed with molybdenum (Mo), chromium (Cr), tungsten (W), silicon (Si), and a combination of these and other elements to produce a large family of alloys. Some of the Ni alloys are highly popular (e.g., Hastelloy C-276, Inconel 600), but other alloys have more of a boutique or niche manufacturing end use. The list of Ni alloys is being revised continuously since almost every other year a new alloy appears in the market. Most, if not all, of the newer alloys are modifications of previously existing alloys, and their development and commercialization are result of research at three or four international primary metal producers of Ni alloys. Little or no research and development on Ni alloys is carried out outside the primary metal producers. The commercially produced Ni alloys in general are classified into two large groups: (1) corrosion-resistant alloys (CRAs) targeted for wet or condensed system applications and (2) superalloys or high-temperature alloys (HTAs) targeted for applications in dry or gaseous corrosion systems. The not well-defined temperature boundary of application between the CRA and the HTA is approximately 1000°F (538°C). This classification does not preclude that CRA may be used at temperatures higher than 538°C and vice versa. Generally, the CRAs are mostly selected for their capacity to resist corrosion in a given environment, and less importance may be given to their mechanical strength. However, most HTAs need to play a dual role, that is, besides their capacity to withstand the aggressiveness of the high-temperature corrosive environment, HTAs also need to keep significant strength at high temperatures. In many instances, for example, near and above 1000°C, alloy selection is dominated by how strong the alloy is in this temperature range. Tables 7.1 and 7.2 show, respectively, abridged lists of representative CRA and HTA.

7.2 General Background

Nickel alloys are in general more corrosion resistant than austenitic stainless steels in a given industrial application. Ni alloys are also more expensive than stainless steels mainly because the base metal Ni is more expensive than Fe but also because the Ni alloys may hold a larger variety and amount of alloying elements, for example, Mo, than the stainless steels. Also in many cases some Ni alloys are not readily available from the market and may need to be especially ordered. One of the large advantages of Ni alloys over austenitic stainless steels is the resistance of the Ni alloys to chloride-induced stress corrosion cracking (SCC), which chronically plagues the performance of the austenitic stainless steels in every industrial application. The first commercial nickel alloy was Monel 400 (N04400) introduced into the market in 1905 (Lamb 2000, Mankins and Lamb 2003). The second alloy to be produced commercially was the Ni–molybdenum (Mo series called Hastelloy B, which was developed in the mid-1920s. The highly popular Ni–Cr–Mo alloy series or Hastelloy C was introduced, as a cast product in the 1930s. The wrought version of alloy C, called C-276, was developed in the 1960s using newer melting practices at that time such as argon–oxygen decarburization process, which was used to reduce the amount of carbon and other impurities such as Si in the melt (Agarwal 2000). In 1931, the alloy Inconel 600 (Ni–Cr–Fe) was developed initially for use in the dairy industry and later in the 1950s and 1960s found a crucial application in the nuclear industry (Lamb 2000). The need for HTAs for aircraft applications led to the development of the Nimonic 80 alloys in the 1940s (Mankis and Lamb 1997). Incoloy 800 was developed in the mid-1950s as another important alloy in the Ni–Cr–Fe family, in part to reduce the amount of Ni that was used in alloy 600. Most of the current most popular commercial Ni alloys were developed and accepted by the industry only in the last 50 years. Figure 7.1 shows a schematic representation of how the Ni alloys may have evolved from the pure Ni base element (Ni-200) by continuously adding or partially removing alloying elements. The list of current alloys in Figure 7.1 is highly abridged for practical representational purposes. Also, many popular alloys in the past are now obsolete, and newer alloys are frequently introduced into the market. Figure 7.1 shows that the original alloy 600 appears

TABLE 7.1   Corrosion-Resistant Alloys

Alloy	UNS	Approximate Composition	YS (0.2%)	UTS	ETF (%)	RH	Applications
		Commercial nickel					
Ni-200	N02200	99Ni–0.2Mn–0.2Fe	190	450	50	60 B	Strong caustic
Ni-301[A]	N03301	93Ni–4.5Al–0.6Ti	860	1170	25	35 C	Fasteners, springs
		Ni–Cu alloys					
Monel 400	N04400	67Ni–31.5Cu–1.2Fe	270	540	43	68 B	HF
Monel K-500[A]	N05500	63Ni–30Cu–3Al–0.5Ti	700	1020	28	30 C	Fasteners, springs
		Ni–Mo alloys					
B-2	N10665	72Ni–28Mo	407	902	61	94 B	Hot hydrochloric
Hastelloy B-3	N10675	68.5Ni–28.5Mo–1.5Cr–1.5Fe	400	885	58	NA	Reducing acids
Nimofer 6629 (B-4)	N10629	65Ni–28Mo–4Fe–1Cr–0.3Al	340	755	40	NA	Hydrochloric, sulfuric
		Ni–Cr–Mo alloys					
C-276	N10276	59Ni–16Cr–16Mo–4W–5Fe	347	741	67	89 B	Versatile CPI and pollution control
Inconel 625	N06625	62Ni–21Cr–9Mo–3.7Nb	535	930	45	95 B	Aerospace, pollution control
Hastelloy C-22	N06022	59Ni–22Cr–13Mo–3W–3Fe	365	772	62	89 B	FGD, CPI, nuclear waste
Hastelloy C-2000	N06200	59Ni–23Cr–16Mo–1.6Cu	345	758	68	NA	CPI, oxidizing and reducing; sulfuric
Nicrofer 5923hMo (59)	N06059	59Ni–23Cr–16Mo–1Fe	340	690	40	NA	Oxidizing and reducing acids, CPI
Inconel 686	N06686	46Ni–21Cr–16Mo–4W–5Fe	364	722	71	NA	Oxidizing and reducing acids; CPI
C-22HS[A]	N07022	59Ni–21Cr–17Mo	1390	1590	20	30 C	Oil and gas
		Ni–Cr–Fe alloys					
Inconel 600	N06600	76Ni–15.5Cr–8Fe	275	640	45	75 B	Nuclear power
Inconel 690	N06690	58Ni–29Cr–9Fe	352	703	46	NA	Nuclear power
Inconel 725[A]	N07725	57Ni–21Cr–18Fe–8Mo–3.4Nb–1.4Ti–0.3Al	916	1264	28	36 C	Oil and gas
Inconel X-750[A]	N07750	73Ni–15Cr–7Fe–2.5Ti–1Nb–0.7Al	868	1270	25	36 C	Nuclear power
Incoloy 800	N08800	33Ni–21Cr–40Fe–0.8(Al+Ti)	250	590	NA	83 B	Steam generator tubing
Incoloy 825	N08825	43Ni–21Cr–30Fe–3Mo–2.2Cu–1Ti	338	662	45	85 B	Oil and gas; sulfuric, phosphoric
Incoloy 945[A]	N09945	50Ni–21Cr–18Fe–3Mo–3Nb–1.5Ti–0.3Al–2Cu	920	1194	28	40 C	Oil and gas
Hastelloy G-30	N06030	44Ni–30Cr–15Fe–5Mo–2Cu–2.5W–4Co	317	689	64	NA	Nitric, phosphoric
Nicrofer 3033 (33)	R20033[D]	31Ni–33Cr–32Fe–1.6Mo–0.6Cu–0.4N	380	720	40	NA	Phosphoric acid

YS, yield stress (MPa); UTS, tensile strength (MPa); ETF, elongation to failure; RH, Rockwell hardness; CPI, chemical process industry; A, thermally treated; NA, not available; D, UNS starts with an R because it classified as Cr-based alloy. Hastelloy is a trademark of Haynes International Inc., Inconel is a trademark of Special Metals Corporation, and Nicrofer is a trademark of Krupp VDM.

TABLE 7.2   HTAs Listed by Ascending UNS Number

Alloy	UNS Number	Approximate Composition	Common High-Temperature Use
Nimonic 75	N06075	78Ni–20Cr–0.4Ti	Gas turbines, heat treatment
Haynes 230	N06230	57Ni–22C2–14W–2Mo–0.3Al–0.02La	Gas turbines, superheater tubes
Inconel 600	N06600	75Ni–16Cr–9Fe	Furnace components
Inconel 601	N06601	60Ni–23Cr–15Fe–1.4Al–0.3Ti	Furnace and heat treatment components, combustion chambers
Hastelloy X	N06602	47Ni–22Cr–18Fe–9Mo–1.5Co–0.6W	Combustion chambers, heat treatment components
Inconel 617	N06617	55Ni–22Cr–12Co–9Mo–1Al	Gas turbine combustion cans, furnace components
Inconel 625	N06625	62Ni–21Cr–9Mo–3.7Nb	Aerospace, pollution control equipment
Haynes 214	N07214	75Ni–16Cr–4.5Al–3Fe–0.01Y	Specialized heat treatment, turbine parts
Inconel 718	N07718	53Ni–19Cr–18Fe–5Nb–3Mo–1Ti–0.5Al	Gas turbines, rocket engines, nuclear applications
Inconel X-750	N07750	72Ni–16Cr–7Fe–2.5Ti–1Nb–0.6Al	Gas turbine components, pressure vessels, applications in nuclear reactors
Incoloy 800HT	N08811	31Ni–45Fe–21Cr–0.4Al–0.6Ti	Industrial furnaces, carburizing equipment
Haynes 242	N10242	65Ni–25Mo–8Cr	Turbine seal rings, fasteners
Haynes HR-160	N12160	37Ni–29Co–28Cr–2.75Si–2Fe	Thermocouple shields, calciner components
Haynes 188	R30188	22Ni–39Co–22Cr–14W–3Fe–0.03La	Aerospace, combustion cans in gas turbine engines

The cobalt-based alloy 188 is also provided for reference purposes.

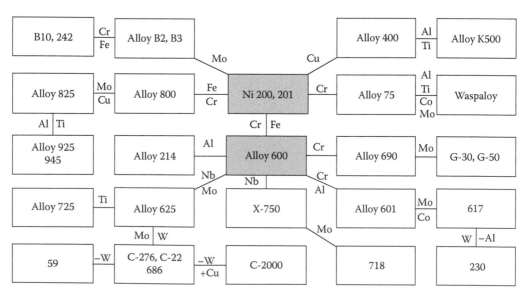

FIGURE 7.1   The large number of Ni alloys mostly evolved from Ni-200 and alloy 600.

to have been modified more times than any of the other Ni alloys, creating a large subfamily of alloys both in the CRA and the HTA families. The main alloying element for Ni alloys is Cr (Figure 7.1), which is added because it forms a chromium oxide film (Cr_2O_3) on the surface of the component and provides protection against further environmental degradation. The protection of chromium oxide is effective both in aqueous environments and in high-temperature gaseous environments. Two other common alloying elements used in CRA and HTA are Mo and W (Figure 7.1). In the CRA, Mo and W provide resistance to general corrosion in reducing acidic conditions, and in HTA these two elements provide additional mechanical strength for high-temperature applications. Other elements added in small amounts to mainly HTA are aluminum (Al), titanium (Ti), and niobium or columbium (Nb or Cb), which are used to form second-phase particles that provide resistance to mechanical deformation at high temperatures.

7.3 Application

As mentioned earlier, the extensive family of Ni alloys (Tables 7.1 and 7.2 and Figure 7.1) may be divided broadly in two groups: (1) CRAs and (2) HTAs or superalloys. These alloys were developed to fill a need in the industry, where highly corrosive streams and high-temperature environments exist. As the newer alloys improve in their resistance to the applied environments, the environments in the industry become progressively more aggressive pushing the boundaries of the resistance of these alloys. That is, there is a continuous feedback relationship between alloy development and field performance.

Both the Ni CRA and HTA alloys were mostly developed for use in areas where the less-expensive iron-based stainless steels would not perform well or in areas where a corroded stainless steel part could not be replaced cost-effectively. The main applications for the CRA are in the area of hot chloride solutions and hot acids. Austenitic stainless steels are chronically susceptible to SCC in hot chloride solutions, and Ni alloys are practically immune to chloride cracking by virtue of their high nickel content (Copson curve). Other use is in hot acids where Ni alloys find an application because they can dissolve larger amounts of beneficial alloying elements than iron. For example, the Ni alloy B-3 is highly resistant to corrosion in hot hydrochloric acid because it contains approximately 29% Mo (Table 7.1).

7.3.1 Uses

Tables 7.1 and 7.2 list some important uses for the CRA and HTA, respectively. The list of uses and applications in Tables 7.1 and 7.2 is highly condensed. The types of degradation that the CRA alloys may undergo in service can generally be classified in three large groups: (1) general or uniform corrosion, (2) localized corrosion (pitting and crevice corrosion), and (3) environmentally assisted cracking (EAC) or SCC. When selecting a material for an application, all the three main modes of degradation should be considered. Some CRAs are targeted for highly specific environments since no other alloy performs as well in those conditions. For example, in highly caustic solutions (>50% NaOH at temperatures higher than 100°C), the best alloy is pure Ni (Ni-200 or Ni-201), and for hot reducing hydrochloric acid environments, the best alloys are B-2 or B-3. Also, for wet nonoxidizing hydrofluoric acid applications, the best alloy is Monel 400. Other alloys such as C-276 were designed to be multipurpose alloys, that is, they perform reasonable well in multiple environments but are not the best one for each separate condition. Alloy C-276 will perform relatively well in caustic solutions (59% Ni), hydrochloric acid (16% Mo), and hydrofluoric acid but may not be the first choice for each separate environment. A multipurpose versatile alloy such as the C family can be used to fabricate a multipurpose vessel or heat exchanger that may be used interchangeably in several streams; however, if the vessel will be used only to handle hot caustic solutions, the alloy of choice for fabrication will be Ni-200.

A similar criterion selection for alloys is used for the high-temperature applications. In general, the high-temperature environments are divided based on the predominant type of degradation that the alloy might experience, centered on the aggressive element in the environment causing the damage.

The main modes of high-temperature degradation are (1) oxidation, (2) carburization and metal dusting, (3) nitridation, (4) halogen corrosion, (5) sulfidation, (6) molten salt corrosion, (7) ash deposit corrosion, and (8) molten metal corrosion (Lai 1990).

7.3.2 Challenges

Ni alloys are in general more corrosion resistant than austenitic stainless steels, both the standard 300 series and the super austenitic 6Mo steels. The challenges that the Ni alloys face are that they are not as popular in the industries in general, since design and even materials and corrosion engineers are not familiar with the several families of Ni alloys and their intended applications. It is commonly found that corrosion engineers in the industry feel uncomfortable dealing with Ni alloys since many times they seem confused about the difference between a B-type alloy and a C-type alloy or which Ni alloy is heat treatable and which one is not. Sometimes Ni alloys have the reputation of being exotic and expensive, that they are difficult to find in the right form (availability off the shelf), and that they are difficult to machine or fabricate. Ni alloys are not common in most industrial applications. For example, considering one field such as oil and gas exploration and production, about 85% of the metals used are carbon steels and low-alloy steel. That is, only 15% of the metals used are labeled as CRAs, which include a huge spectrum of alloys from martensitic 13% Cr stainless steel, to duplex stainless steel, to austenitic 18/8 stainless steels, to nickel-based alloys. Of this 15% segment, over 90% of the alloys labeled as corrosion resistant by the oil and gas used are iron-based alloys (martensitic, duplex, and austenitic stainless), which means that the volume of nickel alloys used in the oil and gas industry is only in the area of 1%–2%. That is, in general this industry prefers to recommend an alloy that is more available and cheaper and replace it when it fails.

7.4 General Properties

Ni alloys are generally produced by melting in electric arc or induction furnaces (Heubner 1998). The melt is then transferred from the furnace to an argon–oxygen decarburizing (AOD) vessel or to a vacuum oxygen decarburizer where the carbon content can be easily reduced to levels lower than 0.03%. Subsequent additions of aluminum and lime may help reducing any oxidized chromium and at the same time these additions act as entrappers for the unwanted sulfur in the melt. The melt is then poured into ingots that later may be vacuum electro-slag remelted (ESR). More sulfur and other impurities are removed from the ingot through the slow ESR process under the cover of a molten salt slag. The higher purity ingots that result from the ESR process are homogenized for about 24 h at temperatures that vary depending on the type of alloy (e.g., amount and type of alloying elements), but it could be as high as 1200°C. After homogenization, the ingots are forged or cogged to transform the original cast microstructure into a more homogeneous wrought microstructure. The forging temperature and the amount of thickness reduction passes during forging depend on the type of alloy. Annealing or soaking is generally required in between reduction passes to avoid cracking, since most Ni alloys contain a high amount of alloying elements that tend to precipitate or segregate as the temperature drops below a certain limit. Depending on the final desired product, for example, the alloy may be hot rolled into plates and then cold rolled into sheet product. A final solution annealing of the sheet may be required before the product leaves the mill. The final annealing produces recrystallization in the material, and the final grain size will depend on the amount of deformation introduced during forging and also the time and temperature of the annealing treatment. The last anneal temperature will also depend on the type of the alloy; it is generally in the order of 850°C for alloys such as 400 and could be approximately 1100°C for alloys such as C-276.

Ni alloys are austenitic materials and therefore have similar fabrication characteristics as the austenitic stainless steels. Ni alloys have high ductility and may be formed by deep drawing. Ni alloys possess great toughness and significant work-hardening properties. These alloys can be joined readily through welding. Many Ni alloys are used as over-alloyed materials to join austenitic stainless steels and as weld overlays on less corrosion resistant alloys such as carbon steel. It is sometimes reported that Ni alloys are difficult to machine since they are gummy and difficult to cut. Variables that should be taken into account to minimize machinability problems include feed rate, cutting depth, machine horsepower and rigidity (no vibration), tool size, tool, and speed. Some Ni alloys such as Ni-200 and alloy 400 are better machined in the work-hardened condition (Nickel Institute 2010). The machinability of Ni alloys improves in the following order: Ni-200 > 400 > 600 > C-276, that is, when the alloys contain a higher amount of alloying elements (such as C-276), the machining is more difficult than for almost pure Ni (such as Ni-200). For the age-hardenable alloys (such as alloy 718), it is recommended to do a rough machining before the strengthening thermal treatment and then finish machining after aging since a size contraction of about 0.07% takes place upon thermal aging, which should be taking into consideration during rough machining (Nickel Institute 2010).

All the Ni alloys' primary metal producers offer detailed data sheets containing the general chemical, physical, and mechanical characteristics as well as their corrosion behavior, fabrication characteristics, joining, and thermal treatments. The data sheets also contain the list of applicable specifications for plate, tube, and welding products, for example, including ASTM, ASME, ISO, DIN, VdTüV, and AWS.

7.4.1 Mechanical

Figure 7.2 shows comparatively the yield strength for non-cold-worked, non-thermally aged CRA. The values of stainless steel type 316 and 304 are added for comparative purposes. Figure 7.2 shows that the yield strength for most of the Ni alloys is in the band from 200 to 400 MPa. It is obvious that the addition of alloying elements increases the yield strength of Ni alloys. For example, the yield strength of Ni-200 is doubled by adding 28% Mo to produce alloy B-2. The yield strength of all the C family of alloys (C-276, C-22, C-2000, 59, and 686) is approximately the same. Figure 7.2 also shows that the yield strength of most of the Ni alloys is similar to that of the popular austenitic stainless steels. The yield strength of the alloys listed in Figure 7.2 can be increased by cold work only, that is, except for alloy 625, they are

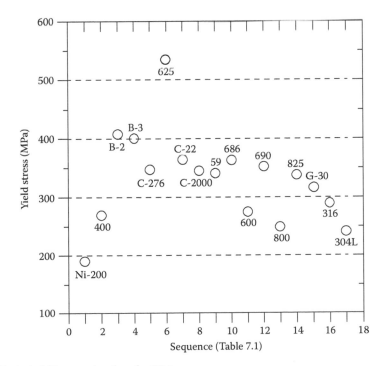

FIGURE 7.2 Typical yield strength values for CRAs.

not precipitation hardenable. Table 7.1 shows that the yield strength of precipitation-hardenable Ni-301 and K-500 is approximately 3–4 times higher than for the nonprecipitation-hardenable cousins (e.g., Ni-200 and 400).

One of the requirements for HTA is that they possess enough strength for the desired high-temperature application. This can be measured as the value of the yield strength as a function of the temperature or as the resistance to creep at temperature under a given applied stress. The primary metal producer's data sheet for each alloy also includes a large variety of other information such as the effect of long-term exposure at higher temperature on mechanical properties such as the impact energy.

7.4.2 Physical

The typical physical properties of the Ni alloys such as density, melting point or melting temperature range, specific heat, electrical resistivity, and thermal conductivity can readily be found in the product specification sheets for each alloy from the primary metal producers.

7.4.3 Thermal

See Section 7.4.2.

7.5 Structure

Ni is fcc and has an austenitic or gamma (γ) microstructure. Ni can dissolve a large amount of alloying elements while still maintaining its austenitic microstructure (Heubner 1998). Some Ni alloys (e.g., 625, 718 or 945) rely on the precipitation of particles of second phases to provide mechanical strength. These second phases are called gamma prime and gamma double prime. Other alloys such as C-22HS may rely on a long-range ordering for their strengthening mechanism.

7.5.1 Atomic or Molecular Bonding

Ni alloys are solid solutions of most metals. A few alloys may contain the metalloid Si and nonmetallic impurities such as C, S, and N. The atoms share free electrons through a metallic bonding, which provides with the specific characteristics such as mechanical ductility and thermal and electrical conductivity.

7.5.2 Microstructure

Ni alloys are solid solutions with an fcc or gamma microstructure. All the CRAs and most of the HTAs consist only of a single fcc phase. Some alloys will have randomly dispersed nonmetallic inclusions such as oxides, metal sulfides, or silicates. These inclusions were formed in the original melt and do not change during annealing studies. A few HTAs under especial heat-treated condition may have a small volume fraction (less than 1%) of a secondary strengthening phase that could be the result of precipitation (such as a carbide or gamma prime) or long-range ordering (such as in C-22HS).

7.6 Degradation Principles for Material (Chemical, Electrochemical, and/or Photochemical)

Like most of the other alloys, nickel alloys are susceptible to degradation since the elements in the alloy have a favorable free energy to return to their mineral natural state (as oxides, sulfides, salts, etc.). As the main alloying element, nickel is less reactive than iron; therefore, nickel alloys are more stable than, for example, a standard steel. The main mode of degradation is electrochemical, and little or no chemical or photochemical degradation would occur. Ni alloys are high-end alloys or CRAs; therefore, their natural degradation may be extremely slow because both the thermodynamic driving force for degradation and the kinetics of the corrosion processes are very slow. Because of its high corrosion resistance and degradation stability, alloy C-22 (Table 7.1) was a candidate for fabricating the container of the nuclear waste repository in the Yucca Mountain project (Gordon 2002), where it was expected to be stable for thousands of years.

7.6.1 Thermodynamics

Most engineering nickel alloys have three or more alloying elements, each with different nobility. Nickel in itself has a standard electromotive force reduction potential of −0.250 V in the standard hydrogen electrode (SHE) scale, where aluminum is −1.662 V and gold is +1.498 V (Jones 1996). An electromotive force of −0.250 V says that nickel can be oxidized by the hydrogen ion even though the driving force is not large. Nickel is more noble than iron (−0.447) but less noble than copper (+0.342). The reduction electromotive force for other typical alloying elements is Cr^{3+} to Cr = −0.744, MoO_2 to Mo = −0.15, and WO_2 to W = −0.12. Some nickel alloys such as Ni-200, 400, B-2, and B-3 may rely on their basic thermodynamic stability to protect against corrosion since they do not form protective passive films in most industrial applications; however, most of the alloys containing Cr offer corrosion resistance by the development of a passive film on the surface that acts as a barrier against the environment.

7.6.2 Kinetics

Nickel alloys, as other industrial engineering alloys such as stainless steels, have widespread applications as CRAs mainly due to its kinetic behavior rather than to pure thermodynamics properties. For example, Ni–Mo alloys such as B-2 have low corrosion rates in highly aggressive hot and reducing hydrochloric acid because the i_0 discharge current for hydrogen on Mo metal is very low. That is, the addition of Mo to the Ni alloy reduces the reduction current of H^+ to H_2, and therefore, the oxidation current (and corrosion rate) of the metal is also reduced (Lizlovs and Bond 1971). Another typical kinetic response of the alloying elements is on the effect of Cr by the formation of Cr_2O_3 rich film on the surface of the alloy, therefore protecting against further degradation. Several Ni alloys contain 15% or higher content of Cr, therefore imparting passivation in most industrial applications. The passive effect of chromium oxide is not only protective in aqueous condensed systems but also in high-temperature gaseous applications.

7.7 Degradation of Specific Systems (Various Alloys, Compositions, etc.)

Since there are two groups of nickel alloys (CRA and HTA), their degradation mechanism will be discussed separately, mainly because the corrosion degradation processes are different.

7.7.1 Corrosion-Resistant Alloys

Table 7.1 shows a condensed list of nickel alloys used for corrosion-resistant applications. Nickel alloys in general are more resistant to corrosion than the common austenitic stainless steels. Also the family or variety of nickel alloys is larger and therefore targeted to more specific applications than the stainless steels. For example, Ni-200 is almost used exclusively for caustic solutions because it is the best alloy for that type of environment and also because Ni-200 is not as good as other alloys for acidic solutions. Other alloys such as C-276 are more versatile, that is, they behave rather well in several types of environments; but it may not be the best alloy for a specific environment. Therefore, the degradation mechanism and corrosion behavior of each alloy group will be discussed separately, and in each group the three types of degradation (general corrosion, localized corrosion, and SCC) will also be discussed separately.

7.7.1.1 Commercially Pure Nickel

Commercially pure nickel (such as Ni-200 in Table 7.1) is used almost exclusively for highly caustic solutions at temperatures in the order of 200°C of higher. There is no better material for hot 50% caustic soda (NaOH) solution than commercially pure nickel. The corrosion rate of Ni-200 in 50% NaOH is negligible up to the boiling point; therefore, Ni-200 can be used for all concentrations of caustic soda and caustic potash at all temperatures up to the molten state. The resistance is due to the formation of a black nickel oxide film on the surface that protects the alloy for extended periods of time (Special Metals 2010). Figure 7.3 shows the corrosion rate by weight (mass) loss of several commercial engineering alloys in boiling 50% NaOH. It is evident that the corrosion rate of the alloys decreases as the amount of Ni in the alloys increases. It should also be mentioned that weight loss may not be the most appropriate way to rate the corrosion performance in caustic solutions since some alloys may suffer intergranular attack (e.g., 600) and others may suffer dealloying (e.g., C-276), reducing further their performance compared to Ni-200. Some alloys such as C-276 may also suffer SCC in hot caustic solutions.

Ni-200 is also used as an atmospheric material both in industrial and marine environments where the metal may develop a pseudo protective salt-based film on the surface (generally nickel sulfate) (Special Metals 2010). Ni-200 also has excellent corrosion resistance in presence of flowing seawater but may suffer under-deposit localized corrosion in presence of stagnant seawater (Special Metals 2010). Ni-200 resists corrosion by cold reducing acids such as sulfuric, hydrochloric, and phosphoric. If the acids are aerated or contain oxidizing salts (such as cupric and ferric ions) and the temperature is increased above ambient, the corrosion rate of Ni-200 may increase significantly. Ni-200 has good resistance to organic acids, especially fatty acids if aeration of the acidic solutions is not high. Ni-200 can also be used in presence of nonoxidizing salts such as chlorides and sulfates.

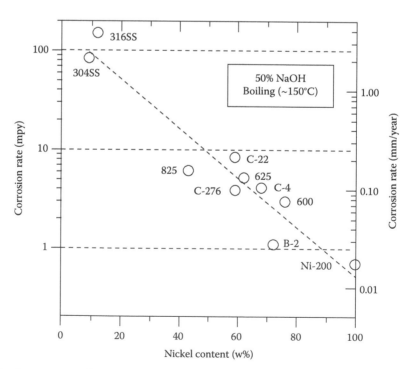

FIGURE 7.3 Corrosion rate by weight loss of commercial alloys in boiling 50% NaOH. The corrosion rate decreases as the Ni content in the alloy increases.

Ni-200 is not known to suffer pitting corrosion but may be susceptible to under-deposit corrosion (crevice corrosion), for example, in presence of stagnant seawater. Ni-200 does not suffer SCC in presence of hot chloride salts like the austenitic stainless steels. Ni-200 could be susceptible to hydrogen embrittlement under specific conditions, for example, if Ni-200 is cold worked and galvanically coupled to carbon steel in acidic media. For example, when Ni-200 specimens were deformed at 6.7×10^{-5} s^{-1} in air, the reduction in area was approximately 70%, but when the straining was carried out in 0.1 M N H_2SO_4 solution at −1000 mV SCE, the reduction in area was approximately 35%. In air, the fracture mode of Ni-200 was dimpled, but in the H_2SO_4 solution the fracture mode was a mixture of quasi cleavage and intergranular cracking showing the effects of hydrogen-induced cracking (Lee and Latanision 1987).

7.7.1.2 Nickel–Copper Alloys

Nickel–copper alloys (e.g., alloy 400, Table 7.1) are used widely for corrosion resistance in the chemical process industry and in marine applications. Typical uses include valves, pumps, and equipment such as vessels and piping handling chemicals. Alloy 400 is resistant to corrosion in reducing acids such as sulfuric and hydrochloric at near ambient temperature, and it is more resistant to oxidizing acids than higher copper alloys. Alloy 400 is resistant to corrosion in flowing seawater, but it may suffer localized corrosion in stagnant seawater. As Ni-200 is used almost exclusively for caustic solutions, alloy 400 is the workhorse alloy to handle hydrofluoric acid, that is, currently alloy 400 is the best alloy in the market for hydrofluoric acid application. However, even alloy 400 may be attacked by hydrofluoric acid in the presence of oxygen or other oxidizing species such as cupric ions. Figure 7.4 shows the cross section of two coupons tested in the liquid phase of 20% HF at 93°C. The amount of oxygen in the system is not known, but it is assumed to be low. The specimen of 20Cb-3 was exposed for 168 h and the specimen of alloy 400 was exposed for 336 h (Rebak et al. 2001). Alloy 400 did not experience evident corrosion or internal penetration, while the stainless steel specimen suffered severe corrosion, including external attack, cracking, and the formation of voids in the internally damaged area. However, when alloy 400 was tested in the same 20% HF at 93°C environment in the vapor phase, it suffered severe intergranular attack (Rebak et al. 2001).

As in the case of Ni-200, alloy 400 does not seem to be highly susceptible to SCC probably because of its low mechanical strength and high ductility (Table 7.1). Alloy 400 was found to be susceptible to SCC in acidic solutions containing mercury salts, in liquid mercury, in hydrofluoric acid, and in fluosilicic acid (The International Nickel Company, 1968). The age-hardenable alloy K-500 is used in seawater application mainly as shafts and bolting due to its high strength; however K-500 may be susceptible to

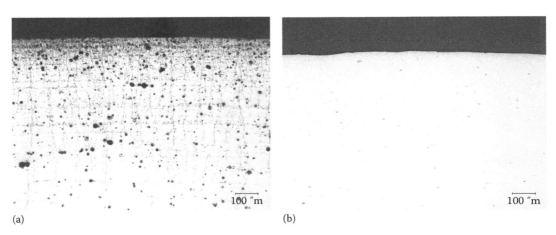

(a)   (b)

FIGURE 7.4 Cross section of corroded coupons after exposure to the liquid phase of 20% HF at 93°C. Magnification x100. (a) Alloy 20Cb-3, 168 h and (b) Alloy 400, 336 h.

HE (Pound 1998). Alloy K-500 has also been used extensively for sour service in oil field applications (Krishnan et al. 2009). Observations of several oil well field cracking failures in K-500 were reported, mainly following an intergranular cracking mode. These failures may be attributed to a hydrogen embrittlement mechanism probably due to cathodic protection or because the alloy was coupled to a less corrosion-resistant carbon steel (Krishnan et al. 2009).

7.7.1.3 Nickel–Molybdenum Alloys

Nickel–molybdenum alloys or B-type alloys (e.g., B-2 and B-3 in Table 7.1) were especially designed to handle hot reducing acids such as hydrochloric at all concentrations and temperatures. Hydrochloric acid is one of the most aggressive industrial acids, and only few materials (like glass and tantalum) can handle the exposure to this acid at boiling and higher temperatures. The B-type alloys have an excellent corrosion resistance in hot hydrochloric acid, mainly because of the effect of Mo that retards the hydrogen cathodic evolution reaction, and therefore, the oxidation reaction is also remarkably low. Figure 7.5 shows the corrosion rate of several commercial alloys in boiling 5% HCl as a function of the Mo content in the alloy, and it is clear that as the amount of Mo increases, the corrosion rate decreases approximately three orders of magnitude. B-2 and B-3 alloys can also be used to fabricate equipment for sulfuric, formic, acetic, phosphoric, and other nonoxidizing acid service. B-2 and B-3 is not recommended to be used in nitric acid or reducing acids contaminated with oxidizing metallic ions such as cupric or ferric. B-type alloys also have high corrosion rates in aerated acids such as HCl and H_2SO_4.

Since B-type alloys do not form a passive film on the surface, they are not susceptible to the classical pitting corrosion. However they may be susceptible to localized attack if coupled locally to a passive metal or graphite, by a mechanism of galvanic corrosion. When alloy B-2 and, to a lesser extent, alloy B-3 are exposed to temperatures in the range 550°C–850°C, they lose ductility due to a solid phase transformation that forms ordered intermetallic phases such as Ni_4Mo. The precipitation of these ordered phases changes the deformation mechanisms of the alloys making them susceptible to EAC such as hydrogen embrittlement (Agarwal et al. 1994, James et al. 1996). The ordered phases can also form in the heat-affected zones (HAZs) of welds rendering it susceptible to environmental cracking. For example, it was reported that B-2 alloy failed by intergranular SCC in the HAZ when exposed to organic

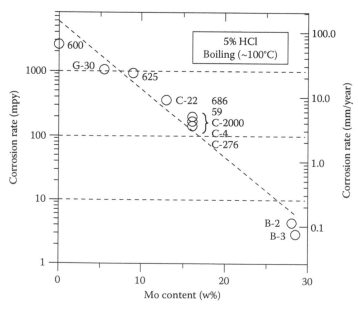

FIGURE 7.5 Corrosion rate by weight loss of commercial alloys in boiling 5% HCl. The corrosion rate decreases as the Mo content in the alloy increases.

solvents containing traces of sulfuric acid at 120°C (Takizawa and Sekine 1985). Laboratory tests were carried out to determine the susceptibility to EAC of B-type alloys in acidic solutions (Nakahara and Shoji 1996). At cathodic potentials (100 and 400 mV below the free corrosion potential), intergranular cracking was reported only for the aged (sensitized) alloys. Since the amount of intergranular brittle cracking increased at the lower applied cathodic potential, this environmentally induced cracking was attributed to hydrogen embrittlement (Nakahara and Shoji 1996).

7.7.1.4 Nickel–Chromium–Molybdenum Alloys

Nickel–chromium–molybdenum is one of the largest families of nickel alloys especially designed to contain all the beneficial alloying elements at their maximum possible concentration. Nickel is there to protect against caustic solutions and chloride cracking, molybdenum is there to give low corrosion rate in reducing hot acids such as hydrochloric, and chromium is there to offer protection against oxidizing acids or oxidizing salts such as cupric and ferric. The first and still most widely used wrought alloy in this family is C-276 (Table 7.1). C-276 is the alloy known popularly as "Hastelloy." Other alloys in the series are C-22, C-2000, 59, and 686 (Table 7.1). C-4 was once a common alloy but now it may be difficult to find in the market. Alloy 625 sometimes is included in the C family, but it may not have enough Mo (9%) to qualify as a C alloy. Ni–Cr–Mo alloys have the appropriate elements to offer outstanding resistance to localized corrosion such as pitting corrosion and crevice corrosion. One parameter that many use in several industries (e.g., chemical process industry, oil and gas) to rate the resistance of an alloy to localized corrosion is the pitting resistance equivalent number (PREN):

$$PREN = Cr + 3.3\left(Mo + \frac{W}{2}\right) + 16N \tag{7.1}$$

where the symbols represent the weight percent of each element in the alloy. Since the solubility of nitrogen (N) in nickel alloys is much lower than in stainless steels, its contribution to the PREN is much lower in nickel alloys than in stainless steels. The element W is added to Equation 7.1 even though it has beneficial effect only in the presence of Mo (Rebak and Crook 1999). Some authors prefer not to add W to the PREN equation (Agarwal 2000). Table 7.3 shows the typical PREN values for CRAs that are used in environments that could cause localized corrosion or SCC. The use of PREN to rank Ni alloys is highly resisted by some scientists since it was initially developed for stainless steels, and they argue that the comparison is not fair since Ni can dissolve more alloying elements than iron. Table 7.3 shows that the highest values of PREN are 75–80 for the Ni–Cr–Mo alloys.

Ni–Cr–Mo alloys are widely used in the pharmaceutical and chemical processing industry to fabricate multipurpose components that can be used sequentially with many different aggressive media. Such components include heat exchangers, reaction vessels, evaporators, and transfer piping. C-276 is an ideal alloy for corrosive media where the conditions are not well known or may suffer frequent "upset" conditions, that is, C-276 is a forgiving alloy if the acid concentration or temperature may increase temporarily or a contaminant may appear in the stream. Ni–Cr–Mo alloys such as C-276 are used to handle a variety of the most severely corrosive media including sulfuric acid, hydrochloric acid, phosphoric acid, organic acids, wet chlorine gas, hypochlorite, and chlorine solutions (Special Metals 2010). C-276 is also recommended for seawater service even under creviced or stagnant conditions where other alloys may suffer severe localized attack. C-276 is also a popular alloy in pollution control equipment such as flue gas desulfurization (FGD) systems, where corrosive gases resulting from the burning of dirty fossil fuels are washed with water and other chemicals to avoid their release to the environment. C-276 is used to construct scrubbers, liners, and ducting. In many FGD systems, the solution not only is acidic but also contains a large amount of chlorides, which may promote localized corrosion such as pitting corrosion or crevice corrosion. C-276 has also been for decades one of the most CRAs for handling the most aggressive corrosive conditions in downhole applications such as tubing in the oil and gas industry.

TABLE 7.3 PREN Number for Selected CRAs

Alloy	Cr	Mo	W, N	PREN
304	19	0	—	19
316	18	2.5	—	26
2205	22	3	0.14 N	34
2507	25	4	0.27 N	43
AL6XN	20	6.3	0.22 N	44
C-276	16	16	4 W	75
625	21	9		51
C-22	22	13	3 W	70
C-2000	23	16		76
59	23	16		76
686	21	16	4 W	80
C-22HS	21	17		77
600	16			16
690	29			29
725	21	8		47
X-750	15			15
800	21			21
825	21	3		31
945	21	3		31
G-30	30	5		47
33	33	1.6		38

Currently, newer high-strength alloys such as C-22HS and 945 may replace some of the applications of C-276 since this alloy can only be made stronger by cold working. One of the advantages of C-276 is that the alloy does not require heat treatment after fabrication such as welding. Newer and improved Ni–Cr–Mo alloys such as C-2000, 59, and 686 find similar applications as C-276.

Ni–Cr–Mo alloys such as C-276, C-22, 59, 686, and C-2000 are not especially susceptible to pitting corrosion but may be prone to crevice corrosion under anodic polarization in high-temperature concentrated chloride solutions under tight occluded conditions, where crevice corrosion develops by the local formation of hydrochloric acid. The crevice corrosion behavior of C-22 was extensively studied for several years at many universities and national laboratories as part of the Yucca Mountain project in the United States. It is now understood that the crevice susceptibility of C-22 (and other Ni–Cr–Mo alloys) depends strongly on several variables such as temperature, chloride concentration, applied potential, and presence of inhibitors such as nitrate in the solution (Rebak 2005, Carranza 2008).

7.7.1.5 Nickel–Chromium–Iron Alloys

Nickel–chromium–iron alloys include a large and diverse group of materials from the very popular alloy 600 (also called commonly "Inconel"), alloy 800, and the G family of alloys. Some of these alloys, for example, G-30 and 825, may also contain some Mo. Alloy 600 may be the nickel alloy that has the largest sale volume in the world since it is used widely in the food industry, the nuclear industry, and the chemical process industry, for thermocouple sheaths, and even for high-temperature applications such as in heat treatment. Figure 7.6 shows the corrosion rate of several commercial alloys in boiling 10% nitric acid. Nitric acid is oxidizing; therefore, alloys such as 200, 400, and B-2 do not perform well since they do not contain Cr. Also, Mo is a detrimental alloying element for nitric acid service. However other alloys such as 825 and G-30 have excellent corrosion resistance because they develop a protective chromium oxide film on the surface. Figures 7.5 and 7.6 show clearly poor corrosion behavior of G-30 in

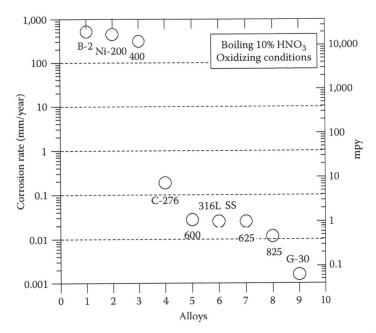

FIGURE 7.6 Corrosion rate by weight loss of commercial alloys in boiling 10% HNO₃. The corrosion rate decreases as the Cr content in the alloy increases.

HCl (reducing acid) and clearly good behavior in nitric acid (oxidizing), just the opposite of the behavior for B-2 alloy. In both figures, Ni is the base element but the good corrosion resistance is provided by the alloying addition, Mo in the case of HCl and Cr in the case of nitric acid.

The Ni–Cr–Fe alloy group is not as resistant to localized corrosion and environmental cracking as the Ni–Cr–Mo group (Agarwal 2000, Rebak 2000a). Table 7.3 shows that the PREN numbers of some of the Ni–Cr–Fe alloys are similar to those of some stainless steels. Alloy 600 has been found to suffer SCC in high-temperature pure water (>300°C) both in service and in the laboratory. Due to its importance for the nuclear industry, the stress cracking of alloys 600 and 690 in pure water and in caustic solutions has been extensively researched in the last three decades (Szklarska-Smialowska and Rebak 1996, Staehle and Gorman 2003, 2004), and thousands of technical papers have been published in this subject. The susceptibility to cracking of alloys 600 and 690 depends strongly on environmental factors such as temperature, level of tensile stresses, deformation rate, presence of hydrogen gas, solution pH and electrochemical potential, and metallurgical factors, such as presence of minor alloying elements (impurities), the amount of cold work, and heat treatment (intragranular or intergranular carbides). Alloy 690, which has double the amount of chromium in alloy 600, has been found to be more resistant than alloy 600 to high-temperature cracking in pure water and in caustic solutions. Alloy 800 is also used in the nuclear power generation. In steam generator applications, alloy 800 is generally more resistant to cracking than alloy 600, probably because of the intermediate Ni composition in alloy 800 (Staehle and Gorman 2004). Alloy 825 is more resistant to SCC in chloride solutions than 316 stainless steels (S31600) due to the higher content of nickel in alloy 825. Slow strain rate tests and U-bend tests have shown that alloy 825 was susceptible to transgranular SCC in 45% MgCl₂ solutions at temperatures above 146°C. Alloy 825 is used extensively in the oil and gas production in sour wells; however, the performance of other nickel alloys such as C-276 and G-50 (N06950) is still superior of that of 825 (Hibner and Tassen 2000). These nickel alloys are used mainly in the cold-worked condition for increased strength. Environmental factors that may affect the stress-cracking performance of alloy 825 (and other alloys) in oil and gas wells include temperature, amount of chloride, and the presence of hydrogen sulfide gas (Hibner and Tassen 2000). Data on the SCC behavior of G-30 alloy are scarce. It has been reported that G-30 components used in the industrial production of hydrofluoric acid suffered cracking (Rebak 2000b). U-bend specimens of G-30 alloy did not crack after exposure for 500 h in 45% $MgCl_2$ solution at 154°C. It has been found that G-30 as well as other nickel alloys would suffer cracking in the aggressive conditions encountered in super critical water oxidation (SCWO) treatments.

7.7.2 High-Temperature Alloys

Numerous applications of industrial alloys operate at temperatures higher than 500°C. They include coal gasification, refineries, waste incineration, and gas turbines (Gleeson 2000). As mentioned before, besides the requirements of fabricability and corrosion resistance required for the low temperature or CRAs, the HTAs also need to be microstructurally stable (not to form detrimental second phases) and possess enough strength and creep resistance. For high-temperature applications, mostly in gaseous atmospheres, the corroding environments are grouped by the type of degradation that they cause to the alloy in service. The most common degradation processes are (1) oxidation, (2) carburization, (3) sulfidation, (4) nitridation, (5) halogenation, and (6) molten salt and hot corrosion. Oxidation is the most important high-temperature degradation mode of Ni alloys (Lai 1990). In general Cr provides the resistance to oxidizing environments, and the benefit of Cr is enhanced if the alloy also contains iron (Mankins and Lamb 2003). Since Ni does not form stable carbides, high-Ni alloys are resistant to carburization processes. High Ni content is also important for alloy stability (maintaining the initial austenitic structure) in prolonged high-temperature applications and for resistance to nitridation (Mankins and Lamb 2003). Cr is the element that generally gives protection against sulfidation. The sulfidation, carburization, and nitridation attacks tend to be more severe in reducing environments (no oxygen) than in environments that contain oxygen.

7.7.2.1 Oxidation

Oxidation is the most common degradation mechanism in the application of HTAs. The oxidation process is controlled by the partial pressure of oxygen and by the presence of other species such as hydrogen (H_2), water vapor (H_2O), and the carbon monoxide to carbon dioxide ratio (CO/CO_2). The most important variable controlling oxidation is the service temperature, since the oxidation rate increases significantly with the temperature. For example, Eiselstein and Skinner (1954) showed that several stainless steels and Ni alloys are resistant to cyclic oxidation in air at 760°C and 871°C. However when the engineering alloys were tested at 982°C, scaling by oxidation was pronounced for the stainless steels, and only alloys such as Nimonic 80 and Incoloy 800 performed well since they contained more than 35% nickel and in the order of 20% chromium (Eiselstein and Skinner 1954, Lai 1990). Smaller additions of Al to alloys containing Cr improve their oxidation response, since Al needs a lower activity of oxygen to form oxide than Cr (Lai 1990). Alloys such as 601 and 214 (Table 7.2) rely on the Cr_2O_3 and Al_2O_3 dual mechanism of protection, especially at temperatures higher than 1000°C when Cr_2O_3 converts into the volatile CrO_3 compound. It is also claimed that the good performance of alloys such as 214 is due to the presence not only of aluminum but also of rare earth yttrium (Y), which may act as seeds or pegs to avoid oxide spallation from the metal surface (Elliott 1990). Testing for oxidation resistance in the laboratory should always include cycling temperatures, since this simulates better actual service conditions (Lai 1990). Figure 7.7 shows the average affected metal depth by oxidation at 980°C, 1095°C, and 1150°C for several engineering HTA (Table 7.2) exposed in a tube furnace to flowing air for 1008 h. Data for stainless steels are also included for reference. For each alloy, as the temperature increased, the penetration damage by oxidation increased. For the Ni alloys, the largest damage by oxidation was for the alloys 625 and HR-160, and the lowest damage was for alloy 214. It is likely that the good behavior of alloy 214 was due to the formation of an alumina scale under the chromia scale, which protected the alloy from rapid oxidation attack (Elliott 1990). The high oxidation rate of stainless steel 316 at T > 1000°C could be

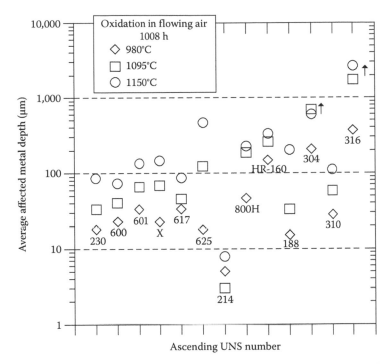

FIGURE 7.7 Oxidation penetration damage for engineering HTAs exposed to flowing air at three different temperatures. (From Haynes International Inc., *Tech Brief H-3022C, High Temperature Alloys*, Haynes International Inc., Kokomo, IN, 2009.)

due to the presence of Mo, which may promote the catastrophic oxidation of the 316SS by the formation of unstable MoO_3 compound (Elliott 1990).

7.7.2.2 Sulfidation, Carburization, and Nitridation

Sulfur is generally present as a contaminant in fossil fuels, mainly in oil and in coal. Sulfur could be present in high-temperature applications in the presence of oxygen as sulfur oxides (SO_2 and SO_3) or in the absence of oxygen as hydrogen sulfide (H_2S). It is generally accepted that when oxygen is present in the system, the HTA would form a protective oxide scale on the surface and therefore is more resistant to sulfur attack than in reducing conditions (Lai 1990). Especially in the absence of a protective oxide scale, sulfur can diffuse into the metal and react with the alloying elements forming sulfides. The *sulfidation* reaction is especially detrimental when the sulfide formed is liquid, such as in the case of nickel sulfide (Ni_3S_2), which has a relatively low melting point of 635°C (Lai 1990). Phase stability diagrams have been constructed to predict regions of stability of oxide phases, mixed oxide and sulfide phases, and the more detrimental sulfide phases based on thermodynamic calculations according the partial pressures of oxygen, hydrogen sulfide + hydrogen, and sulfur (Shifler 2003). The addition of silicon to Cr-containing alloys provides a similar benefit as does Al, that is, Si may provide a "glazing" effect protecting against attacks such as carburization and sulfidation (Elliott 1990).

Carburization is a form of nickel alloy degradation that is generally different from the other elements (e.g., oxygen and sulfur) since the material does not waste away. Carburization occurs by diffusion of carbon into the metal resulting in the formation of carbides that alter the chemical and mechanical properties of that alloy (Grabke 1998). Carburization is generally controlled by the oxygen and carbon activity in the atmosphere (Shifler 2003). Carburization may be partially inhibited in the presence of a high oxygen activity. Carbon mainly diffuses into the nickel alloy following intergranular paths and reacting especially with elements that form stable carbides such as chromium, niobium, tantalum, molybdenum, tungsten, and titanium. The most common ones in high-temperature nickel alloys are the chromium carbides of which there are three forms: $Cr_{23}C_6$, Cr_7C_3, and Cr_3C_2, and they may form at different depths from the surface. Where the activity of carbon is the lowest, the formation of $Cr_{23}C_6$ is favored; and at the surface, where the activity of carbon is the highest, the formation of Cr_3C_2 may be favored (Lai 2007). If the nickel alloy forms a protective Cr_2O_3 oxide on the surface, this will act as a barrier for carbon diffusion (Lai 2007). High-nickel alloys are more resistant to carburization than Fe–Cr–Ni or stainless steel alloys since the solubility and diffusivity of carbon in nickel alloys is lower than in stainless steels. Silicon provides additional resistance to carburization only in environments where SiO_2 is allowed to form, that is, if SiO_2 does not form on the surface of the alloy, Si has no effect on carburization resistance (Lai 2007). Another beneficial element in resisting carburization is aluminum, since alloys containing only small amounts of Al such as 601 and 617 proved to be more carburization resistant than alloys that contain no Al (Lai 2007). Alloys 214, X, and 601 were tested side by side for carburization resistance at 980°C for 55 h in Ar–5%H_2–5%CO–5%CH_4 ($a_C = 1.0$, $p_{O_2} = 9 \times 10^{-22}$ atm), and cross sections of these coupons after testing showed that alloy 214 was clearly more resistant to carburization than either X or 601, probably because 214 contained 4.5% Al (Table 7.2).

Metal dusting is a special case of catastrophic carburization when the activity of carbon in the mostly reducing atmosphere is much higher than one and the attack proceeds mainly on the surface of the alloy creating uneven cavities or pits (Klarstrom et al. 2001, Lai 2007). Metal dusting generally occurs in the temperature range of 430°C–900°C (Lai 2007) and is manifested as a dust on the surface containing carbon and metal particles (Klarstrom et al. 2001). Coupons of several nickel alloys including 214, 230, HR-160, 601, and 800H (Table 7.2) were exposed for up to 10,000 h in a flowing gas mixture composed of 49% CO, 49% H_2, and 2% H_2O with $a_C = 18.9$ at a temperature of 650°C (Klarstrom et al. 2001). Severe metal dusting was reported for the 800H alloy, and as the amount of iron in the alloy decreased and the amount of nickel increased, the resistance to dusting increased. High chromium, molybdenum, and tungsten (such as in alloy 230) were cited as beneficial alloying elements in the resistance to dusting; however, the best dusting resistance was by HR-160, and the authors cited the presence of silicon in the alloy as the main contributing factor (Klarstrom et al. 2001). Surprisingly, the behavior of alloy 214 (with 4.5% Al) was not as good as that of HR-160 (Lai 2007).

Nitridation is a mechanism of degradation similar to carburization; however, it is not as pervasive as carburization. The nickel alloys that are commonly used to avoid nitridation degradation include alloys 600, 188, 230, and 214 (Gleeson 2000) (Table 7.2). The main beneficial element to combat nitridation is nickel itself (Gleeson 2000).

7.7.2.3 Hot Corrosion (Molten Salt) and Halogen Attack (Chloridation)

Hot corrosion is from a molten alkali sulfate salt attack of the nickel alloy exposed to high-temperature gases (Gleeson 2000, Shifler 2003, Shifler 2006). This type of attack may be exacerbated in the presence of chloride salts that dissolve the protective oxides from the surface of the component. Generally, two types of molten sulfate attack are identified: Type I occurs in the temperature range of 850°C–950°C where the activity of sodium oxide is an important factor in the destruction of the protective oxide. Type II occurs at the lower temperature of 650°C–750°C where the attack progresses by uneven pitting of the surface of the nickel alloy (Rapp 1986, Shifler 2006). The attack by alkaline molten sulfates is generally exacerbated by eutectic mixtures of sulfate salts that remain liquid at temperatures well below the melting temperature of the individual sulfates (Shifler 2006). Chromium is the best alloying element to protect against hot corrosion, but sometimes a high-chromium alloy is not enough for protection (Gleeson 2000).

Chloride salts are ubiquitous in many industries where high-temperature nickel alloys are used, such as in the processing (burning) of municipal waste. Some chloride salts such as zinc chloride ($ZnCl_2$) have melting points below 300°C, and when they form on the surface of the component, they tend to wash away the protective oxide scales (Shifler 2003).

7.8 Degradation Protection

7.8.1 Surface Treatments

Nickel alloys are seldom used with a special surface treatment. The typical passivating treatment that is performed on stainless steels, for example, after fabrication, is not needed for Ni alloys. Even for aggressive conditions such as in the chemical process industry, Ni alloys are generally used in the as-fabricated (machined) conditions. Also, nickel alloys such as Ni–Cr–Mo are used in the as-welded conditions, without post-weld heat treatment. Similarly, corrosion-resistant Ni alloys are not used in surface-hardened conditions such as from carburizing or nitriding treatments. In certain applications such as for internal components in reactors, Ni alloys such as weld metals of alloy 182/82/52 (which belong to the family of alloys 600 and 690) may be surface treated by peening (shot, water jet, or laser) to impart compressive stresses on the surface that may retard the onset of environmentally assisted cracking. The surface peening of nuclear reactor components is a rather new treatment that may be applied for replacement parts of the older reactors or for the newly constructed reactors.

7.8.2 Coatings

Coatings are not used for low-temperature corrosion-resistant Ni alloys. However, coatings may be used for high-temperature Ni alloys (superalloys) that may operate at temperatures higher than 900°C such as in turbine blades (Gleeson 2000). At the highest operating temperatures, the base structural alloys do not contain sufficient beneficial alloying elements such as Cr to offer a long-term protection. These coatings could be overlay welded or diffusion bonded and are generally aimed at protecting against oxidation and hot corrosion (Ravi 2003). One protective coating application method is pack cementation, where elements such as Cr, Al, or Si are deposited on the surface of the alloy to a thicknesses less than 100 μm. Other deposition methods include thermal spray or physical vapor deposition by which a protective coating of Ni, Cr, Al, and Y could be concurrently formed as a layer containing approximately 15%–25% Cr, 10%–15% Al, and 0.2%–0.5% Y in a matrix of Ni (Gleeson 2000). Additionally, for turbine applications, for example, a second layer may be applied over the metallic bond coating of NiCrAlY (Shifler 2006). This second layer is not metallic and is called a thermal barrier coating (TBC), consisting mainly of zirconia (ZrO_2) stabilized by approximately 8% of yttria (Y_2O_3) (Shifler 2006).

7.9 Specific Standardized Tests

There are no specific standards and tests geared solely toward evaluating Ni alloys—in general, when Ni alloys are tested for mechanical and corrosion resistance, the methods used are the same as those used for other alloys such as austenitic stainless steels.

7.10 Summary

Ni alloys are produced by dissolving alloying elements into a Ni gamma matrix. The commercially available Ni alloys may contain from over 99% Ni to less than 40% Ni. Since Ni can dissolve a large volume fraction and variety of other metals, Ni alloys can be tailored for highly specific applications. The most common alloying element for Ni is Cr. Other alloying elements include Mo, Fe, Co, Cu, W, Si, Nb, Ti, and Al. Ni alloys are considered some of the most CRAs available in the commercial market. The use of Ni alloys in industrial applications may be divided into (low temperature) CRAs and HTAs. Within each category, large families of Ni alloys exist. For example, CRAs may be targeted for use in caustic environments or where reducing acids or oxidizing acids are present; the specific application determines the selection of the alloying elements. Other alloys such as the Ni–Cr–Mo family contain a reasonable amount of all the beneficial alloying elements so that they could be considered multipurpose alloys.

While CRAs are mostly designed for their capacity to resist corrosion, HTAs need to have sufficient strength, thermal stability, and corrosion resistance in the environment of that application. High-temperature degradation mechanisms may include oxidation, sulfidation, carburization, hot corrosion, and chloridation.

Acknowledgments

I am grateful to my many professors, coworkers, and mentors both at universities and national laboratories and especially in the industry. I also want to acknowledge the patience of my husband Bill Wickline.

Nomenclature and Units

All self-described in the text—international units.

References

Agarwal, D.C. 2000. Nickel and nickel alloys, in R. Winston Revie, Ed., *Uhlig's Corrosion Handbook*, John Wiley & Sons Inc., New York, pp. 831–851.

Agarwal, D.C., U. Heubner, M. Kohler, and W. Herda. 1994. UNS N10629: A new Ni-28% Mo alloy. *Materials Performance*, 33(10), 64–68.

Carranza, R.M. 2008. The crevice corrosion of Alloy 22 in the Yucca Mountain Nuclear waste repository. *JOM*, 60(1), 58–65.

Eiselstein, H.L. and E.N. Skinner. 1954. *The Effect of Composition on the Scaling of Iron-Chromium-Nickel Alloys Subjected to Cyclic Temperature Conditions*, ASTM STP 165, West Conshohocken, PA, p. 162.

Elliott, P. 1990. *Practical Guide to High-Temperature Alloys*, NiDI Series 10056, Nickel Institute, Toronto, Ontario, Canada.

Gleeson, B. 2000. High-temperature corrosion of metallic alloys and coatings, in M. Schütze, Ed., *Corrosion and Environmental Degradation*, Wiley-VCH, Weinheim, Germany, pp. 174–228.

Gordon, G.M. 2002. F.N. Speller award Lecture: Corrosion considerations related to permanent disposal of high-level radioactive waste. *Corrosion*, 58(10), 811–825.

Grabke, H.J. 1998. *Carburization: A High Temperature Corrosion Phenomenon*, MTI 52, Materials Technology Institute of the Chemical Process Industries Inc., St. Louis, MO.

Haynes International Inc., 2009. *Tech Brief H-3022C, High Temperature Alloys*, Haynes International Inc., Kokomo, IN.

Heubner, U. 1998. *Nickel Alloys*, Marcel Dekker Inc., New York.

Hibner, E.L. and C.S. Tassen. 2000. Corrosion resistant OCTG's for a range of sour gas service conditions, Paper 00149, Corrosion/2000, NACE International, Houston, TX.

James, M.M., D.L. Klarstrom, and B.J. Saldanha. 1996. Stress corrosion cracking of nickel-molybdenum alloys, Paper 432, Corrosion/96, NACE International, Houston, TX.

Jones, D.A. 1996. *Principles and Prevention of Corrosion*, 2nd edn., Prentice Hall Inc., Upper Saddle River, NJ.

Klarstrom, D.L., H.J. Grabke, and L.D. Paul. 2001. The metal dusting behavior of several high temperature nickel based alloys, Paper 01379, Corrosion/2001, NACE International, Houston, TX.

Krishnan, K., J. Rooker, and G.B. Chitwood. 2009. Case-history of environmental cracking failures with alloy K-500 for downhole completion tools, Paper 09080, Corrosion/2009, NACE International, Houston, TX.

Lai, G.Y. 1990. *High-Temperature Corrosion of Engineering Alloys*, ASM International, Metals Park, OH.

Lai, G.Y. 2007. *High-Temperature Corrosion and Materials Applications*, ASM International, Metals Park, OH.

Lamb, S. 2000. *CASTI Handbook of Stainless Steels & Nickel Alloys*, CASTI Publishing Inc., Edmonton, AL.

Lee, T.S.F. and R.M. Latanision. 1987. Effects of grain boundary segregation and precipitation on the hydrogen susceptibility of nickel. *Metallurgical and Materials Transactions A*, 18A(9), 1653–1662.

Lizlovs, E.A. and A.P. Bond. 1971. Anodic polarization behavior of 25% chromium ferritic stainless steels. *Journal of the Electrochemical Society*, 118, 22–28.

Mankins, W.L. and S. Lamb. 2003. *ASM Handbook of Nickel and Nickel Alloys*, Vol. 2, ASM International, Metals Park, OH.

Nakahara, M. and T. Shoji. 1996. Stress corrosion cracking susceptibility of nickel-molybdenum alloys by slow strain rate and immersion testing. *Corrosion*, 52(8), 634–642.

Nickel Institute. 2010. nickelinstitute.org.

Pound, B.G. 1998. Effect of heat treatment on hydrogen trapping in alloy K-500. *Corrosion*, 54(12), 988–995.

Rapp, R.A. 1986. Chemistry and electrochemistry of hot corrosion of metals. *Material Science and Engineering*, 87(1987), 319–327.

Ravi, V. 2003. Pack cementation coatings, in S.D. Cramer and B.S. Covino, Jr., Eds., *ASM Handbook Volume 13A: Corrosion: Fundamentals, Testing, and Protection*, ASM International, Materials Park, OH, pp. 763–771.

Rebak, R.B. 2000a. Corrosion of non-ferrous alloys. I. nickel-, cobalt-, copper, zirconium- and titanium-based alloys, in R.W. Cahn, P. Haasen, and E.J. Kramer, Eds., *Corrosion and Environmental Degradation*, Vol. II, Wiley-VCH, Weinheim, Germany, p. 69.

Rebak, R.B. 2000b. Environmentally assisted cracking in the chemical process industry: Stress corrosion cracking of iron, nickel, and cobalt based alloys in chloride and wet HF services, in R. Kane, Ed., *Environmentally Assisted Cracking: Predictive Methods for Risk Assessment and Evaluation of Materials, Equipment and Structures*, ASTM STP 1401, ASTM, West Conshohocken, PA, p. 289.

Rebak, R.B. 2005. Factors affecting the crevice corrosion susceptibility of Alloy 22, Paper 05610 in Corrosion/2005 conference and exposition, NACE International, Houston, TX.

Rebak, R.B. and P. Crook. 1999. Improved pitting and crevice corrosion resistance of nickel and cobalt based alloys, in R.G. Kelly, G.S. Frankel, P.M. Natishan, and R.C. Newman, Eds., *Critical Factors in Localized Corrosion III*, Vol. 98-17, The Electrochemical Society, Pennington, NJ, p. 289.

Rebak, R.B., J.R. Dillman, P. Crook, and C.V.V. Shawber. 2001. Corrosion behavior of nickel alloys in wet hydrofluoric acid. *Materials and Corrosion*, 52(4), 289–297.

Shifler, D.A. 2003. High-temperature gaseous corrosion testing, in S.D. Cramer and B.S. Covino Jr. Eds., *ASM Handbook of Volume 13A: Corrosion: Fundamentals, Testing, and Protection*, ASM International, Material Park, OH, pp. 650–682.

Shifler, D.A. 2006. Factors that influence the performance of high-temperature coatings, Paper 06475, S.D. Cramer and B.S. Covino, Jr., Eds., *Corrosion/2006*, NACE, Houston, TX.

Special Metals. 2010. *Technical Information on Specific Alloys*, Special Metals Corporation, Huntington, WV.

Staehle, R.W. and J.A. Gorman. 2003. Quantitative assessment of submodes of stress corrosion cracking on the secondary side of steam generator tubing in pressurized water reactors: Part 1. *Corrosion*, 59, 931–994.

Staehle, R.W. and J.A. Gorman. 2004. Quantitative assessment of submodes of stress corrosion cracking on the secondary side of steam generator tubing in pressurized water reactors: Part 2. *Corrosion*, 60, 5–63.

Szklarska-Smialowska, Z. and R.B. Rebak. 1996. Stress corrosion cracking of alloy 600 in high-temperature aqueous solutions: Influencing factors, mechanisms, and models, in R.W. Staehle, J.A. Gorman and A.R. McIlree, Eds., *Control of Corrosion on the Secondary Side of Steam Generators*, NACE, Houston, TX, pp. 223–257.

Takizawa, Y. and Sekine, I. 1985. Stress corrosion cracking phenomena on Ni-Mo alloys in high temperature non aqueous solutions, Paper 355, Corrosion/85, NACE, Houston, TX.

The International Nickel Company. 1968. *Corrosion Engineering Bulletin CEB-5*, Inco, New York.

Further Readings

Cramer, S.D. and B.S. Covino Jr. 2003–2006. *Corrosion ASM Metals Handbook*, Vol. 13 (A, B and C), ASM International, Materials Park, OH.

Lamb, S. 2000. *CASTI Handbook of Stainless Steels & Nickel Alloys*, CASTI Publications, Edmonton, AL.

Schütze, M. 2000. *Corrosion and Environmental Degradation*, Wiley-VCH, Weinheim, Germany.

Winston Revie, R. 2000. *Uhlig's Corrosion Handbook*, The Electrochemical Society, John Wiley & Sons, New York.

Web pages of Nickel Institute, Haynes International, Special Metals, Krupp-VDM, Allegheny Technologies and other primary metal producers.

Crystalline Alloys: Copper

Marc A. Edwards
*Virginia Polytechnic
Institute and State
University*

James H. Michel
*Copper Development
Association Inc.*

8.1 Introduction

Copper and copper alloys are used in a multitude of applications in architecture, electrical, electronics, telecommunications, and plumbing tube and fittings. They have excellent corrosion resistance in most natural environments. However, under certain environmental conditions, copper and its alloys are impacted by issues associated with corrosion and materials degradation, resulting in reduced product lifetime and system failures. This chapter provides an overview of the current understanding related to environmental degradation of copper—failure modes, mechanisms, and mitigation strategies.

8.1.1 Copper and Copper Alloys in History and Antiquity

The numerous uses and applications of copper during antiquity hinted at its critical importance to modern civilization, technology, and engineering. Copper was the first metal utilized by mankind in relatively large quantities. The Sumerians used copper for artwork 5000–6000 years ago, and the first copper tubes to transport water were fashioned nearly 3000 years ago (Smith 1965). Copper was perhaps the first disinfectant used to treat drinking water as evidenced by Indian medical instructions from 2000 BC that recommended that water be "treated by boiling and 'by dipping into it a piece of hot copper' and then filtering it (Baker 1948)." The Romans later mastered the use of copper and its alloys for pumps, valves, and other intricate plumbing devices.

8.2 Background

8.2.1 Current Worldwide Demand and Utilization of Copper

Copper production and worldwide use has trended upward in the past century. Some recent declines have occurred during the great recession of 2008. In recent years, copper prices have ranged from $1.35 to $4.25 per pound. Worldwide mine production for 2007 was 15.4 million metric tons, and secondary and recycled sources were 2.7 million metric tons, which together account for a total world refinery capacity of 18 million metric tons. Chile accounts for approximately one-third of world production, followed by Peru and the United States as other major copper-producing nations. Current reserves indicate land-based resources exceed 3 billion metric tons, with more than 700 million metric tons in potentially mineable deep-sea nodules. Recycling of copper scrap accounts for about 31% or greater of U.S. supply.

8.3 Application

8.3.1 Uses

Copper and its alloys have unique combinations of conductivity (both thermal and electrical), durability, and corrosion resistance that make it highly desirable for numerous specialized applications (Tables 8.1 and 8.2). The most significant use for copper and its alloys occurs in the electrical and electronics fields, as the major use of copper is electrical wire for buildings, telephone, and communications cables. Other significant electronic uses include interface cables, motor windings, transformers, grounding, and lightning protection.

Heat exchangers make extensive use of copper's high conductivity in applications that include automotive radiators, offshore platforms and shipboard steam, and water systems. The excellent hydraulic capabilities and relatively high corrosion resistance of copper tube are exploited for use in water piping, hydraulic brake tubing, plumbing fittings and fixtures, and geothermal heat pump applications. Alloys' resistant to seawater and brackish water corrosion finds application in desalination and marine water transmission.

Architectural usage of copper has grown over the years, as copper roofing with characteristic patina represents an obvious application, but its use for interior and exterior wall cladding is also becoming more commonplace. Attractive and strong brass screws and fasteners are also used in both architectural and marine applications, and certain bronze bearings are used in machinery with specialized anti-seize characteristics. A final representative and well-established use takes advantage of the antifouling properties of copper, including use in boats and intake screens.

8.3.2 Challenges and Opportunities

The profile of copper utilization is unlikely to change dramatically in the years ahead, aside from minor adjustments arising as copper and competing materials fluctuate in price. Three issues associated with

TABLE 8.1 Copper and Copper Alloys, General Uses

Alloy UNS No./Common Name	Content	Properties
C10000–C15599 Copper	Min Cu = 93%	Excellent conductivity and ductility
C15600–C19599 High-copper alloys	Cu 96%–99.3%	Greater strength without significant reduction in conductivity
C20500–C28299 Brasses	Copper and zinc	Fair electrical conductivity, excellent forming and drawing, good strength
C30000 series Leaded brasses	Cu, 32%–39% Zn, 1%–3% Pb	Excellent machining
C40000 series Tin brasses	Cu, Zn, 0.5%–2% Sn	Good corrosion resistance combined with strength
C50000 series Phosphor bonzes	Cu, Sn, P, sometimes Zn/Pb	Excellent tensile, fatigue strength, corrosion resistance
C60000 series Aluminum bronzes	Cu, 2%–13% Al	Good strength and formability
C60000 series Silicon bronzes	Cu, Si	Suitable for welding
C60000 series Misc Cu–Zn Alloys	Cu, Zn, sometimes Al, Co, Mn, Ni	Wide range of properties, good corrosion resistance
C70000 series Cupronickels	Cu, Ni	Good forming qualities, high strength, excellent corrosion resistance
C70000 series Nickel silvers	Cu, Zn, Ni	Tarnish resistant, oxidation resistant, high strength
C70000 series Leaded nickel silvers	Cu, Zn, Ni, Pb	Good machining properties

Source: Compiled from information in Mendenhall, J.H., *Understanding Copper Alloys: The Manufacture and Use of Copper and Copper Alloy Sheet and Strip*, John Wiley & Sons, New York, 1977.

TABLE 8.2 Properties of Five Representative Wrought Copper Alloys

Alloy UNS No. Name	Tensile Strength MPa[a]	Thermal Expansion Coefficient (10^{-6} °C)	Conductivity %IACS[b]	Thermal Conductivity (Wcm/cm² °C)	Melting Point °C
C11000 Copper	290	16.8	101	3.94	1065
C26000 Cartridge brass	427	11.1	28	1.21	915
C51000 Phosphor bronze	469	9.9	15	0.71	950
C70600 Copper–nickel 10%	448	9.5	9	0.46	1100
C75200 Nickel–silver, 65-18	510	9.0	6	0.33	1070

Source: Data compiled from CDA, *CDA UNS Standard Designations for Wrought and Cast Copper and Copper Alloys*, Accessed June 10, 2010 at http://www.copper.org/resources/properties/standard-designations/introduction.html, 2010.

[a] 1/2 hard temper.

[b] Relative conductivity compared to International Annealed Copper Standard (IACS).

environmental degradation are becoming increasingly important including antimicrobial properties, environmental regulation, and copper pitting.

The antimicrobial property of copper: As evidenced by its historic practical application as one of the first disinfectants in water treatment, copper is microbiocidal to many bacteria, viruses and fungi, and microorganisms. Recent serendipitous and purposeful discovery of copper control of pathogen growth for *Legionella pneumophila* in water systems and general microbial growth on surfaces is numerous (e.g., Kuhn 1983, Rogers et al. 1991) and has given rise to registration of antimicrobial properties with the U.S. Environmental Protection Agency (EPA) (U.S. EPA 2008). At present registration includes over 450 copper alloys as antimicrobial materials, with human health claims that the materials kill >99.9% of six harmful human pathogens within 2 h contact including *methicillin-resistant Staphylococcus aureus* (MRSA), *vancomycin-resistant enterococcus faecalis* (VRE), *Enterobacter aerogenes*, *E. coli* O157:H7, *Staphylococcus aureus*, and *Pseudomonas aeruginosa*. Many of these benefits of copper are derived from the slow and steady environmental degradation (i.e., corrosion) of the underlying metal during its inter-action with the environment, raising the ironic reality that for copper, at least some environmental degradation is desirable in some instances.

Environmental regulations: The same concerns that have created excitement over use of copper in anti-microbial products have also given rise to increasingly stringent environmental regulation. Although copper is an essential trace micronutrient for humans, human exposure to high levels of copper via drinking water is not recommended, and copper levels are regulated to less than 1–3 ppm in potable water in many countries (Boulay and Edwards 2000). Because copper has toxic effects at much lower levels for microorganisms, resulting regulations in marine or fresh surface water and sewage dis-charge are lower than those allowed for human exposure and are in the range of 0.004–0.200 ppm (Boulay and Edwards 2000). The net result of these regulations is to increase the importance of con-trolling the degradation and release rates of copper in the environment, such that regulated con-centrations are not exceeded. As mentioned earlier, in some situations this creates an interesting balancing act, because antimicrobial uses harness benefits from achieving degradation and copper release rates *above* certain thresholds.

Nonuniform corrosion: There is rising concern about incidence of pinhole leaks from copper pitting corrosion. Surveys and research funded by the Water Research Foundation and the U.S. EPA recently demonstrated that copper did not leak with higher frequency than plastic domestic plumbing systems (Scardina 2008). But 8.1% of homeowners nationally had experienced at least one pinhole leak, and "outbreaks" have been reported in which as many as 50% of homes in some communities are affected by pinholes in just a few years. The total cost to building owners was estimated at $967 million annually in the United States. There is concern that new regulations promulgated by the U.S. EPA, such as mandat-ing removal of natural organic matter (NOM) and changing disinfectant practices, may increase the likelihood of pinhole leak attack (Edwards et al. 1994a). However, said regulations can also decrease the rate of certain types of attack.

8.4 General Properties

Hundreds of copper alloys are currently in use and still more are under development (CDA 2010). The most important classes of copper alloys and their key properties highlight the versatility of the metal and its suitability to a wide range of uses (Table 8.1). Impurities such as phosphorus, tin, selenium, tellurium, and arsenic decrease electrical conductivity; however, in controlled amounts these same elements can provide other benefits. For example, phosphorus improves weldability, tin improves corrosion resistance, and tellurium improves machinability. Also, inhibition of dezincification cor-rosion in alpha brasses can be improved by trace addition of arsenic, phosphorus, tin, or antimony (Joseph 1999).

8.4.1 Mechanical, Physical, Thermal, and Electrical Properties

The specific alloy also controls the color, durability, corrosion resistance, and the physical properties of the final product. Most copper alloys vary from the shiny familiar appearance of a new penny to distinct gold or silvery colors. Mechanical strength can be increased and thermal expansion coefficients decreased by 70%–80% higher relative to pure copper, among five representative wrought copper alloys (Table 8.2). Conductivity and thermal conductivity of alloys can be as low as 6%–8% of that present in pure copper, and melting point is only slightly impacted. Oftentimes, higher strengths result in trade-offs in electrical and thermal conductivity.

8.5 Structure

8.5.1 Atomic or Molecular Bonding

Copper has an atomic number of 29 and an atomic weight of 63.54. An individual copper atom contains a nucleus with 29 positively charged protons and 34 or 36 neutrons. The nucleus is surrounded by an electron cloud comprised of 29 negatively charged electrons arranged in orbitals as described by the notation $1s^2 2s^2 2p^6 3s^2 3p^6 3d^{10} 4s^1$. The first ionization potential relates to the 4s electron and is relatively low at 7.724 eV, meaning the "cuprous" ion; Cu^+ is formed relatively easily compared to other noble metals such as gold. Copper in the "cupric" valence state, Cu^{2+}, is also easily formed since the second ionization potential (for one of the 3d state electrons) is only slightly higher than for cuprous (Joseph 1999).

8.5.2 Microstructure

When molten copper cools, the atoms become arranged in an orderly 3D face-centered cubic pattern on individual cubic unit cells. These cells have one atom at each corner and one centered in each face (Mendenhall 1977). Various amounts of other elements, such as zinc, lead, tin, or nickel, can be alloyed with copper to change tensile strength, hardness, wear resistance, machinability, or corrosion resistance. These elements typically replace copper in the same lattice positions (Mendenhall 1977).

8.5.3 Phases of Brass

When zinc is alloyed with copper, different phases can result depending on the amount of zinc added. Alpha brass occurs when zinc is uniformly dissolved throughout the alloy, as can occur at zinc content below 38%. When zinc content is between 38% and 45%, a combination of alpha and beta brass will form. The two phases have different properties because beta brass has a body-centered cubic lattice arrangement. That is, the unit cell differs from copper's face-centered structure since it has atoms at the eight corners and one in the center of the cube body. As a result the alpha brass is ductile and readily worked at room temperatures, while beta brass is harder and more brittle (Mendenhall 1977).

8.6 Degradation Principles for Materials

8.6.1 Thermodynamics

Copper corrosion is relatively well understood in pure water, in dilute potable water systems, and in marine environments. In pure water three key oxidation states are stable including Cu^0, Cu^{+1}, and Cu^{+2}, which have four corresponding solid phases including Cu^0 (metallic copper), cuprous oxide (i.e., Cu_2O), and two cupric phases that include more soluble cupric hydroxide [$Cu(OH)_2$] and less soluble tenorite [CuO]. A representative redox or Pourbaix-type corrosion diagram (Figure 8.1) illustrates that copper is immune in water with relatively low-redox potentials (or dissolved oxygen concentrations), is passive at higher-redox potentials if protective cuprous or cupric oxides form, but can be subject to severe corrosion at higher-redox potentials at pH below about 7.0.

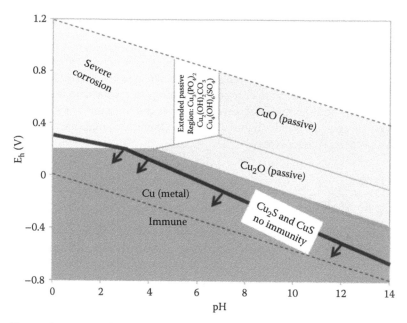

FIGURE 8.1 Thermodynamics and corrosion performance diagram for copper metal at room temperature in water. Passive regions drawn based on <10^{-4} M soluble copper species. Extended passive region based on levels of sulfate, carbonate, and phosphate is encountered in typical potable water applications and can vary depending on the precise species, aging, and other factors. (From Edwards, M. et al., *Water Sci. Technol.*, 1(3), 25, 2001.) Diagram based on data presented in Pourbaix (1966), Duby (1977), Schock et al. (1995), and Edwards et al. (2002). E_h—volts versus standard hydrogen electrode.

In potable water, brackish water, and seawater, a variety of cupric salts with sulfate, chloride, carbonate, and phosphate can form in the pH range of 5–7, which can dramatically extend the range of satisfactory environmental performance for copper and its alloys. Conversely, the immune region of copper in water at very low levels of dissolved oxygen and other oxidants can be wiped out in the presence of hydrogen sulfide species, despite the relatively low solubility of copper sulfide species (Sequeira 1995, Jacobs 1998, Jacobs and Edwards 2000).

On this basis, environmental degradation of copper and its alloys can be controlled by (1) removing oxygen and oxidants to achieve immunity; (2) adjusting pH and alkalinity to achieve passivation or extended passivation through formation of cupric oxide, cupric hydroxide, and cupric carbonate species; or (3) addition of phosphate or other corrosion inhibitors to achieve satisfactory performance at an extended pH range (down to as low as pH of 4–5 with very high doses). Control of microbial activity to prevent formation of sulfide is also important, but this is generally not difficult in relatively pure waters with low content of organic matter (Syrett 1981, Sequeira 1995).

8.6.2 Kinetics

In most applications in which copper products are exposed to the environment, satisfactory performance is generally achieved, unless the concentration of certain reactive constituents exceeds satisfactory ranges. In addition to the thermodynamic impacts described in the preceding section, certain parameters can strongly control the rate of attack.

Effect of oxygen and other oxidants: Oxygen is the ubiquitous "natural" oxidant driving copper alloy corrosion. In potable water, chlorine and chloramine have much stronger oxidizing potential and can create copper corrosion rate orders of magnitude higher for a duration of at least hours, if not indefinitely (Reiber 1991), but in the longer term low levels of free chlorine can be highly beneficial and can mitigate certain problems including those thought to be associated with microbial growth (Edwards and Ferguson 1993, Edwards et al. 2000, Boulay and Edwards 2001).

Effect of pH: As per the prediction of the Pourbaix diagram (Figure 8.1), pHs below about 7.0 can dramatically accelerate copper corrosion versus those encountered at more typical pH (Reiber 1989). The interplay between pH and certain anions is also of high interest, especially that of pH and carbonate species. Specifically, at pHs below about 7.5, the presence of higher carbonate markedly increases copper corrosion rates (Edwards et al. 1994b), perhaps due to the formation of soluble cupric carbonate complexes (Schock et al. 1995, Edwards et al. 1996). The formation potential of cupric carbonate complexes is linearly related to the concentration of dissolved CO_2 in the water, which might explain the success of indices containing dissolved CO_2 in explaining the presence of certain copper corrosion problems (Edwards et al. 1996). At higher pHs, higher carbonate can markedly decrease copper corrosion rates, presumably due to formation of protective malachite scales (Edwards et al. 1994b).

Effect of ammonia: The potential role of ammonia in control of copper corrosion rates is complex. Ammonia only weakly binds cupric species but can form relatively strong complexes with cuprous ions, which potentially creates adverse consequences for corrosion (AWWARF 1996). However, a key emerging issue associated with ammonia is bio-corrosion due to acid-producing nitrifying bacteria (Zhang et al. 2008, 2009a), which can create problems due to lower pH and also because the nitrate/nitrite end products are themselves corrosive in at least some circumstances (Edwards et al. 2004). This issue is deserving of increased scrutiny, given recent trends to add ammonia at levels of about 1 mg/L to potable water during chloramine disinfection.

Effect of temperature: The tendency of corrosion reaction rates to increase at higher temperature is counterbalanced to a large degree, by the fact that solubility and dissolution rates of most copper-passivating scales also decrease at higher temperature. Thus, dissolution rates and corrosion rates for uniform copper corrosion are often practically observed to increase at lower temperatures (Edwards et al. 1996, Boulay and Edwards 2001). Moreover, practical observation of service failures and simple beaker testing indicates that highly protective tenorite (CuO) tends to form more rapidly at higher temperature (Hidmi and Edwards 1999, Edwards et al. 2001) versus less protective and much more soluble cupric hydroxide [$Cu(OH)_2$]. Some practical problems with nonuniform copper corrosion have been reported when temperature exceeds thresholds that favor brochantite formation [$Cu_4(OH)_2SO_4$], because the formation of this scale on top of pits tends to support rapid pitting in hot waters (Mattson and Fredrikksson 1968, Edwards et al. 1994a).

Effect of sulfide: The catalytic role of sulfide in corrosion of copper and its alloys is extensive and was recently reviewed from the perspective of copper pitting and alloy corrosion (Jacobs et al. 1998, Jacobs and Edwards 2000). Sulfides not only attack copper when it would be otherwise immune (Figure 8.1) but can also increase rates of attack in the presence of oxygen by orders of magnitude. Because sulfides originate from microbial reactions in many systems, it is believed that control of these microbes from free chlorine may explain benefits from the presence of this strong oxidizing agent in many systems (e.g., Edwards et al. 2000). Sulfides and sulfate-reducing bacteria (SRB) are also believed responsible for attack on copper and its alloy in fresh and salt waters with high microbial growth potential (Syrett 1977, 1981, Syrett and Wing 1980).

8.7 Degradation of Specific Systems

8.7.1 Copper and Its Alloys in Moist Gas Environments

Exposure of copper and its alloys to the atmosphere generally results in formation of protective layers with a characteristically pleasing blue–green patina. The specific compounds, rate of formation, and characteristics of the patina vary somewhat dependent on the specifics of the local atmosphere including salinity, humidity, and pollutants such as sulfur dioxide and sulfide (Vernon 1931, 1932). However, the practical outcome of exposures to a wide range of conditions including polluted urban, unpolluted rural, and seashore environments shows little significant variation (Tracy et al. 1956). As described in the sections that immediately follow, there are some specialized instances in which atmospheric corrosion leads to serious corrosion problems for copper and its alloys including formicary corrosion and stress corrosion cracking.

8.7.1.1 Formicary Corrosion

If copper is exposed to an atmosphere with sufficient moisture, oxygen from air, and a source of organic acid, the copper metal can be aggressively attacked via formation of tunnels through the wall of the pipe, which are visual analogues to holes formed throughout an "ant-nest" colony. Hence, this type of attack is termed formicary or ant-nest corrosion (Corbett et al. 2000). Since moisture and air are nearly ubiquitous and relatively hard to avoid in at least some circumstances, the key to controlling this problem is to focus on removing organic vapors and residuals that can serve to form organic acids. At present it has been demonstrated that carboxylic acids, chlorinated solvents, soldering flux, and a variety of lubricants can serve as a source of the organic acid (Miyafuji et al. 1995, Corbett and Elliot 2000, Corbett and Severance 2005). ASTM standards exist that call for specified levels of cleanliness on new copper refrigeration materials (Corbett et al. 2000), but the presence of organics in the environment after installation is also cause for concern.

8.7.1.2 Stress Cracking, Intergranular or Season Cracking of Brass

Copper alloys can be subject to mechanical and electrochemical failure via intergranular corrosion, stress corrosion cracking, or season cracking (Thompson and Tracy 1949, Pugh et al. 1966) and dezincification (Lucey 1965). Brass failure from these phenomena can be controlled by the alloy (including trace constituents such as arsenic at levels of 0.02% and zinc levels), how it is produced in terms of residual stresses created in the metal, and how it is installed including mechanical stress and atmospheric or water chemistry. Alloys with about 20% zinc and below are generally immune to stress corrosion cracking. Ammonia is especially prone to causing stress corrosion cracking and intergranular attack (e.g., Guo et al. 2002), but other nitrogen species and other constituents can be involved (Giordano et al. 1997, Rao and Nair 1998). Sources of the nitrogen species can include cement materials and insulation (McDougal and Stevenson 2005). Documented failures from chloramine for brass, via intergranular attack, have also been documented in potable water (Ingleson et al. 1949, Anonymous 1951, Larson et al. 1956). Dependent on the aggressivity of the water and the alloy, the metal can essentially disintegrate with individual grains falling into the water from the device (Sundberg et al. 2003).

8.7.2 Copper Alloys in Aqueous Environments: Uniform Corrosion

When copper is first exposed to a moist aqueous environment, it rapidly forms Cu_2O and $Cu(OH)_2$ scale layers on its surface. Within the constraints mentioned previously in the practical Pourbaix diagram for copper, if corrosion is uniform, the rate of corrosion is of little concern. However, since some benefits of copper products are reliant on leaching of low levels of soluble Cu^{+2} to the water, and many environmental concerns are associated with leaching of higher levels of Cu^{+2}, it is therefore important to understand how much copper leaches to water under what circumstances and when.

It is now understood that soluble copper concentrations in relatively new copper plumbing are controlled by cupric hydroxide equilibrium under at least some circumstances (Werner et al. 1994, Schock et al. 1995, Edwards et al. 1996, Lagos et al. 2001). Over the pH and temperature range most commonly encountered in drinking water distribution and for atmospheric corrosion, equilibration with $Cu(OH)_2$ can lead to maximum soluble copper concentrations of about 2–10 mg/L (Figure 8.2). These levels are of regulatory concern, from both the perspective of the 1.3 mg/L EPA action limit for copper at the consumers' tap and very stringent wastewater discharge levels of 0.005–0.020 mg/L in many localities.

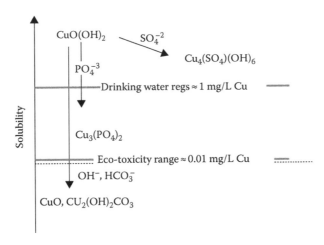

FIGURE 8.2 Initial control of copper solubility by $Cu(OH)_2$ and transitions to less soluble species. (After Edwards, M. et al., *Water Sci. Technol.*, 1(3), 25, 2001.)

As the copper material is further exposed to the environment and the $Cu(OH)_2$ ages, its solubility can be reduced by formation of less soluble solids. Dependent on circumstance, these species can include basic cupric sulfate, basic cupric carbonate (malachite), cupric phosphate, and tenorite species (Hidmi and Edwards 1999). In many circumstances very insoluble tenorite or malachite solid scale (rust) can form in a water of minutes to hours, resulting in soluble levels of copper that are very low from regulatory perspectives (Figure 8.2). In other situations the presence of NOM can "poison" formation of the less soluble scales, which can maintain high levels of copper associated with $Cu(OH)_2$ indefinitely (Edwards et al. 1999, 2001). Progressive formation of malachite over a pipe surface has been elegantly tracked on pipe surfaces in many case studies and has been shown to reduce solubility (Merkel and Pehkonen 2006). The release of copper can be a complex function of stagnation time in some circumstances, with a peak concentration associated with total loss of oxygen from the water and precipitation of released copper on the wall of the pipe if oxygen should disappear (Werner et al. 1994). Similar trends have been associated with copper release and subsequent re-precipitation on growing malachite crystals (Merkel and Pehkonen 2006). Atmospheric corrosion of copper roofs shows similar trends with metal release, with a stable lower level of copper release after an initial time period (Wallinder et al. 2002).

8.7.3 Copper Alloys in Aqueous Environments: Nonuniform Corrosion

In aqueous environments, copper and its alloys can be subject to nonuniform corrosion through a variety of unusual circumstances. Of these problems copper pitting, brass dezincification, and flow-induced (erosion) corrosion are most frequently encountered, although other unusual types of nonuniform attack arising from thermogalvanic, flux, and other factors can also be significant in some circumstances. Each of these is addressed in the sections that follow.

8.7.3.1 Copper Pinhole Leaks and Pitting Corrosion

Premature failure of copper tube from pitting corrosion can culminate in formation of pinhole leaks (Figure 8.3). Under normal circumstances a copper tube can last 60 years or more, even in modestly corrosive waters, but under conditions favoring nonuniform corrosion, failures can occur in as little as a few weeks to a few years (Figure 8.4). Due to the high value of the existing copper potable water infrastructure installed in buildings throughout the United States (probably a net present value on the order of 0.5–1 trillion dollars), extensive research has recently been conducted to better understand the mechanisms behind the problem and its causes (Edwards 2004, Edwards and Abhijeet 2004, Edwards et al. 2009, Scardina et al. 2008).

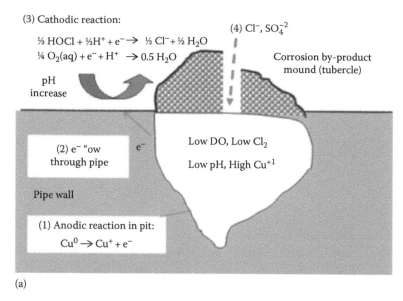

(a)

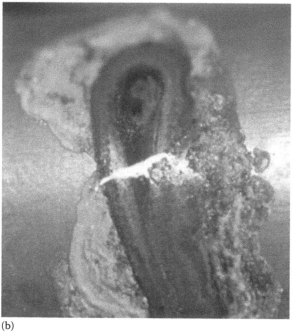

(b)

FIGURE 8.3 Schematic of key reactions in copper pitting (a) and resulting pinhole leak on exterior of pipe (b).

Copper pitting corrosion is now understood to propagate via a classic large-cathode and small-anode mechanism (Figure 8.6), with free chlorine or oxygen as the primary oxidant (Sosa et al. 1999). Considerable progress has recently been made in understanding the role of water chemistry in either initiating or propagating pitting corrosion of copper, especially for a type of attack occurring in certain relatively high-pH water chemistries, which was demonstrated to create fully penetrating pinholes in type M copper tube in just months to a year (Marshall et al. 2003, Rushing and Edwards 2004a, Marshall and Edwards 2005). Other combinations of water chemistry are also believed to cause serious attack in copper tube and can involve sulfides and microbial pitting agents, under deposit attack, and aggressive flux (Table 8.3).

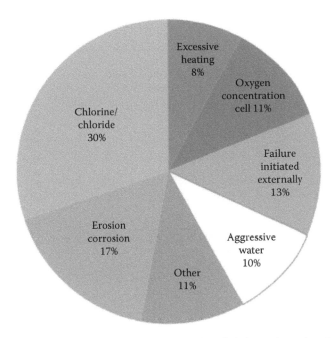

FIGURE 8.4 Recent U.S. experience regarding copper pinhole leak failures. (Raw data from Scardina, P. et al., Assessment of non-uniform corrosion in copper piping, Final Report to the Water Research Foundation, Denver, CO, ISBN: 978-1-60573-017-2, 2008.)

TABLE 8.3 Factors Known and Suspected to Cause Copper Pitting

Suspected Pitting Mechanism	Cause	Characteristic Water	Possible Factors in Water Reducing the Frequency of Failure	Characteristic Location of Worst Pitting
Sulfide attack	Direct corrosion of sulfide	Any raw water with sulfides or SRB	NOM, high pH, molybdenum, phosphate	Near colonies of SRB
High-pH Cl_2	Nonuniform corrosion	pH > 8.5	NOM, phosphate	Pipes in frequent flow and at locations with highest residual
Cl_2–Al high pH	Aluminum deposits cause nonuniform attack	pH > 7.8 Cl_2 > 2.5 ppm Cold water	NOM, phosphate	Pipes in frequent flow and at locations with highest residual
Chloramine–Al	Same as previously mentioned causes	Same as previously mentioned causes, but seems to be hot water	Lower temperatures, lower velocity of water in pipes	Both hot and cold, but hot water recirculation lines worst impacted
Under deposit corrosion	Concentration cells caused by varied oxidant access to surface	Lines or mounds of deposits other than copper	Filter particles from water, reduced oxidant concentration	Deposits often preferentially found on pipe bottom or outside of bends
Other microbial	Concentration cells caused by bacteria	Low alkalinity and poorly buffered water, lack of disinfectants	Low concentrations of AOC, phosphate, fixed N or K limit growth	Farthest reaches of treatment plant
Excessive and aggressive flux	Petroleum-based flux	Not determined	Use modern flux compliant with code	Under or near flux

TABLE 8.4 Guide to Identification of Copper Corrosion Problems and a Summary of Exacerbating and Ameliorating Factors

Characteristic	Uniform Corrosion	Type I Pitting (Cold Water)	Type II Pitting (Hot Water)	Type III Pitting (Soft Water)	High-pH and Chlorine-Induced Pitting
Pit shape	No pits	Deep and narrow	Narrower than type I	Wide and shallow	Wide and shallow
Manifestation of problem	Elevated copper in new buildings	Pipe failure	Pipe failure	Blue water from particulate copper	Pipe failure
Scale or pit morphology	Tarnished copper surface	Malachite, other basic copper salts	Underlying Cu_2O with overlying brochantite, malachite	Underlying Cu_2O with overlying brochantite, some malachite	Same as type II or type III
Water qualities	Most	High alkalinity and low pH; high sulfate; SRB activity	Hot waters, pH below 7.2, high sulfate relative to bicarbonate	Soft waters, pH above 8.0	pH > 8.2, in waters with relatively high chlorine and/or frequent flow
Initiating factors	Nonaggressive water	Stagnation early in pipe life, deposits within pipe including carbon films	Higher temperatures, high chlorine residuals, aluminum	Stagnation early in pipe life, pH's above 8.0; alum coagulation	High chlorine residuals, low alkalinity, frequent flow, Al or Fe particle deposits
Ameliorating factors/ treatments	Higher pH and phosphate inhibitors	NOM; increase pH	Lower temperature, raise pH and alkalinity	NOM; avoid stagnation early in pipe life	Increased phosphate, silicate, NOM, alkalinity

Source: Modified from Edwards, M. et al., *J. Am. Water Works Assoc.*, 86(7), 74, 1994a.

A recent survey suggested that current pinhole leak failures from high chlorine and higher pH have become increasingly important, and classic "type 1" failures associated with lower-pH and higher-alkalinity waters have become much less important (Table 8.4) since the subject was thoroughly reviewed a few decades ago (Edwards et al. 1994, Scardina et al. 2008). This is likely because the pH of U.S. drinking water is being raised markedly for compliance with EPA regulations or because phosphate corrosion inhibitors have been dosed to water supplies.

Certainly, many factors can contribute to formation of pinhole leaks including manufacturing defects, installation and construction issues, nuances of water exposure, and aggressive water (Campbell 1979, Edwards et al. 1994a). Working with the high-pH and high-chlorine water that has been proven to cause pitting by Marshall et al. (2003), Marshall and Edwards (2005), Cong et al. (2007, 2009), and Cong and Scully (2010a,b) have provided in-depth mechanistic insights regarding critical pitting potentials that induce pitting, as well as specific levels of chloride and sulfate that initiate pitting as a function of pH. Lytle and Schock (2008) also independently determined that high-pH and low-alkalinity waters with some chlorine (but without aluminum) could create significant nonuniform corrosion and pitting on copper tubing, and in the process they better defined the range of what may be considered "aggressive" water chemistries (Lytle et al. 2008). Due to these investigations, much more is also understood about factors such as alkalinity, chloride, sulfate, and pH relative to copper pitting in potable waters. Finally, a sequence of researchers at Virginia Tech (i.e., Marshall and Edwards 2006, Lattyak 2007, Edwards and Parks 2008, Custalow 2009) narrowed in on various chemical and physiochemical factors, such as flow velocity and frequency, which can contribute to very rapid pitting attack in this water, as well as silica, phosphate, and other inhibitor doses that can stop pitting. These contributions are reviewed in more detail in a later section.

8.7.3.2 Brass Dezincification

Dezincification is a dealloying process, which arises from selective leaching of zinc from brass (a Cu–Zn alloy), and can produce failure of the metal fitting or choking of flow due to formation of voluminous zinc-containing white precipitates termed meringue (Figure 8.5). Brass failures in building plumbing systems resulting from dezincification corrosion can be expensive (Sarver et al. 2010), and there are also health implications due to increased leaching of lead to water with high dezincification propensity (Kimbrough 2001, Triantafyllidou and Edwards 2007). If dezincification is localized to parts of the brass (i.e., plug dezincification), it can produce regions of spongy brittle copper, which appear as red patches on the brass surface. Layer dezincification results in a more uniform attack on the brass surface.

In the United States, problems with dezincification corrosion were once believed to be solved by use of low-zinc or dezincification-resistant (DZR) brass alloys, but there has been a recent increase in these types of failures due to use of high-zinc brass fittings in plastic plumbing systems (Sarver et al. 2010). The magnitude of dezincification problems is controlled by the alloy, chemistry of the water supply, and physical factors. Specifically, brasses with zinc content below about 15% are generally considered resistant to dezincification in practice, whereas brass with higher zinc contents is susceptible (Oliphant 1978, Sarver 2010). For alpha brasses containing 15%–30% zinc, addition of traces of arsenic and tin can render the alloy resistant (Karpagavalli and Balasubramaniam 2007).

Even with alloys that are considered highly susceptible to dezincification, the water chemistry and physical conditions of exposure can control the extent and type of corrosion problems (Turner 1961). It has long been understood that elevated chloride in water can increase the tendency for all types of dezincification problems, and it has recently been demonstrated that carbonate species (i.e., alkalinity) ameliorates attack, whereas hardness plays little role (Zhang 2009, Zhang et al. 2009b). Benefits from phosphate inhibitors seem to be short-lived (Zhang et al. 2009b). Other influential factors include pH, temperature, aeration, disinfectant type and concentration, other anions, and the chemical makeup of surface films or scales (Sarver et al. 2010).

When the same brass components are exposed to the same water in given plumbing system, attack can be highly variable from device to device, due to other critical physical and environmental factors that contribute to degradation. Specifically, it is suspected that flow velocity, galvanic connections between metallic plumbings subject to differential flow, and galvanic connections to copper can sometimes

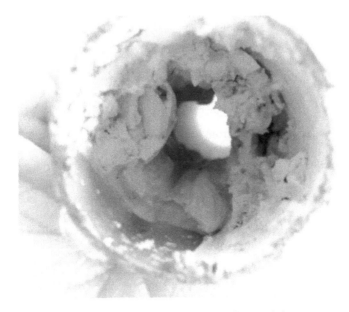

FIGURE 8.5 Voluminous meringue deposits choke flow in dezincification failure.

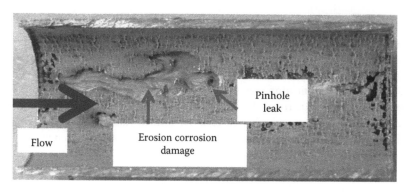

FIGURE 8.6 Illustration of a type of service failure attributed to high-velocity water flow.

control the rate of attack (Kelly et al. 1980, Sarver et al. 2010). Brass is often slightly anodic relative to copper under conditions encountered in the field (Lucey 1973, Nielson and Rislund 1973), and therefore connections to copper can accelerate attack in some circumstances.

8.7.3.3 Erosion Corrosion and Cavitation

Erosion corrosion of copper and its alloys has been attributed to a localized wearing away of the pipe material due to rapidly flowing corrosive liquid (Knutsson et al. 1972) or accelerated attack due to localized high velocity and excessive turbulence (Myers 2005) at flow rates exceeding 2 m/s (Figure 8.6). These explanations have obvious relevance to the practical factors that caused the damage, but do not provide detailed insights to mechanisms or solutions. A recent thorough literature review and associated fundamental research investigation (Coyne 2009, Coyne et al. 2009) isolated and identified key modes of attack hypothesized to contribute to flow-assisted failures including (1) particle impingement, (2) gaseous cavitation and bubble impingement, (3) vaporous cavitation, (4) differential concentration cell attack, (5) flow-assisted pitting, (6) flow electrification, and (7) multiple flow factors in combination (Table 8.5). Each of these is discussed briefly in the sections that follow.

Particle and bubble impingement: The notion that sediment or hard particles might be entrained in flow and impinge on and degrade the surface of copper and its alloys was described by Knutsson et al. (1972). Although this is a well-known attack mode on metal surfaces subject to gaseous flow (Hutchings 1983), recent attempts to reproduce rapid attack on copper even in waters with very high levels of sand and velocities >3 m/s were unsuccessful. Likewise, formation of air bubbles from gaseous cavitation (Novak et al. 2005) and abrasion against copper tube has also been speculated to create rapid attack (Obrecht and Quill 1960a–f, 1961, Myers et al. 1972, Sakamoto et al. 1995), but testing of copper in solutions with bubbly flow did not result in significant problems.

Gaseous cavitation: Cavitation resulting from pressure drops below the boiling point of water creates a micro-jet of water with pressures sufficient to physically gouge and damage metallic copper (During 1997, Novak 2005, Novak et al. 2005). Although classic fluid mechanics does not predict gaseous cavitation at velocities below about 10 m/s, Chan et al. (2002) gathered practical data suggesting it was occurring in Hong Kong potable water lines and that copper is highly susceptible to damage. Coyne et al. (2009) confirmed this hypothesis and further demonstrated that thicker walled tube can have much higher resistance to this failure mode and that the rate of attack from cavitation cannot be reduced by manipulating water chemistry or adding inhibitors.

Differential velocity concentration cell: A specific case of erosion corrosion was described by Obrecht and Quill (1960a–f, 1961) and Murakami et al. (2003), who suggested that high velocity can remove protective surface films (scales) that leave bare copper metal fully exposed to the full corrosiveness of the bulk water. Bengough and May (1924) believed that high-velocity jets impinging against

TABLE 8.5 Summary of Flow-Induced and Flow-Assisted Degradation Mechanisms for Copper and Its Alloys

Type of Attack	Attack Mode	Recent Test Results[a]	References
Particle impingement	Hard sediment and sand physically abrades copper surface at specific point	Unable to reproduce damage using very high concentrations of sand impinging on copper	Cohen and Lyman (1972), Knutsson et al. (1972), Benchaita et al. (1983), Landrum (1990)
Gaseous cavitation and bubble impingement	Abrasion from bubble impingement	Nearly no serious attack from air bubble flow or impingement. Gas bubbles only form at pressures below about 15 psi	Obrecht and Quill (1960f), Myers and Obrecht (1972), Sakamoto et al. (1995), Novak (2005), Novak et al. (2005)
Vaporous cavitation	Implosion of water vapor (>10,000 atm) and severe damage	Very severe attack and failure proven, especially for thin-walled tube	During (1997), Chan et al. (2002), Novak (2005)
Differential velocity Concentration cell attack	Area exposed to high-flow anode to area exposed to lower flow rate in some waters	Sections of copper exposed to high flow subject to severe attack, short term, but not long term. Low levels of Cl_2 inhibit attack	Obrecht and Quill (1960a–f), Murakami et al. (2003)
Flow-assisted pitting	Enhanced transport of oxidants under conditions favoring nonuniform corrosion accelerates pitting	High-pH and high-Cl_2 pitting attack dramatically increased with higher flow rates	Lattyak (2007), Slabaugh and Edwards (2008), Custalow (2009), Custalow et al. (2009)
Flow electrification	Voltage created along flow path via flow creates attack at front of pipe	Can be very significant in relatively pure water without phosphate, silica, or NOM	Edwards et al. (2007), Slabaugh and Edwards (2008)
Multiple factors in combination	Differential concentration cell assisted by removal of protective scale	A test in seawater suggested this mode of attack can be very severe in some water chemistries	Coyne (2009) and Coyne et al. (2009)

[a] Coyne (2009) and Coyne et al. (2009).

copper specimens not only prevented a protective surface scale from forming but also accelerated the corrosion rate. Furthermore, they speculated that the jets also caused separation of the anodic and cathodic regions of the copper, driving accelerated corrosion reaction via formation of a concentration cell as described later in this work. Examination of this hypothesis demonstrated that copper surfaces exposed to high flow were highly anodic in some situations and subject to very high >60 $\mu A/cm^2$ corrosion rates, but that in normal situations the protective scale was not removed even at water velocities >7 m/s (Coyne et al. 2009). Low levels of free chlorine inhibited the attack because in its presence, copper surfaces subject to high flow were cathodic. Thus, a differential velocity concentration cell would be a viable mechanism of rapid attack, if other flow factors remove the protective scale layers (Table 8.5).

Flow-assisted pitting: Certain types of nonuniform pitting, including that from waters of relatively high pH and high Cl_2, are dramatically accelerated at higher flow rates (Slabough et al. 2008, Custalow et al. 2009). In some situations this flow-assisted attack can be confused with attack directly induced by flow.

Flow electrification: In very low-conductivity solutions, copper can be subject to flow electrification, which can create DC electrical currents along the wall of a pipe due to fluid flow past a charged double layer on the pipe surface (Varga and Dunne 1985, Lattyak 2007). Flow electrification is recognized as a significant problem in nonaqueous oil and gas flow application for iron and other metals and can be an issue for copper in very pure waters without phosphate, silica, or NOM (Edwards et al. 2007).

Multiple factors in combination: Hodgkiess and Vassilou (2005) investigated the effect of high-velocity jets impinging perpendicularly against copper–nickel alloys while submerged in 3.5% sodium chloride solution. Weight loss of the plates subjected to jets increased with velocity due to a combination of electrochemical corrosion and mechanical erosion. Coyne et al. (2009) verified that suspected combined effects of cavitation and differential velocity concentration cell drove attack on pure copper in synthetic seawater, recreating failures similar to those observed in service. Addition of trace chlorine to the water mitigated the attack, consistent with predictions based on work with differential velocity concentrations cells.

8.7.3.4 Thermogalvanic Corrosion

Copper and its alloys can be subject to nonuniform corrosion resulting from gradients in temperature along the copper metal. As was the case for erosion corrosion, the water chemistry can determine whether the cooler section of copper is relatively anodic or cathodic versus the warmer section of copper tube. Failures accelerated by differential temperature have been reported under relatively acidic, deaerated, and continuous flow conditions (Berry 1946, Buffington 1947) and in relatively high-salinity tubes with large (>100°C) temperature gradients (Boden 1971). However, performance concerns were also documented from much smaller temperature gradients in aerated potable water in building plumbing systems, due not only to nonuniform corrosion but also from convection currents arising from differences in water density with temperature (Rushing and Edwards 2004b).

8.8 Degradation Protection

Aside from properly selecting the appropriate alloy suitable to the corrosive environments and to avoid associated degradation (Table 8.1), there are two general approaches that can be used to mitigate corrosion of copper and its alloys. The first involves application or removal of surface coatings, and the second involves either natural or chemical corrosion inhibitor.

8.8.1 Surface Treatments and Coatings

Acceptable performance of a copper alloy component in its environment, dependent on circumstance, can require manipulation of conductivity, heat transfer, copper leaching, and corrosivity. Coatings have been developed and used over the years to improve performance.

Removal of deleterious films: Some manufacturing processes can create carbon surface films on finished products that are cathodic to the copper metal and which can increase susceptibility to pitting (Campbell 1950, 1979). Certain manufacturers have taken steps to prevent formation of these films or to actively remove them, with some clear benefits, although the extent to which such films contribute to modern pitting problems is in doubt (Campbell 1979, Smith and Francis 1990, Edwards et al. 1994a).

Tin, epoxy, and plastic coatings: In recent years, "hybrid" materials using copper alloys have emerged that can dramatically expand the range of conductivity/heat transfer applications and environments, in which copper products can provide outstanding performance (Figure 8.7). Plastics, epoxies, and metal coatings, such as tin, are available that can coat either the outer or inner surface. Some of these coatings can even be applied in situ, after copper has failed, and restore the integrity of the copper plumbing system.

8.8.2 Control of Corrosion

Corrosion problems of copper alloys can often be controlled by dosing of corrosion inhibitors or by manipulation of water chemistry. For example, about 58% of potable water utilities in the United States reported using corrosion inhibitors such as phosphates, at an average annual cost of $1.16 per customer

FIGURE 8.7 Plastic and epoxy coatings can protect copper alloys from corrosive environments on either the inside or the outside of the product.

(Scardina et al. 2008). All other utilities are still required to "optimize" corrosion control by adjustment of pH and alkalinity to desirable ranges.

Copper dissolution: Most natural environments accumulate significant concentrations of carbonate species from weathering of minerals, microbial activity, and exposure to the atmosphere. Laboratory experiments determined that for relatively new plumbing, copper release was a linear function of alkalinity if all other factors were equal. The adverse effects are attributable to complexation reactions (i.e., $Cu^{+2} + CO_3^{-2} \rightarrow CuCO_3$), and simplification of the relevant chemical equilibria produces equations of the following form:

$$\text{Maximum Soluble Cu (mg/L)} = 0.83 + 0.015 \, [\text{alkalinity}] \text{ at pH 7.0 for } Cu(OH)_2$$

$$\text{Maximum Soluble Cu (mg/L)} = 0.58 + 0.0013 \, [\text{alkalinity}] \text{ at pH 8.0 for } Cu(OH)_2$$

where alkalinity (i.e., approximately the concentration of bicarbonate in the water) is in terms of mg/L as $CaCO_3$. These predictions have been in reasonable agreement with practical data collected for water utilities (Figure 8.8). By adjusting the pH and alkalinity to appropriate ranges, the extent of copper corrosion can often be controlled (Edwards et al. 1996).

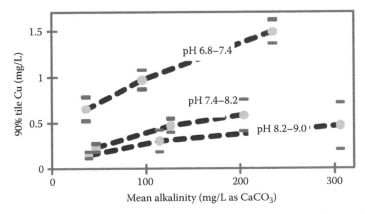

FIGURE 8.8 Copper concentrations reported to the U.S. EPA in relatively new plumbing are linear functions of alkalinity. Data previously shown based on monitoring at over 1000 U.S. water utilities. (After Edwards, M. et al., *J. Am. Water Works Assoc.*, 88(3), 81, 1996.)

Dezincification: Dezincification problems can also be controlled by adjusting pH and alkalinity. Although brass corrosion rates generally decrease at higher pH and lower alkalinity, buildup of meringue dezincification is maximized at about pH 8.3. Thus, meringue dezincification can be addressed by increasing alkalinity or avoiding the pH range of 7.8–8.8 (Sarver et al. 2010).

Copper pitting: There are many manifestations of copper pitting, and all are a function of pH. Types of copper pitting that generally occur in lower-pH and higher-alkalinity waters (Cohen et al. 1972) have been virtually wiped out in the United States due to increased pHs in recent years. Type 3 copper pitting that requires a minimum pH of 8.5 to initiate is problematic in very low-alkalinity waters without free chlorine residuals (Edwards et al. 2000).

Erosion corrosion: Erosion and flow-induced corrosion problems are also strong functions of pH. The original investigations into the problem (Obrecht and Quill 1960a–f) discovered that the rate of attack was greatly accelerated at lower pHs. However, flow-enhanced pitting attack occurring in high-pH and low-alkalinity waters with high chlorine actually becomes much more serious near pH of 9.2 (Marshall and Edwards 2005).

8.8.2.1 Effect of Natural and Artificial Inhibitors

There are a variety of natural and artificial inhibitors in use (McNeill and Edwards 2002, Antonijevic and Petrovic 2008) that can greatly extend the range of environments in which copper products are utilized.

Phosphate: Approximately 50% of all water utilities are currently using an orthophosphate or phosphate corrosion inhibitor (McNeill and Edwards 2002). These inhibitors work best in the pH range of 6.5–7.5 to inhibit copper corrosion. Polyphosphates do not decrease copper corrosion or copper leaching, but tend to increase such problems (Schock et al. 1995, Edwards et al. 2002). Recent research has demonstrated that phosphates can strongly inhibit certain types of copper pitting corrosion (Marshall and Edwards 2006, Lattyak and Edwards 2007, Sarver and Edwards 2010).

Silica: Silica occurs naturally in many waters in the range of 0.5–40 mg/L as Si. Silica levels above 5 mg/L were found to almost completely inhibit pitting corrosion of copper due to high pH and high chlorine levels (Sarver and Edwards 2010). However, silica in surface deposits has often been suspected to contribute to other pinhole leak and pitting problems, although no such link has ever been conclusively proven (Edwards et al. 1994a).

Natural organic matter: Natural waters often contain NOM in the range of 1–20 mg/L, and these inhibitors have long been associated with reduced incidence of copper pinhole corrosion (Campbell 1954, 1979, Edwards et al. 1994a). Levels as low as 0.5 mg/L of humic substances can completely eliminate pitting corrosion of high-pH and high-chlorine waters (Sarver and Edwards 2010). NOM can also influence corrosion of brass alloys, in general, by reducing corrosion rates (Korshin et al. 2000). But higher NOM can sometimes dramatically worsen corrosion of copper by maintaining very high-solubility cupric hydroxide and other compounds (Edwards and Boulay 2001, Edwards et al. 2001). Thus, there is concern that removing NOM from water supplies due to concern over formation of carcinogens in the presence of chlorine might worsen pitting, while in some cases it can reduce copper release.

Nonpotable water inhibitors: In nonpotable closed loop applications, it is possible to use a wider range of inhibitors for copper and its alloys. Effective inorganic inhibitors include chromate, molybdate, and tetraborate, whereas classes of organic inhibitors include azoles, amines, and amino acids (Antonijevic and Petrovic 2008).

8.9 Specific Standardized Tests

Standard test methods exist for determining the dezincification propensity of specific brass alloys under highly aggressive conditions. The general idea behind development of these methods is that brasses that are found to be "resistant" under extreme circumstances should be free from dezincification problems in service.

Standard methods also exist for determining the pitting propensity of copper in specific water(s) of interest and for evaluating observed pitting corrosion. These methods may be used to systematically test the effects of various waters on copper pitting and/or characterize pitting corrosion observed on copper in service.

8.9.1 Standard Test Methods for Brass Dezincification Corrosion

There is currently no ASTM standard test method to determine the dezincification propensity of brass alloys. However, several national/international standards (e.g., Australian Standard (AS) 2345, British-adopted European Standards BS EN 12163 [1998], 12165 [1999], and 12420 [1999]), as well as an ISO Standard (i.e., ISO 6509, 1981), do exist. ISO 6509, *Corrosion of Metals and Alloys—Determination of Dezincification Resistance of Brass*, was adopted in 1981 (Nielsen and Rislund 1973, Nielsen 1985), and it calls for brass test specimens to be exposed to 1% cupric chloride solution at 75°C for 24 h. After exposure to the extreme dezincifying conditions, the specimen is examined metallographically to determine the depth of dezincification corrosion. ISO 6509 does not specify acceptance limits with respect to dezincification depth (AWWARF 1996), but similar standards often do. For example, BS 2872 (1989) (recently replaced by BS EN 12420) utilizes an identical test protocol to ISO 6509 and specified that DZR forged brass specimens must not exhibit dezincification depths exceeding 100 μm (perpendicular to the exposed surface).

Additionally, National Science Foundation (NSF)/ANSI 14, *Plastic Piping System Components and Related Materials*, was updated in 2009 to include requirements on brass dezincification corrosion (NSF 2010). Using the ISO 6509 test protocol, brass specimens are considered DZR if the maximum depth of dezincification does not exceed 200 μm. By 2011, all NSF listed brass valves and fittings must be DZR.

8.9.2 ASTM Standards for Copper Pitting Corrosion

There are currently two relevant standards for pitting corrosion of copper in potable water systems (American Society of Metals 1979): ASTM 2688 and ASTM G46. ASTM 2688, *Standard Test Method(s) for Corrosivity of Water in the Absence of Heat Transfer (Weight Loss Method)*, pertains to an actual method that may be used to subject copper specimens to aggressive water(s). Specimens are mounted in a test apparatus and exposed to flowing water of interest for a measured period of time and then removed from the apparatus and examined for weight loss and pitting corrosion. From quantitative characterizations of these, the rates of both general and pitting corrosion, respectively, can be determined. Rates are typically specified as thickness of copper removed (general corrosion) or depth of copper penetrated (pitting) per time (e.g., mils/year). The most recent revision of the standard (see ASTM D2688-05, 2005) mentions flat copper coupon specimens, but the original (see ASTM D2688-83, 1983, test method C) allowed for exposure of short lengths of actual copper pipe in order to more closely approximate real service conditions.

ASTM G46, *Standard Guide for Examination and Evaluation of Pitting Corrosion*, pertains to identification and characterization of pitting corrosion. It includes standard methods for metallographic inspection, evaluation of pit morphology, quantitative analysis of pitting depth and density, and qualitative analyses of pitting via visual inspection.

8.9.3 Other Desirable Testing Standards

Although no U.S. standards currently exist that regulate use of DZR brass in potable water, other countries including Australia, New Zealand, Great Britain, Sweden, and Scotland have adopted standards or bylaws that require or encourage their use (Sarver et al. 2010).

There is also a significant need for standardized testing, to gauge the dezincification propensity of water to brass, given that water companies and building managers can control the extent of dezincification problems by adjusting pH, alkalinity, phosphate, and other treatments.

There are standard test apparatus to investigate dezincification and stress corrosion cracking susceptibility of particular alloys (Lucey 1965, ASTM G37-90 1994).

8.10 Summary

Copper and its alloys are rightly considered noble metals, with a wide range of environmental applications that take advantage of its unique antimicrobial, physical, and metallurgical properties. In some environments accelerated degradation can occur, which can be counteracted, in most cases, by using corrosion-resistant alloys or changing the water chemistry. In other unusual situations resulting from exposure to water with very low pH, extremely high water velocities (>2 m/s), high levels of hydrogen sulfide or SRB activity, and other unusual circumstances, hybrid materials or other metals should be used.

Acknowledgments

We acknowledge the financial support of the Copper Development Association (CDA) and the NSF under grant CBET-0933246. The views expressed herein are those of the authors and do not necessarily reflect the views of the NSF or the CDA.

Nomenclature and Units

EPA   Environmental Protection Agency
ppm   Part per million

References

American Society of Metals. 1979. *Source Book on Copper and Copper Alloys*, American Society of Metals, Materials Park, OH.

Anonymous. 1951. Review of current investigation no. 8: Waterworks fittings. Research Group Report. *Journal of the Institution of Water Engineers* (Great Britain), 5, 700.

Antonijevic, M.M. and M.B. Petrovic. 2008. Copper corrosion inhibitors: A review. *International Journal of Electrochemical Science*, 3, 1–28.

AS 2345. 2006. *Dezincification Resistance of Copper Alloys*. Australian Standards, Sydney, New South Wales, Australia.

ASTM D2688-05. 2005. *Standard Test Method for Corrosivity of Water in the Absence of Heat Transfer (Weight Loss Method)*. American Society for Testing and Materials, West Conshohocken, PA.

ASTM D2688-83. 1983. *Standard Test Methods for Corrosivity of Water in the Absence of Heat Transfer (Weight Loss Method)*. American Society for Testing and Materials, West Conshohocken, PA.

ASTM G37-90. 1994. *Standard Practice for Use of Mattsson's Solution of pH 7.2 to Evaluate the Stress Corrosion Cracking Susceptibility of Copper-Zinc Alloys*, American Society for Testing and Materials, West Conshohocken, PA, pp. 129–131.

AWWARF. 1996. Copper alloys and solders, in R. Oliphant and M. Schock, Eds., *Internal Corrosion of Water Distribution Systems*, 2nd edn., American Water Works Association, Denver, CO, pp. 269–312.

Baker, M.N. 1948. *The Quest for Pure Water*, The American Water Works Association, New York.

Benchaita, M.T., P. Griffith, and E. Rabinowicz. 1983. Erosion of metallic plate by solid particles entrained in a liquid jet. *Journal of Engineering for Industry*, 105, 215–222.

Bengough, G.D. and R. May. 1924. Seventh report to the corrosion research committee of the institute of metals. *The Journal of the Institute of Metals*, 32, 90–269.

Berry, N.E. 1946. Thermogalvanic corrosion. *Corrosion*, 2(11), 261–267.

Boden, P.J. 1971. Corrosion of Cu and Cu-base alloys under conditions of boiling heat transfer—I. Corrosion of Cu. *Corrosion Science*, 11, 353–362.

Boulay, N. and M. Edwards. 2000. Copper in the urban water cycle. *Critical Reviews in Environmental Science and Technology*, 30(3), 297–326.

Boulay, N. and M. Edwards. 2001. Role of temperature, chlorine and organic matter in copper corrosion by-product release to soft water. *Water Research*, 35(3), 683–690.

BS 2872. 1989. *Copper & Copper Alloys: Forging Stock & Forgings*. British Standards, London, U.K.

BS EN 12163. 1998. *Copper and Copper Alloys—Rod for General Purposes*. British Standards, London, U.K.

BS EN 12165. 1998. *Copper and Copper Alloys—Wrought and Unwrought Forging Stock*. British Standards, London, U.K.

BS EN 12420. 1999. *Copper and Copper Alloys: Forgings*. British Standards, London, U.K.

Buffington, R.M. 1947. Thermogalvanic corrosion II. *Corrosion*, 3(12), 613–631.

Campbell, H.S. 1950. Pitting corrosion in copper water pipes caused by films of carbonaceous material produced during manufacture. *Journal of the Institute of Metals*, 77, 345–366.

Campbell, H.S. 1954. A natural inhibitor of pitting corrosion of copper in tap-waters. *Journal of Applied Chemistry*, 4, 633–647.

Campbell, H.S. 1979. A review: Pitting corrosion of copper and its alloys, in *Localized Corrosion*, NACE, Houston, TX.

CDA. 2010. *CDA UNS Standard Designations for Wrought and Cast Copper and Copper Alloys*. Accessed June 10, 2010 at http://www.copper.org/resources/properties/standard-designations/introduction.html

Chan, W.M., F.T. Cheng, and W.K. Chow. 2002. Susceptibility of materials to cavitation erosion in Hong Kong. *Journal of the American Water Works Association*, 94(8), 76–84.

Cohen, A. and W.S. Lyman. 1972. Service experience with copper plumbing tube. *Materials Protection and Performance*, 11(2), 48–53.

Cong, H., F. Bocher, N.D. Budiansky, M.F. Hurley, and J.R. Scully. 2007. Use of multi-electrode arrays to advance the understanding of selected corrosion phenomenon. *Journal of ASTM International*, 4(10).

Cong, H., H.T. Michels, and J.R. Scully. 2009. Passivity and pit stability behavior of copper as a function of selected drinking water chemistry variables. *Journal of the Electrochemical Society*, 56(1), C16–C27.

Cong, H. and J.R. Scully. 2010a. Effect of chlorine concentration and resultant cathodic half-cell kinetics on natural pitting of copper as a function of water chemistry. *Journal of the Electrochemical Society*, 157(5), C200–C211.

Cong, H. and J.R. Scully. 2010b. The use of coupled multi-electrode arrays to elucidate the pH dependence of copper as a function of water chemistry. *Journal of the Electrochemical Society*, 157(1), C36–C46.

Corbett, R.A. and P. Elliot. 2000. Ant-nest corrosion: Digging the tunnels, in *NACE Corrosion Conference Paper 00646*, Houston, TX.

Corbett, R.A. and D. Severance. 2005. Development of a reproducible method to determine the mechanism and effect of organic acids and other contaminants on the corrosion of aluminum-finned copper-tube heat exchange coils. Final Report to the air conditioning and refrigeration technology institute, Arlington, VA.

Coyne, J. 2009. Flow induced failures of copper drinking water tube. MS thesis, Virginia Tech, Blacksburg, VA.

Coyne, J., M. Edwards, and P. Scardina. 2009. Flow induced failures of copper drinking water tube. Report to International Copper Association, New York.

Custalow, B. 2009. Influences of water chemistry and flow conditions on non-uniform corrosion in copper tube. MS thesis, Virginia Tech, Blacksburg, VA.

Custalow, B., E. Sarver, and M. Edwards. 2009. Factors influencing copper pitting induced by high pH and high chlorine, in *Proceedings of the 2009 AWWA Annual Conference*, San Diego, CA.

Duby, P. 1977. *The Thermodynamic Properties of Aqueous Inorganic Copper Systems*. INCRA Monograph IV, International Copper Research Association, Inc., New York.

During, E.D.D. 1997. *Corrosion Atlas*, Elsevier Science, Amsterdam, the Netherlands.

Edwards, M. 2004. Controlling corrosion in drinking water distribution systems: A grand challenge for the 21st century. *Water Science and Technology*, 49(2), 1–8.

Edwards, M. and D. Abhijeet. 2004. Role of chlorine and chloramine in corrosion of lead-bearing plumbing materials. *Journal of the American Water Works Association*, 96(10), 69–81.

Edwards, M. and N. Boulay. 2001. Organic matter and copper corrosion by-product release: A mechanistic study. *Corrosion Science*, 43(1), 1–18.

Edwards, M. and J.F. Ferguson. 1993. Accelerated testing of copper corrosion. *Journal of the American Water Works Association*, 85(10), 105–113.

Edwards, M., J.F. Ferguson, and S. Reiber. 1994a. The pitting corrosion of copper. *Journal of the American Water Works Association*, 86(7), 74–90.

Edwards, M., L. Hidmi, and D. Gladwell. 2002. Phosphate inhibition of soluble copper corrosion by-product release. *Corrosion Science*, 44, 1057–1071.

Edwards, M., S. Jacobs, and D. Dodrill. 1999. Desktop guidance for mitigating Pb and Cu corrosion by-products. *Journal of the American Water Works Association*, 91(5), 66–77.

Edwards, M., S. Jacobs, and R. Taylor. 2000. The blue water phenomenon. *Journal of the American Water Works Association*, 92(7), 72–82.

Edwards, M., R. Lattyak, P. Scardina, and C. Strock. 2007. Flow electrification and non-uniform copper corrosion in potable water systems. Final Report to Copper Development Association, New York.

Edwards, M., T. Meyer, and J. Rehring. 1994b. Effect of selected anions on copper corrosion rates. *Journal of the American Water Works Association*, 86(12), 73–81.

Edwards, M. and J. Parks. 2008. Secondary effects of implementing arsenic removal treatment-focus on corrosion and microbial growth. *Journal of the American Water Works Association*, 100(12), 108–121.

Edwards, M., K. Powers, L. Hidmi, and M.R. Schock. 2001. The role of ageing in copper corrosion by-product release. *Water Science and Technology*, 1(3), 25–32.

Edwards, M., J.C. Rushing, S. Kvech, and S. Reiber. 2004. Assessing copper pinhole leaks in residential plumbing (*Water Science and Technology*, 49(2), 83–90.), in S. Parsons, R. Stuetz, B. Jefferson, and M. Edwards, Eds., *Scaling and Corrosion in Water and Wastewater Systems*, IWA Publishing, London, U.K.

Edwards, M., M.R. Schock, and T.E. Meyer. 1996. Alkalinity, pH, and copper corrosion by-product release. *Journal of the American Water Works Association*, 88(3), 81–94.

Edwards, M., P. Scardina, R. Taylor, and N. Goodman. 2009. Non-uniform corrosion in copper piping—monitoring techniques. Final Report to the Water Research Foundation, Denver, CO, ISBN: 978-1-60573-054-7.

Giordano, C.M., G.S. Duffo, and J.R. Galvele. 1997. The effect of Cu^{+2} concentration on the stress corrosion cracking susceptibility of α-brass in cupric nitrate solutions. *Corrosion Science*, 39(10–11), 1915–1923.

Guo, X., W. Chu, K. Gao, and L. Qiao. 2002. Relativity between corrosion-induced stress and stress corrosion cracking of brass in an ammonia solution. *Journal of University of Science and Technology Beijing*, 9(6), 431.

Hidmi, L. and M. Edwards. 1999. Role of temperature and pH in $Cu(OH)_2$ solubility. *Environmental Science and Technology*, 33(15), 2607–2610.

Hodgkiess, T. and G. Vassilou. 2005. Complexities in the erosion corrosion of copper-nickel alloys in saline water. *Desalination*, 183, 235–247.

Hutchings, I.M. 1983. *Monograph on the Erosion of Materials by Solid Particle Impact*, Materials Technology Institute of the Chemical Process Industries, Publication No. 10, St. Louis, MO.

Ingleson, H., A.M. Sage, and R. Wilkinson. 1949. The effect of the chlorination of drinking water on brass fittings. *Journal of the Institution of Water Engineers* (Great Britain), 3, 81.

ISO 6509. 1981. *Corrosion of Metals and Alloys—Determination of Dezincification Resistance of Brass*. International Organization for Standardization, Geneva, Switzerland.

Jacobs, S. and M. Edwards. 2000. Sulfide scale catalysis of copper corrosion. *Water Research*, 34(10), 2798–2808.

Jacobs, S., S. Reiber, and M. Edwards. 1998. Sulfide-induced copper corrosion. *Journal of the American Water Works Association*, 90(7), 62–73.

Joseph, G. 1999. *Copper: Its Trade, Manufacture, Use, and Environmental Status*, ASM International, Materials Park, OH.

Karpagavalli, R. and R. Balasubramaniam. 2007. Development of novel brasses to resist dezincification. *Corrosion Science*, 49(3), 963–979.

Kelly, G., G.R. Lebsanft, and J.J. Venning. 1980. Effect of flow on dezincification, in *Proceedings of Australasian Corrosion Association Conference*, Adelaide, South Australia, Australia, vol. 1, pp. H.3.1–H3.14.

Kimbrough, D.E. 2001. Corrosion and the LCR monitoring program. *Journal of the American Water Works Association*, 93, 81–91.

Knutsson, L., E. Mattsson, and B.E. Ramberg. 1972. Erosion corrosion in copper water tubing. *British Corrosion Journal*, 7, 208–211.

Korshin, G.V., J.F. Ferguson, and A.N. Lancaster. 2000. Influence of natural organic matter on the corrosion of leaded brass in potable water. Behavior of the lead phase. *Corrosion Science*, 42(1), 53–66.

Kuhn, P.J. 1983. *Diagnostic Medicine*, Medical Economics Company Inc., Oradell, NJ.

Lagos, G.E., C.A. Cuadrado, and M.V. Letelier. 2001. Aging of copper pipes in drinking water. *Journal of the American Water Works Association*, 93(12), 94–103.

Landrum, J.R. 1990. *Fundamentals of Designing for Corrosion Control*, National Association of Corrosion, Houston, TX.

Larson, T.E., R.M. King, and L. Henley. 1956. Corrosion of brass by chloramine. *JAWWA*, 48(1), 84–88.

Lattyak, R. 2007. Non-uniform copper corrosion in potable water: Theory and practice. MS thesis, Virginia Tech, Blacksburg, VA.

Lattyak, R. and M. Edwards. 2007. Role of zinc and phosphate in mitigation of copper pitting, Presented at the *AWWA Annual Conference*, Toronto, Ontario, Canada.

Lucey, V.F. 1965. The mechanism of dezincification and the effect of arsenic. *British Corrosion Journal*, 1(1), 53–59.

Lucey, V.F. 1973. Relationship between water composition and dezincification of duplex brass, in *Proceedings of British Non-Ferrous Metals Research Association Seminar on Dezincification in Brass Fittings for Water Services*, Copenhagen, Denmark, pp. 1–42.

Lytle, D. and M. Schock. 2008. Pitting corrosion of copper in waters with high pH and low alkalinity. *Journal of the American Water Works Association*, 100(3), 115–128.

Marshall, B. and M. Edwards. 2005. Copper pinhole leak development in the presence of $Al(OH)_3$ and free chlorine, in *Proceedings of the AWWA Annual Conference*, San Francisco, CA, 16pp.

Marshall, B. and M. Edwards. 2006. Phosphate inhibition of copper pitting corrosion, Presented at the *2006 AWWA Water Quality Technology Conference*, Denver, CO, 13pp, November 2006.

Marshall, B.J., J.C. Rushing, and M. Edwards. 2003. Confirming the role of aluminum solids in copper pitting corrosion, in *Proceedings of the American Water Works Association National Conference*, Anaheim, CA, T-7-2, 13pp.

Mattsson, E. and A.M. Fredrikksson. 1968. Pitting corrosion in copper tubes—Cause of corrosion and counter-measures. *British Corrosion Journal*, 3, 246–257.

McDougal, J.L. and M.E. Stevenson. 2005. Stress-corrosion cracking in copper refrigerant tubing. *Journal of Failure Analysis and Prevention*, 5(1), 13–16.

McNeill, L.S. and M. Edwards. 2002. Phosphate inhibitor use at US utilities. *Journal of the American Water Works Association*, 94(7), 57–63.

Mendenhall, J.H. 1977. *Understanding Copper Alloys: The Manufacture and Use of Copper and Copper Alloy Sheet and Strip*, John Wiley & Sons, New York.

Merkel, T.H. and S.O. Pehkonen. 2006. Copper corrosion in domestic drinking water installations—Scientific background and mechanistic understanding of general corrosion. *Corrosion Engineering Science and Technology*, 4(1), 21–37.

Miyafuji, M., R. Ozaki, A. Tsuchiya, T. Kuroda, and K. Minamoto. 1995. *Journal of the Japan Copper and Brass Research Association*, 34, 159–167.

Murakami, M., K. Sugita, A. Yabuki, and M. Matsumura. 2003. Mechanism of so-called erosion corrosion and flow velocity difference corrosion of pure copper. *Corrosion Engineering*, 52(3), 155–159.

Myers, J.R. 2005. Copper tube corrosion in domestic-water systems. *Heating, Piping and Air Conditioning Engineering*, 6, 22–31.

Myers, J.R. and M.F. Obrecht. 1972. Potable water systems: Recognition of the cause vital to minimizing corrosion. *Materials Protection and Performance*, 11(4), 41–46.

NSF—National Sanitation Foundation International. 2010. *NSF/ANSI Standard 14 Update—Dezincification and Stress Corrosion Cracking (SCC) Resistance*. Accessed June 14, 2010 at http://www.nsf.org/business/newsroom/articles/plumbing_1004_didyouknow.asp

Nielsen, K. 1985. Corrosion testing in potable water, in G.S. Haynes and R. Baboian, Eds., *Laboratory Corrosion Tests and Standards*, ASTM STP 866, American Society for Testing and Materials, West Conshohocken, PA, pp. 169–183.

Nielsen, K. and E. Rislund. 1973. Comparative study of dezincification tests. *British Corrosion Journal*, 8, 106–116.

Novak, J.A. 2005. Cavitation and bubble formation in water distribution systems. MS thesis, Virginia Tech, Blacksburg, VA.

Novak, J., P. Scardina, and M. Edwards. 2005. Cavitation and bubble formation in distribution systems, in *Proceedings of the AWWA Annual Conference*, San Francisco, CA, 20pp.

Obrecht, M.F. and L.L. Quill. 1960a. How temperature, treatment, and velocity of potable water affect corrosion of copper and it as alloys in heat exchanges and piping systems. *Heating, Piping and Air Conditioning*, 32(1), 165–169.

Obrecht, M.F. and L.L. Quill. 1960b. How temperature, treatment, and velocity of potable water affect corrosion of copper and it as alloys; cupro-nickel, admiralty tubes resist corrosion better. *Heating, Piping and Air Conditioning*, 32(9) 125–133.

Obrecht, M.F. and L.L. Quill. 1960c. How temperature, treatment, and velocity of potable water affect corrosion of copper and it as alloys; different softened waters have broad corrosive effects on copper tubing. *Heating, Piping and Air Conditioning*, 32(7), 115–122.

Obrecht, M.F. and L.L. Quill. 1960d. How temperature, treatment, and velocity of potable water affect corrosion of copper and it as alloys; monitoring system reveals effects of different operating conditions. *Heating, Piping and Air Conditioning*, 32(4) 131–137.

Obrecht, M.F. and L.L. Quill. 1960e. How temperature, treatment, and velocity of potable water affect corrosion of copper and it as alloys; tests show effects of water quality at various temperatures, velocities. *Heating, Piping and Air Conditioning*, 32(5), 105–113.

Obrecht, M.F. and L.L. Quill. 1960f. How temperature, treatment, and velocity of potable water affect corrosion of copper and it as alloys; what is corrosion? *Heating, Piping and Air Conditioning*, 32(3), 109–116.

Obrecht, M.F. and L.L. Quill. 1961. How temperature, treatment, and velocity of potable water affect corrosion of copper and it as alloys. *Heating, Piping and Air Conditioning*, 33(4), 129–134.

Oliphant, R. 1978. *Dezincification by Potable Water of Domestic Plumbing Fittings: Measurement and Control*, Water Research Centre, Medmenham, U.K.

Pourbaix, M. 1966. *Atlas of Electrochemical Equilibria in Aqueous Solutions*, Pergamon Press, London, U.K.

Pugh, E.N., W.G. Montague, and A.R.C. Westwood. 1966. Stress corrosion cracking of copper. *Corrosion Science*, 6, 345–346.

Rao, T.S. and K.V.K. Nair. 1998. Microbiologically influenced stress corrosion cracking failure of admiralty brass condenser tubes in a nuclear power plant cooled by freshwater. *Corrosion Science*, 40(11), 1821–1836.

Reiber, S.H. 1989. Copper plumbing surfaces: An electrochemical study. *Journal of the American Water Works Association*, 81(7), 114–122.

Rogers, J., J.V. Lee, P.J. Dennis, and C.W. Keevil. 1991. Continuous culture biofilm model for the survival and growth of *Legionella pneumophila* and associated protozoa in potable waters, in *U.K. Symposium on Health-Related Water Microbiology*, Glasgow, U.K.

Rushing, J.C. and M. Edwards. 2004a. Effect of aluminum solids and free Cl_2 on copper pitting. *Corrosion Science*, 46(12), 3069–3088.

Rushing, J.C. and M. Edwards. 2004b. The role of temperature gradients in copper pipe corrosion. *Corrosion Science*, 46, 1883–1894.

Sakamoto, A., T. Yamasaki, and M. Matsumura. 1995. Erosion-corrosion tests on copper alloys for tap water use. *Wear*, 186–187(8), 538–554.

Sarver, E.A. and M.A. Edwards. 2010. Effects of water chemistry and some physical factors on copper pitting corrosion in potable water systems, Presented at the *AWWA ACE*, Chicago, IL.

Sarver, E., Y. Zhang, and M. Edwards. 2010. Review of brass dezincification corrosion in potable water systems. *Corrosion Reviews*, 28(3), 155–196.

Scardina, P., M. Edwards, D. Bosch et al. 2008. Assessment of non-uniform corrosion in copper piping. Final Report to the Water Research Foundation, Denver, CO, ISBN: 978-1-60573-017-2.

Schock, M.R., D.A. Lytle, and J.A. Clement. 1995. *Effect of pH, DIC, Orthophosphate and Sulfate on Drinking Water Cuprosolvency*, USEPA, Washington, DC, EPA/600/R-95/085.

Sequeira, C.A.C. 1995. Inorganic, physicochemical, and microbial aspects of copper corrosion: Literature survey. *British Corrosion Journal*, 30(2), 137–153.

Slabaugh, R. and M. Edwards. 2008. The impacts of flow electrification on copper corrosion in potable water systems, Presented at the *American Water Works Association Annual Conference*, Atlanta, GA.

Smith, B.W. 1965. *Sixty Centuries of Copper*, United Kingdom Copper Association, London, U.K.

Smith, S. and R. Francis. 1990. Use of electrochemical current noise to detect initiation of pitting conditions on copper tubes. *British Corrosion Journal*, 25(4), 285–291.

Sosa, M., S. Patel, and M. Edwards. 1999. Concentration cells and pitting corrosion of copper. *Corrosion*, 55(11), 1069–1076.

Sundberg, R., R. Holm, S. Hertzman, B. Hutchinson, and E. Lindh-Ulmgren. 2003. Dezincification (DA) and intergranular corrosion (IGA) of brass-influence of composition and heat treatment. *Metall*, 57, 721–731.

Syrett, B.C. 1977. Accelerated corrosion of copper in flowing pure water contaminated with oxygen and sulfide. *Corrosion*, 33, 257–262.

Syrett, B.C. 1981. The mechanism of accelerated corrosion of copper-nickel alloys in sulphide-polluted seawater. *Corrosion Science*, 21, 187–209.

Syrett, B.C. and S.S. Wing. 1980. Effect of flow on corrosion of copper-nickel alloys in aerated sea water and in sulfide-polluted sea water. *Corrosion*, 36, 73–85.

Thompson, D.H. and A.W. Tracy. 1949. Influence of composition on the stress-corrosion cracking of some copper-base alloys. *Metals Transactions*, 185, 100–109.

Tracy, A.W., D.H. Thompson, and D.H. Freeman. 1956. The atmospheric corrosion of copper—Results of 20-year tests, in *Symposium on Atmospheric Corrosion of Non-Ferrous Metals*, ASTM STP 175, Philadelphia, PA, 1956, pp. 67–87.

Triantafyllidou, S. and M. Edwards. 2007. Critical evaluation of the NSF 61 section 9 test water for lead. *Journal of the American Water Works Association*, 99, 133–143.

Turner, M.E.D. 1961. The influence of water composition on the dezincification of duplex brass fittings. *Proceedings of the Society for Water Treatment and Examination*, 10, 162–178.

U.S. EPA. 2008. *EPA Registers Copper-Containing Alloy Products*. Accessed June 11, 2010 at http://www.epa.gov/opp00001/factsheets/copper-alloy-products.htm

Varga, I.K. and L.J. Dunne. 1985. Streaming potential cells for the study of erosion-corrosion caused by liquid flow. *Journal of Physics D: Applied Physics*, 18, 211–220.

Vernon, W.H.J. 1931. A laboratory study of the atmospheric corrosion of metals, I. The corrosion of copper in certain synthetic atmospheres with particular reference to the influence of sulphur dioxide in air of various relative humidities. *Transactions of the Faraday Society*, 27, 255–277.

Vernon, W.H.J. 1932. The open-air corrosion of copper, Part III. Artificial production of green patina, *Journal of the Institution of Metals*, 49, 153–161.

Wallinder, O., T. Korpinen, R. Sundberg, and C. Leygraf. 2002. Atmospheric corrosion of naturally pre-patinated copper roofs in Singapore and Stockholm--Runoff rates and corrosion products formation, in H.E. Townsend, Ed., *Outdoor Atmospheric Corrosion*, ASTM STP 1421, American Society for Testing and Materials International, West Conshohocken, PA, pp. 230–244.

Werner, W., H.J. Gorb, and H. Sontheimer. 1994. Untersechuzen zur Flachekorrion in Trinkwasserteitijen aus Kupler. *Wasser. Abwassen.*, 135(2), 92–103.

Zhang, Y. 2009. Dezincification and brass lead leaching in premise plumbing systems: Effects of alloy, physical conditions and water chemistry. MS thesis, Virginia Tech, Blacksburg, VA.

Zhang, Y., N. Love, and M. Edwards. 2009a. Nitrification in drinking water systems. *Critical Reviews in Environmental Science and Technology*, 39, 1–62.

Zhang, Y., E. Sarver, and M. Edwards. 2009b. Effects of water chemistry on dezincification corrosion of brass plumbing components, Presentation at the *American Water Works Association Water Quality Technology Conference*, Seattle, WA.

Zhang, Y., S. Triantafyllidou, and M. Edwards. 2008. Effect of nitrification on GAC filtration on copper and lead leaching in home plumbing systems. *ASCE Journal Environmental Engineering*, 134(7), 521–530.

9

Crystalline Alloys: Zinc

Xiaoge Gregory
Zhang
Zinnovate Inc.

9.1 Introduction and Background

Zinc is a basic element for the healthy and efficient function of human society. It is essential for the normal healthy growth and reproduction of plants, animals, and humans and is commonly used as additives in fertilizers, foods, and medicines (International Zinc Association). Zinc is also the metal used in galvanizing, which is the most cost-effective method for corrosion protection of steel. Nearly half of the zinc is used for this purpose. Through galvanizing, metal spraying, sacrificial anodes, zinc-dust paints, and other methods, zinc-protected steels are widely used in automobiles, building structures, reinforced concrete, roofing, and other domestic and industrial structures. Zinc is also commonly used in many products in various ways and forms of alloying, casting, plating, rolling, sheet, wire, fiber, powder, etc. (Zhang 1996). More recently, zinc in different forms is seen as a unique energy carrier material for materialization of electricity that could potentially be used in a wide range of energy storage and power applications.

In terms of total world consumption, zinc ranks fourth among the metals in worldwide production and consumption, behind only iron, aluminum, and copper, although it is only 23 among the elements in relative abundance in the earth's crust, amounting to 0.013%, compared with aluminum's 8.13% and iron's 5.0%. The widespread use of galvanized steel products owes to the high corrosion resistance of zinc in natural environments, particularly atmospheric environments, and in large measure to its unique property of providing sacrificial protection to steel at places where the zinc coating is damaged. No other coating possesses high corrosion resistance and sacrificial protection at the same time. Also, as a corrosion protection material, degradation and its understanding are particularly important for the effective application of galvanized steel.

9.2 Applications

The applications of zinc can be divided into six major categories: (1) coatings, (2) casting alloys, (3) alloying element in brass and other alloys, (4) wrought zinc alloys, (5) zinc oxide, and (6) zinc chemicals. Cast zinc products are mainly produced by the die-casting process, in which liquid metal is forced under pressure into a cooled metallic die where it solidifies almost instantaneously to produce a fine-grained product. Die cast products are used for automotive parts, household appliances and fixtures, office and computer equipment, and building hardware (Zhang 1996).

Rolled zinc products are in the form of sheet, strip, foil, plate, rod, and wire, with a variety of compositions. Rolled zinc sheet is widely used in building, in the form of roofing, cladding, gutters, rainwater pipes, and flashings. Zinc dust and zinc powder are made by condensation of zinc vapor or by atomization of molten zinc with a jet of air or an inert gas or water. Zinc dust and powder are used mainly as reagents for producing chemicals, in metal refining processes, as a component for making zinc-rich paints, and as an active material for zinc batteries.

Zinc coatings can be produced by hot-dipping, electroplating, mechanical bonding, and thermal spraying (metalizing). Hot-dip galvanizing is the most widely used coating method. Typical applications for zinc and its alloy-coated steel sheet products cover a wide range in the construction, automobile, utility, and appliance industries. Among all coated steel products, continuous hot-dip zinc-coated steel sheet has the widest range of applications and is the largest in terms of tonnage produced and consumed. The electroplated zinc-based coatings are applied primarily on automotive bodies and home appliances where they have advantages in uniform coating thickness and excellent surface characteristics for subsequent painting, but certain disadvantages in cost (Zhang 1996).

Like all products, zinc products face constant challenges in the market place as the needs in society for quality, efficiency, new applications, and other added value and benefit continuously emerge. Despite the wide application of galvanizing, 88% of steel is still not galvanized. The benefit of galvanizing is not always recognized particularly in some developing countries. Also, in some aggressive environments such as in the splash zone by the sea, conventional galvanized coatings do not last and more corrosion-resistant coatings are needed for long-term protection of steel. In the automotive industry, improving fuel efficiency necessitates the use of lighter steel car body and use of high-strength steels. How to produce high-quality zinc coating for high-strength steels is currently a challenge in the galvanizing industry. Another challenge that has been facing the galvanizing industry for a long time is to replace the environmentally challenged chromating surface treatment process. There is also a trend in the industry to develop coating systems that can offer extra surface functions such as heat absorption or reflection, anti-electromagnetic waves, and anti-finger prints, in addition to corrosion resistance with zinc coating and appearance with painting. For application in aggressive environments such as highway bridges and structures immersed in sea water, more corrosion-resistant rebar are needed and a duplex coating system, a polymer coating on top of galvanized rebar, is currently under development.

9.3 General Properties

9.3.1 Physical Properties

Zinc is a silvery blue–gray metal with a relatively low melting point (419.5°C) and boiling point (907°C). Some basic physical properties of zinc are shown in Table 9.1. Zinc crystal has a close-packed hexagonal structure. The axial ratio c/a is 1.856, which is considerably greater than the theoretical value, 1.633, for

TABLE 9.1 Physical Properties of Zinc

Atomic number	30
Atomic weight	65.38
Density	7.14 g cm^{-3}
Melting point	419.5°C
Boiling point, 1 atm.	907°C
Resistivity, 20°C	5.96 Ω cm
Thermal conductivity, 18°C	113 W m^{-1} K^{-1}

the system. Although each zinc atom has 12 near neighbors, 6 are at a distance of 0.2664 nm and the other 6 are at 0.2907 nm. Thus, the bonds between the atoms in the hexagonal basal layers are appreciably stronger than those between the layers. This accounts for much of the deformation behavior and anisotropy of the zinc crystal.

The grain structure in a polycrystalline zinc product has preferred orientations depending on the casting and mechanical working conditions: For cast products the <0001> direction is perpendicular to the axis of the cast columnar crystals; for wire the (0001) plane is parallel to the axis of drawn wire; and for sheet the (0001) plane is parallel to the rolling plane and the <11$\bar{2}$0> direction is parallel to rolling direction for sheet rolled at 20°C.

9.3.2 Mechanical Properties

Zinc is a relatively soft metal. The strength and hardness of unalloyed zinc are greater than those of tin or lead, but appreciably less than those of aluminum or copper. The pure metal cannot be used in stressed applications due to its low creep resistance. Except when very pure, zinc is brittle at ordinary temperatures, but is ductile at about 100°C.

Since atoms in the basal plane are closer to each other than those between adjoining layers, bonding between basal planes is relatively weak, and when under stress the lattice tends to first slip along this plane. At higher temperature, slip may also occur along the (1010) plane. Another major deformation mode of zinc crystal is twinning, which tends to occur along one of the (10$\bar{1}$2) pyramidal planes.

Due to its low melting point, pure zinc recrystallizes rapidly after deformation at room temperature because of the high mobility of the atoms within the lattice. Thus, zinc cannot be work-hardened at room temperature. Zinc has low resistance to creep due to grain boundary migration. The temperature for recrystallization and the creep resistance can be increased through alloying.

9.4 Degradation Principles

9.4.1 Electrochemical Characteristics

The electrochemical properties that are important to the degradation performance of zinc coatings for corrosion protection of steel are its favorable position in the electromotive force series (EMF), fast dissolution/deposition kinetics, large overpotential for hydrogen evolution, and the formation of a porous tenacious corrosion product film. The favorable position in the EMF series (lower than iron but not too low) allows zinc to act as a sacrificial anode when galvanically coupled with steel; the fast reaction kinetics allow a large galvanic protection distance (PD); the large overpotential for hydrogen reaction is the reason that zinc is stable in aqueous environments; and the formation of a porous tenacious corrosion product film provides zinc with high corrosion resistance without losing the effect of galvanic protection.

During the corrosion process zinc is oxidized with simultaneous reduction of hydrogen ions or dissolved oxygen in the electrolyte. The oxidation follows the reaction

$$\text{Zn} = \text{Zn}^{2+} + 2e^- \quad E^0 = -0.763 + 0.0295 \log[\text{Zn}^{2+}] \quad V_{\text{SHE}} \tag{9.1}$$

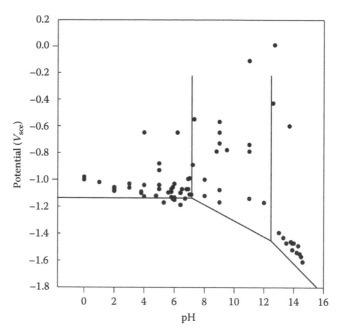

FIGURE 9.1 Corrosion potentials experimentally determined in various solutions with respect to pH; the solid line indicates the reversible potential that is calculated from the Nernst equation assuming 10^{-4} M Zn^{2+} in the solution.

As the standard potential of this reaction is $-0.763V_{SHE}$, zinc is 0.315 V more negative than iron, which is the thermodynamic foundation of using zinc for sacrificial protection of steel. The zinc ion has a radius of 0.74–0.83 Å and a hydration number of 10–12. Owing to the electronic configuration of the outer electronic orbital of the zinc atom, zinc ions typically form tetrahedrally coordinated complexes in solution.

Figure 9.1 shows the Pourbaix diagram of zinc in aqueous solutions. The solid lines define the stability regions of the different solid and dissolved zinc species and, thus, the condition for zinc corrosion and passivation. Also plotted in the figure are the corrosion potentials experimentally determined in solutions of various pH values. The corrosion potential in slightly alkaline solutions, from pH of about 8–12, in which ZnO is the stable form, can be much higher than the reversible value due to the formation of a surface oxide, which, depending on the specific conditions, results in various degrees of passivation.

9.4.2 Oxidation of Zinc

Zinc dissolves readily near its equilibrium potential with the formation of zinc divalent ions. It has a rather high exchange current density in the order of 0.1 A cm^{-2}. In acidic solutions, the dissolution product is simply Zn^{2+}. In alkaline solutions, the predominant zinc species has been identified as the tetrahedral $Zn(OH)_4^{2-}$. Dissolution may occur on a bare surface or on a surface covered by a solid film. Solid films formed during Zn dissolution may have different compositions and various degrees of compactness that significantly affects the dissolution process.

The mechanism of dissolution is different for acidic and alkaline solutions and for complexing and noncomplexing solutions. In acidic and noncomplexing neutral solutions, the Zn/Zn^{2+} electrode reaction has been found to occur in two consecutive charge-transfer steps:

$$Zn = Zn_{ads}^+ + e^- \tag{9.2}$$

$$Zn^+ = Zn^{2+} + e^- \quad \text{r.d.s.} \tag{9.3}$$

where "r.d.s." denotes the rate-determining step.

9.4.3 Hydrogen and Oxygen Reactions

The hydrogen and oxygen reactions are important in the corrosion process, particularly in natural environments where water and air are the major components. The hydrogen evolution can be expressed with the following equations:

$$\text{Acid} \quad H_3O^+ + e^- = H_{ads} + H_2O \quad \text{r.d.s.} \tag{9.4}$$

$$\text{Alkaline} \quad H_2O + e = H_{ads} + OH^- \quad \text{r.d.s.} \tag{9.5}$$

$$2H_{ads} = H_2 \tag{9.6}$$

The exchange current density for hydrogen evolution on a zinc surface is of the order of 10^{-9} A cm^{-2}, which is eight orders of magnitude smaller than that of zinc reaction. This slow kinetics for hydrogen reaction is the reason why zinc is rather stable in water and aqueous solutions.

Different processes may be involved at different overpotentials and in different electrolytes. In solutions with pH values between 3 and 12, hydrogen evolution may be controlled by nonactivation steps. Diffusion of protons to the surface has been found to be the rate-determining step at low overpotentials in slightly acid solutions of pH 3.5–6. In near-neutral and slightly alkaline solutions, hydrogen reduction is affected by the formation of a surface oxide film.

Oxygen reduction is, besides hydrogen evolution, another important cathodic reaction in the corrosion process. On zinc, the reduction reaction has been found to be controlled by different processes in different pH ranges. Within pH 4–6 at corrosion potential, oxygen reduction on zinc is diffusion controlled. In the pH range 6–11, on the other hand, it is controlled by the processes inside the passive film.

When the thickness of the electrolyte on the electrode surface is close to or smaller than that of the diffusion layer, the oxygen reduction rate increases significantly. The reduction current greatly increases with decreasing the electrolyte thickness when it is thinner than 100 μm. The increased reduction current density in thinner electrolytes is due to the reduction of the diffusion layer thickness and to the self-mixing effect in thin electrolytes induced by evaporation. The dependence of oxygen reduction on the thickness of the electrolyte is very important to atmospheric corrosion processes.

9.4.4 Corrosion Products and Passivation

Insoluble zinc corrosion products are produced on the surface as a result of the interaction between the metal substrate and the environment. The morphology of corrosion products can vary greatly depending on the conditions.

Many zinc compounds have been identified in the corrosion products formed in different types of atmospheres. However, for a specific atmosphere, only certain compounds dominate. Generally, among the zinc compounds, oxides, hydroxides, and carbonates are most often found in insoluble corrosion products (Odnevall 1994).

The formation sequence of the major zinc compounds found in corrosion products in atmospheres can be generally illustrated in Figure 9.2. Upon surface wetting during exposure, the zinc surface is covered quickly with zinc hydroxide, which is gradually converted to zinc oxide and carbonate. Within 1 month of exposure, almost all major zinc compounds can be detected in the corrosion products. In the

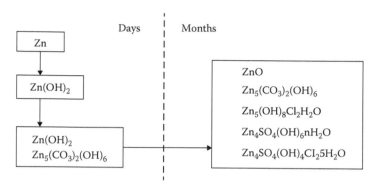

FIGURE 9.2 General sequence of formation of major zinc compounds in the corrosion products formed in atmospheric corrosion environments.

more severe atmospheres, such as marine and industrial atmospheres, the formation of chloride and sulfate compounds can be very rapid, occurring within 1 day. As corrosion continues, the various zinc compounds generally increase in quantity, but may also disappear due to transformation into different compounds depending on the specific environmental factors.

In atmospheric environments, the corrosion of zinc proceeds through the formation of solid corrosion products to dissolution of the corrosion products. In the event of rain, the rainwater that is collected by and run over a zinc surface plus the dissolved/particulate substances including zinc is called runoff. In general, the amount of dissolved and particulate zinc in runoff water relative to the amount of corrosion is about 30%–60% during relatively short exposure and 50%–90% in long exposure. The amount of runoff is affected by many environmental factors such as the amount and intensity of rain, pH, pollutants in the air, and wet and drying patterns.

Zinc surfaces may become passivated when the surface is covered with a compact layer of oxide, which virtually stops any further oxidation. According to the potential–pH diagram shown in Figure 9.1, passivation of a zinc surface does not occur in acidic solutions without the presence of film-forming agents. In slightly alkaline solutions, passivation of zinc is thermodynamically possible through the formation of zinc oxides or hydroxides.

Passivation may also occur due to the formation of zinc salt films under certain conditions. For example, zinc surfaces are readily passivated in chromate solution due to the formation of a composite film of zinc and chromium compounds or in a phosphate solution due to the formation of zinc phosphate film. These two processes are important for the surface treatment of zinc and its alloys.

9.4.5 Galvanic Action

Galvanic action, electrochemical interaction between two metals, occurs when two dissimilar metals immersed in an electrolyte are electrically connected. Figure 9.3 is an illustration for a galvanic couple covered with a thin layer of electrolyte, which is typically encountered in atmospheric environments. The electrochemical interaction between the more noble metal 1 and less noble metal 2 results in a current, galvanic current, flowing from metal 2 to metal 1 through the electrolyte, which is responsible for the galvanic corrosion of metal 2. Galvanic action plays a particularly important role in the performance and degradation process of galvanized steel, which is essentially a galvanic couple.

Many factors can affect the galvanic action for a given metal couple, which may be described by two main groups that are associated with the material and that are associated with the environment. Generally, the material properties of each metal in a given environmental condition are reflected in the potential and surface activity. The difference in the potential and activity between the two metals determines the polarity of the couple and the extent of the galvanic action. Both the potential and the surface activity depend on whether there is a surface film and how compact the film is. When the potential of

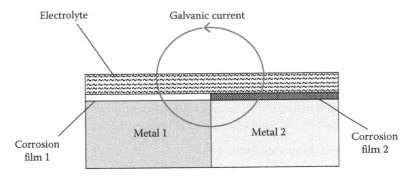

FIGURE 9.3 Schematic illustration of the elements involved in galvanic action.

metal is greatly different from its thermodynamic value under certain situation, a reversed polarity from that indicated by the thermodynamic potential values of the two coupled metals may occur. The direction of galvanic current is reversed when polarity reversal occurs.

The electrolyte affects galvanic action mainly through two aspects: the surface condition of each metal and the conductance of electrolyte. Electrolyte conductance is a very important factor because it determines the extent of galvanic corrosion across the anode surface. Galvanic current is small when electrolyte conductance is low, e.g., either the thickness of electrolyte layer is low or conductivity is low. Also, when conductivity is high, as in sea water, the galvanic corrosion of the anodic metal is distributed uniformly across the surface. As the conductivity decreases, galvanic corrosion becomes concentrated in a narrow region near the junction.

9.5 Degradation of Specific Systems

9.5.1 Aspects of Corrosion of Galvanized Steel

9.5.1.1 Distribution of Corrosion Activities

The overall corrosion behavior of zinc-coated steel with the coating partially removed under a thin moisture layer (e.g., atmospheric environments) is schematically represented in Figure 9.4. There are five regions across the surface. On the zinc side, the surface area away from the zinc/steel boundary experiences only normal corrosion (no galvanic corrosion) and the narrow region at the boundary corrodes galvanically. On the steel side, there are three regions: a region next to the zinc, measured by PD,

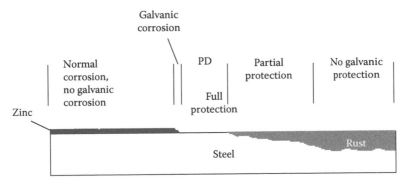

FIGURE 9.4 Schematic illustration of the different regions on a partially zinc-coated steel surface: (1) normal corrosion only; (2) galvanic corrosion mainly; (3) full cathodic protection; (4) partial cathodic protection; and (5) normal corrosion with no cathodic protection. (From Zhang, X.G., *Corrosion*, 56, 139, 2000.)

in which the steel is fully cathodically protected; a region where the steel is partially protected; and a region further away from the zinc where the steel is not cathodically protected and corrodes normally.

The excellent corrosion resistance of zinc-coated steel, whether by pure zinc or zinc alloy coating, is attributed to two principal effects: barrier protection resulting from the high corrosion resistance of zinc coatings and cathodic protection resulting from the galvanic action between the zinc and the steel. Barrier protection is the primary protection mechanism since the majority of the surface area of a galvanized product is covered with the zinc coating, while galvanic protection is secondary and occurs only at places where the zinc coating is not present or has been mechanically removed.

9.5.1.2 Corrosion Rate of Zinc Coatings

The high corrosion resistance of zinc coatings in atmospheric environments is associated with the compactness and tenacity of the corrosion products formed on the surface of zinc. During atmospheric corrosion, the corrosion products formed initially are loosely attached to the surface, but gradually become more adherent and denser, resulting from the wetting and drying cycles of the weathering process. After the formation of this corrosion product layer, further corrosion can proceed only within the pores where the zinc surface is not sealed by the corrosion products, while the rest of the surface area, which is sealed by the corrosion products, is protected from corrosion. This is a dynamic process. With time, some pores become sealed by newly formed corrosion products while some pores are opened due to dissolution of the corrosion products.

The corrosion resistance of zinc can be simplistically described by a model illustrated in Figure 9.5. According to this model, the low corrosion rate observed in atmospheric environments can be described by the following equation:

$$R = \frac{ra}{A} \tag{9.7}$$

where
R is the observed corrosion rate, averaged over the entire surface
r is the actual corrosion rate on an active zinc surface unsealed by corrosion products
a is the area of active zinc surface
A is the area of the entire surface

Because a is small compared to A, the observed rate R is low even though the actual corrosion rate r within the pores may be much higher. Thus, the more compact the corrosion product layer, the smaller the active surface area within the pores and the smaller the observed corrosion rate.

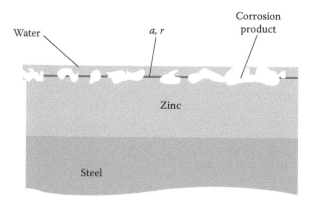

FIGURE 9.5   Schematic illustration of the corrosion mechanism of galvanized steel.

The value of *a*, and thus the observed corrosion rate of zinc, depends on the physical and chemical natures of zinc corrosion products, which depend on environmental conditions and change over time. The more compact and continuous the corrosion product layer, the smaller the active surface area within the pores and the smaller the observed corrosion rate. Thus, the performance of galvanized steel in various environments depends largely on the tendency to form compact and tenacious corrosion products that have a small surface area *a*. The corrosion rates will be high under conditions where tenacious and compact corrosion products cannot form. For example, wet storage stain formed under continuous wetting is loose, having a large *a*, and is thus not protective.

9.5.1.3 Galvanic Corrosion of Zinc Coating

Galvanic corrosion of galvanized steel occurs in areas where the coating is damaged and the steel beneath is exposed, such as at cuts or at scratches. At these areas, the exposed steel is cathodically protected while the surrounding zinc coating is galvanically corroded. However, in most cases for galvanized steel, the amount of coating loss due to galvanic corrosion, compared to normal corrosion, is small because the exposed areas of bare steel are usually too small to cause significant corrosion of the relatively much larger zinc surface area. As a result, the area averaged atmospheric corrosion rate of galvanized zinc coatings, including galvanic and normal corrosion, is usually very similar to that of uncoupled zinc.

Experimental results indicate that the difference in corrosion potentials for uncoupled metals is not a reliable indicator of the extent of galvanic corrosion. For example, the amount of zinc corrosion is larger when zinc is coupled with mild steel than with copper, although the potential difference between zinc and steel is smaller than that between zinc and copper. Other factors, such as reaction kinetics and formation of corrosion products, are more important in determining the rate of galvanic corrosion.

9.5.1.4 Galvanic Protection of Steel

The galvanic corrosion of zinc coating results in galvanic protection of steel. The throwing power of galvanic action, or PD, is illustrated in Figure 9.4. The typical PD for zinc-coated steel with exposed bare steel surface such as at cut edges is typically 1 mm in atmospheric environments. It depends strongly on environmental conditions. As shown in Figure 9.6, the PD in 100% relative humidity was close to zero since the steel surface was almost fully covered with red rust. The very small value of PD was due to the

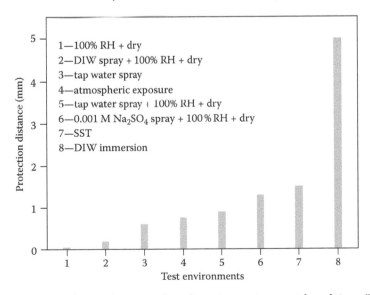

FIGURE 9.6 PD of a planar steel/zinc galvanic couple under various environmental conditions. (From Zhang, X.G., *Corrosion*, 56, 139, 2000.)

extremely high resistance of the very thin moisture layer formed on the surface under the humid condition. A relatively large PD, about 5 mm, was observed under a full immersion test in deionized water (DIW). PD increased from 0.2 mm under a cyclic spray with DIW to 0.6 mm with tap water, indicating the effect of increased conductivity of the water.

9.5.1.5 Polarity Reversal

Polarity reversal of galvanized steel can occur in hot water and diluted solutions, but does not occur in distilled water up to 65°C. Also, without the presence of oxygen, it does not occur in hot water (Zhang 1996). The presence of certain ionic species is often responsible for the reversal. In hot water containing oxygen, the presence of sulfates and chlorides decrease, whereas bicarbonates and nitrates increase the probability of reversal. In the absence of oxygen, zinc is found to be anodic to the steel. Additions of even small amounts (up to 20 ppm) of calcium salts or silicates can also decrease the probability of reversals. The pH of the solutions in which reversal occurs is usually slightly basic.

Polarity reversal observed in hot water and solutions is primarily due to the ennoblement of zinc because the potential of the steel is relatively unaffected by changes in the temperature. The general behavior is as follows: (1) Ennoblement of zinc only occurs in certain waters and solutions; it occurs readily in the presence of bicarbonate and less readily, or not at all, in the presence of chloride or sulfate. (2) The presence of oxygen is necessary for the ennoblement. (3) For a given solution, the tendency for ennoblement increases with increasing temperature.

9.5.2 Corrosion in Atmospheric Environments

Atmospheric corrosion is the most encountered type of corrosion by zinc products as galvanized steel is mostly used in atmospheric environments. Atmospheric corrosion is a complex process involving a large number of interacting and constantly varying factors, such as weather conditions, air pollutants, and material conditions. The combined effect of these factors results in a great variation in corrosion rates, as shown in Figure 9.7 The corrosion rate of zinc in atmospheric environments may vary from as low as about 0.1 μm/year in indoor environments to as high as more than 10 μm/year in some industrial or marine environments, about two orders of magnitude in range. The corrosion rate is lowest in dry, clean atmospheres and highest in wet, industrial atmospheres. Locations near sea water are subject to salt spray and, hence, the corrosion rate can be much higher.

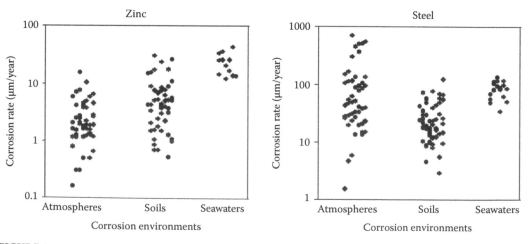

FIGURE 9.7 Corrosion rates of zinc and steel in atmosphere, soil, and sea water. (From Zhang, X.G., *Corrosion*, 55, 787, 1999.)

Except for the initial years, the corrosion loss of zinc is generally observed to be almost linear with respect to time. The corrosion rate in the initial period of exposure tends to be higher than that after several years of exposure. For a given atmosphere, the yearly average corrosion rate may vary because atmospheric conditions, such as the amount of rain or the pollution level, change from year to year.

The most corrosive pollutant in air is sulfur dioxide and high corrosion rates of zinc tend to be found in some industrial areas and highly polluted cities. Other air pollutants, such as NO_x, have a relatively less significant effect on the corrosion of zinc largely due to the much lower content of these species in the air. The corrosion rate of zinc in many cities around the world have been found to decrease over time as the level of pollution in many developed countries has been considerably reduced over the years because of environmental awareness and regulations.

Relative humidity, amount of rain, temperature, and other climatic factors, such as wind and solar radiation, may also affect condensation and the rate of drying, as well as the amount of contaminants and corrosion products retained on the surface. The initial climatic conditions at the time of exposure exert marked effects on the corrosion of zinc. Long-lasting rainfall or a relative humidity at or near 100% during the first days tends to cause a higher corrosion rate.

Many other factors, namely, microscopic ones, such as distance from the ground, orientation of the samples, rain shielding, and distance to local contaminant sources, may also significantly affect the corrosion rate. The size, shape, and orientation of test samples may affect the corrosion rate of zinc considerably. The corrosion rate is higher on the skyward surface than on the groundward surface, even though the wetting time is longer on the groundward surface. This may be attributed to the effect of rain and the retention of larger amounts of pollutants on the skyward surface. Sheltering, which prevents rain from falling on the exposed metal surface, reduced the corrosion rate by two to four times in noncoastal environments. The corrosion rate under sheltered condition can be higher than nonsheltered near the sea coast because deposited surface sea salt has the effect of prolonging the time wetness and rain washes away the salt under nonsheltered condition (Johansson and Gullman 1995).

Corrosion rates many times higher than those typical of an inland location can be found near the seacoast, where the major pollutants in air are chloride salts. In general, at distances greater than 1 km from the seashore, the corrosion rate of zinc is close to that measured inland. The corrosion rate of zinc in a highway environment experienced by automobiles and highway structures due to splashed water that may contain sand, minerals, and de-icing salts is comparable to that of a relatively severe marine atmospheric environment.

In an indoor atmosphere, the corrosion rate of zinc is very low, typically below 0.1 µm/year, making galvanized steel a long-lasting material for indoor applications such as residential housing. Generally, in an indoor environment, the zinc surface only darkens over time as a visible tarnish film forms slowly over the surface. The appearance and the degree of corrosive attack are related to the relative humidity. Significant corrosion only occurs when frequent or sustained moisture condensation occurs on the surface.

9.5.3 In Waters and Solutions

The typical corrosion rate of zinc in distilled water varies over a wide range between 15 and 150 µm/year (Zhang 1996) and depends strongly on the amount of dissolved oxygen and carbon dioxide. The corrosion rate of zinc in distilled water increases only slightly with temperature up to about 50°C, then increases quickly with temperature, reaching a maximum at about 65°C, before decreasing. In general, the corrosion rate of zinc is lower in hard water than in soft water or distilled water. Flowing water causes more corrosion than still water.

The corrosion rate of zinc in seawater may vary between 20 and 70 µm/year, depending on location, length of exposure, type of zinc, etc. It is generally much higher at the beginning of exposure, but tends to decrease with time.

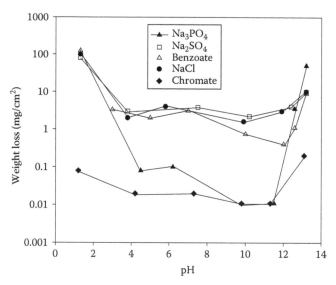

FIGURE 9.8 Effect of pH on weight loss of zinc in 0.1 N solutions of different salts after 4 weeks. (From Lorking, K.F., The corrosion of zinc, Department of Supply, Australian Defence Scientific Service, Metallurgy Report No. 67, Melbourne, Victoria, Australia, May 1967.)

In solutions, the corrosion processes of zinc are greatly influenced by the nature of the anions present. In neutral solutions, with chemical agents that are not electrochemically reactive and that do not form insoluble salts or complex ions with zinc, the corrosion rate of zinc is not very different from that in distilled water. In the absence of reducing or passivating agents, the corrosion of zinc in aqueous solutions is determined primarily by the pH of the solutions. The corrosion rate of zinc in water of pH 6–12 is relatively low primarily due to the formation of protective corrosion products on the surface of the zinc. At pH values lower than 6 or higher than 12, the corrosion rate increases substantially. Figure 9.8 shows the corrosion loss measured for various chemical solutions as a function of pH. The pH dependence of the corrosion rates of zinc in sulfate, chloride, and benzoate is similar to that in water, i.e., relatively low in solutions near neutral or slightly alkaline and high in acidic or strong alkaline solutions. In phosphate solutions, corrosion is inhibited between pH 4 and 12 due to the formation of a less soluble and more protective zinc phosphate film. In chromate solutions, the corrosion of zinc is inhibited almost in the entire pH range of 1–13 due to the formation of a passive chromate-incorporated chromium/zinc oxide film.

9.5.4 In Soils

The corrosion rate of zinc drastically varies from soil to soil because soil may vary over a wide range in chemical and physical properties. For example, the pH of soil may vary from as low as 2.6 to as high as 10.2, and the resistance from several tens of ohms to nearly 100 kΩ. Also, since soil is a highly inhomogeneous environment, both microscopically (e.g., at the dimension of a clay particle) and macroscopically (e.g., at the dimension of a rock), corrosion in soil is seldom uniform across the metal surface (Romanoff 1957).

The factors that may affect the corrosion of zinc and galvanized steel in soils are numerous. There correlation between the corrosion behavior and the various factors in soil is rather poor. In general, the corrosion rate tends to be lower in soils with very high resistivity. Poorly and very poorly aerated soils tend to be more corrosive to zinc. Soils of fair to good aeration, but containing high concentrations of chlorides and sulfates, tend to induce deep pitting. Also, muddy clay and peat (as compared to sand) are, in general, more corrosive to zinc.

9.5.5 Organic Solvents

Corrosion of zinc in organic solvents may greatly vary depending on the specific type and structure of the organic compound. It can be much higher in some solvents while much lower in others than that in water. For example, the corrosion rate of zinc is higher in methanol than in ethanol. The presence of salt or acid generally increases the corrosion rate of zinc in the solvent solution.

Viscosity is a major factor in determining the corrosion rate of zinc in organic solvents containing small amounts of acid. Other factors such as solubility of zinc in solvents, dissociation level of the dissolved acid, or electrolytic conductivity have been found to have only marginal effect on the corrosion rate (Zhang 1996).

9.5.6 Gaseous Environments

Dry air has little effect on zinc at room temperatures, but oxidation occurs rapidly above 200°C. In hydrogen sulfide gas, the corrosion rate of zinc is low when compared to other common metal alloys. The corrosion rate of zinc in a moving mixture of SO_2, water vapor, and air is rather high, but decreases with exposure. When zinc is enclosed with other materials such as woods, which may release vapors in the enclosure, direct contact with these materials causes more corrosion than merely being exposed in the vapor released by these materials.

9.5.7 Alloying

The effect of alloying elements on the atmospheric corrosion performance of zinc is complex. Some elements may be beneficial in one situation while harmful in another. Some elements have little effect on atmospheric corrosion but may enhance corrosion when another element is present.

Among the alloying elements, aluminum is of particular importance as it is widely used in making zinc alloys. When present in small quantities, e.g., 0.3%, aluminum reduces the atmospheric corrosion resistance. With more than 1% aluminum, the atmospheric corrosion resistance of zinc coatings increases with aluminum content. Two major commercial zinc/aluminum alloy coatings, Galfan and Galvalume, have been developed for more corrosion-resistant steel sheets. In general, Galfan is about two times more corrosion resistant than a galvanized coating, and Galvalume is two to four times more corrosion resistant than a galvanized coating. The corrosion of zinc/aluminum alloys proceeds through several stages. First, the zinc preferentially dissolves, leaving an aluminum-rich porous structure. After the zinc is depleted in the coating, depending on the compactness of the remaining structure and the type of atmosphere, red rust may start to form. More recently, more corrosion-resistant Al–Mg–Zn alloy-coated steel sheets have been developed for specialized applications (Tanaka et al. 2001, Tsujimura et al. 2001).

Many different zinc alloy and composite coatings have been explored for automotive applications. Some of the key properties of these products, besides corrosion resistance, are weldability, formability, bondability, and paintability. In general, zinc and zinc alloy coatings substantially reduce the blistering width compared to cold-rolled steel.

9.5.8 Corrosion Forms

The most common form of corrosion encountered by zinc and its alloys is in general corrosion or uniform corrosion. Pitting corrosion and intergranular corrosion are less common for zinc and its alloys; they may occur under certain specific material/environmental conditions. In the galvanizing industry, "wet storage stain" is a particular form of general corrosion that is often encountered during storage and transportation (American Galvanizers Association 1984).

9.5.8.1 Pitting Corrosion

Pitting is not a common form of corrosion in zinc applications. In atmospheric environments, pitting corrosion has been seldom reported as the main cause of failure of zinc products. In soil, pitting may result from the nonuniform nature of corrosion in this environment, and the extent varies significantly depending on the chemical composition and texture of the soil. Pitting may occur in distilled water under an immersed condition, which is believed to be associated with the high resistivity of distilled water.

Pitting can be encountered in hot water and can be a serious problem for galvanized steel hot water tanks. In hot soft water, pitting corrosion is likely to lead to rapid penetration of galvanized coatings because of the reversal of polarity for zinc/steel galvanic couples in hot water. In hot hard water, the corrosion is likely to be stifled by the deposition of a protective scale, depending on the heating method.

9.5.8.2 Intergranular Corrosion

Intergranular corrosion has been observed to occur on zinc alloys in warm and moist conditions. Zinc of high purity is not susceptible to intergranular corrosion. The presence of aluminum as an alloying element or impurity is found to cause intergranular corrosion. It is observed to occur in a concentration range between 0.03% and 50% Al and is most severe around 0.2% Al as shown in Figure 9.9 (Devillers 1974). Aluminum precipitates at the grain boundaries and is responsible for the increased corrosion rate at the grain boundaries.

Temperature and pH are the most significant of the environmental factors. The intergranular corrosion rate sharply increases with temperature above 70°C. Between pH 5 and 10, the corrosion penetration rate is almost constant. Below pH 5, it decreases with decreasing pH. At pH values above 10, it increases drastically with increasing pH.

As a result of intergranular corrosion, the strength of zinc alloys can be drastically reduced. For example, for an alloy containing more than 0.05% Al after 10 days of exposure in 95°C water vapor, there is practically no mechanical strength left (Devillers 1974).

9.5.8.3 Wet Storage Stain

"Wet storage stain" is a term used to describe the zinc corrosion products formed on galvanized steel surfaces during periods of storage and transportation (American Galvanizers Association 1984). It is voluminous, white, powdery, and bulky and is formed when closely packed galvanized articles are stored under damp and poorly ventilated conditions. The crevices formed between the articles can attract and absorb moisture and retain the wetness more readily than the surface area exposed to the open air.

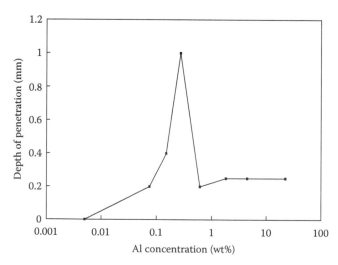

FIGURE 9.9 Average depth of intergranular corrosion penetration of Zn–Al alloys as a function of aluminum concentration in water vapor at 95°C for 10 days. (From Devillers, L.P., *The mechanism of aqueous intergranular corrosion in zinc–aluminium alloys*, PhD thesis, University of Waterloo, Waterloo, Ontario, Canada, 1974.)

The moisture necessary for the formation of "wet storage stain" may result from direct exposure to rain or seawater or from condensation caused by atmospheric temperature changes. Close packing can result in moisture being retained by capillary action between the surfaces in contact because drying is delayed by the lack of circulating air. Under sustained wetting, a fluffy "white rust" is formed. Due to this loose nature, it has little barrier effect on the access of solution to the zinc metal and also prolongs the time of wetness.

"Wet storage stain" discolors the galvanized steel surface and, in some situations, can seriously affect the cosmetic appearance of the galvanized steel articles. However, it is generally not harmful to the long-term corrosion performance. "Wet storage stain" can be prevented by properly stacking, and storing galvanized products under dry and ventilated conditions, and by the use of surface treatments (American Galvanizers Association 1984).

9.6 Degradation Protection

The surface of zinc can be treated to increase the corrosion resistance. Surface treatments are commonly applied in the galvanizing industry mainly for providing some protection of galvanized products during the period of storage and transportation when the products may encounter surface wetness caused by condensation and rain. The surface of zinc can be treated by applying a surface layer of oil, or organic and inorganic solutions, by anodization, by electroplating, and by painting.

The process most commonly used has been chromating. The chromating process involves dipping in a solution containing chromate, resulting in the formation of a conversion coating consisting of chromium oxide and chromate. Chromate treatment is very effective in delaying the onset of corrosion on a zinc surface and in preventing the formation of wet storage stain. However, due to environmental concerns, the use of chromating is becoming increasingly limited and various kinds of non- or low-chromate-containing solution treatment processes have been developed and applied in industry (Carlsson et al. 2001, Yoshimi et al. 2001).

Phosphating is another surface treatment extensively used for surface treatment of zinc and its alloys. Since phosphating significantly alters the surface appearance, it is used less as a surface finishing process and more as a pretreatment for painting (Whitaker and Fry 1958).

Anodization is a process to produce a passive film, either oxide or salt, of certain thickness and properties. Anodic coatings of varying colors, from white to gray to black, can be produced in aqueous solutions of sodium hydroxide and sodium carbonate. Anodization is not commonly practiced since it is a more complex process compared to solution dipping type of treatments.

Plating a more corrosion-resistant metal or alloy layer on the zinc surface can also be used to increase the service life of zinc products. Copper, nickel, and chromium coatings are commonly plated on the surface of zinc die-casting articles for surface appearance and improved corrosion resistance. Care must be taken not to breach the surface coating, however, since that could result in accelerated corrosion of the zinc substrate by galvanic action.

Painting is widely used for increasing corrosion resistance and aesthetic appearance in both coil and other product forms. Painted zinc-coated steel is commonly referred to as a duplex system. Depending on the type of paint formulation and the application process, the corrosion performance greatly varies. Painted galvanized steels, due to their superior corrosion performance and versatility in surface finishing, have been widely used in automotive bodies, appliances, wall panels, building roofs, etc. In particular, the use of painted galvanized steel, starting late in the 1980s, has greatly improved the corrosion life of automotive bodies and is responsible for the virtual disappearance of rusty cars. The thin layer of zinc coating effectively delays the occurrence of red rust on painted car bodies from a few years to more than 10 years. More recently, more complex paint systems are used to provide added functional properties such as light and heat absorption and reflection, finger print resistance, and anti-electromagnetic and anti-slip effects.

In general, corrosion of painted products starts at places where the paint continuity is damaged. The process begins with corrosion of the coating or with paint delamination, followed by corrosion of the substrate and, with time, leads to perforation of the steel. For cold-rolled steel, the corrosion products built up at the corrosion front and may mechanically delaminate the paint. Delamination can occur at different interfaces in a paint-coating-steel system, depending on the material and environmental conditions. The causes of the delamination at the corrosion front can be physical, electrochemical, mechanical, or combinations thereof.

9.7 Corrosion Testing and Specific Standardized Tests

9.7.1 Corrosion Tests

Properly testing the corrosion behavior of zinc-coated steel products for various applications has been an important issue in the galvanizing industry. Various types of corrosion tests have been developed for assessing the corrosion performance of zinc products depending on the application and purpose. Some are industry-wide standardized tests and some are only used by individual companies. The following corrosion tests are commonly used in the steel industry for evaluating the corrosion performance of zinc and zinc-coated steel products:

- Salt Spray Test (ASTM B117). This standard describes the apparatus, procedure, and conditions required to create and maintain the salt spray (fog) test environment. This is the earliest and most widely used standard corrosion test.
- Atmospheric Corrosion Test (ASTM G50-76). This practice defines conditions for exposure of metals and alloys to the weather.
- QUV Weatherometer Test (ASTM G53). This test procedure has been used as a tool to evaluate the performance of polymeric materials under UV light exposure that is encountered in outdoor environment.
- Prohesion Test (ASTM G85 A5). It is a variation of ASTM B117 when a different or more corrosive environment than the salt fog described in ASTM B117 is desired.
- Humidity Test (ASTM D4585). This test is primarily intended to determine the water resistance of coatings by controlled condensation.
- Immersion Corrosion Test (ASTM G31-72). This standard describes accepted procedures and factors that influence laboratory immersion corrosion tests.
- Laboratory Cyclic Corrosion Test (SAE J2334). This is a test procedure to simulate the corrosion performance in an automotive environment. Results from this test can provide good correlation to field condition for painted zinc and its alloy-coated steels.

For application in the automotive industry, testing is typically done on painted panels, while for application in the construction market, testing is typically done on unpainted panels.

9.7.2 Relevance of Corrosion Testing Results

It is important to assess the relevance of the test before selection of a particular corrosion test. The relevance of a corrosion test for an application in a specific environment can be measured using the ratio of the corrosion rate of steel to that of zinc. The principle for the ratio measurement is based on the premise that a relevant test will accelerate corrosion similarly for different materials, such as steel and zinc (Zhang et al. 1998).

Figure 9.10 shows the corrosion rates and corrosion rate ratios obtained for a number of different laboratory tests. Tests 1–6 (with corrosion ratios of less than 5) are not relevant tests for evaluating the performance of galvanized steel in atmospheric environments since the corrosion ratios of steel to zinc in atmospheric environments are higher than 20 (Zhang 1999). In particular, the ASTM Standard Salt Spray Test B117, which gives a ratio of less than 2, is not a relevant test. Here the use

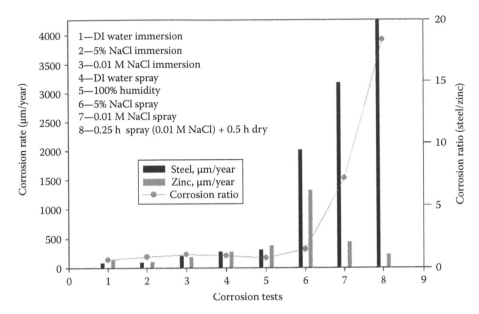

FIGURE 9.10 Corrosion rates of steel and zinc and the corrosion rate ratio between the two materials in different corrosion tests.

of concentrated salt and the lack of cyclic drying are the main factors for lack of relevance of B117 to actual atmospheric performance.

The fundamental reason that corrosion ratios vary with testing conditions is due to the different effects that those conditions have on the formation of corrosion products on the two metals. The rust on steel (due to its porous nature, lack of adherence to the surface, and catalytic effect for cathodic reactions) has little protective effect on the steel surface beneath. On the other hand, the corrosion products formed on zinc surfaces in most natural environments are compact and tenacious and, thus, provide protection to the zinc surface. The effect of periodic drying is to increase the compactness of the corrosion products of zinc but not those of steel.

There is no intrinsically bad test or good test in a general sense, and every test can be a good test if it generates results relevant to the specific application conditions. Every corrosion test has a set of conditions to generate specific corrosion phenomena for a specific metal alloy. Continuous salt spray may provide a close simulation of the condition in a spray zone of a seacoast, but is not a good test for corrosion in atmospheric environments. For atmospheric environments, dilute salt spray, in addition to cyclic drying, produces more realistic results.

9.8 Summary

Environmental degradation is the most important issue for the application of zinc as half of the world zinc production is used for corrosion protection of steel in the form of coatings. The effectiveness of zinc coatings owns to its unique properties: high corrosion resistance in natural environments such that the coatings serve as a good barrier to the environments, and provide sacrificial protection at places where the zinc coating is damaged. No other coatings, metal or nonmetal, have such properties, which is why zinc coatings through galvanizing has become the most important method for corrosion protection of steel.

The detailed degradation behavior and mechanisms of zinc coatings in various specific environments are as complex as the environments themselves, with wide range of corrosion forms and corrosion rates. Owing to the widespread and long time use of zinc-coated steels, a tremendous amount of research work has been done on the various aspects of environmental degradation of zinc coatings, resulting in a large body of technical information distributed in numerous publications and documents. This chapter is only very short synapses of the most pertinent information on the environmental degradation of zinc coatings.

The effectiveness of corrosion protection by zinc coatings can be further enhanced through alloying, painting, or multilayer composite coating systems. New zinc coating systems with better resistance to environmental degradation will allow more steel products be protected under more environmental conditions and for longer times.

References

American Galvanizers Association. 1984. Wet Storage Stain, Brochure.

Carlsson, P., U. Bexell, and S.E. Hornstrom. 2001. Corrosion behavior of Aluzink with different passivation treatment, in *Proceedings, Galvatech'01*, Brussels, Belgium, p.670.

Devillers, L.P. 1974. The mechanism of aqueous intergranular corrosion in zinc–aluminium alloys, PhD thesis, University of Waterloo, Waterloo, Ontario, Canada.

International Zinc Association, http://www.iza.com (accessed June 10, 2013).

Johansson, E. and J. Gullman. 1995. Corrosion study of carbon steel and zinc, in W.W. Kirk and H.H. Lawson, Eds., *Atmospheric Corrosion, STP 1239*, ASTM, Philadelphia, PA, p.240.

Lorking, K.F. 1967. The corrosion of zinc, Department of Supply, Australian Defence Scientific Service, Metallurgy Report No. 67, Melbourne, Victoria, Australia, May 1967.

Odnevall, I. 1994. Atmospheric corrosion of field exposed zinc, PhD thesis, Royal Institute of Technology, Stockholm, Sweden.

Romanoff, M. 1957. *Underground Corrosion*, U.S. National Bureau of Standards Circular 579, Issued April 1957.

Tanaka, S., K. Honda, A. Takahashi, Y. Morimoto, M. Kurosaki, H. Shindou, K. Nishimura, and M. Sugiyama. 2001. The performance of Zn–Al–Mg–Si hot-dip galvanized steel sheet, in *Proceedings, Galvatech'01*, Brussels, Belgium, p.153.

Tsujimura, T., A. Komatsu, and A. Andoh. 2001. Influence of Mg content in coating layer and coating structure on corrosion resistance of hot-dip Zn–Al–Mg alloy coated steel sheet, in *Proceedings, Galvatech'01*, Brussels, Belgium, p.145.

Whitaker, M. and H. Fry. 1958. The anodic oxidation of zinc and zinc–aluminium alloys in a sodium hydroxide-sodium carbonate solution, The British Non-Ferrous Metals Research Association, Report No. A.1172, February 1958.

Yoshimi, N., S. Ando, A. Matsuzaki, T. Kubota, and M. Yamashita. 2001. Newly developed chromium-free thin organic coated steel sheets with excellent corrosion resistance—Trend in coated steel sheets for electrical appliances, in *Proceedings, Galvatech'01*, Brussels, Belgium, p.655.

Zhang, X.G. 1996. *Corrosion and Electrochemistry of Zinc*, Plenum, New York.

Zhang, X.G. 1999. Corrosion ratios of steel to zinc in natural corrosion environments. *Corrosion*, 55, 787 (NACE International, Houston, TX).

Zhang, X.G. 2000. Galvanic protection distance of zinc-coated steels under various environmental conditions. *Corrosion*, 56, 139 (NACE International, Houston, TX).

Zhang, X.G., J. Hwang, and W.K. Wu. 1998. A corrosion testing of steel and zinc, in *Proceedings, Galvatech'98*, Tokyo, Japan, p.410.

Further Readings

Porter, F.C. 1991. *Zinc Handbook: Properties, Processing, and Use in Design*, CRC Press, Boca Raton, FL.

Porter, F.C. 1994. *Corrosion Resistance of Zinc and Zinc Alloys*, CRC Press, Boca Raton, FL.

Zhang, X.G. 1996. *Corrosion and Electrochemistry of Zinc*, Plenum, New York.

Zhang, X.G. 2005. Corrosion of zinc and zinc alloys, in *Corrosion: Materials, ASM Handbook*, Vol. 13B, ASM International, Materials Park, OH, p.402.

10

Nanostructured Alloys

R.K. Singh Raman
Monash University

10.1 Introduction

Single or multiphase polycrystalline solids with grain size typically less than 100 nm are known as nanocrystalline (nc) materials. These materials can be zero dimensional (clusters), one dimensional (lamellar), two dimensional (filamentary), or three dimensional (equiaxed particles) (Suryanarayana and Froes 1992, Siegel 1994). However, a fine microstructural feature (such as a precipitate), having only one of the dimensions less than 100 nm, may still be called nanocrystalline. Nanocrystalline material, a generic term, finds similar expressions in different contexts, such as nanocrystals, nanostructured materials, nanophase materials, nanometer-sized crystalline solids, or solids with nanometer-sized features. In this chapter, the term "nanocrystalline" has been used exclusively for metallic materials with grain size less than 100 nm. Because of the extremely fine grain size, a remarkably high volume of the nc materials is composed of interfaces (grain boundaries and triple points) (Schaefer 1993, Koch et al. 2005). As a result, those properties that depend on the grain size and grain boundaries may be considerably different for nc metals and/alloys and their conventional microcrystalline (mc) counterparts. For example, nc metals and alloys exhibit increased mechanical strength, enhanced diffusivity, and higher specific heat and electrical resistivity. Though ferrous systems (primarily iron and iron–chromium systems) are the major materials covered in this chapter, the reported literature on the corrosion resistance of other systems have also been reviewed.

10.2 Applications (Uses and Challenges)

Attempts are being made for exploiting the attractive physical properties of nc metallic materials. However, several of such applications will require the materials to demonstrate acceptable levels of resistance to environmental degradation. A proper understanding of environmental-assisted degradation of nc metallic materials metals/alloys is particularly important, since grain size and grain boundaries are known to influence corrosion processes. Corrosion resistance of nc metals and alloys has received very limited research attention. However, nc metals and alloys have been reported to exhibit different oxidation/corrosion resistance than their mc counterparts (Kirchheim et al. 1992, Rofagha et al. 1993, Singh Raman et al. 2010). It is also emphasized that besides the interest in investigating the role of nc structure in corrosion, another aspect is the possibility of exploiting the enhanced grain boundary phenomenon (such as diffusion) for the purpose of developing corrosion-resistant alloys with considerably less alloying contents (Singh Raman et al. 2010). In this context, developing commercially viable materials processing routes for producing and retaining nc structure of corrosion-resistant alloys is a major challenge. Therefore, this chapter also provides an overview of the challenge in synthesis and the attempts for circumventing these challenges.

10.3 Structure and Mechanical Properties of Nanocrystalline Metals

Though there are variations of a few orders of magnitude in their grain size, the structure and dimensions of grain boundaries in nc and mc materials have been suggested to be similar (Gleiter 1989, Siegel 1991, 1993a,b). However, based on thermodynamic properties of nc materials, Fecht (1992) suggested the grain boundary energy to be considerably greater in the case of nc iron (produced by ball milling) as compared to the same alloy composition in the mc state. Several other researchers believe grain boundaries in nc state to be more disordered than in the conventional mc materials (Horváth et al. 1987, Mütschele and Kirchheim 1987, Zhu et al. 1987, Haubold et al. 1989, Wallner et al. 1989). Though the "grains" and "grain boundaries" in nc state have been visualized in a way that is entirely different from the traditional concept of large grains separated by considerably thin boundaries in the case of conventional mc materials, in this chapter the terms "grains" and "grain boundaries" will continue to be used in the traditional sense. However, it may be important to reflect that the structure of nc materials is visualized as consisting of two components: a crystalline component (CC), which is formed by small equiaxed single crystals and the intercrystalline component (IC) (Gleiter 1989, 2000), as shown in Figure 10.1. The IC component, which is less close packed, forms a network and surrounds the CC crystallites. The fraction of IC increases with decreasing size of crystallites and may even exceed CC fraction (when the crystallites are of sizes below a critical size).

As described earlier, because of their remarkably fine grain size, the grain boundaries constitute a remarkably large volume fraction of nc materials. As a result, those mechanical, electrical, magnetic, and chemical properties that profoundly depend on grain size and grain-boundary phenomena are also remarkably different for nc materials as compared to their mc counterparts. The basic degradation and mechanical behavior associated with grain size/grain boundary phenomena are particularly relevant in the context of this chapter. While the degradation behavior associated with grain size/grain boundary will be discussed in detail in Sections 10.5 through 10.7, the influence of nc structure on basic mechanical properties (such as hardness and strength) is briefly discussed here.

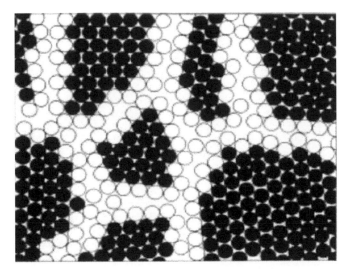

FIGURE 10.1 Model of a nanostructured material: filled circles represent crystalline component atoms whereas the open circles represent intercrystalline component atoms. (From Gleiter, H., *Prog. Mater. Sci.*, 33(4), 223, 1989; Gleiter, H., *Acta Mater.*, 48(1), 1, 2000.)

Yield strength (σ) of conventional mc materials increases linearly with decrease in grain diameter (d), following the Hall–Petch relationship (Hall 1951, Petch 1953)

$$\sigma = \sigma_0 + k_0 d^{-1/2} \tag{10.1}$$

where
  σ_0 is the yield strength due to phenomenon within the grains (i.e., without any contribution from grain boundaries in hurdling dislocation motion)
  k_0 is associated with a measure of resistance due to the grain boundary

Hardness of conventional mc materials shows a similar dependence on grain size.

The experimental attempts to investigate mechanical properties of nc materials are largely limited to hardness testing (e.g., nano-indentation technique), apparently because of the difficulties in producing artifact-free specimens of the desired size for other conventional tests (e.g., tensile testing). However, by deriving yield strength from Vickers hardness (H_v) data ($H_v = 3\sigma$ [Taboor 1951]), the validity of Hall–Petch relationship has been investigated for a few nc metals and alloys. For example, hardness of nc Fe is reported to be several times greater than the mc Fe (Jang and Koch 1990, Le Brun et al. 1992, Fougere et al. 1995, Michalski 1997, Savader et al. 1997, Malow and Koch 1998a,b, Korznikova et al. 1999, Khan et al. 2000, Hernando et al. 2003, Jang and Atzmon 2003, Lee et al. 2005, Rodríguez-Baracaldo et al. 2007). The Hall–Petch behavior is reported to be followed over a considerable range of nanometric grain size. However, a considerable departure is reported in the extremely fine grain size regime (~10 nm), as suggested in Figure 10.2 (Kumar et al. 2003). In fact, below critical nc grain sizes, hardness of a few materials (e.g., Fe [Khan et al. 2000], Cu and Pd [Suryanarayana 2002]) has been found to decrease with grain size (and the phenomenon is called "Inverse Hall–Petch relationship" [Palumbo et al. 1990, Kim and Okazaki 1992, Loffler and Weissmuller 1995, Khan et al. 2000, Suryanarayana 2002, Jang and Atzmon 2003]).

Though there seems to be no concrete consensus on the reason for the departure from the Hall–Petch behavior in the case of nc materials, a few formidable mechanisms have been proposed. In conventional mc materials, the pile-up of dislocations at grain boundaries is the mechanism for hardening due to grain refinement and the Hall–Petch behavior. This mechanism is suggested (Nieh and Wadsworth 1991) to operate also in nc materials but only until the grain size of the nc materials

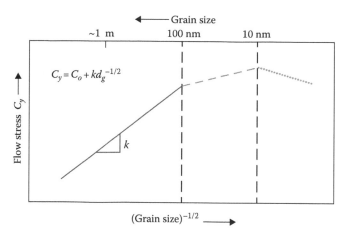

FIGURE 10.2 Dependence of the strength of materials on grain size in the micrometric and nanometric regimes. (From Kumar, K.S. et al., *Acta Mater.*, 51(19), 5743, 2003.)

is large enough to sustain the dislocation pile-up. Other researchers have also suggested that the dislocation source may not operate in the extremely fine-grained materials (Gryaznov et al. 1990, Nieh and Wadsworth 1991). Though a few mechanisms have been proposed for the mechanical behavior of nc materials (Meyers and Chawla 1984, Chokshi et al. 1989, Gryaznov et al. 1990, Nieh and Wadsworth 1991, Suryanarayana et al. 1992), it is fair to conclude that a consensus mechanistic understanding is yet to be reached. Since a detailed treatment is outside the scope of this chapter, the interested readers should read an article by Kumar et al. (2003) and Meyers et al. (2006) for an elaborate mechanistic description.

10.4 Thermal Stability and Synthesis of Nanocrystalline Metals and Alloys

10.4.1 General Principle

Because grain boundaries (described also as IC area as in Figure 10.1) constitute a remarkably high-volume fraction of nc materials, the interfacial energy of such materials is generally considerably higher. As a result, nc materials are exceedingly susceptible to thermally assisted grain growth (Gleiter 1989, 2000), which decreases the total energy of the system. Therefore, it is often essential to take into account their thermal instability while processing such materials. Thermal instability may also be a factor before considering any elevated temperature engineering application of nc materials. In fact, some of the low melting engineering metals (viz., Sn, Pb, Al, and Mg) can undergo considerable grain growth even at normal room temperatures (Birringer 1989, Gleiter 2000). However, metals with high melting points may possess thermal stability at considerably higher temperatures. For example, iron and iron-based alloys resist grain growth up to higher temperatures, 400°C–600°C (Gleiter 1989, Moelle and Fecht 1995, Malow and Koch 1997, Perez et al. 1997, 1998, Bonetti et al. 1999, Natter et al. 2000, Gupta et al. 2008). Similar behavior has been reported for other metals (e.g., Co [Song et al. 2006]). However, there is some inconsistency in the literature on the temperature at which sudden grain growth may kick in for the same material. In their study on grain growth of pulsed electrodeposited nc Fe, Natter et al. (2000) reported a sudden grain growth at 410°C; whereas Malow and Koch (1997) did not find such behavior until a temperature range of 500°C–530°C. This inconsistency is generally attributed to the presence of different levels of impurities in the nc materials that may come inherently from processing route employed in different studies (Moelle and Fecht 1995, Malow and Koch 1997, Bonetti et al. 1999, Omura et al. 1999, Natter et al. 2000). The presence of impurities may enhance thermal stability because of classical reasons such as pinning of the grain boundaries including that due to the formation of second-phase particles (Perez et al. 1997, Song et al. 2006). Other factors have also been reported to impede grain growth, viz., grain boundary segregation (Eckert et al. 1993), solute drag (Knauth et al. 1993), and chemical ordering (Gao and Fultz 1994, Bansal et al. 1995). It is interesting that the role of impurities/alloying elements in impeding grain growth may provide a recipe for developing routes for processing nc materials (alloys) at elevated temperatures.

10.4.2 Synthesis of Fe and Fe-Based Nanocrystalline Materials

Several techniques have been employed for producing nc solids in powder and thin film forms. Table 10.1 lists such techniques for iron and iron-based alloys. However, these techniques can be used also for producing other nc solids.

The techniques of inert gas condensation, pulsed plasma deposition, chemical vapor condensation, and sputtering have exclusively been used for processing thin films or small amounts of nc materials, whereas it is necessary to produce/process bulk samples for corrosion or mechanical testing. Electrodeposition (Natter et al. 2000a,b, Karimpoor et al. 2003) and severe plastic deformation (Fecht et al. 1990, Daróczi et al. 1993, Sinha and Collins 1994, Bonetti et al. 1995, 1999, Moelle and Fecht 1995, Malow and Koch 1997, 1998a,b, Szabó et al. 1997, Elkedim et al. 1998, 1999, Malow et al. 1998, Tian and Atzmon 1999, Khan et al. 2000, Sobczak et al. 2001, Koch 2003, Cheng et al. 2005) have been recognized as the two relatively successful routes for processing nc materials in bulk quantities. Pulsed electrodeposition has been employed successfully for processing nc materials in bulk (Karimpoor et al. 2003), most notably Ni–Fe and Ni–Co alloys (Seo et al. 2005, Hibbard 2006, Hibbard et al. 2006). However, synthesis of nc alloys or metals by electrodeposition often requires use of additives for the purpose of biasing nucleation over growth of the depositing grains. These additives are believed (Koch et al. 2005) to remain in the material as impurities and may cause poor mechanical properties (such as embrittlement), typically observed in nc electrodeposits.

Among the plastic deformation techniques, advanced ball milling has produced artifact-free nc powders (Youssef et al. 2005, 2006). However, the ball-milled powders need to be compacted. Groza (2007) has reviewed various techniques employed for compaction of different nc materials, viz., high pressure/lower temperature compaction, *in situ* consolidation (Youssef et al. 2004), hot compaction (Elkedim et al. 2002), and explosive compaction (Guruswamy et al. 2000). However, compaction

TABLE 10.1 Techniques Employed for Producing Nanocrystalline Iron

Technique	Grain Size (nm)	References
Inert gas condensation	7–31	Tanimoto et al. (1999), Trapp et al. (1995), Hernando et al. (2003), Fougere et al. (1995), Segers et al. (1999)
High-energy ball milling	6–30	Malow and Koch (1997, 1998a), Khan et al. (2000), Moelle and Fecht (1995), Fecht et al. (1990), Daróczi et al. (1993), Sinha and Collins (1994), Bonetti et al. (1995, 1999), Malow et al. (1998), Szabó et al. 1997, Elkedim et al. (1998, 1999), Tian and Atzmon (1999), Sobczak et al. (2001)
Pulsed electrodeposition	16	Natter et al. (2000a,b)
Pulsed plasma deposition	2	Michalski (1997)
Chemical vapor condensation	10	Lee et al. (2005)
Sputtering	10–20	Hernando et al. (2003)
Severe plastic deformation (cold rolled + torsion)	50	Korznikova et al. (1999)

of nc Fe or Fe-based alloys may be a nontrivial task. The difficulties arise due to the restrictions on plastic deformation posed by the body-centered cubic (BCC) structure causing high hardness. Such restrictions necessitate consolidation at high pressures/temperatures. For example, pure iron with an average grain size of 10 nm has a hardness of 10 GPa (Siegel 1997). Plastic deformation, a necessary condition for effective compaction, requires the applied pressure to be in excess of the yield strength and approximately one-third of the hardness (i.e., 3.5 GPa). The necessary plastic flow, high densification, and inter-particle bonding can be achieved by compaction/sintering at high temperatures. However, processing at excessively high temperatures will commonly lead to grain growth, and loss of nc structure.

The author and his coworkers have successfully processed nc Fe–Cr alloy discs with close-to-theoretical density and without excessive grain growth (Gupta et al. 2008). Nc powder of Fe–10 wt% Cr alloy was produced by ball milling, and the powder was compacted into discs by employing a suitable combination of prior thermal softening and compaction at a moderate temperature and pressure. However, it was essential to establish a suitable temperature regime for softening without causing an excessive grain growth. For this purpose, the ball-milled alloy powders were annealed at 500°C, 600°C, and 700°C in a forming gas atmosphere for different durations, and average grain size of the powder was determined by the well-established x-ray diffraction (XRD) technique (De Keijser et al. 1982), after various intermittent durations. Figure 10.3 shows change in XRD patterns of Fe–10Cr alloy powder with increasing time of annealing at 600°C. As shown in Figure 10.3, the diffraction line broadening of Bragg reflection peaks decreases as a result of the grain growth, which increases with the annealing time (Figure 10.4). This pattern was consistent also at 500°C and 700°C. However, grain growth is a strong function of temperature, as suggested from the increasing intensity of the initial grain growth with temperature (Figure 10.4). Initial grain growth is rapid at each of the three temperatures. The as-milled and annealed powder samples were compacted into pellets under a uniaxial pressure. It was necessary to reduce hardness of the powders, before subjecting them to a pressure compaction. Based on the grain growth data shown in Figure 10.4, a prior annealing at 600°C for 30 min was selected with a view to prior softening of the powder without any excessive grain growth. The prior annealing at 600°C for 30 min though would result in some grain growth (as suggested in Figure 10.4), the grain size of the

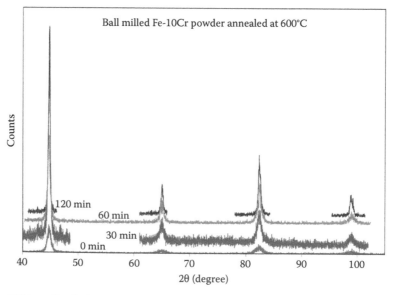

FIGURE 10.3 XRD profiles for ball-milled nanocrystalline Fe–10Cr alloy annealed at 600°C for different times. (From Singh Raman, R.K. et al., *Philos. Mag.*, 90(23), 3233, 2010.)

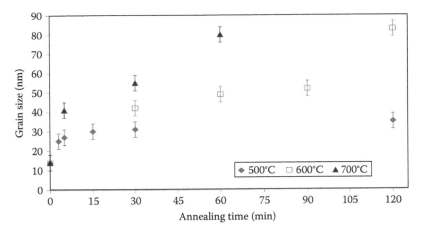

FIGURE 10.4 Grain growth in Fe–10Cr nanocrystalline alloy at 500°C, 600°C, and 700°C. (From Singh Raman, R.K. et al., *Philos. Mag.*, 90(23), 3233, 2010.)

alloy was still found to be 42 nm, which is well within the nc range. Most importantly, because of the softening caused by the prior annealing, it was possible to compact the powder into pellets (diameter = 12 mm, thickness = 1.5 mm) at room temperature at a pressure of 3 GPa. Compacted pellets were sintered for 1 h at 600°C, which further improved the density (close to 100% of the theoretical density). Though the sintering caused some further grain growth, the grain size of the sintered pellets was determined to be 52 ± 4 nm.

10.5 Environmental Degradation of Nanocrystalline Metals and Alloys (Corrosion)

Most conventional metals suffer corrosion as a result of their inherent thermodynamic instability in their elemental form. However, the progress of corrosion is largely governed by one or combination of the following phenomena:

1. Electrochemical nonhomogeneities at the metal surface
2. Diffusion of elemental/ionic species
   a. In the metal substrate
   b. Through the layer of corrosion product or through the electrolytic environment immediately adjacent to the metal surface

The nc structure can remarkably influence the nature and/or degree of both electrochemical nonhomogeneities at the surface and diffusion in the bulk of metals. Besides that, there may be other influences due to the nc structure (the degree of which will vary from metal to metal). For example, the structure (such as grain size) and associated property (such as mechanical property) of the corrosion films developed on nc substrate may be considerably different from those on a mc substrate. Thus, the nc structure of the metal influencing the structure of the corrosion films can also indirectly influence diffusion through the layer of corrosion film.

It is relevant to note that diffusion in the metal and through the corrosion film is the predominating phenomenon in the case of degradation of metals due to gaseous corrosion at elevated temperatures. On the other hand, electrochemical nonhomogeneities will essentially have a predominating influence on electrochemical corrosion at low temperatures.

10.6 Electrochemical Degradation of Nanocrystalline Materials

10.6.1 General Principles

Given that the grain boundaries and triple junctions (i.e., high-energy regions with a much greater degree of disorder) are nearly always anodic (Maurer et al. 1984, Yamashita et al. 1991, Kirchheim et al. 1992, Barbucci et al. 1999, Aledresse and Alfantazi 2004), the corrosion rate of nc materials in most simplistic terms would be expected to be considerably higher than that of mc materials of similar chemical composition, and this view is often supported in the literature (Rofagha et al. 1991, 1993, Kirchheim et al. 1992, Barbucci et al. 1999, Kim et al. 2003, Aledresse and Alfantazi 2004). However, the corrosion resistance of some other materials is reported to be superior in their nc form. Such contrasting observations of the influence of nanostructure may depend on the coupled environmental–materials interactions for that system in specific environments. In the case of electrochemical corrosion of the pure metals, the anodic behavior of grain boundaries and the associated anode–cathode area ratio are the governing factors, whereas the effect of alloying elements would be an additional compounding factor for the case of nc alloys. Early fundamental studies on the role of nanostructure in electrochemical corrosion were carried out on Co (Palumbo et al. 1997, Vinogradov et al. 1999, Jung and Alfantazi 2006, Wang et al. 2007), Cu (Thorpe et al. 1988, Vinogradov et al. 1999, Cheng et al. 2001, Jung and Alfantazi 2006), Ni (Rofagha et al. 1991, Tang et al. 1995, Lu et al. 2000, Liu et al. 2007), and Ni-based binaries (Rofagha et al. 1993, Gonzalez et al. 1996, Kim et al. 2003, Ghosh et al. 2006). The extent of localized corrosion at grain boundaries is reported to decrease in the case of some nc metals (viz., Cu and Co [Vinogradov et al. 1999, Kim et al. 2003, Jung and Alfantazi 2006]), in comparison with their mc counterparts; these observations can be attributed to either (1) decrease in the difference in the electrochemical potentials of grains and grain boundaries or (2) the greater anode-to-cathode area ratio in nc materials.

10.6.2 Electrochemical Degradation of Fe–Cr Alloys

A few studies (Tong et al. 1995, Wang and Li 2002, Kwok et al. 2006, Lu et al. 2006, Meng et al. 2006, Ye et al. 2006) have been carried out to compare the electrochemical corrosion of nc and mc iron–chromium alloys.

Electrochemical corrosion resistance of a nc surface of 316 stainless steel developed by surface mechanical attrition treatment was found to be considerably inferior to the mc unmodified bulk (Figure 10.5). This behavior is attributed to the considerable increase in the "fast diffusion channels" for ions, i.e., grain boundaries and triple junctions in the nc material (Tong et al. 1995). Grain boundary corrosion is suggested to be enhanced in mc materials as a result of the much smaller anode-to-cathode area ratio (than in the nc material). Figure 10.6 presents microscopic evidence that there is greater grain boundary corrosion on the mc surface.

Also, in another example, an nc surface of 304-stainless steel, developed upon sandblasting, showed a considerably inferior corrosion resistance as compared to the mc substrate (Ye et al. 2006). However, when the sandblasted surface was annealed prior to electrochemical testing, that surface demonstrated considerably improved corrosion resistance, to the extent that the annealed surface even had a corrosion resistance superior to that for the mc substrate.

In another study, the grain refinement of stainless steels to a nanometric level is reported to improve the corrosion resistance (Meng et al. 2006), as suggested by the considerably extended passivation stage for the nc sample in the polarization plots in Figure 10.7. In this study, the nc thin films (coatings) on the 309 stainless steel were developed by DC magnetron sputtering where the untreated mc bulk of 309 steel was used for comparison purposes. Authors have attributed this behavior to the greater chromium diffusion in the nc structure.

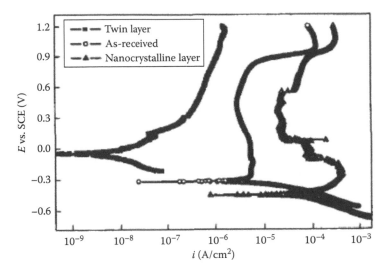

FIGURE 10.5 Potentiodynamic polarization curves for the nanocrystalline and microcrystalline surface of 316 stainless steel in 0.05 M H_2SO_4 + 0.25 M Na_2SO_4. (From Lu, A.Q. et al., *Acta Metall. Sin. (English Letters)*, 19(3), 183, 2006.)

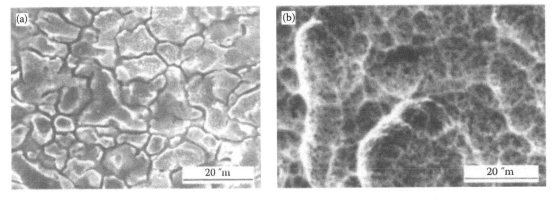

FIGURE 10.6 SEM micrographs showing corrosion product morphology of 316 stainless steel samples after subjecting them to 0.05 M H_2SO_4 + 0.25 M Na_2SO_4 solution at a potential of −900 m V_{SCE}: (a) microcrystalline steel and (b) nanocrystalline steel. (From Lu, A.Q. et al., *Acta Metall. Sin. (English Letters)*, 19(3), 183, 2006.)

It is interesting to note that the processing routes employed for developing the nc structure in steels in the examples described earlier are considerably different from those employed for producing cast stainless steels samples used in these studies for the purpose of comparison with corresponding mc alloys. In fact, in one of these studies (Tong et al. 1995), the crystal structure of the nc thin films was described to be ferritic whereas that for the mc alloy was austenitic. As a result, the two materials indeed show distinctly different polarization behaviors (the latter producing two anodic peaks as opposed to the only one peak for the former [see Figure 10.7]). Moreover, the nc surfaces used for these corrosion studies were in the form of thin surface films. Therefore, it may be appropriate to suggest that the effects of nc structure on corrosion, as claimed in at least two of these studies (Meng et al. 2006, Ye et al. 2006), could also be associated with factors other than the nc structure, such as the levels of the defects present in the nc alloys as well as the difference in density and phase compositions in the nc and mc samples (since each kind of sample was prepared by a different processing route). In fact, one of the studies possibly validates this view since the defective nc structure developed upon sandblasting also

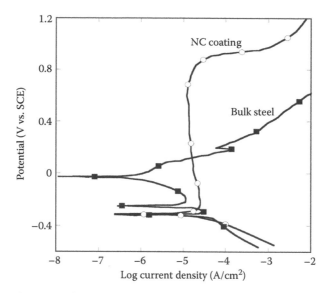

FIGURE 10.7 Potentiodynamic polarization plots of nanocrystalline coating of 309 stainless steel (produced by sputtering) and the bulk microcrystalline steel in a NaCl solution. (Adapted from Ye, W. et al., *Electrochim. Acta*, 51(21), 4426, 2006.)

had a considerably inferior corrosion resistance in comparison to that for the mc substrate, whereas an annealing step before corrosion testing apparently removed these defects and improved its corrosion resistance so that it was now superior to the corrosion resistance of the mc substrate. Hence, it is argued that for making an appropriate comparison of the role of nc vis-à-vis mc structure on the electro-chemical corrosion properties of Fe–Cr alloys, it would be necessary that samples with two distinctly different grain sizes must be processed by the same/similar route. The author and his co-worker have recently carried out a study on the corrosion behavior of nc and mc Fe–Cr alloys prepared by the same processing route.

In that study (Gupta et al. 2010), discs of nc Fe–10Cr and Fe–20Cr alloys were prepared by high-energy ball milling followed by pre-annealing, compaction, and sintering (described in Section 10.4.1). To prepare the mc alloy specimens, the discs of Fe–10Cr and Fe–20Cr alloys were annealed at 800°C, then compacted under a uniaxial pressure of 2 GPa and then sintered at 840°C for 3 h. The grain size of the nc alloys as determined by the XRD technique was 53 nm (±4 nm), whereas the grain size of the mc material was 1.5 µm, as determined by optical microscopy. Potentiodynamic polarization tests of these nc and mc Fe–10Cr and Fe–20Cr alloy samples in 0.5 M H_2SO_4 have produced remarkably reproducible results. However, results suggest that the improvement in the corrosion resistance due to nc structure was considerably less than those reported in the earlier literature (Meng et al. 2006). In fact, the observed improvement due to nc structure was still much less significant, especially in the case of Fe–20Cr alloy, implying that the considerable reported improvement in electrochemical corro-sion resistance as a result of the nc structure, as claimed in that earlier study (Meng et al. 2006), could possibly be ascribed to the factors other than just the nc structure (such as the difference in processing routes used for nc and mc materials). In fact, in that earlier study (Meng et al. 2006), the "faster diffusion of the Cr" was the mechanism suggested to be responsible for the claimed improvement in corrosion resistance of the alloy in its nc form. However, this view of faster diffusion of Cr from the internal bulk region to the interface where the corrosion film forms does not seem plausible because Cr has a very low diffusivity at room temperature. Note that the grain boundary diffusion coefficient of Cr in BCC Fe at room temperature is 4.1×10^{-40} m^2/s and lattice diffusion coefficient is even lower, i.e., 9.4×10^{-48} m^2/s (these diffusion coefficients have been derived by extrapolation of the reported diffusion data at higher temperatures [Wang et al. 2003]). The corresponding diffusion coefficient for nc alloys with a grain size of 10 nm (assuming grain boundary width to be 1 nm) has been calculated to be 1.2×10^{-40} m²/s and that for a fine grain material with grain size of 5 μm is 8×10^{-43} m²/s. Thus with these considerably lower diffusion coefficients, any effective transport of Cr atoms from the bulk to corrosion film/metal interface (and therefore Cr enrichment of corrosion film) does not seem feasible.

10.7 Oxidation Resistance of Nanocrystalline Metals/Alloys

During oxidation, binary alloys with certain alloying elements (viz., Cr, Al, and Si) can form a continuous layer of the chromia, alumina, or silica, conferring substantial oxidation resistance. This has formed the basis of the development of common oxidation resistance alloys, such as stainless steels. Formation of a continuous layer of surface oxide is called external oxidation (Kofstad 1988). If, on the other hand, the inward flux of oxygen exceeds the outward flux of solute during oxidation process, isolated oxide particles form in the subsurface. This phenomenon is called internal oxidation (Kofstad 1988).

For external oxidation and formation of a continuous layer of chromia, alumina, or silica, a critical concentration of solute is required, which can be calculated by Wagner's treatment (Wagner 1952) for various systems/conditions. Critical amount of a solute for such a transition depends directly on its diffusivity in the alloy, besides other factors (viz., concentration of solute element, diffusivity in the oxide scale, temperature). The extremely fine grain size and the high-volume fraction of grain boundaries of nc materials (Gleiter 1989) can cause an extraordinary increase in diffusivity, and nc structures may have beneficial effects in the development of the protective oxide layer. For example, oxidation resistances of an iron–aluminide and an Fe–B–Si alloy in the nc state are reported (Tong et al. 1995, El Kedim et al. 2004) to be superior than in their mc state. This behavior is attributed to Al and Si, the well-known protective oxide film formers, being the predominantly diffusing species respectively in the two alloys, and the nanostructure facilitating their diffusion and expedited formation of protective films (of Al/Si oxide). Since a detailed treatment is outside the scope of this chapter, the interested readers should read an article by Singh Raman et al. (2010) for an elaborate mechanistic description.

10.8 Degradation Protection and Specific Standardized Tests

Since the study of nc alloys is a relatively recent endeavor in the Materials Science field, much R&D activity has been directed at exploring the properties of this new class of materials and synthesizing laboratory quantities of these materials. Thus not much applied technology efforts have been directed at developing specific surface treatments and coatings for providing degradation protection.

Similarly, aside from generic standard laboratory and industrial corrosion property measurement protocols such as the potentiodynamic polarization scans (referred to earlier), no specific standardized tests relevant to measuring the properties of nanostructured alloys have been proposed and accepted by the usual agencies that develop and validate such standardized testing procedures.

10.9 Summary

As a result of their extremely fine grain size, nc metals and alloys possess remarkably different properties than mc metals and alloys. However, because such nc materials have poor thermal stability and are highly susceptible to grain growth, their large-scale processing remains a challenge.

Though there are few studies on the aqueous corrosion of nanostructured metals and alloys, the nc structure is reported to have beneficial influence on corrosion resistance for some systems but may have a deleterious influence in the case of some others.

Most studies have suggested that nc stainless steels do possess superior aqueous corrosion resistances than their mc counterparts. However, the reported explanation/mechanism of such improvement is either not clear or unconvincing.

Nc alloys containing those alloying elements that can form protective oxide scale (viz., Cr, Si, and Al) have shown superior resistance to degradation due to gaseous corrosion at high temperatures. This behavior has been convincingly attributed to the much greater diffusivity of chromium in the nc alloy, which enables the nc alloy to develop a protective layer of chromium oxide, even with only 10 wt% Cr in the alloy (that would be an insufficient Cr content in its mc form).

Acknowledgments

The motivation for accomplishing this chapter came from an experimental and theoretical work of recent years on corrosion/oxidation of nc alloys in the author's group. This program received funding support from Australian Research Council's Discovery Projects scheme, and has benefitted immensely from a collaboration with Prof Carl Koch (North Carolina State University) for synthesis of the nc alloys as well as from the dedicated work of past and present graduate students, Dr. Rajeev Gupta and Dr. B. V. Mahesh.

References

Aledresse, A. and Alfantazi, A. 2004. A study on the corrosion behavior of nanostructured electrodeposited cobalt. *Journal of Material Science*, 39, 1523–1526.

Bansal, C., Gao, Z.Q. et al. 1995. Grain growth and chemical ordering in $(Fe,Mn)_3Si$. *Nanostructured Materials*, 5, 327–336.

Barbucci, A., Farne, G. et al. 1998. Corrosion behaviour of nanocrystalline Cu90Ni10 alloy in neutral solution containing chlorides. *Corrosion Science*, 41, 463–475.

Birringer, R. 1989. Nanocrystalline materials. *Materials Science and Engineering A*, 117, 33–43.

Bonetti, E., L. Del Bianco et al. 1999. Thermal evolution of ball milled nanocrystalline iron. *Nanostructured Materials*, 12(5–8), 685–688.

Bonetti, E., G. Scipione et al. 1995. A study of nanocrystalline iron and aluminium metals and Fe_3Al intermetallic by mechanical alloying. *Journal of Materials Science*, 30(9), 2220–2226.

Cheng, S., E. Ma et al. 2005. Tensile properties of in situ consolidated nanocrystalline Cu. *Acta Materialia*, 53(5), 1521–1533.

Cheng, D., V.L. Tellkamp et al. 2001. Corrosion properties of nanocrystalline Co–Cr coatings. *Annals of Biomedical Engineering*, 29(9), 803–809.

Chokshi, A., Rosen, A. et al. 1989. On the validity of the Hall–Petch relationship in nanocrystalline materials *Scripta Materialia*, 23, 1679.

Daróczi, L., D.L. Beke et al. 1993. Production and magnetic properties of nanocrystalline Fe and Ni. *Nanostructured Materials*, 2(5), 515–525.

De Keijser, T.H., J.I. Langford et al. 1982. Use of the Voigt function in a single-line method for the analysis of x-ray diffraction line broadening. *Journal of Applied Crystallography*, 15(3), 308–314.

Eckert, J., J.C. Holzer et al. 1993. Thermal-stability and grain-growth behavior of mechanically alloyed nanocrystalline Fe–Cu alloys. *Journal of Applied Physics*, 73(1), 131–141.

El Kedim, O., S. Paris et al. 2004. Electrochemical behavior of nanocrystalline iron aluminide obtained by mechanically activated field activated pressure assisted synthesis. *Materials Science and Engineering A*, 369(1–2), 49–55.

Elkedim, O., H.S. Cao et al. 1998. Preparation of nanocrystalline copper by hot and cold compaction: Characterization of mechanical and electrochemical properties. *Materials Science Forum*, 269–272, 843–848.

Elkedim, O., H.S. Cao et al. 1999. Hardness and corrosion performance of nanocrystalline iron powder prepared by ball milling: Comparison with different iron obtained by conventional methods. *Materials Science Forum*, 312–314, 635–640.

Elkedim, O., H.S. Cao et al. 2002. Preparation and corrosion behavior of nanocrystalline iron gradient materials produced by powder processing. *Journal of Materials Processing Technology*, 121(2–3), 383–389.

Fecht, H.J. 1992. Synthesis and properties of nanocrystalline metals and alloys prepared by mechanical attrition. *Nanostructured Materials*, 1(2), 125–130.

Fecht, H., E. Hellstern et al. 1990. Nanocrystalline metals prepared by high-energy ball milling. *Metallurgical and Materials Transactions A*, 21(9), 2333–2337.

Fougere, G.E., J.R. Weertman et al. 1995. Processing and mechanical behavior of nanocrystalline Fe. *Nanostructured Materials*, 5(2), 127–134.

Gao, Z.Q. and Fultz, B. 1994. Inter-dependence of grain growth, Nb segregation, and chemical ordering in Fe-Si-Nb Nanocrystals. *Nanostructured Materials*, 4, 939–947.

Ghosh, S.K., G.K. Dey et al. 2006. Improved pitting corrosion behaviour of electrodeposited nanocrystalline Ni–Cu alloys in 3.0 wt.% NaCl solution. *Journal of Alloys and Compounds*, 426(1–2), 235–243.

Gleiter, H. 1989. Nanocrystalline materials. *Progress in Materials Science*, 33(4), 223–315.

Gleiter, H. 2000. Nanostructured materials: Basic concepts and microstructure. *Acta Materialia*, 48(1), 1–29.

Gonzalez, F., A.M. Brennenstuhl et al. 1996. Electrodeposited nanostructured nickel for in-situ nuclear steam generator repair. *Materials Science Forum*, 225–227, 831–836.

Groza, J. Nanocrystalline powder consolidation methods. In *Nanostructured Materials: Processing, Properties, and Applications*, C.C. Koch, Ed., William Andrew Publishing, Norwich, NY, 2007.

Gryaznov, V.G., V.A. Solov'ev et al. 1990. The peculiarities of initial stages of deformation in nanocrystalline materials (NCMs). *Scripta Metallurgica et Materialia*, 24(8), 1529–1534.

Gupta, R.K. 2010. Synthesis and corrosion behaviour of nanocrystalline Fe-Cr alloys, PhD thesis, Monash University, Melbourne, Victoria, Australia.

Gupta, R., R.K. Singh Raman et al. 2008. Grain growth behaviour and consolidation of ball-milled nanocrystalline Fe-10Cr alloy. *Materials Science and Engineering A*, 494(1–2), 253–256.

Guruswamy, S., R.L. Michael, N. Srisukhumbowornchai, K.M. Michael, and P.T. Joseph. 2000. Processing of Terfenol D alloy based magnetostrictive composites by dynamic compaction. *IEEE Transactions*, 36, 3219–3222.

Hall, E.O. 1951. The deformation and ageing of mild steel: II. Characteristics of the Lüders deformation. *Proceedings of the Royal Society of London*, B64, 747.

Haubold, T., R. Birringer et al. 1989. Exafs studies of nanocrystalline materials exhibiting a new solid state structure with randomly arranged atoms. *Physics Letters A*, 135(8–9), 461–466.

Hernando, A., A. Cebollada, J.L. Menendez, and F. Briones. 2003. Electronic transport in nanocrystalline iron: A low T magnetoresistance effect. *Journal of Magnetism and Magnetic Materials*, 262, 1.

Hibbard, G.D. 2006. Processing and fabrication of advanced materials, in *Proceedings of the 15th International Symposium (MS&T)*, Cincinnati, OH, p. 297.

Hibbard, G.D., K.T. Aust et al. 2006. Thermal stability of electrodeposited nanocrystalline Ni–Co alloys. *Materials Science and Engineering: A*, 433(1–2), 195–202.

Horváth, J., R. Birringer et al. 1987. Diffusion in nanocrystalline material. *Solid State Communications*, 62(5), 319–322.

Jang, D. and M. Atzmon. 2003. Grain-size dependence of plastic deformation in nanocrystalline Fe. *Journal of Applied Physics*, 93(11), 9282–9286.

Jang, J.S.C. and C.C. Koch. 1990. The hall-petch relationship in nanocrystalline iron produced by ball milling. *Scripta Metallurgica et Materialia*, 24(8), 1599–1604.

Jung, H. and A. Alfantazi. 2006. An electrochemical impedance spectroscopy and polarization study of nanocrystalline Co and Co-P alloy in 0.1 M H2SO4 solution. *Electrochimica Acta*, 51(8–9), 1806–1814.

Karimpoor, A.A., U. Erb et al. 2003. High strength nanocrystalline cobalt with high tensile ductility. *Scripta Materialia*, 49(7), 651–656.

Khan, A.S., H. Zhang et al. 2000. Mechanical response and modeling of fully compacted nanocrystalline iron and copper. *International Journal of Plasticity*, 16(12), 1459–1476.

Kim, S.H., K.T. Aust et al. 2003. A comparison of the corrosion behaviour of polycrystalline and nanocrystalline cobalt. *Scripta Materialia*, 48(9), 1379–1384.

Kim, K. and K. Okazaki. 1992. Nano-crystalline consolidation of MA powders by EDC. *Materials Science Forum*, 88, 521–528.

Kirchheim, R., X.Y. Huang et al. 1992. Free energy of active atoms in grain boundaries of nanocrystalline copper, nickel and palladium. *Nanostructured Materials*, 1(2), 167–172.

Knauth, P., A. Charaï et al. 1993. Grain growth of pure nickel and of a Ni–Si solid solution studied by differential scanning calorimetry on nanometer-sized crystals. *Scripta Metallurgica et Materialia*, 28(3), 325–330.

Koch, C.C. 2003. Top-down synthesis of nanostructured materials: Mechanical and thermal processing methods. *Reviews on Advanced Materials Science*, 5(2), 91–99.

Koch, C.C., K.M. Youssef et al. 2005. Breakthroughs in optimization of mechanical properties of nanostructured metals and alloys. *Advanced Engineering Materials*, 7(9), 787–794.

Kofstad, P. 1988. *High Temperature Corrosion*, Elsevier Applied Science and Publishers Ltd., New York.

Korznikova, G.F., A.V. Korznikov et al. 1999. Magnetic hysteric properties and saturation magnetization of nanocrystalline iron. *Journal of Magnetism and Magnetic Materials*, 196–197, 207–208.

Kumar, K.S., H. Van Swygenhoven et al. 2003. Mechanical behavior of nanocrystalline metals and alloys. *Acta Materialia*, 51(19), 5743–5774.

Kwok, C.T., F.T. Cheng et al. 2006. Corrosion characteristics of nanostructured layer on 316L stainless steel fabricated by cavitation-annealing. *Materials Letters*, 60(19), 2419–2422.

Le Brun, P., E. Gaffet et al. 1992. Structure and properties of Cu, Ni and Fe powders milled in a planetary ball mill. *Scripta Metallurgica et Materialia*, 26(11), 1743–1748.

Lee, D.W., J.H. Yu et al. 2005. Nanocrystalline iron particles synthesized by chemical vapor condensation without chilling. *Materials Letters*, 59(17), 2124–2127.

Liu, L., Y. Li et al. 2007. Influence of micro-structure on corrosion behavior of a Ni-based superalloy in 3.5% NaCl. *Electrochimica Acta*, 52(25), 7193–7202.

Loffler, J. and J. Weissmuller. 1995. Grain-boundary atomic structure in nanocrystalline palladium from x-ray atomic distribution functions. *Physical Review B*, 52(10), 7076.

Lu, L., L.B. Wang et al. 2000. Comparison of the thermal stability between electro-deposited and cold-rolled nanocrystalline copper samples. *Materials Science and Engineering A*, 286(1), 125–129.

Lu, A.Q., Y. Zhang et al. 2006. Effect of nanocrystalline and thin boundaries on corrosion behavior of 316L stainless steel using SMAT. *Acta Metallurgica Sinica (English Letters)*, 19(3), 183–189.

Malow, T.R. and C.C. Koch. 1997. Grain growth in nanocrystalline iron prepared by mechanical attrition. *Acta Materialia*, 45(5), 2177–2186.

Malow, T. and C. Koch 1998a. Mechanical properties, ductility, and grain size of nanocrystalline iron produced by mechanical attrition. *Metallurgical and Materials Transactions A*, 29(9), 2285–2295.

Malow, T.R. and C.C. Koch. 1998b. Mechanical properties in tension of mechanically attrited nanocrystalline iron by the use of the miniaturized disk bend test. *Acta Materialia*, 46(18), 6459–6473.

Malow, T.R., C.C. Koch et al. 1998. Compressive mechanical behavior of nanocrystalline Fe investigated with an automated ball indentation technique. *Materials Science and Engineering A*, 252(1), 36–43.

Maurer, R., U. Erb et al. 1984. Intercrystalline corrosion: Factors controlling the enhanced corrosion near grain boundaries. *Materials Science and Engineering*, 63(2), L13–L15.

Meng, G., Y. Li et al. 2006. The corrosion behavior of Fe-10Cr nanocrystalline coating. *Electrochimica Acta*, 51(20), 4277–4284.

Meyers, M. and Chawla, K. *Mechanical Metallurgy: Principles and Applications*, Prentice-Hall, Englewood Cliffs, NJ, 1984.

Meyers, M.A., A. Mishra et al. 2006. Mechanical properties of nanocrystalline materials. *Progress in Materials Science*, 51(4), 427–556.

Michalski, A. 1997. Nanocrystalline iron layers produced by the pulse plasma method. *Nanostructured Materials*, 8(6), 725–730.

Moelle, C.H. and H.J. Fecht. 1995. Thermal stability of nanocrystalline iron prepared by mechanical attrition. *Nanostructured Materials*, 6(1–4), 421–424.

Mütschele, T. and R. Kirchheim. 1987. Hydrogen as a probe for the average thickness of a grain boundary. *Scripta Metallurgica*, 21(8), 1101–1104.

Natter, H., M. Schmelzer et al. 2000a. Grain-growth kinetics of nanocrystalline iron studied in situ by synchrotron real-time x-ray diffraction. *Journal of Physical Chemistry B*, 104(11), 2467–2476.

Natter, H., M. Schmelzer et al. 2000b. In-situ x-ray crystallite growth study on nanocrystalline Fe. *Materials Science Forum*, 343–346, 689–694.

Nieh, T.G. and J. Wadsworth. 1991. Hall-petch relation in nanocrystalline solids. *Scripta Metallurgica et Materialia*, 25(4), 955–958.

Omura, K., H. Miura et al. 1999. Preparation of high nitrogen Cr–Ni and Cr–Mn stainless steel powders by mechanical alloying and their compositional dependence of austenitizing. *Materials Science Forum*, 318–320, 723–732.

Palumbo, G., Erb, U. et al. 1990. Triple line disclination effects on the mechanical behaviour of materials *Scripta Metallurgica et Materialia*, 24, 2347.

Palumbo, G., F. Gonzalez et al. 1997. In-situ nuclear steam generator repair using electrodeposited nanocrystalline nickel. *Nanostructured Materials*, 9(1–8), 737–746.

Perez, R.J., H.G. Jiang et al. 1997. Grain size stability of nanocrystalline cryomilled Fe-3 wt.% Al alloy. *Nanostructured Materials*, 9(1–8), 71–74.

Perez, R., H. Jiang et al. 1998. Grain growth of nanocrystalline cryomilled Fe–Al powders. *Metallurgical and Materials Transactions A*, 29(10), 2469–2475.

Petch, N.J. 1953. The cleavage strength of polycrystals. *Journal of Iron and Steel Institute*, 174, 25.

Rodríguez-Baracaldo, R., J. Benito et al. 2007. Mechanical response of nanocrystalline steel obtained by mechanical attrition. *Journal of Materials Science*, 42(5), 1757–1764.

Rofagha, R., U. Erb et al. 1993. The effects of grain size and phosphorus on the corrosion of nanocrystalline Ni–P alloys. *Nanostructured Materials*, 2(1), 1–10.

Rofagha, R., R. Langer et al. 1991. The corrosion behaviour of nanocrystalline nickel. *Scripta Metallurgica et Materialia*, 25(12), 2867–2872.

Savader, J.B., M.R. Scanlon et al. 1997. Nanoindentation study of sputtered nanocrystalline iron thin films. *Scripta Materialia*, 36(1), 29–34.

Schaefer, H.-E. 1993. Interfaces and physical properties of nanostructured solids. In: *Mechanical Properties and Deformation Behavior of Materials Having Ultra-Fine Microstructures*, M. Nastasi, D. Parkin, and H. Gleiter, Eds., Springer Netherlands, 233, 81–106.

Segers, D., S. Van Petegem et al. 1999. Positron annihilation study of nanocrystalline iron. *Nanostructured Materials*, 12(5–8), 1059–1062.

Seo, J.H., J.K. Kim et al. 2005. Textures and grain growth in nanocrystalline Fe–Ni alloys. *Materials Science Forum*, 475–479, 3483–3488.

Siegel, R.W. 1991. Cluster-assembled nanophase materials. *Annual Review of Materials Science*, 21(1), 559–578.

Siegel, R.W. 1993a. Nanostructured materials—Mind over matter. *Nanostructured Materials*, 3(1–6), 1–18.

Siegel, R.W. 1993b. Synthesis and properties of nanophase materials. *Materials Science and Engineering A*, 168(2), 189–197.

Siegel, R.W. 1994. Nanostructured materials—Mind over matter. *Nanostructured Materials*, 4(1), 121–138.

Siegel, R.W. 1997. Mechanical properties of nanophase materials. *Materials Science Forum*, 235–238, 851–860.

Singh Raman, R.K., R.K. Gupta et al. 2010. Resistance of nanocrystalline vis-á-vis microcrystalline Fe–Cr alloys to environmental degradation and challenges to their synthesis. *Philosophical Magazine*, 90(23), 3233–3260.

Sinha, P. and G.S. Collins. 1994. Mössbauer and PAC studies of nanocrystalline Fe. *Hyperfine Interactions*, 92(1), 949–953.

Sobczak, E., Y. Swilem et al. 2001. X-ray absorption studies of Fe-based nanocrystalline alloys. *Journal of Alloys and Compounds*, 328(1–2), 57–63.

Song, X., J. Zhang et al. 2006. Correlation of thermodynamics and grain growth kinetics in nanocrystalline metals. *Acta Materialia*, 54(20), 5541–5550.

Suryanarayana, C. 2002. The structure and properties of nanocrystalline materials: Issues and concerns. *JOM, Springer-Verlag*, 54, 24–27.

Suryanarayana, C. and F. Froes. 1992. The structure and mechanical properties of metallic nanocrystals. *Metallurgical and Materials Transactions A*, 23(4), 1071–1081.

Suryanarayana, C., D. Mukhopadhyay et al. 1992. Grain size effects in nanocrystalline materials. *Journal of Materials Research*, 7, 2114–2118.

Szabó, S., D.L. Beke et al. 1997. Correlation between the grain-shape and magnetic properties in nanocrystalline iron. *Nanostructured Materials*, 9(1–8), 527–530.

Taboor, D. 1951. *The Hardness of the Metals*, Clarendon Press, Oxford, U.K., 95pp.

Tang, P.T., T. Watanabet et al. 1995. Improved corrosion resistance of pulse plated nickel through crystallisation control. *Journal of Applied Electrochemistry*, 25(4), 347–352.

Tanimoto, H., P. Farber et al. 1999. Self-diffusion in high-density nanocrystalline Fe. *Nanostructured Materials*, 12(5–8), 681–684.

Thorpe, S.J., B. Ramaswami et al. 1988. Corrosion and auger studies of a nickel-base metal-metalloid glass. *Journal of the Electrochemical Society*, 135, 2162–2170.

Tian, H.H. and M. Atzmon. 1999. Kinetics of microstructure evolution in nanocrystalline Fe powder during mechanical attrition. *Acta Materialia*, 47(4), 1255–1261.

Tong, H.Y., F.G. Shi et al. 1995. Enhanced oxidation resistance of nanocrystalline FeBSi materials. *Scripta Metallurgica et Materialia*, 32(4), 511–516.

Trapp, S., C.T. Limbach et al. 1995. Enhanced compressibility and pressure-induced structural changes of nanocrystalline iron: In situ Mössbauer spectroscopy. *Physical Review Letters*, 75(20), 3760.

Vinogradov, A., T. Mimaki et al. 1999. On corrosion of ultra-fine grained copper produced by equichannel angular pressing. *Materials Science Forum*, 312–314, 641–646.

Wagner, C.J. 1952. Theoretical analysis of the diffusion processes determining the oxidation rates of alloys, *Journal of Electrochemical Society*, 369, 99.

Wallner, G., E. Jorra, H. Franz, J. Peisl, R. Birringer, T. Gleiter, T. Haubold, and W. Petry. 1989. Small angle scattering from nanocrystalline Pd. *Materials Research Society Symposium Proceedings*, 132, 149.

Wang, X.Y. and D.Y. Li. 2002. Mechanical and electrochemical behavior of nanocrystalline surface of 304 stainless steel. *Electrochimica Acta*, 47(24), 3939–3947.

Wang, L., Y. Lin et al. 2007. Electrochemical corrosion behavior of nanocrystalline Co coatings explained by higher grain boundary density. *Electrochimica Acta*, 52(13), 4342–4350.

Wang, Z.B., N.R. Tao et al. 2003. Diffusion of chromium in nanocrystalline iron produced by means of surface mechanical attrition treatment. *Acta Materialia*, 51(14), 4319–4329.

Yamashita, M., T. Mimaki et al. 1991. Stress corrosion cracking of [110] and [100] tilt boundaries of α-Cu–Al alloy. *Philosophical Magazine A*, 63(4), 707–726.

Ye, W., Y. Li et al. 2006. Effects of nanocrystallization on the corrosion behavior of 309 stainless steel. *Electrochimica Acta*, 51(21), 4426–4432.

Youssef, K.M., R.O. Scattergood et al. 2004. Ultratough nanocrystalline copper with a narrow grain size distribution. *Applied Physics Letters*, 85(6), 929–931.

Youssef, K.M., R.O. Scattergood et al. 2005. Ultrahigh strength and high ductility of bulk nanocrystalline copper. *Applied Physics Letters*, 87(9), 1–3.

Youssef, K.M., R.O. Scattergood et al. 2006. Nanocrystalline Al–Mg alloy with ultrahigh strength and good ductility. *Scripta Materialia*, 54(2), 251–256.

Zhu, X., R. Birringer et al. 1987. X-ray diffraction studies of the structure of nanometer-sized crystalline materials. *Physical Review B*, 35(17), 9085.

List of Relevant Publications

Gleiter, H. 1989. *Progress in Materials Science*, 33, 223.

Gupta, R.K. 2010. Synthesis and corrosion behaviour of nanocrystalline Fe–Cr alloys, PhD thesis, Monash University, Melbourne, Victoria, Australia.

Koch, C.C., K.M. Youseff, R.O. Scattergood, and K.L. Murty. 2005. *Advanced Engineering Materials*, 7, 787.

Kumar, K.S., H.V. Swygenhoven, and S. Suresh. 2003. *Acta Materialia*, 51, 5743.

Meyers, M.A., A. Mishra, and D.J. Benson. 2006. *Progress in Materials Science*, 51, 427.

Rofagha, R., U. Erb, D. Ostrander, G. Palumbo, and K. Aust. 1993. *NanoStructured Materials*, 2, 1.

Singh Raman, R.K., R.K. Gupta, and C.C. Koch. 2010. *Philosophical Magazine*, 90, 3233.

11

Amorphous Alloys

Koji Hashimoto
Tohoku University

11.1 Introduction (Ideal Structure of Amorphous Alloys)

Ordinary metals have been found in the form of oxides and/or salts in nature. Thus, in nature where oxygen and water exist, the oxidized state is chemically more stable than the metallic state, and the metallic state is formed by reduction from the oxidized state. However, metallic materials are generally separated from oxygen and water in nature due to coverage by an air-formed oxide film, and their oxidation is significantly slower because oxidation is controlled by outward diffusion of cations and inward diffusion of oxygen through the film. This is called spontaneous passivity. If the air-formed film is broken by the attack of given environments, corrosive dissolution and/or rusting of materials occur. Thus, metallic materials other than precious metals should be and can be practically used only in the spontaneously passive state. However, the uniform passive film with uniform protectiveness cannot be formed on heterogeneous metallic materials.

In crystalline materials, heterogeneities in the form of second phases, precipitates, segregates, grain boundaries, dislocations, and stacking faults exist. The corrosion resistance expected from homogeneous solid solution alloys, where the solute atoms are homogeneously distributed, has never been realized in heterogeneous crystalline metallic materials. Any phases with lower concentrations of corrosion-resistant solute elements will be corroded preferentially, and sometimes a chemically heterogeneous site will act as the cathode and stimulate dissolution of the matrix. Consequently, corrosive degradation of crystalline materials generally occurs based on the heterogeneity.

In contrast, amorphous alloys are composed of a single phase of homogenous solid solution without any physical and chemical heterogeneities. Thus, amorphous alloys with sufficient concentrations of corrosion-resistant elements show superior corrosion resistance that has never been found in any crystalline metallic alloys. This interpretation has been first made in 1974 when the extremely high corrosion resistance of amorphous Fe–Cr–P–C alloys was found (Naka et al. 1974). It can still be said that almost all superior chemical properties as well as physical properties of amorphous alloys are based on its homogenous single-phase nature. The material of such a composition showing the synergistic effects of all constituents cannot be found even in the form of a single crystal.

Consequently, the homogenous single-phase nature of amorphous alloys guarantees their high corrosion resistance due to the formation of a uniform passive film without defects and their high resistance to depassivation. Furthermore, because of their similarity to a uniformly random structure in frozen liquid, the solid solubility of alloy solutes are far higher than the solid solubility limit of crystalline stable alloys. Thus, the synergistic effects of high concentrations of alloy constituents which can never exist in crystalline metallic materials can be obtained.

11.2 Background

11.2.1 High Corrosion Resistance due to Homogeneous Nature of Amorphous Alloys

The remarkable corrosion resistance of amorphous alloys was first reported by the astonishing findings that there was no mass loss of an amorphous Fe–8at%Cr–13at%P–7at%C alloy immersed for 168 h in 10% $FeCl_3$ at 60°C, whereas type 316 steel disappeared leaving the oxide film (Naka et al. 1974, 1976). (*Note:* Alloy formulae for amorphous alloys in this chapter are expressed in at%). This result showed not only the first finding of extremely high corrosion resistance of the amorphous alloy but also the fact that type 316 steel will not be corroded in 10% $FeCl_3$ at 60°C, if its oxide film, which remained without dissolution in this solution, were able to cover uniformly the steel without having any heterogeneities.

11.2.2 Enhanced Passive Film Formation due to the Metastable Nature of Amorphous Alloys

Passivation of alloys containing corrosion-resistant solute elements occurs as a result of formation of a passive film in which these corrosion-resistant solute elements are highly concentrated. When the chromium content of crystalline binary Fe–Cr alloys is increased above 13%, these alloys are classified as stainless steels with well-known high corrosion resistance. As shown in Figure 11.1, stainless steels are characterized by the significant enrichment of chromium in the passive film, whereas the cationic composition of air-formed films on binary Fe–Cr alloys is almost the same as the alloy composition (Asami et al. 1978). In other words, anodic polarization in the passive region of the Fe–Cr alloys resulted in an enrichment of chromium in the film, and passivation took place because of the higher chemical stability of the chromium-enriched film. In this connection, the chromium content in the air-formed film on an amorphous Fe–10Cr–13P–7C alloy is almost the same as that in the passive film on the Fe–20wt%Cr crystalline alloy (Asami et al. 1976). Thus, for example, the amorphous Fe–10Cr–13P–7C alloy is spontaneously passive without suffering active dissolution and pitting corrosion up to oxygen evolution even in 2 M HCl as shown in Figure 11.2; also, immersion in 1 M HCl for 168 h gives rise to the enrichment of chromium of up to 97% of the cations as shown in Figure 11.1. In general, the amorphous alloys are characterized by the fact that those corrosion-resistant solute elements with a high affinity to oxygen and a high chemical stability in the oxidized state tend to be concentrated in the film during exposure to air and/or corrosive solutions. This has been interpreted in terms of the higher chemical reactivity due to the metastable nature of amorphous alloys. If an element is able to form a stable oxidized solid, the high activity of the alloy enhances the enrichment of the oxidized elements in the film; in contrast, if another solute or matrix element is unstable in its oxide or oxyhydroxide form, the high activity of the alloy accelerates the dissolution for that unstable element into aqueous environments. Therefore, those amorphous alloys containing sufficient concentrations of corrosion-resistant solute elements are characterized by a rapid formation of uniform corrosion-resistant passive film where those corrosion-resistant elements are highly concentrated. Furthermore, whenever that film is mechanically broken, repassivation in the breached area rapidly occurs unless that alloy itself is massively fractured by a shear stress or hydrogen embrittlement failure mechanism.

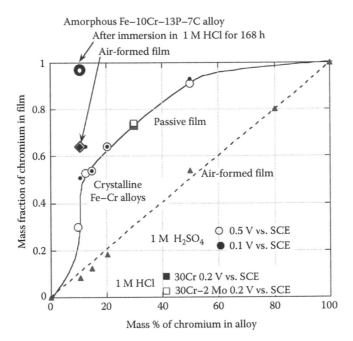

FIGURE 11.1 Mass fraction of chromium in films formed on crystalline Fe–Cr alloys polarized for 1 h in 0.5 M H$_2$SO$_4$, crystalline Fe–30Cr and Fe–30Cr–2Mo alloys polarized for 1 h in 1 M HCl, and amorphous Fe–10Cr–13P–7C alloys exposed to air and immersed for 168 h in 1 M HCl.

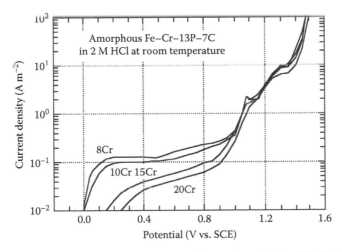

FIGURE 11.2 Potentiodynamic anodic polarization curves of amorphous Fe–Cr–13P–7C alloys in 2 M HCl.

11.2.3 Difficulty in Producing and Processing Amorphous Alloys

In spite of superior characteristics, the preparation of amorphous alloys is restricted. The formation of solid amorphous alloy requires that the randomly distributed atoms in the liquid state are prevented during solidification from moving into those atomic lattice positions that result in a crystalline structure. Thus, for the basic studies of amorphous alloys, specimen preparation is commonly carried out by rapid quenching from the liquid state to form solid amorphous alloy ribbons, fibers, or powders.

Since corrosive degradation is the surface phenomenon, the protective surface coating of conventional metallic materials with corrosion-resistant amorphous alloys has been attempted. Laser and electron beam processing was one of expected methods for preparation of amorphous surface alloys (Yoshioka et al. 1987, Kumagai et al. 1988). It has, however, been found that laser and electron beam processing of amorphous surface alloys was the most difficult technology to form amorphous alloys. For the formation of a linear amorphous alloy in the surface, linearly forming molten alloy by a scanning beam was easily quenched by thermal conduction to the substrate metal to extract the heat from the solidifying surface alloy. However, this procedure should be repeated along the previously formed linear amorphous alloy to cover the entire surface by amorphous surface alloy. Thus, the previously formed amorphous alloy is inevitably heated by the linear scan of a beam on the neighbor alloy surface, and this procedure often induces the crystallization of the previously formed amorphous alloy. Consequently, the formation of a uniform amorphous surface alloy layer requires alloy selection from a very limited number of alloy systems that have a high glass-forming ability. In addition, the rapidly quenched surface alloy layer that is bonded to the underlying solid substrate is subject to very high tensile residual stresses. These stresses are substantially high to bend the specimen and often induce cracking of the amorphous surface alloys thus formed. Therefore, laser and electron beam processing for the formation of amorphous alloy coating seriously limits its application.

Sputter deposition was the widely used method for preparation of amorphous alloys and once was thought to be suitable for production of amorphous surface alloys. However, from a corrosion point of view, the main problem of all kinds of surface coating is always the presence of physical defects. Sputter deposition is not an exception. Even if highly corrosion-resistant materials were chosen for coating, perfect coating without defects was impossible. Thus, the corrosion rate has generally been determined by the density of physical defects in the coated materials regardless of high corrosion resistance of coated materials themselves. Consequently, the idea of the corrosion-resistant coating of amorphous alloys was abandoned, and further effort was directed at producing corrosion-resistant amorphous alloys in a bulk form.

However, even if amorphous bulk alloys are prepared, welding of amorphous alloys is not possible because heating during welding induces the metastable amorphous phase to transform into multiple crystalline phases with a consequent loss of almost all kinds of superiority of the uniform single phase. Because of the sophisticated nature for the production and processing of amorphous bulk alloys and because welding cannot be done, most attempts to directly replace crystalline metallic alloys with comparable amorphous alloys are not suitable; thus, the practical application of amorphous alloys is restricted to only using them when the specific beneficial properties are not realized in any crystalline materials. The typical example of such environments for corrosion-resistant amorphous alloys is concentrated hydrochloric acids.

11.2.4 Production of Amorphous Bulk Alloys

The prerequisite for retaining the randomly distributed array of atoms present in that liquid state in order to form an amorphous solid structure is to prevent the movement of those constituent atoms into crystalline structure during solidification. Thus, superior characteristics of amorphous alloys have mostly been studied using specimens prepared by rapid quenching from the liquid state or sputter deposition. However, since about 1994, consolidation of amorphous alloy powders of a specific composition enabled to form amorphous "bulk" alloys (Kato et al. 1994, Kawamura et al. 1995). Although a Pd–30Cu–10Ni–20P alloy had been solidified in an amorphous alloy rod of 75 mm diameter at the cooling rate of about 0.1 K s^{-1} from the liquid state (Inoue et al. 1996), production of engineering amorphous alloys by slow cooling is difficult. On the other hand, the temperature increase by heating for the most of traditional amorphous alloys leads to the direct crystallization, but there are some alloys of selected compositions which transform to the supercooled liquid state just below the crystallization temperature. If the supercooled liquid is stable enough to process the alloys, consolidation of amorphous alloy powders in the supercooled liquid state enables to form the amorphous alloys in any desirable bulk shapes. The stabilization of the supercooled liquid state requires lower mobility of atoms due to strong interaction among constituent atoms. Higher-dense, random-packed atomic configurations in the amorphous solid are also effective for the formation of amorphous bulk alloys. Thus, the three empirical rules have been reported (Inoue 1999): (1) multicomponent systems consisting of three or more elements, (2) significant difference in atomic size ratios above about 12% among the main three constituent elements, and (3) high negative heats of mixing among the main three constituent elements.

The presence of the supercooled liquid state can be detected by thermal analysis. When an amorphous alloy specimen having the supercooled liquid state is heated at a constant heating rate of 20 K min^{-1}, the endothermic behavior corresponding to the formation of the supercooled liquid is observed up to the appearance of an exothermic peak for crystallization. The lower temperature limit of the endothermic behavior is called the glass transition temperature, Tg, because cooling of the supercooled liquid under Tg results in transformation to the amorphous (glass) solid. The lower bottom of the exothermic peak corresponds to the crystallization temperature, Tx. In the temperature range between Tg and Tx, this alloy is in the supercooled liquid state. If the temperature interval $\Delta T = Tx - Tg$ is approximately 50 K or higher, that is, 2.5 min or longer heating time after reaching Tg at the heating rate of 20 K min^{-1}, it is possible to form an amorphous alloy plate by consolidation of the amorphous alloy powders using a heated twin-roller type rolling mill after quickly preheating these amorphous alloy powders into the supercooled liquid state.

The first attempt for consolidation of corrosion-resistant amorphous alloy powders was performed for Ni–10Cr–5Nb–16P–4B alloy, for the HCl dew point corrosion test in waste incineration environments (Habazaki et al. 2001). The thermal analysis of the melt-spun ribbon-shaped Ni–10Cr–5Nb–16P–4B alloy at a heating rate of 20 K min^{-1} revealed that there was a supercooled liquid state between Tg of 676 K and Tx of 740 K. The gas-atomized Ni–10Cr–5Nb–16P–4B alloy powder was consolidated as follows: (a) about 50 g of powder was sealed in a stainless steel tube of 15 mm inner diameter in vacuum; (b) the stainless steel tube specimen was cold-rolled to the thickness of 7 mm; (c) the cold-rolled specimen was quickly heated to 708 K in the supercooled liquid state of this amorphous alloy in a previously heated furnace; and (d) immediately warm-rolled at 708 K to the total thickness of 4 mm. After removing the sheath, the alloy specimen of about 2 mm in thickness was obtained. The amorphous structure of the bulk alloy specimen was confirmed by x-ray diffraction as shown in Figure 11.3. A high-resolution

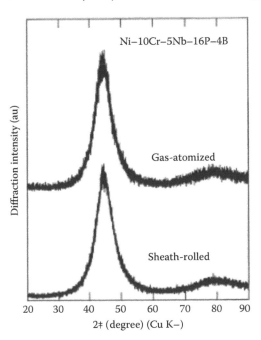

FIGURE 11.3 X-ray diffraction patterns of gas-atomized powders and sheath-rolled plate of Ni–10Cr–5Nb–16P–4B alloy.

transmission electron microscopy image of ultramicrotomed section of the specimen revealed maze patterns typical of the amorphous structure. The corrosion rate in 6 M HCl at 303 K for this sheath-rolled specimen was 1.7×10^{-2} mm year^{-1}, whereas that of an as-cast amorphous alloy specimen was less than 1×10^{-3} mm year^{-1}. The lower corrosion resistance of that sheath-rolled specimen was attributed to the presence of crystallized powder phase in the amorphous matrix.

The performance of the sheath-rolled amorphous Ni–10Cr–5Nb–16P–4B alloy specimen was examined by flue gas exposure in a waste incinerator (Hashimoto et al. 2001). The amorphous alloy sheet showed no detectable mass loss corrosion after exposure for 20 days at 393 and 433 K; in contrast, the mass losses of Type 316L stainless steel and Alloy B samples after exposure at 433 K were more than 80 and 360 mg cm^{-2} year^{-1}, respectively.

11.3 Corrosion-Resistant Amorphous Alloys

For the purpose of choosing the solute elements effective in enhancing the corrosion resistance of homogeneous alloys in concentrated hydrochloric acids, sputter deposition was used for preparation of binary single-phase alloys consisting only of corrosion-resistant elements such as amorphous Cr–Ti (Kim et al. 1993c), Cr–Zr (Kim et al. 1993a), Cr–Nb (Kim et al. 1993b), Cr–Ta (Kim et al. 1993b), and Mo–Zr (Park et al. 1995) alloys and bcc single-phase Mo–Ti (Park et al. 1996a), Mo–Nb (Park et al. 1996b), Mo–Ta (Park et al. 1996c), and Mo–Cr (Park et al. 1995) alloys. The corrosion rates of crystalline and amorphous Cr-valve metal alloys in concentrated hydrochloric acids are shown in Figure 11.4. The corrosion rates of those amorphous binary chromium-valve metal alloys are clearly lower than those of individual component elements of the alloys, and the corrosion rate decreases as the chromium content of the alloy increases despite of the fact that the corrosion rate of pure chromium metal is far higher than those for these valve metals. Among all metallic materials, the amorphous Cr–Ta alloys are the only ones that do not experience corrosion in 12 M HCl, since these alloys are all spontaneously passive in concentrated hydrochloric acids.

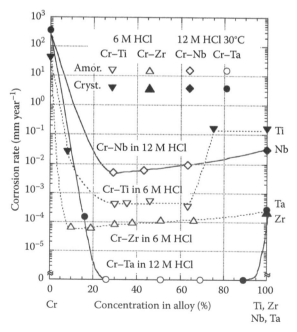

FIGURE 11.4 Corrosion rates of sputter-deposited Cr–Ti and Cr–Zr alloys in 6 M HCl and Cr–Nb and Cr–Ta alloys in 12 M HCl.

The surface characterization was performed using x-ray photoelectron spectroscopy for Cr–Ti alloys (Li et al. 1997a), Cr–Zr alloys (Li et al. 1997b), Cr–Nb alloys (Li et al. 1998a), and Cr–Ta alloys (Kim et al. 1994, Li et al. 1998b). In general, the composition of the spontaneously formed passive film was not largely different from that of the air-formed film; this indicates that spontaneous passivation occurs because the air-formed film has a high stability in concentrated hydrochloric acids. Anions in the films on Cr-valve metal alloys were O^{2-} and OH^-, where the relative amount changed with cationic composition of the films from $CrO_{1-x}(OH)_{1+2x}$ to about $ZrO_{1.4}(OH)_{1.2}$ or $TaO_2(OH)$, although the O^{2-} content was higher in the inner part of the film near the film/alloy interface whereas the OH^- content was higher in the outer part of the film near the film/environment interface. The tendency of charge transfer from Cr^{3+} ions to valve metal cations in the films was noted (e.g., Li et al. 1998b). The electronic interaction between Cr^{3+} ions and valve metal cations in the film indicates that the film is not a heterogeneous mixture of two oxyhydroxides but rather consists of a homogeneous double oxyhydroxides, such as $Cr_{1-x}Ta_xO_y(OH)_{3+2x-2y}$. Because the stability of $Cr_{1-x}Ta_xO_y(OH)_{3+2x-2y}$ is significantly higher than that of $TaO_2(OH)$, those amorphous Cr–Ta alloys did not suffer corrosion when exposed to concentrated hydrochloric acids in contrast to pure tantalum metal that dissolved at the rate of about 2.8 mm year^{-1} in 12 M HCl. In this manner, the synergistic stability of multiple oxides and oxyhydroxides was often higher than the stability of oxides and oxyhydroxides of the individual component elements. Thus, we expect enhanced corrosion resistance for such newly designed homogeneous alloys, even though the stability constants of nonstoichiometric multiple oxides and oxyhydroxides have not been determined.

No concentration gradient of cations was detected in the air-formed film. However, with increasing time of immersion in 12 M HCl, after a week there was a slight decrease in the chromium content near the exterior of the film formed on Cr–Nb alloys (Li et al. 1998a) and Cr–Ta alloys (Li et al. 1998b). In contrast, a slight decrease in the zirconium content near the exterior of the film formed on Cr–Zr alloys in 6 M HCl was observed (Li et al. 1997b).

Figure 11.5 shows the corrosion rates of a series of Mo-corrosion-resistant element alloys in 12 M HCl for Mo–Ti alloys (Park et al. 1996a), Mo–Zr alloys (Park et al. 1995), Mo–Nb alloys (Park et al. 1996b), Mo–Ta alloys (Park et al. 1996c), and Mo–Cr alloys (Park et al. 1995) in 12 M HCl. In contrast to the

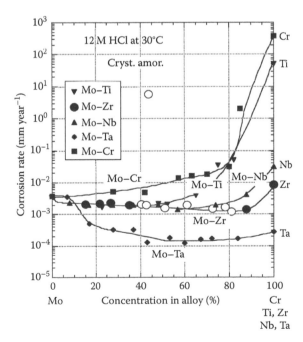

FIGURE 11.5 Corrosion rates of sputter-deposited Mo–Ti, Mo–Zr, Mo–Nb, Mo–Ta, and Mo–Cr alloys in 12 M HCl.

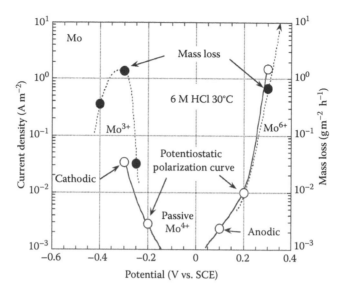

FIGURE 11.6   Potentiostatic polarization and dissolution curves of molybdenum metal in 6 M HCl.

Cr-valve metal alloys, the formation of Mo-valve metal alloys resulted in only slight decrease in their corrosion rates in 12 M HCl. The affinity of valve metals to oxygen was significantly higher than that of molybdenum. Thus, preferential oxidation of valve metals by air exposure led to the formation of the film enriched with valve metals. Nevertheless, the molybdenum addition to valve metals resulted in a slight decrease in the corrosion rate.

Molybdenum is a unique element. Figure 11.6 shows potentiostatic polarization and dissolution curves of molybdenum in 6 M HCl (Habazaki et al. 1992). Molybdenum is passive forming a Mo^{4+} oxide or oxyhydroxide film at the potentials higher than the hydrogen equilibrium potential; but at the potentials higher than about 0.15 V vs. SCE, molybdenum experiences transpassive dissolution forming molybdate ions. The open circuit potential of Mo-valve metal alloys was at about 0.1–0.2 V vs. SCE. Thus, molybdenum near the surface of the film exposed directly to 12 M HCl was dissolved as a molybdate, but the Mo^{4+} was protected by the overlaying valve metal oxyhydroxide film, and thus found in the inner part of the film. The XPS analysis revealed no electronic interaction between molybdenum ions and valve metal cations, and higher concentrations of molybdenum particularly Mo^{4+} in the inner part of the film. Therefore, the spontaneously formed passive film on the Mo-valve metal alloys consisted of bilayer structure of the outer valve metal oxyhydroxide and the inner Mo^{4+} oxide. The presence of the inner Mo^{4+} oxide layer adjacent to the alloys is responsible for their higher corrosion resistance in comparison with the corrosion resistance of the component valve metals. The role of molybdenum in enhancing the corrosion resistance will be further discussed in the section of the amorphous bulk alloys.

It was found from these studies that chromium, tantalum, and molybdenum were effective solute elements in preventing corrosion of amorphous alloys in concentrated hydrochloric acids.

11.4 Amorphous Bulk Alloys Resistant to 12 M HCl

Production of amorphous alloy powders and consolidation in the supercooled liquid state are suitable for the production of engineering materials of amorphous bulk alloys, but not suitable for alloy design because of time-consuming, sophisticated, and expensive procedure. On the other hand, production of amorphous bulk alloy by consolidation in the supercooled liquid state is empirically possible, if the glass-forming ability is high enough to form an amorphous alloy rod of 1 mm diameter by copper-mold casting, and if the temperature interval ΔT is 50 K or higher at the heating rate of 20 K min^{-1}. Thus, conventional examination of amorphous bulk alloy formation was performed by copper-mold casting and thermal analysis of melt-spun amorphous alloy ribbon.

The concentration of HCl azeotrope is lower than 6 M HCl, and the corrosive HCl environments with higher concentrations than 6 M HCl are not usual. The objective was to identify metallic materials resistant to 12 M HCl, which can be used widely in HCl environments to which no conventional crystalline materials are resistant. Sputter-deposited amorphous binary Cr–Ta alloys suffered no corrosion in 12 M HCl. However, binary Cr–Ta alloys cannot be solidified as amorphous bulk alloys. Because a Ni–40Nb alloy had relatively high glass-forming ability, enhancement of corrosion resistance was attempted by substituting some niobium with chromium and tantalum; also, some addition of phosphorus was made to further enhance the glass-forming ability of the alloys. Analogous binaries are known to have significantly high negative heats of mixing: where the heats of mixing for Nb–P, Ta–P, Mo–P, Cr–P, Ni–Nb, Ni–Ta, Ni–P are −81, −81, −45, −41, −30, −29, and −26 kJ (mole of atoms)$^{-1}$, respectively, although those of Cr–Ta, Cr–Nb, Ni–Cr, and Ni–Mo are all lower at only −7 kJ (mole of atoms)$^{-1}$ (de Boer et al. 1988). Preparation of amorphous Ni–(0–5)Cr–(0–12)Mo–(36–x)Ta–xNb–(4–7)P alloy rods was attempted by copper-mold casting. All these alloys were amorphous after melt spinning and showed a wide supercooled liquid region at relatively high temperatures, because of the presence of tantalum. For instance, using thermal analysis at the heating rate of 20 K min^{-1}, the supercooled liquid region of Ni–4Cr–1Mo–22Ta–14Nb–7P alloy was about 100 K (between T_g = 840 K and T_x = 940 K). Thus, if the amorphous alloy rods of 1 mm or thicker diameter are prepared by copper-mold casting, it would be possible to form any desirable bulk shapes of the amorphous alloys by consolidation of those amorphous alloy powders in the supercooled liquid state.

Figure 11.7 shows the compositions of Ni–Cr–Mo–22Ta–14Nb–4(7)P alloy rods of 1 mm diameter made by copper-mold casting, which could be solidified as amorphous single-phase rods. In those alloys whose main components were nickel, tantalum, and niobium, the addition of both chromium and molybdenum was necessary to form the amorphous single-phase structure. Nevertheless, the chromium content of those alloys that were able to be solidified in their amorphous single-phase state was seriously limited. The molybdenum content to be able to form amorphous single phase was also limited, but was to some extent wider than that of chromium. The highest chromium content was realized in the Ni–4Cr–1Mo–22Ta–14Nb–7P alloy, although relative amounts of tantalum and niobium could be interchanged almost freely to retain these casting as amorphous single-phase rods.

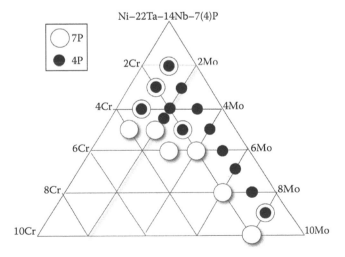

FIGURE 11.7 Compositions of copper-mold cast Ni–Cr–Mo–22Ta–14b–4(7)P alloys identified as an amorphous single phase in the form of rods of 1 mm diameter.

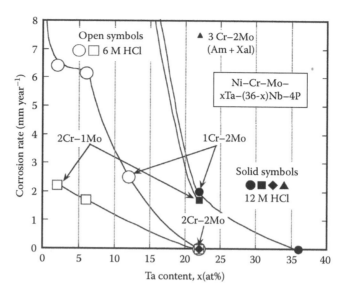

FIGURE 11.8 Corrosion rates of copper-mold cast amorphous Ni–Cr–Mo–xTa–(36-x)Nb–4P alloy rods in 6 and 12 M HCl and that of copper-mold cast Ni–3Cr–2Mo–22Ta–4Nb–4P alloy rod consisting of crystalline phase with the amorphous matrix in 12 M HCl.

Figure 11.8 shows the corrosion rates in 6 and 12 M HCl (Shinomiya et al. 2009). Even in amorphous single-phase alloys, the addition of sufficient chromium, molybdenum, and tantalum was necessary to assure no detectable corrosion in 12 M HCl. A smaller increase in the chromium content was more effective than an equivalent increase in molybdenum content. For instance, Ni–2Cr–2Mo–22Ta–14Nb–4P alloy was immune to corrosion in 12 M HCl; but if the chromium content was dropped to 1 at%, the tantalum content had to be increased to 36 at% for the same level of immunity in 12 M HCl, that is, Ni–1Cr–2Mo–36Ta–4P alloy. Although the increase in the chromium content was effective in enhancing the corrosion resistance, "excess" additions of chromium promoted the formation of crystalline phases within the amorphous matrix with a consequent increase in the corrosion rate as can be seen by the higher corrosion rate of the Ni–3Cr–2Mo–22Ta–14Nb–4P alloy.

Even if zero corrosion mass loss was observed in concentrated hydrochloric acids, potentiodynamic polarization curves measured in 6 M HCl clearly differentiated the beneficial effects of increases in chromium and molybdenum contents as shown in Figure 11.9. These alloys are spontaneously passive without showing active region, and the open circuit potentials are higher than 0 V vs. Ag/AgCl. In Ni–Cr–Mo–22Ta–14Nb–4P alloys, an increase in the chromium content from 1Cr–2Mo to 2Cr–2Mo clearly decreases the anodic current, and an increase in molybdenum content from 2Cr–2Mo to 2Cr–3Mo further decreases the anodic current. However, even if chromium is enhancing the corrosion resistance, the "excess" addition of chromium leads to the formation of crystalline phases within the amorphous matrix, which also leads to an increase in the anodic current; an example is the Ni–3Cr–2Mo–22Ta–14Nb–4P alloy.

Figure 11.10 shows the analytical results obtained by x-ray photoelectron spectroscopy, XPS, of the roles of alloying elements on corrosion performance. The Ni–4Cr–1Mo–24Ta–12Nb–7P alloy was immune to corrosion in both 6 and 12 M HCl. The upper right portion shows the cationic fraction in the surface films and the lower right portion the atomic fraction in the underlying alloy surface. Because no nickel was found in the surface film, the alloy composition without nickel is shown in the upper left portion and the nominal alloy composition is shown in the lower left portion. Air exposure before immersion results in preferential oxidation of chromium, tantalum, and niobium where the chromium enrichment in the air-formed film is remarkably high. After immersion in hydrochloric

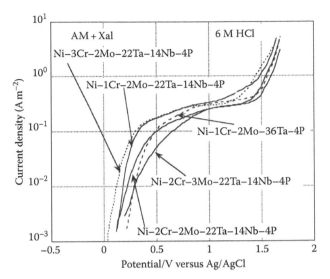

FIGURE 11.9 Potentiodynamic anodic polarization curves of copper-mold cast amorphous Ni–Cr–Mo–Ta–Nb–4P alloy rods and Ni–3Cr–2Mo–22Ta–14Nb–4P alloy rod consisting of crystalline phase with the amorphous matrix in 6 M HCl.

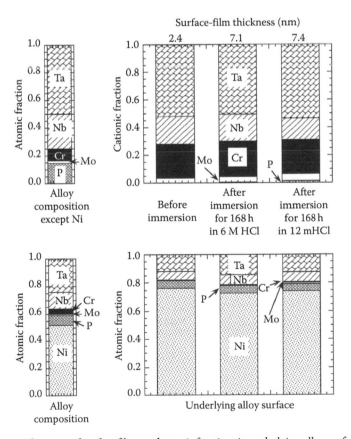

FIGURE 11.10 Cationic fractions of surface films and atomic fractions in underlying alloy surfaces for copper-mold cast amorphous Ni–4Cr–1Mo–24Ta–12Nb–7P alloy rods before and after immersion for 168 h in 6 and 12 M HCl.

acids, the surface film composition is not largely changed from the air-formed film. This indicates that the air-formed film itself is stable and protective in concentrated hydrochloric acids. Thus, spontaneous passivation occurs. This alloy is able to thicken the protective film depending upon the aggressiveness of the environments as shown in the top of Figure 11.10. In this connection, Figure 11.11 shows the impact of film mass (thickness) and the cation content in the film on the "no corrosion" phenomena. Because the chromium and molybdenum contents in the Ni–2Cr–1Mo–24Ta–12Nb–7P alloy are not sufficiently high, the alloy is corroded in 12 M HCl although no mass loss is observed in 6 M HCl. The alloy can increase the thickness of the protective film to resist against corrosion in 6 M HCl. However, the corrosion resistance of the alloy is not sufficient in 12 M HCl where it cannot increase the thickness of the protective film.

As shown in Figure 11.4, zero corrosion mass loss is found only when the double oxyhydroxide film of chromium and tantalum is formed, whereas the double oxyhydroxide film of chromium and niobium is not highly stable and less protective, particularly in 12 M HCl. Consequently, although niobium is found in the film, the contribution of niobium to the high corrosion resistance in the concentrated hydrochloric acids is low and the formation of triple oxyhydroxide particularly containing chromium and tantalum is responsible for no dissolution in 12 M HCl.

Another important fact is that the enhancement of the protectiveness of the film depending upon the increase in the aggressiveness of hydrochloric acids is associated with the increasing the amount of oxidized molybdenum in the film. Angle-resolved XPS gives the information on the distribution of elements as a function of depth below the surface. When the electron take-off angle in XPS is low, the signal from the inner part of the film must travel a longer distance before leaving the specimen and hence be weaker. Consequently, at the lower the take-off angles of electron, the strongest intensity signal is from the outer part of the surface film. Figure 11.12 shows the analytical results of x-ray

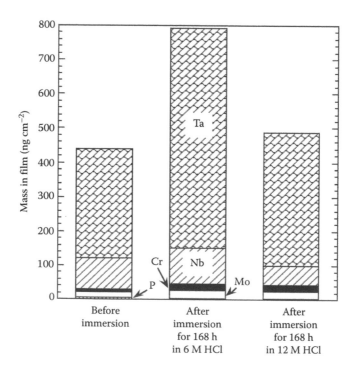

FIGURE 11.11 Masses of cations in the surface films on copper-mold cast amorphous Ni–2Cr–1Mo–22Ta–14Nb–4P alloy rod before and after immersion for 168 h in 6 and 12 M HCl.

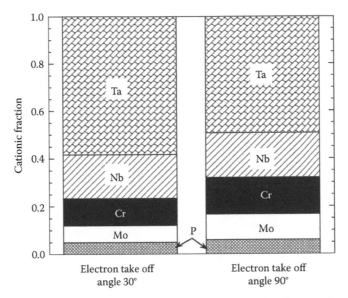

FIGURE 11.12 Analytical results of x-ray photoelectron spectra measured at electron take-off angles of 30° and 90° for cationic fractions in the surface film on copper-mold cast amorphous Ni–2Cr–8Mo–22Ta–14Nb–7P alloy rod after immersion for 336 h in 12 M HCl.

photoelectron spectra measured at two different electron take-off angles taken from the same specimen. The tantalum content is higher for the results analyzed using the data measured at the 30° electron take-off angle, whereas molybdenum content is higher in the analytical results obtained using the data measured at the 90° electron take-off angle. The chromium content shows a similar trend as that for molybdenum. The slight chromium deficiency in the exterior region of the film is similar to that found for the $Cr_{1-x}Ta_xO_y(OH)_{3+2x-2y}$ film formed on the binary Cr–Ta alloys in 12 M HCl (Li et al. 1998b). The affinity of molybdenum for oxygen is significantly lower than those of chromium, tantalum, and niobium. After preferential oxidation of chromium, tantalum, and niobium, the concentration of molybdenum under the film at the underlying alloy surface becomes high enough so that oxidation of molybdenum can occur, forming molybdenum oxide mostly located under the chromium-, tantalum-, and niobium-enriched outer portion of the film. The analytical results of angle-resolved XPS for molybdenum are shown in Figure 11.13. Molybdenum, particularly Mo^{4+}, is richer in the inner part of the film. Oxidation of molybdenum under the overlaying triple oxyhydroxide film of chromium, tantalum, and niobium will form MoO_2 near the underlying alloy surface. If molybdenum is oxidized to the high-valence Mo^{6+} on the outer film surface, the Mo^{6+} would be dissolved in the form of molybdate.

Consequently, the high chemical stability of the triple oxyhydroxide of chromium, tantalum, and niobium is protective without dissolving in 12 M HCl and is responsible for zero corrosion mass loss. The enhanced corrosion resistance caused by molybdenum is due to the formation of a MoO_2 film under the triple oxyhydroxide film of chromium, tantalum, and niobium where the MoO_2 film acts as the diffusion barrier suppressing the outward diffusion of cations and inward diffusion of oxygen.

In this manner, the passive film formed on the alloys with chromium, molybdenum, tantalum, and niobium consists of the bilayer structure where the outer triple (of chromium, tantalum, and niobium) oxyhydroxide portion of the film is stable even in 12 M HCl and the inner MoO_2 portion of the film acts as the diffusion barrier that suppresses the corrosive dissolution and oxidation of the metal alloy substrate.

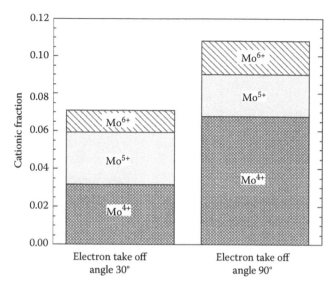

FIGURE 11.13 Analytical results of x-ray photoelectron spectra measured at electron take-off angles of 30° and 90° for fractions of molybdenum ions in the surface film on copper-mold cast amorphous Ni–2Cr–8Mo–22Ta–14Nb–7P alloy rod after immersion for 336 h in 12 M HCl.

11.5 Summary

Amorphous alloys have many attractive characteristics, including extremely high corrosion resistance if the sufficient amounts of corrosion-resistant elements are added. The superiority of amorphous alloys is based on the homogeneous single-phase nature without any chemical and physical heterogeneities. However, for the formation of the amorphous structure, the movement of constituent atoms to coordinate themselves into crystalline structure during solidification must be prevented. Heating during welding will destroy the superior characteristics by introducing the heterogeneities due to crystallization. Consequently, the application of corrosion-resistant amorphous alloys should be limited to the very aggressive environments where any conventional crystalline metallic materials cannot be used. Concentrated hydrochloric acids are examples of such environments. Some amorphous bulk alloys showed zero corrosion mass loss due to spontaneous passivation even in 12 M HCl, although the alloy compositions suitable for forming corrosion-resistant amorphous bulk alloys are limited. The technology for producing amorphous bulk alloys in desirable shapes was also interpreted.

References

Asami, K., K. Hashimoto, T. Masumoto, and S. Shimodaira. 1976. ESCA study of the passive film on an extremely corrosion resistant amorphous iron alloy. *Corrosion Science*, 16, 909–914.

Asami, K., K. Hashimoto, and S. Shimodaira. 1978. An XPS study of the passivity of a series of iron-chromium alloys in sulfuric acid. *Corrosion Science*, 18, 151–160.

de Boer, F.R., R. Boom, W.C.M. Mattens, A.R. Miedema, and A.K. Niessen. 1988. *Cohesion in Metals: Transition Metal Alloys, Cohesion and Structure*, Vol. 1, North-Holland, Amsterdam, the Netherlands.

Habazaki, H., A. Kawashima, K. Asami, and K. Hashimoto. 1992. The corrosion behavior of amorphous Fe-Cr-Mo-P-C and Fe-Cr-W-P-C alloys in 6 M HCl. *Corrosion Science*, 33, 225–236.

Habazaki, H., H. Ukai, K. Izumiya, and K. Hashimoto. 2001. Corrosion behaviour of amorphous Ni-Cr-Nb-P-B bulk alloys in 6 M HCl. *Materials Science and Engineering A*, 318, 77–86.

Hashimoto, K., H. Katagiri, H. Habazaki et al. 2001. Extremely corrosion-resistant bulk amorphous alloys. *Materials Science Forum*, 377, 1–8.

Inoue, A. 1999. Stabilization and high strain-rate superplasticity of metallic supercooled liquid. *Materials Science and Engineering A*, 267, 171–183.

Inoue, A., N. Nishiyama, and T. Masuda. 1996. Preparation of bulk glassy $Pd_{40}Ni_{10}Cu_{30}P_{20}$ Alloy of 40 mm in diameter by water quenching. *Materials Transactions, JIM*, 37, 181–184.

Kato, A., T. Suganuma, H. Horikiri, Y. Kawamura, A. Inoue, and T. Masumoto. 1994. Consolidation and mechanical properties of atomized Mg-based amorphous powders. *Materials Science and Engineering A*, 179/A180, 112–117.

Kawamura, Y., H. Kato, A. Inoue, and T. Masumoto. 1995. Full strength compacts by extrusion of glassy metal powder at the supercooled liquid state. *Applied Physics Letters*, 67, 2008–2010.

Kim, J.H., E. Akiyama, H. Habazaki, A. Kawashima, K. Asami, and K. Hashimoto. 1993a. The corrosion behavior of sputter-deposited amorphous chromium-zirconium alloys in 6 M HCl solution. *Corrosion Science*, 34, 1817–1827.

Kim, J.H., E. Akiyama, H. Habazaki, A. Kawashima, K. Asami, and K. Hashimoto. 1993b. The corrosion behavior of sputter-deposited amorphous Cr-Nb and Cr-Ta alloys in 12 M HCl solution. *Corrosion Science*, 34, 1947–1955.

Kim, J.H., E. Akiyama, H. Habazaki, A. Kawashima, K. Asami, and K. Hashimoto. 1994. An XPS study of the corrosion behavior of sputter-deposited amorphous Cr-Nb and Cr-Ta alloys in 12 M HCl solution. *Corrosion Science*, 36, 511–523.

Kim, J.H., E. Akiyama, H. Yoshioka et al. 1993c. The corrosion behavior of sputter-deposited amorphous titanium- chromium alloys in 1 M and 6 M HCl solutions. *Corrosion Science*, 34, 975–988.

Kumagai, N., Y. Samata, S. Jikihara, A. Kawashima, K. Asami, and K. Hashimoto. 1988. Laser and electron beam processing of electrodes consisting of amorphous Ni-valve metal-platinum group metal surface alloys on valve metals. *Materials Science and Engineering*, 99, 489–492.

Li, X.-Y., E. Akiyama, H. Habazaki, A. Kawashima, K. Asami, and K. Hashimoto. 1997a. Spontaneously passivated films on sputter-deposited Cr-Ti alloys in 6 M HCl solution. *Corrosion Science*, 39, 935–948.

Li, X.-Y., E. Akiyama, H. Habazaki, A. Kawashima, K. Asami, and K. Hashimoto. 1997b. An XPS study of passive films on corrosion-resistant Cr-Zr alloys prepared by sputter deposition. *Corrosion Science*, 39, 1365–1380.

Li, X.-Y., E. Akiyama, H. Habazaki, A. Kawashima, K. Asami, and K. Hashimoto. 1998a. An XPS study of passive films on sputter-deposited Cr-Nb alloys in 12 M HCl solution. *Corrosion Science*, 40, 821–838.

Li, X.-Y., E. Akiyama, H. Habazaki, A. Kawashima, K. Asami, and K. Hashimoto. 1998b. Electrochemical and XPS studies of the passivation behavior of sputter-deposited amorphous Cr-Ta alloys in 12 M HCl. *Corrosion Science*, 40, 1587–1604.

Naka, M., K. Hashimoto, and T. Masumoto. 1974. Corrosion resistance of amorphous alloys with chromium [in Japanese]. *Journal of the Japan Institute of Metals*, 38, 835–841.

Naka, M., K. Hashimoto, and T. Masumoto. 1976. High corrosion resistance of chromium-bearing amorphous iron alloys in neutral and acidic solutions containing chloride. *Corrosion*, 32, 146–151.

Park, P.Y., E. Akiyama, H. Habazaki, A. Kawashima, K. Asami, and K. Hashimoto. 1995a. The corrosion behavior of sputter-deposited amorphous Mo-Zr alloys in 12 M HCl. *Corrosion Science*, 37, 307–320.

Park, P.Y., E. Akiyama, H. Habazaki, A. Kawashima, K. Asami, and K. Hashimoto. 1996a. The corrosion behavior of sputter-deposited Mo-Ti alloys in concentrated hydrochloric acid. *Corrosion Science*, 38, 1649–1667.

Park, P.Y., E. Akiyama, H. Habazaki, A. Kawashima, K. Asami, and K. Hashimoto. 1996b. The corrosion behavior of sputter-deposited Mo-Nb alloys in 12 M HCl solution. *Corrosion Science*, 38, 1731–1750.

Park, P.-Y., E. Akiyama, A. Kawashima, K. Asami, and K. Hashimoto. 1995b. The corrosion behavior of sputter-deposited Cr-Mo alloys in 12 M HCl solution. *Corrosion Science*, 37, 1843–1860.

Park, P.-Y., E. Akiyama, A. Kawashima, K. Asami, and K. Hashimoto. 1996c. The corrosion behavior of sputter-deposited Mo-Ta alloys in 12 M HCl solution. *Corrosion Science*, 38, 397–411.

Shinomiya, H., Z. Kato, K. Enari, and K. Hashimoto. 2009. Effects of corrosion-resistant elements on the corrosion resistance of amorphous bulk Ni-Cr-Mo-Ta-Nb-4P alloys in concentrated hydrochloric acids. *ECS Transactions*, 16(32), 9–18.

Yoshioka, H., K. Asami, A. Kawashima, and K. Hashimoto. 1987. Laser processed corrosion-resistant amorphous Ni-Cr-P-B surface alloys on a mild steel. *Corrosion Science*, 27, 981–995.

12

Metal Matrix Composites

Lloyd H. Hihara
University of Hawaii
at Manoa

12.1 Introduction

Metal matrix composites (MMCs) are metals that are usually reinforced with particles or fibers to achieve a combination of properties that are not normally available from existing metallic alloys. The achievement of superior properties such as high specific stiffness and strength is generally realized by the rule of mixtures (ROMs) (Piggott 1980), where MMC properties are proportional, to some degree, to that of the relative volume percentages of the reinforcement material and matrix alloy phases. While the ROMs can be used with some success in predicting certain mechanical and physical properties, corrosion resistance is normally compromised and less than that of the individual MMC constituents.

Reinforcements generally are in the form of continuous fibers (F) and monofilaments (MFs) or discontinuous particles (P) and whiskers (Wh); these can be metallic (e.g., tungsten), nonmetallic (e.g., carbon or boron), or ceramic (e.g., silicon carbide) and may be conductive (e.g., carbon), semiconductive (e.g., silicon), or insulative (e.g., alumina). When reinforcements are continuous, the MMCs are often referred to as continuously reinforced (CR), and when they are discontinuous, the MMCs are referred to as discontinuously reinforced (DR). Designation notation for all types of MMCs has not been unified, and for consistency in this document, the following sequential notation system will be used: matrix alloy/reinforcement material/reinforcement volume percent and form-heat treatment. The later designations can be left out if only making reference to the type of MMC so that, for example, Al6061/SiC/45P-T6 MMC has been simplified to Al6061/SiC MMC.

12.2 Background

The development of MMCs having continuous or discontinuous reinforcements began in the 1960s. Some examples sorted by metal matrix and then reinforcement species were aluminum (Al) (Burte et al. 1966, Hill and Stuhrke 1968), nickel (Ni) (Burte et al. 1966), steel (Burte et al. 1966), copper (Cu) (Burte et al. 1966), and titanium (Ti) (Burte et al. 1966, Galasso and Pinto 1969) all reinforced with boron (B) MF, copper (Kelly and Tyson 1965) reinforced with tungsten (W) MF, and aluminum reinforced with discontinuous alumina (Al_2O_3) (Brenner 1962a,b, Sutton and Chorne 1963, Sutton 1966) and silicon carbide (SiC) whiskers (Divecha et al. 1969). In the 1970s, aluminum reinforced with graphite (Gr) fibers (Harrigan 1991) or SiC particles (Harrigan 1991) was developed. A concerted effort to develop MMCs, however, only materialized beginning in the 1980s when the United States focused on using MMCs for the Strategic Defense Initiative and the National Aerospace Plane and Japan incorporated MMCs into automotive components (Evans et al. 2003). Research was also conducted in Europe (Evans et al. 2003), Russia, and other countries. Today, aluminum MMCs are the most widely available (Harrigan 1991). Owing to manufacturing complexity and the high cost of continuous reinforcements, CR MMC structures are usually very expensive compared to DR MMCs. CR MMCs generally have enhanced properties along the reinforcement axis and exhibit anisotropic behavior, whereas DR MMCs generally have lower property enhancements, but tend to have more isotropic properties, and are cheaper and thus are more readily available and used.

The earliest articles on the corrosion of MMCs were primarily those for Al/B/MF MMCs that were published in the late 1960s. In the 1970s, articles were published on the corrosion of Al/Gr/F MMCs, Al/Al$_2$O$_3$/F MMCs, as well as magnesium (Mg) matrix MMCs. In the 1980s, literature appeared on the corrosion of Al/SiC MMCs, as well as lead (Pb), depleted uranium (DU), and stainless steel (SS) MMCs.

12.3 Application

12.3.1 Uses

MMCs are generally used in specialized, "high cost/value added" applications where their combination of unique properties gives them a comparative advantage over other structural materials. MMCs are often tailored to increase stiffness (Weeton et al. 1987), strength (Weeton et al. 1987), and thermal conductivity (Park and Foster 1990), as well as to reduce density (Weeton et al. 1987), coefficient of thermal expansion (CTE) (Harrigan 1991), friction (Liu et al. 1991), and wear (Lim et al. 1991). Their use has been predominantly in the automotive and aerospace industries as high-performance structural components and in the electronics industry as thermal management components (Evans et al. 2003). MMCs are also used in other niche markets such as in the recreation industry where high-performance materials are valued (Evans et al. 2003).

12.3.1.1 Automotive

Structural-grade DR Al/SiC/P MMCs have been used in the automotive sector instead of conventional metal alloys for components such as engine pistons, connecting rods, rear-wheel driveshafts, brake calipers, cylinder liners, pushrods, rocker arms, valve guides, and wheels due to properties such as lower weight, higher-temperature tensile and fatigue strengths, higher moduli, higher wear resistance, lower coefficient of friction, lower CTE, and higher thermal conductivity to enhance vehicle performance (DWA Technologies Inc. 2007). Structural-grade CF Al/Al$_2$O$_3$/F MMCs have been used to replace some cast-iron automotive components such as engine blocks (Saffil 2002), pistons (Saffil 2002), and push-rods (3M 2002) where properties such as lower weight, lower thermal conductivity and expansion, good wear resistance, improved high-temperature tensile and fatigue strengths, and high-damping capacity are desired.

12.3.1.2 Aerospace

The earliest successful application of CF MMCs was the use of Al/B/MF MMC structural tubes on the space shuttle that resulted in a 44% reduction in weight over aluminum alloy components specified in the original design (Buck and Suplinskas 1987). Antenna booms in the Hubble Space Telescope (Harrigan 1991) were fabricated from Al/Gr/F MMC based on its high stiffness and a near-zero CTE property. DR Al/SiC/P MMCs have been used on aircraft for fan exit guide vanes in turbine engines, ventral fins, helicopter blade sleeves, and fuel access covers (ALMMC 2007, DWA Aluminum Composites 2010).

12.3.1.3 Electronic

The electronic-grade MMCs usually contain high levels of thermally conductive reinforcements (e.g., high-purity green SiC and graphite) to dissipate heat. For example, Al6092/SiC/44P MMCs (DWA Technologies Inc. 2007) and Cu/Gr MMCs (Marcus et al. 1987, JW Composites LC 2010, Metal Matrix Cast Composites LLC 2010) have been used for electronic packaging, heat sinks, and sliding electrical contacts. For another DR MMC application, a pure Al/Si/43P MMC (with better machinability than MMCs reinforced with harder reinforcements) was used as electronic packaging material owing to its lower CTE, which better matched that of its GaAs electronic components.

12.3.2 Challenges

One of the greatest challenges of MMCs is the absence of an inventory of commercially available off-the-shelf (COTS) MMCs with known/standardized properties because each MMC, generally fabricated by either liquid-state or solid-state processing (Chawla and Chawla 2006), has properties that are strongly dependent on its processing-history specifics. For liquid-state processing, various methods such as casting, squeeze casting, and spraying have been employed, whereas in solid-state processing, powder metallurgical technology and diffusion bonding have been used successfully. Other more exotic methods including vapor-state processing have also been developed. In addition to processing-history property variations, much variability in MMC properties can be attributed to differences in the reinforcement-phase properties due to purity and macroscopic (shape) and crystallographic forms, as well as to differences in the concentration and distribution of the reinforcements within the matrix. Thermomechanical processing histories are also important because an MMC cannot simply be "resolutionized" to reset its microstructure to original or desired conditions as in conventional metal alloys. Thermomechanical processing of an MMC may fracture reinforcements, redistribute reinforcements, and cause interphase reactions between reinforcement and matrix that cannot be reversed.

12.4 Physical Properties

Depending on the type and distribution/orientation of the reinforcements within the MMC microstructure, MMCs can have either anisotropic or isotropic mechanical properties. CR MMCs primarily have anisotropic properties as the reinforcement fibers or MFs are usually aligned uniaxially. DR MMCs generally have more isotropic properties as it is usually desirable to have reinforcement particles randomly scattered/oriented throughout the MMC microstructure. Some bulk composite properties can be estimated to some degree of accuracy with simple models such as the ROMs, while other properties require more sophisticated models. In the models in the succeeding text, the subscripts MMC, m, r, and f are used to represent those of the MMC, matrix, reinforcement, and fiber, respectively.

12.4.1 Density

The density of the MMC (ρ_{MMC}) can be estimated (Equation 12.1) by the ROM using original densities of the matrix (ρ_m) and reinforcement (ρ_r) phases and their respective volume fractions, v_m and v_r (Chawla and Chawla 2006):

$$\rho_{MMC} = \rho_m v_m + \rho_f v_f = \rho_m(1-v_f) + \rho_f v_f \qquad (12.1)$$

since

$$v_m + v_f = 1 \qquad (12.2)$$

12.4.2 Mechanical Properties

12.4.2.1 Elastic Modulus

For a CR MMC that is uniaxially reinforced, the upper bound of the elastic modulus in the longitudinal ($E_{CR\,MMC\text{-}L}$) direction (i.e., in the same direction along which the fibers are oriented) is estimated (Equation 12.3) to be as follows (Ashby and Jones 1996):

$$E_{CR\,MMC\text{-}L} = E_m v_m + E_f v_f = E_m(1-v_f) + E_f v_f \qquad (12.3)$$

The lower bound of the elastic modulus for a CR MMC in the transverse ($E_{CR\,MMC\text{-}T}$) direction (i.e., in the direction perpendicular to axis of the fibers) is estimated (Equation 12.4) to be (Ashby and Jones 1996)

$$E_{CR\,MMC\text{-}T} = \frac{1}{(v_m/E_m) + (v_f/E_f)} = \frac{1}{((1-v_f)/E_m) + (v_f/E_f)} \qquad (12.4)$$

For a particulate DR MMC, which normally has isotropic properties, the elastic modulus is bounded by Equations 12.3 and 12.4 and usually resides nearer to the lower bound (Equation 12.4) (Ashby and Jones 1996).

12.4.2.2 Strength

For a CR MMC that is uniaxially reinforced, the upper bound of the ultimate tensile strength in the longitudinal direction $\left(\sigma^{TS}_{CR\,MMC\text{-}L}\right)$ is estimated to be as follows:

$$\sigma^{TS}_{CR\,MMC\text{-}L} = v_f \sigma^{frac}_f + (1-v_f)\sigma^{yield}_m \qquad (12.5)$$

where
σ^{frac}_f is the fracture strength of the fiber
σ^{yield}_m is the yield strength of the matrix at the strain value where fiber fracture occurs

For a CR MMC that is uniaxially reinforced, the transverse strength $\sigma^{TS}_{CR\,MMC\text{-}T}$ is much lower than the value for the longitudinal direction since the fibers do not directly contribute to the transverse MMC strength. Similarly, even with similar reinforcement volume fractions and the same matrix, DR MMCs have much lower strengths than the equivalent CR MMC in the longitudinal direction; however, DR MMCs have the advantage of having more isotropic properties.

12.4.3 Thermal Properties

12.4.3.1 Thermal Expansion

The CTEs of the CR MMCs (α_{MMC}) were estimated using the values for the matrix (α_m) and fiber (α_f), the elastic moduli (E_m and E_f), volume fractions (v_m and v_f), and Poisson's ratios (v_m and v_f) (Schapery 1968). In the longitudinal direction, the CTE $\alpha_{CR\,MMC\text{-}L}$ for uniaxially reinforced CR MMCs when the fibers and matrix have similar Poison's ratios was estimated to be

$$\alpha_{CR\,MMC\text{-}L} = \frac{(\alpha_m E_m v_m + \alpha_f E_f v_f)}{E_m v_m + E_f v_f} \tag{12.6}$$

In the transverse direction, the CTE $\alpha_{CR\,MMC\text{-}T}$ for uniaxially reinforced CR MMCs was estimated to be

$$\alpha_{CR\,MMC\text{-}T} = (1 + v_m)\alpha_m v_m + (1 + v_f)\alpha_f v_f - \alpha_{CR\,MMC\text{-}L}(v_m v_m + v_f v_f) \tag{12.7}$$

For an isotropic DR MMC, the CTE is estimated as follows (Turner 1946):

$$\alpha_{DR\,MMC} = \frac{\alpha_m v_m K_m + \alpha_r v_r K_r}{v_m K_m + v_r K_r} \tag{12.8}$$

where K_m and K_r are the bulk moduli of the matrix and reinforcement, respectively.

12.4.3.2 Thermal Conductivity

The thermal conductivity of the MMC (k_{MMC}) can be estimated by the values of the matrix (k_m) and reinforcement (k_r) and volume fraction of the matrix and the reinforcements.

For a CF MMC in the longitudinal direction, the expression for the thermal conductivity is as follows (Behrens 1968):

$$k_{CFMMC\text{-}L} = k_{f\text{-}L} v_f + k_m v_m \tag{12.9}$$

where $k_{f\text{-}L}$ is the thermal conductivity of the fiber in the longitudinal direction.

In the transverse direction, the thermal conductivity (Chawla and Chawla 2006) is

$$k_{CFMMC\text{-}T} = \frac{k_{f\text{-}T} k_m}{k_{f\text{-}T} v_f + k_m v_m} \tag{12.10}$$

where $k_{f\text{-}T}$ is the thermal conductivity of the fiber in the transverse direction.

For DR MMCs, a method to estimate the thermal conductivity can be found in the reference by Rayleigh (Rayleigh 1892).

12.5 Microstructure

CR MMCs are usually reinforced with fibers having diameters of approximately 10 μm (Figure 12.1a) or with MF having diameters of approximately 150 μm (Figure 12.1b). DR MMCs are reinforced with particulate reinforcements that have volume percents ranging from ~15% to 25% for structural MMCs (Figure 12.2a) and values greater than 30% for electronic-grade MMCs (Figure 12.2b). The particle-size distribution can be relatively uniform in structural MMCs (Figure 12.2a) but is highly

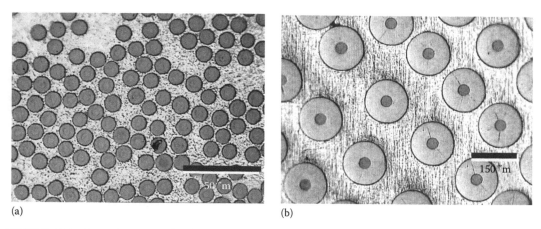

(a)   (b)

FIGURE 12.1 Optical micrographs of transverse sections of CR MMCs: (a) pure Al/Gr/50F MMC, (b) Al6061/SiC/48MF MMC. (Photo courtesy of Shruti Tiwari.)

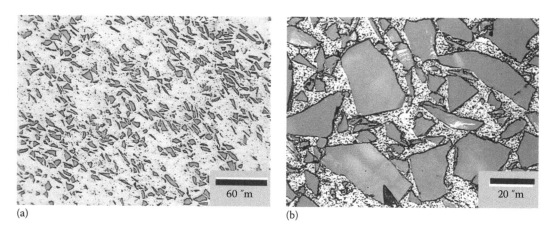

(a)   (b)

FIGURE 12.2 Optical micrograph of DR MMCs: (a) Al 6092/B$_4$C/20P-T6 with a nominally uniform particle-size distribution, (b) Al 6092/SiC/50P-T6 with highly varied particle-size distribution. (Photo courtesy of George Hawthorn.)

variable in electronic-grade MMCs (Figure 12.2b) in order to attain higher reinforcement volume fractions. The overall microstructure of the MMC depends on (a) the type, relative amount, and morphology of the reinforcement phase as well as the matrix system used; (b) the processing route (e.g., liquid or solid state); (c) any subsequent heat treatments employed; and (d) other relevant processing and material factors. In addition to the reinforcement and matrix present in the microstructure, additional interphases may be formed by the reaction between the reinforcement and matrix during processing.

12.5.1 Reinforcements

Typical reinforcements that have been used in MMCs are boron, carbon or graphite, silicon, silicon carbide, boron carbide, titanium diboride, alumina, and muscovite. Their physical and mechanical properties are discussed in the succeeding text and summarized in Table 12.1.

Pure alumina is an insulator, and its morphology is usually in the form of particles and fibers. The particles have a typical size of 1–50 μm, and the fibers are usually 10 μm in diameter.

TABLE 12.1 Reinforcement Properties

Material	Notes	Size or Diameter (μm)	Density (g/cm³)	Electric Resistivity (Ω cm)	Tensile Strength (GPa)	Elastic Modulus (GPa)	CTE (μ/K)	Refs.
Alumina	Particles, whiskers	1–50	3.8	$>10^{14}$		430	7	Bolz and Tuve (1973), Clyne and Withers (1995)
	Saffil® fiber	3	3.5		1.5	285	7.7	Clyne and Withers (1995)
	DuPont FP fiber	20	4.0		1.3–2.1	380	8.3	Clyne and Withers (1995)
Boron				6.7×10^5				Greenwood and Earnshaw (1984)
	W-cored MF	50–100	2.6	2×10^{-1}	2.7–7.0	400	5.0	Hihara (1997), Clyne and Withers (1995)
Boron carbide	Particles, whiskers	1–50	2.5	10^0	2.1	480		Clyne and Withers (1995), Yamada et al. (2003)
Carbon, graphite	Pitch-based	5–13	2.0	2.5×10^{-4}– 7.5×10^{-4}	2.0	380–690	−1.4	Clyne and Withers (1995), Weeton et al. (1987)
	PAN-based high strength	7	1.9		4.8	230	−1.2	Clyne and Withers (1995)
	PAN-based high modulus	7	1.9		2.4	412	−1.2	Clyne and Withers (1995)
Muscovite mica	$KAl_3Si_3O_{10}(OH)_2$	<70		10^{13}–10^{17}				Clauser (1963), Nath et al. (1980)
Silicon		5–50		10^{-2}–10^5				Bolz and Tuve (1973), Hihara (2005b)
Silicon carbide	Particles, whiskers	1–50 (p) 0.1–1 (w)	3.2	10^{-5}–10^{13}		450–700	4	Ichinose (1987), Clyne and Withers (1995)
	C-cored MF	150	3.0	$\approx 10^{-2}$	3.45	400	—	Textron Specialty Materials (1990), Hihara (1997)
Titanium diboride	Particles, whiskers	44	4.5			350–570	8.1	Clyne and Withers (1995), Covino and Alman (2002)

Boron reinforcements are usually in the MF form and consist of a sheath of polycrystalline boron with a core of tungsten with a thin surrounding tungsten boride interaction phase that forms while the boron is being chemically vapor deposited onto a resistance-heated continuous tungsten filament during processing (Adler and Hammond 1969, Tsirlin 1985). Although pure boron is an insulator and cannot support electrical currents, the resistivity of B MF is many orders of magnitude lower than that of pure boron due to tungsten and tungsten borides in the core (Adler and Hammond 1969, Tsirlin 1985). B MF has a density higher than that of pure B due to the W core. Typical B MFs have a diameter of approximately 100 μm (Weeton 1969).

Boron carbide (B_4C) is a semiconductor (SC) with a complex icosahedral crystal structure (Greenwood and Earnshaw 1984). B_4C, the third hardest material (behind diamond and cubic boron nitride) known to man (Suri et al. 2010), has a density approximately 20% lower than SiC and also has good neutron adsorption properties (Greenwood and Earnshaw 1984), making it attractive for the nuclear industry. B_4C reinforcements are usually in the form of particles, with a typical size of 1–50 μm.

Carbon and graphite fibers, which are conductors, can be derived from various precursors such as pitch or polyacrylonitrile (PAN) (Chawla and Chawla 2006). Graphite fibers typically have a core of radially oriented basal graphite planes that is surrounded circumferentially by an outer layer of basal planes; thus, the microstructure in the plane perpendicular to the fiber axis has a spoke-like structure surrounded by an onion skin structure (Diefendorf 1984). The elastic modulus of carbon–graphite fibers can usually be increased by processing at higher temperatures to increase the degree of graphitization; this, however, can also lead to lower strength due to higher residual stresses. Graphite fibers also have negative CTE and hence have been used to fabricate low CTE MMCs (Harrigan 1991). Continuous and chopped carbon fibers are used in MMCs. The diameter of the carbon and graphite fibers is approximately 10 μm.

Muscovite $(KAl_3Si_3O_{10}(OH)_2)$ mica is an insulator and has been used as particulate reinforcement to impart antiseizure properties to MMCs (Nath et al. 1980). Mica exhibits cleavage along the basal plane similar to graphite and hence can impart antifriction properties. Typical particle sizes are less than 70 μm (Nath et al. 1980).

Silicon is an SC and is used in electronic-grade particulate DR aluminum MMCs. The hardness of pure Si is approximately half that of SiC, and hence, Al/Si MMCs can be machined by conventional means. The typical Si particle size is in the range of 5–50 μm (Hihara 2005b).

Lower-purity, black particulate SiC is used in structural-grade DR MMCs, and high-purity, green particulate SiC is used in electronic-grade DR MMCs owing to the higher thermal conductivity of the green high-purity SiC. The electrical resistivity of SiC, which is an SC, depends on its purity and thus can span approximately 18 orders of magnitude (Ichinose 1987). Therefore, in the unpure state, the resistivity of SiC can be relatively rather low. Some types of SiC MFs also have carbon cores and carbon-rich surfaces resulting in very low resistivities (Hihara 1997). SiC reinforcements are available in particulate, whisker, fiber, and MF forms. SiC particles can range in sizes from ~10 to 50 μm, whiskers have typical diameters of 0.1–1 μm with an aspect ratio of 10–20, fibers have diameters of approximately 10 μm, and MFs have diameters of approximately 150 μm.

Titanium diboride (TiB_2) has a very low resistivity that is comparable to that of metals. It has been used in the particle form with a typical size of 44 μm (Covino and Alman 2002).

12.5.2 Matrix

The resulting microstructure of the MMC matrix can depend on processing conditions and routes (e.g., liquid- or solid-state processing). When MMCs are formed by a casting process, the last melt to solidify is frequently enriched in solute that often solidifies around the reinforcements as solute-rich zones (Clyne and Withers 1995). In contrast, such solute-rich zones can be eliminated by using solid-state processing (Clyne and Withers 1995), for example, in diffusion-bonded MMCs, where metallic foils and their reinforcements are consolidated under heat and pressure. There is a risk, however, of incomplete matrix/matrix and matrix/reinforcement bonding if processing conditions are not optimized, as was evident in some Al/B/MF MMCs (Sedriks et al. 1971, Bakulin et al. 1978).

12.5.3 Interphase

During the fabrication processing of MMCs, thermally activated reactions between certain reinforcement and matrix phases may lead to the formation of an interphase at the reinforcement–matrix interface. The extent of interphase formation can depend on the processing temperature as well as the composition of the matrix that affects the chemical activities of the reacting species. Table 12.2 shows some interphase reaction products that can form in a variety of MMC systems.

TABLE 12.2 MMC Systems and Possible Interphases

Matrix	Reinforcement	Interphase or Reaction Product	Reference
Aluminum	Al_2O_3	None	Chawla and Chawla (2006)
	B	AlB_2	Chawla and Chawla (2006)
	B_4C	Al_4C_3	Grytsiv and Rogl (2004)
	C	Al_4C_3	Iseki et al. (1984), Chawla and Chawla (2006)
	SiC	Al_4C_3	Iseki et al. (1984), Chawla and Chawla (2006)
	ZrO_2	$ZrAl_3$	Chawla and Chawla (2006)
Magnesium	Al_2O_3	$MgO, MgAl_2O_4, Al$	Clyne and Withers (1995), Chawla and Chawla (2006)
	SiC	Mg_2Si, C	Clyne and Withers (1995)
	SiO_2	Mg_2Si, MgO	Clyne and Withers (1995)
Copper	C	None	Chawla and Chawla (2006)
	W	None	Chawla and Chawla (2006)
Titanium	Al_2O_3	TiO_2, Al	Clyne and Withers (1995)
	B_4C	TiB, TiC	Clyne and Withers (1995)
	C	TiC	Blackwood et al. (2000)
	SiC	TiC, Ti_5Si_3	Chawla and Chawla (2006)
	TiB_2	TiB	Covino and Alman (2002)
Nickel	SiC	Ni_2Si, C	Clyne and Withers (1995)

12.6 Degradation Principles for MMCs (Electrochemical, Chemical, and/or Photochemical)

12.6.1 Thermodynamics

The thermodynamic driving force for an MMC to corrode is generally similar to that of a monolithic matrix alloy. In deaerated environments, the thermodynamic driving force $\left(E_{cell}^{deaerated}\right)$

$$E_{cell}^{deaerated} = E_{H^+/H_2} - E_{M^{n+}/M} \tag{12.11}$$

is the potential difference between the equilibrium potential of proton reduction (E_{H^+/H_2})

$$E_{H^+/H_2} = E_{H^+/H_2}^{\circ} - \frac{RT}{nF}\ln\frac{p_{H_2}}{a_{H^+}^2} \tag{12.12}$$

based on the proton reduction reaction

$$2H^+ + 2e^- \rightarrow H_2 \tag{12.13}$$

and the equilibrium potential of the matrix alloy $(E_{M^{n+}/M})$

$$E_{M^{n+}/M} = E_{M^{n+}/M}^{\circ} - \frac{RT}{nF}\ln\frac{a_M}{a_{M^{n+}}} \tag{12.14}$$

based on the matrix metal dissolution reaction

$$M \rightarrow M^{n+} + ne^-$$ (12.15)

In aerated environments, the thermodynamic driving force $\left(E_{cell}^{aerated}\right)$ is the potential difference

$$E_{cell}^{aerated} = E_{O_2/OH^-} - E_{M^{n+}/M}$$ (12.16)

between the equilibrium potential of oxygen reduction (E_{O_2/OH^-})

$$E_{O_2/OH^-} = E_{O_2/OH^-}^\circ - \frac{RT}{nF} \ln \frac{a_{OH^-}^4}{p_{O_2}}$$ (12.17)

based on the oxygen reduction reaction

$$O_2 + 4e^- + 2H_2O \rightarrow 4OH^-$$ (12.18)

and the equilibrium potential of the matrix alloy $E_{M^{n+}/M}$.

For corrosion to occur, the cell potential E_{cell} of the corrosion reaction must be positive, which corresponds to a decrease in Gibbs free energy. In Equations 12.12, 12.14, and 12.17, n is the number of electrons transferred in the reaction, R is the universal gas constant, T is temperature in K, F is the Faraday, a_i is the activity of species i, and p_i is the partial pressure of gas i. The value of E_{cell} only indicates that the reaction is thermodynamically possible or impossible. Whether the MMC corrodes differently from its monolithic matrix alloy is dependent on the ability of the reinforcements or interphases to support the cathodic reactions or undergo anodic dissolution. Therefore, knowledge of kinetics is required to estimate the rates of the corrosion reaction.

12.6.2 Kinetics

In the study of corrosion, polarization diagrams (Gellings 1976) are used to determine the rates of metal dissolution, oxygen reduction, and proton reduction. The thermodynamic driving force for the electrochemical reaction is measured as potential on the vertical axis of the polarization diagram. The kinetics of the electrochemical reaction is measured as current on the horizontal axis of the polarization diagram. Background information of polarization diagrams that are used to describe metallic corrosion is covered in Chapter 1. Since MMCs are comprised of heterogeneous constituents, their electrochemical or galvanic interaction will govern the MMC corrosion rate, which can be studied using polarization diagrams.

12.6.2.1 Galvanic Corrosion between the Matrix and Reinforcements

When two dissimilar electrically conductive materials are coupled together and placed in an electrolyte, the potential of the more noble material (more positive electrode potential) and that of the more active material (more negative electrode potential) will equilibrate to the galvanic potential (E_{Galv}), resulting in the change of the normal corrosion behavior of each constituent material. This phenomenon is known as galvanic corrosion (Wesley and Brown 1948) and is a primary concern regarding the corrosion behavior of MMCs that have either conductive or semiconductive reinforcements and interphases. Reinforcement resistivities are tabulated in Table 12.2. In many cases, the reinforcements (e.g., graphite, SiC, and TiB_2) are relatively inert and noble and become the cathode, while the metal matrix that is often an active metal (e.g., Al or Mg) becomes the anode, resulting in corrosion behavior unlike that of the monolithic matrix alloy. The use of polarization diagrams and the mixed-potential theory (Fontana and Greene 1978), however, enables one to predict the galvanic tendencies in an MMC. Several scenarios are discussed later for MMCs reinforced with conductive or semiconductive reinforcements. To keep the explanations simple, it will be assumed that the anodic and cathodic curves of the matrix metal are identical in both deaerated and aerated environments. The intersection of the cathodic curve of the reinforcement and the anodic curve of the matrix metal identifies the galvanic potential (E_{Galv}) and the galvanic current (I_{Galv}).

Scenario 1—Galvanic corrosion increases with the presence of dissolved oxygen

The polarization curve (Figure 12.3a) of an active metal is plotted with the cathodic curves of a noble reinforcement in deaerated and aerated environments. The dissolution rate for the matrix metal will increase from its normal corrosion current I_{Corr} in the uncoupled state to a higher value $I_{Galv}^{Deaerated}$ when it is coupled to the reinforcement in a deaerated environment and to an even larger value $I_{Galv}^{Aerated}$ in an aerated environment. The galvanic potential also increases from $E_{Galv}^{Deaerated}$ in the deaerated environment to a more positive value $E_{Galv}^{Aerated}$ in the aerated environment.

Scenario 2—Galvanic corrosion does not increase with the presence of dissolved oxygen

The corrosion current of an active metal (Figure 12.3b) will increase from the normal corrosion current I_{Corr} to a value of $I_{Galv}^{Aerated, Deaerated}$ that is essentially identical for deaerated and aerated environments. The galvanic potential will be similar for both deaerated and aerated environments. This situation can arise if the diffusion-limited, oxygen reduction current is less than $I_{Galv}^{Aerated, Deaerated}$.

Scenario 3—Galvanic corrosion is limited by passivation

The galvanic-corrosion current $I_{Galv}^{Passive}$ of a passive metal (Figure 12.4a) is identical to the normal corrosion current I_{Corr} due to passivation of the matrix metal. The galvanic potential, however, of the passivated matrix metal $E_{Galv}^{Passive}$ will be more positive than the E_{Corr} of the monolithic matrix metal. If the matrix metal is susceptible to pitting, for example, in a halide-containing environment, the matrix metal can pit when it is coupled to the reinforcement, resulting in a galvanic current $I_{Galv}^{Pitting}$ higher than

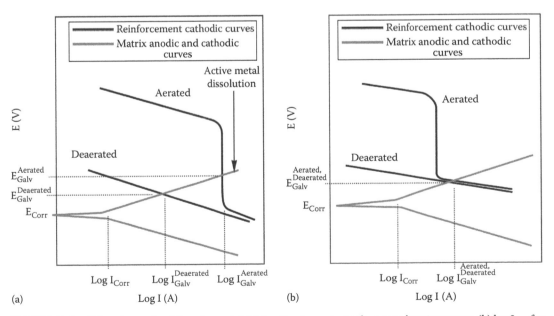

(a)   (b)

FIGURE 12.3 Galvanic couple with active metal. (a) Log I_{Galv} is greater in the aerated environment, (b) log I_{Galv} for both deaerated and aerated environments is approximately equal.

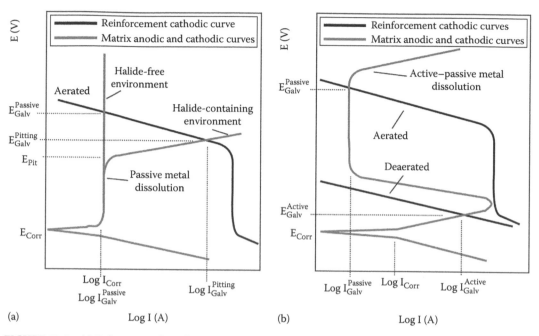

FIGURE 12.4   (a) Galvanic couple with passive metal. Log I_{Corr} and log I_{Galv} are identical in the halide-free environment. Log I_{Galv} increases in the halide-containing environment. (b) Galvanic couple with active–passive metal. Log I_{Galv} is lower in the aerated environment due to anodic protection.

the normal I_{Corr} of the monolithic matrix metal and a galvanic potential $E_{Galv}^{Pitting}$ higher than the normal E_{Corr} of the monolithic matrix metal.

Scenario 4—Galvanic corrosion decreases in the presence of dissolved oxygen

This interesting phenomenon could occur when an active–passive metal is galvanically coupled to a more noble reinforcement. The normal I_{Corr} of the monolithic matrix metal (Figure 12.4b) will increase to I_{Galv}^{Active} in a deaerated environment, but then it will decrease to $I_{Galv}^{Passive}$ in an aerated environment as oxygen reduction on the reinforcement phase polarizes the metal to $E_{Galv}^{Passive}$ into the passive regime, which anodically protects the metal.

Scenario 5—Galvanic corrosion decreases or increases under illumination

If the MMC reinforcements are SCs, illumination can suppress galvanic currents if the reinforcement is an n-type SC or can cause accelerated galvanic corrosion if the reinforcement is a p-type SC.

An n-type SC is photo-anodic and can promote photooxidation reactions under illumination, such as the oxidation of water. Hence, in an MMC, an n-type SC reinforcement phase could polarize the MMC from its galvanic potential in the dark E_{Galv}^{Dark} to more negative potentials $E^{Illuminated}$ inducing cathodic protection with lower matrix dissolution currents $I_{Dissolution}^{Illuminated}$ (Figure 12.5a) (Ding and Hihara 2008b). Accordingly, anodic current densities on Al 6092/Al$_2$O$_3$/20 P-T6 MMC electrodes immersed in air-exposed 0.5 M Na$_2$SO$_4$ solutions increased sharply with illumination, which was attributed to photo-anodic currents due to water oxidation on n-type TiO$_2$ particles in the MMC that were likely introduced with the Al$_2$O$_3$ reinforcements (Ding and Hihara 2008b). The open-circuit potentials also decreased with illumination, indicating that the n-type TiO$_2$ induced cathodic protection. Outdoor exposures of these Al 6092/Al$_2$O$_3$/20 P-T6 MMCs showed that corrosion products were thinner on the topside of specimens exposed to sunlight as compared to the backside of the specimens not exposed to sunlight (Adler et al. 2005).

A p-type SC is photocathodic and under illumination promotes photoreduction reactions, such as proton or oxygen reduction. Hence, a p-type SC could polarize the MMC in the dark from E_{Galv}^{Dark} to

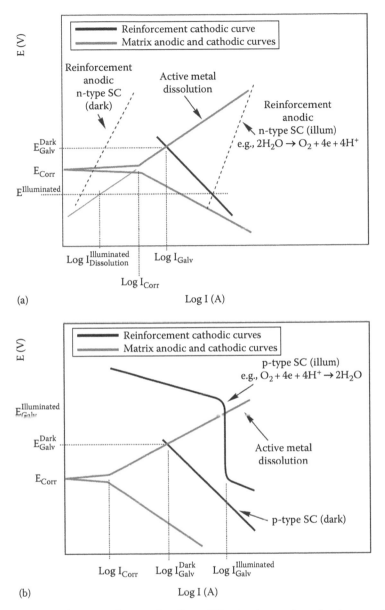

FIGURE 12.5 Polarization diagrams showing the effect of illumination on the galvanic-corrosion behavior of a metal alloy coupled to an SC. (a) For n-type SC, illumination stimulates the oxidation of water, causing the galvanic potential to drop resulting in lower metal dissolution rates. (b) For p-type SC, illumination increases the cathodic current resulting in higher galvanic-corrosion rates.

more positive galvanic potentials under illumination $E_{Galv}^{Illuminated}$, causing the cathodic current to increase to $I_{Galv}^{Illuminated}$ (Figure 12.5b) (Ding and Hihara 2008b). Interestingly, MMCs containing p-type SCs had thicker corrosion products on the sunlit surfaces as apposed to the shaded surfaces (Adler et al. 2005).

These examples demonstrate various mechanisms of MMC corrosion based on the known electrochemical behavior of their constituents. The utilization of polarization diagrams and the mixed-potential theory can be helpful in predicting obvious corrosion concerns that may surface in the actual MMC and help to engineer MMCs with acceptable corrosion resistance. The use of the mixed-potential theory, however, cannot replace an experimental study of the actual corroding

MMC to understand all aspects of the corrosion behavior because unexpected behavior may result from specific processing-related issues.

12.6.3 Processing and Microstructural Effects on MMC Corrosion

Sometimes, the corrosion behavior of an actual MMC can be different from that predicted using the mixed-potential theory, especially if only the matrix and reinforcement phases are considered. Other factors such as the formation of interphases, segregation of alloying elements, mismatch in CTE of reinforcement and matrix, and impurities introduced into the MMC microstructure during processing may also affect corrosion behavior.

12.6.3.1 Interphase Formation

Interphase formation may be accelerated when the processing temperature is too high or when insufficient isolation (e.g., when an inert coating barrier on the reinforcement is absent or too thin to prevent interdiffusion) is present between the reactive reinforcement and matrix phases. Some interphase compounds can become an extra tertiary phase that can significantly alter MMC corrosion behavior.

For example, in Al/B/MF MMCs (Pohlman 1978), galvanic currents could not be measured between aluminum and virgin B MFs but were measured between aluminum and B MFs that were extracted from the MMC. In this case, the extracted B MF had an external 4 μm thick interphase layer of aluminum boride that formed during MMC processing. This aluminum boride coating was an efficient cathode, whereas the virgin B MF was not. Similarly, in aluminum MMC reinforced with carbon-containing reinforcements such as graphite and carbon (Becher 1963), SiC (Iseki et al. 1984), and B_4C (Grytsiv and Rogl 2004), an aluminum carbide (Al_4C_3) interphase layer can form at elevated temperatures during processing. That Al_4C_3 could hydrolyze in the presence of moisture forming methane and aluminum hydroxide. The rate of hydrolysis for hot-pressed, porous Al_4C_3 (78% of theoretical density) in pure water at 30°C was measured to be approximately 1% per hour (Hihara 1989). In some Al/Gr MMC corrosion studies, the hydrolysis of this Al_4C_3 did result in methane generation (Portnoi et al. 1981, Buonanno 1992) and fissuring at the graphite fiber–aluminum interfaces (Buonanno 1992).

12.6.3.2 Segregation of Alloying Elements

The presence of the reinforcement phases in MMCs can affect the precipitation morphology of normal intermetallic phases that are found in certain monolithic matrix alloys. For example, intermetallic phases may form preferentially around the reinforcements by solute rejection during solidification (Mortensen et al. 1988). These intermetallics have intrinsic electrochemical properties (e.g., corrosion potentials, pitting potentials, and corrosion current densities [Birbilis and Buchheit 2005]) and, hence, can have an effect on MMC corrosion behavior. Intermetallics that have more noble potentials than the matrix can accelerate dissolution of the matrix by galvanic action, whereas intermetallics that are more active than the matrix may go into dissolution, leaving fissures that can lead to pitting in the MMC. Hence, variations in the corrosion behavior between the monolithic matrix alloy and the MMC may become evident if the presence of the reinforcement significantly alters the intermetallic morphology (e.g., preferred precipitation along reinforcement–matrix boundaries in MMCs, compared to standalone precipitation that occurs in the monolithic alloy).

In Al/Al_2O_3 MMCs, localized corrosion along fiber–matrix interfaces (Bruun and Nielsen 1991) was attributed to the dissolution of Al_8Mg_5 and Mg_2Si intermetallics, and pit formation (Yang and Metzger 1981) was attributed to the dissolution of $MgAl_3$.

12.6.3.3 Mismatch in Reinforcement and Matrix CTE

If there is a large mismatch in the CTE between the reinforcement and matrix, the crystallographic structure of the matrix may be thermally disrupted by the formation of dislocations (Arsenault 1991) during any heating and cooling cycles occurring during processing and subsequent heat treatments.

It is known that the process of cold working (to intentionally generate high dislocation densities) used to increase yield strengths in metals can alter the corrosion behavior of metals such as steel (Uhlig and Revie 1985) and aluminum (Butler and Ison 1978) and therefore similarly may also affect MMCs (Hihara 1997). This effect has been suggested as the cause for localized corrosion observed near the SiC–Al interface in some Al/SiC MMCs (Yao and Zhu 1998, Ahmad et al. 2000).

12.6.3.4 Microstructural Contamination

The processing of MMCs is generally very different from that of their monolithic matrix alloys, and hence, contaminants not usually associated with the monolithic alloy processing may enter MMC microstructures. In specific cases, reinforcements may be coated with compounds to provide barrier protection or to enhance wetting by the matrix alloy during processing. For example, in Al MMCs reinforced with Ni-coated graphite fibers, the Ni coating dissolves resulting in the formation of Al–Ni intermetallics (i.e., Al_3Ni, Al_3Ni_2, and $AlNi$) (Wielage and Dorner 1999). Also, in Al MMC reinforced with TiB_2-coated graphite fibers, chlorides originating from $TiCl_4$ and BCl_3 processing gasses (Harrigan and Flowers 1979) contaminate the MMC microstructure, forming microstructural chlorides that can induce pitting (Hihara and Latanision 1991) even in chloride-free environments. A polarization diagram of the actual Al6061/Gr/50F-T6 MMC was compared to those generated from monolithic Al6061-T6 and graphite fibers (in a mixed-electrode diagram) exposed to 0.5 M Na_2SO_4 using the mixed-potential theory (Hihara and Latanision 1991). The mixed-electrode diagram shows that Al6061-T6 passivates during the anodic scan, whereas the anodic polarization diagram of the actual Al6061/Gr/50F-T6 MMC shows that pitting is induced at approximately $-0.6V_{SCE}$. Buonanno (1992) later verified that the anodic polarization diagram of chloride-free Al/Gr MMCs traces that of the mixed-electrode diagram.

12.7 Corrosion of MMC Systems

The corrosion behavior of aluminum, magnesium, titanium, copper, SS, lead, DU, and zinc MMCs will be presented.

12.7.1 Aluminum MMCs

Aluminum is a reactive metal but generally has good resistance to corrosion in chloride-free, near-neutral solutions (Deltombe et al. 1974) due to the formation and stability of a passive oxide film. In acidic and basic solutions, however, the passive oxide film is not stable, resulting in high corrosion rates (Deltombe et al. 1974). High corrosion rates are also caused by aggressive ions such as chlorides that cause pitting of the aluminum surface, and the susceptibility of pitting increases (lowering of E_{Pit}) with increasing halide concentration (Galvele 1978). However, in order for aluminum to pit in the open-circuit condition, it must be polarized to potentials noble to E_{Pit} by a cathodic reaction. Hence, one of the primary concerns for Al MMCs in corroding environments is that galvanic action between the matrix and any conducting or semiconducting reinforcements (where that phase can become an efficient cathode) would accelerate corrosion. When reinforcements are insulating, galvanic action is not a concern unless the formation of an "active" interphase or induced precipitation of intermetallics in the matrix can become cathodic sites. The present caveat is that conductive (e.g., graphite), semiconductive (e.g., SiC, Si, B_4C), and insulative (e.g., Al_2O_3) reinforcements are being used in Al MMCs.

12.7.1.1 Aluminum/Graphite MMCs

Al/Gr MMCs use chopped or continuous fiber reinforcements and are susceptible to severe galvanic corrosion in aerated chloride-containing solutions (Figure 12.6) since graphite is an efficient cathode. The main cathodic reaction is oxygen rather than proton reduction (Dull et al. 1977); hence, galvanic-corrosion rates in deaerated chloride solutions (Figure 12.7) are negligible. Some Al/Gr MMCs may also contain the Al_4C_3 interphase that forms during processing, and the presence of

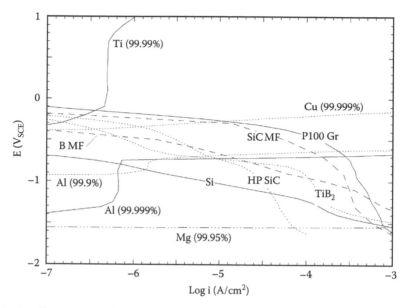

FIGURE 12.6 A collection of anodic polarization diagram of ultrapure Al (99.999%) (Hihara and Latanision 1992), pure Al (99.9%) (Lin 1995), ultrapure Ti (99.99%) (Tamirisa 1993, Hihara and Tamirisa 1995), ultrapure Cu (99.999%), and pure Mg (99.95%) (Kondepudi 1992) exposed in deaerated 3.15 wt% NaCl at 30°C and cathodic polarization diagrams of P100 Gr (ends exposed) (Hihara and Latanision 1992), HP SiC (Hihara and Latanision 1992), SiC MF (ends exposed) (Hihara and Latanision 1994), Si (Lin 1995), TiB$_2$ (Hihara and Latanision 1992), and B MF (ends exposed) (Hihara 1997) exposed to aerated 3.15 wt% NaCl at 30°C.

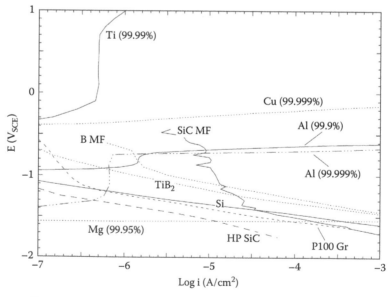

FIGURE 12.7 A collection of anodic polarization diagram of ultrapure Al (99.999%) (Hihara and Latanision 1992), pure Al (99.9%) (Lin 1995), ultrapure Ti (99.99%) (Tamirisa 1993, Hihara and Tamirisa 1995), ultrapure Cu (99.999%), and pure Mg (99.95%) (Kondepudi 1992) and cathodic polarization diagrams of P100 Gr (ends exposed) (Hihara and Latanision 1992), HP SiC (Hihara and Latanision 1992), SiC MF (ends exposed) (Hihara and Latanision 1994), Si (Lin 1995), TiB$_2$ (Hihara and Latanision 1992), and B MF (ends exposed) (Hihara 1997) exposed to deaerated 3.15 wt% NaCl at 30°C.

Al_4C_3 is a problem since it readily decomposes in water (Hihara 1989) to produce CH_4 and aluminum hydroxide $Al(OH)_3$ (Portnoi et al. 1981).

Due to this reactivity of Al/Gr MMCs in water, they are more suited for use in dry environments.

12.7.1.2 Aluminum/Silicon Carbide MMCs

Al/SiC MMCs are perhaps the most widely available and utilized. Silicon carbide particles, whiskers, fibers, or MFs are used in these MMCs. Since SiC is an SC, it can have a large range of electrical resistivity values dependent on its impurity species and concentration. When resistivities are low, SiC can serve as a cathode for proton and oxygen reduction, leading to galvanic action. Hence, large variations in Al/SiC MMC corrosion behavior are often related to the intrinsic properties and the impurity content of each type of SiC reinforcements used.

The cathodic polarization diagrams of HP SiC and SiC MF (with carbon cores exposed) in aerated 3.15 wt% NaCl (Figure 12.6) demonstrate the variation in galvanic action when reinforcements were coupled to an equal area of aluminum. For actual MMCs in aerated, chloride-containing environments, the corrosion rate of the particulate- and whisker-reinforced Al/SiC MMCs is higher than that of its monolithic matrix alloy (Metzger and Fishman 1983, Sun et al. 1991, Modi et al. 1992, Hawthorn 2004) and also increases as the SiC content increases (Lore and Wolf 1981, Hawthorn 2004). In various 90-day humidity-chamber corrosion tests (Hawthorn and Hihara 2004), Al6092/SiC/P-T6 MMCs showed a progressive increase in the corrosion rate as the SiC content was increased from 5, 10, 20, 40 to 50 vol%. For MMCs that had equal amounts of SiC reinforcement (i.e., 50 vol%), the corrosion rate was noticeably lower for the MMC reinforced with high-purity, high-resistivity green SiC compared to a comparable MMC that was reinforced with low-purity, low-resistivity black SiC (Hawthorn and Hihara 2004). The black SiC is likely to support more cathodic currents leading to higher galvanic action and corrosion rates in comparison to the green SiC with higher resistivity. The large variation in resistivity of the SiC that is due to differing impurity levels may be the reason for conflicting results being presented in the literature. For example, no obvious evidence of galvanic corrosion was found in Al6061/SiC MMCs with 13–26 vol% SiC particles (Griffiths and Turnbull 1994).

The corrosion of the MMC also depends on the anodic behavior of the metal matrix. However, in Al/SiC MMCs, the presence of the SiC particles did not significantly affect the passive current densities (Trzaskoma et al. 1983, Golledge et al. 1985, Hawthorn 2004, Hawthorn and Hihara 2004) and pitting potentials (Trzaskoma et al. 1983, Aylor and Moran 1985, Coleman et al. 1994, Griffiths and Turnbull 1994, Monticelli et al. 1995, Nunes and Ramanathan 1995, Roper and Attwood 1995, Shimizu et al. 1995, Kiourtsidis et al. 1999, Hawthorn and Hihara 2004). During anodic polarization in 0.1 N NaCl aqueous solutions, however, the pits on Al/SiC/Wh MMCs were much smaller but of higher spatial density (Trzaskoma 1990) compared to the pitting characteristics observed on wrought and powder-compacted monolithic alloys. Those pits did not nucleate at SiC particles but rather nucleated at intermetallic particles that were also smaller and more numerous compared to those present in the monolithic matrix alloys (Trzaskoma 1990). The presence of SiC whiskers (Trzaskoma 1990) and particles (Griffiths and Turnbull 1994) can enhance the precipitation of some intermetallic phases.

The presence of the SiC reinforcements can also have other effects on the microstructure and corrosion behavior of these MMCs. Corrosion could also be affected by (1) segregation of alloying elements to the SiC–Al interface (Garrard 1994), (2) the generation of higher dislocation densities due to a mismatch between the CTE of SiC and aluminum (Hihara 1997, Yao and Zhu 1998, Ahmad et al. 2000), (3) differences in void content (Paciej and Agarwala 1988), and (4) any agglomeration of SiC particles (Bhat et al. 1991). Thermomechanical processing can be beneficial to the corrosion resistance of MMCs. For example, certain solution heat treatments and high extrusion ratios improved the corrosion resistance of an Al7091/SiC/20 P MMC (Paciej and Agarwala 1988). The number of voids and agglomerates of SiC particles (Bhat et al. 1991) were reduced by extrusion, improving the corrosion resistance of the original as-cast MMCs. A finer and more homogenous distribution of secondary phases at the T4 temper in comparison to the O and F tempers (Ahmad and Abdul Aleem 1996) also resulted in improved corrosion resistance.

The formation of Al_4C_3 at matrix–reinforcement interfaces has been identified as another source of corrosion for MMCs reinforced with SiC particles (Park and Lucas 1997) and SiC Nicalon™ fibers (Coleman et al. 1994). In the open-circuit condition, pits have been observed to initiate at SiC–Al interfaces and could be caused by the hydrolysis of Al_4C_3 at the SiC–Al interface (Hihara et al. 2005).

The formation of micro-crevices resulting from the network of reinforcement particles in surface relief due to matrix dissolution accentuates localized acidification at anodic sites and alkalinization at cathodic sites. In Al6092/SiC/P-T6 MMCs (Hihara et al. 2004, Ding and Hihara 2005), the localized acidification and alkalinization results in accelerated corrosion since the aluminum matrix loses its ability to passivate in either acidic or alkaline environments.

12.7.1.3 Aluminum/Boron MMCs

Al/B/MF MMCs are usually fabricated by diffusion bonding B MFs between aluminum foils (Schwartz 1984). Although pure boron is an insulator and should not induce galvanic action, B MF generally contains tungsten and tungsten borides in its core (Tsirlin 1985), making its resistivity many orders of magnitude lower than that of pure boron. Accordingly, electrode sites at the exposed ends of the virgin B MF will support cathodic currents (Figures 12.6 and 12.7) and therefore can induce galvanic corrosion. In addition, galvanic currents could also be measured between aluminum and extracted B MF (from the matrix) that contained a 4 µm thick interphase layer of aluminum boride that formed during processing (Pohlman 1978). The corrosion rate of Al/B/MF MMCs also increased with the B MF content (Pohlman 1978), indicating higher galvanic action due to more B MF cathodic sites. Preferential sites of corrosion were also noticed at diffusion-bonded, B MF–foil and foil–foil interface regions (Sedriks et al. 1971, Bakulin et al. 1978, Pohlman 1978).

12.7.1.4 Aluminum/Alumina MMCs

The Al_2O_3 reinforcements are usually in the form of particles, short fibers, or continuous fibers. Galvanic action is not expected between Al_2O_3 and aluminum since Al_2O_3 is an insulator.

The presence of the Al_2O_3 reinforcements usually do not significantly affect pitting potentials (Coleman et al. 1994, Monticelli et al. 1995, Nunes and Ramanathan 1995, Roper and Attwood 1995, Shimizu et al. 1995, Fang et al. 1999, Hihara et al. 2005) in chloride solutions. Passive current densities below the pitting potential have been reported to increase with Al_2O_3 content (Fang et al. 1999). In other instances, however, passive current densities for a particulate Al6092/Al_2O_3/20P-T6 MMC were consistent with other types of MMCs in sodium sulfate solutions and in chloride solutions (under the pitting potential) (Hawthorn and Hihara 2004).

In Al/Al_2O_3 MMCs, corrosion initiation usually occurs at intermetallic particles or contaminants introduced in the microstructure. Al/Al_2O_3 MMCs with pure Al matrices, therefore, usually have excellent corrosion resistance due to low amounts of intermetallic particles that otherwise could serve as cathodic sites (Zhu 2008). For example, only very slight corrosion damage was observed on a pure Al/Al_2O_3/50F MMC exposed to marine atmosphere 0.5 mile from the coastline for an 11-month period (Hihara 2005b).

The segregation of alloying elements and the presence of intermetallics near reinforcements may contribute to localized corrosion in those regions. Preferential corrosion near fibers (Yang and Metzger 1981, Agarwala 1982, Bertolini et al. 1999, Zhu 2008) and particles (Nunes and Ramanathan 1995, Bertolini et al. 1999, DeSalazar et al. 1999) has been noticed in chloride-containing solutions. In an Al (2 wt% Mg)/Al_2O_3 MMC, Fe and high levels of Mg ($\approx$10 wt%) were detected near fibers (Yang and Metzger 1981). It was suspected that the presence of Mg originated from Mg_2Al_3, which is rapidly attacked at low potentials (Yang and Metzger 1981). In other Al/Al_2O_3 MMCs (Bruun and Nielsen 1991), Al_8Mg_5 and Mg_2Si intermetallics have been reported to induce corrosion. Precipitates, intermetallics, and even residual contaminants from the reinforcements have been found to act as cathodic sites: In Al–2 wt% Cu/Al_2O_3/50F MMCs, corrosion initiated at copper-rich precipitates on the fiber–matrix interface (Zhu 2008). In Al6092/Al_2O_3/20P-T6 MMCs, Fe–Si–Al intermetallics and low-resistivity

Ti oxide or suboxide particles (likely introduced with the alumina reinforcement) were identified as cathodic sites that induce corrosion (Ding and Hihara 2008a).

12.7.1.5 Aluminum/Mica MMCs

Experimental particulate MMCs reinforced with muscovite ($KAl_3Si_3O_{10}(OH)_2$) mica particles (less than ≈70 µm) (Nath et al. 1980) were developed by casting of various aluminum alloys (Nath and Namboodhiri 1988, 1989) for applications where antifriction, seizure resistance, and high-damping capacity are important (Rohatgi et al. 1986). Galvanic corrosion between mica and aluminum is not a concern since muscovite is an insulator with resistivities ranging from ~10^{13} to 10^{17} Ω cm (Clauser 1963). There are other issues with mica, however, that can lead to degradation. Although muscovite is insoluble in cold water (Weast 1986), it has also been reported to absorb moisture and then swell (Nath and Namboodhiri 1988), and accordingly, exfoliated mica particles have been observed in Al/mica MMCs (Nath and Namboodhiri 1988). In 3.5 wt% NaCl solutions, Al/mica MMCs had pitting potentials ~20–30 mV lower than the monolithic matrix alloys. In addition, intermetallics were preferentially attacked, regions around and away from mica particles pitted, and mica–aluminum interfaces corroded (Nath and Namboodhiri 1988).

12.7.1.6 Stress-Corrosion Cracking Al MMCs

There have been only relatively few studies on stress-corrosion cracking and corrosion fatigue of DR and CR Al MMCs.

12.7.1.6.1 Discontinuously Reinforced MMCs

Stress-corrosion cracking studies for alternate exposure and immersion in NaCl solutions have been conducted on aluminum MMCs reinforced with Al_2O_3 particles (Monticelli et al. 1997) and SiC particles (Monticelli et al. 1997, Yao 1999, Kiourtsidis and Skolianos 2000) and whiskers (Monticelli et al. 1997). The Al2024/Al_2O_3/P MMC was susceptible to stress-corrosion cracking while subjected to three-point beam bending and alternate exposure or continuous immersions in a NaCl solution (Monticelli et al. 1997). Under the same conditions, however, the 6061 Al MMCs reinforced with SiC particles and SiC whiskers were not susceptible to stress-corrosion cracking (Monticelli et al. 1997). Similarly, Al2024/SiC/P MMCs were not prone to stress-corrosion cracking under constant strain at 75% of ultimate tensile strength while exposed to an aerated NaCl solution (Kiourtsidis and Skolianos 2000). Slow strain rate tension testing of Al2024/SiC/P MMCs in 3.5 wt% NaCl at room temperature indicated that the MMC lost up to 10% of failure strength compared to air-exposed specimens (Yao 1999).

12.7.1.6.2 Continuously Reinforced MMCs

Stress-corrosion cracking studies for samples in an immersed aqueous state have been conducted on aluminum MMCs reinforced with unidirectional, graphite fibers (Davis et al. 1982), boron MFs (Sedriks et al. 1971), and Nextel 440 (Al_2O_3, SiO_2, B_2O_3) fibers (Berkeley et al. 1998). Al6061/Gr/F MMCs were stressed parallel to the fiber axis in natural seawater. Failure was stress dependent at high stress levels and occurred in less than 100 h. At lower stresses, failure was primarily caused by extensive corrosion and therefore was relatively independent of stress levels. Al2024/B/MF MMCs stressed parallel to the fiber axis at 80% fracture strength in a NaCl solution did not fail in 1000 h but failed after 500 h when H_2O_2 was added to the NaCl solution. Extensive intergranular matrix corrosion and broken filaments at random sites were observed. The monolithic matrix alloy failed within 10 h under similar conditions. For Al2024/B/MF MMCs stressed perpendicular to the fiber axis at 90% yield strength in NaCl and NaCl with H_2O_2 solutions, failure occurred by intergranular matrix corrosion and separation at diffusion-bonded, fiber–matrix interfaces. Failure times decreased with increasing B MF content; therefore, the presence of the MF was deleterious when stresses were perpendicular to the fiber axis. For the Al6061/Nextel/F MMCs, specimens were exposed to a pH 2 NaCl solution in the stressed and unstressed states (Berkeley et al. 1998). The composite strength was measured prior to and after exposure to assess damage. The prevailing mode of failure was attributed to the presence of extensive corrosion along the fiber–matrix interface and not stress-corrosion cracking.

12.7.1.7 Corrosion Fatigue in Al MMCs

Corrosion-fatigue studies have been conducted on Al MMCs reinforced with graphite fibers (Davis et al. 1982), SiC whiskers (Hasson et al. 1984, Yau and Mayer 1986, Jones 1991, Minoshima et al. 1998), and SiC particles (Hasson et al. 1984, Buck and Thompson 1991). Processing conditions and type of reinforcement affect corrosion-fatigue behavior. Unnotched Al6061/Gr/F MMCs were exposed to natural seawater and stressed parallel to the fiber axis. These MMCs were processed with either silica (SiO_2)-coated or TiB_2-coated graphite fibers. For a given stress amplitude, the MMC with TiB_2-coated fibers had the longest corrosion-fatigue life, followed by the MMC with the SiO_2-coated fibers and the monolithic matrix alloy. At low stress amplitudes corresponding to longer exposure times, the MMC with the SiO_2-coated fibers suffered premature failure due to extensive corrosion. In Al/SiC MMCs, fatigue crack rates of compact tension specimens are usually higher when in NaCl solutions as compared to air (Yau and Mayer 1986) or argon (Buck and Thompson 1991). Loading frequency affects corrosion-fatigue crack rates (Buck and Thompson 1991), but no consistent trends were observed. Fatigue (Yau and Mayer 1986) and corrosion-fatigue (Buck and Thompson 1991) crack rates are influenced by loading and extrusion or rolling direction. The nucleation of a crack was also observed at the bottom of a corrosion pit (Minoshima et al. 1998). The shape of the reinforcement constituent may also have significant effects on stress-corrosion and corrosion-fatigue behavior, based on modeling that considers crack-tip strain rate (Jones 1991). The model predicts that crack rates are reduced by increasing the reinforcement length-to-diameter ratio, which implies that MMCs reinforced with whiskers are more resistant to stress-corrosion and corrosion fatigue than those reinforced with particles. This analytical implication is in agreement with experimental results where Al6061/SiC/Wh MMCs were found to have longer corrosion-fatigue lives than Al6061/SiC/P MMCs in salt-ladened moist air (Hasson et al. 1984).

12.7.2 Magnesium MMCs

Of the structural metals, magnesium is the lightest (density of only 1.7 g/cm³) but most active in the electromotive series (Uhlig and Revie 1985). Therefore, Mg is attractive for lightweight MMCs but can be very susceptible to corrosion, particularly if it is reinforced with more noble reinforcement constituents. Interestingly, ultrapure Mg has a very low rate of corrosion (0.25 mm/year in seawater), but commercially pure Mg corrodes at about 100–500 times faster due to impurities (Uhlig and Revie 1985). For example, iron, nickel, copper, and cobalt, which have low hydrogen overvoltages, can significantly accelerate the corrosion rate of Mg (Butler and Ison 1978). Another characteristic of Mg is that its corrosion rate is generally not affected significantly by dissolved oxygen (Uhlig and Revie 1985) since the primary cathodic reaction in Mg corrosion is proton reduction. This behavior, however, may not translate to Mg MMCs if the reinforcements are catalytic to oxygen reduction. Hence, Mg MMC corrosion rates may significantly increase with aeration, whereas those of monolithic alloys generally do not. Corrosion studies have been conducted on Mg MMCs reinforced with graphite, SiC particles, SiC MF, B MF, and Al_2O_3 fibers.

12.7.2.1 Magnesium/Graphite MMCs

Since graphite is a conductor and catalytic to oxygen reduction, the corrosion rate of Mg/Gr MMCs should increase with aeration, as depicted in the polarizations diagrams showing anodic curves of pure Mg (in deaerated 3.15 wt% NaCl) with cathodic curves of pitch-based graphite fibers in aerated (Figure 12.6) and deaerated (Figure 12.7) 3.15 wt% NaCl. The galvanic corrosion of Mg is also cathodically controlled, and therefore, galvanic-corrosion rates should increase with increasing area fraction of cathodic reinforcement. Actual Mg/Gr MMCs immersed in air-exposed 0.001 N NaCl corroded severely within 5 days (Trzaskoma 1986). The open-circuit potential and corrosion rate

FIGURE 12.8 Exfoliated Mg AZ91C/Gr/12.7F MMCs with AZ31B Mg skins after a 25-year exposure period in an air-conditioned environment.

of Mg AZ91C/Gr/40P MMC was 0.3 V more noble and 40 times greater, respectively, than that of the monolithic matrix alloy in a deaerated 50 ppm chloride solution (Czyrklis 1985). Even in an air-conditioned environment with relatively dry air, a Mg AZ91C/Gr/12.7F MMC with AZ31B Mg skins exfoliated over a 20-year period (Figure 12.8).

12.7.2.2 Magnesium/Silicon Carbide MMCs

As a reinforcement for MMCs, SiC can be used in the form of particles, fibers, or MFs. Since SiC is an SC where its resistivity is highly dependent on its purity, galvanic corrosion with Mg can sometimes be a potential concern. Anodic polarization diagrams of pure magnesium with cathodic polarization diagrams of HP SiC and SiC MF exposed in aerated (Figure 12.6) and deaerated 3.15 wt% (Figure 12.7) show that galvanic-corrosion rates (as determined by the mixed-potential theory) are greater in aerated solutions due to oxygen reduction (Hihara and Kondepudi 1993, 1994). In addition, the galvanic-corrosion rates are much higher when Mg is coupled to SiC MF with carbon cores and surfaces in comparison to when Mg is coupled to HP SiC that has a much higher resistivity. Studies on a model MMC consisting of high-purity magnesium and well-separated SiC particles exposed to 3.5 wt% NaCl also did not show evidence of galvanic corrosion between the particles and matrix (Nunez-Lopez et al. 1996). In salt spray tests (Nunez-Lopez et al. 1995), Mg ZE41A/SiC/12P MMCs with SiC particles ranging in size up to approximately 20 μm did not show preferential attack between the SiC particles and the surrounding matrix. Instead, macroscopic anodic and cathodic regions encompassing many particles developed. Anodic regions spread over the MMC surface much more rapidly than on the monolithic alloy, and the local corrosion rates were approximately three times greater for the MMC. The authors (Nunez-Lopez et al. 1995) speculated that the higher corrosion rates on these MMCs could have been caused by iron contamination of the magnesium matrix during its processing in a steel crucible.

12.7.2.3 Magnesium/Boron MMCs

Although pure boron is an insulator (Stroganova and Timonova 1978, Timonova et al. 1980) and galvanic corrosion with Mg is not a concern, MMCs are usually reinforced with tungsten-cored B MFs that have lower resistivity than pure boron due to the presence of tungsten cores and the associated formation of tungsten borides (Tsirlin 1985). In NaCl solutions, galvanic currents could be measured between virgin B MF (tungsten cores either shielded or exposed) and Mg (Stroganova and Timonova 1978) or a Mg alloy (Timonova et al. 1980) where the galvanic currents were higher whenever the tungsten cores were exposed (Stroganova and Timonova 1978, Timonova et al. 1980) since pure tungsten is an effective cathode (Timonova et al. 1980). Galvanic current densities were even higher and increased approximately five times whenever Mg was coupled to B MF extracted from the matrix (Timonova et al. 1980). Corrosion rates of Mg alloy (MA2-1)/B/MF MMCs exposed in 0.005 and 0.5 N NaCl solutions were 12.5 and 81.7 g/m² day, respectively, which are about six times the values of the monolithic matrix alloy in similar environments.

12.7.2.4 Magnesium/Alumina MMCs

Since Al_2O_3 is an insulator, galvanic corrosion should not be expected with magnesium. The corrosion rates of a Mg AZ91C/Al_2O_3/F MMC (Levy and Czyrklis 1981), however, was approximately 100 times greater than that of the monolithic matrix alloy Mg AZ91C in 3.5 wt% NaCl at 25°C but similar to that of the matrix alloy in distilled water at 20°C. Interestingly, the higher corrosion rates in the 3.5 wt% NaCl and the similar corrosion rates in distilled water are behaviors that could be expected from a galvanic couple. When the electrolyte has low ohmic resistance such as in the case for 3.5 wt% NaCl, the galvanic-corrosion rate would be high, but when the ohmic resistance is high such as the case for distilled water, the anode and cathode constituents will effectively be decoupled and corrode at their normal uncoupled rates. Although galvanic corrosion is not expected between magnesium and Al_2O_3, there is a possibility that conducting interphases or precipitates may have formed due to the presence of the Al_2O_3 fibers. For example, the open-circuit potential of Mg AZ91C/Al_2O_3/F MMC was more noble than that of the matrix alloy in a 50 ppm Cl^- solution (Czyrklis 1983), which could indicate that more noble precipitates or interphases were present.

12.7.3 Titanium MMCs

Titanium is one of the most corrosion-resistant metals due to the formation of a protective surface oxide film. Its pitting potential in 3.15 wt% NaCl at 30°C has been measured to be approximately $12V_{SCE}$. Hence, under normal conditions, galvanic corrosion between titanium and conducting reinforcements should be inconsequential. Titanium, with a density of 4.5 g/cm^3, is less active than aluminum and magnesium in the electromotive series and is one of the most noble metals in the galvanic series due to its ability to passivate. Titanium MMCs are being considered for aerospace, automotive, sporting goods, cutlery, and biomedical applications.

12.7.3.1 Titanium/Graphite MMCs

Porous titanium/titanium carbide (TiC) and Gr/P MMCs (Blackwood et al. 2000) were fabricated by heat-treating porous titanium/Gr/P MMCs to react the Ti with graphite. Polarization tests were conducted in 0.9 wt% NaCl and lactated Ringer's solution (for simulated in vitro use) that is isotonic with blood and contains sodium chloride, sodium lactate, calcium chloride, and potassium chloride. The anodic polarization current densities of these MMCs were significantly higher than that of pure monolithic titanium that passivated. The authors attributed the higher corrosion rates of the Ti/TiC, Gr/P MMC to its porosity, which may have prevented complete passivation of the titanium matrix. However, another possibility for the higher current densities in the MMC could be the oxidation of graphite to CO_2 during anodic polarization.

12.7.3.2 Titanium/Silicon Carbide MMCs

Corrosion studies were conducted on titanium alloy Ti–15V–3Cr–3Sn–3Al (Ti 15-3)/SiC/MF MMCs (Hihara and Tamirisa 1995) and titanium aluminide (α_2-Ti$_3$Al) (14 wt% aluminum, 21 wt% niobium, balance titanium)/SiC/MF MMCs (Saffarian and Warren 1998). The corrosion behavior of Ti 15-3/SiC/MF MMC was investigated in 3.15 wt% NaCl. Polarization diagrams of the actual Ti 15-3/SiC/MF MMC were compared to those generated from monolithic Ti 15-3 and SiC MF using the mixed-potential theory and showed very good agreement indicating that interphases that could affect the electrochemical behavior of the MMC did not form (Hihara and Tamirisa 1995). The results showed that the matrix passivated while the carbon cores of the SiC MF were oxidized and consumed and the carbon-rich outer surface of the SiC MF flaked and peeled, likely forming CO_2 similar to that observed for graphite fibers under anodic polarization (Hihara and Latanision 1991). In aerated 3.15 wt% NaCl, the galvanic current density between the Ti 15-3 and the SiC MF cannot exceed that of the passive current density of Ti 15-3, and accordingly, the galvanic current densities measured using the zero-resistance ammeter technique confirmed that galvanic currents between Ti 15-3 and SiC MF are negligible. Somewhat similar corrosion behavior was observed for the α_2-Ti$_3$Al/SiC/MF MMC (Saffarian and Warren 1998), with the exception that the α_2-Ti$_3$Al matrix is less resistant to pitting compared to that of Ti 15-3. During anodic polarization, the α_2-Ti$_3$Al/SiC/MF MMC pitted at approximately $1V_{SCE}$ in 0.5 N NaCl, which was approximately 0.5 V less than that of the monolithic matrix alloy. Some matrix pitting and crevice corrosion around the SiC MFs were also observed after anodic polarization. The galvanic current density of the α_2-Ti$_3$Al/SiC/MF MMC was negligible and limited to the passive current density of the α_2-Ti$_3$Al matrix (Saffarian and Warren 1998).

12.7.3.3 Titanium/Titanium Carbide and Titanium/Titanium Diboride MMCs

Pure titanium/titanium carbide (Ti/TiC/P) and pure titanium/titanium diboride (Ti/TiB$_2$/P) MMCs, containing 2.5, 5, 10, or 20 vol% of reinforcement, were fabricated by cold isostatic pressing followed by sintering (Covino and Alman 2002). Interphase products were not identified at Ti–TiC interfaces; however, a TiB interphase was identified at Ti–TiB$_2$ interfaces after processing. The MMCs were anodically polarized in deaerated 2 wt% HCl in the temperature range of 50°C–90°C. For pure monolithic Ti, the passive current density was approximately 10^{-5} A/cm^2 throughout the temperature range. The dissolution current density of the Ti/TiC/P and Ti/TiB$_2$/P MMCs, however, generally increased with increasing temperature and reinforcement content, with maximum values approximately 20 and 100 times, respectively, higher than that of pure titanium. For both types of MMCs, the titanium matrices were virtually uncorroded; however, the TiC particles showed some signs of degradation, and the TiB$_2$ particles and TiB interphase were significantly corroded.

12.7.4 Copper MMCs

Copper is generally thought to have relatively good corrosion resistance because it is one of the most noble metals in the electromotive force series (i.e., only Hg, Pd, Pt, and Au have more noble potentials). In the galvanic series, however, Cu–Ni alloys, titanium, and passivated super alloys and some SSs are more noble than copper. Copper is relatively heavy, with a density of 8.96 g/cm^3. Reinforcements for copper MMCs are typically chosen to impart strength and stiffness, reduce weight, enhance thermal and electrical properties, improve machinability, and enhance wear resistance. Initial studies (Aylor 1987) were conducted on a wide variety of experimental copper and copper alloy MMCs reinforced with graphite, SiC, TiC, silicon nitride, boron carbide, and Al$_2$O$_3$ for marine applications. The MMCs generally showed corrosion behavior that was similar to that of the monolithic alloys, although in some cases, the corrosion rates were higher for the MMCs. Other studies have focused on copper MMCs for electronic, thermal, and tribological applications.

12.7.4.1 Copper/Graphite MMCs

Graphite is conductive and relatively inert, and its potential is noble to that of copper. The corrosion behavior of pure copper/graphite MMCs reinforced with 1.2–40 vol% particles and 50 vol% fibers were investigated in deaerated and aerated 3.5 wt% NaCl solutions (Sun et al. 1993). The corrosion potential of the particulate-reinforced MMCs became more noble as the graphite content was increased for both deaerated and aerated solutions. In the aerated solution, the corrosion potential of the MMC reinforced with 50% fiber was similar to that reinforced with 40% particles; however, in the deaerated solution, the MMC reinforced with the fibers was significantly more active than the particulate-reinforced MMC. The similarity in the corrosion potential in the aerated solution for the fiber-reinforced and particulate-reinforced MMCs is likely since oxygen reduction is normally diffusion limited and relatively independent of the substrate as long as it is conductive, whereas in deaerated solutions, hydrogen evolution kinetics depend on the substrate and could be different on the fibers versus the particles due to different degrees of graphitization. The implications of the corrosion potential results earlier are that the corrosion rates in aerated chloride solutions will likely increase primarily with the graphite content independent of form (i.e., fiber vs. particulate), whereas in deaerated solutions, the corrosion rates will depend on both the graphite content and form (i.e., fiber vs. particulate).

12.7.4.2 Copper/Silicon Carbide MMCs

SiC is an SC and its resistivity is dependent on its purity, and therefore, SiC of low purity can be conductive and promote galvanic action, whereas SiC of high purity will be insulative and not support galvanic action. The electrical and electrochemical properties of SiC can also be affected by light and affect corrosion differently depending on whether the SiC is n-type or p-type (see Section 12.6.2.1). The corrosion behavior of pure copper/SiC MMCs reinforced with 0, 5, 10, and 20 vol% particles was examined in a 5 wt% NaCl solution (Lee et al. 1999). The porosity in the MMCs ranged from 2.2% to 3.5% and generally increased with increasing SiC content. Interestingly, the corrosion potentials became more active, and the corrosion current densities increased with increasing SiC content, which indicated that the SiC was not serving as efficient cathodes but was enhancing dissolution rates of the matrix. Accordingly, upon inspection of the MMC electrodes, significant corrosion was observed at SiC–copper interfaces. Hence, the decrease in corrosion potential with increasing SiC content is likely to have been caused by an increase in anodic sites at voids and SiC–copper interfaces.

12.7.4.3 Copper/Alumina MMCs

Alumina is an insulator and cannot promote galvanic corrosion. The corrosion behavior of copper MMCs reinforced with 2.7 vol% Al_2O_3 was examined in deaerated and aerated 3.5 wt% NaCl (Sun and Wheat 1993). The corrosion rates of the MMCs and monolithic copper were comparable, and the corrosion potentials of the MMC were only 0.01–0.02 V more active than that of monolithic pure copper.

12.7.5 Stainless Steel MMCs

SSs are iron–chromium alloys that contain a minimum of approximately 12 wt% Cr needed for passivation. SSs often contain other alloying elements such as nickel and molybdenum to improve corrosion resistance. In the galvanic series, SSs occupy the most noble positions when in the passive state but also drop down to potentials lower than brass when in the active state (LaQue 1948). Ferritic 434L SS/Al_2O_3/P MMCs have been developed for potential application in chemical processing plants, turbine blades, and heat exchanger tubes (Mukherjee and Upadhyaya 1985, Mukherjee et al. 1985a,b), and austenitic 316L SS/Al_2O_3/P and 316L SS/Y_2O_3/P MMCs have been investigated for enhanced strength and wear resistance (Velasco et al. 1997).

12.7.5.1 Stainless Steel/Alumina MMCs

Alumina is an insulator and cannot promote galvanic corrosion. Sintered 434L SS/Al_2O_3/P MMCs and sintered 434L SS alloy (without Al_2O_3 particles) were examined (Mukherjee et al. 1985a,b, Mukherjee and Upadhyaya 1985). The volume percent of Al_2O_3 particles ranged from 0% to 8%. The effect of small amounts of titanium and niobium alloying elements on corrosion resistance was also investigated. There was no strong correlation between Al_2O_3 content and corrosion behavior in 1 N H_2SO_4 (Mukherjee et al. 1985a,b, Mukherjee and Upadhyaya 1985), and the passive current densities for almost all materials were high and within an order of magnitude of 1 mA/cm². In the 5 wt% NaCl solutions at the open-circuit potential, i_{Corr} of the MMCs was less than 10 μA/cm² (Mukherjee et al. 1985a); however, upon polarization, all materials displayed active corrosion behavior.

Sintered 316L SS/Al_2O_3/P MMCs (Velasco et al. 1997) were reinforced with 3–5 wt% Al_2O_3 particles and additions of 2 wt% chromium diboride (CrB_2) or 1 wt% boron nitride (BN) were used as sintering aids. The MMCs were porous and their densities ranged from 86% to 96% of the theoretical value. Unreinforced 316L SS specimens were also fabricated using powder metallurgy without sintering aids, resulting in specimens with 85% of theoretical density. Less porosity was present in the reinforced MMCs as compared to the unreinforced pure 316L SS specimen. The test samples were immersed in

10 wt% sulfuric acid (H_2SO_4) at room temperature for 24 h, 1 wt% hydrochloric acid (HCl) at room temperature for 24 h, and boiling 10 wt% nitric acid (HNO_3) for 8 h. The unreinforced 316L SS specimens passivated in the 10 wt% H_2SO_4 solution, whereas the corrosion rate of the MMC generally increased with increasing Al_2O_3 content to a maximum value of approximately 4 mm/year. The MMCs had lower corrosion rates than the unreinforced 316L SS specimen in the 1 wt% HCl solution but higher corrosion rates than the unreinforced specimens in the boiling nitric acid solution. There was no strong correlation between Al_2O_3 content in the MMCs and the corrosion rates in 1 wt% HCl and boiling 10 wt% HNO_3 solutions.

12.7.5.2 Stainless Steel/Yttria MMCs

Yttria (Y_2O_3) is an insulator and galvanic effects are not expected. Sintered 316L SS/Y_2O_3/P MMCs (Velasco et al. 1997) were also reinforced with 3–5 wt% Y_2O_3, and additions of 2 wt% chromium diboride (CrB_2) or 1 wt% BN were used as sintering aids. The Y_2O_3 MMCs were sintered to 88%–96% of theoretical density. In sulfuric, hydrochloric, and nitric acid solutions, the Y_2O_3-reinforced MMCs exhibited less corrosion resistance as compared to the Al_2O_3-reinforced MMCs. The Y_2O_3 particles also showed better bonding to the matrix (as compared to the Al_2O_3 particles) probably forming a complex $YCrO_3$ interphase oxide. It is possible that the formation of the reaction layer around the Y_2O_3 particles may have depleted chromium from the matrix, resulting in lower corrosion resistance than the Al_2O_3-reinforced MMCs.

12.7.6 Lead MMCs

Lead has a potential of $-0.126V_{SHE}$ in the emf series and occupies an intermediate position in the sgalvanic series in between brasses and active SSs. Lead is a heavy metal with a density of 11.4 g/cm^3. At room temperature, lead is already at approximately 0.5 of its melting temperature and is relatively soft. Hence, the interest in lead MMCs is usually for increased strengths. Although lead can also be strengthened by alloying with arsenic, antimony, or calcium, its corrosion resistance will be sacrificed by these solute additions.

The corrosion behavior of pure lead MMCs reinforced with Al_2O_3, SiC, glass–quartz, or carbon fibers of various volume percents was studied (Dacres et al. 1981, 1983, Viala et al. 1985) for potential use as positive electrode grids in lead–acid batteries. Pure lead has very good corrosion resistance in lead–acid battery environments (consisting of sulfuric acid solutions) but is heavy and lacks sufficient mechanical strength. The experimental pure lead MMCs were therefore developed to increase strength and reduce weight while retaining the corrosion resistance of pure lead (Dacres et al. 1981, 1983, Viala et al. 1985). The MMCs were anodically polarized at 1.226 V (vs. mercury/mercurous sulfate reference electrode) in sulfuric acid solutions (of 1.285 specific gravity) at 50°C, 60°C, and/or 70°C. At 1.226 V, lead and water are oxidized to lead dioxide (PbO_2) and molecular oxygen (O_2), respectively (Burbank 1959, Burbank et al. 1971). About one-third of the total anodic current is consumed in the oxidation of lead under these conditions (Dacres et al. 1983). In continuous fiber-reinforced lead MMCs, the wetting or intimate contact between the lead matrix and the fibers is very important in preventing the electrolyte from seeping into the bulk of the MMC leading to internal crevice corrosion (Dacres et al. 1983). In some Pb/Al_2O_3 MMCs, poor bonding between the fibers and the matrix allowed electrolyte ingress into the fiber–matrix interfaces, which resulted in accelerated corrosion (Dacres et al. 1981) and swelling of the MMC due to entrapment of the corrosion products (Dacres et al. 1983). The Pb/SiC/F MMCs also did not have good corrosion resistance (Dacres et al. 1983, Viala et al. 1985), and the composites were destroyed (Viala et al. 1985) by swelling. The SiC fibers (which did not have carbon cores) were stable. The corrosion resistance of the Pb/glass–quartz/F MMCs was also not satisfactory. The Pb/graphite/F MMCs also degraded due to the oxidation of the graphite fibers (Viala et al. 1985).

Particulate lead-alloy (i.e., Pb 80%, 20% Sb) MMC was reinforced with 1–5 wt% zircon ($ZrSiO_4$) particles (Seah et al. 1997b) to improve strength. Since zircon is an insulator, it should not induce galvanic corrosion of the lead alloy. Exposure of the MMCs in a 1 N NaCl solution (Seah et al. 1997b) over a 72 h period, however, showed that the corrosion rate (as gauged by weight loss measurements) increased with increasing zircon content.

12.7.7 Depleted Uranium MMCs

Uranium is an active metal with an equilibrium potential of $-1.80V_{SHE}$ in the electromotive force series (Uhlig and Revie 1985). DU consists primarily of isotope U-238 and is the by-product of refining natural uranium to obtain nuclear fuel that is isotope U-235. DU is a very heavy metal with a density of 18.9 g/cm^3. Depleted uranium/tungsten fiber (DU/W/F) MMCs were developed to create high-density materials.

Since tungsten is a metal and electrically conductive, galvanic corrosion of DU is a concern. In air-exposed 3.5 wt% NaCl solutions at room temperature (Trzaskoma 1982), the open-circuit potential of tungsten fiber ($-0.25V_{SCE}$) is noble to that of the DU alloy ($-0.80V_{SCE}$), and the galvanic-corrosion current density measured between equal areas of tungsten fibers and the DU alloy is equal to about 4×10^{-5} A/cm^2. The open-circuit potentials of the DU/W/F MMC and galvanic couples consisting of tungsten fiber and DU alloy of equal areas are $-0.78V_{SCE}$ and $-0.77V_{SCE}$, respectively, and fall between those of tungsten fibers and the DU alloy. In a 30-day exposure test in the NaCl solution, the DU/W/F MMC lost 43.56 mg/cm^2, which was about 1.3 times that of the monolithic DU alloy.

12.7.8 Zinc MMCs

Zinc is an active metal and has an equilibrium potential of $-0.763V_{SHE}$ (Uhlig and Revie 1985) and is more active than aluminum and plain-carbon steel in the galvanic series (Uhlig and Revie 1985). Zinc has a density of 7.14 g/cm^3, and its alloys are known to have excellent wear and bearing characteristics (Smith 1993). A zinc-based ZA-27 casting alloy was reinforced with 1, 3, and 5 wt% graphite particles ranging in sizes from 100 to 150 μm for potential use as a bearing material (Seah et al. 1997a). Since graphite is a noble conductor, galvanic corrosion of Zn is possible if moisture is present. The zinc MMCs were resistant to corrosion in SAE 40 grade lubricant that had been in service for 6 months in an internal combustion engine. The lack of corrosion was likely due to low levels or the absence of moisture in the lubricant. The MMC corroded in a 1 N HCl electrolyte.

12.8 Degradation Protection of MMCs

Corrosion protection of MMCs is usually more difficult to achieve versus their monolithic matrix alloys due to their heterogeneous microstructures. The reinforcements may induce galvanic corrosion of the matrix, dissolve and go into solution, and/or disrupt the passive film on the MMC surface. Hence, it is anticipated that experimentation will be needed when selecting and then verifying corrosion protection strategies. Some options for attaining corrosion protection could be the use of claddings, inhibitors, or coatings.

Claddings are protective metal layers bonded to the surface of the MMC. The resulting corrosion properties would be that of the clad metal, and its durability would depend on the cladding thickness and the attainment of integral clad metal/MMC bonds. Corrosion problems could arise if the cladding is breached or becomes debonded.

Inhibitors could be effective if the MMCs are used in closed systems. Proven inhibitors for the monolithic matrix alloys may or may not be effective for the MMC, and a testing program to find suitable inhibitors for a particular application is highly recommended. If the MMC contains conducting reinforcements, cathodic inhibitors may be effective. In any case, it may be worthwhile to use a mixture of cathodic and anodic inhibitors to achieve optimal results.

The application of nonmetallic coatings may be the most practical solution if the MMC lacks inherent corrosion resistance. The use of impervious, inhibitive, or cathodically protective coatings will depend on the application and substrate. Coating systems often include a pretreatment, a primer, and a topcoat (generally an organic based paint). The pretreatment can be a wash to improve adhesion or a conversion coating that reacts with the metallic substrate to form a layer that can also improve adhesion and impart barrier as well as corrosion-inhibitive properties. The primer may also contain corrosion inhibitors and improve the adhesion of the topcoat. Testing and verification will be necessary because a proven coating system for a monolithic matrix alloy may not be as effective for the MMC of that alloy. The degree of adhesion and wettability between the coating and the reinforcement or differences in the electrochemical properties of the alloy and MMC may affect the coating effectiveness.

Other problems could arise from techniques such as anodization if the MMC reinforcements interfere with that process. For example, if an aluminum MMC is anodized, the reinforcements could impede the growth of a continuous passive aluminum oxide film (Greene 1992), or the reinforcements can be damaged by oxidation (e.g., graphite fibers oxidize to CO_2 [Hihara 1989]). Various studies on the corrosion protection of MMCs utilizing organic coatings, inorganic coatings, anodization, and chemical conversion coatings have been conducted, and the results have been summarized in Table 12.3. The studies have generally shown that the MMC receives the best protection from coatings that completely shield them from the environment.

12.9 Specific Standardized Tests

Standardized corrosion tests have been primarily designed for monolithic alloys but can often be used to assess the corrosion performance of MMCs. The evaluation of the corrosion damage, however, is generally more difficult for MMCs compared to monolithic alloys. The corrosion assessment of CR MMC can be particularly difficult because the fibers are often left in surface relief as the matrix corrodes, making it difficult to accurately assess weight loss and depth of penetration metrics. Corrosion assessment of DR MMCs is less problematic since the reinforcement particles dislodge when the matrix recedes during corrosion. Some useful corrosion assessment procedures and testing techniques for specific types of corrosion that can be applied to MMCs with careful data interpretation are summarized in the succeeding text. A more thorough coverage of testing procedures for MMCs are covered in the literature (Hihara 2005a,b).

12.9.1 Uniform Corrosion

For MMCs that corrode uniformly, corrosion assessments can be made by using gravimetric, metallographic, and electrochemical techniques. For DR MMCs, uniform corrosion assessment is likely to be similar to that for monolithic alloys since reinforcement particles become dislodged as the matrix recedes. For CR MMCs, since fibers are often left in relief as the matrix corrodes, their residual presence can interfere with gravimetric and depth of penetration measurements.

Procedures for gravimetric evaluation can be found in the American Society for Testing and Materials (ASTM) G1 Standard Practice for Preparing, Cleaning, and Evaluating Corrosion Test Specimens. Since this standard was developed for monolithic materials, results for MMCs should be carefully analyzed to determine if the data are meaningful. For example, corrosion of some CR MMCs is sometimes accompanied by weight gain and swelling that is due to the entrapment of corrosion products in the MMC microstructure.

Metallographic analyses can be used when reinforcements left in surface relief due to corrosion can interfere with depth of penetration measurements. Specimen cross sections may be mounted and polished and then analyzed to identify the true penetration depth of the corroding matrix using a microscope and a calibrated eyepiece. Details for determining penetration depth are available in ASTM G46 Standard Guide for Examination and Evaluation of Pitting Corrosion.

If a sample corrodes uniformly, corrosion rates can also be measured electrochemically using procedures in ASTM G102 Standard Practice for Calculation of Corrosion Rates and Related Information

TABLE 12.3 Summary of Corrosion Protection Studies on MMCs

MMC Type	Substrate Protected	Coating/Treatment	Environment	Outcome	Reference
Al/Gr	Al/Gr MMC	Ti and Ni cladding	Marine	Delamination from exposed edges	Payer and Sullivan (1976)
	Al/Gr MMC	Inorganic diamond-like coating	NaCl solution	Short-term protection	Wielage et al. (2000)
	Surface Al foils on MMC	Organic coatings	Marine and NaCl solution	Protection	Payer and Sullivan (1976), Mansfeld and Jeanjaquet (1986), Lin, Shih et al. (1992)
	Surface Al foils on MMC	CVD and PVD inorganic coatings	Marine	No protection	Payer and Sullivan (1976)
	Surface Al foils on MMC	Electroplated Ni coating (without defects)	Marine	Protection	Payer and Sullivan (1976)
	Surface Al foils on MMC	Electroless Ni coating (with defects)	Marine	Accelerated corrosion	Aylor and Kain (1983), Aylor et al. (1984)
	Surface Al foils on MMC	Electrodeposited Al/Mn on electroless Ni coating	Marine	Protection if panel edges were sealed to prevent Al/Gr exposure	Payer and Sullivan (1976)
	Surface Al foils on MMC	Anodization with dichromate sealing	Marine	Protection if panel edges sealed	Aylor and Kain (1983)
	Surface Al foils on MMC	Chromate/phosphate conversion coating	Marine	Protection if edges sealed with epoxy	Aylor and Kain (1983)
	Surface Al foils on MMC	Chemical passivation with $CeCl_3$	NaCl solutions	Delayed pitting on surface foils	Hinton et al. (1986), Mansfeld et al. (1989)
Al/SiC	Al/SiC MMC	Organic epoxy coating	Marine and NaCl solution	Protection	Aylor and Kain (1983), Lin, Shih et al. (1992)
	Al/SiC MMC	Plasma sprayed alumina coating	Marine	Protection	Aylor and Kain (1983)
	Al/SiC MMC	Flame sprayed Al coating	Marine	Protection	Aylor and Kain (1983)
	Al/SiC MMC	Anodization	Marine and NaCl solution	Various levels or protection	Crowe et al. (1981), Trzaskoma and McCafferty (1986), Lin, Greene et al. (1992), Chen and Mansfeld (1997), Hou and Chung (1997), He et al. (2011)
	Al/SiC MMC	Chemical passivation with $CeCl_3$	NaCl solution	Limited protection	Mansfeld et al. (1989), Ahmad and Aleem (2002)
Al/Al_2O_3	Al/Al_2O_3 MMC	Chemical passivation with $CeCl_3$	NaCl solution	Limited protection	Hamdy et al. (2002), Traverso et al. (2002)
Al/Al18B4O33	Al/Al18B4O33 MMC	Chemical passivation with $CeCl_3$	NaCl solution	Limited protection	Hu et al. (2006)
Al/AlN	Al/AlN	Anodization	NaCl solution	Limited protection	Hou and Chung (1997)
Mg/TiB	Mg/TiB MMC	Acrylamide treatment	NaCl solution	Limited protection	Parthiban et al. (2010)

from Electrochemical Measurements. Note that when determining MMC corrosion rates electrochemically, the corrosion currents generally originate from the matrix alloy and not the reinforcement; hence, if corrosion currents are later converted to depth of penetration values, the reinforcement volume fraction, geometry, and orientation must be taken into consideration.

12.9.2 Pitting Corrosion

The extent of pitting can be determined using ASTM G46 Standard Guide for Examination and Evaluation of Pitting Corrosion. Pitting potentials for MMCs can be determined by generating anodic polarization diagrams. Refer to ASTM G5 Standard Reference Test Method for Making Potentiostatic and Potentiodynamic Anodic Polarization Measurements to ensure that the electrochemical setup is working properly. Electrochemical impedance spectroscopy has also been used to detect the onset of pitting in some MMCs (Mansfeld et al. 1990).

12.9.3 Stress Corrosion

Studies on stress-corrosion behavior of MMCs have been limited. The type of tests that can be performed on an MMC is dependent on the form of the MMC and whether the MMC can be successfully precracked. Some types of MMCs are only produced in the form of relatively thin panels (e.g., less than about 0.25 cm thick), ribbons, and wires, whereas other types are available as ingots, bars, or tubes. Precracking CR MMCs can be problematic since the cracks do not always grow perpendicular to the direction of applied stress (Chawla 1991).

A variety of stress-corrosion testing methods, using unnotched, notched, and precracked specimens are well documented (Sedriks 1990). The following ASTM standards are also helpful in developing a stress-corrosion testing program: ASTM E399 Standard Test Method for Linear-Elastic Plane-Strain Fracture Toughness of Metallic Materials, G30 Standard Practice for Making and Using U-Bend Stress-Corrosion Test Specimens, G38 Standard Practice for Making and Using C-Ring Stress-Corrosion Test Specimens, G39 Standard Practice for Preparation and Use of Bent-Beam Stress-Corrosion Test Specimens, and G49 Standard Practice for Preparation and Use of Direct Tension Stress-Corrosion Test Specimens.

12.9.4 Determination of Corrosion Behavior Using Polarization Techniques

The anodic and cathodic behavior of MMCs can be determined using polarization measurements. Refer to ASTM G5 Standard Reference Test Method for Making Potentiostatic and Potentiodynamic Anodic Polarization Measurements for conducting polarization tests. To identify variations in corrosion behavior of the MMC compared to its monolithic matrix alloy, it is recommended that polarization diagrams of the MMC and monolithic matrix alloy be generated and compared. If the form of the reinforcements allows consolidation into a bulk electrode, polarization diagrams of that reinforcement material can also aid in the understanding of MMC degradation. As an example, conductive graphite fibers can be made into a bulk polymer–matrix composite that can be fabricated into an electrode. Some detailed examples comparing the polarization diagrams of the MMCs with those of the matrix alloy and reinforcements can be found in the literature (Hihara 2005b).

12.9.5 Galvanic Corrosion between Matrix and Reinforcement Constituents under Immersed Conditions

The zero-resistance ammeter technique described in ASTM G71 Standard Practice for Conducting and Evaluating Galvanic Corrosion Tests in Electrolytes can be used to measure galvanic-corrosion currents between the monolithic matrix alloy and electrically conductive reinforcement constituents.

This technique does not apply when the MMC has electrically insulating reinforcements. If virgin reinforcement materials are used, the effects of subsequent processing steps such as interphase formation cannot be determined. It is normally possible to make electrodes from continuous fibers; however, if the reinforcements are particulates, consolidated bulk specimens usually need to be used. This could introduce some error since the properties of these bulk-sized reinforcements could be substantially different from that of the actual particulates particularly if during consolidation their impurity levels are different or changed. For many reinforcements that are SCs, this caveat applies since their electrical and electrochemical properties can be very sensitive to impurity levels. Techniques using polarization diagrams of the MMCs and their constituents that can be used to estimate galvanic currents between the matrix and reinforcement phases are discussed elsewhere (Hihara 2005a,b).

12.9.6 Loss in Mechanical Properties

Corrosion damage of MMCs can also be indirectly assessed by measuring any loss in strength after environmental exposure. Such changes in strength are measured by comparing properties from virgin samples to those that have been exposed to corrosive environments. These mechanical tests combined with fractographic analysis of failure surfaces can be useful for assessing any corrosion damage in MMCs that penetrates deep into the bulk (e.g., through reinforcement–matrix interfaces).

12.10 Summary

MMC components are being utilized in defense as well as commercial applications in aerospace, aircraft, marine and land vehicles, electronics, etc. MMCs have been developed by reinforcing metals with fibers or particulates to create materials with superior performance characteristics (e.g., specific strength and stiffness and thermal conductivity) or to tailor physical parameters such as CTEs that are not achievable with conventional monolithic alloys. The resultant heterogeneous microstructure, however, usually compromises the corrosion resistance of MMCs in comparison to their monolithic matrix alloys. The corrosion properties of each MMC can be strongly affected by the electrical resistivity of the reinforcements as well as their catalytic properties for oxygen reduction and/or hydrogen evolution that can promote galvanic corrosion of the matrix. Electrically conductive reinforcements such as graphite often strongly promote galvanic corrosion of matrices that are not in the passive state. If reinforcements are p-type SCs, exposure to illumination could induce galvanic corrosion of the matrix, while n-types under illumination could suppress galvanic corrosion. Electrically insulative reinforcements such as alumina cannot induce galvanic corrosion of the matrix; however, the corrosion behavior of the MMC in comparison to the monolithic matrix alloy may still be different. Whether the reinforcements are conductive, semiconductive, or insulative, they can affect the precipitation of intermetallics and induce higher dislocation densities in the matrix phase and also might react with the matrix to form interphases. All of these phenomena may alter the corrosion behavior of MMCs in comparison to that of the monolithic matrix alloy. Corrosion initiation in MMCs may also be more sensitive to concurrent physical damage such as debonding at the reinforcement–matrix interface leading to the formation of minute crevices that may enhance corrosion initiation. Variations in the quality (e.g., purity), size, and morphological form (e.g., fiber vs. particulate) of the reinforcements; the manufacturing technique for producing the MMC (e.g., powder metallurgy vs. casting) as well as in-process parameter changes during thermomechanical processing; and other factors could all alter corrosion behavior. Hence, it is not prudent to assume that any specific class of MMCs (e.g., Al/SiC MMCs) will have consistent corrosion behavior. Corrosion protection of MMCs is more challenging than that of their monolithic matrix alloys due primarily to the heterogeneous MMC microstructure. Options are claddings, surface treatments, and coatings. A thorough testing and verification program is recommended whenever corrosion protection strategies are selected for a particular MMC application. Standardized corrosion tests that are applied to MMCs usually have been those that were developed for monolithic alloys. The assessment of corrosion damage for MMCs is generally more challenging compared to that of monolithic alloys when MMC reinforcements are left in surface relief as the matrix corrodes and when corrosion penetrates deep into the MMC bulk along continuous fiber–matrix interfaces. In these cases, corrosion assessment such as depth of penetration and weight loss measurements becomes more difficult to evaluate and interpret. In short, MMCs often have a superior combination of properties not achievable in conventional monolithic alloys, but their limits of chemical stability should be thoroughly investigated before they are incorporated and used in critical applications.

Acknowledgments

The author is very grateful to Ms. Shruti Tiwari and Mr. George Hawthorn for the MMC micrographs and Dr. Suraj P. Rawal of Lockheed Martin Space Systems for the Mg MMC specimen.

References

3M. 2002. Metal matrix composites, www.3m.com/market/industrial/mmc/(last accessed in 2002).

Adler, R.P.I. and M.L. Hammond. 1969. Macrostructural integrity of vapor-deposited boron and silicon carbide filament. *Applied Physics Letters*, 14(11), 354–358.

Adler, R.P.I., D.J. Snoha et al. 2005. Characterization of environmentally exposed aluminum metal matrix composite corrosion products as a function of volume fraction and reinforcement specie, paper 06T029, in *Tri Service Corrosion Conference*, Orlando, FL, 2005.

Agarwala, V.S. May 1982. Corrosion behavior of aluminum/alumina composite. Abstract no. 15. In *Extended Abstracts*, Vol. 82-1, The Electrochemical Society, Montreal, Quebec, Canada.

Ahmad, Z. and B.J. Abdul Aleem. 1996. Effect of temper on seawater corrosion of an aluminum-silicon carbide composite alloy. *Corrosion*, 52(11), 857–864.

Ahmad, Z. and B.J.A. Aleem. 2002. Degradation of aluminum metal matrix composites in salt water and its control. *Materials and Design*, 23, 173–180.

Ahmad, Z., P.T. Paulette et al. 2000. Mechanism of localized corrosion of aluminum-silicon carbide composites in a chloride containing environment. *Journal of Materials Science*, 35, 2573–2579.

ALMMC. October 2007. Aluminum Metal-Matrix Composites Consortium, www.almmc.com

Arsenault, R.J. 1991. Strengthening of metal matrix composites due to dislocation generation through CTE mismatch. In R.K. Everett and R.J. Arsenault, Eds., *Metal Matrix Composites: Mechanisms and Properties*, Academic Press, New York, p. 79.

Ashby, M.F. and D.R.H. Jones. 1996. *Engineering Materials I*, Butterworth Heinemann, Oxford, U.K.

Aylor, D.M., R.J. Ferrara et al. 1984. Marine corrosion and protection for graphite/aluminum metal matrix composites. *Materials Performance*, 23(7), 32–38.

Aylor, D.M. 1987. Corrosion of metal matrix composites. In L.J. Korb and D.L. Olson, Eds., *Metals Handbook, Corrosion*, 9th edn., ASM International, Metals Park, OH, pp. 859–863.

Aylor, D.M. and R.M. Kain. 1983. Assessing the corrosion resistance of metal matrix composite materials in marine environments. In Vinson, J.R. and M. Taya, Eds., *Recent Advances in Composites in the United States and Japan, ASTM STP 864*, American Society for Testing and Materials, Philadelphia, PA, pp. 632–647.

Aylor, D.M. and P.J. Moran. 1985. Effect of reinforcement on the pitting behavior of aluminum-base metal matrix composites. *Journal of the Electrochemical Society*, 132, 1277–1281.

Bakulin, A.V., V.V. Ivanov et al. 1978. Corrosive behavior of boron-aluminum composite material. *Zaschita Metallov*, 14(1), 102–104.

Becher, H.J. 1963. Section 15. Aluminum. In G. Brauer, Ed., *Handbook of Preparative Inorganic Chemistry*, Vol. 1., Academic Press, New York, p. 832.

Behrens, E. 1968. Thermal conductivity of composite materials. *Journal of Composite Materials*, 2, 2.

Berkeley, D.W., H.E.M. Sallam et al. 1998. The effect of pH on the mechanism of corrosion and stress corrosion and degradation of mechanical properties of AA6061 and Nextel 440 fiber-reinforced AA6061 composites. *Corrosion Science*, 40(No. 2/3), 141–153.

Bertolini, L., M.F. Brunella et al. 1999. Corrosion behavior of a particulate metal-matrix composite. *Corrosion*, 55(4), 422–431.

Bhat, M.S., M.K. Surappa et al. 1991. Corrosion behaviour of silicon carbide particle reinforced 6061/Al alloy composites. *Journal Materials Science*, 26(18), 4991–4996.

Birbilis, N. and R.G. Buchheit. 2005. Electrochemical characteristics of intermetallic phases in aluminum alloys: An experimental survey and discussion. *Journal of the Electrochemical Society*, 152(4), B140–B151.

Blackwood, D.J., A.W.C. Chua et al. 2000. Corrosion behaviour of porous titanium-graphite composites designed for surgical implants. *Corrosion Science*, 42, 481–503.

Bolz, R.E. and G.L. Tuve. 1973. *CRC Handbook of Tables for Applied Engineering Science*, CRC Press, Boca Raton, FL, 262–264.

Brenner, S.S. 1962a. Mechanical behavior of sapphire whiskers at elevated temperatures. *Journal of Applied Physics*, 33, 33.

Brenner, S.S. 1962b. Whisker-reinforced metals. *Journal of Metals*, 14(11), 808.

Bruun, N.K. and K. Nielsen. 1991. *Metal Matrix Composites—Processing, Microstructure and Properties: 12th Risø International Symposium on Materials and Science*, Roskilde, Denmark.

Buck, M.E. and R.J. Suplinskas. 1987. *Engineered Materials Handbook on Composites*, Vol. 1, ASM International, Metals Park, OH, pp. 851–857.

Buck, R.F. and A.W. Thompson. 1991. Environmental fatigue in Al-SiC composites. In R.H. Jones and R.E. Ricker, Eds., *Environmental Effects on Advanced Materials*, The Minerals, Metals, and Materials Society, Pennsylvania, PA, pp. 297–313.

Buonanno, M.A. 1992. The effect of processing conditions and chemistry on the electrochemistry of graphite and aluminum metal matrix composites. PhD thesis, Massachusetts Institute of Technology, Cambridge, MA.

Burbank, J. 1959. The anodic oxides of lead. *Journal of the Electrochemical Society*, 106, 369.

Burbank, J., A.C. Simon et al. 1971. The lead-acid cell. In P. Delahay, Ed., *Advances in Electrochemistry and Electrochemical Engineering*, Vol. VIII, Wiley Interscience, New York, p. 157.

Burte, H.M., F.R. Bonanno et al. 1966. Metal matrix composite materials. In *Orientation Effects in the Mechanical Behavior of Anisotropic Structural Materials*, ASTM Special Technical Publication No. 405, American Society for Testing and Materials, Seattle, WA.

Butler, G. and H.C.K. Ison. 1978. *Corrosion and Its Prevention in Waters*, Robert E. Krieger Publishing, New York.

Chawla, K.K. 1991. In R.K. Everett and R.J. Arsenault, Eds., *Metal Matrix Composites: Mechanisms and Properties*, Academic Press, New York, pp. 235–253.

Chawla, K.K. and N. Chawla. 2006. *Metal Matrix Composites*, Springer, New York.

Chen, C. and F. Mansfeld. 1997. Corrosion protection of an Al 6092/SiCp metal matrix composite. *Corrosion Science*, 39(6), 1075.

Clauser, H.R. 1963. *The Encyclopedia of Engineering Materials and Processes*, Reinhold Publishing Corporation, New York.

Clyne, T.W. and P.J. Withers. 1995. *An Introduction to Metal Matrix Composites*, Cambridge University Press, Cambridge, U.K.

Coleman, S.L., V.D. Scott et al. 1994. Corrosion behaviour of aluminium-based metal matrix composites. *Journal of Materials Science*, 29, 2826–2834.

Covino, B.S., Jr. and D.E. Alman. 2002. Corrosion of titanium matrix composites, in *Proceedings of the 15th International Corrosion Congress*, Madrid, Spain, Viajes Iberia Congresos.

Crowe, C.R., D.G. Simons, and M.D. Brown. 1981. An investigation of anodized SiC/Al metal matrix composites. Abstract No. 151. In *Extended Abstracts Volume 81-2*, The Electrochemical Society, Denver, CO.

Czyrklis, W.F. 1983. Corrosion evaluation of metal matrix composite FP/Mg AZ91C, in *Tri-Service Corrosion Conference*, U.S. Naval Academy, Annapolis, MD, 1983.

Czyrklis, W.F. 1985. *Conference Proceedings of Corrosion 85*, Paper No. 196, National Association of Corrosion Engineers, Boston, MA.

Dacres, C.M., S.M. Reamer et al. 1981. Anodic corrosion of fiber reinforced lead composites for use in large lead-acid batteries. *Journal of the Electrochemical Society*, 128, 2060–2064.

Dacres, C.M., R.A. Sutula et al. 1983. A comparison of procedures used in assessing the anodic corrosion of metal matrix composites and lead alloys for use in lead-acid batteries. *Journal of the Electrochemical Society*, 130, 981–985.

Davis, D.A., M.G. Vassilaros et al. 1982. Corrosion fatigue and stress corrosion characteristics of metal matrix composites in seawater. *Materials Performance*, 21(3), 38–42.

Deltombe, E., C. Vanleugenhaghe et al. 1974. Aluminium. In M. Pourbaix, Ed., *Atlas of Electrochemical Equilibria in Aqueous Solutions*, National Association of Corrosion Engineers, Houston, TX, pp. 168–176.

DeSalazar, J.M.G., A. Ureña et al. 1999. Corrosion behaviour of AA6061 and AA7005 reinforced with Al_2O_3 particles in aerated 3.5% chloride solutions: Potentiodynamic measurements and microstructure evaluation. *Corrosion Science*, 41, 529–545.

Diefendorf, R.J. 1984. Carbon fiber structure, Report # NSWC MP 84-258, in R.N. Lee, Ed., *Critical Issues in Materials Technology Workshop on Transverse Strength in Carbon-Fiber/Aluminum Composites*, Naval Surface Weapons Center, White Oak, Silver Spring, MD.

Ding, H. and L.H. Hihara. 2005. Localized corrosion currents and pH profile over B_4C, SiC and Al_2O_3 reinforced 6092 aluminum composites I. In 0.5 M Na_2SO_4. *Journal of the Electrochemical Society*, 152(4), B161–B167.

Ding, H. and L.H. Hihara. 2008a. Effect of embedded titanium-containing particles on the corrosion of particulate alumina reinforced aluminum-matrix composite. *ECS Transactions*, 11(15), 935.

Ding, H. and L.H. Hihara. 2008b. A "Photochemical Corrosion Diode" model depicting galvanic corrosion in metal-matrix composites containing semiconducting constituents. *ECS Transactions*, 11(18), 41.

Divecha, A.P., P. Lare et al. 1969. Silicon carbide whisker metal matrix composites, AFML-TR-69-7.

Dull, D.L., W.C.J. Harrigan et al. 1977. The effect of matrix and fiber composition on mechanical strength and corrosion behavior of graphite-aluminum composites. Final Report. The Aerospace Corporation, El Segundo, CA.

DWA Aluminum Composites. June 2010. www.dwa-dra.com/ (accessed July 3, 2013).

DWA Technologies Inc. October 2007. http://dwatechnologies.com/ (accessed July 3, 2013).

Evans, A., C. San Marchi et al. 2003. *Metal Matrix Composites in Industry: An Introduction and a Survey*, Kluwer Academic Publishers, Dordrecht, the Netherlands.

Fang, C.-K., C.C. Huang et al. 1999. Synergistic effects of wear and corrosion for Al_2O_3 particulate-reinforced 6061 aluminum matrix composites. *Metallurgical and Materials Transactions A*, 30A, 643–651.

Fontana, M.G. and N.D. Greene. 1978. *Corrosion Engineering*, McGraw-Hill, Inc., New York.

Galasso, F. and J. Pinto. 1969. Compatibility of silicon carbide coated boron fibers in a titanium matrix. *Fibre Science and Technology*, 2(2), 89–95.

Galvele, J.R. 1978. Present state of understanding of the breakdown of passivity and repassivation. In R.P. Frankenthal and J. Kruger, Eds., *Passivity of Metals*, The Electrochemical Society, Inc., Princeton, NJ, pp. 285–327.

Garrard, W.N.C. 1994. The corrosion behaviour of aluminum-silicon carbide composites in aerated 3.5% sodium chloride. *Corrosion Science*, 36(5), 837–851.

Gellings, P.J. 1976. *Introduction to Corrosion Prevention and Control for Engineers*, Delft University Press, Delft, the Netherlands.

Golledge, S.L., J. Kruger et al. October 1985. *Extended Abstracts*, Vol. 85-2, The Electrochemical Society, Las Vegas, NV.

Greene, H.J. 1992. Evaluation of corrosion protection methods for aluminum metal matrix composites. PhD thesis, University of Southern California, Los Angeles, CA.

Greenwood, N.N. and A. Earnshaw. 1984. *Chemistry of the Elements*, Pergamon Press, Ltd., Oxford, U.K.

Griffiths, A.J. and A. Turnbull. 1994. An investigation of the electrochemical polarisation behaviour of 6061 aluminum metal matrix composites. *Corrosion Science*, 36(1), 23–35.

Grytsiv, A. and P. Rogl. 2004. Aluminum–boron–carbon. In G. Effenberg and S. Ilyenko, Eds., *Light Metal Systems. Part 1: Selected Systems from Ag-Al-Cu to Al-Cu-Er*, Vol. 11A1, Springer, Berlin, Germany.

Hamdy, A.S., A.M. Beccaria et al. 2002. Corrosion protection of aluminum metal-matrix composites by cerium conversion coatings. *Surface and Interface Analysis*, 34, 171–175.

Harrigan, W.C.J. 1991. Metal matrix composites. In R.K. Everett and R.J. Arsenault, Eds., *Metal Matrix Composites: Processing and Interfaces*, Academic Press, Boston, MA, pp. 1–16.

Harrigan, W.C.J. and R.H. Flowers. 1979. Graphite-metal composites titanium-boron vapor deposit method of manufacture. In J.A. Cornie and F.W. Crossman, Eds., *Failure Modes in Composites IV*, The Metallurgical Society of AIME, Warrendale, PA, pp. 319–335.

Hasson, D.F., C.R. Crowe et al. 1984. Fatigue and corrosion fatigue of discontinuous SiC/Al metal matrix composites. In J.G. Early, T.R. Shives, and J.H. Smith, Eds., *Failure Mechanisms in High Performance Materials*, Cambridge University Press, Cambridge, U.K., pp. 147–156.

Hawthorn, G.A. 2004. Outdoor and laboratory corrosion studies of aluminum-metal matrix composites. MS thesis, Department of Mechanical Engineering, University of Hawaii at Manoa, Honolulu, HI.

Hawthorn, G.A. and L.H. Hihara. 2004. Out-door & laboratory corrosion studies of aluminum metal-matrix composites, in *U.S. Army Corrosion Summit*, Cocoa Beach, FL.

He, C., D. Lou et al. 2011. Corrosion protection and formation mechanism of anodic coating on SiCp/Al metal matrix composite. *Thin Solid Films*, 519, 4759.

Hihara, L.H. 1989. Corrosion of aluminum-matrix composites. PhD thesis, Massachusetts Institute of Technology, Cambridge, MA.

Hihara, L.H. 1997. Corrosion of aluminum-matrix composites. *Corrosion Reviews*, 15(3–4), 361–386.

Hihara, L.H. 2005a. Metal-matrix composites (Chapter 58). In R. Baboian, Ed., *Corrosion Tests and Standards: Application and Interpretation*, ASTM International, Materials Park, OH, pp. 637–655.

Hihara, L.H. 2005b. Corrosion of metal-matrix composites, in S.D. Cramer and J.B.S. Covino, Eds., *ASM Handbook, Corrosion: Materials*, Vol. 13B, ASM International, Materials Park, OH, pp. 538–539.

Hihara, L.H., T.S. Devarajan et al. 2005. Corrosion initiation and propagation in particulate aluminum-matrix composites, in *Tri-Service Corrosion Conference*, Orlando, FL.

Hihara, L.H., H. Ding et al. 2004. Corrosion-initiation sites on aluminum metal-matrix composites, in *U.S. Army Corrosion Summit*, Cocoa Beach, FL.

Hihara, L.H. and P.K. Kondepudi. 1993. The galvanic corrosion of silicon carbide monofilament/ZE41 magnesium metal-matrix composite in 0.5 M sodium nitrate. *Corrosion Science*, 34, 1761–1772.

Hihara, L.H. and P.K. Kondepudi. 1994. Galvanic corrosion between SiC monofilament and magnesium in NaCl, Na2SO4 and NaNO3 solutions for application to metal-matrix composites. *Corrosion Science*, 36, 1585–1595.

Hihara, L.H. and R.M. Latanision. 1991. Localized corrosion induced in graphite/aluminum metal-matrix composites by residual microstructural chloride. *Corrosion*, 47, 335–341.

Hihara, L.H. and R.M. Latanision. 1992. Galvanic-corrosion of aluminum matrix composites. *Corrosion*, 48(7), 546–552.

Hihara, L.H. and R.M. Latanision. 1994. Corrosion of metal-matrix composites. *International Materials Review*, 39, 245.

Hihara, L.H. and C. Tamirisa. 1995. Corrosion of SiC monofilament/Ti-15-3-3-3 metal-matrix composites in 3.15 wt.% NaCl. *Materials Science and Engineering A*, 198, 119–125.

Hill, R.J. and W.F. Stuhrke. 1968. The preparation and properties of cast boron-aluminum composites. *Fibre Science and Technology*, 1(1), 25–42.

Hinton, B.R.W., D.R. Arnott et al. 1986. Cerium conversion coatings for the corrosion protection of aluminium. *Materials Forum*, 9, 162.

Hou, J. and D.D.L. Chung. 1997. Corrosion protection of aluminum-matrix aluminum nitride and silicon carbide composites by anodization. *Journal of Materials Science*, 32, 3113–3121.

Hu, J., X.H. Zhao, S.W. Tang, and M.R. Sun. 2006. Corrosion protection of aluminum borate whisker reinforced AA6061Composite by cerium oxide-based conversion coating. *Surface & Coatings Technology*, 201, 3814.

Ichinose, N. 1987. *Introduction to Fine Ceramics*, John Wiley & Sons, Ltd., New York.

Iseki, T., T. Kameda et al. 1984. Interfacial reactions between silicon carbide (SiC) and aluminum during joining. *Journal of Materials Science*, 19, 1692–1698.

Jones, R.H. 1991. Stress corrosion cracking of metal matrix composites: Modeling and experiment. In R.H. Jones and R.E. Ricker, Eds., *Environmental Effects on Advanced Materials*, The Minerals, Metals and Materials Society, Warrendale, PA, pp. 283–295.

JW Composites LC. 2010. http://www.jwcomposites.com/products.htm (accessed July 3, 2013).

Kelly, A. and W.R. Tyson. 1965. Tensile properties of fibre-reinforced metals: Copper/tungsten and copper/molybdenum. *Journal of the Mechanics and Physics of Solids*, 13(6), 329–338.

Kiourtsidis, G.E. and S.M. Skolianos. 2000. Stress corrosion behavior of aluminum alloy 2024/silicon carbide particles (SiC_p) metal matrix composites. *Corrosion*, 56(6), 646–653.

Kiourtsidis, G.E., S.M. Skolianos et al. 1999. A study on pitting behaviour of $AA2024/SiC_p$ composites using the double cycle polarization technique. *Corrosion Science*, 41, 1185–1203.

Kondepudi, P.K. 1992. Corrosion behavior of magnesium matrix composites. M.S. Thesis. Department of Mechanical Engineering, University of Hawaii at Manoa, Honolulu, HI.

LaQue, F.L. 1948. Behavior of metals and alloys in sea water. In H.H. Uhlig, Ed., *Corrosion Handbook*, Wiley, New York, p. 416.

Lee, Y.-F., S.-L. Lee et al. 1999. Wear and corrosion behaviors of SiC_p reinforced copper matrix composite formed by hot pressing. *Scandinavian Journal of Metallurgy*, 28, 9–16.

Levy, M. and W.F. Czyrklis. October 1981. *Extended Abstracts*, Vol. 81-2, The Electrochemical Society, Denver, CO.

Lim, T., C.S. Lee et al. 1991. *Conference Proceedings, ICCM/8*, Honolulu, HI, Society for the Advancement of Material and Process Engineering (SAMPE), Covina, CA.

Lin, S., H. Greene et al. 1992. Corrosion protection of Al/SiC metal matrix composites by anodizing. *Corrosion*, 48(1), 61–67.

Lin, S., H. Shih et al. 1992. Corrosion protection of aluminum alloys and metal matrix composites by polymer coatings. *Corrosion Science*, 33(9), 1331–1349.

Liu, Y., P.K. Rohatgi et al. 1991. *Conference Proceedings, ICCM/8*, Honolulu, HI, Society for the Advancement of Material and Process Engineering (SAMPE), Covina, CA.

Lin, Z.J. 1995. Corrosion study of silicon-aluminum metal-matrix composites. M.S. Thesis. Department of Mechanical Engineering, University of Hawaii at Manoa, Honolulu, HI.

Lore, K.D. and J.S. Wolf. 1981. *Extended Abstracts*, Vol. 81-2, The Electrochemical Society, Denver, CO.

Mansfeld, F. and S.L. Jeanjaquet. 1986. The evaluation of corrosion protection measures for metal matrix composites. *Corrosion Science*, 26, 727–734.

Mansfeld, F., S. Lin et al. 1989. Corrosion protection of aluminum alloys and aluminum-based metal matrix composites by chemical passivation. *Corrosion*, 45, 615–630.

Mansfeld, F., S. Lin et al. 1990. Pitting and passivation of aluminum alloys and aluminum-based metal matrix composites. *Journal of the Electrochemical Society*, 137(1), 78–82.

Marcus, H.L., W.F. Weldon et al. 1987. Technical report contract number N62269-85-C0222. University of Texas, Austin, TX.

Metal Matrix Cast Composites LLC. 2010. http://www.mmccinc.com/thermal.htm

Metzger, M. and S.G. Fishman. 1983. Corrosion of aluminum-matrix composites. *Industrial and Engineering Chemistry. Product Research and Development*, 22, 296–302.

Minoshima, K., I. Nagashima et al. 1998. Corrosion fatigue fracture behaviour of a SiC whisker-aluminum matrix composite under combined tension-torsion loading. *Fatigue and Fracture of Engineering Materials and Structures*, 21, 1435–1446.

Modi, O.P., M. Saxena et al. 1992. Corrosion behaviour of squeeze-cast aluminum alloy-silicon carbide composites. *Journal of Materials Science*, 27, 3897–3902.

Monticelli, C., F. Zucchi et al. 1995. Application of electrochemical noise analysis to study the corrosion behavior of aluminum composites. *Journal of the Electrochemical Society*, 142(2), 405–410.

Monticelli, C., F. Zucchi et al. 1997. Stress corrosion cracking behaviour of some aluminum-based metal matrix composites. *Corrosion Science*, 39(10–11), 1949–1963.

Mortensen, A., J.A. Cornie et al. 1988. Solidification processing of metal-matrix composites. *Journal of Metals*, 40, 12.

Mukherjee, S.K., A. Kumar et al. 1985a. Anodic polarization study of sintered 434L ferritic stainless steel and its particulate composites. Anodic polarization study of sintered 434L ferritic stainless steel and its particulate composites. *Powder Metallurgy International*, 17, 172–175.

Mukherjee, S.K., A. Kumar et al. 1985b. Effect of iron phosphide (Fe2P) additions on corrosion behavior of ferritic stainless steel composite compacts containing 6 vol. % aluminum oxide. *British Corrosion Journal*, 20, 41–44.

Mukherjee, S.K. and G.S. Upadhyaya. 1985. Corrosion behavior of sintered 434L ferritic stainless steel-alumina composites containing early transition metals. *Materials Chemistry and Physics*, 12, 419–435.

Nath, D., R.T. Bhat et al. 1980. Preparation of cast aluminum alloy-mica particle composites. *Journal of Materials Science*, 15, 1241–1251.

Nath, D. and T.K. Namboodhiri. 1988. Corrosion of an aluminum alloy-mica particulate composite in 3.5% sodium chloride. *Composites*, 19, 237–243.

Nath, D. and T.K. Namboodhiri. 1989. Some corrosion characteristics of aluminum-mica particulate composites. *Corrosion Science*, 29, 1215–1229.

Nunes, P.C.R. and L.V. Ramanathan. 1995. Corrosion behavior of alumina-aluminum and silicon carbide-aluminum metal-matrix composites. *Corrosion*, 51(8), 610–617.

Nunez-Lopez, C.A., H. Habazaki et al. 1996. An investigation of microgalvanic corrosion using a model magnesium-silicon carbide metal matrix composite. *Corrosion Science*, 38(10), 1721–1729.

Nunez-Lopez, C.A., P. Skeldon et al. 1995. The corrosion behaviour of Mg alloy ZC71/SiC$_p$ metal matrix composite. *Corrosion Science*, 37(5), 689–708.

Paciej, R.C. and V.S. Agarwala. 1988. Influence of processing variables on the corrosion susceptibility of metal-matrix composites. *Corrosion*, 44, 680–684.

Park, G.B. and D.A. Foster. 1990 July. *International Technical Conference Proceedings, SUR/FIN'90*, American Electroplaters and Surface Finishers Society, Inc., Boston, MA.

Park, J.K. and J.P. Lucas. 1997. Moisture effect on SiC$_p$/6061 Al MMC: Dissolution of interfacial Al$_4$C$_3$. *Scripta Materialia*, 37(4), 511–516.

Parthiban, G.T., D. Malarkodi et al. 2010. Corrosion protection by acrylamide treatment for magnesium alloy metal matrix composite (MMC) reinforced with titanium boride. *Surface Engineering*, 26(5), 378.

Payer, J.H. and P.G. Sullivan. 1976. Corrosion protection methods for graphite fiber reinforced aluminum alloys, *Bicentennial of Materials, 8th National Sample Technical Conference*, Vol. 8, Society for the Advancement of Material and Process Engineering, Seattle, WA, October 12–14, pp. 343–352.

Piggott, M.R. 1980. *Load Bearing Fibre Composites*, Pergamon Press, New York.

Pohlman, S.L. 1978. Corrosion and electrochemical behavior of boron/aluminum composites. *Corrosion*, 34, 156–159.

Portnoi, K.I., N.I. Timofeeva et al. 1981. Effect of the content of a carbide phase on properties of carbon-aluminum. *Poroshkovaya Metallurgiya*, (No. 2 (218)), 45–49.

Rayleigh, L. 1892. On the influence of obstacles arranged in rectangular order upon the properties of the medium. *Philosophical Magazines*, 34, 481–502.

Rohatgi, P.K., R. Asthana et al. 1986. Solidification, structures, and properties of cast metal-ceramic particle composites. *International Metals Review*, 31, 115.

Roper, G.W. and P.A. Attwood. 1995. Corrosion behaviour of aluminum matrix composites. *Journal of Materials Science*, 30, 898–903.

Saffarian, H.M. and G.W. Warren. 1998. Aqueous corrosion study of α_2-Ti$_3$Al/SiC composites. *Corrosion*, 54(11), 877–886.

Saffil. May 2002. Saffil. www.saffil.com (accessed July 3, 2013).

Schapery, R.A. 1968. Thermal expansion coefficients of composite materials based on energy principles. *Journal of Composite Materials*, 2(3), 380–404.

Schwartz, M.M. 1984. *Composite Materials Handbook*, McGraw-Hill, Inc., New York.

Seah, K.H.W., S.C. Sharma et al. 1997a. Corrosion characteristics of ZA-27-graphite particulate composites. *Corrosion Science*, 39(1), 1–7.

Seah, K.H.W., S.C. Sharma et al. 1997b. Corrosion behaviour of lead alloy/zircon particulate composites. *Corrosion Science*, 39(8), 1443–1449.

Sedriks, A.J. 1990. *Stress Corrosion Cracking Test Methods*, National Association of Corrosion Engineers, Houston, TX.

Sedriks, A.J., J.A. Green et al. 1971. Corrosion behavior of aluminum-boron composites in aqueous chloride solutions. *Metallurgical Transactions*, 2, 871–875.

Shimizu, Y., T. Nishimura et al. 1995. Corrosion resistance of Al-based metal matrix composites. *Materials Science and Engineering A*, 198, 113–118.

Smith, W. 1993. *Structure and Properties of Engineering Alloys*, McGraw-Hill, New York.

Stroganova, V.F. and M.A. Timonova. 1978. *Metallovedenie i Termicheskaya Obrabotka Metallov*, 44–46.

Sun, H., E.Y. Koo et al. 1991. Corrosion behavior of SiC$_p$/6061 Al metal matrix composites. *Corrosion*, 47(10), 741–753.

Sun, H., J.E. Orth et al. September 1993. Corrosion behavior of copper-based metal-matrix composites. *Journal of Metals*, 45, 36–41.

Sun, H. and H.G. Wheat. 1993. Corrosion study of Al$_2$O$_3$ dispersion strengthened Cu metal matrix composites in NaCl solutions. *Journal of Materials Science*, 28, 5435–5442.

Suri, A.K., C. Subramanian et al. 2010. Synthesis and consolidation of boron carbide: A review. *International Materials Reviews*, 55(1), 4–40.

Sutton, W.H. August 1966. Whisker composite materials—A prospectus for the aerospace designer. *Astronautics and Aeronautics*, 4, 46.

Sutton, W.H. and J. Chorne. 1963. Development of high-strength, heat-resistant alloys by whisker reinforcement. *ASM Metals Engineering Quarterly*, 3(1), 44.

Tamirisa, C. 1993. Corrosion behavior of silicon-carbide reinforced titanium 15-3 metal matrix composite in 3.15 wt% NaCl. M.S. Thesis. Department of Mechanical Engineering, University of Hawaii at Manoa, Honolulu, HI.

Traverso, P., R. Spiniello et al. 2002. Corrosion inhibition of Al 6061 T6/Al2O3p 10% (v/v) composite in 3.5% NaCl solution with addition of cerium (III) chloride. *Surface and Interface Analysis*, 34:185–188.

Timonova, M.A., G.I. Spiryakina et al. 1980. Corrosion resistance and electrochemical properties of a magnesium matrix composite material. *Metallovedenie i Termicheskaya Obrabotka Metallov*, 11, 33–35.

Trzaskoma, P.P. 1982. Corrosion rates and electrochemical studies of a depleted uranium alloy tungsten fiber metal matrix composite. *Journal of the Electrochemical Society*, 129, 1398–1402.

Trzaskoma, P.P. 1986. Corrosion behavior of a graphite fiber/magnesium metal matrix composite in aqueous chloride solution. *Corrosion*, 42, 609–613.

Trzaskoma, P.P. 1990. Pit morphology of aluminum alloy and silicon carbide/aluminum alloy metal matrix composites. *Corrosion*, 46, 402–409.

Trzaskoma, P.P., E. McCafferty et al. 1983. Corrosion behavior of silicon carbide/aluminum metal matrix composites. *Journal of the Electrochemical Society*, 130, 1804–1809.

Trzaskoma, P.P. and E. McCafferty. 1986. The effect of anodic coatings on the pitting of silicon carbide/aluminum metal matrix composites. In Alwitt, R.S. and G.E. Thompson, Eds., *Aluminum Surface Treatment Technology*, The Electrochemical Society, Denver, CO, pp. 171–180.

Tsirlin, A.M. 1985. Boron filaments. In W. Watt and B.V. Perov, Eds., *Strong Fibres (Handbook of Composites, Vol. 1)*, Elsevier Science Publishers B.V., Amsterdam, the Netherlands, pp. 155–199.

Turner, P.S. 1946. Thermal-expansion stresses in reinforced plastics. *Journal of Research of the National Bureau of Standards*, 37, 239.

Uhlig, H.H. and R.W. Revie. 1985. *Corrosion and Corrosion Control*, John Wiley & Sons, Inc., New York.

Velasco, F., N. Anton et al. 1997. Mechanical and corrosion behaviour of powder metallurgy stainless steel based metal matrix composites. *Materials Science and Technology*, 13(10), 847–851.

Viala, J.C., M. El Morabit et al. 1985. Mechanical properties and corrosion behavior of lead-silicon carbide fiber and lead-carbon fiber composites made by electrodeposition. *Materials Chemistry and Physics*, 13, 393–408.

Weast, R.C. 1986. *CRC Handbook of Chemistry and Physics*, CRC Press, Boca Raton, FL.

Weeton, J.W. 1969. Fiber-metal matrix composites fabrication, applications, mechanical properties and powder metallurgy. *Machine Design*, 41, 142–156.

Weeton, J.W., D.M. Peters et al. 1987. *Guide to Composite Materials*, American Society for Metals, Metals Park, OH.

Wesley, W.A. and R.H. Brown. 1948. Fundamental behavior of galvanic couples, in H.H. Uhlig, Ed., *The Corrosion Handbook*, John Wiley & Sons, Inc., New York, pp. 481–495.

Wielage, B. and A. Dorner. 1999. Corrosion studies on aluminum reinforced with uncoated and coated carbon fibres. *Composites Science and Technology*, 59, 1239–1245.

Wielage, B., A. Dorner et al. 2000. Corrosion protection of carbon fibre reinforcced aluminum composite by diamondlike carbon coatings. *Materials Science and Technology*, 16, 344–348.

Yamada, S., K. Hirao et al. 2003. Mechanical and electrical properties of B4C-CrB2 ceramics fabricated by liquid phase sintering. *Ceramics International*, 29, 299.

Yang, J.Y. and M. Metzger. October 1981. *Extended Abstracts*, Abstract No. 155, Vol. 81-2, The Electrochemical Society, Denver, CO.

Yao, H.-Y. 1999. Effect of particulate reinforcing on stress corrosion cracking performance of a $SiC_p/2024$ aluminum matrix composite. *Journal of Composite Materials*, 33(11), 962–970.

Yao, H.-Y. and R.-Z. Zhu. 1998. Interfacial preferential dissolution on silicon carbide particulate/aluminum composites. *Corrosion*, 54(7), 499–503.

Yau, S.S. and G. Mayer. 1986. Fatigue of metal matrix composite materials. *Materials Science and Engineering*, 82, 45–47.

Zhu, J. 2008. Corrosion of continuous alumina fiber reinforced aluminum-matrix composites. PhD thesis. Department of Mechanical Engineering, University of Hawaii at Manoa, Honolulu, HI.

Further Readings

Aylor, D.M. 1987. Corrosion of metal matrix composites, in *Metals Handbook, Corrosion*, 9th edn., ASM International, Metals Park, OH, pp. 859–863.

Hihara, L.H. 2005. Corrosion of metal-matrix composites, in S.D. Cramer and J.B.S. Covino, Eds., *ASM Handbook, Corrosion: Materials*, Vol. 13B, ASM International, Materials Park, OH, p. 526.

Hihara, L.H. 2005. Metal-matrix composites, chapter 58, in R. Baboian, Ed., *Corrosion Tests and Standards: Application and Interpretation*, ASTM International, Materials Park, OH, p. 637.

Jones, R.H. 2001. Metal matrix composites, in R.H. Jones, Ed., *Environmental Effects on Engineered Materials*, Marcel Dekker, Inc., pp. 379–390.

Lucas, K.A. and H. Clarke. 1993. *Corrosion of Aluminum-Based Metal Matrix Composites*, Research Studies Press, Ltd., England, U.K.

Trzaskoma, P.P. 1991. Corrosion, in R.K. Everett and R.J. Arsenault, Eds., *Metal Matrix Composites: Mechanisms and Properties*, Academic Press, New York, pp. 383–404.

II

Polymers

13

Forms of Polymer Degradation: Overview

Margaret Roylance
*U.S. Army Natick Soldier
Research, Development
and Engineering Center*

David Roylance
*Massachusetts Institute
of Technology*

13.1 Introduction

13.1.1 Usage of Polymeric Materials

Although polymers have existed naturally on Earth since well before the dawn of humans, the ubiquitous role they play in today's society is much more recent. The modern plastics industry is often dated from the mid-nineteenth century, with John Hyatt's invention of celluloid (a synthetic modification of natural cellulose). The first wholly synthetic polymer was phenolic, invented by Leo Baekeland by condensing phenol with formaldehyde in 1906. In the decades following 1930, industrial research by Wallace Carothers and others produced a great outpouring of new polymers: nylon, polyethylene terephthalate, polystyrene, and many others. The bar graph in Figure 13.1 shows how polymers now rank alongside metals, ceramics, and natural materials in terms of annual tonnage (a graph using volume rather than weight would show polymers even more prominently, due to their low density).

13.1.2 Nature of Polymeric Materials

Most commercial polymers are long-chain hydrocarbon molecules, containing thousands of carbon atoms bonded covalently along the length of the chain. (Some important polymers [such as nylon] have noncarbon elements such as nitrogen or oxygen in the chain, and one [polydimethyl siloxane] has no carbons along the chain). Figure 13.2 shows the composition of polyethylene. The single covalent bonds

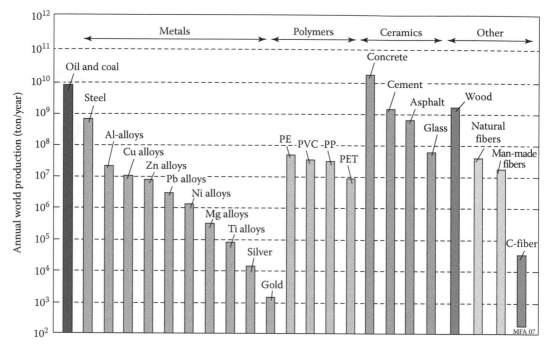

FIGURE 13.1 Graph of materials production ca 2008. (From Ashby, M.F., *CES EduPack*, University of Cambridge/ Granta, Cambridge, U.K., 2008.)

FIGURE 13.2 "Space-filling" model of a portion of the polyethylene molecule. Black, carbon; white, hydrogen.

in the chain are very strong and stable, accounting for polymers' well-known chemical stability. There is also easy molecular rotation around these bonds, leading to substantial flexibility in the chain.

The high bond strength exists from carbon to carbon along the chain, but the lateral bonding to adjacent molecules or nearby portions of the same molecule is much weaker. Since mechanical loads applied to a bulk polymer act on both the weak and the strong bonds, polymers are much weaker and more compliant than metals and ceramics that have strong bonds in all directions. But if the molecules in a polymer are oriented along a common direction as in a fiber, the material can exhibit very high strength and stiffness in that direction. (This is true in the polyaramid Kevlar®, for instance.) Such a material is much weaker in its transverse direction, however; that is, such a material is *anisotropic*.

The combination of strong covalent bonds within an individual polymer molecule and weaker secondary bonds between neighboring molecules produces the useful ability in some polymers to be molded into complex shapes at relatively low processing temperatures. *Amorphous* polymers are those that do not form ordered crystalline structures, and they can be processed at temperatures above their so-called glass transition temperature (T_g). Above T_g, individual amorphous polymer molecules are free to undergo large-scale motion past one another, providing molecular mobility required for molding without scission of main-chain bonds and resultant thermal degradation. Some polymers do form ordered crystallites, and these *crystalline* polymers must be heated above their crystalline melting point (T_m) for molding.

13.1.3 Classes of Polymeric Materials

Polymers exist in a wide variety of classifications and functions. There are some 30 major chemical classes of polymers, most with many subcategories of molecular weights, fillers, chemical modifications, and other features. They are classified by function (commodity, engineering, and specialty), industrial sectors (plastics, textiles, rubbers, adhesives), morphology (crystalline versus amorphous), or processability (thermoplastic versus thermoset). Polymers are *thermoplastics* in terms of processability if they can be melted and remolded repeatedly. Polymers may also be *thermosets* such as vulcanized rubber or cured epoxy—these materials can be set to a given shape only once; they cannot be recycled after that other than being ground and used as fillers in other materials.

Unlike metals, most polymers are not available as alloys or solid solutions of one another. Most polymers are immiscible in each other, so they cannot be mixed (an alloy of polystyrene and polyphenylene oxide is an exception). Although classical alloys are uncommon, different monomers may be copolymerized in a single main-chain molecule. Copolymers may be formed randomly or by alternating sequences of individual monomers. These are called block copolymers. Block copolymers will often form complex phase-segregated morphologies and may retain certain physical and mechanical properties characteristic of each separate starting polymer. Phase-separated polymer blends may also be used to tailor polymer properties. Polystyrene is a brittle plastic, but it can be toughened significantly by blending it with a well-chosen size distribution of rubber particles. Advanced composites can also be created by strengthening certain polymers (notably polyesters, phenolics, and epoxies) with strong, stiff fibers (notably glass, aramid, or carbon).

13.2 Mechanisms of Environmental Degradation in Polymers

Depending upon the environment, polymers are generally more durable than metals because the covalent bonds of which polymer main chains are composed are inherently stable. Polymers do not rust, but they can exhibit environmental degradation through a variety of mechanisms. These include thermal and photolytic depolymerization, moisture absorption and hydrolysis, microbial degradation, and flammability.

13.2.1 Thermal Depolymerization

Although long-chain single-bonded polymer molecules are chemically stable, they can be depolymerized back to the low molecular weight state if sufficient thermochemical driving energy is available. This degradation is usually negligible at ambient temperatures, but in some polymers it may be a problem during processing. One important example is a condensation polymer such as nylon 66. During polymerization, nylon 66 converts two monomers—adipic acid and hexamethylene diamine—to three products: nylon polymer, water, and heat. At sufficiently high extents of reaction, as the concentration of product rises, the probability of back reaction to monomer increases, lowering the yield of polymer. This is handled industrially by removing water and keeping the temperature low during polymerization. But later, when the polymer is melted during extrusion or injection molding, especially if water is present, conditions will be right for depolymerization. This degradation mechanism—a consequence of Le Chatelier's principle—is avoided by making sure the polymer resin is dried before melt processing.

13.2.2 Photolytic Oxidation

The photon energy in solar radiation is the most damaging component of the outdoor environment, serving to initiate a wide variety of chemical changes in polymeric materials (Trozzolo 1972). According to the Planck–Einstein law $E = hv = hc/\lambda$ (where E, h, $v \cdot c$, and λ are energy, Planck's constant, frequency, and wavelength, respectively), the energy contained in a photon rises as the wavelength decreases. Although the sun emits radiation over a wide range of wavelengths extending from below 100 nm to over 3000 nm, the Earth's atmosphere prevents radiation of wavelength less than approximately 290 nm from reaching the surface. Figure 13.3 shows the spectral distribution of surface sunlight in the ultraviolet (UV) range and compares the solar spectrum with the energy received from an S-1 mercury sunlamp at a distance of 220 mm. The energy of UV photons is comparable with that of the dissociation energies of polymeric covalent bonds, which lie in the range of approximately 290–460 kJ/mol. Such photons have the capability of altering the polymer's chemical structure.

In order to induce chemical change, the photon must first be absorbed by the material, and a material totally transparent in the UV range will not exhibit photoinitiated corrosion. Many chemical entities often found in polymers, however, have characteristic UV absorptions. The carbonyl group, for instance, has a broad absorption peak at approximately 300 nm due to the excitation of a nonbonding electron into the π^* molecular orbital. Such UV-absorbing groups may be present naturally in the polymer, or they may be introduced adventitiously by any of several means, such as oxidation during fabrication, polymerization anomalies, and introduction of various additives. The energy contained in photon-excited high-energy orbitals may be dissipated harmlessly in the form of heat, but a certain fraction of these excited states may relax by initiating chemical change. It is this latter process that produces deterioration.

The deterioration process is frequently some form of photoinitiated oxidation. Typically, the process is initiated when a sufficiently energetic photon strips away a labile proton (H•) from the polymer (RH), leaving behind a free radical (R•):

$$RH \xrightarrow{\ hv\ } R\bullet + H\bullet \tag{13.1}$$

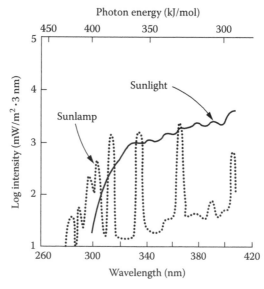

FIGURE 13.3 Typical spectrum of surface solar intensity and emission spectrum from S-1 mercury sunlamp. (From Hirt, R.C. and Searle, N.Z., *Weatherability of Plastic Materials*, Kamal, M.R., Ed., Wiley-Interscience, New York, 1967, p. 61.)

The radical is highly unstable and reacts easily with atmospheric oxygen to form a peroxy radical, which further reacts to form hydroperoxide:

$$R\bullet + O_2 \rightarrow ROOH \qquad (13.2)$$

The oxidation process may now be repeated, so that the initial abstraction by the photon serves to set up a chain reaction that may include thousands of steps. The hydroperoxides formed are generally felt to be the direct cause of degradation in polymer properties, as they decompose via mechanisms that cause polymer chain scission and/or cross-linking.

Scission and cross-linking may be regarded as competitive mechanisms, with one or the other dominating in a particular situation. Cross-linking may also be deleterious, in which it reduces molecular mobility and leads to the sort of brittleness often seen in elastomers that have been exposed to sunlight.

Another deleterious result may be the production of chromophoric chemical species. Such groups may impart an unacceptable discoloration to the polymer if they absorb in the visible range of light, and an autocatalytic UV degradation may be established if UV-absorbing chromophores are produced, which in turn serve to capture more UV photons. These groups, however, may also serve as convenient means of monitoring the extent of the deterioration process, since they are often amenable to quantitative analysis by suitable spectroscopic techniques.

13.2.3 Moisture Absorption

Although polymers do not rust in the presence of environmental moisture, most polymers absorb some moisture through diffusion in high-humidity environments. The extent of moisture absorption in a given polymer will be a function of its hydrophilicity or chemical compatibility with water. Absorbed water is generally less damaging to polymeric materials than is sunlight, but its effects may be important in certain cases. Its role is usually to act as a plasticizer, decreasing the glass transition or softening temperature and yield strength of the polymer. This latter effect has caused concern for polymer matrix composites employed in advanced aircraft, where absorbed moisture may lower the resin matrix T_g to the point that the material is unable to withstand the severe aerothermal loadings encountered (Hawkins 1972). Cases in which water causes permanent degradation by hydrolysis of chemical bonds are not common, and the plasticizing effects can generally be reversed upon drying the material. Rainfall, however, may act in concert with photoinitiated oxidation to produce erosion of the material, serving to wash away the embrittled surface layer so as to expose new material to direct sunlight.

Water is of special concern in the case of glass-fiber-reinforced polymer matrix composites, since glass is known to be subject to permanent hydrolytic damage, especially when simultaneously exposed to stress. Water causes significant reductions in the strengths of fiberglass composites, and it has been common practice in the marine industry to design to "wet" rather than dry strength limits. The nature of the water-induced damage is not well understood, since its effect is often but not always reversible on drying (Hertz 1973). The difference between wet and dry values has been reduced by the development of surface treatments applied to the fibers before impregnation. These surface treatments are considered by many to stabilize fiber–matrix adhesion by forming chemical bonds between the glass fibers and organic resin, although others argue that the treatment acts primarily to shield locally defective regions of the fiber from the corrosive effects of the water.

13.2.4 Outmigration of Diluents and Unreacted Monomers

Polymers can be filled and modified in a great many ways and this versatility is an important key to their commercial success. Fillers and additives used in polymer products include plasticizers and reactive diluents, antioxidants, UV stabilizers, colorants, flame retardants, mechanical reinforcements, and inert fillers to reduce costs. Plasticizers are important fillers, often used in polyvinyl chloride (PVC) to impart flexibility to what is inherently a rigid plastic. The characteristic "plastic" odor often detected in children's toy stores is due to the outmigration of the phthalate plasticizer from PVC. This outmigration is a degradation mechanism, as loss of plasticizer reduces the material's flexibility. Outmigration of diluents occurs by diffusion, the same mechanism that allows environmental moisture to migrate into the material, increasing its flexibility and decreasing its high temperature stability.

Most materials can pose certain safety problems, during their manufacture, use, or disposal. As mentioned earlier, polymers are low-energy stable compounds, generally inert and unreactive. As such, they pose few dangers to human health or the environment. However, the monomers from which they are made are sometimes toxic, and not all of these chemicals are converted to the polymeric state after polymerization. These materials can migrate from the bulk of the material to the surface and into the environment. Most consumers are now aware that a constituent of polycarbonate, bisphenol-A, is thought to produce certain health problems, especially in infants. This worry has impacted consumer acceptance of what has long been considered a versatile plastic of extremely high quality.

13.2.5 Flammability

Polymers are synthesized from fossil fuels, so it is not surprising that they can burn. One measure of their flammability is the limiting oxygen index (LOI), the percentage of oxygen in the air needed to support combustion. LOI over 23 indicates a resistance to burning, and many polymers have LOI greater than this. For instance, the LOI of PVC is near 30, although, when plasticized, its LOI drops to near 18. Even with a room temperature LOI above 23, however, polymers can contribute to fires when fed by nearby flame, for instance, by burning wood. In addition, the LOI of a given polymer may be temperature dependent, and it may drop as the temperature rises during a fire.

As mentioned earlier, combustion products from burning polymers can be dangerous. Chlorine is released when PVC burns, for instance, and of course chlorine is toxic. However, polymer outgassing products probably are not the main danger in fires. Most fires occur in conditions of limited oxygen, so that substantial carbon monoxide concentrations are developed. CO is such a major threat in fires that polymer products, while of concern, are not usually the main problem.

Even in the absence of unreacted monomers or additives, toxic chemicals can be released if the polymer degrades, for instance, in a fire.

13.2.6 Physical Aging and Other Mechanical Effects

Physical aging occurs in polymers due to a variety of processes. Over long periods of time, small-scale rearrangement of short segments of the polymer main chain and side chains may lead to changes in the physical properties of the material such as increasing density or changes in dimensions under stress. In some polymers, under stresses applied during use, these molecular rearrangements can lead to other more pronounced changes such as *crazing*. Crazing (Kambour 1973) is a degradation mechanism prevalent in amorphous plastics such as polystyrene and polycarbonate. Crazes look like cracks but are actually internal debond regions in which a void is spanned by fibrils of polymer connecting the uncrazed material on either side of the material. Figures 13.4 and 13.5 show the appearance and morphology of crazes.

These crazes grow under the influence of tensile stresses and are facilitated by factors such as absorbed diluents that increase molecular mobility. Diluents most effective in promoting crazing are those whose solubility parameters are close to that of the polymer itself; acetone in polystyrene or polycarbonate is a dramatic example.

Crazes differ from true cracks in that the fibrils are capable of supporting mechanical stress, and a crazed specimen may exhibit a substantial fraction of the failure stress of the uncrazed polymer. However, as the load is maintained or increased, the fibrils may rupture, so that crazing can be a precursor to brittle fracture. A phenomenon known as *environmental stress cracking* (ESC) is related to craze breakdown.

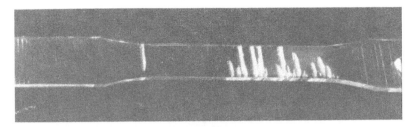

FIGURE 13.4 Crazes in a tensile specimen. (From Kambour, R., *J. Polym. Sci. Macromol. Rev.*, 7, 154, 1973.)

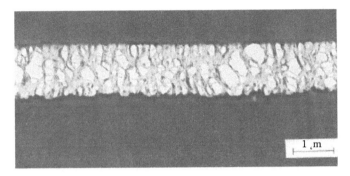

FIGURE 13.5 Micrograph of a craze, showing the fibrils spanning the void region within the craze. (From Kambour, R., *J. Polym. Sci. Macromol. Rev.*, 7, 154, 1973.)

When a compatible environmental agent diffuses along an internal crack by capillary action, it can plasticize the material in the high-stress region at the tip of the crack. This facilitates crazing at the crack tip, which under continued stress experiences fibril rupture and crack extension. Eventually the crack extends through the entire specimen, leading to a brittle-appearing fracture. Both stress and environment are thus involved in this process, which accounts for a large fraction of failures in polymeric articles.

Creep and creep rupture are other mechanical processes that occur in polymers during use. These time-dependent degradation and failure processes reflect the underlying molecular mobility of the materials and may be exacerbated by elevated temperature and humidity. Like all the other mechanisms of polymer degradation discussed earlier, they are thermally activated processes.

13.2.7 Kinetics of Degradation

As described earlier, there are very many distinct mechanisms of material degradation and a kinetic scheme could conceivably be developed for each of them. But as a simple general approach, degradation may be seen as a kinetic process in which the chemical bonds providing the material's strength are gradually consumed, or a second material diffuses in or out of the bulk of the material.

As an example that contains a reasonable overall picture of such a process, we might propose a first-order mechanism for main-chain bond scission in which the number of unbroken bonds decreases at a rate proportional to the number of unbroken bonds remaining:

$$\frac{dn}{dt} = -Kn \rightarrow \frac{dn}{n} = -Kdt \rightarrow n = n_0 e^{-Kt} \tag{13.3}$$

where
  n is the fraction of unbroken bonds remaining
  K is a rate constant for the process

In such a process, the number of unbroken bonds falls to zero only at $t \to \infty$, and clearly total failure will occur well before that. Perhaps a reasonable scaling law would take the lifetime-to-failure t_f to scale with the *average* time $\langle t \rangle$ for a bond scission, which can be computed as

$$t_f \approx \langle t \rangle = \frac{\int n \cdot t \, dt}{\int n \, dt} = \frac{1}{K} \tag{13.4}$$

Treating the process as thermally activated and stress-aided, we can write the rate constant K as an Arrhenius–Eyring expression:

$$K = K_0 \exp \frac{-(E^* - \psi V^*)}{kT} \tag{13.5}$$

where
  E^* and V^* are an activation energy and volume
  ψ is the mechanical stress on the bond

Determining ψ is nontrivial, as ψ obviously varies over the distribution of bonds and is dependent on the material's microstructure. But as another approximation, we might take the atomic stress to scale with the externally applied stress, giving

$$\psi \approx \sigma \to t_f = t_0 \exp \frac{(E^* - \sigma V^*)}{kT} \tag{13.6}$$

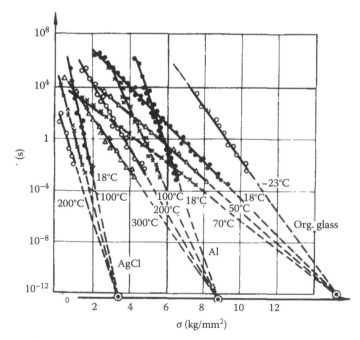

FIGURE 13.6 Time and temperature dependence of the creep-rupture lifetime of solids. Org. glass is PMMA. (From Zhurkov, S.N., *Int. J. Fract. Mech.*, 1, 311, 1965.)

where $t_0 = 1/K_0$. A relation similar to this was proposed by Zhurkov (1965), who conducted a large number of important and innovative studies of the physics of thermomechanical degradation. He argued that mechanical fracture is in fact a thermal degradation process, in which an applied stress acts to lower the energy barrier to thermofluctuational bond dissociation. Figure 13.6 shows this relation to describe the time to failure in a wide variety of materials, including ceramics, metals, and polymers.

13.2.8 Bacterial and Fungal Degradation

Attack by microorganisms is another possible mechanism of environmental degradation in polymers. Microbial attack is most problematic in high-humidity environments, since growth of microbial species on a polymeric material requires the presence of environmental moisture. Bacterial and fungal growth has been observed on some solid polymers, but it is most commonly observed in textiles. Microbial attack can degrade the properties of a range of textile materials such as fibers, fabrics, and webbings. Even in relatively dry environments, if the material is buried underground or worn as part of a garment, sufficient environmental moisture may be present to facilitate microbial attack.

13.3 Protection against Environmental Degradation in Polymers

Physical or chemical stabilization against environmental degradation in polymers may be achieved by blocking any of the steps in the deterioration process. Since the effects of UV radiation are the most serious threat to environmental durability, a number of techniques have been developed to counter them. UV radiation may be excluded by various coatings or paints. Fillers such as carbon black may also be used as screening agents, or the radiation may be absorbed harmlessly by chemical agents that dissipate the photon energy without chemical change. Radical scavengers may be employed to terminate chain radicals and halt the propagation steps, and various deactivators are available that serve to stabilize the hydroperoxide groups formed during photolytic oxidation (see Equations 13.1 and 13.2). The reader is referred elsewhere (e.g., Hawkins 1972) for a more complete treatment of stabilization.

13.3.1 Paints and Coatings

Organic paints and other coatings have been used successfully to protect polymeric materials from UV radiation, but these conventional surface treatments cannot exclude diffusion of environmental moisture into the bulk of the polymer. Since absorption of moisture into the material does not result in main-chain bond scission, this rarely causes degradation leading to failure. However, moisture absorption can have serious consequences in the use of polymer matrix composites in high-performance applications such as aircraft structures. Strength and stiffness values used to design aircraft structures from these materials must be measured under "hot wet" conditions to account for the decreased properties the materials will exhibit if those conditions are encountered in flight (MIL-HDBK-17F 2002).

13.3.2 Novel Surface Treatments

Novel surface treatments are being developed to provide greatly enhanced hydrophobicity and/or oleophobicity to the surface of polymer films or fabrics (Lafuma and Quere 2003, Tuteja et al. 2007). Such surfaces could provide some level of protection against moisture sorption and subsequent diffusion, protection against microbial attack, and even the potential for self-cleaning. These treatments are based on the behavior of some naturally occurring surfaces such as that of the lotus leaf. Observing these natural surfaces provides clues to creating superhydrophobic/superoleophobic surfaces, that is, surfaces that have water and oil contact angles greater than 150°. Since the wettability of a solid surface is determined by two parameters, surface energy (γ) and geometrical roughness, the combination of these two parameters can be used for the development of superhydrophobic and superoleophobic surfaces. Development of these technologies is still at an early stage, and the durability of the surface treatments is often insufficient for commercial use. Great strides have been made in this area in the past 5–10 years, however, and this type of surface treatment may be used more extensively in the future.

There are also a range of antimicrobial surface treatments that have been used to prevent the growth of bacterial and fungal organisms on both naturally occurring and synthetic polymers. These include formulations based upon cationic silver or copper (Russell et al. 1994, Tumer et al. 1999), and quaternary ammonium compounds or QACs (Fabian et al. 1997). In contrast to the action of antibiotics that kill microbial species by interruption of metabolic pathways within the cell, these treatments appear to work primarily by disruption of the cellular membrane, although some effects on cell metabolism may occur.

Another novel approach to providing antimicrobial properties to polymers is the incorporation or attachment of antimicrobial peptides (AMPs) to polymeric surfaces (Arcidiacono et al. 2006, Strauss et al. 2010). AMPs are small peptides containing anywhere from 2 to 50 amino acids and are found in the immune systems of all classes of life. They are potent naturally occurring antibiotics. Significant effort is focused currently on investigation of the mechanism of antimicrobial activity in AMPs and their potential use in antimicrobial surface treatments.

13.4 Summary

Materials engineers usually want to eliminate or slow degradation, since that leads to longer lifetimes and increased safety in their designed products. Polymers are attractive materials candidates in this regard, since they do not exhibit the rusting so troublesome in ferrous materials. On the other hand, environmental activists often criticize plastic products exactly because they are so stable, which contributes to litter that is very slow to disappear. We have outlined in the previous discussion how polymers, while free of rusting, exhibit their own degradation mechanisms, perhaps the most pervasive being photolytic bond scission leading to discoloration and loss of mechanical strength and stiffness. Polymeric materials, especially in the form of textiles, are also subject to microbial attack that is not present in metals. The materials engineer and systems designer must be aware of these features affecting the durability of polymers, considering them appropriately in product design and taking steps to stabilize the material against these mechanisms.

References

Arcidiacono, S., C.M. Mello, and K. Senecal. 2006. Incorporation of antimicrobial peptides into polymeric films and coatings, in *Proceedings of the International Nonwovens Technical Conference*, Houston, TX, September 25–28, 2006.

Ashby, M.F. 2008. *CES EduPack*, University of Cambridge/Granta, Cambridge, U.K.

Fabian, J., T. October, A. Cherestes, and R. Engel. 1997. Polycations: Syntheses of polyammonium strings as antibacterial agents. *SYNLETT*, 8, 1007.

Hawkins, W.L., Ed. 1972. *Polymer Stabilization*, Wiley-Interscience, New York.

Hertz, J. 1973. General Dynamics Corp. Report GDCA-DBG73-005.

Hirt, R.C. and N.Z. Searle. 1967. Energy characteristics of outdoor and indoor exposure sources and their relation to the weatherability of plastics. In M.R. Kamal, Ed., *Weatherability of Plastic Materials*, Wiley-Interscience, New York, p. 61.

Kambour, R. 1973. A review of crazing and fracture in thermoplastics. *Journal of Polymer Science: Macromolecular Reviews*, 7, 154.

Lafuma, A. and D. Quere. 2003. Superhydrophobic states. *Nature Materials*, 2, 457.

MIL-HDBK-17F. 2002. Lamina, laminate, and special form characterization. In *The Composite Materials Handbook*, Vol. 1, ASTM International, West Conshohocken, PA, Chapter 6, pp. 6–29.

Russell, A.D., F.R.C. Path, and W.B. Hugo. 1994. Antimicrobial activity and action of silver. *Progress in Medicinal Chemistry*, 31, 351.

Strauss, J., A. Kadilak, C. Cronin, C.M. Mello, and T.A. Camesano. 2010. Binding, inactivation, and adhesion forces between antimicrobial peptide cecropin P1 and pathogenic *E. coli*. *Colloids and Surfaces B: Biointerfaces*, 75, 156.

Trozzolo, A.M. 1972. Stabilization against oxidative photodegradation. In W.L. Hawkins, Ed., *Polymer Stabilization*, Wiley-Interscience, New York, p. 159.

Tumer, M., H. Koksal, S. Serin, and M. Digrak. 1999. Antimicrobial activity studies of mononuclear and binuclear mixed-ligand copper(II) complexes derived from Schiff base ligands and 1,10-phenanthroline. *Transition Metal Chemistry*, 24(1), 13.

Tuteja, A., W. Choi, M. Ma, J.M. Mabry, S.A. Mazella, G.C. Rutledge, G.H. McKinley, and R.E. Cohen. 2007. Designing superoleophobic surfaces. *Science*, 318, 1618.

Zhurkov, S.N. 1965. Kinetic concept of the strength of solids. *International Journal of Fracture Mechanics*, 1, 311.

14

Thermoplastic Polymers

Kent R. Miller
The University of Akron

Xiaojiang Wang
The University of Akron

Mark D. Soucek
The University of Akron

14.1 Introduction

Polymers can be found virtually everywhere and their potential applications are almost limitless. Polymers can be used as replacements for other materials such as wood, metal, and glass. In addition, polymers are fueling their own unique uses because many polymers have properties that cannot be harnessed from any other source. A polymer can be classified into one of three groups: thermoplastic, elastomer, or thermoset. Thermoplastics are polymers that melt upon application of heat and solidify when cooled. Elastomers are rubbery polymers that are easily stretched and can recover to their original dimensions. Finally, thermosets are polymers that are irreversibly cured into a rigid material. Out of these three groups, thermoplastics have the widest range of properties that can easily be tuned for various applications. The extensive properties that can be obtained by a thermoplastic are available due to the large variety of synthetic methods available (radical, emulsion, ionic, ring opening, etc.). Not only can preexisting thermoplastics be made through various routes, but new thermoplastics are continually being developed to replace other materials, improve properties over another polymer, and perform tasks that no other material can. The days of an all-purpose polymer are getting shorter and shorter; smart thermoplastic polymers are being produced that have one specific use and are extremely good at their job. Surface-active, biocompatible, conductive (thermal, ion, electrically), and magnetic thermoplastic polymers are just a few examples of unique polymers that are being created. Because of vast properties and uses of thermoplastics, the following discussion on degradation will focus on thermoplastic polymers.

With the production of more and more new thermoplastics, the work required to characterize each one and understand how each will react to their surrounding environment is vastly increasing. For a thermoplastic to be useful and have a long lifetime, its surrounding environment cannot easily degrade it, unless designed to do so. It is because of all the new thermoplastics being produced, along with the ever-increasing market of thermoplastic polymer products, that a good understanding of polymer degradation is required. Although each system will behave differently, it is important to understand the basic concepts of degradation so that the breakdown of a thermoplastic polymer system can be expected, planned for, and hopefully prevented.

14.2 Background

The existence of various types of thermoplastics, each with a unique structure, allows for a wide range of products to be produced. Each of these products has a different manufacturing history, which when coupled with the different structures of the polymers means that the degradation path of a polymer is a complex process that does not follow one set path. Not only do structure and the manufacturing process affect degradation, but environmental conditions also have to be taken into account. As for the initial onset of degradation, this is usually caused by the presence of impurities (leftovers from processing steps or additives) or weak linkages along the polymer backbone such as a head-to-head junction.

As the degradation process occurs, side products are formed and the original properties of the system may change. The chemical structure changes via bonds being broken and/or formed. Since most degradation takes place on the surface, the air–polymer interface will develop different properties from the bulk. The surface can start to become brittle and discolor and then crack. The mechanical properties of the polymer will decline as degradation continues. Furthermore, any additives that have been incorporated into the system can be degraded and lost causing additional breakdown of the polymer and deterioration of properties (Allen and Edge 1992).

In most thermoplastic polymer products, degradation is highly unwanted as it causes product performance to decrease over time, physical appearance diminishes, and ultimately the product will fail. To help overcome degradation problems, stabilizers are usually incorporated into the polymer to protect it. Stabilizers work by accepting destructive species present in the polymer system, preventing attack on the polymer chain. Due to proliferation of radiation, stabilizers that absorb outside radiation and dissipate the energy without damaging the polymer are widely used. Radical acceptors are also highly used because many degradation pathways involve radical intermediates (Emanuel and Buchachenko 1987, Pritchard 1998, Hamid 2000, Zweifel et al. 2009).

It is important to realize that there do exist several applications where degradation of polymers is required and/or beneficial. In drug delivery applications, medication is encapsulated by a polymer and inserted into the body. The polymer responds to the biological environment, expanding and contracting to allow the medication to diffuse out of the capsule. Once all the medication has been released, the leftover polymer will degrade into biologically safe products that can be easily passed through the body (Galaev and Mattiasson 2008). Biodegradable polymers are also very important for environmental protection and waste management. Once a material is disposed, it is beneficial for that product to be capable of breaking down into safe by-products that can be incorporated back into the environment without any adverse affects. This is very important when one considers the tremendous volume of polymer products disposed every day (Scott 2002, Bastioli and Rapra Technology Limited 2005).

Due to the fact that polymer degradation is dependent on so many factors, it is beyond the scope of this chapter to provide all the information necessary to understand and predict how every type of polymer will degrade in a given system. The purpose of this chapter is to educate the reader on several degradation mechanisms and factors that influence degradation. While reading this chapter, it is important to keep in mind that multiple degradation pathways can be happening simultaneously and degradation is not restricted to singular methods. In addition, degradation is highly contingent upon the polymer and the environmental system it is used in. By understanding the degradation pathways available and how each is initiated will help in predicting how a polymer will degrade in a given system. However, successful prediction of degradation in new and unique systems is very difficult due to numerous factors that arise in each system along with the fact that impurities are almost always present that can lead to unexpected degradation pathways.

14.3 Polymer Structure

There are many types of thermoplastics currently available and more and more are being developed each day. Numerous properties are achieved by thermoplastics through variation in their bonds, structure, side chains, and molecular weight (MW). A change in one or all of these properties will completely change the characteristics of that polymer such as the mechanical, optical, and degradation properties. The degradation effects on a polymer through changes to the structure will be considered here. An important consideration is that some changes to a polymer system, such as MW, will affect degradation differently depending on the polymer being used.

14.3.1 Chemical Bonds and Bond Substitution

The bonds that are present in a polymer are very important to the degradation process. This is because degradation implies bonds, in either the backbone or side chain, being broken. The ease with which a bond is broken depends on its energy; weaker bonds are broken first. The main input of energy to begin degradation comes from the application of heat and/or ultraviolet (UV) energy. For example, poly(tetrafluoroethylene) (PTFE) has much better thermal stability than polyethylene (PE) because C–F bonds have relatively higher dissociation energy than C–H bonds. Common polymer bonds and their bond dissociation energies are shown in Table 14.1 (Bovey and Winslow 1979, Noggle 1996, Mark 2007).

The substitution of a bond also affects how that bond will react. Bonds that are more substituted can homolytically cleave easier to form a free radical. This is because higher substitution results in a more stabilized free radical. Tertiary radicals (3°) are more stable than secondary radicals (2°), which are more stable than primary radicals (1°) (Carey and Sundberg 2007; Figure 14.1).

Due to the stability of substituted free radicals, polymers that are more highly branched may more easily form free radicals and degrade. This can be seen in the fact that highly branched low-density

TABLE 14.1 Typical Bond Dissociation Energies

Bond	Energy (kJ/mol)
C–C	284–368
C–N	293–343
C–Cl	326
C–O	350–389
$CH_3C(O)$–H	368
ROO–H	377
C–H	381–410
N–H	390
C–F	452
C=C	615
C=N	615
C≡C	812
C≡N	891

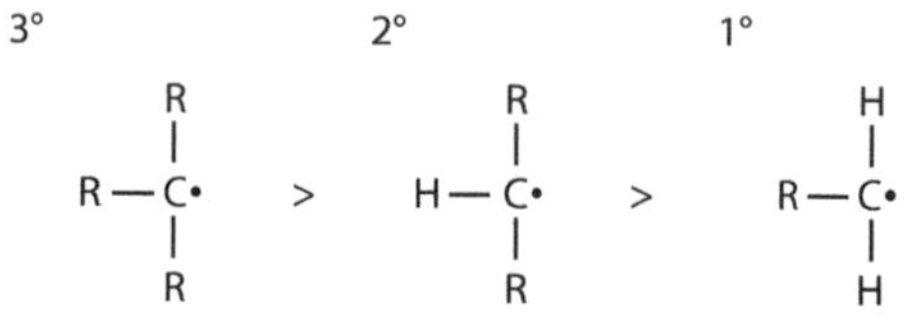

FIGURE 14.1 Free radical stability.

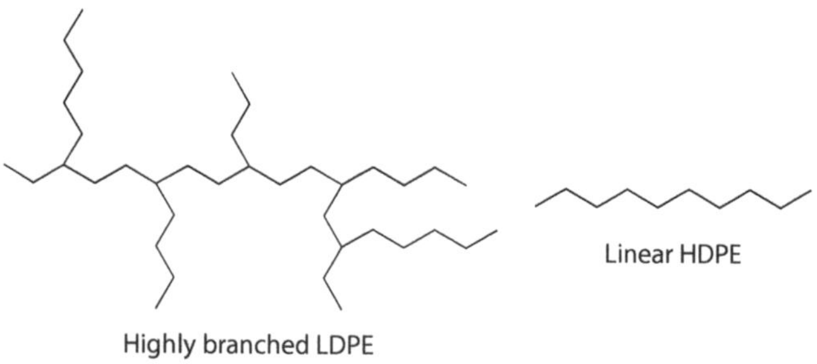

FIGURE 14.2 Branched versus linear PE.

polyethylene (LDPE) offers less resistance to thermo-oxidative degradation than the more linear high-density polyethylene (HDPE) (Wright 2001; Figure 14.2).

Additionally, branched LDPE is more susceptible to xenon arc weathering degradation than linear low-density polyethylene (LLDPE). This is due to the fact that LDPE has a relatively larger number of tertiary sites that allow free radicals to be developed more readily (Roy et al. 2010). Polylactides (PLAs) have also been shown to undergo increased enzymatic degradation and alkaline hydrolysis as the number of molecular branches of PLA is increased (Numata et al. 2007). On the other hand, for stress-induced degradation during turbulent flows, linear high MW polymers are prone to undergo chain scission more readily due to the buildup of stresses from fluid drag (Xue et al. 2005). Increasing the branching or grafting side chains to the polymeric backbone has been shown to enhance shear stability against chain scission (Agarwal et al. 1982).

For star-shaped polymers, the influence of arm length and number on enzymatic degradation of star-shaped polymers has been studied. The star-shaped poly(ε-caprolactone) (PCL) with the same MW but different arm numbers and length (one to five arms, more arm number means shorter arm length) was prepared for degradation mechanism study. The results showed that the biodegradation rate of star-shaped PCL with one to three arms increased with increasing arm number. It was explained that the decrease of the arm length leads to a decrease of crystallinity and lamellar thickness, resulting in the increase of the enzymatic biodegradation rate. However, further increase of arm number from three to five resulted in the decrease of the biodegradation rate. It was explained that the central cores of star-shaped PCLs play the dominant role in determining biodegradation behavior. The chain mobility of each arm of the star-shaped PCL was strongly limited by the central core and decreases significantly when arm number exceeds three, resulting in the decrease of the degradation rate (Xie et al. 2008).

14.3.2 Isomerism

Isomerization is the conversion of one molecule to another molecule that possesses the exact same number and type of atoms but differs in structure or configuration from the original molecule. The effect of isomerization on degradation is dependent on the polymer being used. Different isomers determine the extent of degradation, because the mobility, strength, and interactions of the polymer will all change for different isomers. A polymer can be hard and crystalline or soft and noncrystalline by just going from one isomer to another. Not only does this change the physical properties of the polymer, but the

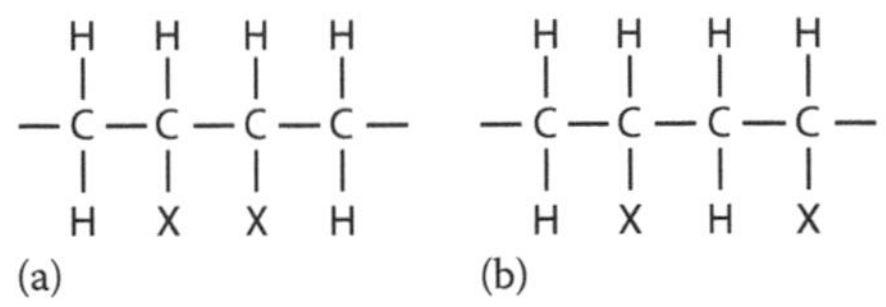

FIGURE 14.3 Configurations of vinyl polymers: (a) head-to-head, (b) head-to-tail.

degradation properties are also affected. The exact effect on degradation cannot be generalized for all types of polymers; this is because property changes upon isomerization are highly dependent on the chemical composition of the polymer. Three types of isomerism encountered with polymer are directional, geometrical, and stereochemical:

1. *Directional isomerism*: When a radical attacks an asymmetric vinyl monomer, two types of addition are possible: *head-to-head* or *head-to-tail*, as illustrated in Figure 14.3. The *head-to-tail* configuration predominates in most polymers because high steric hindrance of large side groups does not permit like ends to be next to each other.
2. *Geometrical isomerism*: Geometrical isomerism exists in polymers that contain double bonds in the polymer backbone. Such polymers have distinct structural isomers, that is, *cis* and *trans* structures. For instance, polymerization of diene monomers such as butadiene can produce polymers with varying geometrical structures (Figure 14.4).
3. *Stereochemical isomerism*: In stereoisomers, atoms are linked in the same order in a *head-to-tail* configuration but differ in their spatial arrangement. Especially for asymmetric monomers, the orientation of each monomer added to the growing chain is described by the term tacticity. Three different arrangements have been characterized: isotactic, syndiotactic, and atactic, as depicted in Figure 14.5. Isotactic polymers have the side groups on one side of the chain, syndiotactic ones have the side groups alternating, and atactic ones have the side groups arranged randomly.

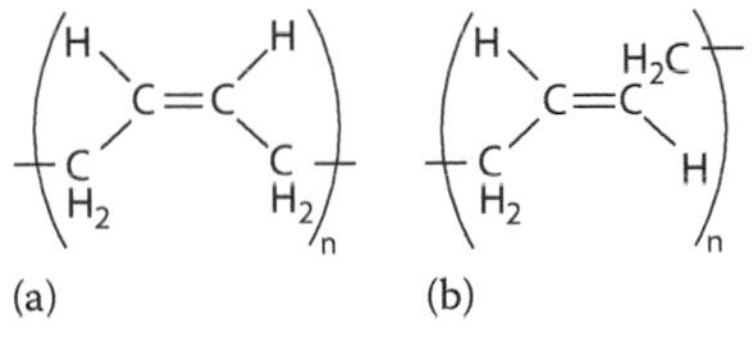

FIGURE 14.4 Geometrical isomerism of polybutadiene: (a) *cis*, (b) *trans*.

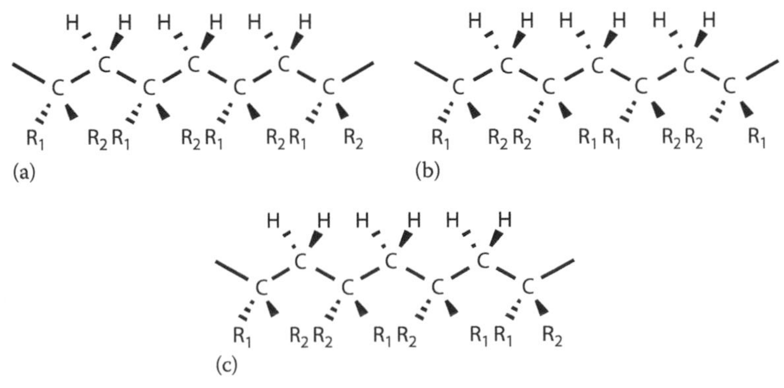

FIGURE 14.5 Stereoisomerism of vinyl polymers: (a) isotactic, (b) syndiotactic, (c) atactic.

Effects of directional isomerism: For the polystyrene (PS) system, it was found that PS containing the head-to-head unit is prone to thermal decomposition, compared with PS containing no head-to-head units (Howell et al. 2003, 2006, Howell 2007, Howell and Chaiwong 2009). The fully head-to-tail PS prepared by nitroxyl-mediated polymerization is much more thermally stable than the corresponding PS prepared by conventional radical polymerization (Howell et al. 2003). This was attributed to thermal stresses that promote homolysis of the bond linking the head-to-head components (Howell and Chaiwong 2009). The study of the thermal degradation of poly(methyl methacrylate) also suggested that head-to-head units are weak links and prone to thermal decomposition (Holland and Hay 2002).

Effects of geometrical isomerism: For a 1,4-polybutadiene system, the polybutadiene with *cis* geometrical structure is soft, can crystallize upon pronounced stretching, and has a low melting point. On the other hand, the polybutadiene with *trans* geometrical structure is hard, can crystallize readily upon stretching, and has a relatively high melting point (Stivala 1980). Gel permeation chromatography (GPC) and Fourier transform infrared spectroscopy (FTIR) analyses indicated that the rate of enzymatic degradation of the 1,4-polybutadiene is not influenced by the configurations of double bonds, *cis* and *trans* forms (Choi et al. 2009). In another study, mass chromatograph analyses indicated that the amount of the volatile products formed during thermal degradation is influenced by the *cis/trans* ratio of 1,4-polybutadienes. With increasing *trans* content, the relative quantities of volatile products of 4-vinyl-1-cyclohexene and 1,3-butadiene decreased, while cyclopentene and 1,3-cyclohexadiene increased (Tamura and Gillham 1978).

Effects of stereochemical isomerism: It has been found that the tacticity affects the thermal degradation and, therefore, the thermal stability of polymers (Millán et al. 1975, Suzuki et al. 2005, Chen et al. 2007). Taking PS as an example, it has three tactic forms: isotactic (iPS), syndiotactic (sPS), atactic (aPS), as depicted in Figure 14.6.

Chen et al. (2007) studied the thermal degradation of PS in three tactic forms using thermogravimetric analysis (TGA). To rule out the effects from factors other than tacticity, the experiments were performed on PS with similar MW and under a similar set of conditions. It was found that iPS degraded at a higher temperature compared to sPS and aPS. Meanwhile, the thermal stability of sPS was similar to that of aPS. The results were interpreted by two approaches, the activation energy and intramolecular hindrance. Isoconversional kinetic analysis of nonisothermal TGA data showed higher activation energy for degradation of iPS than for aPS and sPS. Another contribution to better thermal stability of iPS was the hindered molecular mobility that could come from the intramolecular confinement. This was proved by studying the glass transition dynamics of iPS and aPS samples. The theoretical simulations (Brückner et al. 2002, Allegra 2004) also suggested that the neighboring phenyl groups within a same polymer chain interacted stronger in iPS than in sPS. Consequently, iPS had to overcome a larger rotational energy barrier or, in other words, larger steric hindrance to degrade.

14.3.3 Molecular Weight (MW)

MW has a great effect on polymer degradation. The effects of MW on the degradation of PLA (L/D) 96/4 copolymers during melt spinning were investigated. The results showed that higher MW PLA degraded faster than that of low MW PLA. It was explained that the high MW PLA chains were subject to higher shear stress and high melt temperatures, whereas the low MW PLA was not. This is because low MW PLA has a much lower melt viscosity and therefore is affected very little by melt processing (Paakinaho et al. 2009). Low MW polymers have shorter chain lengths, meaning the chains have low entanglement, and can be separated more easily.

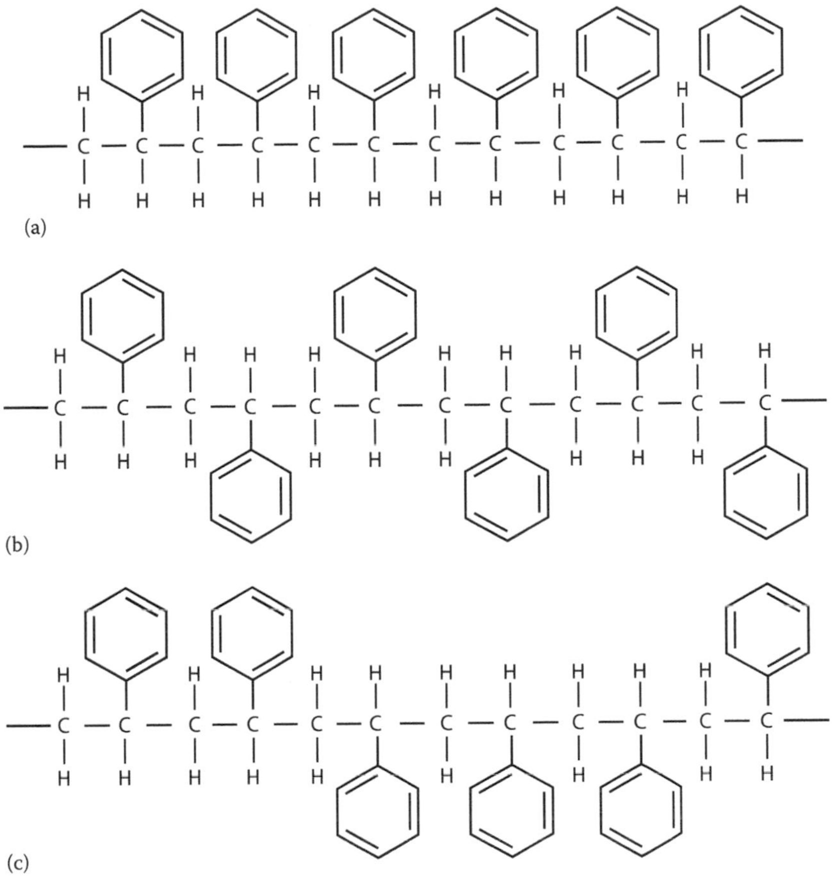

(a)

(b)

(c)

FIGURE 14.6 Stereoisomerism of PS: (a) iPS, (b) sPS, (c) aPS.

The effect of MWs on the thermal degradation kinetics of PS dissolved in mineral oil has been investigated at 275°C. The results showed that the random-scission rate coefficient dependence on MW is linear at low MWs but second order at higher MWs (Madras et al. 1997). One possible explanation is that long chains have a greater tendency to form tight curves than short chains. The resulting bond-angle strain weakens the C–C bond strength, causing a higher C–C scission rate (Vollhardt and Schore 1994).

In biomedical applications, degradation with a controllable rate is an important requirement for polymer constructs. A good approach to regulate the polymer degradation rate involves controlling the molecular weight distribution (MWD), which can be achieved by adjusting the ratio of high to low MW of polymer (Lu et al. 2009). For example, controlling the bimodal MWD of chitosan allows one to regulate degradation of chitosan (Kong et al. 2002).

14.3.4 Copolymer Sequences

We have limited the discussion of polymer microstructure to homopolymers obtained by polymerization of a single monomer. Besides these types of polymers, two or more different monomers can react to form copolymers. The comonomers may be combined at random, in regular alternation, or in block or graft structures (see Figure 14.7; Sperling 2006). Aside from the comonomer sequence and method of attachment, copolymer microstructure is also subject to the effects of stereo- and regio-sequence as

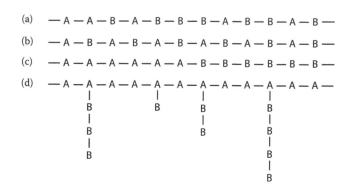

FIGURE 14.7   Types of copolymers: (a) random, (b) alternating, (c) block, (d) graft.

discussed previously. The physical and chemical properties of copolymer materials can be regarded as a combination of the properties of the two polymers in the "mixture."

The alternative, block, and random copolymers, even with the same composition, usually show different thermal stability (Zhou 2009). The thermal degradation of three polyethersulfone (PES) and polyetheretherethersulfone (PEES) copolymers with the same composition but different sequence distribution was investigated. The results showed that the average apparent activation energies of the isothermal degradations of the various copolymers were different from each other. The alternative copolymer had the highest, the block copolymer had the medium, and random copolymer had the lowest apparent activation energies. It was explained that the alternative copolymer had a more normalized chain than random or block copolymers. So the thermal stability of three copolymers decreased according to the following order: alternative > block > random (Zhou et al. 2007, Zhou 2009). Another biodegradation study of the copolymer poly(1,5-dioxepan-2-one) with PLA showed similar results; the block copolymer was more stable than the corresponding random copolymer (Watanabe et al. 2004).

14.3.5 Polymer Morphology

Thermoplastic polymers that are crystalline have higher thermal, oxidative, and chemical stabilities over ones that are amorphous. This is because crystalline polymers require higher energy inputs to overcome their high inter- and intramolecular forces. The degradation of amorphous polymers occurs quicker due to the faster diffusion rate of molecules, such as oxygen, in the amorphous polymer. For semicrystalline polymers (spherulites contained in an amorphous matrix), degradation mainly occurs in the amorphous sections. Oxygen diffuses throughout the amorphous regions, causing oxidative damage at the spherulite boundaries. Because of this process, higher crystalline materials will be damaged much more due to oxidation of the spherulite boundaries than that of lower crystalline materials since much of the adhesive force will be lost. This can be seen in the oxidation of HDPE and LDPE. HDPE, with its higher crystallinity and lower number of tertiary sites, oxidizes slower than LDPE, but the physical degradation by small amounts of oxidation is much greater in HDPE than in LDPE (Scott 2002).

In the crystalline regions, the degradation rate of biomaterials was affected by the lamellar crystal size, crystal morphology, and crystal structure. Single crystals have been used as a very useful model substrate to investigate the degradation mechanism in the crystalline region. The study of enzymatic degradation of several aliphatic polyester single crystals showed that degradation occurred mainly from the crystal edges rather than at the crystal surfaces (Iwata and Doi 1999). The study of enzymatic degradation of poly(3-hydroxybutyrate) (P3HB) showed that at a constant overall degree of crystallinity, the degradation rate decreases with an increase in the average size of the crystals (Tomasi et al. 1996). Another study showed that crystalline structure plays an important role in the degradation of P3HB. The results showed that both the crystallinity and the lamellar thickness of P3HP samples decrease roughly in the order of β-form ≥ γ-form > δ-form, and the degradation rates also decrease in the same order.

It indicated that the crystalline structure is one of the most important factors (more important than the crystallinity and lamellar thickness in this case) in controlling the enzymatic degradation of P3HP (Zhu et al. 2008).

14.4 Degradation Principles for Material

14.4.1 Thermodynamics

As with any chemical reaction, polymeric degradation is driven by the change in free energy, ΔG. For a reaction to occur spontaneously, ΔG must be less than 0. The change in the Gibbs free energy is defined as

$$\Delta G^\circ = \Delta H^\circ - T\Delta S^\circ = -RT \ln K \tag{14.1}$$

where
  ΔH is the change in enthalpy
  T is the temperature (°K)
  ΔS is the change in entropy
  K is the equilibrium constant

Since most polymer degradation is endothermic ($\Delta H > 0$), ΔG becomes negative whenever there is an increase in entropy caused by the breakup of the polymer chain into smaller chains and monomer units.

The activation energy for polymer degradation is largely dependent on the strength of the chemical bonds making up the polymer chain. As a polymer system heats up, the vibration energy increases and can cause bond breakage. Therefore, bonds with higher dissociative energies provide greater thermal stability. The activation energy for polymer degradation, which is the energy required to break a single bond, is usually between 165 and 420 kJ/mol. This energy corresponds to radiation wavelengths of 720–280 nm. Based on these degradation energies, it can easily be seen that near UV radiation, 300–400 nm wavelengths, given off by sunlight or artificial sources, can easily lead to photochemical degradation (Krevelen and Te Nijenhuis 2009).

Since temperature change plays an important role in degradation, the ceiling temperature is an important parameter to know. The ceiling temperature is the temperature at which the rate of polymerization equals the rate of depolymerization (Scheme 14.1).

P_n is a polymer chain with n number of monomers, M is a monomer, k_p is the rate of polymerization, and k_d is the rate of depolymerization. Initially the rate of polymerization is highest, but as the temperature is raised, the rate of depolymerization increases. At equilibrium, the ΔG for the reaction will be zero. The ceiling temperature is calculated by setting $\Delta G = 0$ and defining the equilibrium constant, K, as k_p/k_d in Equation 14.1 and solving for T:

$$K = \frac{k_p}{k_d} = \frac{[P_{n+1}]}{[P_n][M]} = \frac{1}{[M]_c} \tag{14.2}$$

$$T_c = \frac{\Delta H^\circ}{\Delta S^\circ + R\ln[M]_c} \tag{14.3}$$

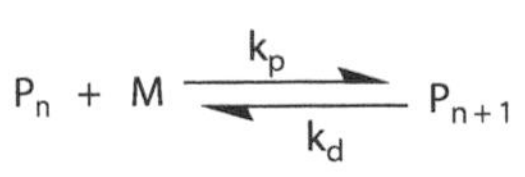

$$P_n + M \underset{k_d}{\overset{k_p}{\rightleftharpoons}} P_{n+1}$$

SCHEME 14.1 Polymerization reaction.

and

$$[M]_c = \exp\left(\frac{\Delta H^\circ}{RT_c} - \frac{\Delta S^\circ}{R}\right) \tag{14.4}$$

where
   $[M]_c$ is the equilibrium monomer concentration
   T_c is the ceiling temperature

Inspection of Equations 14.3 and 14.4 shows that at a set monomer concentration, there exists a certain temperature, T_c, where polymerization will not occur. If an active degradation site is produced, the polymer chain will readily depolymerize to a monomer concentration of $[M]_c$ for the given temperature. It is important to note that other polymer degradation pathways exist below the ceiling temperature besides just depolymerization.

14.4.2 Kinetics

Polymer degradation can happen through many mechanisms; examples include the following: chemical, thermal, photolytic, biodegradation, electrochemical, and mechanical. Degradation is usually a mixture of the various types of mechanisms as opposed to only one. A common degradation method is one that is initiated by an active species, giving rise to a bimolecular reaction (Scheme 14.2).

The first stage of the reaction is the approach of the reacting species A and B. The second is the formation of a contact pair between the two reacting species (A–B) that then produces the products C and D. The kinetic rate equations are

$$-\frac{d[A]}{dt} = -\frac{d[B]}{dt} = k_1[A][B] - k_{-1}[A-B] \tag{14.5}$$

$$-\frac{d[A-B]}{dt} = k_{-1}[A-B] + k_2[A-B] - k_1[A][B] \tag{14.6}$$

If the first stage, species approach, is the limiting stage, then the reaction is diffusion limited. On the other hand, if the contact pair is the limiting stage, the reaction is kinetically limited. By far the most prevalent active species that cause degradation are free radicals. Free radical degradation is a diffusion-limited process.

Naturally occurring peroxides are one of the major species that form and react with free radicals, giving rise to oxidative degradation of the polymer. Polymer radicals can also react with singlet oxygen to form peroxy radicals, causing further degradation of the polymer. The pathways for oxidative degradation are shown in Scheme 14.3.

The initial formation of radicals from a polymer chain, PH, is generally accomplished through the adsorption of energy (thermal, photo, or radiation based) by a polymer. The absorbed energy generates an excited species that can decompose into radicals. Additionally, any peroxides present in the polymer matrix can easily absorb energy and decompose into radicals. This type of oxidative degradation results in lower MW polymer chains, due to chain scission, as well as the production of ketones and

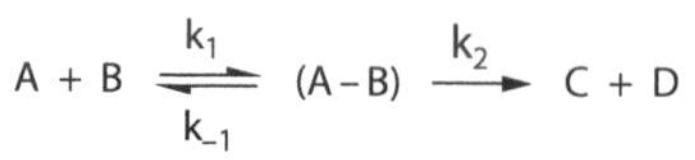

$$A + B \underset{k_{-1}}{\overset{k_1}{\rightleftharpoons}} (A-B) \xrightarrow{k_2} C + D$$

SCHEME 14.2   Bimolecular reaction.

Initiation

$$PH \longrightarrow P\cdot + H\cdot$$

$$POOH \longrightarrow P\cdot + \cdot OOH$$

$$POOH \longrightarrow PO\cdot + \cdot OH$$

Propagation

$$P\cdot + O_2 \longrightarrow POO\cdot$$

$$POO\cdot + PH \longrightarrow POOH + P\cdot$$

$$PO\cdot(\cdot OH) + PH \longrightarrow POH\,(H_2O) + P\cdot$$

Termination

$$2POO\cdot \longrightarrow POOP + O_2$$

$$2P\cdot \longrightarrow PP$$

$$POO\cdot + H\cdot \longrightarrow POOH$$

$$2POO\cdot \longrightarrow \text{Ketones and alcohols}$$

SCHEME 14.3 Oxidative degradation reactions.

alcohols (Hamid et al. 1992). Owing to the damaging effects of oxidation, stabilizers are usually added to polymer systems to suppress oxidative degradation.

Since most polymers are exposed to UV radiation sources, photoinduced degradation is a big problem. Photoinduced degradation begins when a molecule absorbs radiation energy. Upon absorbing energy, a molecule enters an excited state. This short-lived state will dissipate its excess energy through one of many pathways. Pathways include but are not limited to conversion to thermal energy, conversion between states, dissipation of radiation, intermolecular rearrangement, and free radical formation. The lifetime in an excited state is given by

$$\tau = \frac{1}{\sum_i k_i} \tag{14.7}$$

where k_i is the rate constant of any pathway for energy dissipation (Guillet 1985). The chromophores that contribute the most to photodegradation are usually impurities that contain carbonyl groups, peroxides, metallic species, aromatics, and/or polymer–oxygen complexes (Osawa 1992). Once these chromophores are excited, dissipation into free radicals can happen, leading to photooxidation. See aforementioned oxidative degradation for reaction pathways.

Besides radiation, degradation can be initiated via an electric field. As electric current transverses a polymer film, it causes the polymer to heat up due to the electrical resistance of the polymer. Kinetic energy is transferred from the electrons as they collide with the material; this is known as Joule heating (Dissado and Fothergill 1992, Plawsky 2001, Karniadakis et al. 2005). Ohm's law gives the conversion of electrical energy into heat (Q), Joule heating:

$$Q = \frac{V^2}{R} \tag{14.8}$$

where
  V is the voltage
  R is the resistance of the polymer

In addition to initiating degradation, the passage of electric current through a polymer can increase the degradation rate initiated by other means. This is especially damaging to a polymer if hydrolysis is happening due to the enhanced dissociation of neutral species into ionic ones (Dissado and Fothergill 1992).

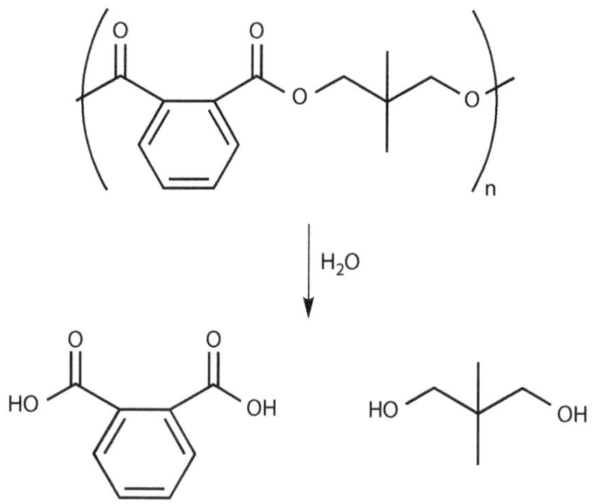

FIGURE 14.8 Hydrolysis of a polyester.

Hydrolysis is the process of a molecule reacting with water (H^+ and ^-OH) and forming two different molecular species. Hydrolysis affects many polymers due to the presence of hydrolyzable bonds in many polymer backbones, such as polyesters, polycarbonates, polyureas, polyurethanes (PUs), polyamides, and polyethers. The reaction mechanism for hydrolytic degradation of a polyester is shown in Figure 14.8.

Although hydrolysis can occur quickly in low MW systems, it is generally a slow process. This is because the attack of hydrolyzable bonds usually takes place on the surface of the bulk polymer due to, in most cases, the hydrophobic nature of organic polymer chains. Subsequent diffusion of an aqueous solution from the polymer surface into the bulk is very sluggish (Grassie and Scott 1985). For polymers with very low water diffusion coefficients, degradation can be assumed to only take place on the surface and the hydrolytic degradation rate (Chu et al. 1997) can be given as follows:

$$\frac{dn}{dt} = K\left(\frac{C_p C_{cat}}{Z}\right)S \tag{14.9}$$

where
    n is the number of bonds capable of hydrolytic scission
    K is the rate constant of hydrolytic scissionable bonds
    C_p is the surface concentration of the hydrolytic scissionable bonds
    C_{cat} is the concentration of any catalyzing agent on the surface
    Z is dependent on the macromolecular packing
    S is the surface area of the polymer

The hydrolysis rate constant is greatly affected by the pH of the aqueous solution where the reaction takes place. Both high and low pH can increase the rate of hydrolysis depending on the behavior of the catalyst in acidic/basic media. Furthermore, the polymer itself can influence the hydrolysis as a function of pH. Low pH can give rise to autocatalysis (e.g., as with PLA) in which monomers are produced from the polymer chain with carboxylic end groups causing the pH to decrease more, further increasing the reaction rate (Göpferich 1996).

An alternative degradation pathway is biodegradation. Biological microorganisms such as bacteria and fungi can eat away at a polymeric material, leading to biodegradation. Additionally, biological organisms can eat additives present in a polymer (e.g., stabilizers) permitting the polymer to degrade much easier through other pathways. Biodegradation is also the process of enzyme-catalyzed

$$A + E \rightleftharpoons E \cdot A \longrightarrow P + E$$

SCHEME 14.4 Active intermediate formation during enzyme reaction.

degradation. Enzymes, which are very common in biological organisms, are protein or protein-like high MW molecules that enable a reaction to occur at a highly accelerated rate. Activation energy is lowered by the enzyme due to the alternate reaction pathway it provides. During the course of the reaction, the enzyme (E) first forms a complex, active intermediate with the reactant (A) and then forms the product (P); the enzyme is not destroyed during this process (Scheme 14.4).

Working through the rate equations of the aforementioned reaction with the assumption that the enzyme concentration is constant and all enzymes molecules are participating in the reaction, the Michaelis–Menten equation is obtained:

$$\frac{dP}{dt} = \frac{V[A]}{K_m + [A]} \tag{14.10}$$

where
  V is the maximum rate of the reaction
  K_m is the Michaelis constant

It is important to note that enzymes are bond specific; enzymes only participate in one type of reaction. However, for the reaction that an enzyme does take place in, the rate is drastically increased by several orders of magnitude. Enzymes will continue to enable reactions until they are denatured, lose an active site, or unfold.

14.5 Degradation of Specific Systems

An exhaustive survey on the degradation behavior of the multitude polymer families is beyond the scope of this chapter. Instead, both familiar and representative polymers are selected for discussion on polymer degradation behavior.

14.5.1 One Polymeric Material System

PE is a widely used thermoplastic polymer and holds a special position as a packaging film because of its desirable mechanical properties and reasonable cost. The nature of PE degradation has been investigated many times in the past decades. PE is degraded by environmental factors such as thermal, oxidative, and photochemical. Besides environmental conditions, the degradation process depends significantly on the variables such as the extent and type of branching, crystalline structure, and MW of PE.

14.5.1.1 Thermal Degradation

Thermal degradation includes thermal decomposition if oxygen is not involved and thermal oxidation if oxygen is involved. There are three types of thermal degradation processes: random scission, chain depolymerization, and substituent reactions. Thermal decomposition of PE typically starts with a random scission. During random scission, the polymer chain breaks down at random points along the chain (Mark 2007).

The PE thermal degradation mechanism in the absence of oxygen is shown in Scheme 14.5 (Bockhorn et al. 1999, Moldoveanu 2005, Ito and Nagai 2008, Vinu et al. 2010). The PE thermal degradation starts with random scission to generate highly reactive polymeric radicals that are surrounded by secondary hydrogens. The propagation reaction can take place by a combination of intramolecular hydrogen transfers with the generation of more stable polymeric radical chains, intermolecular hydrogen transfer with

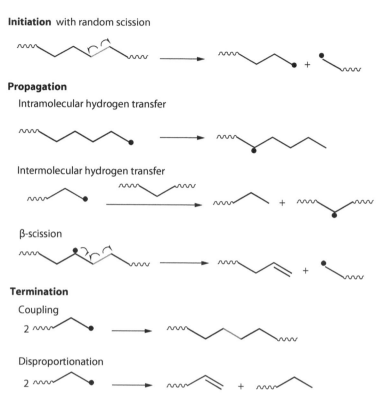

Initiation with random scission

Propagation

Intramolecular hydrogen transfer

Intermolecular hydrogen transfer

β-scission

Termination

Coupling

Disproportionation

SCHEME 14.5   Radical mechanism of the thermal degradation of PE in the absence of oxygen.

the generation of new polymeric radical chains and stable molecules, or β-scission with the generation of new polymeric radical chains and α-olefins. Termination reactions can also take place to generate stable molecules such as alkanes and α-olefins by a combination of coupling and disproportionation (Moldoveanu 2005).

14.5.1.2 Thermo-Oxidative Degradation

The study of PE's thermal decomposition is important in relation to the polymers' resistance to heating (Dolezal et al. 2001). The typical reactions during the PE thermo-oxidative degradation process are shown in Scheme 14.6 (Bamford and Tipper 1975, Gugumus 1997, 2002a,b, 2005).

Similar to thermal decomposition, the initiation reaction in the thermo-oxidative degradation is also the generation of the polymeric radicals. The polymeric radicals are then converted into highly active polymeric peroxy radicals by reaction with oxygen. These polymeric peroxy radicals have been detected by electron spin resonance (ESR) (Chien and Boss 1967). The polymeric peroxy radical attacks a hydrogen atom in another polymer molecule, leading to intermolecular hydrogen atom abstraction. Hydrogen abstraction, especially intramolecular hydrogen abstraction, by peroxy radicals is by far the dominant propagation reaction during the PE thermo-oxidative degradation process (Gugumus 2002a,b). The hemolytic decomposition of the polymer hydroperoxides generates polymeric alkoxy radicals. The polymeric alkoxy radicals can split by the β-scission process, which plays an important role in aldehyde group and in alkyl radical formation at the end of the chain (Adams 1970, Adams and Goodrich 1970). The carbonyl groups have been detected by infrared (IR) spectroscopy (Satoto et al. 1997). The polymeric alkoxy radicals (PO·) react with other molecules (PH) in the polymer and abstract hydrogen atoms, leading to the formation of hydroxyl groups (POH) and new macroradicals (P·). The hydroxyl groups have also been detected by IR spectroscopy (Heacock et al. 1968, Bamford

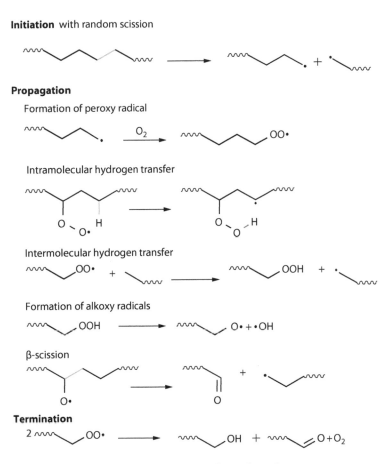

Initiation with random scission

Propagation
Formation of peroxy radical

Intramolecular hydrogen transfer

Intermolecular hydrogen transfer

Formation of alkoxy radicals

β-scission

Termination

SCHEME 14.6 Typical reactions during the PE thermo-oxidative degradation process.

and Tipper 1975). Chain termination is usually the reaction of two polymeric peroxy radicals with each other, if the oxygen concentration is sufficient for rapid transformation of alkyl radicals into polymeric peroxy radicals (Gugumus 2002a,b).

14.5.1.3 Photooxidative Degradation

The mechanisms of PE photooxidative degradation have been previously outlined (Gugumus 1990). Under normal conditions, photooxidative and thermo-oxidative degradations are similar. The main differences are the initiation steps and the fact that thermo-oxidative degradation occurs throughout the bulk of the polymer, whereas photooxidative degradation occurs only on the surface (Singh and Sharma 2008).

Direct photolysis of the C–C and C–H bond is impossible, because solar radiation cannot be absorbed by pure PE and does not have enough energy to break the C–C or C–H bond of PE (Wypych 2003). Initiation may be caused by the following reasons (Bamford and Tipper 1975, Wypych 2003, Ito and Nagai 2008, Singh and Sharma 2008):

1. *Catalyst residue*: The transition metal (Ti) as a polymerization catalyst may remain in the PE and generate radicals (Scheme 14.7).
2. *Carbonyl groups*: Carbonyl groups are formed due to the oxidation of PE during synthesis or processing. The PE polymer chain containing the carbonyl group can absorb UV and generate radicals via Norrish type I and II reactions (Scheme 14.8).

$$TiCl_4 \xrightarrow{h\upsilon} TiCl_3 + Cl\cdot$$

$$Cl\cdot + PH \xrightarrow{h\upsilon} P\cdot + ClH_2$$

SCHEME 14.7 Radical generation by Ti during photodegradation processes.

Norrish type I (α-scission)

Norrish type II (β-scission)

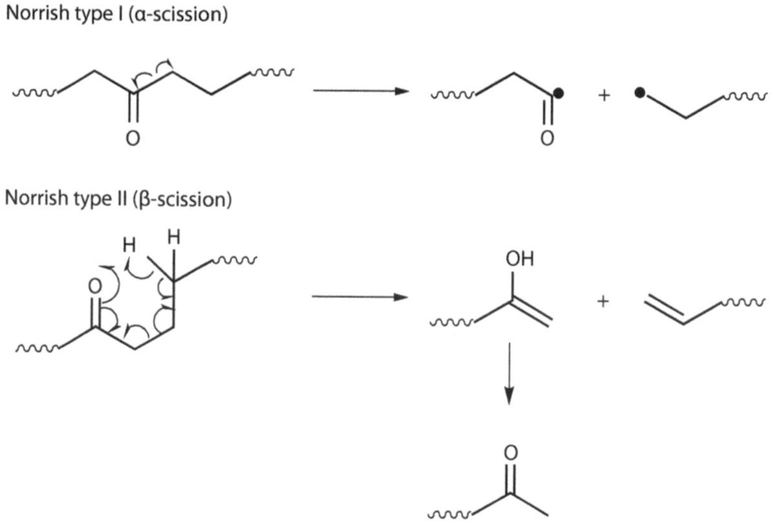

SCHEME 14.8 Radical generation via Norrish-type reaction during photodegradation processes.

14.5.2 Mixture of Polymeric Materials

As polymer blends have been widely used, their degradation has been studied to understand the relationship between polymer blend miscibility and degradation (Utracki 2002). For most blended systems, the total degradation rate is between the degradation rates of each polymer component. A free radical is first generated in the less stable polymer that can cause the more stable polymer to degrade (Wypych 2003). In some cases, the polymer blend has worse stability than the sum of the components; polypropylene (PP)/PE blend is an example (Sadrmohaghegh et al. 1981). In another case, polymer blends have better stability than the sum of the components, such as polyvinyl chloride (PVC)/PU blend (Osawa et al. 1994).

14.5.3 Interaction with Pigments, Dyes, and Fillers

Pigments, dyes, and fillers are capable of changing the properties of the polymer phase, which is why it is important to know how the polymer will react when these are added. Either pigments, dyes, and fillers can catalyze polymer degradation or degradation can be catalyzed by any impurities present. Degradation from pigments, dyes, and fillers can be due to the fact that these additives are able to change the thermal stability of the polymer, absorb any stabilizers present such as antioxidants, and provide nucleation sites for polymer crystallization. Conversely, pigments, dyes, and fillers can help stabilize polymers from degradation by absorbing UV, moisture, and heat. Mechanical properties can also be greatly altered. But unlike fillers, pigments/dyes are usually added in such small quantities that the mechanical properties of the polymer are not significantly impacted (Rothon and Rapra Technology Limited 2003).

Pigments and dyes are mainly used to provide color to a polymer product to make it cosmetically more appealing. Besides just affecting the visual aspect of the polymer, pigments, and dyes, the photostability of the polymer product itself is affected. Effects can be positive, screening of UV light, or negative, acceleration/catalyzing photochemical degradation. The main factors influencing how the added pigment/dye will react with the polymeric system are the particle size distribution, surface treatment of particles (in the case of pigments), and any bonding that may be present between the polymer and the pigment/dye.

When dyes are used in a polymer, an important property to consider is the lightfastness of the dye. Lightfastness is the fading resistance that a dye possesses. Typically, aggregation of dyes results in higher lightfastness over that of fully dispersed dyes. Additionally, lightfastness can be reduced due to the presence of chromophores left over from the production of the polymeric product. This is counteracted by a higher dye concentration, so the chromophore activity will be quenched. Apart from dye fading, photo-tendering is another important concern. Phototendering is when polymer degradation is accelerated or sensitized as a result of added dye or pigment. Two mechanisms have been developed that describe the photodegradation/oxidation of a polymer system with added dye or pigment. One mechanism is based on a reaction with oxygen and the other involves hydrogen abstraction (Allen 1994).

In the first mechanism, proposed by Egerton (1947), radiation is absorbed by the dye/pigment propelling it into an excited state. This excited state, usually the triplet state, is quenched by ground-state oxygen, producing singlet oxygen. The singlet oxygen then goes on to react with water or the polymer to form peroxides that will result in additional oxidative degradation of the polymer. For this reaction, see Scheme 14.9; D represents the dye/pigment and P the polymer.

For the second mechanism, proposed by Bamford and Dewar (1949), the excited dye/pigment molecule abstracts hydrogen from the polymer chain or water. A free radical is produced on the polymer chain that can react with oxygen to form a peroxy radical. This will then abstract hydrogen, forming a free radical on the polymer chain, and so on. The photodegradation mechanism that each dye/pigment follows is dependent upon its chemical composition (Scheme 14.10).

Photodegradation and fading of dyes are very important due to the wide use of dyes in the textile industry where the dyes provide color to polymer fibers. Polyester, a widely used fiber, is often dyed with an azobenzene dye. Azobenzene compounds absorb UV light and undergo electronic $n \rightarrow \pi^*$ and $\pi \rightarrow \pi^*$ transitions. Since the low energy transition of azobenzene is forbidden, an electron-releasing group such as diethylamino can be added to make this transition highly favorable. Additionally, these dyes are quite stable because the main photodeactivation pathway from an excited state is *trans–cis* isomerization.

$$D \longrightarrow {}^1D^* \longrightarrow D^{3*}$$
$$D^{3*} + O_2 \longrightarrow D + {}^1O_2$$
$${}^1O_2 + PH \longrightarrow \text{Oxidation products}$$
$${}^1O_2 + 2H_2O \longrightarrow 2H_2O_2$$
$$H_2O_2 + PH \longrightarrow \text{Oxidation products}$$

SCHEME 14.9 Photodegradation of dye/pigment with production of singlet oxygen.

$$D \longrightarrow D^*$$
$$D^* + PH \longrightarrow DH\cdot + P\cdot$$
$$P\cdot + O_2 \longrightarrow PO_2\cdot$$
$$PO_2\cdot + PH \longrightarrow PO_2H + P\cdot$$
$$PO_2\cdot + DH \longrightarrow PO_2H + D$$
$$D^* + DH^- \longrightarrow D\cdot^- + HO\cdot$$
$$HO\cdot + PH \longrightarrow H_2O + P\cdot$$

SCHEME 14.10 Photodegradation of dye/pigment with hydrogen abstraction.

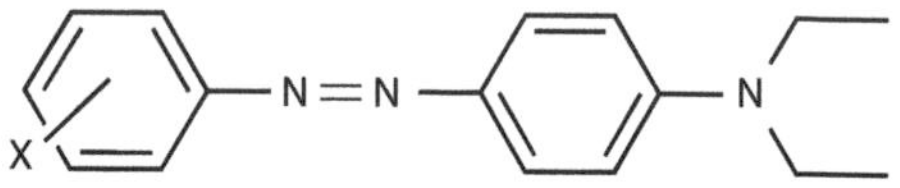

FIGURE 14.9 4-Diethylaminoazobenzene.

As such, substituted 4-diethylaminoazobenzene (Figure 14.9) dyes are commonly used. The photodegradation of the 4-diethylaminoazobenzene is due to the photoreduction of the azo group that will yield hydrozo derivatives and substituted anilines (Caronna et al. 2001).

Determining the photodegradation rate of dyes requires taking multiple factors into account: solvent present, environmental conditions, and dye interactions with the polymer. For substituted 4-diethylaminoazobenzene, the solvent is very important. Methanol and n-hexane both allow the dye to remain stable, whereas acetone causes the dyes to photodegrade. Photodegradation begins due to the photoinduced transfer of an electron from the azo dye to acetone (Scheme 14.11). From here, an ethyl group is lost from the N,N-diethylamino group. This product can further degrade to give lower MW species by additional electron transfer. Since the photodegradation involves free radicals, the presence of molecular oxygen acts as a strong inhibitor. Molecular oxygen can quench the excited state of the dye; this results in the formation of singlet oxygen while the dye returns to its ground state. Photodegradation half-lives for substituted 4-diethylaminoazobenzene under oxygen

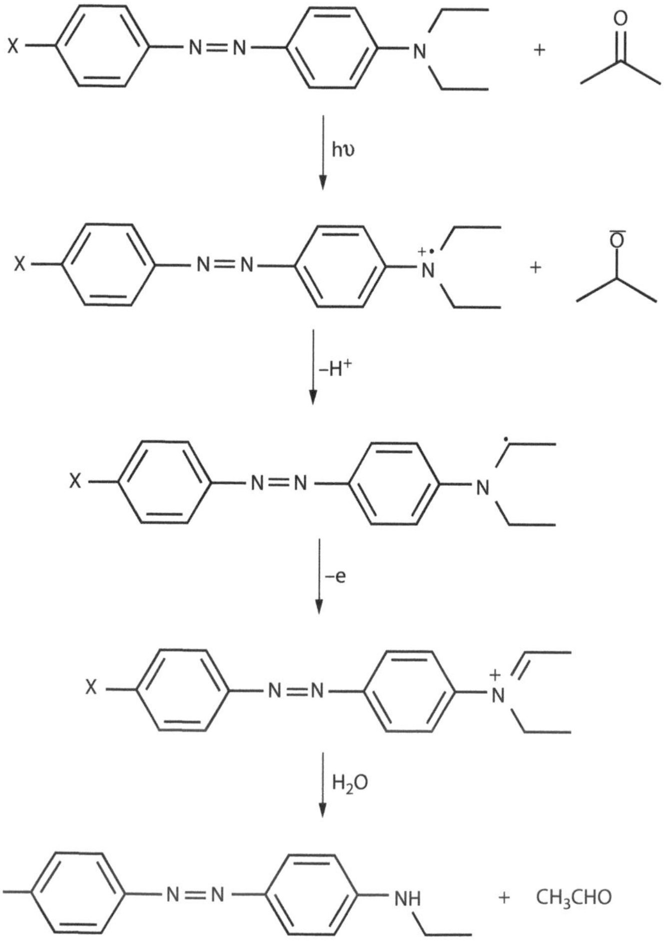

SCHEME 14.11 Photodegradation of 4-dithylaminoazobenzene in acetone.

TABLE 14.2 Photodegradation Half-Lives and % Degradation of Substituted 4-Dethylaminoazobenzene in Acetone and Solid Crystalline Form

	Under O_2	Under N_2
Photodegradation Half-Lives of Dyes in Acetone		
Compound (X)	$t_{1/2}$ min	$t_{1/2}$ min
H	90	70
p-OCH$_3$	90	12
p-NO$_2$	290	15
p-COOH	110	65
o-NO$_2$	70	70
Photodegradation Percentage in Solid Crystalline Form		
Compound (X)	% Photodegradation	% Photodegradation
H	4	9
p-OCH$_3$	38	58
p-NO$_2$	14	25
p-COOH	27	36
o-NO$_2$	5	9

Source: Caronna, T. et al., *Dyes Pigm.*, 49(2), 127, 2001.

and nitrogen are shown in Table 14.2. It can be seen that the presence of molecular oxygen will indeed inhibit photodegradation both in acetone and in crystalline forms (Caronna et al. 2001).

A widely used pigment is titanium dioxide, TiO_2. Like many pigments, TiO_2 strongly absorbs UV radiation. This is beneficial because it shields the polymer from the damaging effects of UV. Conversely, when TiO_2 absorbs UV, it can react with oxygen and water leading to the formation of oxidants. The oxidants will degrade the polymer and cause a coating to chalk. Chalking is the degradation of the outer surface of polymer and subsequent rubbing off of exposed and unbound pigment particles. The reaction in Scheme 14.12 shows the formation of oxidants from TiO_2.

In the previous reaction scheme, (e/p) represents a separated electron/hole pair, e is an electron, and p is a hole (Wicks 2007). Since TiO_2 reacts with singlet oxygen, this reaction mechanism follows that proposed by Egerton.

To alleviate the harmful effects of TiO_2 and keep the beneficial aspects of it, various surface treatments and/or coatings have been used on TiO_2 particles to prevent enhanced polymer degradation due to that pigment. It has been found that using untreated TiO_2 pigment resulted in large amounts of pitting and stripping of the polymer surface due to the high photoreactivity of TiO_2. For TiO_2 pigments surface treated with Al_2O_3 and Al_2O_3/SiO_2, the rate of oxidative degradation was much less (Watson et al. 2008). The surface treatment decreases the amount of conversion of excited-state TiO_2 into an oxidant. Since surface treatment or coating of TiO_2 decreases the oxidative degradation effect of the pigment, it is rare to find industrial pigments where the TiO_2 particles are uncoated.

$$TiO_2 \xrightarrow{h\upsilon} TiO_2^* {}_{(e/p)}$$

$$TiO_2^* {}_{(e/p)} + O_2 \longrightarrow TiO_{2(p)} + O_2{}^{\cdot-}$$

$$TiO_{2(p)} + H_2O \longrightarrow TiO_2 + H^+ + HO\cdot$$

$$H^+ + O_2{}^{\cdot-} \longrightarrow HOO\cdot$$

$$2HOO \longrightarrow H_2O_2 + O_2$$

$$TiO_2^* {}_{(e/p)} + H_2O_2 \longrightarrow TiO_2 + 2HO\cdot$$

SCHEME 14.12 Oxidant formation from irradiation of TiO_2.

Fillers are added to a multitude of polymers to help improve mechanical properties and decrease the costs of the product, since fillers are generally much cheaper than the polymer. Additionally, fillers are added to alter barrier, optical, thermal, electrical, and other properties. Although there are many benefits of adding filler, fillers can sometimes lead to enhanced breakdown and degradation of the polymer. The extent of degradation is dependent on the type of filler, the type of polymer, and the environment.

With the incorporation of fillers, stabilization/degradation can be quite similar or very different from that for pigments; it all depends on the chemical makeup of the filler. One commonly used filler in polymer systems is clay. There are many types of clays and each one reacts differently with different polymers. In one study (Zhou and Xanthos 2010), the hydrolytic degradation of PLA as a function of clay type was investigated. The clays chosen were cationic montmorillonite (MMT), a smectite cationic clay with negatively charged layers and cations in the interlayers, and anionic hydrotalcite (HT), an anionic clay with positively charged layers and anions in the interlayers. It was found that both types of clay resulted in a lower hydrolytic degradation rate. This was due to the ability of the clays to absorb excess water present in the polymer system. Furthermore, these alkaline clays help keep the pH higher than that of unfilled PLA. This reduces the autocatalytic effect of PLA hydrolysis; chain degradation causes a decrease in pH due to the formation of polymer chains with carboxylic end groups, resulting in an increase in the hydrolytic degradation rate.

In another study (Yang et al. 2005), the photodegradation of HDPE with various inorganic fillers was investigated. By measuring the formation of carboxylic acid groups, the main oxidative product, the extent of photooxidation of each system was determined. It was found that $CaCO_3$, talc, and wollastonite had little to no effect on the photooxidation of HDPE as there were virtually no oxidative products formed; these fillers act as stabilizers. Conversely, black mica, mica, kaolin, and diatomite all lead to the formation of oxidative products with kaolin and diatomite producing the most. The photooxidation of HDPE by the fillers is related to the UV reflectance of the filler. Fillers that reflect UV light can help protect the polymer, while those that absorb UV light can act as photooxidative promoters. $CaCO_3$, talc, and wollastonite all had high UV reflectance or low absorption, while the others had low reflectance or high absorbance. Wollastonite was the lowest with 18.7% UV absorbance and kaolin was the highest with 78.5% UV absorbance. These results show that, indeed, fillers with the most UV absorption will promote photooxidation, while those with little to no UV absorption can act as photooxidative stabilizers.

Although the aforementioned additives are widely used, there are numerous other pigments and fillers available. Each one has its own benefits and drawbacks depending on the polymer chain structure, use of product, and the surrounding environmental conditions. Since there are so many variations in pigments and fillers, it is important to evaluate each one individually to determine their stabilization and/or degradation effects.

14.5.4 Interaction of Polymers with Substrates

Adequate adhesion between the polymer, particularly coatings, and the substrate is critical to prevent the failure of coatings or exposure of the substrate to aggressive environments. Consequently, substrate corrosion takes place whenever there is lack of protection by the coating. There are two types of interaction between the polymer and substrate, namely, physical and chemical interaction. Physical interaction or mechanical interaction refers to the penetration of polymers into the microscopic or even macroscopic crevices of the rough substrate surface. Chemical interaction includes primary bonds (covalent and ionic) and secondary bonds (van der Waals forces, dipole interaction, and dispersion forces) (Sørensen et al. 2009).

Metals or metal alloys are widely used in high-rise buildings, bridge constructions, automobiles, appliances, tools, pipes, railroad tracks, etc., due to their high structural strength per unit mass. However, metal substrates are usually inclined to oxidize and finally corrode away. Hence, coatings are necessary for corrosion protection of many of these structural metals. Most metallic surfaces are covered with a thin oxide layer after environmental exposure. Besides, these metal surfaces are frequently contaminated by oil, which reduces the surface tension. Therefore, surface cleaning and pretreatments are desirable before applying a coating (Wicks 2007).

In general, the organic coatings cling to the metallic substrates via hydrogen bonding. As previously mentioned, the metal surface is usually covered with a thin layer of oxides with polar groups (i.e., hydroxyl groups), which allow organic coatings to form strong hydrogen bonding with these oxides. Much stronger adhesion can be obtained if the coating reacts with the metal as in the case when conversion pretreatments based on chromating and phosphating are used. The deposition of a conversion layer on the metal may passivate that surface and enhance the adhesion for subsequent applications of coatings. The adhesion can also be improved by means of coupling agents that form stronger covalent bonds between the coating and the substrate.

Other than metals, other regularly used substrates are glasses and plastics. In order to enhance adhesion of coatings to glass substrates, reactive silanes containing trialkoxysilyl group attached to a hydrocarbon chain and another functional group, such as amine, epoxy, or vinyl, can be used. Regarding the plastic substrates, the main problem is their low surface free energies that cause dewetting of coatings. In order to attain satisfactory adhesion to plastics, surface treatments are usually required to increase its surface energy. This can be done by oxidizing the surface to generate polar groups like hydroxyl, carboxylic acid, and ketone groups. The presence of these groups not only increases the surface free energy so that wetting is more likely with a broader range of coating materials but also provides hydrogen-bond acceptor and donor groups for interacting with the coatings (Wicks 2007).

14.6 Degradation Protection

To help protect a polymer from undergoing degradation, stabilizers are added. There are many types of stabilizers; each has a specific function in order to slow or stop degradation from taking place. Depending on the type of stabilizer, the polymer can be protected from various forms and stages of degradation. Some stabilizers, such as UV and heat absorbers, help prevent degradation from even beginning. While others, like antioxidants absorb or reduce propagating species, free radicals, into benign species. In general practice, a variety of stabilizers are added to a system to provide the most protection possible. Although a mix of stabilizers is needed, the actual amount added is usually less than 1%. Various types of stabilizers and their mechanisms will be discussed later; it is important to note that the classification of stabilizers used is for ease of presentation and that numerous stabilizers provide multiple stabilization mechanisms.

14.6.1 Radiation/Light Stabilization

Protection against radiation is very important to ensure a long lifetime for a polymer product. Additives for radiation protection can be classified into four subcategories: (1) UV screener, (2) UV absorber, (3) excited-state quencher, and (4) free radical scavenger (Pritchard 1998). The first category, UV screener, is usually not a specific stabilizer. Instead, UV screeners are additives used in a system to impact properties other than stabilization. Pigments, used to add color to system, are the main type of additives that can act as UV screeners. For a pigment to be an active UV screener, it should have a high UV reflectance to prevent light from penetrating the surface. Pigments such as titanium dioxide, chromium oxide, and ultramarine blue are all good screeners.

Although absorbers and quenchers have separate protection methods, they are generally used together because many stabilizers in these classes have dual roles as both an absorber and a quencher. Both absorber and quencher stabilizers work to prevent degradation from the beginning; usually these are also classified as UV stabilizers. An absorber will absorb incoming UV radiation and convert it to thermal energy. Quenchers return a molecule in the excited state back to the ground state through the emission of thermal energy as well. To be efficient, absorbers must have a high absorption in the same wavelength region as those for the polymer and impurities present in that system. Quenchers should have strong absorption in the wavelength region where the excited states of the polymer and impurities emit radiation. This overlap is important to prevent the polymer and/or impurities from initiating degradation; this, however, is the ideal case. Degradation is never really fully stopped. Stabilizers can only decrease the rate at which degradation happens. A very effective stabilizer acts as both an absorber and a quencher.

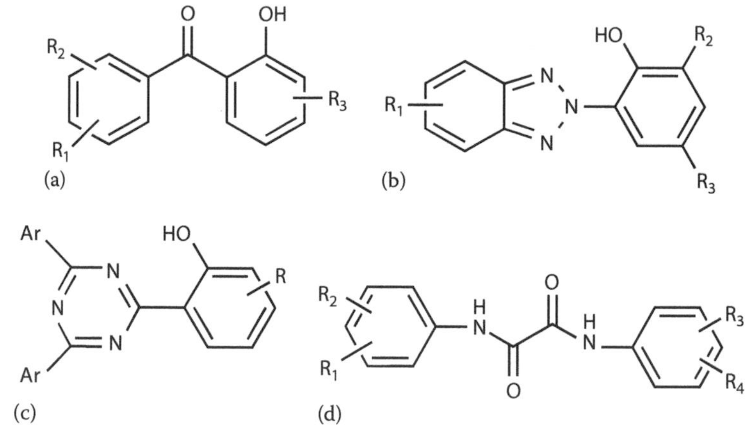

FIGURE 14.10 UV stabilizers: (a) benzophenones, (b) benzotriazoles, (c) *s*-triazines, (d) oxalanilides.

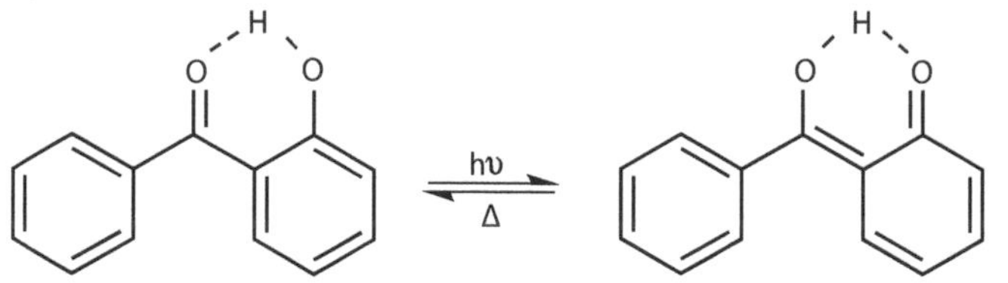

SCHEME 14.13 UV energy dissipation through hydrogen transfer.

There are many UV stabilizers available, but most can be classified into one of four classes: benzo-phenones, benzotriazoles, *s*-triazines, and oxalanilides (Figure 14.10).

UV energy is absorbed by the stabilizers and converted to heat by means of intermolecular rear-rangement, hydrogen transfer, or *cis–trans* isomerization (Pritchard 1998, Wicks 2007). For example, 2-hydroxyvenzophenone will absorb UV energy and dissipate this energy by undergoing hydrogen transfer through an intermediate six-membered ring, as shown in Scheme 14.13.

The fourth category, scavenging of free radicals, is typically accomplished with metal complexes and hindered piperidine compounds. Hindered piperidine compounds are the most efficient light stabilizers and as such are widely used. Metal complexes are not as efficient as hindered piperidine compounds but are still in use. Metal complexes will only be covered in this section because hindered piperidine com-pounds will be covered later in the antioxidant section. Metal complexes are multifunctional stabilizers that can decompose hydrogen peroxide in a dark reaction, scavenge free radicals, act as UV absorbers, and/or quench excited states/singlet oxygen. All metal complexes do not perform each of the aforemen-tioned stabilization mechanisms; the structure of the metal complex will determine which ones and how many stabilization mechanisms it participates in (Pritchard 1998). Figure 14.11 shows an example of a metal complex.

If a sulfur donor ligand is present, the metal complex shows marked quenching of singlet oxygen. Additionally, sulfur plays an important role in the decomposition of hydrogen peroxides: sulfur is oxi-dized to form sulfur oxides that undergo further oxidation to produce sulfur dioxide.

14.6.2 Antioxidants

The use of antioxidants is very important in maintaining the life of a product and preventing degrada-tion from taking place during thermal processing. This is because free radicals are easily produced from hydrogen peroxides present in the material. Hydrogen peroxides are quite unstable and will form

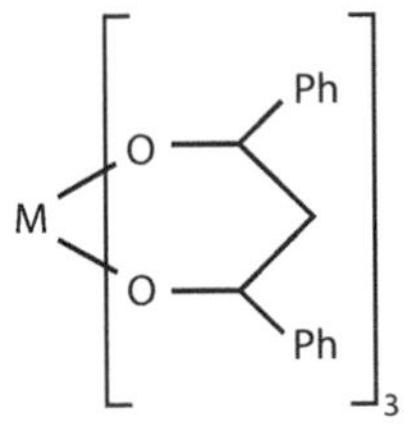

FIGURE 14.11 Tris(dibenzoylmethanato)chelate; M represents Cr or Fe.

free radicals relatively easily in the presence of heat, light, and metal ions. If left alone, the formed free radicals will break down the polymer via autoxidation. Since peroxides are generated during the oxidation of the polymer, the chain reaction does not self-terminate through radical coupling but instead is autoaccelerated as more and more peroxides are generated, causing more free radicals to be formed (Pritchard 1998, Lutz and Grossman 2001).

Antioxidants can be separated into two classes depending on how the oxidation process is hindered: chain breaking or preventative. Chain-breaking antioxidants work to remove propagating free radicals. This can be done by electron or hydrogen transfer to reduce the free radicals, ROO, to ROOH. Common chain-breaking antioxidants include hindered phenols and aromatic amines. Preventative antioxidants work to inhibit the formation of free radicals. This is done by decomposing peroxides through a non-radical mechanism. Common preventative antioxidants are phosphate esters and sulfur compounds that decompose peroxides into alcohol and other side products, depending on the chemical composition of the antioxidant (Pospíšil and Klemchuk 1989, Lutz and Grossman 2001). Figure 14.12 shows example structures for chain-breaking and preventative antioxidants.

As mentioned earlier, hindered piperidine compounds are excellent light stabilizers. Additionally, these compounds act as excellent antioxidants and are commonly referred to as hindered amine light stabilizers (HALS). There are various HALS, and most are based off of substituted 2,2,6,6-tetramethyl-piperidine. HALS act like metal complexes as well as function as chain-breaking antioxidants; however, HALS do not absorb UV. The success of HALS is derived from the stability of the nitroxy radical that forms in the presence of UV and oxygen. Once formed, the nitroxy radical can react with carbon-centered radicals to produce hydroxylamines and ethers that in turn react with peroxy radicals. This regenerates the nitroxy radical and terminates the peroxy radical. The reactions of HALS, based on 2,2,6,6-tetramethylpiperidine, are shown in Scheme 14.14, and reactions 2 and 3 make up the Denisov cycle (Pospíšil and Klemchuk 1989, Al-Malaika et al. 1999, Bieleman 2000, Chanda and Roy 2007).

To get the most degradation protection with HALS, UV absorbers should also be added. The UV absorbers will reduce the rate of radical formation, while the HALS decreases the oxidative degradation rate due to radicals.

Due to the various mechanisms of degradation, it can easily be seen that multiple additives are needed to ensure proper degradation protection. There exists multiple stabilizer additives and each has its own advantages and disadvantages. Stabilizers can play a specific role, such as a UV absorber, or have multiple roles, such as metal complexes. Regardless of how many roles a single stabilizer can perform, the best

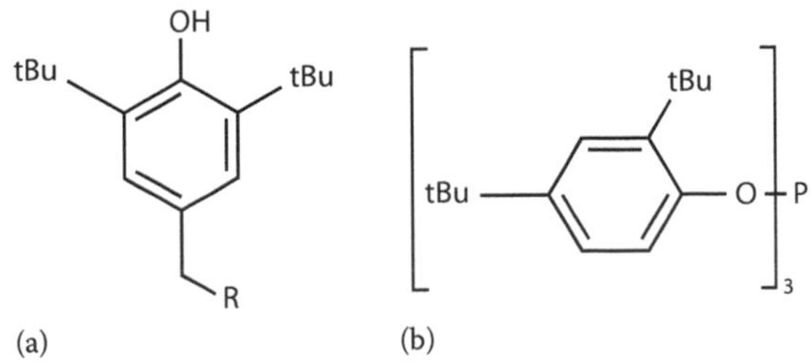

FIGURE 14.12 Example structures of antioxidants: (a) chain-breaking antioxidant, (b) preventative antioxidant.

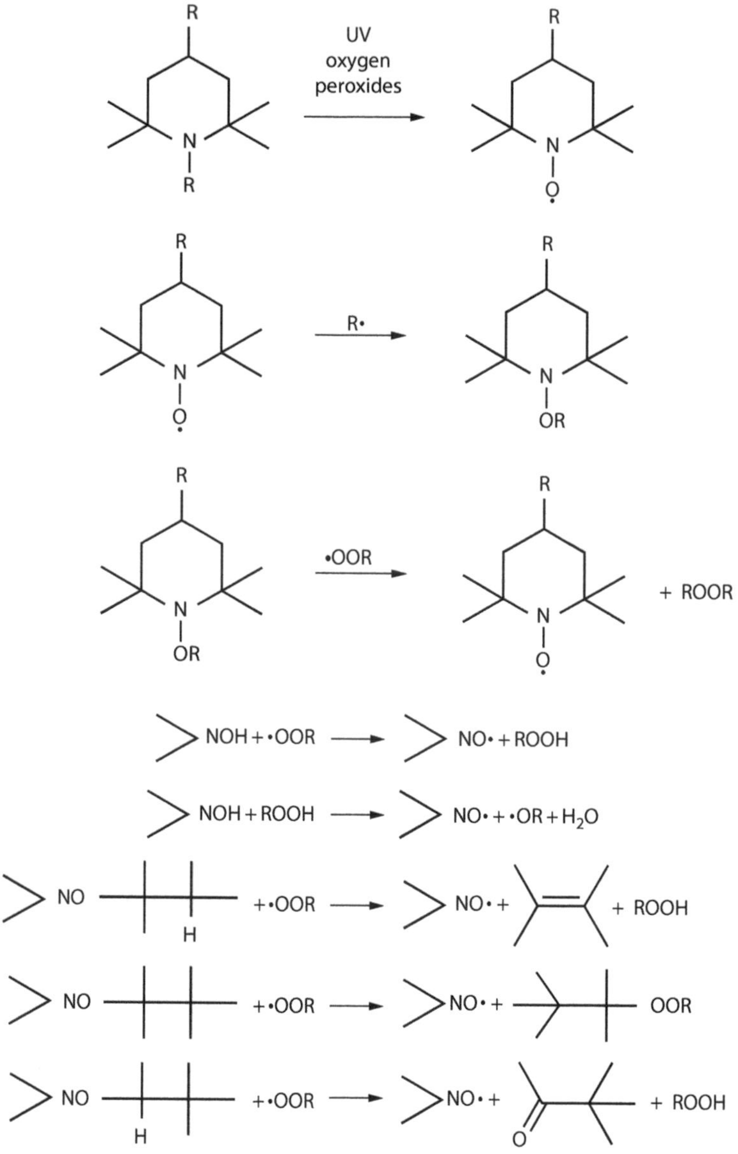

SCHEME 14.14   Reactions of HALS.

degradation properties are obtained using multiple stabilizers. The choice of stabilizers depends heavily on the polymer, manufacturing process, and exposure environment. By choosing the right combination of stabilizers, the lifetime of a polymer product can be greatly improved.

14.7 Specific Standardized Tests

As discussed so far, there are various types of degradation, that is, thermal, hydrolytic, photoinduced/UV, biodegradation, as well as weathering. National and international standard test methods have been developed to evaluate the degradation behavior. The American Society for Testing and Materials (ASTM) standards and methods are widely employed by industrial and academic researchers. Table 14.3

TABLE 14.3 ASTM Standards and Methods for Evaluating Various Types
of Degradation

Type of Degradation	ASTM Standards and Methods
Thermal	E2550-07, E1877-00 (2010)
Hydrolytic	D4502-92 (2004), E895-89 (2008)
Photoinduced/UV	D4587-05, D5208-09, D5272-08, G154-06
Bio	D3273-00 (2005), D3274-95 (2002), D3456-86 (2008), D4939-89 (2007), D4610-98 (2009), D5988-03, D5988-03, D6954-04 E1279-89 (2008), G21-09, G29-96 (2010)
Weathering	
Natural	D1014-09, D1435-05, D4141-07, D4364-05, D4674-02, G7-05, G24-05, G84-89 (2005), G90-10, G92-86 (2003)
Artificial	D750-06, D822-01 (2006), D1499-05, D2565-99 (2008), D3361-01 (2006), G60-01 (2007), G85-09

is a list of ASTM testing methods for different types of degradation. Please note that this table is not a complete list, but rather it is a representative of tests commonly reported in the literature.

14.8 Summary

Degradation is the process of breaking down a polymer into smaller and smaller segments. This occurs via the absorption of energy from the environment. Once enough energy is absorbed, bond breakage will occur. The resulting products can be active species, such as a free radical, that further attacks the polymer, or it can be an inactive side product, as in the case of hydrolysis. Initiation of polymer degradation is usually caused by the presence of impurities or weak linkages along the polymer backbone.

The rate and extent of degradation is highly contingent upon the polymer and the environmental system that it is exposed to. Not only do different polymers behave differently, but also different structures of the same polymer will exhibit different degradation characteristics. Additionally, multiple degradation pathways can be happening simultaneously for any given system. Protection against degradation is usually accomplished with a mixture of stabilizers to defend against different degradation initiation and propagation mechanisms. Overall, degradation is a complex process that is reliant on the polymer system, process conditions, and the environment.

ASTM Method Titles

ASTM D1014-09 Standard Practice for Conducting Exterior Exposure Tests of Paints and Coatings on Metal Substrates.

ASTM D1435-05 Standard Practice for Outdoor Weathering of Plastics.

ASTM D1499-05 Standard Practice for Filtered Open-Flame Carbon-Arc Exposures of Plastics.

ASTM D2565-99 (2008) Standard Practice for Xenon-Arc Exposure of Plastics Intended for Outdoor Applications.

ASTM D3273-00 (2005) Standard Test Method for Resistance to Growth of Mold on the Surface of Interior Coatings in an Environmental Chamber.

ASTM D3274-95 (2002) Standard Test Method for Evaluating Degree of Surface Disfigurement of Paint Films by Microbial (Fungal or Algal) Growth or Soil and Dirt Accumulation.

ASTM D3361-01 (2006) Standard Practice for Unfiltered Open-Flame Carbon-Arc Exposures of Paint and Related Coatings.

ASTM D3456-86 (2008) Standard Practice for Determining by Exterior Exposure Tests the Susceptibility of Paint Films to Microbiological Attack.

ASTM D4141-07 Standard Practice for Conducting Black Box and Solar Concentrating Exposures of Coatings.

ASTM D4364-05 Standard Practice for Performing Outdoor Accelerated Weathering Tests of Plastics Using Concentrated Sunlight.

ASTM D4502-92 (2004) Standard Test Method for Heat and Moisture Resistance of Wood-Adhesive Joints.

ASTM D4587-05 Standard Practice for Fluorescent UV-Condensation Exposures of Paint and Related Coatings.

ASTM D4610-98 (2009) Standard Guide for Determining the Presence of and Removing Microbial (Fungal or Algal) Growth on Paint and Related Coatings.

ASTM D4674-02 Standard Test Method for Accelerated Testing for Color Stability of Plastics Exposed to Indoor Fluorescent Lighting and Window-Filtered Daylight.

ASTM D4939-89 (2007) Standard Test Method for Subjecting Marine Antifouling Coating to Biofouling and Fluid Shear Forces in Natural Seawater.

ASTM D5208-09 Standard Practice for Fluorescent Ultraviolet (UV) Exposure of Photodegradable Plastics.

ASTM D5272-08 Standard Practice for Outdoor Exposure Testing of Photodegradable Plastics.

ASTM D5988-03 Standard Test Method for Determining Aerobic Biodegradation in Soil of Plastic Materials or Residual Plastic Materials After Composting.

ASTM D6868-03 Standard Specification for Biodegradable Plastics Used as Coatings on Paper and Other Compostable Substrates.

ASTM D6954-04 Standard Guide for Exposing and Testing Plastics that Degrade in the Environment by a Combination of Oxidation and Biodegradation.

ASTM D750-06 Standard Test Method for Rubber Deterioration Using Artificial Weathering Apparatus.

ASTM D822-01 (2006) Standard Practice for Filtered Open-Flame Carbon-Arc Exposures of Paint and Related Coatings.

ASTM E1279-89 (2008) Standard Test Method for Biodegradation By a Shake-Flask Die-Away Method.

ASTM E1877-00 (2010) Standard Practice for Calculating Thermal Endurance of Materials from Thermogravimetric Decomposition Data.

ASTM E2550-07 Standard Test Method for Thermal Stability by Thermogravimetry.

ASTM E895-89 (2008) Standard Practice for Determination of Hydrolysis Rate Constants of Organic Chemicals in Aqueous Solutions.

ASTM G154-06 Standard Practice for Operating Fluorescent Light Apparatus for UV Exposure of Nonmetallic Materials.

ASTM G21-09 Standard Practice for Determining Resistance of Synthetic Polymeric Materials to Fungi.

ASTM G24-05 Standard Practice for Conducting Exposures to Daylight Filtered Through Glass.

ASTM G29-96 (2010) Standard Practice for Determining Algal Resistance of Plastic Films.

ASTM G60-01 (2007) Standard Practice for Conducting Cyclic Humidity Exposures.

ASTM G7-05 Standard Practice for Atmospheric Environmental Exposure Testing of Nonmetallic Materials.

ASTM G84-89 (2005) Standard Practice for Measurement of Time-of-Wetness on Surfaces Exposed to Wetting Conditions as in Atmospheric Corrosion Testing.

ASTM G85-09 Standard Practice for Modified Salt Spray (Fog) Testing.

ASTM G90-10 Standard Practice for Performing Accelerated Outdoor Weathering of Nonmetallic Materials Using Concentrated Natural Sunlight.

ASTM G92-86 (2003) Standard Practice for Characterization of Atmospheric Test Sites.

References

Adams, J.H. 1970. Analysis of nonvolatile oxidation products of polypropylene. 1. Thermal oxidation. *Journal of Polymer Science Part A-1—Polymer Chemistry*, 8(5), 1077–1090.

Adams, J.H. and J.E. Goodrich. 1970. Analysis of nonvolatile oxidation products of polypropylene. 2. Process degradation. *Journal of Polymer Science Part A-1—Polymer Chemistry*, 8(5), 1269–1277.

Agarwal, S.H., R.F. Jenkins, and R.S. Porter. 1982. Molecular characterization and effect of shear on the distribution of long branching in polyvinyl acetate. *Journal of Applied Polymer Science*, 27(1), 113–120.

Al-Malaika, S., A. Golovoy, and C.A. Wilkie. 1999. *Chemistry and Technology of Polymer Additives*, Blackwell Science, Malden, MA.

Allegra, G. 2004. Macromolecular internal viscosity. The role of stereoregularity. *Macromolecular Symposia*, 218(1), 89–100.

Allen, N.S. 1994. Photofading and light stability of dyed and pigmented polymers. *Polymer Degradation and Stability*, 44(3), 357–374.

Allen, N.S. and M. Edge. 1992. *Fundamentals of Polymer Degradation and Stabilisation*, Elsevier Applied Science, New York.

Bamford, C.H. and M.J.S. Dewar. 1949. Absolute velocity constants in the autoxidation of tetrahydronaphthalene. *Nature (London, United Kingdom)*, 163, 215.

Bamford, C.H. and C.F.H. Tipper. 1975. *Degradation of Polymers, Comprehensive Chemical Kinetics*, Elsevier Scientific Pub. Co., New York.

Bastioli, C. and Rapra Technology Limited. 2005. *Handbook of Biodegradable Polymers*, Rapra Technology, Shrewsbury, U.K.

Bieleman, J. 2000. *Additives for Coatings*, Wiley-VCH, New York.

Bockhorn, H., A. Hornung, U. Hornung, and D. Schwaller. 1999. Kinetic study on the thermal degradation of polypropylene and polyethylene. *Journal of Analytical and Applied Pyrolysis*, 48(2), 93–109.

Bovey, F.A. and F.H. Winslow. 1979. *Macromolecules. An Introduction to Polymer Science*, Academic Press, New York.

Brückner, S., G. Allegra, and P. Corradini. 2002. Helix inversions in polypropylene and polystyrene. *Macromolecules*, 35(10), 3928–3936.

Carey, F.A. and R.J. Sundberg. 2007. *Advanced Organic Chemistry*, 5th edn., Springer, New York.

Caronna, T., F. Fontana, B. Marcandalli, and E. Selli. 2001. Photostability of substituted 4-diethylaminoazobenzenes. *Dyes and Pigments*, 49(2), 127–133.

Chanda, M. and S.K. Roy. 2007. *Plastics Technology Handbook*, 4th edn., Plastics Engineering Series, CRC Press/Taylor & Francis Group, Boca Raton, FL.

Chen, K., K. Harris, and S. Vyazovkin. 2007. Tacticity as a factor contributing to the thermal stability of polystyrene. *Macromolecular Chemistry and Physics*, 208(23), 2525–2532.

Chien, J.C.W. and C.R. Boss. 1967. Electron spin resonance spectra of low molecular weight and high molecular weight peroxy radicals. *Journal of the American Chemical Society*, 89(3), 571–575.

Choi, B.H., A. Chudnovsky, R. Paradkar, W. Michie, Z.W. Zhou, and P.M. Cham. 2009. Experimental and theoretical investigation of stress corrosion crack (SCC) growth of polyethylene pipes. *Polymer Degradation and Stability*, 94(5), 859–867.

Chu, C.-C., J.A. Von Fraunhofer, and H.P. Greisler. 1997. *Wound Closure Biomaterials and Devices*, CRC Press, Boca Raton, FL.

Dissado, L.A. and J.C. Fothergill. 1992. *Electrical Degradation and Breakdown in Polymers*, IEE Materials and Devices Series, P. Peregrinus Ltd., London, U.K.

Dolezal, Z., V. Pacakova, and J. Kovarova. 2001. The effects of controlled aging and blending of low- and high-density polyethylenes, polypropylene and polystyrene on their thermal degradation studied by pyrolysis gas chromatography. *Journal of Analytical and Applied Pyrolysis*, 57(2), 177–185.

Egerton, G.S. 1947. The action of light on dyed and undyed cotton. *Journal of the Society of Dyers and Colourists*, 63, 161–171.

Emanuel, N.M. and A.L. Buchachenko. 1987. in C.H.R.I.d. Jonge, Ed., *Chemical Physics of Polymer Degradation and Stabilization, New Concepts in Polymer Science*, VNU Science Press, Utrecht, the Netherlands.

Galaev, I. and B. Mattiasson. 2008. *Smart Polymers: Applications in Biotechnology and Biomedicine*, 2nd edn., CRC Press, Boca Raton, FL.

Göpferich, A. 1996. Mechanisms of polymer degradation and erosion. *Biomaterials*, 17(2), 103–114.

Grassie, N. and G. Scott. 1985. *Polymer Degradation & Stabilisation*, Cambridge University Press, Cambridge, U.K.

Gugumus, F. 1990. Mechanisms of photooxidation of polyolefins. *Angewandte Makromolekulare Chemie*, 176, 27–42.

Gugumus, F. 1997. Thermooxidative degradation of polyolefins in the solid state: Part 5. Kinetics of functional group formation in PE-HD and PE-LLD. *Polymer Degradation and Stability*, 55(1), 21–43.

Gugumus, F. 2002a. Re-examination of the thermal oxidation reactions of polymers. 3. Various reactions in polyethylene and polypropylene. *Polymer Degradation and Stability*, 77(1), 147–155.

Gugumus, F. 2002b. Re-examination of the thermal oxidation reactions of polymers. 2. Thermal oxidation of polyethylene. *Polymer Degradation and Stability*, 76(2), 329–340.

Gugumus, F. 2005. Physico-chemical aspects of polyethylene processing in an open mixer. Part 15: Product yields on bimolecular hydroperoxide decomposition. *Polymer Degradation and Stability*, 89(3), 517–526.

Guillet, J. 1985. *Polymer Photophysics and Photochemistry: An Introduction to the Study of Photoprocesses in Macromolecules*, Cambridge University Press, Cambridge, U.K.

Hamid, S.H. 2000. *Handbook of Polymer Degradation*, 2nd edn., Environmental Science and Pollution Control Series 21, Marcel Dekker, New York.

Hamid, S.H., M.B. Amin, and A.G. Maadhah. 1992. Weathering degradation of polyethylene, in S.H. Hamid, M.B. Amin, and A.G. Maadhah, Eds., *Handbook of Polymer Degradation*, Marcel Dekker, Inc., New York.

Heacock, J., F.B. Mallory, and F.P. Gay. 1968. Photodegradation of polyethylene film. *Journal of Polymer Science Part A-1—Polymer Chemistry*, 6(10PA), 2921–2934.

Holland, B.J. and J.N. Hay. 2002. The effect of polymerisation conditions on the kinetics and mechanisms of thermal degradation of PMMA. *Polymer Degradation and Stability*, 77(3), 435–439.

Howell, B.A. 2007. The utilization of TG/GC/MS in the establishment of the mechanism of poly(styrene) degradation. *Journal of Thermal Analysis and Calorimetry*, 89(2), 393–398.

Howell, B.A. and K. Chaiwong. 2009. Thermal stability of poly(styrene) containing no head-to-head units. *Journal of Thermal Analysis and Calorimetry*, 96(1), 219–223.

Howell, B.A., Y. Cui, K. Chaiwong, and H. Zaho. 2006. Polymerization as a means of stabilization for polystyrene. *Journal of Vinyl & Additive Technology*, 12(4), 198–203.

Howell, B.A., Y.M. Cui, and D.B. Priddy. 2003. Assessment of the thermal degradation characteristics of isomeric poly(styrene)s using TG, TG/MS and TG/GC/MS. *Thermochimica Acta*, 396(1–2), 167–177.

Ito, M. and K. Nagai. 2008. Degradation issues of polymer materials used in railway field. *Polymer Degradation and Stability*, 93(10), 1723–1735.

Iwata, T. and Y. Doi. 1999. Crystal structure and biodegradation of aliphatic polyester crystals. *Macromolecular Chemistry and Physics*, 200(11), 2429–2442.

Karniadakis, G., A. Beşkök, and N.R. Aluru. 2005. *Microflows and Nanoflows: Fundamentals and Simulation, Interdisciplinary Applied Mathematics*, Vol. 29, Springer, New York.

Kong, H.J., K.Y. Lee, and D.J. Mooney. 2002. Decoupling the dependence of rheological/mechanical properties of hydrogels from solids concentration. *Polymer*, 43(23), 6239–6246.

Krevelen, D.W. van, and K. Te Nijenhuis. 2009. *Properties of Polymers: Their Correlation with Chemical Structure: Their Numerical Estimation and Prediction from Additive Group Contributions*, 4th completely revised edn., Elsevier, Amsterdam, the Netherlands.

Lu, G.Y., B.Y. Sheng, G. Wang, Y.J. Wei, Y.D. Gong, and X.F. Zhang. 2009. Controlling the degradation of covalently cross-linked carboxymethyl chitosan utilizing bimodal molecular weight distribution. *Journal of Biomaterials Applications*, 23(5), 435–451.

Lutz, J.T. and R.F. Grossman. 2001. *Polymer Modifiers and Additives, Plastics Engineering*, Marcel Dekker, New York.

Madras, G., G.Y. Chung, J.M. Smith, and B.J. McCoy. 1997. Molecular weight effect on the dynamics of polystyrene degradation. *Industrial & Engineering Chemistry Research*, 36(6), 2019–2024.

Mark, J.E. 2007. *Physical Properties of Polymers Handbook*, 2nd edn., Springer, New York.

Millán, J., E.L. Madruga, and G. Martínez. 1975. The influence of the tacticity on thermal degradation of PVC. II. New results obtained with fractionated polymers. *Die Angewandte Makromolekulare Chemie*, 45(1), 177–184.

Moldoveanu, S. 2005. *Analytical Pyrolysis of Synthetic Organic Polymers, Techniques and Instrumentation in Analytical Chemistry*, Elsevier, Amsterdam, the Netherlands.

Noggle, J.H. 1996. *Physical Chemistry*, 3rd edn., HarperCollins College Publishers, New York.

Numata, K., R.K. Srivastava, A. Finne-Wistrand, A.C. Albertsson, Y. Doi, and H. Abe. 2007. Branched poly(lactide) synthesized by enzymatic polymerization: Effects of molecular branches and stereo-chernistry on enzymatic degradation and alkaline hydrolysis. *Biomacromolecules*, 8(10), 3115–3125.

Osawa, Z. 1992. Photoinduced degradation of polymers, in S.H. Hamid, M.B. Amin, and A.G. Maadhah, Eds., *Handbook of Polymer Degradation*, Marcel Dekker, Inc., New York.

Osawa, Z., T. Sunakami, and Y. Fukuda. 1994. Photodegradation of blends of poly(vinyl chloride) and polyurethane. *Polymer Degradation and Stability*, 43(1), 61–66.

Paakinaho, K., V. Ella, S. Syrjala, and M. Kellomaki. 2009. Melt spinning of poly(L/D)lactide 96/4: Effects of molecular weight and melt processing on hydrolytic degradation. *Polymer Degradation and Stability*, 94(3), 438–442.

Plawsky, J.L. 2001. *Transport Phenomena Fundamentals, Chemical Industries*, Marcel Dekker, New York.

Pospíšil, J. and P.P. Klemchuk. 1989. *Oxidation Inhibition in Organic Materials*, 2 Vol., CRC Press, Boca Raton, FL.

Pritchard, G. 1998. *Plastics Additives: An A-Z Reference*, 1st edn., Polymer Science and Technology Series, Chapman & Hall, London, U.K.

Rothon, R.N. and Rapra Technology Limited. 2003. *Particulate-Filled Polymer Composites*, 2nd edn., Rapra Technology, Shrewsbury, U.K.

Roy, P.K., P. Singh, D. Kumar, and C. Rajagopal. 2010. Manganese stearate initiated photo-oxidative and thermo-oxidative degradation of LDPE, LLDPE and their blends. *Journal of Applied Polymer Science*, 117(1), 524–533.

Sadrmohaghegh, C., G. Scott, and E. Setoudeh. 1981. Effects of reprocessing on polymers. 3. Photooxidation of polyethylene-polypropylene blends. *Polymer Degradation and Stability*, 3(6), 469–476.

Satoto, R., W.S. Subowo, R. Yusiasih, Y. Takane, Y. Watanabe, and T. Hatakeyama. 1997. Weathering of high-density polyethylene in different latitudes. *Polymer Degradation and Stability*, 56(3), 275–279.

Scott, G. 2002. *Degradable Polymers: Principles and Applications*, 2nd edn., Kluwer Academic Publishers, Dordrecht, the Netherlands.

Singh, B. and N. Sharma. 2008. Mechanistic implications of plastic degradation. *Polymer Degradation and Stability*, 93(3), 561–584.

Sørensen, P., S. Kiil, K. Dam-Johansen, and C. Weinell. 2009. Anticorrosive coatings: A review. *Journal of Coatings Technology and Research*, 6(2), 135–176.

Sperling, L.H. 2006. *Introduction to Physical Polymer Science*, 4th edn., Wiley, Hoboken, NJ.

Stivala, S.S. 1980. Structure vs stability in polymer degradation. *Polymer Engineering & Science*, 20(10), 654–661.

Suzuki, S., Y. Nakamura, A.T.M. Kamrul Hasan, B. Liu, M. Terano, and H. Nakatani. 2005. Dependence of tacticity distribution in thermo-oxidative degradation of polypropylene. *Polymer Bulletin*, 54(4), 311–319.

Tamura, S. and J.K. Gillham. 1978. Pyrolysis molecular weight chromatography vapor-phase infrared spectrophotometry—Online system for analysis of polymers. 4. Influence of cis-trans ratio on thermal-degradation of 1,4-polybutadienes. *Journal of Applied Polymer Science*, 22(7), 1867–1884.

Tomasi, G., M. Scandola, B.H. Briese, and D. Jendrossek. 1996. Enzymatic degradation of bacterial poly(3-hydroxybutyrate) by a depolymerase from *Pseudomonas lemoignei*. *Macromolecules*, 29(2), 507–513.

Utracki, L.A. 2002. *Polymer Blends Handbook*, Kluwer Academic Publishers, Dordrecht, the Netherlands.

Vinu, R., A. Marimuthu, and G. Madras. 2010. Enzymatic degradation of poly(soybean oil-g-methyl methacrylate). *Journal of Polymer Engineering*, 30(1), 57–76.

Vollhardt, K.P.C. and N.E. Schore. 1994. *Organic Chemistry*, 2nd edn., W.H. Freeman, New York.

Watanabe, Y., H. Shirahama, and H. Yasuda. 2004. Syntheses of random and block copolymers of lactides with 1.5-dioxepan-2-one and their biodegradability. *Reactive & Functional Polymers*, 59(3), 211–224.

Watson, S., A. Forster, I. Hsiang Tseng, L.-P. Sung, J. Lucas, and A. Forster. 2008. Effects of pigment and its dispersion on UV degradation of epoxy and acrylic urethane polymeric systems. *Materials Research Society Symposium Proceedings*, Vol. 1056E (Nanophase and Nanocomposite Materials V), Paper #1056-HH03-67, National Institute of Standards and Technology, Gaithersburg, MD.

Wicks, Z.W. 2007. *Organic Coatings: Science and Technology*, 3rd edn., Wiley-Interscience, Hoboken, NJ.

Wright, D.C. 2001. *Failure of Plastics and Rubber Products: Causes, Effects, and Case Studies Involving Degradation*, Rapra Technology, Shrewsbury, U.K.

Wypych, G. 2003. *Handbook of Material Weathering*, 3rd edn., ChemTec Pub., Toronto, Ontario, Canada.

Xie, W.Y., N. Jiang, and Z.H. Gan. 2008. Effects of multi-arm structure on crystallization and biodegradation of star-shaped poly(epsilon-caprolactone). *Macromolecular Bioscience*, 8(8), 775–784.

Xue, L., U.S. Agarwal, and P.J. Lemstra. 2005. Shear degradation resistance of star polymers during elongational flow. *Macromolecules*, 38(21), 8825–8832.

Yang, R., J. Yu, Y. Liu, and K. Wang. 2005. Effects of inorganic fillers on the natural photo-oxidation of high-density polyethylene. *Polymer Degradation and Stability*, 88(2), 333–340.

Zhou, X.M. 2009. Thermal behavior of PES/PEES copolymers with different sequence distribution: Comparative study of the kinetics of degradation. *Journal of Applied Polymer Science*, 111(2), 833–838.

Zhou, X.M., Z.M. Gao, and Z.H. Jiang. 2007. Effect of sequence distribution on short-range ordering in poly(ether sulfone) copolymers. *Journal of Applied Polymer Science*, 103(1), 534–537.

Zhou, Q. and M. Xanthos. 2010. Effects of cationic and anionic clays on the hydrolytic degradation of polylactides. *Polymer Engineering & Science*, 50(2), 320–330.

Zhu, B., Y. He, H. Nishida, K. Yazawa, N. Ishii, K. Kasuya, and Y. Inoue. 2008. Crystalline-structure-dependent enzymatic degradation of polymorphic poly(3-hydroxypropionate). *Biomacromolecules*, 9(4), 1221–1228.

Zweifel, H., R.D. Maier, and M. Schiller. 2009. *Plastics Additives Handbook*, 6th edn., Hanser Publications, Cincinnati, OH.

15

Thermosetting Polymers

Drew Pavlacky
North Dakota State University

Chris Vetter
North Dakota State University

Victoria J. Gelling
North Dakota State University

15.1 Introduction

This chapter focuses on the degradation and measurement of degradation of thermosetting polymers (thermosets). It provides information regarding the application, properties, structure, and degradation of thermosets with a focus on polymeric coatings based on thermosets.

15.2 Background

Thermosetting polymers have the distinction of being continuous, 3D networks whose polymeric backbones are comprised exclusively of covalently bonded atoms. This is the key to their mechanical properties and chemical resistance that differs greatly from thermoplastic polymers, which are covalently bonded chains held together by secondary bonding such as van der Waals forces. They are formed primarily through a step growth polymerization method, but chain growth thermosets are possible. Section 15.4 outlines the functionalities that are most commonly used in thermoset polymers and the reactions that they undergo to form cross-linked networks (Pascault et al. 2002).

15.3 Application

15.3.1 Uses

Due to stability with respect to temperature for thermosets in comparison with thermoplastics, thermosets are used for a wide variety of applications such as car parts, wrinkle-resistant shirts, fiberglass, adhesives, plywood, chemical agent-resistant coating (CARC), and composites. The focus of this chapter is primarily on thermosets for coating and paint applications. It is interesting to note that coatings based on traditional thermoplastic polymers require very high solvent concentrations perhaps approaching 80–90 vol% solvent. The environmentally friendly desire for low-solvent coatings has led to an increased use of thermoset polymers due to the lower molecular weights of the thermosetting resins in comparison to the higher molecular weight thermoplastics.

15.3.2 Challenges

One challenge for thermosets is that higher cure temperatures are required. With increasing energy costs, maintaining large ovens at high temperatures can be prohibitively expensive. Also, for thermoset coatings on tempered metal substrates, caution must be used so that the lowest cure temperature possible is used so that the existing temper of the metal is not reduced. Another related issue with thermoset polymers is that of limited shelf or storage life. A compromise must be reached whereby the polymer is stable under normal storage conditions, which may include rather high ambient storage temperatures—just imagine the temperature of a un-air-conditioned semitrailer in Arizona in August—but still does not require extremely high curing temperatures.

15.4 Structure of Thermosetting Polymers

15.4.1 Bonding

15.4.1.1 Examples of Thermoset Chemistry and Structure

One of the most widely used thermoset polymer systems is epoxy resin. *Epoxies* get their name from the three-membered cyclic ethers in their structures that are known as epoxy or oxirane rings. They undergo cross-linking reactions through nucleophilic attack on the least sterically hindered carbon of the epoxy ring that then breaks to form a hydroxyl group on the most sterically hindered carbon in the ring. Amines are often used because primary and secondary amines will undergo this reaction at room temperature. Any nucleophile can be used, however, as long as there is sufficient energy, usually by external heating in the form of thermal energy, for the reaction to take place. For cross-linking to occur in any thermoset, one of the resins, in this case either the epoxy or the cross-linker, must have a functionality of three or greater. If all resins present in the reaction only have a functionality of two, cross-linking will not occur and linear polymer chains will be formed. Figure 15.1 details the reaction between one epoxy group and one secondary amine. Note that here the symbols O, H, and N represent the corresponding elements while R, R′, and R″ are different constituents. This reaction would then have to take place a minimum of three times on one polymer chain for cross-linking to take place (Wicks et al. 1999).

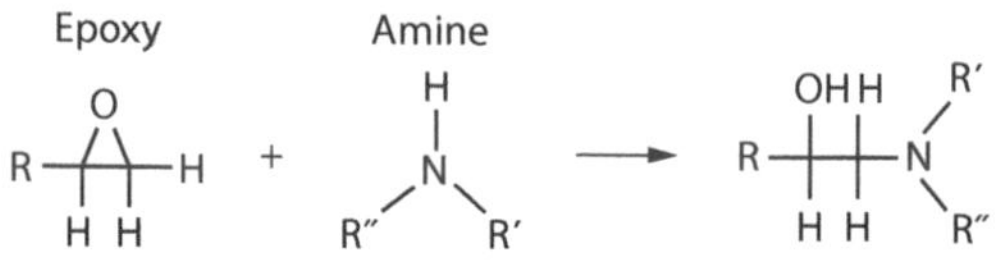

FIGURE 15.1   The cross-linking reaction between an epoxy functional resin and a secondary amine.

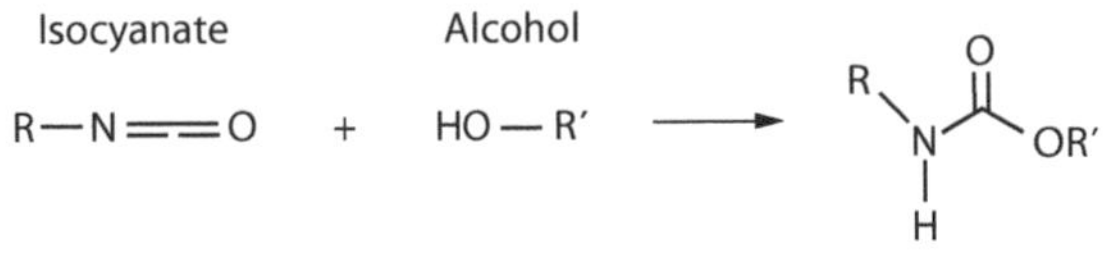

Isocyanate   Alcohol

R—N==O   +   HO—R'   ⟶   [structure: R–N(H)–C(=O)–OR']

FIGURE 15.2   An isocyanate reacting with an alcohol to form a urethane.

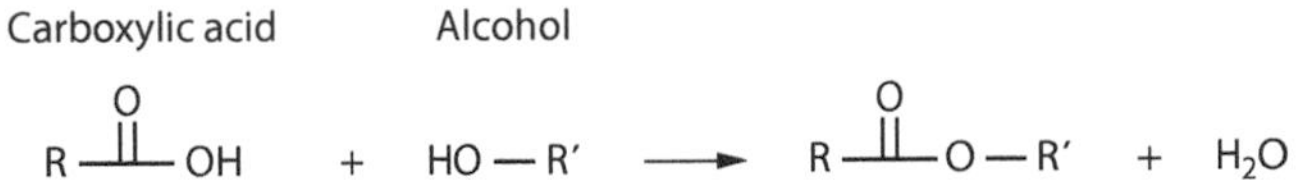

Carboxylic acid   Alcohol

R—C(=O)—OH   +   HO—R'   ⟶   R—C(=O)—O—R'   + H_2O

FIGURE 15.3   The reaction of a carboxylic acid with an alcohol to form a polyester.

Polyurethanes are another group of polymers that are used quite extensively. While epoxies have a wide range of cross-linking chemistries that can be used, polyurethanes can only be synthesized by the reaction of an isocyanate with an alcohol. This reaction can be seen in Figure 15.2. Generally, primary alcohols are more reactive than secondary alcohols that in turn are more reactive than tertiary alcohols. Polyurethanes are able to form intermolecular hydrogen bonds that add to the abrasion resistance of the material. When the polymer is placed under stress, the hydrogen bonds can break to absorb energy and then can reform after the stress has been removed (Wicks et al. 1999).

Polyesters often have hydroxyl (–OH) functionality and are cross-linked with isocyanates to make polyurethanes. The reaction between a carboxylic acid and an alcohol to form a polyester can be used to form cross-links. This reaction can be seen in Figure 15.3. One degradation problem that polyesters have is that they can undergo a hydrolysis reaction with water. Because of this, one must choose the polyols and acids used in polyester synthesis carefully so that the end product has acceptable hydrolytic stability. There must be enough steric hindrance near the ester linkage to inhibit the hydrolysis reaction (Wicks et al. 1999).

Amino resins are the other commonly used functionality for thermoset resins. They are based on urea, melamine, methacrylamide, glycoluril, or other amino compounds. These monomers are reacted with formaldehyde and then alcohols to yield ethers. These compounds can then be cross-linked with polyols in an acid-catalyzed reaction to form thermoset polymers. This cross-linking reaction can be seen in Figure 15.4. For this reaction to result in a thermoset polymer, R″ should be a relatively low molecular weight moiety, and R‴ should be a larger molecular weight moiety with multiple hydroxyl functional groups (Wicks et al. 1999).

So far, this section has focused on the functional groups on the end of polymer branches that allow the formation of thermosetting polymers. The other consideration to take into account is the aliphatic and aromatic content of the chains between the functional groups. Linear aliphatic linkages between functional groups tend to be flexible with carbon-to-carbon bonds that can rotate resulting in a softer more elastic polymer. An example of aliphatic and aromatic linkages can be seen in Figure 15.5. These aliphatic linkages also tend to absorb much less ultraviolet (UV) radiation than aromatic linkages, thus greatly increasing their durability to sunlight. Aromatic linkages, especially in the form of benzene rings within the structure, tend to resist deformation. This results in a harder, more brittle polymer. These linkages tend to absorb UV radiation, and so they have increased degradation rates when that polymer is exposed to sunlight (Strong 2006).

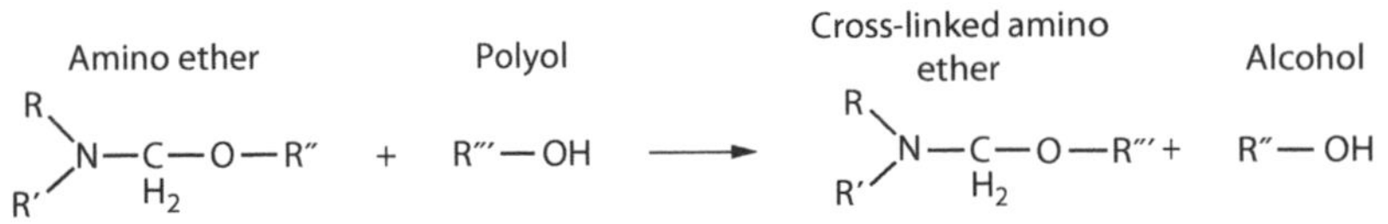

Amino ether   Polyol   Cross-linked amino ether   Alcohol

[structure: (R)(R')N—C(H_2)—O—R″]   +   R‴—OH   ⟶   [structure: (R)(R')N—C(H_2)—O—R‴]   +   R″—OH

FIGURE 15.4   Cross-linking between an amino ether and a polyol.

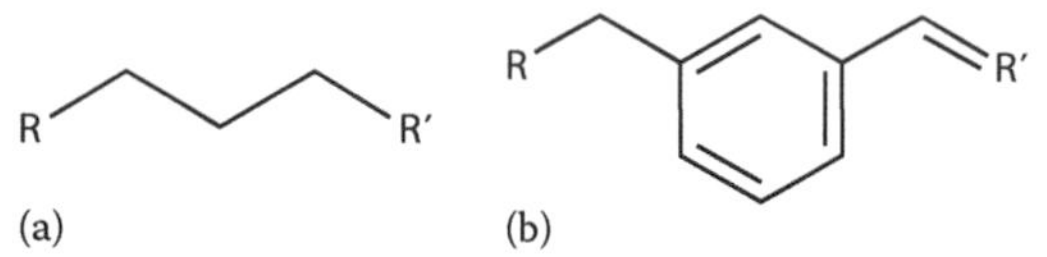

(a)   (b)

FIGURE 15.5   Examples of an aliphatic linkage (a) and an aromatic linkage (b).

15.4.2 Microstructure

Due to cross-linking, it is very difficult for thermosetting polymers to crystallize. Therefore, any microstructural features that can be observed in these polymers are due to phase separation, during the curing process, before the polymer is fully cross-linked, thus limiting the mobility of these polymer chains. This often requires using block copolymers to create different phases or two prepolymers that are not mutually soluble or have dissimilar densities.

15.4.2.1 Examples

One example of how phase separation can be used in coating technology is for self-stratifying coatings. These coatings are composed of at least two prepolymer systems that are not mutually soluble, to a large extent, with one another. Therefore, during the curing process, before there has been a large viscosity increase, the coating will phase separate into stratified layers. The complex mechanisms by which this may happen are discussed by Verkholantsev (1995).

   One application where these types of coatings have been found to be useful is for marine antifouling coatings. In this case, to create a low surface energy coating that marine organisms could not adhere to or could be removed with minimal effort, polydimethyl siloxane (PDMS) was used. PDMS, however, is quite soft so it does not possess the desirable mechanical properties for use as a ship coating. To remedy this, the PDMS must be mixed in with a more polar polyurethane formulation (a polyol mixed with an isocyanate). After the coating is applied, the PDMS will form primarily at the surface of the coating to create a low-energy surface, while the polyurethane would make up the bulk of the coating to provide the required mechanical properties (Pieper et al. 2007).

   These coatings may also be of use in the automotive industry. The many layers of coatings that are applied to automobile bodies add significantly to their cost. Self-stratifying coatings could limit the number of processing steps required to apply these coatings by forming multiple layers during one application process. Indeed, patents for this technology are currently being filed (Baghdachi et al. 2010).

15.5 General Properties

Thermosetting polymers can exhibit a large range of properties depending on several polymer characteristics. The specific molecular structure can cause the inherent properties to be vastly different within a thermoset class because these properties depend on whether the structure contains aromatic (i.e., benzene ring) rings or is aliphatic in nature. For instance, phenolics are usually very hard and brittle when compared to polyesters because phenolics contain a large quantity of aromatic groups with limited chain motion during the application of stress. Meanwhile, unsaturated (i.e., contains carbon–carbon double bonds) polyesters exhibit more chain mobility such that when stressed, the polymeric chains can disentangle more easily to avoid drastic brittle failure.

   The extent of cross-linking of thermoset systems heavily controls the mechanical, physical, thermal, and various other properties of thermosets. The density of cross-links can be approximately predetermined by examining the given polymer and the equivalent weight of the hardener functional group. Depending on the functional group ratios set forth by the manufacturer of the thermoset, the key characteristic in thermoset properties can be established. This key characteristic is known as the cross-link density where the average molecular weight between the cross-links determines how the cross-linked system will act. Figure 15.6 shows a cross-linked epoxy system of varying molecular weight of the

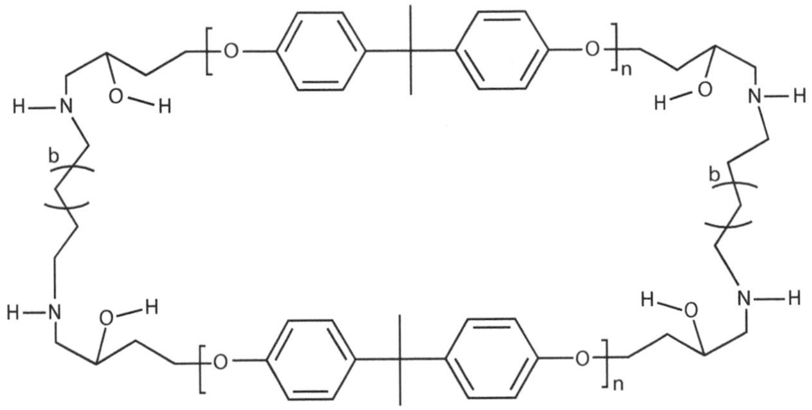

FIGURE 15.6 Diglycidyl ether of bisphenol A (DGEBA) epoxy cross-linked with an amine.

amine cross-linker (remember that the cross-links occur in three dimensions, so the secondary amines can be further cross-linked).

If the amine group has a "b" value of five, the amine cross-linker would be 1,8-diaminooctane, which would be more flexible than if "b" was equal to one (i.e., 1,4-butanediamine). However, the 1,8-diaminooctane cross-linked system has a lower cross-link density and is more susceptible to diffusion of gas or liquids due to the larger interstitial spacing. These are just two simple polymer characteristics that cause properties to vary not only from thermoset to thermoset but also within a given thermosetting polymer category.

15.5.1 Physical Properties

The numerous physical properties of thermosetting polymers are wide ranging since thermosets have a large amount of polymers within this realm. Therefore, the appearance, density, and barrier properties as well as the behavior of some of these properties after degradation shall be covered in Section 15.9.

15.5.1.1 Appearance

One of the major uses for thermosetting polymers is in the area of coatings; therefore, it is important for these polymers to have desirable aesthetics. The main properties that dictate the overall appearance of a coating system are its color and gloss. Color is exhibited by the different wavelengths being absorbed or diffracted by the polymeric system constituents. Color can also be altered with the addition of pigments or dyes. However, for clear coat applications, as the name suggests, the polymer should be colorless or near colorless. For instance, epoxies usually exhibit near colorless appearance due to their chemical structures, but phenolics tend to be more yellow due the highly aromatic nature present within their polymeric backbone.

Color can be measured using a variety of different instrumentation, but spectrophotometers are the instruments that most easily quantify colors by numerical systems under a given light source. Color matching can also be performed with this instrument to determine if a given numerical parameter defined by the user has been attained. Visual color matching can be also performed using the protocol described in the ASTM D 1729 standard.

Gloss is another appearance property that can determine whether a certain thermoset will be used over another. Specular gloss is the measure of luminous reflectance relative to a standard at various viewing angles. This is a measure of how shiny the object appears to the observer and depends on the angle of observance. Gloss is both a representation of the underlying substrate composition as well as the surface finish. The smoother the surface is, the more glossy it appears; inherently most thermosetting

TABLE 15.1   Densities of Common Thermosets

Thermosetting Polymer	Phenolics	Epoxies	Polyurethanes	Polyesters	Amino Resins
Range of density (g/cm^3)^a	1.35–2.13	0.86–2.60	0.40–1.24	0.60–2.00	1.50–1.70
Average density (g/cm^3)^a	1.49	1.36	1.03	1.15	1.55

^{a} Properties derived from numerous samples found on the material database (www.matweb.com).

polymers are glossy in their virgin state but can be deliberately changed during synthesis and manufacturing or manipulated by mechanical means (after curing) to have lower gloss.

15.5.1.2 Density

Thermosets typically have similar densities, but as newer technologies are developed and the criteria for some applications become more stringent, densities of thermosets have become lower than what was conventionally typical. Densities of polymeric resins are usually measured according to ASTM D 792 or ASTM D 1895. Table 15.1 shows the ranges of densities for the five most common types of thermosetting polymers that were presented in Section 15.4. Many applications require low densities so that the operational structure has lower weight. This is especially true in the transportation industry where fuel savings are the main economic driving force for selecting a particular thermoset coating product.

15.5.1.3 Barrier Properties

Many thermosets are used for coating applications, and many coating applications are used specifically as barrier coatings. These coatings provide low diffusivity to water or other atmospheric chemicals so that the underlying substrate, whether it is the steel frame of an automobile or an aluminum soda can, will not be compromised throughout its intended lifetime. Cross-link density is a very critical property for understanding barrier properties and is defined as the number of cross-links per unit volume. The greater the cross-link density, the less interstitial space can exist between polymeric chains for water, air, solvents, etc., to pass through, thus leading to coating systems with lower permeation values. However, extensive cross-linking can negatively affect some mechanical properties as described next in Section 15.5.2. Permeability of polymers is usually measured in units of volume or length per unit area per unit time (i.e., cm^3/cm^2/s or cm/cm^2/s), but several other metrics with different units for permeability are accepted. For example, the permeability of a thermoset polyester is roughly 4×10^{-8} g/m s at 37°C, while at the same nominal temperature, an epoxy system has a permeability of 1×10^{-8} g/m s for water as the moving/diffusing species.

15.5.2 Mechanical Properties

Unlike the monolithic metals with long-range order covered in earlier chapters, the mechanical properties of thermosets are not as reproducible or as consistent due to the presence of disordered polymeric chains. The bulk strength of these polymer systems is usually derived mainly from the strength of the covalent bonds between both the polymer and its corresponding hardener molecules. Thermosets can even have different properties that depend on the manufacturing process used; namely, certain extrusion processes can align molecular chains, whereas casting usually produces randomly oriented polymeric chains. This process-induced morphology along with other variables produces a larger standard deviation in mechanical properties than is typical for metallic materials.

15.5.2.1 Tensile Properties

Tensile properties of thermosets are also more difficult to measure due to their viscoelastic behavior that is also time and applied load dependent. After a slight initial elastic deformation occurs, a combination of strain-rate-dependent viscous and elastic effects influences how the polymer component under external load deforms and approaches its rupture point. In modeling these systems with equivalent mechanical components, a combination of springs (elastic effects) and dashpots (viscous effects) can be used in

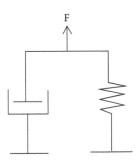

FIGURE 15.7   Voigt or Kelvin model of simple viscoelastic solids.

TABLE 15.2   Mechanical Properties of Thermosetting Polymers

Thermosetting Polymer	Phenolics	Epoxies	Polyurethanes	Polyesters	Amino Resins
Tensile strength (MPa)[a]	50–55	55–130	55–70	35–104	55–70
Elastic modulus (GPa)[a]	2.7–4.1	2.8–4.1	1.5–2.0	2.0–4.4	3.0–4.5
Strain at break (%)[b]	2	6	6	3	1

[a] Properties derived from numerous samples found on the material database (www.matweb.com).
[b] Properties cited in context with Askeland and Phule (2006).

series and/or parallel to simulate the mechanical response of that material. The most widely accepted model for viscoelastic solids is the Voigt or Kelvin model where a spring and dashpot are in parallel (Strong 2006). Figure 15.7 shows this simple model of behavior.

It can be seen from the model that it is a combination of elastic solid (i.e., Hooke's law) and viscous liquid (i.e., Newtonian fluid). Again, this is simple approximation of a system that is obviously more complex than this model, but overall the model helps gain an understanding of the material's behavior under loading. By determining the magnitude for the characteristic parameters for each equivalent mechanical constituent, an understanding of the amount of permanent deformation, sound damping, etc., can be deduced. Table 15.2 lists typical values for the tensile strength, elastic modulus, and strain at failure for common thermosetting polymers. The ranges of values are representative of several commercially available products within each class of thermosets.

It can be seen that most amino resins have a stiffer behavior with the lowest elongations at break. This can be attributed to the presence of cyclic nitrogen containing rings within most of these resins. For instance, melamine formaldehyde has cyclic melamine groups with carbon-to-nitrogen double bonds that act very stiffly. Another factor that contributes to the high modulus and low elongation values is the size and frequency of the side chains containing very large ether groups (Wicks et al. 2007). Phenolics and epoxies also possess cyclic structures in their polymeric backbone making them relatively stiff; however, bisphenol A-derived epoxies have a greater degree of flexibility due to the presence of large molecular weight segments between the benzene groups. Thus, properties can be tailored for any strength, stiffness, or elongation values by selecting the appropriate polymer backbone or hardening agent and the amounts of these precursors. The standard most often used for the testing of polymeric specimens in tension is ASTM D 638 where a dog-bone-shaped specimen is utilized to direct forces into the reduced cross-sectional region so that failure initiates in that region.

15.5.2.2 Compressive Properties

Since thermosets are covalently bonded polymeric chains, when loaded in compression, their polymer chains fail in a buckling column fashion. The bulk specimen can be thought of as consisting of several parallel thin columns supporting a load, and once a few of the columns buckle, the whole structure begins to buckle. Accordingly, most polymers do not have equivalent tensile and compressive values for their strength, stiffness, or elongation values unlike monolithic metals. In general, their compressive properties are lower in magnitude than those for their tensile counterparts. However, thermosets also generally have higher compressive properties than those for thermoplastics because their cross-linking bonds act as stabilizers against buckling. To determine compressive properties, ASTM D 695 can be used, but other standards are available if the thermosetting polymer is in a foamed form.

15.5.2.3 Flexural Properties

Flexural properties reveal a phenomenon not seen in metal or ceramic materials. Since both tensile and compressive modes are present during flexural loading, the compressive side of the specimen usually fails first due to their compressive strengths and stiffness being lower than the corresponding tensile properties. Flexural properties may be of greatest interest to design personnel because they are the one of the most common loading and failure situations for nonreinforced polymers. Car seats, benches, polymeric flooring, etc., are all subjected to such flexural loading. Flexural properties can be measured by following the protocol described in ASTM D 790. Three-point bending or four-point bending can be used, but four-point is more common since there is a constant bending moment between the inner two bearing points. This constant moment produces less measurement error and, thus, greater accuracy for the measured results because the distance from the support to the center points (the dimension measured/used in three-point bending calculations) can be neglected.

15.5.3 Thermal Properties

15.5.3.1 Glass Transition Temperature

The glass transition temperature (T_g) is that temperature at which a polymer transitions from its glassy state into a more pliable rubbery-leather state. Above the T_g, long-chain motion of the polymer backbone allows deformation to occur in rubbery manner, while polymers below their T_g act in a glassy manner because their chains are held in place. Due to the cross-linking that occurs in thermoset chemistry, thermosetting polymers usually have higher glass transition temperatures than thermoplastics. This occurs because their cross-links impede the long-range polymeric chain motion needed for a rubbery deformation. With thermosets, the T_g can be an indicator of how cross-linked the system is and gives insight into the extent that curing has occurred during polymer processing. However, the desired cross-link density is usually a variable that can be changed for its intended application, so large ranges of T_gs are possible for a variety of thermosetting polymers. For instance, epoxy resins have a range of 100°C–250°C, while thermosetting polyesters have a range of 75°C–150°C (Agarwal et al. 2006). The differential scanning calorimetry (DSC) technique is used for measuring the glass transition temperatures of liquid polymers by measuring the heat absorbed by the test sample as the temperature is slowly raised. With DSC, the T_g is determined by a change in the heat flow due to chain motion of twisting, bending, etc., that causes the specimen to absorb energy. Other commonly used methods for evaluating thermosets are dynamic mechanical analysis (DMA) and thermal mechanical analysis (TMA) where a sudden change in deformation under constant load is measured as the temperature is ramped up through a temperature range. ASTM D 7028 is an accepted test method for determining the glass transition temperature with a DMA.

15.5.3.2 Heat Deflection Temperature

The heat deflection temperature (HDT) is a common measure for the formability of polymers under specified loading conditions. The HDT of a polymer is measured by placing a specimen in a three-point bend loading jig and applying a constant force; the HDT is the temperature at which a sample with standardized dimensions has deformed to a given amount. The T_g of the polymer is always greater than the HDT since the T_g is usually measured by placing an infinitesimally small load (relative to DMA and TMA loads) on the specimen, while the standard load for determining the HDT is either 0.46 or 1.80 MPa. The HDT is measured by following the protocol proscribed in ASTM D 648; there, the required dimensions of the sample and required deformation percentage are specified. This measurement gives

TABLE 15.3 HDTs of Thermosetting Polymers

Thermosetting Polymer	Phenolics	Epoxies	Polyurethanes	Polyesters	Amino Resins
Deflection temperature (°C)[a]	160–226	49–243	55–136	47–270	155–270

[a] Properties derived from numerous samples found on the material database (www.matweb.com).

thermoset processors and formers an indication of the maximum temperature that the cured material can be formed at without possibly degrading it. Table 15.3 demonstrates the HDT ranges under the 1.8 MPa loading procedure for the five common thermoset materials being covered in this chapter. Phenolics and amino resins have higher deflection temperatures due to their rigid nature that are due to the highly aromatic and cyclic bonding in their polymer backbone.

15.5.3.3 Coefficient of Thermal Expansion

The coefficient of thermal expansion (CTE) of a material is a measure of the rate of a material contraction or expansion during cooling or heating, respectively. When energy in the form of heat is introduced into a thermosetting polymer, the polymeric chains begin twisting and moving apart to dissipate the energy. This motion causes the material's dimensions to essentially increase in multiple directions since their polymeric chains are in a random orientation. Therefore, two modes of measurement (linear and volume coefficients of thermal expansion) have been identified to understand how the thermoset behaves. ASTM D 696 lists the protocol for measuring linear thermal expansion, while ASTM D 864 is used to measure the volume expansion. Since thermosets are cross-linked, the covalent bonding of the cross-links to the polymeric backbone causes less motion/expansion upon heating. Thus, thermosets usually have smaller coefficients of thermal expansion. For example, epoxies have a linear CTE of $45–65 \times 10^{-6}/°C$, while a high-performance, polyphenylene sulfide has a higher linear CTE of $99 \times 10^{-6}/°C$ (Agarwal et al. 2006).

15.6 Degradation Principles for Thermosets

15.6.1 Thermodynamics

Polymer degradation implies the breaking of chemical bonds within a polymer matrix leading to a change in appearance, mechanical properties, barrier properties, or any other property that is needed for a polymer to perform its function in a given application. A set amount of energy is needed to achieve this. That amount of thermal, chemical, or radiative energy is equivalent to the strength of the chemical bond that holds the polymer together. Table 15.4 shows the amount of energy required to break various chemical bonds also known as the bond dissociation energy (Lide 2003).

It is important to note that the data in Table 15.4 were found strictly for diatomic bonding cases. Additional chemical bonds on either atom would change that bond dissociation energy. These values essentially represent the positive enthalpy change associated with the breaking of that chemical bond. For all cases of degradation, a greater amount of negative enthalpy change associated with the degradation reaction is required to break these covalent bonds. Typically, this takes high temperature, UV radiation, strong oxidants such as ozone, or large amounts of mechanical strain (Schnabel 1981).

While one can try to construct polymers with higher bond strengths in the backbone, these modification effects are not large (less than an order of magnitude) when dealing with typical organic compounds

TABLE 15.4 Bond Dissociation Energies for Various Diatomic Chemical Bonds

Diatomic Bond	C–C	C–N	C–H	C–O	O–H	N–H
Bond dissociation energy (kJ/mol)	610 ± 2.0	748.0 ± 10	338.4 ± 1.2	1076.5 ± 0.4	429.99 ± 0.38	≤ 339

Source: Lide, D., *Handbook of Chemistry and Physics*, 84th edn., CRC Pr I Llc., New York, 2003.

such as thermoset polymers. With the abundance of UV radiation that many thermoset polymers are exposed to outdoors, degradation becomes a kinetic issue and not a thermodynamic one. Because sufficient energy to break chemical bonds will be in abundance in an exterior/outdoors application, these reactions are going to take place; it is just a matter of the rate at which they happen. If that rate can be slowed sufficiently, the coating will last for an acceptable amount of time.

15.6.2 Kinetics

In polymer degradation, there two basic forms of reactions occur: there are chain reactions such as autoxidation and single-step processes or reactions such as the hydrolysis of polyesters. In chain reactions, the reaction rate varies exponentially with the initiation rate, while in single-step processes, the reaction rate varies linearly with the initiation rate. Therefore, if one can suppress the initiation of specific chain reactions in a specific polymer system, the degradation rate of that polymer can be greatly decreased. The other issue to be taken into account is to possibly reduce/eliminate/shield the presence of certain moieties within the polymer that are susceptible to specific degradation pathways (Schnabel 1981). An example of this would be aromatic rings such as a styrene ring or phenolic resins that tend to strongly absorb UV radiation. These moieties tend to form free radicals on surrounding linkages leaving the aromatic ring intact to continue absorbing radiation and initiating degradation. In these cases, it is usually necessary to coat the polymer with a UV blocking layer. Therefore, the more aromatic groups that are present in a polymer, the faster it will be degraded by UV radiation. Usually, the use of these polymers in exterior applications is to be avoided (Rosu et al. 2009).

In the case of polyesters, steric hindrance can be used to suppress hydrolysis. When there are carbons in the six positions around the ester, they tend to crowd the ketone group and prevent its reaction with water. This can drastically reduce the degradation kinetics of polyesters. Anchimeric effects (intramolecular catalysis effects), however, can also play a role in accelerating the hydrolysis if the moieties next to the polyester are not chosen carefully. These effects are greatest for telechelic groups such as glycols, while steric effects dominate on the backbone of polyesters (Johnson et al. 2004).

In the case of thermodegradation, the reaction mechanism and the kinetics of that mechanism are largely dependent on the structure of the polymer and must be determined on a case-by-case basis. Some general rules do apply however. Aromatic rings and double bonds within the polymer backbone tend to resist thermal degradation. In the case of combustion, polymers containing these moieties tend to char, which limits the combustion reaction. Cross-linking also helps produce this effect (Jellinek 1983).

Overall, the mechanisms and kinetics of polymer degradation are quite specific to each polymer structure, but there are some general rules that do apply. These mechanisms have been discussed in general terms for conciseness and to give an overview of thermoset polymers in general. A large amount of work has been done investigating the degradation of each system specifically and should be consulted before using a polymer in a given application.

15.7 Degradation Protection

Degradation can come in many forms whether it is from thermal, UV exposure, oxidation, moisture, etc., so a wide variety of degradation inhibitors have been instituted into thermosetting polymers to decrease the degradation process. The following subsections will cover three of the most common types of degradation protection materials.

15.7.1 Ultraviolet Stabilizers

UV light is present in everyday applications and has a relatively high energy due to its short wavelengths of roughly 10–400 nm. UV light can cause degradation to thermosets in a variety of ways, but the most

$$A \xrightarrow{\text{UV}} A^*$$

$$A^* \longrightarrow A + heat$$

and

$$(\text{Polymer}) \xrightarrow{\text{UV}} (\text{Polymer})^*$$

$$(\text{Polymer})^* + Q \longrightarrow (\text{Polymer}) + Q^*$$

$$Q^* \longrightarrow Q + heat$$

FIGURE 15.8 Reaction schemes of both absorbing and quenching mechanisms of UV stabilization. (From Wicks, Jr. Z. et al., *Organic Coatings Science and Technology*, John Wiley & Sons, Inc., Hoboken, NJ, 2007.)

noted is the creation of free radicals within the polymeric system. In some instances, this free-radical creation can cause polymeric chain scission, which basically breaks the covalent bonds within the chain, or the radical can further induce cross-linking to occur that may make the thermoset very brittle.

UV stabilizers can work in two separate manners, either they absorb the UV radiation or they quench the free-radical species. Both mechanisms require the release of heat to return the absorber or quencher back to its ground state, which allows it to repeat the process of stabilizing the UV radiation. The reaction scheme of both quenchers (Q) and absorbers (A) can be seen in Figure 15.8.

It can be seen that absorbers directly transfer the incoming radiation to thermal energy, while the quenchers find the radical form of the polymer and quench its excitation into the form of heat.

Depending on the polymeric system, some quenchers may act as absorbers; in other systems, because of the nature of stabilization, it is dependent of the polymeric structure. Some of the most common types of quenchers and absorbers are benzophenones, benzotriazoles, triazines, and oxalanilides (Wicks et al. 2007). All these compounds contain aromatic rings, which are known constituents that absorb UV.

15.7.1.1 Pigments

A special section of UV stabilizers are categorized by their colorant nature. These UV absorbers are known as pigments. The most efficient UV absorber used is carbon black. Another well-known pigment that absorbs UV is titanium dioxide specifically in the rutile form. The anatase form also absorbs UV, but the characteristic diameter of the rutile form makes it a more efficient pigment for absorption (Wicks et al. 2007). Another pigment, yellow iron oxide, has been shown to effectively absorb UV radiation as well (Sun et al. 2007). Other pigments that act as absorbers are zinc oxide, magnesium oxide, calcium carbonate, and barium sulfate (Schnabel 1981). Another characteristic that makes carbon black so beneficial is its ability to slow degradation via UV radiation, as well as also to act as an antioxidant.

15.7.2 Antioxidants

Oxidation is a degradation process that can be simply from atmospheric oxygen creating or bonding with free radicals or autoxidation that can occur after the absorption of UV radiation. Peroxides and hydroperoxides are a by-product in the oxidation process and are constituents that will eventually cause polymeric chain scission. However, the oxidation occurs, and it is detrimental to the mechanical, chemical, and thermal properties of the polymer, but this can be counteracted with the use of antioxidants.

Antioxidants provide protection by two related mechanisms. First is the preventative antioxidants that act by interrupting the oxidation process by preventing the oxidation species from reacting with the polymer, which can be seen in Figure 15.9. Two such common antioxidants are sulfide and phosphite

$$\text{PolymerOOH} + (\text{PhO})_3\text{P} \longrightarrow \text{PolymerOH} + (\text{PhO})_3\text{P} = \text{O}$$

FIGURE 15.9 Phosphite antioxidant reaction scheme to create alcohol. (From Wicks, Jr. Z. et al., *Organic Coatings Science and Technology*, John Wiley & Sons, Inc., Hoboken, NJ, 2007.)

$$R_2N - R' + O_2 \xrightarrow{\text{UV}} R_2NO^* + \text{Products}$$

$$R_2NO^* + (\text{Polymer})^* \longrightarrow R_2NOH + R_2NO(\text{Polymer}) + \text{Polymer(minus H)}$$

$$R_2NOH(\text{polymer}) + POO^* \longrightarrow R_2NO^* + \text{PolymerOOH(Polymer)}$$

FIGURE 15.10 HALS reaction to mediate UV radiation. (From Wicks, Jr. Z. et al., *Organic Coatings Science and Technology*, John Wiley & Sons, Inc., Hoboken, NJ, 2007.)

compounds that are initially oxidized to sulfates and phosphates and are relatively harmless in polymeric substrates (Wicks et al. 2007).

Second is a chain-breaking antioxidant that intercepts the autoxidation radical species from propagating, thus causing a peroxide or hydroperoxide to be formed. This is why both preventative and chain-breaking antioxidants must be used together so that the peroxide specimens from the chain-breaking process can be neutralized by the preventative antioxidants. Some widely accepted chain-breaking antioxidants for uses in thermosetting systems include phenolics, aromatic amines, and heterocyclic amines (Pospísil and Nespurek 1995).

15.7.3 Hindered Amine Light Stabilizers

Hindered amine light stabilizers (HALSs) protect against light-initiated degradation as the name suggests but can also act as stabilizers to thermal degradation as well. Most hindered amines have a nitrogen containing ring (i.e., piperidine) that has many pendent groups around that amine, thus sterically hindering the nitrogen. The nitrogen is also attached to another substituent, which can be hydrogen or alkyl groups. When UV light and oxygen are present, oxygen will react with the nitrogen of the HALS to form nitroxyl (R_2NO) radicals, thus kicking out the substituent group on the nitrogen. These nitroxyl radicals reduce the propagation of the degradation in the polymeric system by reacting with the peroxy radicals that form during autoxidation. After reaction with the peroxy radical, another nitroxyl radical is formed that can again repeat the process of subduing the harmful peroxy radicals (Hailant 2008). This reaction scheme can be seen in Figure 15.10.

Some common HALS include derivatives of 2,2,6,6-tetramethylpiperidine that have a substituent hydrogen or alkyl groups, alkanoyl HALS, hydroxylamine ethers, and varieties of these with different molecular weights and pHs (Wicks et al. 2007).

15.8 Specific Standardized Tests

Most thermosets are susceptible to oxidation, UV light, or thermal degradation that leads to a wide variety of standardized tests to understand what element of degradation will cause damage to the thermoset. The American Society of Testing and Materials (ASTM) and the International Standards Organization (ISO) have set forth guidelines for testing protocols to determine if material degradation is occurring. Since both standards organizations use many of the same guidelines, many of the ASTM standards will be covered as opposed to duplicating ISO standards. The following sections will cover commonly utilized standards and how they relate to the degradation of thermoset polymers. The standards to be covered can all be viewed, in summary, and purchased at the website for the ASTM (www.astm.org).

15.8.1 Accelerated Exposure

Environmental exposure subjects materials to a variety of different UV wavelengths at relatively low intensities and can take months to even years to show degradation characteristics. When researching the effects of UV exposure, researchers need results in shorter time frames, so accelerated exposure instruments are commonly used in industry and academia. These exposure facilities incorporate the use of light sources containing wavelengths in the UV spectrum, but the relative intensities are much higher than found in outdoor exposure, thus accelerating the weathering of the specimen. However, the correlation between natural environmental exposure and accelerated exposure is difficult if not impossible; therefore, research is being conducted to obtain a better understanding of accelerated weathering.

ASTM D 1435 is the protocol for "Standard Practice for Outdoor Weathering of Plastics." It involves the use of an outdoor setting to monitor the response of the plastic to UV radiation, temperature changes, and pollutants, among many other factors. However, the use of this standard does not show repeatability from location to location, so it is becoming more accepted to send specimens to be tested to specific locations with nominally standard/consistent exposure conditions. Along the same lines, ASTM D 4364 is the test standard "Standard Practice for Performing Outdoor Accelerated Weathering Tests of Plastics Using Concentrated Sunlight" where a reflection device is used to concentrate the UV, visible, and infrared radiation supplied by the sun onto a plastic sample where the increased flux increases the "natural" weathering rate. ASTM G 154 encompasses the use of UV lamps in testing chambers with higher intensity UV wavelengths to illuminate the sample to increase the degradation speed and thus decrease the amount of exposure time. The standard entitled "Standard Practice for Operating Fluorescent Light Apparatus for UV Exposure of Nonmetallic Materials" can use various different UV lamps including UVA-340, UVB-313, and UVA-351. The numbering of these lamps is significant because it is the desired maximum intensity wavelength, in nanometers, for that lamp. ASTM G 154 also includes the use of moisture and a programmed schedule of differing chamber temperatures to increase the aging rate of the sample. Finally, ASTM D 2565 entitled "Standard Practice for Xenon-Arc Exposure of Plastics Intended for Outdoor Applications" specifies that a xenon-arc lamp be used instead of UV lamps to supply the UV radiation. The xenon arc has a different spectral power distribution than the UVA and UVB lamps covered earlier. Usually, xenon-arc lamp outputs cover the UV, visible, and IR wavelengths radiation spectra that is similar to natural exposure. This corresponds to chamber exposures having different levels of weathering effectiveness on thermoset polymers.

Thermal aging is also another area of concern for many thermosetting polymer applications. If used in an exterior application, temperature ranges can vary drastically dependent on location. ASTM D 6944 titled "Standard Practice for Resistance of Cured Coatings to Thermal Cycling" determines the performance of cured coatings under a variety of specifications. The adherence of coatings to their substrate is measured since this functionality is needed to provide their primary function of protecting the underlying substrate. Also, since their mechanical properties must be retained if the application calls for that performance, these too are evaluated. ASTM D 6944 uses a method of freezing and thawing the coating system to produce internal mechanical stresses due the differences in CTEs between several of the coating constituents to evaluate such thermomechanical causes of coating failure.

15.8.2 Measurement of Degradation

The measurement of specimens to see if degradation has occurred due to weathering is dependent upon the application and the mode of failure, so standardization of these tests is somewhat trivial. Application-driven testing of degraded materials is usually the researcher's responsibility. Comparisons are usually made between virgin material and the weathered material. This can involve the use of ASTM standards specified in Section 15.5 to account for changes to the appearance and mechanical, thermal, electrical, etc., properties of the thermoset. As thermosets are weathered by accelerants, a drop in strength and stiffness of flexural nature are to be expected due to the polymeric chain scission, but in some instances, free-radical polymerized thermosets show a slight increase in properties at the onset of weathering due to the increased cross-linking from free-radical formation (Hammami and Al-Ghuilani 2004). Color and gloss are other properties monitored to determine if weathering will cause an appearance-related effect/change of the thermoset. The ASTM D 4674 "Standard Practice for Accelerated Testing for Color Stability of Plastics Exposed to Indoor Office Environments" uses accelerated weathering to determine if the color change exhibited by indoor polymers is acceptable.

Gloss may also change upon weathering due the surface roughening from polymeric chains moving about due to chain scission (Correia et al. 2006). Again, these are just a small sampling of the acceptable tests to determine if degradation is occurring.

15.8.3 Degradation Measurement of Coatings

Historically, evaluation of coating degradation and the subsequent corrosion processes have been a visual science. A perusal of ASTM methods for corrosion detection of coated surfaces details the emphasis on the visual aspect of corrosion detection. In ASTM B117 (Standard Test Method of Salt Spray Testing), one of the most widely accepted corrosion evaluation methodologies provides no insight for interpreting the results; therefore, many users of ASTM B117 rely on mass loss measurements. Mass loss measurements, while valid especially for uniform corrosion processes, provide very little insight into the overall processes and mechanisms of coating or material failure. Other users of ASTM B117 scribe through the coating in a standardized fashion to expose bare metal. The panels are then exposed to the salt spray chamber where, at timed intervals, the panels are removed and examined for blistering away from the scribe. The scribe is investigated for any undercutting due to corrosion or loss of adhesion. Indeed, ASTM D1654 (Standard Test Method for Evaluation of Painted or Coated Specimens Subjected to Corrosion Environments) details the scribing technique often used for ASTM B117 testing. In ASTM D1654, a rating system is provided where the rating is determined by approximate inches of creepage from the scribe and the percent of area failed (nonscribe area). Advances have been made in the visual assessment of coating condition as noted by ASTM D714 (Standard Test Method for Evaluation Degree of Blistering of Paints). In this standard, a photographic technique is used to remove some of the user error in estimating the uniformity and size of blistering of coated samples. The standard provides pictures of samples with various blistering concentrations such as medium dense and dense blistering.

Electrochemical techniques allow for the determination of changes in material properties that often occur prior to any observable visual changes and provide information regarding corrosion mechanisms. Electrochemical impedance spectroscopy (EIS) and scanning vibrating electrode technique (SVET) provide one with information that is far superior to the historically conventional visual assessments (Scheffey 1986). Now, small changes in capacitance and resistance of a coated sample can be monitored via EIS, and small currents associated with corrosion can be measured via SVET. These changes can be followed by using electrical circuit element modeling to extract further information from the EIS results. Furthermore, EIS characterization of the degradation of coated substrates may allow for the quick determination of the barrier properties of a coating and, therefore, allow for the extraction of a service life prediction.

For example, EIS is a frequency-dependent method where a small voltage perturbation is made to a sample, usually about its naturally occurring corrosion potential (Mansfeld 1995). The corresponding current flow due to the voltage perturbation is measured; a form of Ohm's law (Equation 15.1) is followed letting one calculate the impedance (ohms) as the voltage and current are known:

$$z(\omega) = \frac{v(\omega)}{i(\omega)} \tag{15.1}$$

where
   $z(\omega)$ is the complex impedance and accounts for the relationship between amplitudes of the voltage and current signals as well as the phase shift between them
   $v(\omega)$ and $i(\omega)$ are the voltage perturbation and corresponding current, respectively

EIS measurement results for a coated metal often display data similar to that shown in the Figure 15.11. The graphs on Figure 15.11 are commonly referred to as a Bode plot and phase angle plot. In the graphs,

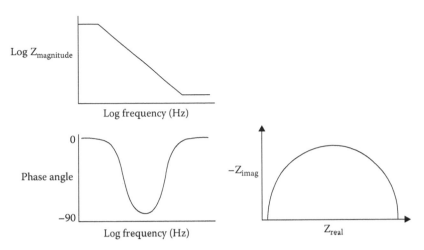

FIGURE 15.11 Schematic of EIS results for metal coated with a thermoset coating.

the $Z_{modulus}$ (ohms) is plotted versus the frequency (Hz), and the phase angle is plotted against the frequency (Hz). The right graph is referred to as a Nyquist plot. In it, the Z_{imag} is plotted versus the Z_{real}. The Zimag is any component of the impedance that is not in phase with the initial perturbation, while the Zreal is the remaining impedance that is in phase with the perturbation. From this graphical form, it is easily observable whether or not that sample is displaying a capacitive behavior, a Randle's cell behavior (a resistor followed by a capacitor and resistor in parallel), or diffusion controlled behavior; many other behaviors are also possible. The behavior of a thermoset-coated metal is depicted in Figure 15.11 (Randle's cell behavior). While EIS as a method for studying corrosion and coatings has matured greatly over the past few decades, its use has only been selectively applied at this point.

Coating systems often utilize thermosetting polymers due to desirable mechanical, thermal, processing, and various other properties. The primary function of coatings is to provide a protective layer from numerous elements and compounds that cause the substrate to degrade. One of the major users for coatings is the corrosion mitigation sector of the corrosion industry. Many common construction materials such as steels and aluminum alloys undergo corrosion readily when an electrolyte containing medium is present. Barrier coatings perform the task of separating the substrate from these types of solutions. The aforementioned nondestructive method, EIS, can be used to measure the integrity of the entire coating and substrate system. It has been shown that water ingression can be monitored by EIS in thermoset composite systems (Davis et al. 2004). Another advantage of EIS over qualitative techniques is the use of modeling of the electrochemical impedance spectra with electrical resistors, capacitor, constant phase elements, etc., to determine specific tendencies. For instance, pore resistance is an element of circuit modeling that corresponds to the ease of solution penetration via pore development resulting from coating aging via either natural or accelerated weathering (Lavaert et al. 2000). Figure 15.12 is an example of the electrochemical impedance spectra for epoxy coatings filled with mica platelets that have been weathered under the ASTM B 117 (constant exposure to a salt spray at 35°C). The pore resistance can be estimated by roughly evaluating the point where the magnitude of the impedance behavior changes from a roughly 45° angle (i.e., capacitive behavior) to horizontal. These results show that the pore resistance drops during each successive testing period until finally the coating has become porous enough that the measurement reveals an almost completely resistive pattern. Again, these are only a sampling of the variety of electrochemical processes used to determine if degradation, either by corrosion, pore formation, etc., is occurring within the specimen. From Figure 15.12, one can determine that while the coating did display high impedance initially, indicating good barrier performance, the coating failed very quickly. After only 9 days of weathering, the coating was providing no protection at all as evidenced by the extremely low impedance values.

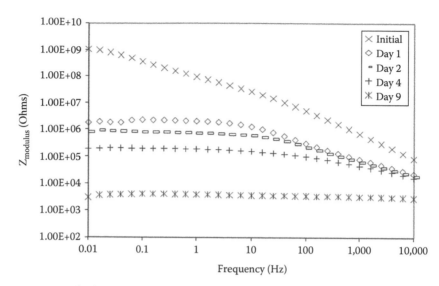

FIGURE 15.12   EIS results for an epoxy coating filled with mica platelets.

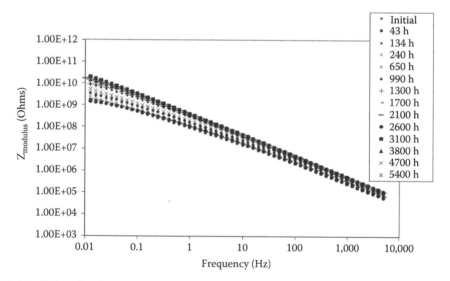

FIGURE 15.13   EIS results of a CARC coating system after accelerated weathering.

In comparison with the aforementioned EIS results, the results in Figure 15.13 for a standard Army CARC coating display extremely durable barrier properties (United States Department of Defense 2005). After 4300 h of weathering in ASTM B117, only one order of magnitude decrease in low-frequency impedance can be observed. CARC coatings are formulated to have extremely high cross-link densities so that these coatings have high chemical resistance and high barrier properties; CARC also provides low observability (camouflage) properties to Department of Defense vehicles and structures.

15.9 Summary

Thermosetting polymers have a wide range of uses, partly from their benefits and advantages in comparison with thermoplastic polymers. Certainly, thermosets have a wide range of properties that require study and measurement via a multitude of ASTM and ISO standards. In addition to these standard methods of measurement, the degradation of thermoset coatings can be studied via electrochemical techniques that provide quantitative results detailing the changes in resistance and capacitance that occur with water ingression and coating degradation.

Acknowledgments

The support of this research by the U.S. Army Research Laboratory under grant No. W911NF-04-2-0029 and W911NF-09-2-0014 is gratefully acknowledged.

References

Agarwal, B., L. Broutman, and K. Chandrashekhara. 2006. *Analysis and Performance of Fiber Composites*, John Wiley & Sons, Inc., Hoboken, NJ.

Askeland, D. and P. Phule. 2006. *The Science and Engineering of Materials*, Thompson, Toronto, Ontario, Canada.

Baghdachi, J., H. Hernandez, and C. Templeman. 2010. Self-stratifying automotive topcoat compositions and processes, United States patent application 20100087596.

Correia, J., S. Cabral-Fonseca, F. Branco et al. 2006. Durability of pultruded glass-fiber-reinforced polyester profiles for structural applications. *Mechanics of Composite Materials*, 42(4), 325–338.

Davis, G., M. Rich, and L. Drzal. 2004. Monitoring moisture uptake and delamination in CFRP-reinforced concrete structures with electrochemical impedance sensors. *Journal of Nondestructive Evaluation*, 23(1), 1–9.

Haillant, O. 2008 Spectroscopic characterization of the stabilising activity of migrating HALS in a pigmented PP/EPR blend. *Polymer Degradation and Stability*, 93(10), 1793–1798.

Hammami, A. and N. Al-Ghuilani. 2004. Durability and environmental degradation of glass-vinylester composites. *Polymer Composites*, 25(6), 609–616.

Jellinek, H. 1983. *Degradation and Stabilization of Polymers (Retroactive Coverage)*, Elsevier, Amsterdam, the Netherlands.

Johnson, A., J. Wegner, and M. Soucek. 2004. Hydrolytic stability of oligoesters: Comparison of steric with anchimeric effects. *European Polymer Journal*, 40, 2773–2781.

Lavaert, V., M. De Cock, M. Moors et al. 2000. Influence of pores on the quality of a silicon polyester coated galvanised steel system. *Progress in Organic Coatings*, 38(3–4), 213–221.

Lide, D. 2003. *Handbook of Chemistry and Physics*, 84th edn., CRC Pr I Llc, New York.

Mansfeld, F. 1995. Use of electrochemical impedance spectroscopy for the study of corrosion protection by polymer coatings. *Journal of Applied Electrochemistry*, 25(3), 187–202.

Pascault, J., H. Sautereau, J. Verdu et al. 2002. *Thermosetting Polymers*, Marcel Dekker AG, New York.

Pieper, R., A. Ekin, D. Webster et al. 2007. Combinatorial approach to study the effect of acrylic polyols composition on the properties of crosslinked siloxane-polyurethane fouling-release coatings. *Journal of Coatings Technology and Research*, 4, 453–461.

Pospísil, J. and S. Nespurek. 1995. Chain-breaking stabilizers in polymers: The current status. *Polymer Degradation and Stability*, 49(1), 99–110.

Rosu, D., L. Rosu, and C. Cascaval. 2009. IR-change and yellowing of polyurethane as a result of UV irradiation. *Polymer Degradation and Stability*, 94, 591–596.

Scheffey, C. 1986. Electric fields and the vibrating probe for the uninitiated. *Progress in Clinical and Biological Research*, 210, xxv–xxxvii.

Schnabel, W. 1981. *Polymer Degradation: Principles and Practical Applications*, Macmillan Publishing Company Inc., New York.

Strong, B. 2006. *Plastics: Materials and Processing*, Pearson Education, Inc., Upper Saddle River, NJ.

Sun, C., B. Zhou, N. Li et al. 2007. Preparation of nanometer yellow iron and its UV absorption capacity. *Journal of University of Science and Technology Beijing—Mineral, Metallurgy, Material*, 14(1), 72–76.

United States Department of Defense. 2005. Coating, aliphatic polyurethane, single component, chemical agent resistant. Military Specification MIL-DTL-53039B, Washington, DC.

Verkholantsev, V. 1995. Heterophase and self-stratifying polymer coatings. *Progress in Organic Coatings*, 26, 31–52.

Wicks Jr., Z., F. Jones, P. Pappas et al. 1999. *Organic Coatings Science and Technology*, Wiley-Interscience, New York.

Wicks Jr., Z., F. Jones, P. Pappas et al. 2007. *Organic Coatings Science and Technology*, John Wiley & Sons, Inc., Hoboken, NJ.

16
Elastomers

Anusuya Choudhury
Indian Institute of Technology,
Kharagpur

Anil K. Bhowmick
Indian Institute of Technology,
Kharagpur

16.1 Introduction

The term "elastomer" refers to any member of a class of polymeric substances that possess elasticity, that is, the ability to regain shape after deformation. Elastomers can be naturally occurring polymers, such as natural rubber (NR), or they can be synthetically produced substances, such as butyl rubber, Thiokol, or neoprene. For a substance to be a useful elastomer, it must possess high molecular weight and flexible polymer chains. Elastomers are the base material for all rubber products, both natural and synthetic, and for many adhesives. Each of the monomers that link to form the polymer is usually composed of carbon, hydrogen, oxygen, and/or silicon. Elastomers are amorphous polymers at temperatures above their glass transition temperature, where considerable segmental motion is possible (Blow and Hepburn 1971, Morton 1973, Mark et al. 2005, Billmeyer 2007). In order to make them useful materials, elastomers are invariably cross-linked.

At ambient temperatures, usually amorphous rubbers are thus relatively soft (E ~ 3 MPa) and deformable. In order to be made into useful rubber products, elastomeric materials must be subjected to various modifications. These include strengthening of the material by cross-linking the polymer chains (for instance, with sulfur atoms by a process known as vulcanization), further strengthening by fillers such as carbon black, and also treatment with chemicals that provide resistance to weathering and chemical attack. For fabrication into adhesives, elastomers are often dissolved in organic solvents and treated with various other additives to improve their application, adhesion, and durability. The elasticity is derived from the ability of the long chains to reconfigure themselves to distribute an applied stress. The covalent cross-linkages ensure that the elastomer will return to its original configuration when the stress is removed. As a result of this extreme flexibility, elastomers can reversibly extend from 5% to 700%, depending on the specific material. Without the cross-linkages or with short, uneasily reconfigured chains, the applied stress would result in permanent deformation. Temperature also affects the demonstrated elasticity of a polymer. Elastomers that have been cooled to a glassy or crystalline phase will have less mobile chains, and consequentially less elasticity, than those manipulated at temperatures higher than the glass transition temperature of that polymer. It is also possible for a polymer to exhibit elasticity that is not due to covalent cross-links but instead for thermodynamic reasons.

Nowadays, rubber is used in diversified applications ranging from toys, garments and undergarments, stationery supplies for students, accessories such as bands, musical instruments, automobile parts, seals, washers, and other utensil components to tires and significant machinery parts in factories. Elastomeric materials are virtually indispensable with regard to their myriad industrial, medical, and consumer applications and continue to make up a considerable portion of the annual production and sales of polymers. Thermoplastic elastomer (TPE) materials also find applications in the automotive and household appliance sectors.

16.2 Applications of Elastomers

The most notable application of rubber occurs for the modern transportation industry, which relies substantially upon rubber tires for on- and off-road propulsion, whether by truck, car, motorcycle, or bicycle, as well as for aircraft. Rubber is the ideal material for tires because of its ability to accomplish multiple critical functions simultaneously: sealing the pressurized cushion of air that softens the ride, providing an extremely flexible and durable membrane to contain this air so that one can realize the benefit of that cushion, and offering high surface friction to give the vehicle traction for propulsion, steering, and braking. In addition, rubbers can be molded into extraordinarily complex configurations and can be bonded to virtually any substrate or reinforcing material to form a composite component, greatly enhancing the engineer's ability to tailor a component function.

However, in all these applications, rubber degrades where the properties change with time and temperature due to the aging process. Rubber degradation is also a function of the specific structure of each polymer. Some rubbers degrade faster than others. In the following paragraphs, applications of a few representative rubbers are given along with their structure.

Choosing the right elastomer for each application is an extremely important material-selection decision (Blow and Hepburn 1971, Morton 1973, Bhowmick et al. 1994, Bhowmick and Stephens 2001, Billmeyer 2007, Bhowmick 2008).

16.2.1 Natural Rubber

The use of NR is widespread, ranging from household to industrial products, where it enters the production stream at the intermediate stage or as final products. Despite the competition of synthetic compounds, NR continues to hold an important place in tire consumption. In particular, its superior tear strength and excellent degradation resistance during heat up makes it better suited for high-performance tires used on racing cars, trucks and buses, and aircraft. Multiple uses of NR in hoses, footwear, battery boxes, foam mattresses, balloons, toys, etc., are well known. In addition to this, NR now finds extensive use in soil stabilization and in vibration absorption and as an asphalt filler for road construction. A variety of NR-based engineering products have been developed for use in these fields.

16.2.2 Styrene–Butadiene Rubber

This elastomer is the first competitor of NR. It is widely used in different fields of applications like pneumatic tires, shoe heels and soles, gaskets, and chewing gum. Styrene–butadiene rubber (SBR) latex (emulsion) is widely used in coated papers and is one of the most cost-effective resins to bind pigmented coatings. It is also used in building applications, as a sealing and binding agent, and as an alternative to polyvinyl acetate (PVA). Though it is more expensive, it offers better durability, reduced shrinkage, and increased flexibility, as well as resistance to emulsification in damp conditions. SBR can be used to "tank" damp rooms or surfaces, a process in which the rubber is painted onto the entire surface forming a continuous, seamless damp-proof liner. Additionally, it is used in some rubber cutting boards.

16.2.3 Ethylene Propylene Diene Methylene Rubber

Ethylene propylene diene methylene (EPDM) rubber is used in vibrators and seals; glass-run channels; radiator, garden, and appliance hoses; tubing; washers; belts; electrical insulation; and speaker cone surrounds. EPDM has good resistance to ozone aging and weathering, water and steam, dilute acids, alkalies, ketones, alcohol, as well as fire-retardant hydraulic fluids and hydraulic brake fluids. However, this elastomer is susceptible to degradation in contact with mineral oils and greases.

16.2.4 Nitrile Rubber

This rubber is used in healthcare industry for the preparation of non-latex gloves, automotive transmission belts, hoses, O-rings, gaskets, oil seals, V-belts, synthetic leather, printer's roller, and cable jacketing. Nitrile-butadiene rubber (NBR) in its latex form is used in the preparation of adhesives and pigment binder. NBR is a general-purpose elastomer for hydraulic and pneumatic systems and is used as a seal energizer for low pressure applications (e.g., pneumatic). It has excellent resistance to mineral oils, greases, water, and hydrocarbon fuels, but not to automotive brake fluid. NBR also has excellent compression set resistance and good elastic properties.

16.2.5 Polyurethane

Polyurethane (PU) products are used in the form of foams accounting for over three-quarters of its global consumption. PU is a premium material for superior sealing performance for use in general hydraulic applications (except it is not suitable for hot water applications), and it has excellent mechanical properties such as high tensile strength, abrasion resistance, tear strength, and extrusion resistance.

16.2.6 Silicone Rubber

Silicone rubbers are widely used in industry, and there are multiple formulations. Silicone rubber has great cold flexibility as well as high thermal resistance. It has excellent dielectric properties and very good resistance to attack by oxygen, ozone, and sunlight.

16.2.7 Fluorinated Rubber

Fluorinated rubber has excellent resistance to mineral oils, hot water, steam, fuels, alcohol, hot oils, and aliphatic and aromatic hydrocarbons; however, it has poor resistance to methanol, ketones, and esters. It further exhibits excellent chemical resistance for use in harsh environments such as phosphate esters.

16.2.8 Hydrogenated Nitrile Rubber

Hydrogenated nitrile rubber (HNBR) has improved wear and extrusion resistance over standard NBR. HNBR has outstanding resistance to heat, ozone, weathering, mineral oils, crude oils with amines, fuels, greases, aliphatic hydrocarbons, and industrial chemicals. It has very good mechanical properties and has an extended high-temperature range (it can be used up to 150°C).

16.2.9 Polychloroprene Rubber

The chief characteristics of this elastomer are its flame retardancy and excellent aging properties in ozone and weathering environments along with resistance to abrasion and flex fatigue. It has good resistance to mineral oils with high aniline points, as well as greases, refrigerants, and water.

16.3 General Properties of Elastomers

A wide range of variations of physical properties like tensile strength, elongation percentage, and abrasion resistance can be found among different commercial rubber products made of the same basic elastomer—and these are the factors that usually greatly affect the quality and durability of rubber seals, cushions, and other rubber products. Other factors like chemical resistance and use temperature range also vary but normally to a lesser degree, making the following information useful for the purpose of pointing the designer/user in the right general direction as far as the selection of rubber compounds is concerned (Tables 16.1 and 16.2). The tables in the succeeding text provide general information for various common elastomeric compounds. Basic elastomers are mixed with a variety of chemicals and ingredients to obtain desired physical properties (Blow and Hepburn 1971, Morton 1973, Bhowmick et al. 1994, Bhowmick and Stephens 2001, Bhowmick, 2008).

16.4 Structure of Elastomers

Most elastomers are hydrocarbons; that is, they are composed principally of carbon and hydrogen and their compounds. Some occur naturally, for example, polyisoprene, which is the latex of the rubber tree and is processed into NR. In 1826, English chemist Michael Faraday (1791–1867) analyzed NR and found it to have the empirical (simplest) formula C_5H_8, along with 2%–4% protein and 1%–4% acetone-soluble materials (resins, fatty acids, and sterols). The molecular weights of rubber molecules range from 50,000 to 3,000,000. Sixty percent of these molecules have molecular weights greater than 1,300,000. The repeating unit in NR has the *cis* configuration (with chain extensions on the same side of the ethylene double bond), which is essential for elasticity. The cis double bonds in the hydrocarbon chain provide planar segments that stiffen, but do not straighten the chain. At 2%–3% cross-linking, a useful soft rubber, which no longer suffers stickiness and brittleness problems on heating and cooling, is obtained. The following illustration (Figure 16.1) shows a cross-linked section of amorphous rubber. The more highly ordered chains in the stretched conformation are entropically unstable and return to their original coiled state when allowed to relax. The molecular structure of various rubbers is given in Table 16.1 (Blow and Hepburn 1971, Morton 1973, Bhowmick et al. 1994, Bhowmick and Stephens 2001, Billmeyer 2007, Bhowmick 2008).

16.5 Principles of Polymer Degradation

Elastomers are processed above their softening temperature. Conditions such as high temperature, presence of oxygen, and substantial shear stress can cause chemical reactions to occur, which in turn cause the onset of degradation. Thermal decomposition is a chemical process in which solid material generates volatile fragments that can burn above the solid material. Elastomers can break down thermally by an oxidative process or by the action of heat. In the presence of oxidants, the decomposition process is accelerated (Slusarski 1984). There are several ways in which polymers decompose: random-chain scission, end-chain scission, chain stripping, and cross-linking. The result of these chemical reactions is the buildup or breakdown of the polymer chain that considerably influences the liquid- and solid-state properties of the polymer. Hence, in all studies that involve melt conditioning

TABLE 16.1 Structure and General Properties of Elastomers

Common Name(s)	Designation and Structure	Composition	General Properties	General Chemical Resistance	
				Resistant to	Attacked by
Natural, gum rubber	NR	Isoprene, natural	Excellent physical properties including abrasion and low-temperature resistance Poor resistance to petroleum-based fluids	Most moderate chemicals, wet or dry, organic acids, alcohols, ketones, aldehydes	Ozone, strong acids, fats, oils, greases, most hydrocarbons
SBR (Buna-S)	SBR	Styrene–butadiene	Good physical properties and abrasion resistance to petroleum-based fluids	Most moderate chemicals, wet or dry, organic acids, alcohols, ketones, aldehydes	Ozone, strong acids, fats, oils, greases, most hydrocarbons
EPDM (EP rubber)	EPDM, EPM	Ethylene propylene diene; ethylene propylene	Excellent ozone, chemical, and aging resistance Poor resistance to petroleum-based fluids	Animal and vegetable oils, ozone, strong and oxidizing chemicals	Mineral oils and solvents, aromatic hydrocarbons

(continued)

TABLE 16.1 (continued)   Structure and General Properties of Elastomers

Common Name(s)	Designation	Composition	General Properties	General Chemical Resistance	
				Resistant to	Attacked by
Neoprene	CR	Chloroprene	Good weathering resistance Flame retarding Moderate resistance to petroleum-based fluids	Moderate chemicals and acids, ozone, oils, fats, greases, many oils, and solvents	Strong oxidizing acids, esters, ketones, chlorinated, aromatic, and nitro-hydrocarbon
Buna-N (Nitrile)	NBR	Nitrile–butadiene	Excellent resistance to petroleum-based fluids Good physical properties	Many hydrocarbons, fats, oils, greases, hydraulic fluids, chemicals	Ozone (except PVC blends), ketones, esters, aldehydes, chlorinated and nitro-hydrocarbons
Silicone	Q, Si	Polysiloxane	Excellent high- and low-temperature properties Fair physical properties	Moderate or oxidizing chemicals, ozone, concentrated sodium hydroxide	Many solvents, oils, concentrated acids, dilute sodium hydroxide
Butyl	IIR	Isobutene–isoprene	Very good weathering resistance Excellent dielectric properties Low permeability to air Good physical properties Poor resistance to petroleum-based fluids	Animal and vegetable fats, oils, greases, ozone, strong and oxidizing chemicals	Petroleum, solvents, coal tar solvents, aromatic hydrocarbons

Material	Abbreviation	Monomers / structure	Properties	Resistant to	Attacked by
Urethane	AU, EU	Polyurethane	Good aging and excellent abrasion, tear, and solvent resistance Poor high-temperature properties	Ozone, hydrocarbons, moderate chemicals, fats, oils, greases	Concentrated acids, ketones, esters, chlorinated and nitro-hydrocarbons
Viton, fluoroelastomer	FKM	Vinylidene fluoride, hexafluoropropylene, tetrafluoroethylene	Excellent oil and air resistance both at low and high temperatures Very good chemical resistance	All aliphatic, aromatic, and halogenated hydrocarbons, acids, animal and vegetable oils	Ketones, low molecular weight esters and nitro-containing compounds
ECH, Hydrin, Herchlor	ECO	Ethylene oxide chloromethyl-oxirane	Good low-temperature properties Excellent oil and ozone resistance Fair flame resistance Low permeability to gases	Similar to nitrile with ozone resistance	Ketones, esters, aldehydes, chlorinated, and nitro hydro

FKM: $-(CF_2\text{-}CH_2)x-(CF\text{-}CF_2\text{-})y-(CF_2\text{-}CF_2)z$, CF_3

TABLE 16.2 Physical Properties of Elastomers

Physical Properties	Polyisoprene (NR)	Styrene–Butadiene (SBR)	Ethylene Propylene (EPDM)	Chloroprene (Neoprene)	Acrylonitrile–Butadiene (Nitrile)	Polyester/Polyether Urethane (Urethane)	Polysiloxane (Silicone)
Specific gravity	0.93	0.94	0.86	1.23	1.00	1.05–1.25	0.95–1.20
Durometer, Shore A	30–100	40–100	30–90	40–95	30–90	55–100	25–90
Tensile strength	E	G–F	VG	VG	VG	E	G–F
Elongation	VG–G	G	G	G	G	VG–G	E–VG
Compression set	G	G	G	G–F	G	E–G	E–G
Heat resistance	F	G–F	E–VG	G–F	G	G–F	E
Resilience or rebound	E	G–F	G	VG	G–F	E–F	G
Impact	E	E	G	G	F	E–G	G–P
Abrasion	E	E–G	E–G	E–G	E–G	E	F–P
Tear	E	F	G–F	G–F	G–F	E	F–P
Cut growth	E	G	G	G	G	E–G	F–P
Flame resistance	P	P	P	G	P	F–P	G–F
Gas impermeability	F	F	G–F	G–F	G	P–F	G–F
Weather resistance	F–P	F	E	VG–G	G–F	E–G	E
Temperature range, °F	−50° to 225°	−50° to 225°	−20° to 350°	−50° to 225°	−40° to 275°	−50° to 250°	−150° to 550°

Sources: Rubber Manufacturers Association (U.S.), *Sheet Rubber Handbook—Gasket and Packing Materials*, 1968; Blow, C.M. and Hepburn, C., *Rubber Technology & Manufacture*, Butterworths, London, U.K., 1971; Morton, M., *Rubber Technology*, Van Nostrand Reinhold, New York, 1973.

E, excellent; VG, very good; G, good; F, fair; P, poor.

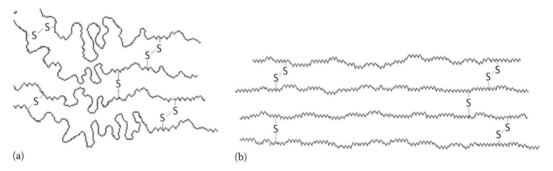

(a)   (b)

FIGURE 16.1 (a) Unstretched rubber and (b) stretched rubber.

(either in a melt blender or extruder), care should be given to prevent degradation of those polymers during the conditioning process.

The most common types of degradation occur through chemical reactions at the molecular level. Degradation resulting from various reactions with oxygen (especially at high temperature) is the most important mechanism that influences the liquid- and solid-state properties of polymer. Even during processing in enclosed equipment, sufficient oxygen is still present for oxidation reactions to take place. Degradation by heat alone (thermal degradation or thermolysis) is important; however, in a restricted oxygen atmosphere, these reactions may take place with thermal oxidation.

Most commercial polymers are manufactured by processes involving chain polymerization, polyaddition, or polycondensation reactions. These processes are generally controlled to produce individual polymer molecules with defined

- Molecular weight (or molecular weight distribution)
- Degree of branching
- Composition

Once the initial product of these processes is exposed to further shear stress, heat, light, air, water, radiation, or mechanical loading, chemical reactions start in the polymer, which have the net result of changing the chemical composition and the molecular weight of the polymer.

These reactions, in turn, lead to a change in the physical and optical properties of the polymer. In practice, any change of the polymer properties relative to the initial, desirable properties is called degradation. In this sense, "degradation" is a generic term for any number of reactions that are possible in a polymer.

16.5.1 Degradation in Presence of Flames

Fire typically occurs from an ignition source, such as a burning match or spark, meeting a flammable substance. In order for the solid to burn, it must be volatilized. Volatilization occurs when heat breaks down the polymer strands within the material creating chemical fragments that vaporize, which is necessary because combustion is almost exclusively a gas-phase phenomenon. Polymer decomposition begins at the solid phase and continues through the liquid (molten) and gas phases (Weil 1974, Camino and Costa 1988, Encyclopedia of Polymer Science and Technology 1993). The decomposition creates low molecular weight compounds that eventually enter the gas phase. Heat produced from the combustion causes further decomposition and volatilization that in turn produces further combustion. As a material is exposed to fire and its temperature begins to rise, several things start to happen. In thermoplastics, the polymer chains acquire vibrational energy with the bonds becoming more active until the glass transition temperature is reached, gradually transforming the solid to a highly viscous liquid. Upon further heating, the vibrationally excited bonds begin to break as a result of which fragments can escape to the gas phase. The simplest bond cleavage is random scission, where the backbone of the polymer is cleaved irregularly along the chain. This results in a rapid reduction of both average molecular weight as well as the melt viscosity. Upon further decomposition, a gaseous species would be generated. If the polymer backbone is a highly stable species or does not contribute to the gas yield, fragmentation of the pendant group can occur.

16.5.2 Rubber Degradation: Some Examples

Extensive studies regarding degradation of rubber and rubber blends have been found in the literature during the past two decades and very useful information has been drawn from these studies.

Bhattacharjee et al. (1991) studied the low-temperature degradation of NBR and HNBR thermogravimetrically and found that the molecular weight of NBR first increased and then decreased after prolonged aging. In the case of HNBR, the molecular weight first decreased and then increased. They attributed these results to the formation of carbonyl and carbolic functionalities. NBR undergoes gel formation with aging that is not observed in highly saturated NBR even after prolonged aging times. Infrared (IR) spectroscopic studies indicate that both the saturated and unsaturated NBRs generate carbonyl and ester functionalities on thermal oxidation. Electron spectroscopy for chemical analysis shows that the oxidation in saturated polymers takes place through the nitrile group, which is converted to the imide group. Degradation of saturated and unsaturated NBRs at high temperature (30°C–800°C) under nitrogen atmosphere by thermogravimetric analysis (TGA) shows that the temperature at which the degradation is maximum is much higher in the case of HNBR. The reaction kinetics indicates that the degradation is first order in the case of hydrogenated polymers. The activation energy for the degradation is 350 and 175 kJ mol^{-1} for HBNR and non-HNBR, respectively.

De Sarkar et al. (1999) studied the degradation of SBR at high temperatures having different levels of unsaturation under anaerobic and aerobic conditions using thermogravimetric analysis (TGA), differential scanning calorimetry (DSC), infrared spectroscopy (IR), and nuclear magnetic resonance (NMR) spectroscopy. The results were anomalous in nature in the presence of air due to cross-linking and oxidation of rubber. Isothermal data, IR, and NMR results revealed that thermal isomerization, cyclization, oxidation, depolymerization, and chain scission processes occurred during the course of degradation (De Sarkar et al. 1999).

Garbarczyk et al. (2002) characterized aged NBR by NMR spectroscopy and found that aging proceeds mainly via additional cross-linking. Scission of the polymer chains does not appear to contribute much to aging. Manoj and De (1998) studied the self-cross-linking of polyvinyl chloride (PVC)–NBR blends during processing at elevated temperatures. The extent of the reaction depends on the blend composition, processing temperature, shear rates, fill factor, and the presence of a PVC stabilizer. The reaction is found to involve hydrolysis of the nitrile group by hydrochloric acid. Varghese et al. (2001) studied the thermal behavior of another blend: NBR/poly(ethylene-co-vinyl-acetate) blend by thermogravimetry. The effects of blend ratio, different cross-linking systems (sulfur, peroxide, and mixed), various fillers (silica, clay, and carbon black), and filler loading on the thermal properties were evaluated in this study. It was found that the initial decomposition temperature increased with the addition of NBR to ethylene vinyl acetate (EVA). Among the various cross-linking system studied, the peroxide-cured system showed the highest initial decomposition temperature. This is associated with the high bond-dissociation energy of C–C linkages. The addition of fillers improved the thermal stability of the blend. The thermal aging of these blends was carried out at 50°C and 100°C for 72 h. It was seen that the properties are not affected by the mild aging condition.

Saha (2001) investigated the rheological and morphological characteristics of polychloroprene (CR)/PVC blends. The effect of temperature and mixing speed was studied. The results showed that there exists specific interaction between polar group of PVC and CR. This does not help in mixing and in forming a single phase. Temperature and mixing speed (the mixing speeds were 10, 20, 30, and 40 rpm and the mixing set temperatures were 140°C, 150°C, 160°C, and 170°C) contribute to the formation and perfection of a PCP network structure.

Deuri et al. (1986) studied the aging of a rocket insulator compound based on EPDM, filled with cork, asbestos fiber, and iron oxide, in air and nitrogen atmospheres from 100°C to 180°C for various times. The tensile strength, tear strength, and hardness of the insulator increase with increasing time and temperature of aging. The tensile strength of the samples aged in air under different extensions, however, decreases with increasing aging time. The activation energy for fracture of filled compound under normal aging is 56 kJ mol^{-1}, as compared to 30–36 kJ mol^{-1} for the gum compounds.

Choudhury and Bhowmick (1989) carried out aging of NR–polyethylene TPE composites with various levels of interaction in air at various temperatures and times of aging. Changes in mechanical properties such as tensile strength, modulus, and elongation at break have been studied as a function of composite composition and time and temperature of aging.

Amraee (2009) studied the effect of heating on thermal degradation of elastomers containing butadiene units. The aim of this research was to develop a new method for studying the thermal degradation behavior of elastomers containing butadiene units by isothermal and anisothermal analysis. Accelerated aging during isothermal TGA shows that when BR, SBR, and NBR elastomers are subjected to specific time and temperature conditions in nitrogen atmosphere, an interchain reaction occurs. Experimental observations also reveal that thermal degradation mechanism of these elastomers depends on heating rate and time–temperature history. For these types of elastomers, the effect of time is similar to temperature, and time–temperature superposition principle can be employed for studying the thermal degradation mechanism of elastomer containing butadiene units.

16.5.3 Rubber Degradation in Adverse Environments

16.5.3.1 Oxidative Degradation

Oxidative degradation of organic materials is a natural and continuous process (Figure 16.2). It occurs in varying degrees and at various rates depending upon the environmental stresses placed upon that material. Natural and synthetic rubbers are particularly susceptible to degradative changes resulting from interaction with molecular oxygen. These changes can manifest themselves as drastic swings in tensile strength, hardness, elongation, tack, and adhesion values in a specific elastomer system. Surprisingly, even a small amount of oxygen is sufficient to cause such changes. Detrimental reactions lead to scission of rubber chains as well as cross-linking in various proportions. They also result in introduction of various peroxidic and oxy functional groups at points along the main polymer chain. Obviously, one well-placed polymer scission can result in reduction of the polymer's molecular weight by one-half, while one single cross-link can produce a doubling of the molecular weight leading to aggregation and gel formation. Such changes affect the viscosity and strength properties and ultimately the processability and performance of a product even though the bulk of the material is relatively unchanged.

16.5.3.2 Ozonolysis

Cracks are formed in elastomers by ozone attack. Tiny traces of this gas in air attack double bonds within the rubber chain. NR, SBR, and NBR are sensitive rubbers and undergo rapid degradation by ozone attack. Ozone cracks form in products under tension, but the critical stress is very small. The cracks are always oriented at right angles to the strain axis and thus can appear around the circumference in a bent rubber tube. Such cracks are very dangerous when they occur in fuel pipes as they grow inward from the outside exposed surfaces toward the bore of the pipe, which may allow fuel leakage and can lead to accidental fires. The problem of ozone cracking can be prevented by adding antiozonants to a rubber before vulcanization. Currently, ozone cracks are rarely seen in automobile tire sidewalls due to the use of these additives. However, the problem still does recur in unprotected products such as rubber tubing and seals.

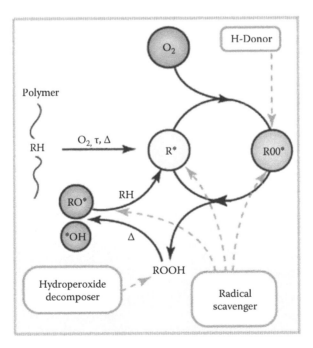

FIGURE 16.2 Oxidative degradation scheme of polymer. (This figure has been taken from the SpecialChem4Adhesives.com that is the SpecialChem website dedicated to adhesives and sealants formulation.)

16.5.3.3 Thermal Degradation

The temperature range in which elastomers exhibit their specific properties is of the order of 400 K. During heating of an elastomer, many physical transformations and chemical reactions can be observed. In principle, a loss of elasticity may take place due to a transition into a glassy state, crystallization, cross-linking, degradation, depolymerization, or destruction (Bhattacharjee et al. 1991). The factors that influence the life of an elastomeric product fall into two basic categories that can be labeled as "product characteristics" (Table 16.3) and "aging processes." Temperature, sunlight, and stresses can accelerate or initiate the impending chemical changes. Aging involves three distinct and potentially co-synchronous mechanistic routes for most (sulfur) vulcanizates. These can be identified as

1. Continuing sulfur chemistry
2. Shelf aging
3. Atmospheric aging

Shelf aging is basically oxidative degradation; however, apart from the obvious influence of oxygen, one must also consider the catalytic effects of heat, light, internal and external stresses or strains, and prooxidant metals. The chemistry of oxidative degradation is not yet completely resolved, but in a simplistic form, oxidation of a sulfur-vulcanized polyolefin such as NR proceeds via at least one chain reaction sequence that introduces C–C and C–O–O–C cross-links between polymer chains as well as C–O–O–C rings within the same polymer chain. Another set of chain reactions between oxygen and the sulfur atoms of cross-links or pendant groups also plays a vital role in oxidation of a polymer. These two sequences of chain reactions can result in both chain scission and formation of additional cross-links. The reactions between sulfur and oxygen can also, eventually, lead to sulfuric acid formation, which is a particular problem with ebonites. There is, therefore, a large range of both sequential (or chain) reactions and competing reactions, and that the one that predominates depends on factors such as the composition of the vulcanizate as well as the influence of heat, light, and metal catalysis. During heat aging, the rate of reaction of oxygen with an elastomer and the rate of diffusion of oxygen into bulk material are balanced. If the temperature is relatively low, it has been postulated that for an unprotected vulcanizate, diffusion predominates, and therefore, there is slow oxidation throughout the product, but as the temperature rises, the rate of oxidation increases much more than the rate of diffusion and so substantial oxidation occurs on the surface and an oxidized (hard) surface skin is formed. As oxidation continues, the chain breakdown may become more significant and the hard surface then softens and turns sticky. To complicate matters further, under certain conditions, this order can be reversed and an initially sticky degraded surface can harden with further oxidation.

TABLE 16.3 Use Temperature Characteristics of Rubber

Elastomer	Characteristics
Natural rubber (NR)	Temperature range of use: −70°C to 100°C
Polybutadiene (BR)	Temperature range of use: −80°C to 90°C
Styrene–butadiene rubber (SBR)	Temperature range of use: −50°C to 100°C
Nitrile–butadiene rubber, acrylonitrile–butadiene rubber (NBR)	Temperature range of use: −40°C to 120°C
Hydrogenated nitrile rubber (HNBR)	Temperature range of use: −40°C to 150°C
Chloroprene rubber (CR)	Temperature range of use: −45°C to 110°C
Ethylene propylene diene (EPDM) rubber	Temperature range of use: −50°C to 150°C
Ethylene vinyl acetate rubber (EVM/EVA)	Temperature range of use: −30°C to 170°C
Butyl rubber (IIR)	Temperature range of use: −40°C to 140°C

16.5.3.4 Degradation in Aggressive Medium

Rubbers are often subjected to very aggressive environments. As many rubbers have reactive sites in the form of double bonds or polar groups or some cross-links, these rubbers have a tendency to react with the environment and eventually degrade.

As mentioned earlier, cracks have been found in many elastomers by ozone attack. The extent of surface degradation is strong enough to affect bulk mechanical properties. The time-dependent chemical degradation of accelerated sulfur-cured EPDM rubber containing 5-ethylidene-2-norbornene as diene in an acidic environment (20% Cr/H_2SO_4) was investigated (Mitra et al. 2006). Changes in mechanical properties were monitored through retention of tensile strength, elongation at break, modulus, and microhardness. The topographical damage at the surface due to the aqueous acid-induced chemical degradation was reported. The results also indicated that the chemical degradation proceeded mainly via hydrolysis of cross-links. The oxygenated species combined with each other on prolonged exposure.

Thermal aging at very high temperature of various rubbers like acrylate rubber, fluororubber, and their blends in presence of various solvents was a subject of study by Kader and Bhowmick (2003).

In an interesting study (Bhowmick and Gent 1983), soft CR vulcanizates were shown to lose their tensile strength and exhibited continuous tear propagation at room temperature under relatively large tear forces. When immersed in solutions of ferric chloride, these CR materials showed rapid tearing, and they tear at significant lower forces than in water or in sodium chloride solutions. Although they swell continuously in water and in salt solutions, the rate of swelling seems far too low to account for the weakening observed. Moreover, the swelling is greater in water, whereas the weakening is specific to ferric chloride solution. The reaction between iron chloride and the CR molecules was the reason for lowering of strength of neoprene compounds.

Abu-Isa (1983) made an interesting study by putting elastomers in gasoline blends. The effects of ethanol/gasoline and methyl-*t*-butyl ether/gasoline mixtures on swelling and tensile properties of selected automotive elastomers were determined. The results showed that in the case of ethanol/gasoline combinations, the elastomers were more severely affected by the mixtures than by the pure components. The ethanol/gasoline mixtures are less severe than the methanol/gasoline mixtures on the elastomers studied.

16.6 Prevention of Degradation by Synthesis of Nanocomposites

The ability of an elastomeric composition to be able to perform at elevated temperatures for an extended period of time is a paramount requirement for a large number of applications. The introduction of few elastomers such as NBR, HNBR, fluoroelastomer (FKM), and hydrogenated SBR has provided elastomers with superior resistance to specific degradation mechanisms even in hot air environments and with good temperature performance up to the range of about 150°C, combined with good oil and fluid resistance (Bhattacharjee et al. 1992, Bender and Campomizzi 2001). Nevertheless, there is still a desire to further extend the upper operating range of elastomers. Thus, the synthesis of polymer nanocomposites ("nanocomposites" are defined here as polymer matrix composites where one of the phases is dispersed in another in nanometric level) has long been a crucial topic for advanced scientific research and industrial development, presenting a variety of potential applications in transportation, construction, and food packaging (Messermith and Giannelis 1994, Burnside and Giannelis 1995, Chen et al. 2001, Li et al. 2001, Joly et al. 2002, Shach et al. 2005). Though many organic or inorganic additives can be incorporated for the reinforcement of the polymer matrix, clay is considered to be an optimal choice because of its small particle size, intercalation properties (from the perspective of modification efficiency), and product cost (Jordan 1949, Ho and Glinka 2003, Shach et al. 2005). Besides economic and environmental factors, the natural abundance of clay and its potential to increase the mechanical strength and chemical resistance does make it a suitable filler for polymer composites. Moreover, it is believed that the superior properties of the elastomer–clay nanocomposites not only occur because of the dispersion of nanoscale clay particles within the elastomeric matrix but are also the result of a strong interaction between the elastomeric matrix and the clay layers. For this chapter, a three-part identification/characterization convention "AB-xy-4" has been adopted, where AB is the elastomeric matrix identifier, xy is the type of treated/untreated particulate filler, and "4" represents the filler volume concentration—typically 4 parts per hundred rubber (phr) or 4% as reported in this chapter.

The clay most often used is montmorillonite (MMT) that belongs to the general family of 2:1 layered silicates. These are composed of a regular stacking of 2D plates like layers bound together with weak interatomic forces. The clays used in this chapter were obtained from Southern Clay Products, Gonzales, Texas, United States. In fact, an MMT nanocomposite was the first successfully demonstrated example of a polymer–clay nanocomposite (Kojima et al. 1995, Lan et al. 1995, Wang and Pinnavaia 1998, Heinemann et al. 1999, Chen et al. 2000, Hasegawan et al. 2000, Fu and Qutubuddin 2001, Rong et al. 2001, Chang et al. 2002, Bokobza et al. 2004, Ganguly et al. 2008, Maji and Bhowmick 2009). Natural Na-MMT is hydrophilic in nature and thus is not compatible with most of the organic polymers. However, the replacement of sodium cations in the interlayer space of MMT with organic cations yields organophilic MMTs, which are compatible with such polymers. The formation of such polymer–filler nanocomposites affects the thermal behavior of the matrix because the presence of well-dispersed nanofillers can cause modification of typical polymer degradation pathways. This concept was first introduced by researchers from Toyota (Okada et al. 1990) who discovered the possibility of building useful nanocomposites from nylon 6 and organophilic clay. Recent studies on these clay nanocomposites reveal that inclusion of this nanoclay phase improves the barrier properties and lowers the coefficient of thermal expansion of these polyimide films. In addition to the improved mechanical properties, heat release rates from cone calorimetric experiments have also been reduced typically by 50% or more for both intercalated and exfoliated materials (Du et al. 2002).

Previously, several reports have shown that the inclusion of nanoclay raises the degradation temperature of those plastics, namely, nylon 6, polylactides, polystyrene, and polypropylene (Jang and Wilkie 2005, Weifu et al. 2006, Golebiewski and Galeski 2007, Kiliaris et al. 2009, Zhou and Xanthos 2009). Bhowmick et al. have done extensive work to improve the thermal stability of elastomers like polybutadiene rubber (BR), acrylonitrile–butadiene rubber (NBR), SBR, fluoroelastomer, HNBR, silicone rubber, and TPEs by incorporating into them different types of nanoclay (Ganguly and Bhowmick 2007, Maiti et al. 2008, Sadhu et al. 2008, Choudhury et al. 2010a,b, Roy and Bhowmick 2010).

The thermal degradation behavior of fluoroelastomer nanocomposites both in nitrogen and oxygen atmospheres and also at high and low temperatures has been studied by Maiti et al. (2008). An upward shift of T_{max} (temperature corresponding to where the maximum degradation occurs) by 15°C and 11°C in the cases of the unmodified MMT (Cloisite® NA+ designated as NA) clay and the modified clay (Cloisite® 20A designated as 20A) filled (both with a 4 phr particulate content) samples, respectively, was observed from this study as represented in Figure 16.3a and b. Also, a significant reduction in the rate of decomposition of these nanocomposites containing unmodified clays as a major degradation step was obtained from this same study. Two non-isothermal kinetic methods (either the Kissinger or the Flynn–Wall–Ozawa technique) and one isothermal kinetic method have been used to study the thermal degradation of fluoroelastomer–clay nanocomposites. The respective activation energies were found to increase with the addition of the nanoclays. Using the Flynn–Wall–Ozawa method, the activation energy was determined to be 169 kJ mol⁻¹ for the unmodified clay-filled sample and is also higher than the value for the unfilled fluoroelastomer (145 kJ mol⁻¹) and the modified clay-filled samples (155 kJ mol⁻¹). This higher thermal stability for the unmodified clay-filled fluoroelastomer in both nitrogen and oxygen atmospheres is attributed to a better polymer–filler interaction. Specifically, it is due to polar–polar interaction between the unmodified clay and the polar fluoroelastomer where the polar hydroxyl groups of the unmodified clay attract the chains of fluoroelastomers having $C^{\delta+}$–$F^{\delta-}$ bonds. In the case of the modified clays, there may be some incompatibility with the matrix due to modification by long-chain aliphatic amines. The better interaction present in the case of the unmodified clay is also

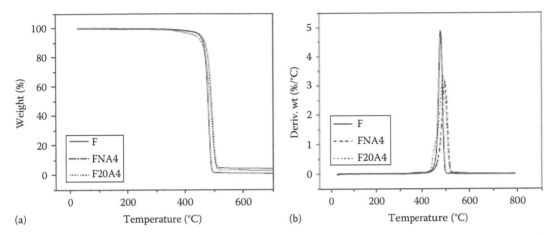

FIGURE 16.3 Typical (a) TG and (b) DTG traces of different fluoroelastomer (F) nanocomposites with 4 phr concentrations of NA and 20A fillers (in N_2).

evident from the higher gel fraction of the corresponding nanocomposite (3.41%) than that of the modified clay-filled sample (2.19%).

The effect of the modified clay on the thermal properties of SBR, NBR, and butadiene rubber (BR) in order to understand their effects on the thermal and thermo-oxidative behaviors of these nanocomposites with special reference to the nature of matrix and clay has been investigated by Sadhu et al. (2008). Figure 16.4a represents the differential thermogravimetric (DTG) curves of unfilled (BR) and unmodified (BRNA4) and modified clay-filled (BRo-MMT4) nanocomposites. As observed from the figure, the decomposition of gum polybutadiene rubber takes place mainly in two steps. The first degradation step is due to the cyclization of the butadiene component, and the second one (where 90% of the weight loss takes place) corresponds to the decomposition of the cyclized polybutadiene structure and limited depolymerization of the first stage. Incorporation of the unmodified nanoclay into the polybutadiene rubber does not cause a major increase in the T_{max} value corresponding to the last degradation step (T_{max} = 459°C). However, there is a significant improvement in thermal stability when the modified clay is present in the 34NBR matrix, which is evident from the higher T_{max} value (T_{max} = 467°C) of the modified clay-filled polybutadiene rubber vulcanizate. There is a significant reduction in the rate of decomposition in the presence of the modified clay, especially in the case of this major degradation step. The rate decreases from 45% min^{-1} for the gum to 37% min^{-1} in the case of the modified clay-filled sample (34NBRo-MMT4). In the case of SBR-based vulcanizate and its nanocomposites, with four phr clay loading (SBRN4 and SBROC4, figure is not shown here), the addition of the modified clay causes

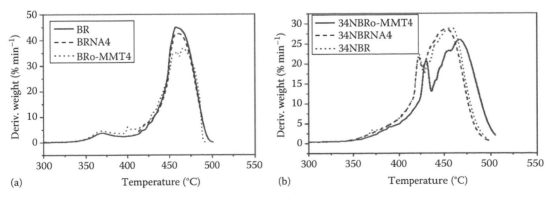

FIGURE 16.4 The DTG curves of (a) BR, (b) 34NBR, and their nanocomposites.

only slight improvement in the major degradation step (448°C–451°C) unlike BR. The rate of degradation of SBR at the T_{max} is again decreased significantly (32%–25% min^{-1}) in the case of the modified clay-filled SBR compared to that of the unmodified clay-filled or unfilled elastomer. For 34NBR and its nanocomposite, filled with the modified and the unmodified clay (34NBRo-MMT4 and 34NBRNA4, respectively), the DTG curves show multiple decomposition steps in nitrogen atmosphere as shown in Figure 16.4b. The presence of the modified clay shifts the T_{max3} to 467°C with a reduction in decomposition rate from 29% to 26% min^{-1}. Now, BR and SBR, being nonpolar in nature, have good interaction with the organically modified filler. The major chain scission step is predominantly influenced by the presence of the clays, as it directly interacts with the main chain. In the case of 34NBR, the polymer chains have H-bonding interaction with the clay along with the van der Waals type of interaction. This, in turn, improves the thermal stability of the nanocomposite.

The effect of various nanofillers on the thermal stability and degradation kinetics of HNBR nanocomposites has been reported by Choudhury et al. (2010a). In that work, five different species of nanofillers have been selected. These are 2D MMT (where there are two organically modified and one unmodified filler, designated as 30B, 15A, and NA, respectively), 1D sepiolite (SP), and zero-dimensional nanosilica (A300). Typical TG and DTG curves in air atmosphere at a heating rate of 20°C min^{-1} for the unfilled HNBR and its nanocomposites are shown in Figure 16.5a (where the HNBR matrix phase is designated as S1 so that the nanocomposite with 4 phr filler loading is designated as S1-F-4, where F represents the filler species). For neat HNBR, there appear three degradation temperatures, whereas for the three nanocomposites, their DTG curves show almost single-step degradation with well-defined initial and final degradation temperature, which may be a result of the scission of the main macromolecular chains as depicted in Figure 16.5a. Shifts in T_{max} values toward higher temperatures by 12°C, 6°C, 4°C, 12°C, and 16°C, respectively, are observed for S1-30B-4, S1-15A-4, S1-NA-4, S1-SP-4, and S1-A300-4, as compared with the neat elastomer (S1). Thus, improvement in thermal stability (T_{max}) is highest for S1-A300-4, followed by S1-30B-4 and S1-SP-4. From the TG curves of the A300, 30B, and SP (Figure 16.5b), it is observed that thermal stability of A300 is much higher than the later two. Also, silica nanofillers contain many surface hydroxyl groups that may interact with the polar groups present in the elastomer. Moreover, silica acts as a heat sink. Its specific heat capacity value of 700 J $(kg K)^{-1}$ is high enough to increase the thermal stability of the corresponding nanocomposites. Thus, thermal stability of nanosilica-filled composites is excellent especially for low filler loading.

The compatibility of the organic polymer matrix with the NA filler is much inferior to nanocomposites containing 30B, 15A, and SP fillers. Thus, S1-NA-4 has inferior thermal stability as compared to the other three clays. Moreover, the intergallery spacing of NA clay phase prior to incorporation into the

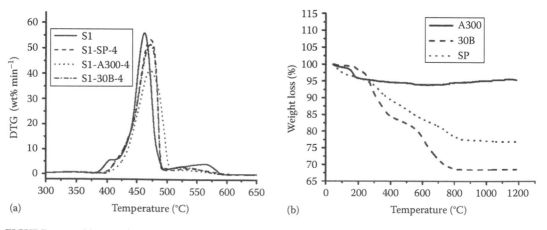

FIGURE 16.5 (a) Typical DTG traces of different S1 nanocomposites in air and (b) weight loss versus temperature of different nanofillers.

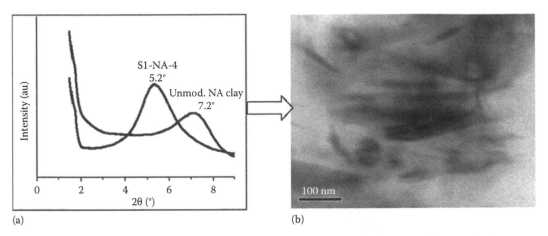

(a)   (b)

FIGURE 16.6 (a) XRD traces of the pure NA clay and the S1-NA-4 and (b) TEM photograph of S1-NA-4 nanocomposite.

nanocomposite is very small (only 1.22 nm, as calculated from the 7.2° x-ray diffraction [XRD] peak, whereas there is a larger intergallery spacing of 1.69 nm for S1-NA-4 nanocomposite as calculated from the comparable 5.2° XRD peak shown in Figure 16.6a). Only a few polymer chains of 34HNBR (with 34% acrylonitrile content), being bulky in nature, can find their ways into such a small gallery space, which results in poor polymer–filler interaction, and this is confirmed from both XRD and transmission electron microscopy (TEM) photographs (Figure 16.6a and b).

In the case of the nanocomposite with 30B (where the intergallery cations are replaced by polar surfactant), its XRD-measured gallery spacing (1.95 nm) is higher than that (1.69 nm) for the NA nanocomposite. Thus, the HNBR molecules have easily entered into the gallery space aiding the breakup of the layer structure resulting in fully exfoliated layers within the matrix that also get well dispersed throughout the bulk and surface of the polymer matrix as confirmed by XRD, TEM, and atomic force microscopy (AFM) photographs (Figure 16.7a). As a consequence this nanocomposite has excellent thermal stability (T_{max} increased by 12°C over neat elastomer). The nanocomposite with 15A (another organically modified MMT that has been modified with long amine chains) thus has a larger intergallery spacing (3.26 nm as obtained from the XRD studies, Figure 16.7b) than that for the 30B nanocomposite. It is known that a longer chain is always thermally less stable than a shorter one. Further, from the XRD, TEM, and AFM photographs (Figure 16.7b), a partially intercalated morphology is obtained for the 15A nanocomposite. It is a well-known fact that exfoliated silicate layers provide better barrier property than an intercalated one (Choudhury et al. 2010a). Here the Cloisite 30B acts as strong barrier to the diffusion of gases such as oxygen and nitrogen, while the intercalated morphology of the 15A nanocomposite makes the barrier property of this corresponding nanocomposite (T_{max} increased by 6°C over the neat elastomer) inferior to that of the 30B nanocomposite. On the other hand, when fibrous or rodlike SP particles are distributed uniformly throughout the matrix (as evident from the TEM and the AFM photographs in Figure 16.7c), this can improve the thermal property of the HNBR matrix, just like 30B because of the high thermal stability of this filler.

The activation energy for degradation of unfilled HNBR and its nanocomposites has been determined by using non-isothermal Flynn–Wall–Ozawa and Kissinger as well as by isothermal methods (Choudhury et al. 2010a). The total activation energy of degradation of the nanocomposite—a sum of the activation energy of matrix degradation, the activation energy of nanofiller degradation, and the energy required to break the interaction between the filler and the matrix—has been found to be much higher than the neat elastomer using these non-isothermal and isothermal methods. Since the degradation activation energy of the nanofiller is highest for nanosilica and the energy required to break polymer–filler interaction is highest for S1-30B-4 and S1-SP-4, the total activation energy was observed to be

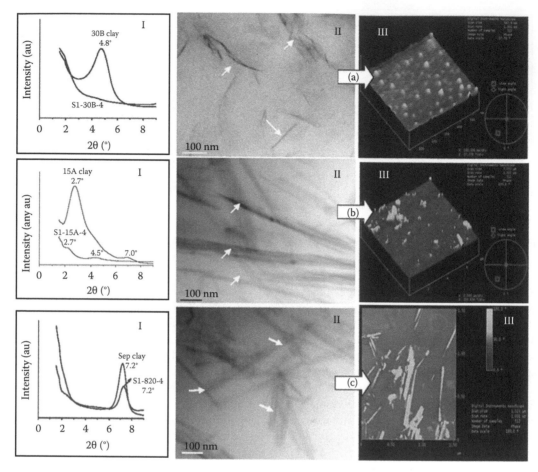

FIGURE 16.7 (I) XRD traces, (II) TEM, and (III) AFM photographs of (a) S1-30B-4, (b) S1-15A-4, and (c) S1-SP-4 nanocomposites.

much higher for S1-A300-4, S1-30B-4, and S1-SP-4. The values obtained are 169, 166, and 168 kJ mol⁻¹, respectively, by the Kissinger method and 162, 163, and 161 kJ mol⁻¹, respectively, by the Flynn–Wall–Ozawa method. The same evaluations for those neat elastomers resulted in activation energy values of 125 and 138 kJ mol⁻¹, respectively, by those methods.

Thermal stability of silicone rubber is higher because of the Si–O bond. However, it could be further improved by nanocomposite formation. This improvement in the thermal degradability of these nanocomposites in comparison to the virgin polymer is reported by Roy and Bhowmick (2010). Inclusion of a nanoscale dispersion of the sepiolite into the in situ generated PDMS matrix leads to an increase in the overall oxidative thermal stability of the elastomeric matrix. For the unfilled polymer, there is a single degradation temperature at 350°C, while for the nanocomposite, tremendous increments of both T_{max} and T_i values were observed.

Ganguly and Bhowmick (2007) reported remarkable enhancements of thermal properties of TPE (sulfonated SEBS)-based nanocomposites. The maximum degradation temperature increased from 415°C for pristine SEBS to 437°C for 3 wt% sulfonated SEBS, and the value increased to 444°C for the nanocomposite. Upon incorporation of MMT clay into pristine SEBS matrix, only agglomerated clusters are formed as shown in the TEM image in Figure 16.8a. Proper dispersion of clay platelets could not be achieved in SEBS–MMT mostly due to incompatibility of hydrophobic polymer with hydrophilic clay particles. After grafting, better distribution of exfoliated MMT clay platelets in the grafted SEBS matrix

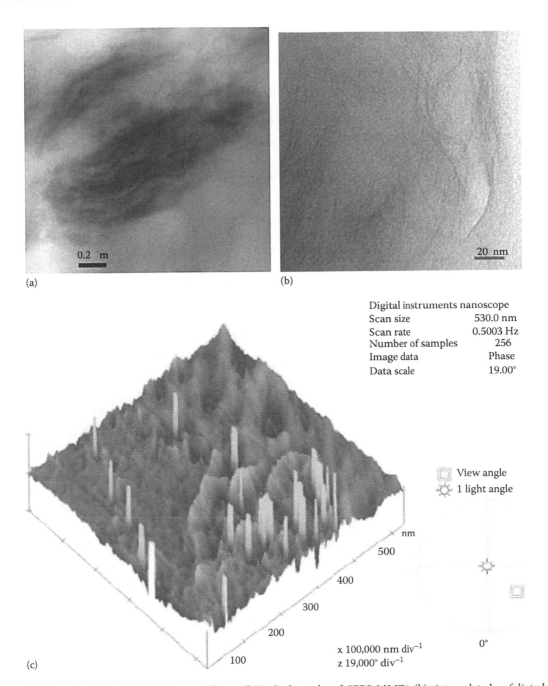

Digital instruments nanoscope
Scan size   530.0 nm
Scan rate   0.5003 Hz
Number of samples   256
Image data   Phase
Data scale   19.00°

View angle
1 light angle

nm
500
400
300
200
100

x 100,000 nm div^{-1}
z 19,000° div^{-1}

0°

(c)

FIGURE 16.8 Bright-field TEM morphology of (a) thick stacks of SEBS-MMT4 (b), intercalated exfoliated S3-SEBS-MMT4, and (c) 3D AFM phase image of S6-SEBS-MMT4 showing fine clay layers (2–6 nm thick) impregnated from SPS domains of the matrix.

is observed (Figure 16.8b and c). It is unique to have such wonderful exfoliation of individual clay layers with the same MMT clay at identical loading of 4 wt% in grafted SEBS matrices. Fine clay layers can be seen in the 3D AFM phase image (Figure 16.8c) for the grafted SEBS nanocomposite. Thus, this morphology appears to enhance the earlier noted improved thermal properties of the grafted SEBS-based nanocomposite.

16.6.1 Mechanisms of Degradation

Two mechanisms have been suggested to account for the reduction in the heat release rate: (1) a barrier mechanism, in which the clay functions as a barrier to mass transfer of the polymer, and (2) a radical trapping mechanism, which occurs due to the presence of iron or other paramagnetic impurities, as a structural component in the clay.

The mechanism of degradation for the fluoroelastomer in terms of two competing net processes has been explained by Maiti et al. (2008):

1. Degradation of the main chain carbon bonds, that is, main chain scission
2. Degradation by the splitting off of an adjacent hydrogen and fluorine, as hydrogen fluoride generating double bonds

It is explained that the degradation mechanism remains the same even after addition of nanoclays.

Choudhury et al. (2010a) have explained the degradation mechanism of HNBR and its nanocomposite with FTIR spectroscopy. For the neat HNBR (S1), the peaks at 1446–1463 cm^{-1} for $-CH_2$ deformation, 990 cm^{-1} for disubstituted double bonds (trans–CH-wagging vibration), 2213 cm^{-1} for acrylonitrile group, and 970 cm^{-1} for CH-wagging absorbance of the hydrogenated polymer were observed, as depicted in Figure 16.9. For the aged sample, a broad absorption at 1760–1720 cm^{-1} due to C = 0 functionality was observed. In order to quantify this increase for the aged sample, the absorbance at 1730 cm^{-1} is divided by absorbance at 1460 cm^{-1} due to CH_2 deformation. It was assumed that the concentration of the $-CH_2$ group remains constant throughout the low-temperature aging process and the attack takes place first on the double bonds. The K value, the ratio of absorbance at 1730–1460 cm^{-1}, was found to be much higher for the neat elastomer than the nanocomposite, which confirmed that the level of oxidation reaction is higher in the case of neat elastomer than the nanocomposites. A similar observation was made when the ratio of absorbance at 1730–2213 cm^{-1} (for the $-CN$ group) was analyzed. An additional peak at 970 cm^{-1}, in the case of S1-aged sample, may be due to the formation of some secondary cyclic alcohol. However, in the case of S1-A300-4 nanocomposite (aged), this peak was absent. Nanofillers having a higher thermal conductivity as well as a greater heat capacity value than that for the monolithic elastomer absorb the heat transmitted from the surroundings and thus retard the direct thermal impact to the polymer backbone. As a result, the temperature at which

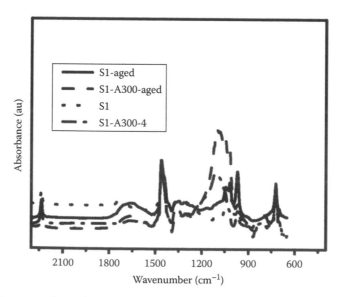

FIGURE 16.9 FTIR spectra of S1 and S1-A300-4 before and after aging at 175°C for 12 h.

major degradation of the neat elastomer takes place is shifted to a higher temperature by approximately 16°C for the nanocomposites. This observation depicts that clay nanofillers form shielding layers on the surface that in turn protects the rubber matrix phase from the attack of oxygen to some extent and inhibits the degradation process.

16.6.2 Aging and Lifetime

Choudhury et al. (2010b) have further addressed the long-term thermo-oxidative stability and useful lifetime of their nanocomposites compared to those of their virgin polymer counterparts. The material was subjected to accelerated heat aging testing at 70°C, 100°C, and 135°C over different time periods. The addition of organically modified clay was found to improve the retention of physico-mechanical properties of the elastomer by disrupting the oxidative process as shown in Figures 16.10 through 16.12. Upon aging at longer times, the properties of neat HNBR were observed to have experienced a drastic reduction in tensile strength and elongation at failure. It can be visualized from Figure 16.10a and b that aging at 70°C, the neat HNBR shows marginal change in properties during the early stages of degradation. An increase in tensile strength is observed up to 48 h of aging (Figure 16.10a), beyond which there is a continuous decrease in properties of the neat elastomer. After 168 h of aging, the tensile strength was reduced by as much as 40% of its original value. These results can be explained as follows: Partial cross-linking of the elastomer backbone at the initial stage of aging takes place (Bhattacharjee et al. 1992,

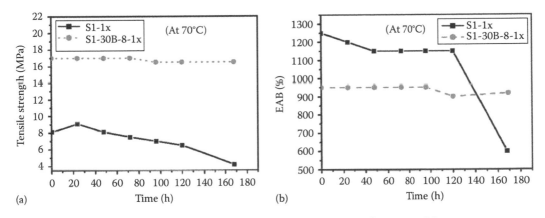

FIGURE 16.10   (a) Tensile strength and (b) elongation at break versus time of aging at 70°C.

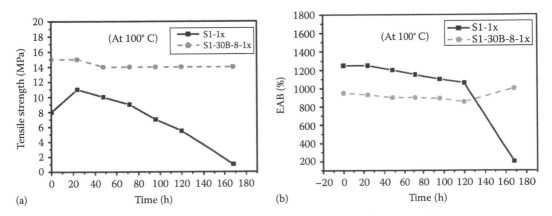

FIGURE 16.11   (a) Tensile strength and (b) elongation at break versus time of aging at 100°C.

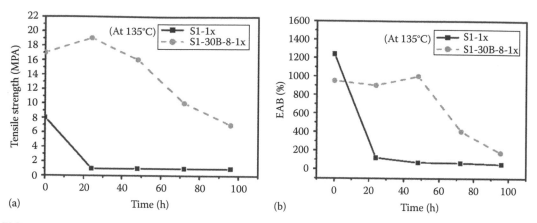

FIGURE 16.12   (a) Tensile strength and (b) elongation at break versus time of aging at 135°C.

Bender and Campomizzi 2001). Upon further aging for prolonged time periods at 70°C, the main chain starts degrading resulting in a drastic reduction of the strength and elongation values, as reflected in Figure 16.10a and b. These are also apparent from the FTIR studies (discussed later). No significant changes in these properties for the comparable nanocomposite were observed under the same conditions. The symbol "1x" in Figures 16.10 through 16.12 means the sample cured with 1 phr (parts per 100 g of rubber) of curing agent.

A similar trend was observed at an aging temperature of 100°C (Figure 16.11a and b). For up to 48 h of aging, the tensile strength of the neat elastomer increased by 30%, while no such change was observed for the properties of the comparable nanocomposite. The overall reduction in properties for the neat elastomer at 100°C after 168 h was 80%, whereas the nanocomposite retained almost all its properties under similar aging conditions.

On aging at 135°C, a remarkable drop in strength and elongation properties for the neat elastomer took place even at the early stages of aging (24 h). In fact, S1-1x (unfilled elastomer vulcanizate) loses its rubberlike properties within 24 h of aging at this temperature, which can be clearly visualized from Figure 16.12a and b. On the other hand, no significant change in properties is observed, at this initial stage of aging, for the comparable nanocomposite. As reflected in Figure 16.12a and b, the nanocomposite retained its properties up to 48 h of aging. However, beyond this period, deterioration in the properties of the nanocomposite did take place. Thus, Figures 16.10 through 16.12 explain the better retention of tensile strength and elongation values at failure for the nanocomposite compared to the comparable neat elastomer after aging in air.

In order to study the nature of these reactions, representative samples were analyzed at 100°C and 135°C, using FTIR spectroscopy. Figure 16.13 represents the FTIR spectra of the unfilled and the clay-filled elastomer vulcanizates before and after aging at 135°C. The S1 vulcanizate before aging shows a sharp peak at 720 cm^{-1} due to the presence of an ethylene moiety. Another peak at 970 cm^{-1} is primarily due to disubstituted double bond. The sharp peaks that appear at 1460 and 2900 cm^{-1}, respectively, correspond to the $-CH_2-$ and C–H groups present in the matrix. The small peak at 1720 cm^{-1} is due to the typical carbonyl range, and finally, that at 2237 cm^{-1} is the characteristic peak of the acrylonitrile group. In the case of the nanocomposite, two sharp peaks are observed at 1032 and 920 cm^{-1}, mainly due to the Si–O vibration. On aging at 135°C for 48 h, a major increase in peak intensity at 1721, 1367, and 1163 cm^{-1} is noted for the unfilled elastomer vulcanizate. The first prominent peak at 1721 cm^{-1} is due to the asymmetric and symmetric stretching vibrations of vinyl carbonyl that is formed from hydroperoxide with loss of a water molecule at this temperature. The mechanism of degradation of HNBR

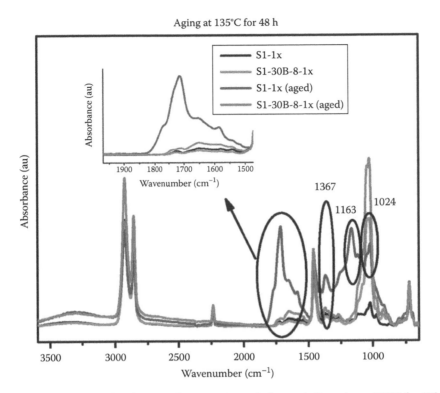

Aging at 135°C for 48 h

FIGURE 16.13 FTIR photography of S1-1x and S1-1x-30B-8-1x before and after aging at 135°C for 48 h.

has been discussed by Bender and Campomizzi (2001). The mechanism shows how vinyl carbonyl is formed from ROOH with loss of a water molecule. It is further clear from the mechanism that oxygen attacks the diene part of the elastomer and causes oxidative degradation. The peak at 1171 and 1017 cm^{-1} is assigned to the asymmetric and symmetric stretching vibration of the ester group (–C–O–C). This implies that during the process of aging (at 135°C), additional ester links are formed in the case of unfilled HNBR. This is in accordance with the mechanism suggested (Bender and Campomizzi 2001). The peak at 1367 cm^{-1} is due to the C–H bending vibration of the saturated carbon chain. The peak at 970 cm^{-1}, however, disappears in the aged unfilled sample due to the utilization of the double bond during the thermo-oxidative reaction. Bender and Campomizzi (2001) also noted that during aging of HNBR, abstraction of the allylic hydrogen leads to the formation of a delocalized radical that is more stable than an alkyl radical. Oxygen molecules as diradical can react with the polymer radicals. On the other hand, the more stable peroxy radical in the presence of oxygen abstracts a hydrogen atom from another polymer chain to form a hydroperoxide and a new peroxy radical at high temperature. Hydroperoxide can undergo number of reactions leading to higher oxidized moieties such as ketones, carboxylic acids, and alcohols. The alcohol then undergoes oxidation and forms ester links. As a consequence, the properties of the unfilled elastomer reduce tremendously at this aging temperature (Figure 16.12a and b). Addition of nanoclay protects the sample from oxidative reaction, and as a result, the formation of ester and carbonyl linkages is inhibited, which has been confirmed from the FTIR spectrum of S1-30B-8 vulcanizate. This is the reason why, on addition of nanoclay, physico-mechanical properties of the elastomer are retained as depicted in Figures 16.10 through 16.12.

The estimated lifetime of a polymer to failure can be defined as the time when the properties of the polymer reach 50% to its original value.

The logarithmic form of the Arrhenius equation with the constant terms combined in B

$$\ln t_i = \frac{E}{RT_i} + B \tag{16.1}$$

where

t$_i$ is the reaction time (min)
E is the activation energy (J mol^{-1})
R is the gas constant (8.314 J mol^{-1} K^{-1})
T$_i$ is the absolute temperature

will be used to quantify these values for estimated lifetimes.

A plot of log (time) versus 1/T gives a straight line with the slope E/R and is known as an "Arrhenius plot." However, over the last few decades, many degradation studies have dealt with the complexity of polymer lifetime prediction and extrapolation. Perhaps the key weakness in testing for the linear Arrhenius behavior is often the limited availability of experimental data. Although there is no exact correspondence between the actual life and predicted life, this is the standard procedure followed by the rubber industry and the methodology for carrying out this procedure is described in the ISO Standard 11346.

Figure 16.14 represents a plot of aging coefficient (ratio of tensile strength of aged samples over the tensile strength of an unaged sample) versus time of aging (in hours).

In order to predict the life of clay-filled as well as the unfilled elastomer vulcanizate, the Arrhenius graph of log (time) against 1/T (Equation 16.1) was plotted. The time taken to reach 50% of the original value (aging coefficient) was determined for each temperature from Figure 16.14 either directly or by extrapolation.

Figure 16.15 represents the plot of log (time) versus 1/T for the unfilled and the nanoclay-filled elastomer vulcanizate. The error in the measurement of log t varies from ±1.5% to ±3.0%. Extrapolating the line, the lifetime of each article at 40°C can be calculated from the slope of the straight line, as shown in Table 16.4.

Cloisite 30B, being an organically modified clay, is quite compatible with organic polymers. The x-ray diffractograms of the unaged nanocomposite vulcanizate and the samples aged at different times

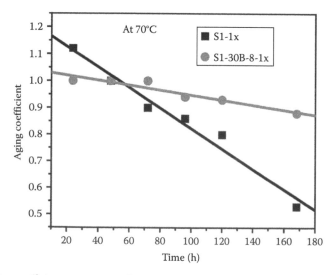

FIGURE 16.14 Aging coefficient versus time of aging at 70°C.

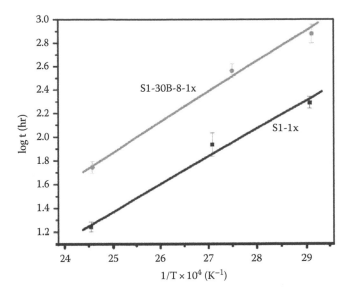

FIGURE 16.15 Arrhenius plot of log t versus 1/T for S1-1x and S1-30B-8-1x.

TABLE 16.4 Lifetime of HNBR and Its Nanocomposite at 40°C

Samples	Lifetime (Years)
S1-1x	15
S1-30B-8-1x	45

and temperatures are shown in Figure 16.16a. No peak is observed for the nanocomposite. This indicates a complete breakdown of the normal layered structure of the clay, resulting from its exfoliation. The absence of any (001) reflection peak in the $2\theta = 2°–10°$ range further confirmed that morphology does not change on aging. It is a conventionally known fact that complete dispersion achieved by an exfoliated system provides superior physico-mechanical properties. Adequate dispersion of the nanoclay within the bulk of the matrix is further confirmed from the TEM photograph of S1-30B-8-1x vulcanizate (Figure 16.16b).

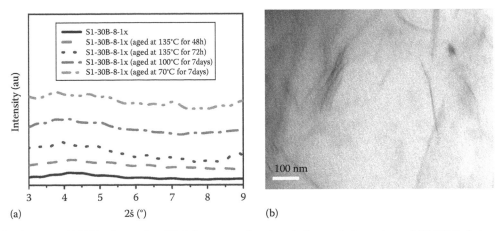

FIGURE 16.16 (a) XRD of nanoclay-filled elastomer vulcanizate before and after aging and (b) TEM photograph of S1-30B-8-1x.

TABLE 16.5   Specific Standardized Tests

Testing Protocol for Physical Properties	ISO or ASTM # and Title
Chemical resistance	ASTM C868-02(2012): *Standard Test Method for Chemical Resistance of Protective Linings*
UV/light resistance	ASTM G7, G24, D1014, D4141—Common standards for outdoor weathering
Heat resistance	ASTM D573: *Standard Test Method for Heat Aging, Heat Resistance, Oven Aging of Rubber*
Flame resistance	ASTM D6413/D6413M-12: *Standard Test Method for Flame Resistance of Textiles*
Moisture resistance	ASTM F1900-98(2012): *Standard Test Method for Water Resistance of Footwear Using a Walking Step Simulator*
Weather resistance	ASTM D1149-07(2012): *Standard Test Methods for Rubber Deterioration—Cracking in an Ozone Controlled Environment*
Useful temperature range °F	ASTM F495-99a(2011): *Standard Test Method for Weight Loss of Gasket Materials upon Exposure to Elevated Temperatures*
Example: extrapolated useful lifetimes and useful temperature range	ISO#11346: Rubber, vulcanized or thermoplastic—estimation of lifetime and maximum temperature of use
Acid and high-temperature resistance	ASTM D6909-10: *Standard Specification for High Temperature and Acid-Resistant Fluorocarbon Terpolymer Elastomer*
Oxygen resistance	ASTM D572-04(2010): *Standard Test Method for Rubber-Deterioration by Heat and Oxygen*

Incorporation of the tallow quaternary ammonium ions expands the gallery gap between the clay layers of 30B. S1, having 34% acrylonitrile content (bulky in size), easily enters the clay gallery space, and finally breaks the layer structure to attain an exfoliated morphology. Moreover, modification of the clay with a polar surfactant further facilitates the interaction between the clay and the polar elastomer. Here, improvement in the properties takes place due to intercalation of polymer chains within the clay layers along with H-bond formation between the CN groups of the rubber and the OH groups present on the clay surface and surfactant.

16.7 Specific Standardized Tests

Along with the development of science and technology, there has been an increased demand for materials with higher-temperature capabilities and excellent thermal performance. Performances during outdoor applications are critical issues for the commercial exploitation of these end products. The useful lifetime of an article, at a given service temperature, is often assessed by determining the time required for a particular property of interest to degrade to a critical value (Bender and Campomizzi 2001). The properties of rubber are subject to changes ultimately to the point where the material is no longer capable of fulfilling its functions. The life of a rubber component is of critical importance in many applications, and hence, it becomes necessary to predict its longevity. Thus, the comparative prediction of the full lifetime of an article is often based on these accelerated aging tests, by conducting characterizations at ordinary time intervals of exposure but where exposure conditions are selected to accelerate aging. In this way, change in bulk properties can be recorded quickly (Table 16.5).

16.8 Conclusions

In this chapter, a combined approach was used to develop an understanding of elastomer degradation and prevention of this degradation by formation of nanocomposites. The study discussed stepwise the applications of several elastomers along with their general and physical properties showing many specific examples like NR, SBR, NBR, and PU. The chapter further highlighted the principles of degradation of elastomers and the attempts of many researchers, who tried various methods to prevent elastomer degradation. Though many inorganic and organic fillers were incorporated into nanocomposites to improve the degradation properties of rubber, nanoclay was found to be the best choice based on its small particle size and the high reinforcement nature of the clay. It was observed that nanofillers form shielding layers on the surface of rubber that in turn protect the rubber from the attack of oxygen to some extent, thus inhibiting the degradation process. This chapter also delineates methods to calculate the lifetime of these elastomers and their nanocomposites.

List of Abbreviation and Symbols

AFM	Atomic force microscopy
BR	Polybutadiene rubber
CR	Chloroprene rubber
ENR	Epoxidized natural rubber
EPDM	Ethylene propylene diene methylene rubber
FKM	Fluorocarbon elastomer
HNBR	Hydrogenated nitrile rubber
34HNBR	Hydrogenated nitrile rubber with 34% acrylonitrile content
NA	Unmodified sodium montmorillonite clay, Cloisite® Na+
NBR	Nitrile rubber
NR	Natural rubber
o-MMT	Organically modified sodium montmorillonite clay
PVA	Polyvinyl acetate
PU	Polyurethane
PCP	Polychloroprene
SBR	Styrene–butadiene rubber
SEBS	Poly(styrene–ethylene-co-butylene–styrene) triblock copolymer
SP	Sepiolite clay
T_i	Temperature for onset of degradation
T_{max}	Temperature at which maximum degradation occurs
TEM	Transmission electron microscopy
TGA	Thermogravimetric analysis
XRD	X-ray diffraction
10A	Cloisite® 10A (a natural montmorillonite modified with dimethyl benzyl, hydrogenated tallow quaternary ammonium salt)
15A	Cloisite® 15A (a natural montmorillonite modified with dimethyl, dihydrogenated tallow quaternary ammonium salt)
20A	Cloisite® 20A (a natural montmorillonite modified with dimethyl, dihydrogenated tallow, quaternary ammonium salt)
30B	Cloisite® 30B (a natural montmorillonite organically modified with methyl, tallow, bis-2-hydroxyethyl, quaternary ammonium salt)
A300	Nanosilica (surface area = 300 [270–330] $m^2\ g^{-1}$, pH = 3.7–4.7, moisture content = 1.5%, and SiO_2 content = 99.8%)

References

Abu-Isa, I.A. 1983. Elastomer—Gasoline blends interactions II: Effects of ethanol/gasoline and methyl-*t*-butyl ether/gasoline mixtures on elastomers. *Rubber Chemistry and Technology*, 56, 169–196.

Amraee, A.I. 2009. The effect of heat history on thermal degradation of elastomers containing butadiene units. *Journal of Applied Polymer Science*, 113, 3896–3900.

Bender, H. and E. Campomizzi. 2001. Improving the heat resistance of hydrogenated nitrile rubber com-
    pounds. Part 1: Aging mechanisms for high saturation rubber compounds. *KGK Kautschuk Gummi
    Kunststoffe*, 54, 14–21.
Bhattacharjee, S., A.K. Bhowmick, and B.N. Avasthi. 1991. Degradation of hydrogenated nitrile rubber.
    Polymer Degradation and Stability, 31, 71–87.
Bhattacharjee, S., A.K. Bhowmick, and B.N. Avasthi. 1992. Preparation of hydrogenated nitrile rubber
    using palladium acetate catalyst: Its characterization and kinetics. *Journal of Polymer Science Part-A:
    Polymer Chemistry*, 30, 471–484.
Bhowmick, A.K. 2008. *Current Topics of Elastomers Research*, Taylor & Francis Inc., Boca Raton, FL.
Bhowmick, A.K. and A.N. Gent. 1983. Strength of neoprene compounds and the effect of salt solutions.
    Rubber Chemistry and Technology, 56, 845–852.
Bhowmick, A.K., M.M. Hall, and H. Benary. 1994. *Rubber Products Manufacturing Technology*, Marcel
    Dekker Inc., New York.
Bhowmick, A.K. and H.L. Stephens. 2001. *Handbook of Elastomers*, Revised Edition, Marcel Dekker Inc.,
    New York.
Billmeyer, F.W. 2007. *Textbook of Polymer Science*, Wiley, New York.
Blow, C.M. and C. Hepburn. 1971. *Rubber Technology & Manufacture*, Butterworths, London, U.K.
Bokobza, L., A. Burr, G. Garnaud, M.Y. Perrin, and S. Pagnotta. 2004. Fibre reinforcement of elasto-
    mers: Nanocomposites based on sepiolite and poly(hydroxyethyl acrylate). *Polymer International*,
    53, 1060–1065.
Burnside, S.D. and E.P. Giannelis. 1995. Synthesis and properties of new poly(dimethyl siloxane) nano-
    composites. *Chemistry of Materials*, 7, 1597–1600.
Camino, G. and L. Costa. 1988. Performance and mechanisms of fire retardants in polymers—A review.
    Polymer Degradation and Stability, 20, 271–294.
Chang, Y.-W., Y. Yang, S. Ryu, and C. Nah. 2002. Preparation and properties of EPDM/organomontmoril-
    lonite hybrid nanocomposites. *Polymer International*, 51, 319–324.
Chen, G., S. Liu, S. Chen, and Z. Qi. 2001. FT-IR spectra, thermal properties, and dispersibility of
    a polystyrene/montmorillonite nanocomposite. *Macromolecular Chemistry and Physics*, 202,
    1189–1193.
Chen, G., S. Liu, S. Zhang, and Z. Qi. 2000. Self-assembly in a polystyrene/montmorillonite nanocompos-
    ite. *Macromolecular Rapid Communications*, 21, 746–749.
Choudhury, N.R. and A.K. Bhowmick. 1989. Ageing of natural rubber-polyethylene thermoplastic elasto-
    meric composites. *Polymer Degradation and Stability*, 25, 39–47.
Choudhury, A., A.K. Bhowmick, C. Ong, and M. Soddemann. 2010a. Effect of various nanofillers on
    thermal stability and degradation kinetics of polymer nanocomposites. *Journal of Nanoscience and
    Nanotechnology*, 10, 5056–5071.
Choudhury, A., M. Soddemann, and A.K. Bhowmick. 2010b. Effect of organo-modified clay on acceler-
    ated aging resistance of hydrogenated nitrile rubber nanocomposites and their lifetime prediction.
    Polymer Degradation and Stability, 95, 2555–2562.
De Sarkar, M., P.P. De, and A.K. Bhowmick. 1999. Influence of hydrogenation and styrene content on
    the unaged and aged properties of styrene-butadiene copolymer. *Journal of Material Science*, 34,
    1741–1747.
Deuri, S., A.K. Bhowmick, and B.N. Avasthi. 1986. Aging of rocket insulator compound based on EPDM.
    Polymer Degradation and Stability, 16, 221–239.
Du, J., J. Zhu, C.A. Wilkie, and J. Wang. 2002. An XPS investigation of thermal degradation and charring
    on PMMA clay nanocomposites. *Polymer Degradation and Stability*, 77(3), 377–381.
Fu, X. and S. Qutubuddin. 2001. Polymer–clay nanocomposites: Exfoliation of organophilic montmoril-
    lonite nanolayers in polystyrene. *Polymer*, 42, 807–813.
Ganguly, A. and A.K. Bhowmick. 2007. Sulfonated Styrene-(ethylene-co-butylene)-styrene/Montmorillonite
    clay nanocomposites: Synthesis, morphology, and properties. *Nanoscale Research Letters*, 3, 36–44.

Ganguly, A., A.K. Bhowmick, and Y. Li. 2008. Insights into montmorillonite nanoclay based ex situ nanocomposites from SEBS and modified SEBS by small-angle X-ray scattering and modulated DSC studies. *Macromolecules*, 41, 6246–6253.

Garbarczyk, M., W. Kuhn, J. Klinowski, and S. Jurga. 2002. Characterization of aged nitrile rubber elastomers by NMR spectroscopy and microimaging. *Polymer*, 43, 3169–3172.

Golebiewski, J. and A. Galeski. 2007. Thermal stability of nanoclay polypropylene composites by simultaneous DSC and TGA. *Composites Science and Technology*, 67, 3442–3447.

Hasegawan, N., H. Okamoto, M. Kato, and A. Usuki. 2000. Preparation and mechanical properties of polypropylene–clay hybrids based on modified polypropylene and organophilic clay. *Journal of Applied Polymer Science*, 78, 1918–1922.

Heinemann, J., P. Reichert, R. Thomann, and R. Mulhaupt. 1999. Polyolefin nanocomposites formed by melt compounding and transition metal catalyzed ethene homo- and copolymerization in the presence of layered silicates. *Macromolecular Rapid Communications*, 20, 423–430.

Ho, D.L. and C.J. Glinka. 2003. Effect of solvent solubility parameters on organoclay dispersions. *Chemistry of Materials*, 15, 1309–1312.

Jang, B.N. and C.A. Wilkie. 2005. The effects of clay on the thermal degradation behavior of poly(styrene-co-acrylonitrile). *Polymer*, 46, 9702–9713.

Joly, S., G. Garnaud, R. Ollitrault, and L. Bokobza. 2002. Organically modified layered silicates as reinforcing fillers for natural rubber. *Chemistry of Materials*, 14, 4202–4208.

Jordan, J.W. 1949. Organophilic bentonites-swelling in organic liquids. *Journal of Physical and Colloid Chemistry*, 53, 294–306.

Kader, M.A. and A.K. Bhowmick. 2003. Thermal ageing, degradation and swelling of acrylate rubber, fluororubber and their blends containing polyfunctional acrylates. *Polymer Degradation and Stability*, 79, 283–295.

Kiliaris, P., C.D. Papaspyrides, and R. Pfaender. 2009. Influence of accelerated aging on clay-reinforced polyamide 6. *Polymer Degradation and Stability*, 94, 389–396.

Kojima, Y., A. Usuki, M. Kawasumi et al. 1995. Novel preferred orientation in injection-molded nylon 6-clay hybrid. *Journal of Polymer Science Part-B: Polymer Physics*, 33, 1039–1045.

Lan, T., P.D. Kaviratna, and T.J. Pinnavaia. 1995. Mechanism of clay tactoid exfoliation in epoxy-clay nanocomposites. *Chemistry of Materials*, 7, 2144–2150.

Li, X., T. Kang, W.-J. Cho, J.-K. Lee, and C.-S. Ha. 2001. Preparation and characterization of Poly(butylene terephthalate)/Organoclay nanocomposites. *Macromolecular Rapid Communications*, 22, 1306–1312.

Maiti, M., S. Mitra, and A.K. Bhowmick. 2008. Effect of nanoclay on high and low temperature degradation of fluoroelastomer. *Polymer Degradation and Stability*, 93, 180–200.

Maji, P.K. and A.K. Bhowmick. 2009. Influence of number of functional groups of hyperbranched polyol on cure kinetics and physical properties of polyurethanes. *Journal of Polymer Science Part-A: Polymer Chemistry*, 47, 731–745.

Manoj, N.R. and P.P. De. 1998. An investigation of the chemical interactions in blends of poly(vinyl chloride) and nitrile rubber during processing. *Polymer*, 39, 733–741.

Mark, H.F. *Encyclopedia of Polymer Science and Technology*. 1993. 4th Edn., Wiley Interscience, New York, pp. 930–996.

Mark, J.E., B. Erman, and F.R. Eirich. 2005. *Science and Technology of Rubber*, 3rd edn., Elsevier, Amsterdam, the Netherlands.

Messermith, P.B. and E.P. Giannelis. 1994. Synthesis and characterization of layered silicate epoxy nanocomposites. *Chemistry of Materials*, 6, 1719–1725.

Mitra, S., A. Ghanbari-Siahkali, P. Kingshott, H.K. Rehmeier, H. Abildgaard, and K. Almdal. 2006. Chemical degradation of crosslinked ethylene-propylene-diene rubber in an acidic environment. Part I. Effect on accelerated sulphur crosslinks. *Polymer Degradation and Stability*, 91, 69–80.

Morton, M. 1973. *Rubber Technology*, Van Nostrand Reinhold, New York.

Okada, O., M. Kawasumi, A. Usuki, Y. Kojima, T. Karauchi, and O. Kamigaito. 1990. Synthesis and properties of nylon 6/clay hybrids. *Proceedings of the Symposium on Materials Research Society*, Vol. 171, p. 45.

Rong, J., Z. Jing, H. Li, and M. Sheng. 2001. A polyethylene nanocomposite prepared via in-situ polymerization. *Macromolecular Rapid Communications*, 22, 329–334.

Roy, N. and A.K. Bhowmick. 2010. Novel in situ polydimethyl siloxane-sepiolite nanocomposites: Structure-property relationship. *Polymer*, 51(23), 5172–5185.

Sadhu, S., R.S. Dey, and A.K. Bhowmick. 2008. Thermal degradation of elastomer based nanocomposites. *Polymer Composites*, 16, 283–293.

Saha, S. 2001. Rheological and morphological characteristics of polyvinylchloride/polychloroprene blends—Effect of temperature and mixing speed. *European Polymer Journal*, 37, 399–410.

Shach, D., G. Fytas, and D. Vlassopolelos. 2005. Structure and dynamics of polymer-grafted clay suspensions. *Langmuir*, 21(1), 19–25.

Slusarski, L. 1984. Thermal stability of elastomers. *Journal of Thermal Analysis*, 29, 905–912.

Varghese, H., S.S. Bhagawan, and S. Thomas. 2001. Thermogravimetric analysis and thermal ageing of crosslinked nitrile rubber/poly(ethylene-co-vinyl acetate) blends. *Journal of Thermal Analysis and Calorimetry*, 63, 749–763.

Wang, Z. and T.J. Pinnavaia. 1998. Hybrid organic–Inorganic nanocomposites: Exfoliation of magadiite nanolayers in an elastomeric epoxy polymer. *Chemistry of Materials*, 10(7), 1820–1826.

Weifu, D., X. Liu, Y. Zhang et al. 2006. Effect of rubber on properties of nylon-6/unmodified clay/rubber nanocomposites. *European Polymer Journal*, 42, 2515.

Weil, E.D. 1974. *New Approaches to Flame Retardancy in Flammability of Solid Plastics*, Technomic Publishing Company, Philadelphia, PA, pp. 1–10.

Zhou, Q. and M. Xanthos. 2009. Nanosize and microsize clay effects on the kinetics of the thermal degradation of polylactides. *Polymer Degradation and Stability*, 94, 327–338.

17

Polymer Matrix Composites

Larry Gintert
Beechcraft Corporation

Chad Ulven
*North Dakota State
University*

Lawrence Coulter
*U.S. Air Force Research
Laboratory*

17.1 Introduction

Composite materials are used in a variety of applications due to various reasons including weight minimization, corrosion and wear resistance, erosion resistance, high-temperature performance, thermal and acoustical insulation, electrical performance, and lifecycle cost reductions. In military applications, aircraft as well as ground vehicles and tactical weapons employ composite materials in order to provide better mobility due to weight reductions offered. Some applications require the low weight and high stiffness achievable only with the use of structural sandwich design approaches. Other applications involving usage in a highly corrosive environment require the use of composite materials in order to maintain system reliability at an affordable cost. The variety of options for materials and processes available to the designer mandate that an understanding of the operational environment, system-level performance including maintenance and repair, and cost and ecological impact are taken into consideration early in the design phase.

The global family of composite material systems has evolved to include not only polymer matrix composites (PMCs) but also metal matrix composites (MMCs) and ceramic matrix composites (CMCs); this discussion focuses only on PMCs. It is assumed that the reader is familiar with the basics of composite material systems involving the combination of different constituent materials to establish an end product that exhibits superior properties to any of the constituents for the particular application. The composite material is generally comprised of constituent matrix and reinforcement materials. The matrix encompasses and provides stability to the reinforcement constituents that are typically in the form of a continuous filament, fiber, or particle. For example, a well-known PMC material is fiberglass/polyester used in many sporting goods applications because of its relatively light weight, high stiffness, and low cost. However, fiberglass/polyester is also used in many industrial applications because of its inherent resistance to corrosion where metallic materials simply are not economical due to the maintenance and repair or replacement costs.

17.2 Background

As composite materials are utilized in high-performance designs, they are typically required to be assembled in a fashion that places them into contact with dissimilar materials with different electrochemical potentials, resulting in the susceptibility to galvanic corrosion. An example of this is an unprotected carbon-reinforced PMC component assembled to aluminum alloy structure such as in some airframe applications. The nobler carbon fiber acts as the cathode, while the more active aluminum acts as the anode when they are in contact and have electrical continuity in the presence of an electrolyte (see Figure 17.1). Thus the aluminum corrodes if not properly designed with suitable isolation methods (Hamm 1994). Jones provides an excellent introduction to the varieties of composite materials, including PMCs as well as other inorganic composite materials, and briefly discusses the degradation mechanisms to consider (Jones 1975).

In addition to the galvanic corrosion concerns of an assembly containing PMCs, the degradation of constituent materials within the PMCs is also a concern in certain environments, such as under extreme humidity or ultraviolet (UV) light conditions. High-humidity or water immersion environments can affect the critical interface between the reinforcement (fiber) and the matrix in a way that is detrimental to material strength. UV exposure can degrade the polymer matrix constituent material if not properly protected, especially for relatively thin sections. Likewise, other environmental considerations are mandated for similar concerns that are not typically considered under "corrosion" considerations.

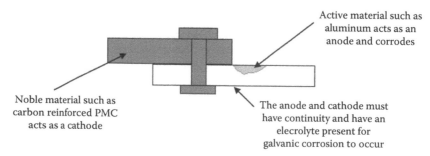

FIGURE 17.1 Example of unprotected carbon-reinforced PMC assembled to aluminum. (Adapted from Gintert, L. and Hihara, L., Corrosion considerations for military applications of composite material systems, *Proceedings of the NACExpo*, NACExpo, Houston, TX, 2001.)

17.3 Application

17.3.1 Uses

Military aircraft have recently adopted the use of PMCs for primary structural applications such as new wing skins for the A-6 Intruder, the wing spars and fuselage components of the F-22 Raptor air dominance fighter (Lockheed Martin Aeronautical Systems 1998), and even more recently Boeing has used PMCs extensively in their new 787 Dreamliner aircraft (Kristoff 2009). Rotary wing aircraft have long used PMCs and composite sandwich construction as main rotor material systems for military aircraft applications. Commercial automotive vehicles have employed PMCs in structural body and chassis components as well as wheels and other parts since the Chevrolet Corvettes began using fiberglass as the primary structural material for its sleekly contoured body panels. Recreational watercraft designs have used fiberglass-reinforced plastic materials of construction for decades largely because of their workability and resistance to corrosion in a marine environment. Likewise, sporting goods such as tennis racquets, skis, golf clubs, and backpacking gear are heavy users of PMCs in order to achieve lightweight requirements while maintaining the desired structural performance. Common industrial applications of PMCs include exhaust fan components and ductwork sections because of their resistance to corrosion in certain chemical environments. More recently, wind turbines and bridge components have been developed and are being incorporated into designs. As PMC materials become more broadly utilized and the design principles are more broadly accepted, the applications will expand.

17.3.2 Challenges

While PMC materials provide properties that enable the systems to achieve the required operational performance levels, they may impose high lifecycle costs driven by maintenance demands if not properly managed. The assembled components operating in the intended environment are critical considerations during design and material/process selection. Sealants, coating systems, fluid exposure, and operational atmospheric conditions are all important factors to consider. The interaction of two faying surfaces of an assembly may produce a by-product that affects other materials in the assembly, so it is important that the full assembly is evaluated to determine true performance. The complete lifecycle of a system needs to consider not only the useful life of the component including maintenance costs but also end-of-life and disposal costs, including disposal of cleaning, sealants, and coatings used during maintenance.

Little is known of the long-term effects of environmental degradation of PMC structures. These structures, which include only a few primary structures, have only so far seen relatively short exposure times. If the industry is expecting a structure to last 30, 40 years, or longer, surveillance and monitoring of the system health needs to be conducted. While PMC materials have been employed in applications such as helicopter main rotor blades and general aviation primary structure for several decades, the aerospace industry has only recently begun detailed inspection of PMC structures that have seen long-term use (Dutta et al. 2004, Tomblin et al. 2010). Since WWII, fiberglass hulls have been employed by the U.S. Navy, but little has been documented about their environmental durability. Other industries have been using PMC materials for a relatively short time. Likewise, studies of longer service lifetimes (over 30 years) need to be conducted to see the real effects of environmental degradation.

17.4 General Properties

PMC materials are often selected during the design phase of an application due to their attractive material properties, including not only the structural performance properties of high stiffness and high strength at minimal weight but also their resistance to corrosion or degradation under certain environmental conditions. Composites are often used due to their ability to modify or enhance surface properties of a component for improved wear resistance, thermal/electrical/noise insulation, or other unique characteristics.

17.4.1 Physical and Mechanical

The main physical and mechanical properties of PMCs with respect to understanding and controlling degradation include density, color, shrinkage, water absorption, ductility, elasticity, and strength. Mechanical and physical properties of PMC materials are affected primarily by the properties of the reinforcement and matrix constituents and the manner in which they are combined during processing. Fiber orientation/configuration and degree of cure can affect properties of pristine materials, but some properties are affected under different environmental exposures, some of which degrade (or "corrode") the material permanently. Many of the physical and mechanical properties of PMCs mentioned are governed by the structure, molecular weight, crystallinity, cross-link density, etc., of the polymer matrices. For instance, most polymers subject to UV degradation need to be protected by a suitable coating system for corrosion protection. Note that some formulations of polymers that are UV resistant have been developed to address this concern.

The density of PMCs is controlled by the volume occupied by and the density of each constituent material. In addition, the quality of the PMC plays a role in the global density. If voids are present in the PMC structure as a result of lack of control during composite processing, the density will be reduced along with the mechanical properties of that PMC. The presence of voids is important to the degradation potential of a PMC. The more voids that are present, the greater the ability for foreign atoms/molecules to diffuse/transport through and collect in the structure, thereby advancing other mechanisms of degradation.

17.4.2 Thermal

Thermal properties of PMCs are governed by the type and respective amounts of matrix and reinforcement materials as well as the architecture (i.e., placement or orientation) of the reinforcement. As with physical and mechanical properties, thermal properties of PMCs can be highly anisotropic. This anisotropy is largely governed by the reinforcement architecture. Important thermal properties to consider in PMCs with respect to degradation include coefficient of thermal expansion (CTE), glass transition temperature (T_g), heat distortion temperature (HDT), and residual thermally induced stress created during processing of PMCs. Especially when coupled with moisture to create hygrothermal effects, the thermal properties of PMCs are very important to understand when trying to limit degradation.

The CTE of a PMC is most often represented by α and is greatly dependent on the type of matrix and reinforcement present, amount of matrix and reinforcement present, and the architecture in which the reinforcement is organized. When the CTE of the reinforcement is lower than that of the matrix, the reinforcement is understood to be constraining the matrix from expansion. Therefore, when unidirectional long fibers are used in a composite where the matrix has a higher CTE than the fibers, the composite will exhibit much lower CTE in the longitudinal direction of the fibers as compared to the transverse direction in the composite. The differences between the extent of expanding and contracting of the constituent materials can induce microscopic strains, which can degrade the PMC properties over time if large cyclic changes in temperature are expected during the application lifetime.

In most PMCs, the T_g is governed by the matrix material. All polymeric materials possess a T_g where the mechanical behavior changes from a glassy/brittle behavior to a leathery/flexible state as a function of increasing temperature. Comparing various T_g across different polymer types, in general elastomers exhibit the lowest T_g, followed by thermoplastics and then thermosets. The atomic structure of the polymer backbone, molecular weight, degree of crystallinity, and degree of cross-linking all contribute to the T_g. Glass transition temperature will play a role in the corrosion resistance of a material by allowing more molecular mobility in the structure as a function of temperature. If the T_g of a PMC has been surpassed during normal or abnormal operation for any period of time, much more mobility by diffusion and transport of atoms/molecules will occur. This increased mobility of foreign atoms/molecules can be very detrimental to the integrity of the PMC structure as other mechanisms of degradation can be easily initiated.

The HDT of PMCs is the temperature at which a specified specimen shape deflects to a certain extent under a specified load. Because of the specificity to shape and loading conditions used, HDT is not considered a material property; however, the HDT of a PMC is usually greatly influenced by the T_g of the polymer matrix. In addition, the HDT of a PMC is important to consider in design as it determines the maximum useful temperature where the material can be used. Typically, the HDT of PMCs will be lower than the T_g of the polymer matrix. Just as with T_g, HDT is highly dependent on the molecular backbone structure, degree of cross-linking, degree of crystallinity, and amount of secondary interaction within the structure. In terms of potential for degradation, a compounding detrimental effect can be observed in PMCs whenever high levels of stress are present in conjunction with high service temperature.

During the processing of PMCs (usually at elevated temperatures), large residual thermal stresses can develop as a result of large differences in CTE between the matrix and reinforcement that can also be directional (non-isotropic) due to the orientation of the reinforcement. When thermosetting polymer matrices are used to process PMCs, the curing reaction between the constituents creates an exotherm that raises the temperature of not only the matrix but also of the reinforcement it encapsulates. Once the structure and bonds between the matrix and reinforcement are established during the exotherm, upon subsequent cooling these phases with different CTEs will experience differential stresses/strains also stretching bonds between the two phases. This state of stretched bonds will cause localized residual stress states between matrix and reinforcement (Parlevliet 2006, Wisnom 2006). In addition, hygrothermal internal stresses in PMCs can be generated from the differences in swelling occurring in the constituents as a result of moisture uptake. In both cases of either residual thermal stresses or hygrothermal stresses, very transient and localized nonuniform stress fields can be produced due to PMC anisotropic properties thereby degrading load-bearing properties (Parlevliet et al. 2007, Benkhedda et al. 2008, Maier and Hofmann 2008).

17.5 Structure

17.5.1 Atomic or Molecular Bonding

The types of atomic or molecular bonding found in PMCs are complex and governed by the types of matrix and reinforcement used as well as any chemical treatments used to couple these two or more material phases. Polymer matrices in a PMC are in general based on forming a continuous network of polymer molecules surrounding the reinforcement. These polymers mainly contain carbon–carbon bonds that are covalently bonded along the backbone of each chain along with several other secondary types of interactions perpendicular to that chain. The reinforcement phase, however, can have covalent, metallic, or ionic bonding depending on its material type. For instance, carbon fiber contains an arrangement of carbon atoms covalently bonded together in the form of hexagonal layers densely packed together because of secondary interactions, whereas glass fibers are amorphous solids composed primarily of silica backbone, which possesses a mixture of covalent and ionic bonding character. Other reinforcement phases may have metallic (for metals) or ionic (e.g., for SiC or other ceramics) bonds. It is the nature of this bonding and atoms involved for both the matrix and reinforcement phases that most influences the resistance of PMCs to degradation within certain environments.

Specifically, it is the structure of the atoms and the strength of their bonds that make up the matrix and reinforcement phases that control not only the mechanical and physical properties of that PMC but also the types of and rate/extent of degradation that occurs in PMCs. The respective atomic/molecular structure dictates the PMC's susceptibility to (1) environmental conditions such as UV, oxidation, and resistance to microorganisms or (2) influences of the chemical solubility, permeability, and electrical currents on that PMC's performance. For instance, nylon reinforced with glass fiber will inherently absorb moisture because of the polar similarities between water molecules and those in nylon's structure. In another case, an epoxy reinforced carbon fiber PMC is known to be sensitive to types of hydraulic fluids used in aircraft; thus, the possibility of contact with hydraulic fluids must be considered in the design and material selection phases.

17.5.2 Microstructure

The microstructure found within PMCs is quite complex because of the large extensive amount of surface area (per unit volume) between the reinforcement and surrounding matrix. The region of contact between the two PMC phases of material is commonly referred to as the interphase and is considered to be an additional phase of material present in the composite as a consequence of the chemical/mechanical interaction between these different materials (see Figure 17.2). The surface area (i.e., boundary layer) between the matrix material and the interphase, as well as between the fiber material and the interphase, is known as interfaces (Jang 1994). In general, the interphase is the region where the strength and integrity of the PMC is developed as a result of good/desirable chemical and mechanical interactions. However, the interphase can also be the location of weakness once interfacial bonding is compromised by different mechanisms of degradation.

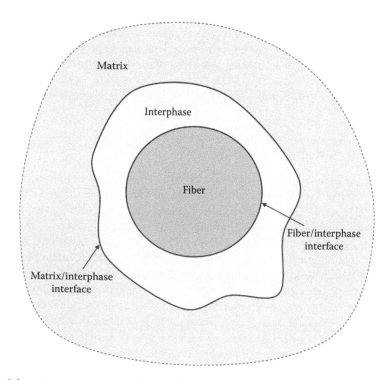

FIGURE 17.2 Schematic microstructure of PMCs illustrating the formation of an interphase and interfaces.

17.5.3 Macrostructure

The macrostructure found within PMCs is also quite complex because of the multitude of different fiber reinforcement architectures that can be selected, for instance, the sequence and stacking of individual plies of fabric reinforcement. Many different reinforcement forms consisting of nonwovens, wovens, and knits are available for PMC design; each possesses benefits and disadvantages in processability, final properties/structural integrity, and resistance to degradation. In addition, the stacking sequence of reinforcement plies can be highly varied resulting in PMCs with property differences in permeability, diffusion, and transport of various foreign atoms/molecules. Interrupting the path of transport abruptly through strategic layering can also bring about the cessation or retardation of certain degradation mechanisms and may also create deleterious stress concentrations.

The design of joints through selective macroscopic layer stacking, where PMCs' components are connected to one another or to other materials to create assemblies, is another important structural aspect to consider in controlling and/or understanding the degradation of PMCs. Joining PMCs to other materials and structures can be accomplished using mechanical fastening, adhesive bonding, or a combination of both. In mechanical fastening, the choice of fastener in combination with the materials being joined is critical in order to limit the effects of galvanic potentials. Similarly for the choice of an adhesive, it must provide an intimate bond to enable proper load transfer without creating chemical free energy differences that may drive certain transport phenomenon that leads to certain degradation mechanisms. Additional discussion regarding the degradation of adhesively bonded structures is provided in Chapter 18 "Adhesive Bonds."

17.6 Degradation Principles for PMCs

The degradation in PMCs can be broadly governed by their susceptibility to environmental weathering, chemical solubility, permeability, and electrical conductance. Each of these conditions can create unique types of deterioration (chemical, electrochemical, and/or photochemical) of the PMC structure. However, depending on the type of matrix material, reinforcement material, and interphase created between the former two, one constituent may or may not play a role in any or all of the degradation mechanisms occurring in certain environments. In general for PMCs, the matrix is designed to protect the structural reinforcement from environmental attack and therefore typically governs the type of degradation that can occur as well as the rate it happens. The following paragraphs describe in detail the mechanism of degradation in PMCs under various adverse conditions and the reasons for their susceptibility in certain environments. Additional discussion regarding the various polymer degradation mechanisms is provided in Chapter 13 "Forms of Polymer Degradation."

17.6.1 UV Degradation

Ultraviolet radiation (i.e., photochemical) induced degradation of PMCs is mainly governed by the susceptibility of the polymer matrix to UV in the near-surface region as the depth of penetration by UV light is rather limited. Also due to the nature of PMC processing techniques, most PMCs have an inherent or intended polymer-rich surface that fully encapsulates the reinforcement. Polymeric matrices will break down under UV exposure through two ways: (1) by adding thermal energy to the polymer and breaking down the structure through thermal degradation and (2) by exciting electrons in the covalent bonds of the polymeric structure (Strong 2006). Through excitation of electrons, those bonds become weaker and easier to break with long-term excitement or additional stress application (i.e., exterior loading of the structure). This occurs when the energy of a specific frequency of light closely matches the covalent bond energy existing between atoms in polymeric structures. If the depth of UV light penetration is sufficient enough to reach the reinforcement of the PMC, a similar type of structural degradation may occur in the interphase and/or fiber depending on the types of bonds present (i.e., primary, secondary) in the interphase and the atomic structure of the reinforcement (fiberglass, carbon fiber, aramid fiber, etc.). However, a very negligible amount of physical and mechanical property loss would be realized if the PMC structure is relatively thick compared to the UV depth of penetration.

Certain polymer matrices are more susceptible to UV radiation than others because of structural differences that ultimately affect the integrity of the entire PMC (Strong 2006). For instance, polyethylene (PE) without a UV stabilizer will degrade at a much higher rate than polymethyl methacrylate. Therefore, some thermoplastics can be extremely sensitive to UV, while some can be very resistant. Since thermosets are in general moderately sensitive to UV without a chemical stabilizer, ultimately most PMCs for use in applications exposed to UV will utilize a chemical stabilizer or absorber such as carbon black and titanium dioxide. UV resistive coatings have also proven to be quite successful with PMCs because the strong chemical bonds that can be created between the polymeric-based coating and polymeric-based composite are more readily matched than when these coatings are applied to metallic substrates.

17.6.2 Thermo-Oxidative Degradation

When PMCs are exposed to high temperatures either repeatedly or for sustained periods through radiative (such as UV), convective, or conductive heating exposures, thermally activated oxidation of certain polymeric structures tends to occur. As with UV degradation, the level of oxidative degradation in polymeric matrices with high-temperature exposures is very dependent on the type of polymer used for the matrix. Oxidation of polymeric structures occurs as the result of thermally rupturing a covalent bond, thereby producing a free radical that can readily combine with oxygen from the surrounding event (Strong 2006). Oxidation of a polymeric structure will produce a localized discontinuity or weakness within the structure, leading to reduced mechanical properties. Similar to the case of UV degradation, the effect of surface oxidation in the polymer matrix of a relatively thick PMC component may only cause a negligible physical and mechanical property loss. However, the thermo-oxidative degradation of polymeric matrices may provide new pathways into PMCs for other agents of degradation such as chemical solubilization, water permeability, and/or microorganism attack to occur.

17.6.3 Microbiological Degradation

Some newer uses of PMCs in applications include outdoor decking products, composting and trash bins, and medical implant devices, where contact with natural products for long durations creates potential degradation by microorganisms (i.e., biodegradation). However, just as with UV and thermo-oxidation mechanisms of PMC deterioration, the class of polymer matrix used will determine the extent of microbiological degradation that will occur within the composite. In addition, the type of polymeric structure, molecular weight, and crystallinity of the polymer matrix also influences the susceptibility to biodegradation (Gu 2003). Synthetic polymeric matrices, such as polyolefins, biodegrade following a two-stage process known as oxo-degradation. In the first stage, synthetic polymers are broken down into oligomers, dimers, monomers, etc., by exoenzymes from microorganisms; this is the primary cause of physical and mechanical property loss (Gu 2003). In the second stage of biodegradation, other bacteria, fungi, enzymes, etc. (see Figure 17.3), start the bioassimilation of these smaller polymer chains, further breaking them down and giving rise to biomass, CO_2, H_2O, and CH_4 production. The speed of biodegradation depends on the temperature, humidity, and number and type of microbes present in the environment surrounding the PMC.

17.6.4 Chemical Solubility Induced Degradation

Due to their designed superior corrosion resistance, PMCs are being utilized in more and more harsh environments to replace traditionally developed metallic counterparts under the same

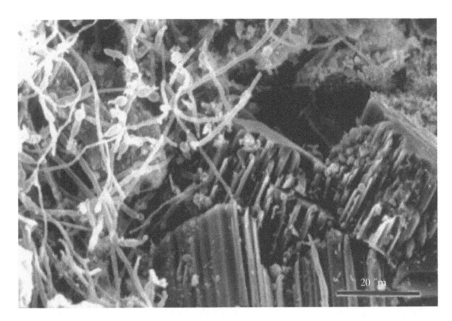

FIGURE 17.3 SEM image showing colonization of bacteria and fungi on surfaces of PMC. (Adapted from Gu, J.D., *Int. Biodeterior. Biodegradation*, 52, 69, 2003.)

exposure conditions. Many of these harsh applications involve containment functions or coating functions to protect a structure from a variety of chemicals. The chemical solubility of a PMC is largely governed by the surrounding polymer matrix, but the interphase and fiber could also play a critical role in providing resistance to chemical solubility and maintenance of physical and mechanical properties; note here that the depth of penetration of solvents is usually greater than that for UV penetration. Chemical attack of polymeric matrices can cause (1) swelling or softening of the structure, (2) dissolving of the entire structure, or (3) reactions with portions of the structure to permanently alter the nature of that PMC by forming new bonds (Strong 2006).

The degree or severity of chemical degradation in a polymer matrix, interphase, or fiber is a function of the chemical nature of the constituent materials. When the chemical nature of the solvent is similar to that of the solute (i.e., matrix, interphase, fiber), degradation is more likely to occur. The greater the chemical affinity between the solvent and the PMC components of the system/structure, the more the damage is likely to occur. As an example, the polarity of water molecules very closely matches that of the structure of polyamides and therefore readily tends to penetrate into the structure, causing swelling and loss of stiffness. If the swelling of the structure does not lead to dissolution or other permanent chemical reactions, it is often referred to plasticization where there is a loss of strength and modulus while there may be increase in elongation and toughness up to a certain point. The effects of plasticization or swelling of a PMC due to chemical solubility can be reversed in certain systems if the solvent can be fully removed and provided no permanent chemical reactions/changes have taken place.

17.6.5 Hygrothermal Degradation

Possibly, the most common form of degradation occurring in PMCs is due to hygrothermal effects (these produce property changes as a function of increasing temperature and humidity conditions). A majority of the polymeric matrices used for structural PMCs are at some level sensitive to moisture especially at high water concentration levels and with high-temperature exposures. Therefore, a large amount of research has been dedicated to the deterioration of physical and mechanical properties in PMCs as a function of hygrothermal effects over the past couple decades (Mallick 1993, Agarwal et al. 2006,

Strong 2008). Many specific studies on hygrothermal effects are summarized in Section 17.7; however, a brief overview of the susceptibility to and the characterization of moisture absorption under high temperatures and in high-humidity environments is provided below.

In general, PMCs are more susceptible to hygrothermal degradation than UV, thermo-oxidation, or microbiological degradation processes because moisture is absorbed not only by the matrix but by the fiber, as well as the interphase, and any porous regions or areas where microcracking or delaminations have already occurred within PMCs. Very similar to the prior description of chemical solubility, uptake of water (i.e., diffusivity) will generally create swelling of the different PMC constituents (i.e., matrix, reinforcement, and interphase). However, each of these constituents will generally uptake different amounts of moisture and at different rates. As a result of these differences in absorption, additional strain fields will be imposed between the matrix and reinforcement creating localized regions of weakness, especially when the structure is already under external load and/or interfacial bonds are swollen.

The moisture uptake of PMCs is governed by the permeability of the composite structure and can be broken into two stages. Initially (i.e., for short time periods), the moisture uptake at elevated temperatures is quite linear and predictable using a simple Fickian diffusion type of model. This initial moisture uptake phase is usually characterized as plasticization of the matrix resulting in the release of internal stresses. However, after prolonged exposure to hygrothermal conditions, permanent damage has been observed to occur in PMCs. Irreversible degradation such as hydrolysis of the matrix and/or interphase and interfacial decohesion between the interphase and the fiber may occur because of osmotic cracking (Bergeret et al. 2009). This type of permanent degradation is most worrisome for PMC structures used in marine or high-humidity/high-temperature environments. The first stage of moisture transport into PMCs is quite reversible, and studies have shown that virgin physical and mechanical properties can be recovered if the structure can be properly desorbed of moisture. However, if moisture is not removed properly or when osmotic cracking has occurred in the PMC structure, large delaminations can occur, leading to gross losses in physical and mechanical properties.

17.6.6 Electrochemical Degradation

In general, PMCs are nonconductive in nature due to the absence of long-range crystalline structure within the polymer matrix that also usually completely surrounds/isolates the reinforcement phase. Oftentimes, most of the reinforcement types used in PMCs such as fiberglass, aramids, and polyethylene are also nonconductive because they inherently have high band gaps. However, carbon or metallic fibers and fillers can be highly conductive, so when they are used as reinforcements in those PMCs, the longitudinal direction of those PMCs can also be conductive. Therefore, although most PMCs are not susceptible to galvanic corrosion, in those specific cases where conductive fibers are used, this degradation potential can become very critical. An example would be in systems/assemblies where carbon fiber-based PMCs are used in direct contact with metallic components such as aluminum. In this situation, even though the carbon fiber may not necessarily be in direct contact with the aluminum, whenever an electrolyte solution is present, such as from atmospheric moisture condensation, these conditions will be sufficient enough to establish a galvanic coupling cell and therefore will induce corrosion (Gebhard et al. 2009). Further detail of this type of degradation is provided in Section 17.7.

17.6.7 Mechanical Degradation

In all of the previously listed degradation mechanisms, one important aspect was neglected, that of the concurrent influence of mechanical loading in conjunction with adverse environments. Many more PMCs are being used in critical structural applications where the maintenance of physical and mechanical properties is of utmost importance for the safety of the personnel or society where they are used. In many cases, stress corrosion occurs as a function of one or more of the degradation mechanisms discussed earlier, while at the same time the structure is being loaded monotonically (i.e., pseudo-static loads, creep), cyclically (i.e., fatigue), or dynamically (i.e., impact). In fact, typically the application of stress to a PMC also undergoing some type of degradation will exacerbate or expedite the degradation processes taking place, leading to an even larger decrement of physical and mechanical properties as well as the reduction in the projected service life of that structure. Further detail regarding stress corrosion of a few specific PMC systems is provided in Section 17.7.

17.7 Degradation of Specific Systems

The degradation of many commonly used PMCs has been studied in a spectrum of environments where they are currently or may be employed. In other words, PMCs have been designed to replace traditional metallic material in many applications because of their improved corrosion resistance resulting in their use in some rather harsh environments. In addition, many PMC components being used to date are part of a larger system or assembly that includes metallic components to which they joined using either adhesive bonds or mechanical fasteners or both. Such joining of dissimilar materials can potentially create galvanic couplings based on the materials used in the PMCs and the other materials to with which they are in contact. This problem becomes very severe whenever the respective electrochemical potentials of all the contacting elements are significantly different.

The following sections briefly describe some case studies for both thermoset-based and thermoplastic-based PMCs. In general, more research has been dedicated to understanding the degradation potential related to thermoset-based PMCs, but recently some research has also investigated thermoplastic-based PMCs. The studies of material systems surveyed here are not intended to be comprehensive but rather provide a broad overview from recent findings for a variety of PMCs experiencing a variety of degradation mechanisms in their respective exposure environments.

17.7.1 Thermoset PMCs

A majority of the degradation studies on PMCs to date have been focused on thermoset matrix-based composites. The following paragraphs summarize recent studies on thermoset PMCs covering a range of aspects for (1) different material configurations, (2) the effects of different processing methods, (3) the influence of fiber architecture, and (4) the degree of resin cure on the degradation potential. The greatest cause of PMC degradation as discussed recently by multiple researchers has been attributed to hygrothermal effects. Therefore, many of the studies surveyed later use hygrothermal conditioning as a means to distinguish between differences in performance where these changes are a function of aspects listed previously.

The study of hygrothermal effects on carbon fiber/epoxy PMCs has recently received much attention (Wan et al. 2005, Wang and Hahn 2007, Du and Jana 2008, Karbhari and Xian 2009, Khan et al. 2010). In a study by Du and Jana (2008), highly conductive graphite/epoxy PMCs suitable for bipolar plates in proton exchange membrane fuel cells were investigated for degradation in such an application. Different boiling solutions (i.e., water, an aqueous sulfuric acid solution, and an aqueous hydrogen peroxide solution) were used to evaluate the diffusivity in these PMCs and any potential degradation. A Fickian behavior was observed, indicating no apparent chemical breakdown of the epoxy matrix. A maximum of 4.22% of water was absorbed after 6 months of immersion, with the diffusivity (10.2×10^{-6} mm^2/s) being the highest in the acid solution. However, no major dimensional, surface appearance, or morphological changes were observed. In addition, very little change in electrical conductivity or mechanical properties were measured. Finally, a slight change in the T_g of the PMC occurred as a result of the plasticization effect on the matrix, but when dried, T_g was fully recoverable.

Another study on carbon/epoxy PMCs was conducted recently by Kahn et al. (2010). This study was focused on the effects of PMC processing techniques on their relative resistance to moisture uptake. Two methods of composite manufacturing were utilized: traditional autoclave and the Quickstep technique (Quickstep Technologies Pty Ltd, www.quickstep.com.au). The Quickstep method uses a relatively high ramp rate compared to traditional autoclave and as a result absorbed slightly more water than autoclaved PMCs because of its higher cross-link density. However, very little difference was found for the reduction of physical and mechanical properties (T_g, flexural, interlaminar shear strength, etc.) for these two different processing techniques. Some permanent damage to the PMC structure occurred as a function of the moisture uptake as assessed by DMTA and FTIR; however, a majority of the degradation was found to be reversible upon drying. In a very similar study (Kootstookos and Burchill 2004), the effect of the degree of cure on the degradation of fiberglass-reinforced vinyl ester PMCs was studied. It was also found that the degree of cure did not affect the durability of fiberglass-reinforced vinyl ester PMCs exposed to alkaline, acidic, or hydrocarbon environments. However, the reinforcement volume fraction of the PMC did, where the resistance to degradation was found to decrease as fiber volume fraction increased.

Karbhari and Xian (2009) investigated the moisture uptake and desorption rates in pultruded carbon fiber/epoxy PMCs subjected to hygrothermal conditioning. They found the initial moisture uptake and desorption could easily be described by a Fickian model. However, for the long-term behavior, the non-Fickian behavior that was observed required a two-stage model (involving a Langmuir dual mode diffusion response) to properly describe the entire moisture absorption behavior. As with the previous two studies, T_g was found to decrease because of the plasticization effect of water being absorbed into the epoxy matrix and where the permanent network/interface deterioration was attributed to elevated temperatures. Other studies on carbon fiber/polyimide PMCs (Han and Nairn 2003) and fiberglass/epoxy PMCs (Yann et al. 2006) investigated hygrothermal aging that produced similar results. A loss in toughness was measured for carbon fiber/polyimide PMCs (Han and Nairn 2003), and a loss in stiffness was exhibited for fiberglass/epoxy PMCs (Yann et al. 2006). Both effects were attributed to the initial plasticization of the matrix and interphase prior to permanent damage from interfacial debonding, polymeric chain scission, etc. (see Figure 17.4).

In a carbon/epoxy PMCs study by Wan et al. (2005), the influence of three-dimensional braided fiber architecture versus unidirectional fiber architecture on the moisture absorption and structural deterioration potential was investigated. These PMCs were vacuum-assisted resin transfer molding (VARTM) processed and hygrothermally aged at 37°C for up to 1700 h. It was shown that the calculated moisture diffusion rates fit a Fickian model for the moisture absorption behavior where the type of fiber architecture influenced the moisture uptake rate, namely, that the 3D fiber-based PMC exhibited a lower

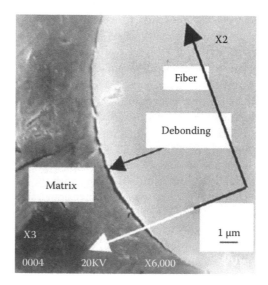

FIGURE 17.4 SEM image of fiber/matrix debonding after hygrothermal conditioning. (Adapted from Yann, R. et al., *Mech. Mater.*, 38, 1143, 2006.)

diffusion rate than the unidirectional fiber-based PMC. A similar work by Boukhoulda et al. (2006) showed that the diffusion mass flow rate was not isotropic and was influenced by the angle between the measured diffusion flow direction relative to the longitudinal fiber orientation in those PMCs. Also they investigated the influence on moisture absorption and subsequent material degradation for E-glass/epoxy and carbon/epoxy PMCs that were hygrothermally aged in varied relative humidity and temperatures for up to 180 h. The moisture diffusion rate was found to be higher along the direction of fibers compared to that perpendicular to the fibers. The maximum moisture absorption value for the E-glass/epoxy was found to occur more quickly than that for the carbon/epoxy PMCs. Finally, it was commented that manufacturing quality of these PMCs had a large influence on absorption with larger-sized and higher-density defects in the PMC structure contributing to greater amounts of water absorption and the corresponding acceleration of the material degradation process.

Wan et al. (2005) also investigated the effects of concurrently applied tensile stresses on PMCs during hygrothermal aging and found increases in the maximum amount of moisture absorbed, whereas applied compressive stresses reduced the total moisture absorbed in these PMCs. It was suggested that the magnitude and sign of externally applied stresses affects the matrix free-volume, the propagation of matrix cracks, and the moisture sorption along fiber/matrix interfaces. In a similar work by Megel et al. (2001), the resistance to stress-corrosion cracking was investigated for pultruded unidirectional E-glass-reinforced polyester, epoxy, or vinyl ester matrix PMCs. Samples loaded in a four-point bend fixture were exposed to a nitric acid solution. Initiation of stress corrosion was found at the E-glass fiber/matrix interfaces located along the surface of these specimens. A very distinct mechanism of crack initiation, subcritical crack extension, and stable crack propagation was observed. The E-glass/vinyl ester PMC had the highest resistance to stress corrosion followed by the E-glass/epoxy and E-glass/polyester PMCs, respectively.

17.7.2 Thermoplastic PMCs

As mentioned previously, advanced engineering grades of thermoplastic-based PMCs are generally more resistant to corrosion than commodity grade thermoplastic-based PMCs, elastomeric-based PMCs, or thermoset-based PMCs. However, more advanced engineering grades of thermoplastic-based PMCs are currently being designed to be used as bearings and antifriction coatings in pumps/hydraulic systems and in biomedical applications where they encounter some rather harsh environments. The following paragraphs provide an overview of various degradation studies recently conducted on thermoplastic PMCs intended to be utilized in these harsh conditions.

Gebhard et al. (2009) investigated the effects of galvanic coupling stainless steel to polyetheretherketone (PEEK) short fiber PMCs reinforced with either polyacrylonitrile (PAN) or pitch-based carbon fibers in an electrolyte solution. Short fiber composite specimens composed of PEEK with PAN-based carbon fibers or PEEK with pitch-based carbon fibers with fiber volume fractions of 10% and 30% were processed with a twin-screw extruder using injection molding to make those specimens. The corrosion of the carbon fibers was observed due to an electrochemical degradation reaction that occurred at the sites where the surface-exposed fibers within the PMC contacted the stainless steel. However, this corrosion only occurred in the 30% volume carbon fiber PMCs regardless of source (i.e., PAN vs. pitch). Regarding the carbon fiber corrosion, the pitch-based carbon fibers corroded at a higher rate than the PAN-based carbon fibers as evident from the greater amount of debonding and fiber cracking (see Figure 17.5).

The work by Pillay et al. (2009) revealed the type of degradation observed in carbon fiber-reinforced polyamide-6 PMCs exposed to UV and moisture independently. Exposure to UV was found to cause yellowing of the samples as well as increases in the crystallinity that was localized to only the near surface of the composite. The UV exposure was found to not significantly alter the bulk flexural or impact properties of this composite because of the limited depth of UV penetration/degradation. Moisture absorption on the other hand was found to have a more profound effect on the mechanical properties. Flexural and impact properties reduced significantly as a function of increasing moisture absorption due to a plasticization effect on the matrix and fiber/matrix interphase. No influence on the melting

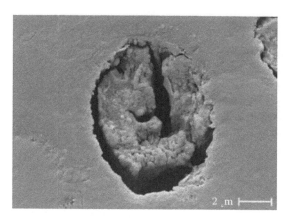

FIGURE 17.5   SEM image of fiber cracking and fiber/matrix debonding caused by galvanic corrosion. (Adapted from Gebhard, A. et al., *Corros. Sci.*, 51, 2524, 2009.)

peaks or crystallinity was observed with increasing amounts of moisture absorption; and once dried, these composites recovered their flexural and impact properties to within 10% of virgin values.

In a study by Pegoretti and Penati (2004), thermoplastic PMCs based on recycled poly(ethylene terephthalate) (rPET) and short glass fibers that were injection mold processed were studied for any changes in structure and properties after hygrothermal aging (at 70°C and 80% or 100% RH). Water uptake was observed to behave in a Fickian manner with diffusivity decreasing as fiber content increased or the relative humidity decreased. In addition, these PMCs absorbed more water than the monolithic polymer itself indicating that other mechanisms such as capillarity and/or transport phenomenon via microcracks were concurrently active. Finally, the T_g of these PMCs were found to decrease as a function of hygrothermal aging time.

In a similar study by Foulc et al. (2005), glass fiber-reinforced PET composites (with 30% glass fiber by weight) were produced by injection molding and subsequently subjected to hygrothermal conditioning by water immersion in an autoclave run at 120°C and 1.6 bars. In this study, however, water absorption was characterized as having a non-Fickian behavior accompanied by a rapid/drastic reduction in mechanical strength. The degradation of these PMC properties was postulated to occur by two mechanisms: one being a reversible plasticization of the matrix and the other being the nonreversible random chain scission caused by hydrolysis of the amorphous phase in the polymer. Finally, interfacial decohesion was observed leading to the formation of cracks and voids that promoted the premature fracture of this PMC.

Many other significant studies evaluating hygrothermal effects on glass fiber-reinforced thermoplastics have been conducted over the past couple decades involving a variety of different matrices: poly (butylenes terephthalate) (Mohd Ishak et al. 2001), polyurethane (Boubakri et al. 2009), polyamide-6,6 (Bergeret et al. 2009), etc. In all of these studies, the diffusivity was found to increase as hygrothermal aging temperature and time increased while the resultant mechanical properties decreased significantly. The most critical aspect for controlling/limiting the rate and extent of degradation in these composites was through the modification of the fiber/matrix interface (Boubakri et al. 2009). Finally, some of the effects of hygrothermal aging were found to be reversible if the composite could be made to simply desorb moisture without causing significant, permanent polymer degradation (Boubakri et al. 2009).

17.7.3 Structural Degradation of PMC Components or at PMC Surfaces (e.g., Faying surfaces, Hybrid configurations, Lightning Strike Systems, Fasteners, Sealants)

Systems or functional structures are seldom made solely of PMC composite materials. Aircraft systems fasteners, sealants, lightning strike mesh, fittings, press fit bushings, etc., are almost always used in carefully assembling the final configuration of such functional structures. For example, to avoid situations

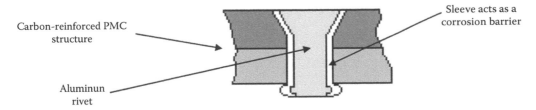

FIGURE 17.6 Sleeved rivet concept for protecting fasteners. (Adapted from Gintert, L. and Hihara, L., Corrosion considerations for military applications of composite material systems, *Proceedings of the NACExpo,* Houston, TX, 2001.)

where a galvanic cell with a large galvanic difference is created (such as between carbon fibers and most metals), great care needs to be exercised in making material selections for these systems and structures. One example is the widespread use of titanium fasteners to join carbon/epoxy PMC components to the remaining structure. Titanium fasteners are known to provide very good resistance to galvanic corrosion that results when other metals (e.g., aluminum) come in contact with carbon fibers. Thus, a common fastening approach to controlling assembly costs involves the use of titanium sleeves in conjunction with using less-expensive aluminum rivets as shown in Figure 17.6. The titanium sleeve provides compatible interfaces to the aluminum rivet and the carbon/epoxy PMC while preventing direct contact between the aluminum and carbon/epoxy PMC. Although titanium is electrically conductive and does not galvanically decouple the aluminum rivet from the carbon/epoxy PMC, the damage resulting from galvanic corrosion is known to be less when titanium serves as an intermediate barrier between aluminum and carbon/epoxy PMCs.

Composite sandwich structures employing thin outer face sheets attached (typically by adhesive bonding) to lightweight core materials that separate both face sheets and thus provide high rigidity and high bending strength properties to that sandwich component are subject to corrosion degradation if the interior core is not properly sealed and protected. Due to the complex nature of this sandwich structure, any penetrations or other leakage paths into the interior can expose the core materials to moisture, thus creating structural problems including corrosion. Figure 17.7 is an example of such damage on a component from a large military aircraft still in service today. The remains of the core in these components were "poured" out of the damaged area because they were so badly corroded. Proper treatment of core materials combined with effective sealing and periodic maintenance is required to prevent this type of failure. Such considerations not only are important for the original design but also apply to the repair and maintenance of sandwich structures. Whenever an anodized metal core material is cut or machined either during assembly or repair operations, the fresh metal exposed on the cut edges of the

FIGURE 17.7 Aluminum honeycomb sandwich structure with core degradation. (Adapted from Gintert, L. et al., Corrosion control for aluminum honeycomb sandwich structures, *Proceedings of the SAMPE,* Long Beach, CA, 2002.)

core must be appropriately treated and restored; otherwise, its original corrosion protection treatment will be severely compromised.

17.8 Degradation Protection

17.8.1 Surface Treatments

Surface treatments could be applied either to the PMC structure itself or to the adjacent piece/component that is in contact with that PMC. A common surface treatment for aerospace fasteners or other inserts is to coat them with some type of barrier such as a sealant or an epoxy primer. These treatments are only partially successful since they tend to wear and degrade during use in service, leaving the metal fastener subject to subsequent attack. When using a fastener with poor corrosion properties, care should be used in selecting and relying on that surface treatment to solely protect the fastener. The reason for this caveat is that significant, rapid reduction in the strength of any fastener can rapidly occur when even small failures occur in that coating.

Another common surface treatment for carbon fiber PMCs is to incorporate a surface ply of fiberglass during lay-up. This ply is sometimes called a scratch ply, sacrificial ply, barrier ply, or veil ply. The use of this fiberglass ply can protect the remaining structure from damage during paint/depaint operations and also can be used to separate any adjacent metal components from contacting the carbon fibers located within the near surface of that PMC component.

17.8.2 Coatings

Coatings for PMC structures can be tremendously important for mitigating environmental degradation. Coatings are usually systems with surface preparations, primers, base coats, top coats, and other application schemes that protect the underlying PMC component. Many PMC matrices are very susceptible to UV damage; even PMCs containing UV inhibitors can be seriously degraded after a very short period of UV exposure. Paints and surface treatments with UV protection should always be used with PMC structures that are intended for outdoor use. These paints and coating systems must also be compatible with the PMC substrate, especially where the use of specific primers and surface preparation techniques are often critical for extending service lives of PMC parts.

17.8.3 Design Features

Design features can be very important in protecting PMC structures from environmental degradation. The materials chosen are important, but any design that does not incorporate proper degradation mitigation features will be doomed to failure. Some considerations to address are as follows:

1. Elimination of water entrapment: Are there proper pathways for water to escape?
2. Separation of dissimilar materials: Is there a good faying surface preparation and separation?
3. Are fasteners and other metal inserts separated or insulated from direct contact with carbon fibers?
4. Do sandwich or hybrid structures have adequate sealing and closeouts that prevent water entry?
5. Is there adequate drainage when rinse water and cleaning products are used?
6. Will maintenance operations affect the characteristics of the part and/or its degradation mechanisms?

17.8.4 Repair and Maintenance Considerations (Lessons Learned)

Maintenance, Repair and Overhaul (MRO) activities should be considered while producing the original design. The lifecycle design of a structure should incorporate whether de-coating and re-coating operations will be needed. If a surface is expected to experience multiple coating/re-coating operations, the possibility that significant degradation and damage to surface and near-surface layers should be considered. Damage from aggressive depainting operations can easily remove enough protective surface plies to render the part unusable.

Any repair done to a structure should incorporate all of the original methods that were intended to mitigate environmental degradation. Thus, if the originally designed structure is painted, any repair patches also must be similarly painted. If bolt or fastener repair operations are performed, the repair instructions should require using the same mitigation strategies, for example, using originally specified sealants, titanium fasteners, etc.

Fasteners and other metal fittings will need periodic inspections over the life of a structure. Since the galvanic problem between carbon and metals is so widespread, any PMC structures will need more frequent inspection of the fasteners themselves with considerations for the possibility of having their periodic replacement.

During any repair of lightning strike systems used in conjunction with PMCs, it is important to maintain the integrity of the corrosion protection scheme; this is particularly sensitive to the materials and processes employed, largely because electrical continuity is required in order for this protection to continue to function. When sealants are employed, it is important to understand the long-term durability properties of these materials as they may shrink and/or crack over time thus allowing leakage paths to form.

Replacing and maintaining existing corrosion mitigation systems for fasteners already in place during removal and replacement operations is also important.

17.9 Specific Standardized Tests

For PMCs, some specific test standards exist for determining the degree and rate for certain types of degradation. Standards based on plastics are mainly utilized for PMCs with regard to measuring their degradation because the performance of the polymer matrix usually controls these PMC properties as well. The common standards used for PMCs include those for (1) UV radiation (ASTM D1435, D4329, and D4364), (2) water absorption (ASTM D570, D5229), and (3) resistance of PMCs to chemical reagents (ASTM D543). The following paragraphs summarize each of these common/standardized methods for determining degradation behavior of various PMCs.

Water absorption standards (ASTM D570 and D5229) are mainly used both for the measurement of moisture uptake in composites and as a standard means of conditioning composite samples for subsequent physical and mechanical property evaluations. ASTM D570 covers the relative rate of absorption of water by plastics after immersion. Thus, many studies on short fiber or filled plastics will utilize this standard as a means for measuring the same property. Also addressed in this standard are the variety of different processing techniques used to mold PMC plastics and measure the rate of water uptake within their respective specimen classes and geometries. ASTM D5229 covers moisture absorption or desorption through the thickness of flat or curved PMC panels based on a single-phase Fickian behavior. As with ASTM D570, ASTM D5229 can be used to "condition" composite specimens in a standard way to attain reproducible conditions for physical and mechanical property evaluations.

A variety of UV radiation exposure standards (ASTM D1435, D4329, and D4364) exist to cover natural, artificial accelerated and accelerated natural exposures of plastics. These standards are often used for PMCs as well because the concentrated, detrimental effects that UV typically has on the polymer matrix are mainly limited to the near surface of the polymer or the PMC composite. ASTM D1435 covers the exposure procedures for simulating natural, long-term exposures of plastics in order to assess their real-time, in-service degradation. ASTM D4329 and D4364 are intended to accelerate the UV degradation process in plastics in order to "condition" samples for other physical and mechanical property evaluations that depend on the extent/duration of UV degradation.

ASTM D543 is typically used to evaluate the resistance of PMCs to different chemical reagents that they may contact while in service. Again, this standard was developed specifically for plastics. However, because the susceptibility of PMCs to chemical reagents is mainly governed by the polymer matrix, this standard is sometimes used for PMCs. This standard uses immersion tests in a wide range of reagents where it is meant only as a guide for comparing the relative resistance of various plastics to typical chemical reagents. The selection of the types and concentrations of reagents, the immersion duration, and the temperatures and/or stresses applied are meant to mimic service exposure conditions; thus, the resultant properties measured should be representative of those in the intended use scenario for that particular material.

In addition to the ASTM Standards described previously for measuring the influence or extent of environmental degradation on various PMCs, some researchers have been investigating nondestructive methods for detecting degradation in situ (Karalekas et al.). Karalekas et al. demonstrated that a fiber Bragg grating (FBG) sensor could be used to measure axial strains in a cylindrical glass fiber-reinforced epoxy composite subjected to hygrothermal aging. Through extensive experimental investigations along with numerical simulations, it was determined that this method was a valid technique for characterizing hygrothermal aging-induced strain as well as determining the relative extent of moisture absorption by its effects on swelling and interfacial damage. Being able to quantify the degradation of PMC composites nondestructively during their use in specific applications is the needed next step in mitigating potentially catastrophic PMC structural failures and improving the confidence for utilizing PMCs in a wider variety of applications.

Wang and Hahn (2007) used atomic force microscopy (AFM) to characterize the interface in carbon fiber PMCs when exposed to hygrothermal treatments. They found that debonding occurs along the interface of unidirectional carbon fibers when conditioned in a 100% humidity chamber at temperatures up to 200°C for up to 1500 h (Figure 17.2). The debonding observed was attributed to matrix swelling caused by moisture absorption, while higher temperatures cause matrix shrinkage. Very precise dimensional changes were measured using AFM. This proves that with its ease of operation and nondestructive nature, AFM is a useful tool for characterizing PMCs exposed to various environments.

17.10 Summary

While PMCs have been used extensively in many different fields during the past several decades, there has been minimal open-source literature regarding their degradation mechanisms until the past decade. Long-term studies or observations have not been well documented or openly shared.

Intelligent design is critical from a corrosion protection standpoint as PMC components are often incorporated into systems/structures with other metallic components, where many of the previously cited corrosion mitigation rules are commonly applicable. PMC materials exhibit degradation in some ways that are different from those for metals, especially when it concerns galvanic corrosion of metals in contact with the carbon fiber-reinforced PMCs. Repair practices introduce new challenges for corrosion protection. Appropriate repairs for coatings, sealants, and protective interface materials need to be engineered/specified into standardized repair designs/procedures with appropriate lifecycle considerations.

In addition to the galvanic corrosion concerns for an assembly containing PMCs and metallic components, the degradation of the constituent materials within PMCs is also a concern. Corrosion can affect the critical interface between the reinforcement (fiber) and the matrix in a way that may be detrimental to material strength. UV exposure can degrade the polymer matrix constituents if not properly protected. Likewise, other environmental influences on PMC performance should be considered since similar concerns may not typically be considered under "normal metallic corrosion conditions." In general, moisture is more damaging to PMC materials than UV exposure due to the limited depth of penetration of UV; whereas, moisture can diffuse through the entire structure and affects the fiber/matrix interface potentially causing major changes in bulk properties.

The usage of PMCs is increasing throughout the world in many industries and with many innovative uses that have not been seen before. Increased usage will mean an increase in the potential for problems related to the unique nature of PMCs. PMCs are very useful and have many advantages, but a proactive approach to corrosion mitigation of these PMC materials is also prudent and necessary. Common, previously applied design wisdom used to state that "corrosion doesn't affect PMCs," but experience and greater knowledge of the physics of interfaces in PMCs shows that there are many, sometimes novel, corrosive type effects. Being forward looking about these novel and common problems will go a long way to assuring the successful usage of these unique PMC materials.

References

Agarwal, B.D. et al. 2006. *Analysis and Performance of Fiber Composites*, 3rd edn., John Wiley & Sons Inc., Hoboken, NJ, pp. 416–431.

Benkhedda, A. et al. 2008. *Composites Structures*, 82, 629–635.

Bergeret, A. et al. 2009. *Polymer Degradation and Stability*, 94, 1315–1324.

Boubakri, A. et al. 2009. *Materials and Design*, 30, 3958–3965.

Boukhoulda, B.F. et al. 2006. *Composite Structures*, 74, 406–418.

Du, L. and S.C. Jana. 2008. *Journal of Power Sources*, 182, 223–229.

Dutta, P.K., C.C. Ryerson, and P. Charles. 2004. Thermal deicing of polymer composite helicopter blades, *Proceedings of the Materials Research Society Fall Symposium*, Boston, MA.

Foulc, M.P. et al. 2005. *Polymer Degradation and Stability*, 89, 461–470.

Gebhard, A. et al. 2009. *Corrosion Science*, 51, 2524–2528.

Gintert, L. and L. Hihara. 2001. Corrosion considerations for military applications of composite material systems, *Proceedings of the NACExpo*, Houston, TX.

Gintert, L., M. Singleton, and W. Powell. 2002. Corrosion control for aluminum honeycomb sandwich structures, *Proceedings of the SAMPE*, Long Beach, CA.

Gu, J.D. 2003. *International Biodeterioration & Biodegradation*, 52, 69–91.

Hamm, C.D. 1994. Corrosion protection measures for CFC/Metal joints of fuel integral tank structures of advanced military aircraft, *Proceedings of the Corrosion Detection and Management of Advanced Airframe Materials*, Seville, Spain, October 5–6.

Han, M.H. and J.A. Nairn. 2003. *Composites Part A*, 34, 979–986.

Jang, B.Z. 1994. *Advanced Polymer Composites*, ASM International, Metals Park, OH, p. 37.

Jones, R.M. 1975. *Mechanics of Composite Materials*, Scripta Book Company, Washington, DC, p. 2.

Karalekas, D. et al. 2009. *Composites Science and Technology*, 69, 507–514.

Karbhari, V.M. and G. Xian. 2009. *Composites Part B*, 40, 41–49.

Khan, L.A. et al. 2010. *Composites Part A*, 41, 942–953.

Kootstookos, A. and P.J. Burchill. 2004. *Composites Part A*, 35, 501–508.

Kristoff, S. 2009. The Boeing 787 Dreamliner. http://mechanical-engineering.suite101.com/article.cfm/the_boeing_787_dreamliner

Lockheed Martin Aeronautical Systems. 1998. F-22 Raptor: Air dominance for the 21st century. *Advanced Materials & Processes*, 153, 23–26.

Maier, G. and F. Hofmann. 2008. *Composites Science and Technology*, 68, 2056–2065.

Mallick, P.K. 1993. *Fiber-Reinforced Composites: Materials, Manufacturing, and Design*, 2nd edn., Marcel Dekker Inc., New York, pp. 307–318.

Megel, M. et al. 2001. *Composites Science and Technology*, 61, 231–246.

Mohd Ishak, Z.A. et al. 2001. *European Polymer Journal*, 37, 1635–1647.

Parlevliet, P.P. et al. 2006. *Composites Part A*, 37, 1847–1857.

Parlevliet, P.P. et al. 2007. *Composites Part A*, 38, 1581–1596.

Pegoretti, A. and A. Penati. 2004. *Polymer Degradation and Stability*, 86, 233–243.

Pillay, S. et al. 2009. *Composites Science and Technology*, 69, 839–846.

Strong, A.B. 2006. *Plastics: Materials and Processing*, 3rd edn., Pearson Prentice Hall, Upper Saddle River, NJ, pp. 152–159.

Strong, A.B. 2008. *Fundamentals of Composites Manufacturing: Materials, Methods, and Applications,* 2nd edn., Society of Manufacturing Engineers, Dearborn, MI, pp. 287–292.

Tomblin, J., L. Salah, and C. Davis. 2010. Aging aircraft evaluation of a Beechcraft starship main wing. *Presented at the 13th Joint FAA/DoD/NASA Aircraft Airworthiness and Sustainment Conference,* Austin, TX, May 10–13.

Wan, Y.Z. et al. 2005. *Composites Part A,* 36, 1102–1109.

Wang, Y. and T.H. Hahn. 2007. *Composites Science and Technology,* 67, 92–101.

Wisnom, M.R. 2006. *Composites Part A,* 37, 522–529.

Yann, R. et al. 2006. *Mechanics of Materials,* 38, 1143–1158.

18

Adhesive Bonds

Guy D. Davis
*Consultant in Materials
Science*

18.1 Introduction

Adhesive bonding is widely used to fasten many materials and fabricate structures. It offers several advantages over mechanical fastening, but as with any process, there are disadvantages as well, as shown in Table 18.1 (Kinloch 1987, Sharpe 1990, Dixon 2005, Pocius 1997).

18.2 Background

There are three principal classifications of the types of bonding that occurs between an adherend and an adhesive. The most common bonding mechanism is electrostatic attraction. Here, oppositely charged polar groups (e.g., H, O, N, Cl-containing moieties) or dipoles of the adhesive and the adherend attract each other when the two materials are in intimate contact. Dispersion forces between transient or induced dipoles can also contribute to bond strength. Covalent bonds between the adhesive and the adherend represent the second bonding mechanism. If mutually attractive chemical groups are available on both sides of the interface, they can react to form very strong and durable bonds. Silane coupling agents are one example of using specific chemically reactive groups to form covalent bonds. The third bonding mechanism is mechanical interlocking or physical bonding. Here, the liquid adhesive flows into "nooks and crannies" of the adherend surface. If the surface morphology is convoluted or complex, once the adhesive hardens or cures, it cannot disengage from the adherend and a strong, durable bond is formed.

Given the topic of this book, the durability of adhesively bond structures is emphasized here. For discussion of other aspects of adhesive bonding, the reader is referred to the sources listed at the end of this chapter. Although obtaining high initial bond strength is relatively easy, maintaining good bond durability in aggressive environments is more challenging. The most important factor controlling bond degradation

TABLE 18.1 Adhesive Bonding: Advantages and Disadvantages by Classification

Classification	Advantages	Disadvantages
Manufacturability	The ability to join materials dissimilar in chemistry, size, or thickness	Rigorous process control is often necessary for high performance
	The ability to join thin sheets and large areas	The adherend surface condition can be critical for high performance
	Concomitant sealing	
	Improved appearance with fewer or no spot welds and fasteners	
	No requirement for mechanical energy during bonding, in many cases, so that shock-sensitive materials can be joined into an assembly	Heat, pressure, jigs, and fixtures may be needed for assembly
	Ability to be automated	
Performance	A reduction in the number of holes and fasteners that can cause stress concentrations in the structure	Bonded structures are usually difficult to disassemble for repair
	Excellent load transfer and a more uniform stress distribution	Joint strength cannot be measured nondestructively
	Improved fatigue resistance	
	Improved corrosion resistance in many cases	
	Improved vibration damping	
	Increased design flexibility, for example, honeycomb structures where a honeycomb core is bonded to two sheets of material on either side	Long-term performance can be difficult to determine from short-term tests
		Operating temperatures can be limited

of most materials is moisture (Kinloch 1987, Davis 2003, Davis et al. 2009). Moisture is pervasive over much of the world and is responsible for the vast majority of bond failures—both in the field during service and in the laboratory during research and development.

The importance of adhesive bond durability will vary depending on the particular application and environment and the consequences of bond failure. One of the most critical cases is military and civilian aircraft, especially those that operate in tropical, coastal, or marine locations. At the other extreme, less critical applications include those subject to low stresses and protected from harsh environments, such as interior furniture in temperate climates or even wrapped gift packages. Increased durability generally requires additional initial cost through the use of more expensive materials or processing. Consequently, it makes little sense for a bond to have a significantly longer lifetime than the item or system of which it is a part. In this chapter, we will concentrate on applications where durability is a critical issue, such as the aerospace industry.

The rate of bond degradation depends on a number of variables that can be grouped into three categories: environment, material, and stress. The environment is dominated by temperature and moisture, but can also include the concentration of aggressive ions, such as chlorides, and the presence of fuels, deicers, and other fluids. Electrochemical potential, either directly applied or created by galvanic couples, and extreme pH can also promote degradation. The material grouping is all-inclusive and includes the adherend, the adhesive, and the interphase between them. Finally, the stresses to which the bond is subject either during or after exposure also influence its lifetime or residual strength.

Each of these factors will be discussed in more detail. We also review the means to enhance durability. Entire books can be and have been written on durability; in this short chapter, we can only touch on the subject. For more details, the reader is referred to the many reviews in the literature (Kinloch 1982, 1983a,

Venables 1984, Marceau and Thrall 1985, Brockmann et al. 1986, Pitrone and Brown 1988, Brinson 1990, Brockmann 1990b, Clearfield et al. 1990, Minford 1991, Boerio et al. 1993, Minford 1993, Davis and Venables 2002, Davis 2003, Park et al. 2010).

18.3 Applications

18.3.1 Uses

Adhesively bonded materials have been reported as early as 70,000 years ago with Neanderthals using natural bitumen or pitch to bond stone points to wooden handles (Koller et al. 2001, Boëda et al. 2008). In some cases, it is suggested that the early stone-age people processed the pitch rather than using it as found. Adams and Fay (2011) have traced the historical use of adhesives from prehistoric times to the twentieth century.

Modern-day uses of adhesive bonding are diverse as adhesives can be developed with a wide variety of properties. The applications of adhesive bonding are widespread: The aerospace industry relies heavily on adhesively bonded construction including honeycomb structures where the honeycomb core is bonded to metallic or composite skins for lightweight yet strong and stiff components. Another aerospace application is solid rocket motors where the rubber-based fuel is bonded to the metallic or composite case. The automobile industry is another major user of adhesive bonding with hem-flange joints (joints between the outer and inner door panels) and windshield attachment being but two examples. Tires require strong bonding between the steel cords and the rubber matrix. The construction industry has numerous applications ranging from plywood and particle board to bonding floorboards to joists for stiffer and quieter construction. Bonding floor and wall tile, as well as wall paper, are other examples in home construction. Furniture and cabinets also exhibit adhesive bonding, either veneer or joints. Packaging is another common example of adhesive bonding ranging from simple sealing to the construction of composite materials, such as metal–plastic laminates. Medical applications include adhesive bandages, patches for drug release, sealing of wounds, and dental crowns, braces, and other orthodontics. Pressure-sensitive adhesive labels are yet another common application.

18.3.2 Challenges

Achieving initial bond strength can usually be obtained with proper choices of adhesives, substrate (adherend) preparation, and joint design. Challenges can arise based on constraints on one or more of these areas or on other issues, including bonding environment, time, or cost.

Greater challenges arise in achieving bond durability in the given operating environment, especially for high-performance bonds. Although high-temperature requirements can present problems for adhesive bonds, moisture is much more common limitation to the lifetime of an adhesive joint and will be the dominant focus of this chapter.

18.4 Environment

As already mentioned, moisture is the bane of most adhesive bonds. It is nearly impossible to keep water from a bond exposed to the outside environment (Comyn 1983). It can readily diffuse through the adhesive or the adherend, if that component is permeable, as typical for an organic matrix composite material. Moisture can also wick or travel along the interface, and it can migrate via capillary action through cracks and crazes in the adhesive. According to Comyn (1983), once moisture is present, it can attack the bond by

- Reversibly altering the adhesive, for example, plasticization
- Swelling the adhesive and inducing concomitant stresses
- Disrupting secondary bonds across the adherend/adhesive interface
- Irreversibly altering the adhesive, for example, hydrolysis, cracking, or crazing
- Hydrating or corroding the adherend surface

The first three of these processes are reversible to one extent or another. Provided that bond degradation has not proceeded too far, if the joint is dried out (which may be a long process), the bond can regain some of its lost strength (Sung 1990, Boerio et al. 1993). There appears to be a critical water concentration, below which either no weakening occurs (Gledhill and Kinloch 1976, Kinloch 1987) or whatever weakening that does occur is reversible (Kinloch 1979, Sung 1990). This critical water concentration is dependent on the materials used in the joint and is likely to be dependent on the temperature and stress as well. At higher moisture levels, some strength may be recovered upon drying, but at a certain point, the failure becomes nearly catastrophic and is beyond recovery.

Upon moisture penetration, the locus of failure commonly switches from cohesive within the adhesive to at or near the interface. Because metal oxide surfaces are polar, the dominant adhesion mechanism is electrostatic attraction. However, these same polar groups also attract water molecules that can disrupt any dispersive bonds across the interface between the adherend and the adhesive. This disruption can be seen thermodynamically by the work of adhesion in an inert medium, W_A, which can be represented as (Kinloch 1987)

$$W_A = \gamma_a + \gamma_s - \gamma_{as} \tag{18.1}$$

where
γ_a and γ_s are the surface free energies of the adhesive and substrate, respectively
γ_{as} is the interfacial free energy

In the presence of a liquid such as water, the work of adhesion, W_{Al} becomes

$$W_{Al} = \gamma_{al} + \gamma_{sl} - \gamma_{as} \tag{18.2}$$

where γ_{al} and γ_{sl} are now the interfacial free energies of the adhesive/liquid and substrate/liquid interfaces, respectively. In an inert environment, the work of adhesion for a bonded system will be positive indicating a stable interface, whereas in the presence of water, the work of adhesion may become negative, indicating an unstable interface that may dissociate. Table 18.2 shows, in fact, that moisture will displace epoxy adhesives from iron (steel), aluminum, and silicon substrates and promote disbanding (Kinloch 1987). In contrast, although moisture weakens epoxy/carbon fiber bonds, these remain thermodynamically stable. Practical experience with both metal and composite joints confirms these predictions (Kinloch 1987).

The data presented in Table 18.2 illustrate the potential disastrous results when relying solely on electrostatic forces or dispersive bonds across the interface between an epoxy adhesive and metals or ceramics. To illustrate this danger, demonstration specimens can be produced that exhibit good initial strength, but fall apart under their own weight when a drop of water is placed at the crack tip. Covalent bonds or mechanical interlocking will resist or prevent bond failure resulting from moisture intrusion into the bondline.

TABLE 18.2 Work of Adhesion for Various Interfaces

Interface	Work of Adhesion (mJ/m²)		Interfacial Debonding after Immersion in Water?
	In Inert Medium	In Water	
Epoxy/ferric oxide (mild steel)	291	−255	Yes
Epoxy/alumina	232	−137	Yes
Epoxy/silica	178	−57	Yes
Epoxy/carbon fiber-reinforced plastic (CFRP)	88–99	22–44	No

Source: Kinloch, A.J., *Adhesion and Adhesives: Science and Technology*, Chapman & Hall, London, U.K., 1987.

18.5 Materials

The adherend, adhesive, and interphase between them are the major factors in determining bond durability. For example, the simple disruption of the dispersive forces already described indicates that joints made with composite adherends will be inherently more stable than those made with metallic adherends. To increase durability, most metallic and many polymeric adherends undergo surface treatments designed to alter the surface chemistry or morphology to promote primary covalent chemical bonds and/or physical bonds (mechanical interlocking) to maximize, supplement, or replace secondary dispersive bonds. These treatments are discussed elsewhere (Kinloch 1983a, 1987, Thrall and Shannon 1985, Wegman 1989, Clearfield et al. 1990, 1991, Critchlow and Brewis 1995, 1996, Pocius 1997, Critchlow et al. 1998, Davis and Venables 2002, Park et al. 2010). An intent of each treatment is to provide interfacial bonding that is resistant to moisture intrusion.

Formation of durable chemical bonds is an obvious means to stabilize the interface and has been demonstrated for phenolic/alumina joints (Lewis et al. 1974), silane and other coupling agents (Miller and Ishida 1991, Owen 2002), and sol–gel coatings (Blohowiak et al. 1996, 1998). Silane coupling agents and sol–gel coatings are discussed later.

For most structural joints using epoxy adhesives and metallic adherends, moisture-resistant chemical bonds are not formed and mechanical interlocking on a microscopic scale is needed between the adhesive/primer and adherend for good durability. In these cases, even if moisture disrupts interfacial chemical bonds, a crack cannot follow the convoluted interface between the polymer and oxide, and the joint remains intact unless this interface or the polymer itself is destroyed.

The scale of the microscopic surface roughness is important to assure good mechanical interlocking and good durability. Although roughness on any scale serves to increase the effective surface area of the adherend and, therefore, to increase the number of primary and secondary bonds with the adhesive/primer, surfaces with features on the order of a few nanometers exhibit superior performance to those with features on the order of micrometers. Several factors contribute to this difference in performance. The larger-scale features are fewer in number and generally are smoother (on a relative scale) so that interlocking is less effective. Depending on the particular treatment used, there may also be loosely bound detritus that prevents bonding to the integral adherend surface (Bishopp et al. 1988). In addition, the larger-scale roughness frequently allows trapped air and surface contaminants to remain at the bottoms of troughs and pores (Bishopp et al. 1988, Davis 1991). These unbonded regions limit joint performance by reducing both chemical and physical bonds and serving as local stress concentrators. In contrast, smaller–scale nanoroughness tends to be more convoluted in morphology and generates strong capillary forces as the primer wets the surface, drawing the polymer into all the "nooks and crannies" of the oxide and displacing trapped air and some contaminants to form a nanocomposite interphase (Davis 1991). Indeed, cross-sectional micrographs show complete filling of the nanopores (Venables 1984, Marceau 1985, Bishopp et al. 1988, Brockmann et al. 1986, Clearfield et al. 1990, 1991).

This dependence on the degree and scale of roughness is illustrated in Figure 18.1, which shows wedge test results of titanium bonds with several surface preparations (Brown 1982). Because the titanium surface is stable under these conditions, differences in the joint performance can be attributed solely to differences in the polymer-to-oxide bonds and correlate very well with adherend roughness. The poorest performing group of pretreatments (Class I: phosphate fluoride [PF] and modified phosphate fluoride [MPF]) produced relatively smooth surfaces. The intermediate group (Class II: Dapcotreat [DA], dry Pasa Jell [DP], liquid Pasa Jell [LP], and Turco [TU]]) exhibited macrorough surfaces with no nanoroughness. They had significant improvements in durability over the smooth adherends, but not as good as the Class III pretreatment (chromic acid anodization [CAA]), which provided a very complex microroughness. Subsequent Class III tests using sodium hydroxide anodization provide further evidence to this correlation (Kennedy et al. 1983, Filbey et al. 1987, Shaffer et al. 1987). Both of these give very good durability performance and exhibit high levels of nanoroughness.

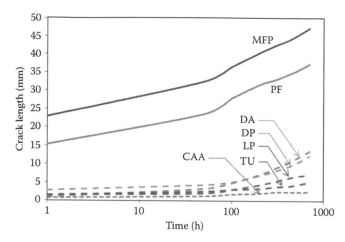

FIGURE 18.1 Wedge test results for Ti adherends with several different surface treatments having differing degrees and scales of roughness—Class I: phosphate fluoride (PF) and modified phosphate fluoride (MPF); Class II: Dapcotreat (DA), dry Pasa Jell (DP), liquid Pasa Jell (LP), and Turco (TU); Class III: chromic acid anodization (CAA). Specimens were exposed to 100% relative humidity at 60°C. (Data from Brown, S.R., *Proceedings of the 27th National SAMPE Symposium*, SAMPE, Azusa, CA, 1982, p. 363.)

18.5.1 Adherends

The adherend often establishes ultimate joint durability. The morphology of its surface determines the degree of physical bonding (mechanical interlocking) with the polymer, and its chemistry, in part, determines the degree and type of chemical bonding. Furthermore, the stability of the adherend and its surface determines the ultimate limit of durability. Once the adherend becomes corroded or otherwise degraded, the bondline is reversibly damaged and the joint fails.

Each material exhibits its own form of degradation and conditions under which the degradation occurs. For aluminum adherends, moisture causes hydration of the surface, that is, the Al_2O_3 that is formed during the surface treatment is transformed into the oxyhydroxide AlOOH (boehmite) or tri-hydroxide $Al(OH)_3$ (bayerite). The transformation to the hydroxide results in an expansion of the interphase (the volume occupied by the hydroxide is larger than that originally occupied by the Al_2O_3). This expansion and the corresponding change in surface morphology induce high stresses at the bondline. These stresses, coupled with the poor mechanical strength of the hydroxide, promote crack propagation near the hydroxide/metal interface.

The rate of hydration of the aluminum oxide depends on a number of factors, including surface chemistry (treatment), presence of hydration/corrosion inhibitors in the primer or applied to the surface, temperature, and the amount of moisture present at the surface or interface. One surface treatment that provides an oxide coating that is inherently hydration resistant is phosphoric acid anodization (PAA) (Marceau 1985). Its stability is due to a layer of phosphate incorporated into the outer Al_2O_3 surface during anodization; only when this phosphate layer goes into solution does the underlying Al_2O_3 hydrate to AlOOH (Davis et al. 1982). The hydration process is illustrated in the surface behavior diagram of Figure 18.2 (Davis et al. 1982, Davis 1986). It shows hydration to occur in three stages: (1) a reversible adsorption of water, (2) slow dissolution of the phosphate layer followed by rapid hydration of the freshly exposed Al_2O_3 to AlOOH, and (3) further hydration of AlOOH to $Al(OH)_3$.

Although the evolution of surface chemistry depicts the hydration of bare surfaces, the same process occurs for buried interfaces within an adhesive bond. This was first demonstrated by using electrochemical impedance spectroscopy (EIS) on an adhesive-covered Forest Products Laboratory etched (FPL, a sodium dichromate and sulfuric acid etch) aluminum adherend immersed in hot water for several months (Davis et al. 1995). EIS, which is commonly used to study paint degradation and substrate

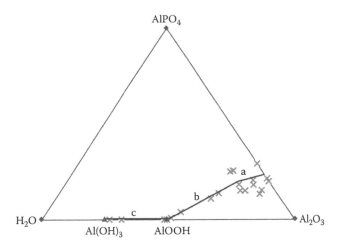

FIGURE 18.2 Surface behavior diagram showing hydration of the PAA Al_2O_3 surface. Hydration occurs in three stages: (a) reversible adsorption of moisture, (b) hydration of the Al_2O_3 to AlOOH, and (c) further hydration to $Al(OH)_3$. The numbers represent hours of exposure to high humidity. (Data from Davis, G.D. et al., *J. Mater. Sci.*, 17, 1807, 1982; Davis, G.D., *Surf. Interface Anal.*, 17, 439, 1991.)

corrosion (Mansfeld et al. 1982, Scully 1989), showed absorption of moisture by the epoxy adhesive and subsequent hydration of the underlying aluminum oxide after 100 days (Figure 18.3). At the end of the experiment, aluminum hydroxide had erupted through the adhesive.

Subsequent investigations showed that identical hydration reactions occurred on both bare aluminum surfaces and bonded surfaces but at very different rates of hydration (Davis et al. 2000). An Arrhenius plot of incubation times prior to hydration of bare and buried FPL surfaces clearly showed that the hydration process exhibited the same energy of activation (~82 kJ/mol) regardless of the bare or covered nature of the surface (Figure 18.4). On the other hand, the rate of hydration varies dramatically, depending on the concentration of moisture available to react at the oxide–polymer interface or the oxide surface. The epoxy-covered surfaces have incubation times (and rate constants) three to four orders of magnitude longer than bare, immersed specimens while bare surfaces exposed to high

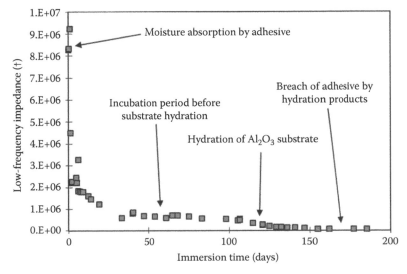

FIGURE 18.3 Low-frequency electrochemical impedance of an epoxy-coated FPL aluminum adherend as a function of immersion time in 50°C water. (Data from Davis, G.D., *Surf. Interface Anal.*, 9, 421, 1986.)

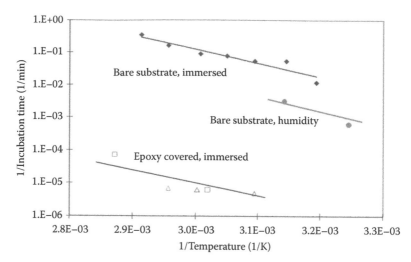

FIGURE 18.4 Arrhenius plot of incubation times prior to hydration of FPL aluminum under various conditions. (From Davis, G.D. et al., *J. Adhes.*, 72, 335, 2000. With additional data.)

humidity exhibit intermediate incubation times. Such findings reflect the limited amount of moisture absorbed by the epoxy and free to react with the oxide.

Steel adherends are also subject to corrosion in moist environments. Unfortunately, no general etch or anodization treatment has been developed that provides superior bond durability (McNamara and Ahearn 1987, Clearfield et al. 1990, 1991). In part, this is due to the lack of a coherent, adherent stable oxide—iron oxides generally do not protect the underlying substrate from the environment. Equally important, the different steel metallurgies react to chemical treatments differently—a procedure that may give good results for one steel alloy may give very poor results for another, similar steel alloy.

Steel bonds are often designed to minimize cost as long as certain performance standards are met (Brockmann 1983). The use of corrosion inhibitors, paints, and sealants often provides suitable protection from the environment. The most common surface treatments are grit blasting or other mechanical abrasion processes that clean the surface and provide a more chemically reactive oxide. Improvements to grit blasting, based on either performance or cost, have been reported for individual steels (David 1973, Devine 1977, Russell et al. 1981, Minford 1983, Pocius et al. 1984, Smith 1984, Trawinski 1984); however, rankings of different treatments commonly vary from researcher to researcher because of different steels or exposure/test conditions.

Deposited coatings often provide better bond durability than native surface treatments. For example, optimized conversion coatings can provide a microscopically rough surface that is resistant to corrosion (Janssen 1980, Chandler 1982, Trawinski et al. 1984, Bishof et al. 1985, Trawinski 1985). They serve to stabilize the surface from degradation and to form physical bonds with the adhesive/primer. Smaller-grain coatings tend to be preferred over larger-grain coatings as they provide better physical bonding and greater resistance to fracture. Again, differences in the adherend metallurgy can cause differences in the coating morphology and chemistry. Nonetheless, such conversion coatings and other deposited coatings provide the best durability for steel bonds.

Joints made with steel and other metals can also be subject to cathodic disbondment if they are immersed in an electrolyte and subjected to a cathodic potential, such as that created when the adherend is in electrical contact with a more electrochemically active metal (Stevenson 1990). Although corrosion of the adherend is suppressed via cathodic protection, the rate of bond failure is increased. After an induction period that depends on the imposed cathodic potential and temperature (Stevenson 1990), interphasal debonding occurs (Boerio et al. 1987, Watts 1988); such disbondment does not occur in the absence of a cathodic potential. Several mechanisms have been proposed to explain this phenomenon.

These include hydrogen evolution at the steel substrate (Stevenson 1990), degradation, and weakening of the polymer in the high pH environment generated at the interface (Hammond et al. 1981, Watts and Castle 1983, Watts 1988), osmotic pressure resulting from lead chlorides formed at the interface by dehydrohalogenation of the polymer (Hammond et al. 1981), and breaking of secondary and primary bonds at the interface (Kinloch 1987).

Bond failure can also occur if the surface is anodic relative to another joint component. An example would be clad aluminum adherends, where a thin layer of pure aluminum overlays the base alloy. Such a surface layer is designed to be more corrosion resistant than the alloy, but to also act as a sacrificial anode should corrosion occur. Although this approach works well for corrosion protection of the substrate material, it can be a disaster for bonded material if the adherend surface/interface corrodes. As a result, American companies tend to use unclad aluminum for bonding and provide other means of corrosion protection, such as painting (Reil 1971, Kinloch 1987). On the other hand, European companies commonly use clad adherends, but with a thicker oxide (CAA) (Fokker 1978, Bell Helicopter 1980, Brockmann et al. 1986, Clearfield et al. 1990, MIL-A-8625C) that provides bondline corrosion protection.

In contrast to aluminum and steel, titanium adherends are stable under conditions of moderately elevated temperatures and humidity. Although moisture has been shown to accelerate the crystallization of the amorphous oxide of titanium adherends anodized in chromic acid (CAA) to anatase (Natan and Venables 1983), the crystallization, along with the resulting morphology change, is very slow relative to the changes observed with aluminum and steel. For moderate conditions, the key requirement for a titanium treatment is a convoluted nanorough surface to promote physical bonding. Critchlow and Brewis extensively reviewed reported results of different surface treatments for titanium (Critchlow and Brewis 1995).

At elevated temperatures, where titanium would need to be chosen over aluminum, another failure mechanism becomes effective—dissolution of oxygen into the titanium substrate (Shaffer et al. 1987, Clearfield et al. 1990). As a result, CAA-treated joints, with their relatively thick oxide, perform poorly when exposed to high temperatures over extended periods (Progar and St. Clair 1986) as the oxide dissolves leaving a defect-riddled interface that fails when stress is applied. Alternative treatments that have been shown effective at high temperatures include sol–gel and plasma-sprayed deposited coatings (Shaffer et al. 1987, Blohowiak et al. 1996, 1998, Clearfield et al. 1990, Cobb et al. 1999, Park et al. 2000). These coatings do not have oxygen that can diffuse into the metal and are stable at elevated temperatures.

Another means by which temperature can influence bond durability is through stresses that develop when different parts of a joint have different coefficients of thermal expansion (CTEs). This consideration is especially important when different classes of materials are being bonded together. Polymer CTEs usually are 10–100 times those of other materials (Comyn 1990). Stresses begin to develop across the interphase once the adhesive cures to a solid (rubbery) state and the joint begins to cool (Comyn 1990). As long as the adhesive is above the glass transition temperature (T_g), it will generally be compliant enough to relax and accommodate these interphasal stresses. However, once T_g is reached, the adhesive is less compliant and stresses begin to build up. Thus, the thermal stresses in a joint will depend on the CTE differences between the substrate(s), adhesive, and any overlying layers or films and on the degree of cooling below T_g. One way to minimize these stresses is to blend low and high CTE polymers to match the CTE of the substrate—a procedure most relevant to polymeric substrates (Charles 1990). Another is to incorporate mineral fillers into the adhesive to reduce its CTE (Comyn 1990). In cases of mismatched adherends, for example, composite to aluminum, a near room-temperature-curing adhesive may be the best solution (Arah et al. 1987, 1989).

18.5.2 Adhesives/Primers

The effects of water and temperature on the adhesive itself are also of utmost importance to the durability of bonded structures. In the presence of moisture, the adhesive can be affected in a number of ways, depending on its chemistry and how rapidly the water permeates through and causes significant

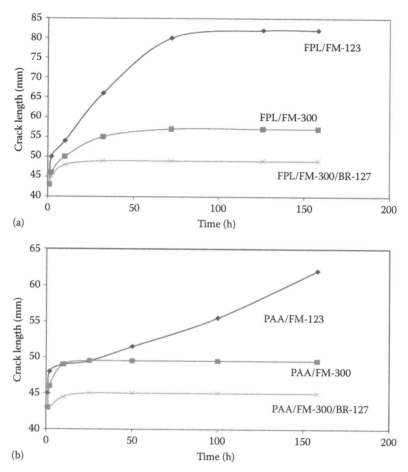

FIGURE 18.5 Wedge test results for (a) FPL and (b) PAA aluminum adherends: FM-123 (moisture-wicking) adhesive, FM-300 (moisture-resistant) adhesive, and BR-127 primer and FM-300.

property changes (Kerr et al. 1967, Bascom 1970, Minford 1983, Jurf and Vinson 1985). The potential efficacy of moisture penetration on the locus of failure of bonded joints has been discussed in the previous section. As expected, elevated temperature conditions tend to degrade joint strength at a faster rate. Figure 18.5 compares the wedge test performance of aluminum bonds formed with a water-wicking adhesive and a water-resistant adhesive.

Of primary importance in moist environments is the plasticization, or softening, of the adhesive, a process that depresses T_g and lowers the modulus and strength of the elastomer (Brewis et al. 1980, 1982, Shalash 1980). Plasticization of the adhesive may also allow disengagement from a nano- or microrough adherend surface to reduce physical bonding and thus reduce joint strength and durability (Davis et al. 1985). On the other hand, it may allow stress relaxation or crack blunting and improve durability (Kinloch 1983b).

Brewis et al. studied the effects of moisture and temperature on the properties of epoxy-aluminum joints by measuring changes in the mechanical strength properties of the soaked adhesive (Brewis et al. 1982). The T_gs of the wet adhesive and relative strengths of wet and dry joints were evaluated for up to 2500 h. They concluded that the joint weakening effect of water was due to plasticization of the adhesive that, in turn, was dependent on the rate of water diffusion within the adhesive.

The softening behavior has also been observed with epoxy adhesive–Al single lap joints exposed to 100% relative humidity at 50°C for 1000 h (Shalash 1980). As shown in Figure 18.6, wet and dry joints

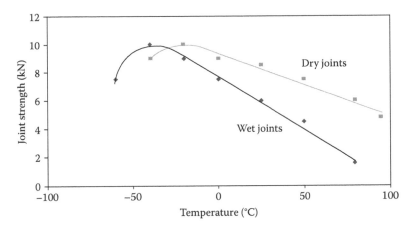

FIGURE 18.6 Strength of wet and dry lap joints with FM-1000 adhesive as a function of temperature: ■, dry joints; ◆, joints preconditioned for 1000 h at 50°C and 100% relative humidity. (Data from Brewis, D.M. et al., *Polymer*, 21, 357, 1980.)

exhibited similar strength–temperature relationships, but with the former being shifted to a lower temperature by 30°C–50°C, a quantity close to the water induced depression of the T_g. Hence, in this case, the T_g depression acts as a shift factor that defines the strength–temperature relationship between the dry and wet adhesives so that at a given temperature a wet joint exhibits lower strength than a dry one.

Water entering a joint can also cause swelling, which tends to introduce stresses to weaken the bonded system. Weitsman has shown that normal stresses resulting from swelling (3%) of an epoxide adhesive are manifested at the edges of the joint; however, after an initial rise, the stress concentration decreases with time, suggesting that they do not contribute to long-term structural weakening (Weitsman 1977). Temperature cycling, especially if the extremes exceed either the boiling or freezing points of water, can be an issue for adhesives with absorbed moisture. The volume expansion of water upon freezing or boiling can cause bondline stresses that lead to bond failure.

As discussed earlier, chemical bonds between the adhesive and adherend help to stabilize the interface and increase joint durability. Aluminum joints formed with phenolic adhesives generally exhibit better durability than those with epoxy adhesives (Kinloch 1983b, 1987, Brockmann 1990a). This is partly attributable to strongly interacting phenolic and aliphatic hydroxyl groups that form stable primary chemical bonds across the interface (Lewis et al. 1974, Knop and Pilato 1985, Tobiason 1990). Nonetheless, epoxy adhesives are more widely used due to their greater toughness and lower temperatures and pressures required during cure.

Silanes and other coupling agents can be applied to various substrates or incorporated into an adhesive/primer to serve as hybrid chemical bridges to increase the bonding between organic adhesive and inorganic adherend surfaces (Plueddemann 1982, 1985, Bascom 1990, Buckley and Schroeder 1990). Such bonding increases the initial bond strength and stabilizes the interface to also increase the durability of the resulting joint. Silane-based primers have been shown to be effective in increasing the environmental resistance of joints prepared from aluminum (Patrick et al. 1981) and titanium (Boerio and Dillingham 1984) alloys. Plueddemann has shown that the resulting interphase can be designed for maximum water resistance by employing hydrophobic resins and coupling agents and by providing a high degree of cross-linking (Plueddemann 1991).

Corrosion-inhibiting adhesive primers are commonly applied onto bonding surfaces soon after the surface treatment (Bascom 1990). Their primary function is to wet the adherend and penetrate the "nooks and crannies" to form both chemical and physical bonds. They also perform other functions essential for durable bonds: creation of a stable surface, prevention of contamination or mechanical damage of surfaces that have been chemically etched or anodized, and corrosion inhibition to the bonded and nonbonded areas of the assembly. Primer systems have historically included chromates, which provide corrosion inhibition. More recent, environmentally safer primers have replaced chromates with other corrosion inhibitors. Figure 18.5 illustrates the improvement in joint performance using a corrosion-preventative primer (Clearfield et al. 1990, 1991).

A hybrid surface treatment/primer sol–gel process has been developed that provides a graded interphase between the metal and the adhesive (Blohowiak et al. 1996, 1998, Hergenrother 2000, Park et al. 2000). The sol–gel coating is typically 0.5–2 μm thick and consists of an inorganic component, an organic component, and, in some cases, a coupling agent. The inorganic component is concentrated at the metal surface (aluminum, titanium, and steel adherends have been used) while the organic component and the coupling agent are concentrated at the adhesive or conventional primer (if used). Both abraded and otherwise cleaned metal surfaces and surfaces with conventional surface treatments, such as described earlier, have been used. In this system, strong covalent bonds that are resistant to moisture attack are formed between the metal and the inorganic component and between the polymer and the organic component. Accordingly, physical bonding (mechanical interlocking) is less important, and these surfaces tend to be smoother than the intentionally nanorough oxides normally grown on the metal surfaces. Excellent bond durability results have been reported under moisture conditions and elevated temperatures (titanium). Because the sol–gel process does not involve toxic materials, such as hexavalent chromium, or strong acids or bases, it has the potential to be an environmentally friendly surface treatment.

18.6 Stresses

The stresses that a joint experiences during environmental exposure also influence its durability, that is, it exhibits either decreased lifetime or decreased residual strength (Cotter 1977, Kinloch 1987, Pitrone and Brown 1988, Boerio et al. 1993). As with moisture, there may be a critical stress level below which failure does not occur (Ripling et al. 1971, Cherry and Thompson 1977) or is not accelerated (depending on the moisture level). The type of stress is also important. For example, cyclic stresses degrade the bond more rapidly than constant stresses (Marceau and Thrall 1985, Pitrone and Brown 1988). Furthermore, tensile or opening stresses are more damaging than shear stresses, which are more damaging than compressive stresses.

The stresses on a joint make primary and secondary chemical bonds, both within the polymer itself and across the polymer–oxide interface, more susceptible to environmental attack by lowering the activation energy for bond breaking (Comyn 1983, Kinloch 1983b, 1987). The stresses can also increase the rate of transport of moisture in the adhesive, possibly via crazing or the formation of microcracks (Comyn 1983, Kinloch 1983b, 1987, Good 1990) or increasing the free volume of the polymer to allow more moisture ingress. Joints subjected to thermal "spikes" or cycling, such as those present in high-speed military aircraft, are particularly vulnerable to this type of aging. Thus, weight gains in bonded composite systems that encountered one to four spikes (0°C–150°C–0°C) per day were proportional to the total number of spikes (Browning 1977, 1978), suggesting that water was entering the microcracks formed during thermal cycling. Additionally, studies of the chemical hydrolysis of epoxides by water (80°C) indicated that although unstressed samples were unaffected for up to 3 months, stressed systems induced hydrolysis of ester groups within days (Antoon et al. 1981, Antoon and Koenig 1981).

The stress at the crack tip of a wedge test specimen, together with the presence of moisture at the tip, serves to make this test specimen more severe than soaked lap shear specimens or similar types and a better evaluation of relative durability. In fact, Boeing correlated the results of wedge tests from actual aircraft components with their in-service durability (Marceau and Thrall 1985, Kuperman and Horton 1990). Wedge test specimens fabricated from components that had exhibited service disbonds showed significant crack growth during the first hour of exposure whereas those fabricated from good components showed no crack growth during this time period. In contrast, lap shear specimens and porta shear specimens all demonstrated high stresses regardless of the service conditions. As a result, wedge tests are commonly used for process control (crack length after approximately 1 h) and research and development (crack growth for several days or weeks).

18.7 Means to Improve Durability

Although the degradation of a bond is likely to be inevitable, there are means to slow down the process, some of which were discussed earlier. For convenience of the discussion, these methods can also be classified as environment, materials, and design related.

18.7.1 Environment

Changing the environment to which a bond is exposed is probably the most effective means of ensuring good durability; bonded structures are not likely to degrade at moderate temperatures and in low humidity. Unfortunately, this usually is not a viable option. However, it may be possible to protect the bond from its external environment, at least for a period of time.

Sealants can be used to slow down moisture ingress from joint edges and seams (Reinhart 1990). If all the bondline edges are sealed, moisture is excluded from the adhesive and the interphase. Proper application is necessary to prevent moisture accumulation and to ensure the absence of an easy path to the interface. One category of sealants, water-displacing corrosion inhibitors, can even creep under existing water films, displacing the moisture and eliminating the corrosive environment. In addition, depending on the modulus of the sealant, it can serve to reduce stresses at the edge arising from a sharp or abrupt end of the joint.

18.7.2 Material

Material selection and preparation are perhaps more feasible options than changing the environment. Although not an economical solution, substitution of titanium for aluminum would solve many moisture-related problems.

Surface preparation is another commonly used means to increase durability. We have already seen that the durability of nanorough surfaces is superior to smooth surfaces or to surfaces with only larger-scale roughness (Figure 18.1) and that hydration-resistant aluminum surfaces provide further improvements (see Section 18.5.1). Figure 18.5 also illustrated this enhancement. For a given adhesive, PAA surfaces that are more hydration resistant show less crack growth than FPL surfaces. This improvement of PAA surfaces over FPL surfaces has also been demonstrated in the field, most notably in Vietnam, where FPL-treated joints suffered a large number of disbonds whereas PAA-treated joints were significantly more reliable (Boerio et al. 1993).

Selection of water-resistant adhesives and/or corrosion-resistant primers is common and was also illustrated in Figure 18.5 (Clearfield et al. 1990, 1991). Here, selection of a water-resistant adhesive (Cytec FM-300) decreased the final crack length in the wedge test by 1–2 cm for both FPL and PAA surfaces compared to a water-wicking adhesive (FM-123). The use of a chromate-containing (corrosion-resistant) primer (Cytec BR-127) further decreased the final crack length (Section 18.5.2).

18.7.3 Joint Design

Proper design of a joint or structure is also necessary to maximize durability. Although moisture cannot be prevented from reaching a bondline, it can be slowed or reduced in quantity. One way is to prevent pooling or other accumulation of water by designing the geometry to promote runoff or including adequate drain holes. Maintenance is then required to ensure that the holes do not become plugged.

Another approach to improve durability involves overdesigning the bond so that the actual stresses experienced are a small fraction of the stresses that the joint is capable of withstanding. Stresses are thereby reduced to below any critical level, and the load can be carried even if moisture creates a disbond over a portion of the joint. Of course, this approach may not be feasible from a cost or weight standpoint. Alternatively, the bond can be designed so that moisture has a long diffusion path to reach a critical area—the same general principle by which sealants work. Also because most adhesive bonds are weakest in tension and tension increases opportunities for moisture diffusion, joints should be designed so that the stresses are compressive or shear in nature.

18.8 Summary

Long-term durability is one of the most critical properties of many adhesive bonds. Although it can be difficult to achieve in aggressive environments, modern materials and processes have proven successful in increasing durability. Moisture is the cause of most environmentally induced bond failures. It can weaken or disrupt secondary (dispersion-force) bonds across the adhesive–adherend interface, especially those involving high-energy surfaces such as metals; as a result, the joint may need to rely solely on primary (covalent or ionic) or physical (mechanical interlocking) bonds. More severe degradation can subsequently occur with hydration or corrosion of the adherend surface. At this point, the joint will fail regardless of the type of bonding at the interface.

Most means of improving durability involve slowing down the degradation mechanisms or providing additional bonding schemes, for example, primary and/or physical bonds that are less susceptible to degradation. Surface preparations that provide physical bonds and a hydration-resistant surface are typical examples. The use of coupling agents, phenolic-based adhesives (with aluminum adherends), and sol–gel treatments are other examples where stable chemical bonds are formed in the interphase and slow down bond degradation.

Nomenclature

Adherend: One of two bodies held together with an adhesive, also known as a substrate.

Adhesive: A material capable of holding two bodies or adherends together by surface attachment.

Interface: A two-dimensional boundary between two materials or phases. The materials or phases may be solid, liquid, or gas.

Joint: A three-dimensional region or boundary between two materials where the local properties are different from those of bulk properties of the two materials.

Primer: A coating applied to a substrate, prior to an adhesive, to improve the performance of the adhesive bond or to protect the substrate from contamination or damage.

Specific ASTM Standardized Tests

D0896 Practice for resistance of adhesive bonds to chemical reagents.
D0897 Test method for tensile properties of adhesive bonds.
D0903 Test method for peel or stripping strength of adhesive bonds.
D0904 Practice for exposure of adhesive specimens to artificial light.
D0905 Test method for strength properties of adhesive bonds in shear by compression loading.
D0906 Test method for strength properties of adhesives in plywood type construction in shear by tension loading.
D0950 Test method for impact strength of adhesive bonds.
D1002 Test method for apparent shear strength of single-lap-joint adhesively bonded metal specimens by tension loading (metal-to-metal).
D1062 Test method for cleavage strength of metal-to-metal adhesive bonds.
D1101 Test methods for integrity of adhesive joints in structural laminated wood products for exterior use.
D1144 Practice for determining strength development of adhesive bonds.

D1151 Practice for effect of moisture and temperature on adhesive bonds.

D1183 Practices for resistance of adhesives to cyclic laboratory aging conditions.

D1184 Test method for flexural strength of adhesive bonded laminated assemblies.

D1780 Practice for conducting creep tests of metal-to-metal adhesives.

D1781 Test method for climbing drum peel for adhesives.

D1828 Practice for atmospheric exposure of adhesive-bonded joints and structures.

D1876 Test method for peel resistance of adhesives (T-peel test).

D1879 Practice for exposure of adhesive specimens to ionizing radiation.

D1995 Test methods for multi-modal strength testing of autohesives (contact adhesives).

D2093 Practice for preparation of surfaces of plastics prior to adhesive bonding.

D2094 Practice for preparation of bar and rod specimens for adhesion tests.

D2095 Test method for tensile strength of adhesives by means of bar and rod specimens.

D2293 Test method for creep properties of adhesives in shear by compression loading (metal-to-metal).

D2294 Test method for creep properties of adhesives in shear by tension loading (metal-to-metal).

D2295 Test method for strength properties of adhesives in shear by tension loading at elevated temperatures (metal-to-metal).

D2339 Test method for strength properties of adhesives in two-ply wood construction in shear by tension loading.

D2557 Test method for tensile-shear strength of adhesives in the subzero temperature range from \-267.8 to \-55|SDC (\-450 to \-67|SDF).

D2559 Specification for adhesives for structural laminated wood products for use under exterior (wet use) exposure conditions.

D2651 Guide for preparation of metal surfaces for adhesive bonding.

D2674 Methods of analysis of sulfochromate etch solution used in surface preparation of aluminum.

D2918 Test method for durability assessment of adhesive joints stressed in peel.

D2919 Test method for determining durability of adhesive joints stressed in shear by tension loading.

D3163 Test method for determining strength of adhesively bonded rigid plastic lap-shear joints in shear by tension loading.

D3164 Test method for strength properties of adhesively bonded plastic lap-shear sandwich joints in shear by tension loading.

D3164 Test method for strength properties of adhesively bonded plastic lap-shear sandwich joints in shear by tension loading.

D3165 Test method for strength properties of adhesives in shear by tension loading of single-lap-joint laminated assemblies.

D3166 Test method for fatigue properties of adhesives in shear by tension loading (metal/metal).

D3167 Test method for floating roller peel resistance of adhesives.

D3310 Test method for determining corrosivity of adhesive materials.

D3433 Test method for fracture strength in cleavage of adhesives in bonded metal joints.

D3434 Test method for multiple-cycle accelerated aging test (automatic boil test) for exterior wet use wood adhesives.

D3482 A Practice for determining electrolytic corrosion of copper by adhesives.

D3528 Test method for strength properties of double lap shear adhesive joints by tension loading.

D3535 Test method for resistance to creep under static loading for structural wood laminating adhesives used under exterior exposure conditions.

D3632 Test method for accelerated aging of adhesive joints by the oxygen-pressure method.

D3658 Test method for determining the torque strength of ultraviolet (UV) light-cured glass/metal adhesive joints.

D3762 Test method for adhesive-bonded surface durability of aluminum (wedge test).

D3807 Test method for strength properties of adhesives in cleavage peel by tension loading (engineering plastics-to-engineering plastics).

D3808 Test method for qualitative determination of adhesion of adhesives to substrates by spot adhesion.

D3929 Test method for evaluating stress cracking of plastics by adhesives using the bent-beam method.

D3931 Test method for determining strength of gap-filling adhesive bonds in shear by compression loading.

D3933 Guide for preparation of aluminum surfaces for structural adhesives bonding (phosphoric acid anodizing).

D3983 Test method for measuring strength and shear modulus of nonrigid adhesives by the thick-adherend tensile-lap specimen.

D4027 Test method for measuring shear properties of structural adhesives by the modified-rail test.

D4499 Test method for heat stability of hot-melt adhesives.

D4501 Test method for shear strength of adhesive bonds between rigid substrates by the block-shear method.

D4502 Test method for heat and moisture resistance of wood-adhesive joints.

D4562 Test method for shear strength of adhesives using pin-and-collar specimen.

D4680 Test method for creep and time to failure of adhesives in static shear by compression loading (wood-to-wood).

D4688 Test method for evaluating structural adhesives for finger jointing lumber.

D4896 Guide for use of adhesive-bonded single lap-joint specimen test results.

D5041 Test method for fracture strength in cleavage of adhesives in bonded joints.

D5573 Practice for classifying failure modes in fiber-reinforced plastic (FRP) joints.

D5574 Test methods for establishing allowable mechanical properties of wood-bonding adhesives for design of structural joints.

D5649 Test method for torque strength of adhesives used on threaded fasteners.

D5656 Test method for thick-adherend metal lap-shear joints for determination of the stress–strain behavior of adhesives in shear by tension loading.

D5824 Test method for determining resistance to delamination of adhesive bonds in overlay-wood core laminates exposed to heat and water.

D5868 Test method for lap shear adhesion for fiber-reinforced plastic (FRP) bonding.

D6004 Test method for determining adhesive shear strength of carpet adhesives.

D6463 Test method for time to failure of pressure sensitive articles under sustained shear loading.

D6862 Test method for 90 degree peel resistance of adhesives.

D7247 Test method for evaluating the shear strength of adhesive bonds in laminated wood products at elevated temperatures.

References

Adams, R.D. and P. Fay. 2011. A history of adhesive bonding from ancient times to today, in *Proceedings of the Annual Meeting Adhesion Society*, The Adhesion Society, Blacksburg, VA.

Antoon, M.K. and J.L. Koenig. 1981. *Journal of Polymer Science, Physics Edition*, 19, 197.

Antoon, M.K., J.L. Koenig, and T. Serafini. 1981. *Journal of Polymer Science, Physics Edition*, 19, 1567.

Arah, C.O., D.K. McNamara, J.S. Ahearn, A. Berrier, and G.D. Davis. 1987. In *Proceedings of the Fifth International Symposium on Structural Adhesive Bonding*, Dover, NJ, p. 440.

Arah, C.O., D.K. McNamara, H.M. Hand, and M.F. Mecklenburg. 1989. *Journal of Adhesion Science and Technology*, 3, 261.

Bascom, W.D. 1970. *Journal of Adhesion*, 2, 161.

Bascom, W.D. 1990. In *Engineered Materials Handbook, Vol. 3: Adhesives and Sealants*, H.F. Brinson, chm., ASM International, Metals Park, OH, p. 254.

Bell Helicopter, Textron, Inc. 1980. Surface preparation of materials for adhesive bonding. Process Specification 4352, Rev J, June.

Bishof, C., A. Bauer, R. Kapeele, and W. Possart. 1985. *International Journal of Adhesion and Adhesives,* 5, 97.

Bishopp, J.A., E.K. Sim, G.E. Thompson, and G.C. Wood. 1988. *Journal of Adhesion,* 26, 237.

Blohowiak, K.Y., K.A. Krienke, J.H. Osborne, and R.B. Greegor. 1998. In *Proceedings of the Workshop on Advanced Metal Finishing Techniques for Aerospace Applications,* Keystone, CO.

Blohowiak, K.Y., J.H. Osborne, K.A. Krienke, and D.F. Sekits. 1996. In *Proceedings of the 28th International SAMPE Tech Conference,* Covina, CA, p. 440.

Boëda, E., S. Bonilauri, J. Connan, D. Jarvie, N. Mercier, M. Tobey, H. Valladas, H. Sakhel, and S. Muhesen. 2008. *Antiquity,* 82(318), 853.

Boerio, F.J., G.D. Davis, J.E. deVries, C.E. Miller, K.L. Mittal, R.L. Opila, and H.K. Yasuda. 1993. *Critical Reviews in Surface Chemistry,* 3, 81.

Boerio, F.J. and R.G. Dillingham. 1984. In K.L. Mittal, Ed., *Adhesive Joints,* Plenum Press, New York, p. 541.

Boerio, F.J., S.J. Hudak, M.A. Miller, and S.G. Hong. 1987. *Journal of Adhesion,* 23, 99.

Brewis, D.M., J. Comyn, and R.J.A. Shalash. 1982. *International Journal of Adhesion and Adhesives,* 2, 215.

Brewis, D.M., J. Comyn, R.J.A. Shalash, and J.L. Tegg. 1980. *Polymer,* 21, 357.

Brinson, H.F., chm. 1990. *Engineered Materials Handbook, Vol. 3: Adhesives and Sealants,* ASM International, Metals Park, OH.

Brockmann, W. 1983. In A.J. Kinloch, Ed., *Durability of Structural Adhesives,* Applied Science, London, U.K., p. 281.

Brockmann, W. 1990a. In H.F. Brinson, chm., *Engineered Materials Handbook, Vol. 3: Adhesives and Sealants,* ASM International, Metals Park, OH, p. 663.

Brockmann, W. 1990b. In L.H. Sharpe and S.E. Wentworth, Eds., *The Science and Technology of Adhesive Bonding,* Gordon and Breach, New York, p. 53.

Brockmann, W., O.D. Hennemann, H. Kollek, and C. Matz. 1986. *International Journal of Adhesion and Adhesives,* 6, 115.

Brown, S.R. 1982. In *Proceedings of the 27th National SAMPE Symposium,* SAMPE, Azusa, CA, p. 363.

Browning, C.E. 1977. In *Proceedings of the 22nd National SAMPE Symposium,* SAMPE, Azusa, CA, p. 365.

Browning, C.E. 1978. *Polymer Engineering and Science,* 18, 16.

Buckley, W.O. and K.J. Schroeder. 1990. In H.F. Brinson, chm., *Engineered Materials Handbook, Vol. 3: Adhesives and Sealants,* ASM International, Metals Park, OH, p. 175.

Chandler, H.E. 1982. *Metal Progress,* 121, 38.

Charles, H.K. 1990. In H.F. Brinson, chm., *Engineered Materials Handbook, Vol. 3: Adhesives and Sealants,* ASM International, Metals Park, OH, p. 579.

Cherry, B.W. and K.W. Thompson. 1977. In K.W. Allen, Ed., *Adhesion-1,* Applied Science, London, U.K., p. 251.

Clearfield, H.M., D.K. McNamara, and G.D. Davis. 1990. In H.F. Brinson, chm., *Engineered Materials Handbook, Vol. 3: Adhesives and Sealants,* ASM International, Metals Park, OH, p. 259.

Clearfield, H.M., D.K. McNamara, and G.D. Davis. 1991. In L.H. Lee, Ed., *Adhesive Bonding,* Plenum, New York, p. 203.

Cobb, T.Q., W.S. Johnson, S.E. Lowther, and T.L. St. Clair. 1999. *Journal of Adhesion,* 71, 115.

Comyn, J., 1983. In A.J. Kinloch, Ed., *Durability of Structural Adhesives,* Applied Science, London, U.K., p. 85.

Comyn, J. 1990. In H.F. Brinson, chm., *Engineered Materials Handbook, Vol. 3: Adhesives and Sealants,* ASM International, Metals Park, OH, p. 616.

Cotter, J.L. 1977. In W.C. Wake, *Developments in Adhesives, Vol. 1,* Applied Science, London, U.K., p. 1.

Critchlow, G.W., K.H. Bedwell, and M.E. Chamberlain. 1998. *Transactions of the Institute of Metal Finishing,* 76, 209.

Critchlow, G.W. and D.M. Brewis. 1995. *International Journal of Adhesion and Adhesives*, 15, 161.

Critchlow, G.W. and D.M. Brewis. 1996. *International Journal of Adhesion and Adhesives*, 16, 255.

David, H.L. 1973. *Metal Deformation*, 19, 27.

Davis, G.D. 1986. *Surface Interface Analysis*, 9, 421.

Davis, G.D. 1991. *Surface Interface Analysis*, 17, 439.

Davis, G.D. 2003. In A. Pizzi and K.L. Mittal, Eds., *Handbook of Adhesive Technology*, 2nd edn., Marcel Dekker, New York, p. 273.

Davis, G.D., J.S. Ahearn, L.J. Matienzo, and J.D. Venables. 1985. *Journal of Materials Science*, 20, 975.

Davis, G.D., L.A. Krebs, L.T. Drzal, M.J. Rich, and P. Askeland. 2000. *Journal of Adhesion*, 72, 335.

Davis, G.D., R.A. Pethrick, and J. Doyle. 2009. *Journal of Adhesion Science and Technology*, 23, 517.

Davis, G.D., T.S. Sun, J.S. Ahearn, and J.D. Venables. 1982. *Journal of Materials Science*, 17, 1807.

Davis, G.D. and J.D. Venables. 2002. In M. Chaudhury and A.V. Pocius, Eds., *Adhesion Science and Engineering 2: Surfaces, Chemistry, and Applications*, Elsevier, Amsterdam, the Netherlands, p. 947.

Davis, G.D., P.L. Whisnant, and J.D. Venables. 1995. *Journal of Adhesion Science and Technology*, 9, 433.

Devine, A.T. 1977. Adhesive bonded steel: Bond durability as related to selected surface treatments, Technical Report ARLCD-TR-77027, ARRADCOM, Dover, NJ.

Dixon, D.G. 2005. In D.E. Packham, Ed., *Handbook of Adhesion*, 2nd edn., Wiley, West Sussex, U.K., p. 40.

Filbey, J.A., J.P. Wightman, and D.J. Polgar. 1987. *Journal of Adhesion*, 20, 283.

Fokker VFW B.V. 1978. Process Specification TH6.7851, the Netherlands, August.

Gledhill, R.A. and A.J. Kinloch. 1976. *Journal of Adhesion*, 8, 11.

Good, G. 1990. In H.F. Brinson, chm., *Engineered Materials Handbook, Vol. 3: Adhesives and Sealants*, ASM International, Metals Park, OH, p. 651.

Hammond, J.S., J.W. Holubka, J.W. DeVries, and R.A. Dickie. 1981. *Corrosion Science*, 21, 239.

Hergenrother, P.M. 2000. *SAMPE Journal*, 36, 30.

Janssen, E. 1980. *Materials Science and Engineering*, 42, 309.

Jurf, R.A. and J.R. Vinson. 1985. *Journal of Materials Science*, 20, 2979.

Kennedy, A.C., R. Kohler, and P. Poole. 1983. *International Journal of Adhesion and Adhesives*, 3, 133.

Kerr, C., N.C. MacDonald, and S.J. Orman. 1967. *Journal of Applied Chemistry*, 17, 62.

Kinloch, A.J. 1979. *Journal of Adhesion*, 10, 193.

Kinloch, A.J. 1982. *Journal of Materials Science*, 17, 617.

Kinloch, A.J., Ed. 1983a. *Durability of Structural Adhesives*, Applied Science, London, U.K.

Kinloch, A.J. 1983b. In A.J. Kinloch, Ed., *Durability of Structural Adhesives*, Applied Science, London, U.K., p. 1.

Kinloch, A.J. 1987. *Adhesion and Adhesives: Science and Technology*, Chapman & Hall, London, U.K.

Knop, A. and L. Pilato. 1985. *Phenolic Resins*, Springer-Verlag, New York, Chapter 11.

Koller, J., U. Baumer, and D. Mania. 2001. *European Journal of Archaeology*, 4(3), 385.

Kuperman, M.H. and R.E. Horton. 1990. In H.F. Brinson, chm., *Engineered Materials Handbook, Vol. 3: Adhesives and Sealants*, ASM International, Metals Park, OH, p. 801.

Lewis, B.F., W.M. Bowser, J.L. Horn, Jr., T. Luu, and W.H. Weinberg. 1974. *Journal of Vacuum Science and Technology*, 11, 262.

Mansfeld, F., M.W. Kendig, and S. Tsai. 1982. *Corrosion*, 38, 478.

Marceau, J.A. 1985. In E.W. Thrall and R.W. Shannon, Eds., *Adhesive Bonding of Aluminum Alloys*, Marcel Dekker, New York, p. 51.

Marceau, J.A. and E.W. Thrall. 1985. In E.W. Thrall and R.W. Shannon, Eds., *Adhesive Bonding of Aluminum Alloys*, Marcel Dekker, New York, p. 177.

McNamara, D.K. and J.S. Ahearn. 1987. *International Metallurgical Reviews*, 32, 292.

MIL-A-8625, Anodic coatings for aluminum and aluminum alloys. USA Military Specification.

Miller, J.D. and H. Ishida. 1991. L.H. Lee, Ed., *Fundamentals of Adhesion*, Plenum Press, New York, p. 291.

Minford, J.D. 1983. In A.J. Kinloch, Ed., *Durability of Structural Adhesives*, Applied Science Publishers, London, U.K., p. 135.

Minford, J.D. 1991. In L.H. Lee, Ed., *Adhesive Bonding*, Plenum, New York, p. 239.

Minford, J.D. 1993. *Handbook of Aluminum Bonding Technology and Data*, Marcel Dekker, New York.

Natan, M. and J.D. Venables. 1983. *Journal of Adhesion*, 15, 125.

Owen, M.J. 2002. In M. Chaudhury and A.V. Pocius, Eds., *Adhesion Science and Engineering 2: Surfaces, Chemistry and Applications*, Elsevier, Amsterdam, the Netherlands, p. 403.

Park, S.Y., W.J. Choi, H.C. Choi, H. Kwon, and S.H. Kim. 2010. *Journal of Adhesion*, 88, 192.

Park, C., S.E. Lowther, J.G. Smith, Jr., J.W. Connell, P.M. Hergenrother, and T.L. St. Clair. 2000. *International Journal of Adhesion and Adhesives*, 20, 457.

Patrick, R.L., J.A. Brown, N.M. Cameron, and W.G. Gehman. 1981. *Applied Polymer Symposium*, 16, 87.

Pitrone, L.R. and S.R. Brown. 1988. In S. Johnson, Ed., *Adhesively Bonded Joints: Testing, Analysis, and Design*, ASTM STP 981, American Society for Testing and Materials, Pennsylvania, PA, p. 289.

Plueddemann, E.P. 1985. In D.M. Brewis and D. Briggs, Eds., *Industrial Adhesion Problems*. Orbital Press, Oxford, U.K., p. 148.

Plueddemann, E.P. 1982. *Silane Coupling Agents*, Plenum Press, New York.

Plueddemann, E.P. 1991. In L.-H. Lee, Eds., *Fundamentals of Adhesion*, Plenum Press, New York, p. 279.

Pocius, A.V. 1997. *Adhesion and Adhesive Technology*, Hanser/Gardner, Cincinnati, OH.

Pocius, A.V., C.J. Almer, R.D. Wald, T.H. Wilson, and B.E. Davidian. 1984. *SAMPE Journal*, 20, 11.

Progar, D.J., and T.L. St. Clair. 1986. *International Journal of Adhesion and Adhesives* 6, 25.

Reil, F.J. 1971. *SAMPE Journal*, 7, 16.

Reinhart, T.J. 1990. In H.F. Brinson, chm., *Engineered Materials Handbook, Vol. 3: Adhesives and Sealants*, ASM International, Metals Park, OH, p. 637.

Ripling, E.J., S. Mostovoy, and C.F. Bersch. 1971. *Journal of Adhesion*, 3, 145.

Russell, W.J., R. Rosty, S. Whalen, J. Zideck, and M.J. Bodnar. 1981. Preliminary study of adhesive bond durability on 4340 steel substrates, Technical Report ARSCD-TR-81020. ARRADCOM, Dover, NJ.

Scully, J.R. 1989. *Journal of the Electrochemical Society*, 136, 979.

Shaffer, D.K., H.M. Clearfield, C.P. Blankenship, Jr., and J.S. Ahearn. 1987. In *Proceedings of the 19th SAMPE Tech Conference*, SAMPE, Azusa, CA, p. 291.

Shalash, R.J.A. 1980. PhD thesis, Leicester Polytechnic, Leicester, U.K.

Sharpe, L.H. 1990. In H.F. Brinson, chm., *Engineered Materials Handbook, Vol. 3: Adhesives and Sealants*, ASM International, Metals Park, OH.

Smith, T. 1984. *Journal of Adhesion*, 17, 1.

Stevenson, A. 1990. In H.F. Brinson, chm., *Engineered Materials Handbook, Vol. 3: Adhesives and Sealants*, ASM International, Metals Park, OH, p. 628.

Sung, N.-H. 1990. In H.F. Brinson, chm., *Engineered Materials Handbook, Vol. 3: Adhesives and Sealants*, ASM International, Metals Park, OH, p. 622.

Thrall, E.W. and R.W. Shannon, Eds. 1985. *Adhesive Bonding of Aluminum Alloys*, Marcel Dekker, New York.

Tobiason, F.L. 1990. In I. Skeist, Ed., *Handbook of Adhesives*, 3rd edn., Van Nostrand Reinhold, New York, Chapter 17.

Trawinski, D. 1984. *SAMPE Quarterly*, 16, 1.

Trawinski, D. 1985. In *Proceedings of the 27th National SAMPE Symposium*, SAMPE, Azusa, CA, p. 1065.

Trawinski, D.L., D.K. McNamara, and J.D. Venables. 1984. *SAMPE Quarterly*, 15, 6.

Venables, J.D. 1984. *Journal of Materials Science*, 19, 2431.

Watts, J.F. 1988. *Surface Interface Analysis*, 12, 497.

Watts, J.F. and J. E. Castle. 1983. *Journal of Materials Science*, 18, 2987.

Wegman, R.F. 1989. *Surface Preparation Techniques for Adhesive Bonding*, Noyes, Park Ridge, NJ.

Weitsman, Y. 1977. *Journal of Composite Materials*, 11, 378.

Further Reading: Select List of Important Relevant Publications

Brinson, H.F., chm. 1990. *Engineered Materials Handbook, Vol. 3: Adhesives and Sealants*, ASM International, Metals Park, OH.

Chaudhury, M. and A.V. Pocius. 2002. *Surfaces, Chemistry and Applications*, Elsevier, Amsterdam, the Netherlands.

Dillard, D.A. and A.V. Pocius. 2002. *The Mechanics of Adhesion*, Elsevier, Amsterdam, the Netherlands.

Kinloch, A.J., Ed. 1983. *Durability of Structural Adhesives*, Applied Science, London, U.K.

Kinloch, A.J. 1987. *Adhesion and Adhesives: Science and Technology*, Chapman & Hall, London, U.K.

Lee, L.H., Ed. 1991a. *Adhesive Bonding*, Plenum, New York.

Lee, L.H., Ed. 1991b. *Fundamentals of Adhesion*, Plenum, New York.

Minford, J.D. 1993. *Handbook of Aluminum Bonding Technology and Data*, Marcel Dekker, New York.

Packham, D.E., Ed. 2005. *Handbook of Adhesion*, 2nd edn., Wiley, West Sussex, U.K.

Pizzi, A. and K.L. Mittal. 2003. *Handbook of Adhesive Technology*, 2nd edn., Marcel Dekker, New York.

Pocius, A.V. 1997. *Adhesion and Adhesive Technology*, Hanser/Gardner, Cincinnati, OH.

Sharpe, L.H. and S.E. Wentworth, Eds. 1990. *The Science and Technology of Adhesive Bonding*, Gordon and Breach, New York.

Skeist, I., Ed. 1990. *Handbook of Adhesives*, 3rd edn., Van Nostrand Reinhold, New York.

Thrall, E.W. and R.W. Shannon, Eds. 1985. *Adhesive Bonding of Aluminum Alloys*, Marcel Dekker, New York.

Wegman, R.F. 1989. *Surface Preparation Techniques for Adhesive Bonding*, Noyes, Park Ridge, NJ.

III

Ceramics and Glassy Materials

19

Ceramics

Dennis W. Readey
*Colorado School of Mines
and
University of Illinois at
Urbana-Champaign*

19.1 Introduction

This chapter focuses on the degradation of engineering ceramics under ambient conditions. A ceramic here is defined as "something useful made from one or more inorganic compounds," which is somewhat more precise than the more-often-used definition that a ceramic is an "inorganic nonmetallic solid" (O'Bannon 1984, Perkins 1984). Under the former definition, minerals and their weathering are excluded from consideration and as are silicate glasses and cements because they are being covered under separate chapters in this volume. "Ambient conditions" are temperatures at or near room temperatures to eliminate the behavior of ceramics at high temperatures where engineering ceramics are frequently used and limit the scope of this discussion to manageable proportions. This eliminates discussion of the degradation of ceramics by high-temperature reactive gases, molten salts, molten metals, slags, and other high-temperature corrosive fluids that covers a wide body of literature, some of which has been summarized in several publications (Blachere and Pettit 1989, Fordham 1990, Tressler and McNallan 1990, Clark and Zoitos 1992, Gogotsi and Lavrenko 1992, Nickel 1994, McCauley 2004, Salem and Fuller 2008). In addition, there are now over 200 volumes of *Ceramic Transactions*, which are published transactions of various sessions at meetings of mainly the American Ceramic Society over the last 25 years, and about 10% of which deal with corrosion and environmental aspects of ceramics: the most recent being *Ceramic Transactions, Volume 22* (Fox et al. 2010). Here, the discussions are limited to the degradation of ceramics in liquid and gaseous environments since this has been the focus of "ambient conditions" research on the degradation of ceramics.

Another boundary condition to be satisfied is the definition of an "engineering ceramic." In general, this would eliminate ceramic artware and other ceramic end items such as tableware and sanitary ware. However, engineering ceramics would include such things as building bricks—and other clay-based materials of construction—abrasives, refractories, glass, and cement, the latter two are already excluded. However, modern engineering ceramics are used for a number of applications where their unique properties are critical for the development of a device, component, or system. These applications are not exclusively structural where only the mechanical properties are important, which is usually the case for metals. Other properties of interest include thermal insulation, thermal conduction, electric insulation, electronic conduction, ionic conduction, optical transmission, and magnetic behavior. Table 19.1 lists a number of ceramic materials, their applications, and the properties of interest for these applications. These are the materials whose stability or degradation is of greatest interest today.

The discussion focuses on solid–fluid interactions leading to degradation or corrosion of ceramics, and "fluids" include both gases and—since the temperatures are near room temperature—aqueous solutions. Some of this is presented in the previous list of general references (Blachere and Pettit 1989, Fordham 1990, Tressler and McNallan 1990, Clark and Zoitos 1992, Gogotsi and Lavrenko 1992, Nickel 1994, McCauley 2004, Salem and Fuller 2008). The goal is not to present a comprehensive list of experimental results for a number of engineering ceramics with or without a critical analysis of what the data imply about the mechanism of degradation. Neither will a grid of engineering ceramics nor their property degradation in a variety of ambient corrosive environments be presented. There are several reasons for this. First, there are limited data available on the behavior of specific systems but a lot of data on many systems. Second, much of the data that do exist for many materials are not necessarily consistent: particularly when trying to imply a mechanism for the degradation, which is the important link to degradation, predictability, and prevention. Third, the degradation of most engineering ceramics—particularly in aqueous solutions—can depend critically on the presence of impurities and second phases that are not always identified. Fourth, the data on the degradation of ceramics are dispersed over a wide variety of quite disparate fields: physical metallurgy, extractive metallurgy, ceramics, geology, nuclear materials, biomaterials, aquatic and environmental science, and medicine, with each field injecting its own perspective, focus, and vocabulary. Fifth, a specific ceramic material, such as silicon carbide (SiC), can be processed in a variety of ways that lead to different microstructures with different levels of residual porosity and second phases whose compositions will vary with the manufacturer—all of which can lead to very different corrosion behavior for a "specific material": there is no such thing as a standard SiC or any other ceramic. Sixth, and finally, it is difficult to summarize and compare results of different investigations since they are frequently carried out in fluids with wide variations in conditions—temperatures from room temperature to hydrothermal solutions with temperatures of several hundred degrees—and no consistencies of solution chemistries. As a result, performing an exhaustive collection and evaluation of the literature on a given material quickly becomes a large undertaking.

TABLE 19.1 Example of Engineering Ceramics

Ceramic Material	Application	Properties
Al_2O_3, MgO, $MgAl_2O_4$, ZrO_2, $MgCr_2O_4$	Thermal insulation Refractories	High melting point (2050°C) "Low" thermal conductivity, k Slag, molten metal resistance
Al_2O_3	Na-vapor lamps	High melting point Corrosion resistance Optical translucency
SiO_2	Fiber optic cables	Optical transparency Chemical, mechanical, and thermal stability
UO_2, PuO_2	Nuclear fuels	High melting point Chemical and mechanical stability Ability to dissolve fission products
Al_2O_3, SiC	Abrasives	Hardness Chemical stability
SiC_w–Al_2O_3, TiC, WC, Si_3N_4	Cutting tools	Hardness Wear resistance (toughness)
Al_2O_3, BeO, AlN	Semiconductor device packages	Electric insulators High thermal conductivity (30–200 W/m-K) Thermal, mechanical, chemical stability
ZrO_2 (+ CaO, MgO, or Y_2O_3 in solid solution)	Metal-forming dies	Hardness Strength (toughness)
ZrO_2 (+ CaO, MgO, or Y_2O_3 in solid solution)	Oxygen sensor	Oxygen ion conductor High melting point Chemical and mechanical stability
ZrO_2 (+ CaO, MgO, or Y_2O_3 in solid solution)	Fuel cell electrolyte	Chemical and mechanical stability
ZrO_2 (+ CaO, MgO, or Y_2O_3 in solid solution)	Thermal barrier coating	Low thermal conductivity (3 W/m-K) Thermal, chemical, mechanical stability
ZrO_2 (+ CaO, MgO, or Y_2O_3 in solid solution), SiC, $MoSi_2$, $LaCrO_3$	Heating elements, high-temperature conductors (fuel cells)	Oxidation resistant (SiO_2) High electric conductivity, σ High melting points
ZnO, SiC	Surge suppressors (varistors)	Both high and low electric conductivity Switchable
ZnO films	TV filters	Piezoelectric
$MgAl_2O_4$	Infrared (IR) windows	IR transparent (up to $\lambda \approx 4.5$ μm)
ZnS, ZnSe, NaCl, KCl	Infrared (IR) windows	IR transparent (up to $\lambda \approx 11$ μm)
SiO_2 glass (+oxides)+ Nd_2O_3	Glass lasers	Transparent Thermal and mechanical stability
$Y_3Al_5O_{12}$ (YAG)	Laser crystals	"
$LiNbO_3$	Optical communications and computers	Optical transparency Index of refraction varies with electric field strength.
$CaWO_3$, YVO_3:Eu	Phosphors (TV, x-ray, lamps, etc.)	Fluorescence Thermal, chemical, mechanical, and radiation stability
$MgCr_2O_4$ (plus solid solution additions)	Temperature sensors (thermistors)	Electronic conductivity Chemical and thermal stability

(*continued*)

TABLE 19.1 (continued)   Example of Engineering Ceramics

Ceramic Material	Application	Properties
$Y_3Fe_5O_{12}$ (yttrium iron garnet, YIG), $MgFe_2O_4$, (magnesium ferrite) (plus solid solutions)	"Soft" magnetic materials for high frequencies: transformers, radar	Magnetic Direction of magnetization easy to change Good electric insulator
$BaFe_{12}O_{19}$ (plus solid solutions)	"Hard" magnetic materials, permanent magnets	Strongly magnetic Hard to switch magnetization Inexpensive
$Na_2O \cdot 11Al_2O_3$ (beta-alumina)	Solid Na–S battery electrolyte	Good ionic (Na^+) conductor Mechanical, chemical, and thermal stability
V_2O_5–$Li_3V_2O_5$, Li_xFePO_4	Battery electrodes	Good ionic and electronic conductivity Ability to charge and discharge
$BaTiO_3$ (plus solid solutions)	Capacitors	High dielectric constant, $k_r \approx 6000$
$BaTiO_3$, $PbTi_xZr_{1-x}O_3$ (plus solid solutions)	Sonar, ultrasonic generators	Piezoelectric (voltage produces strain and vice versa)
Al_2O_3, ZrO_2, $Ca_5(OH,F)(PO_4)_3$ (apatite)	Biomaterials: bone replacement	Biocompatible Corrosion resistant Controlled porosity
$LaMnO_3$, $LaCrO_3$	Fuel cell components Gas separation membrane	High electronic conductivity Oxidation resistant High-T (1000°C) operation High ionic conductivity
$BaCeO_3$	Hydrogen ion conducting Fuel cell electrolyte	Good ionic (H^+) conductivity Poor electronic conductor Reasonable conductivity at low temperatures (700°C)
TiO_2 and other titanates	Paint pigments Solar cells Nuclear waste disposal	Large index of refraction Nontoxic Ideal band gap
$ZnCrO_4$	Corrosion-resisting pigment	Strongly oxidizing Produces protective oxide
SnO_2, Cd_2SnO_4, $CuInO_2$	Transparent conductors (solar cells, deicers, etc.)	Good electronic conductivity Optically transparent in visible

However, a recent excellent review on the degradation of ceramics has attempted just that (Morrell 2010). Nevertheless, the author concludes (Morrell 2010, 2302)

> From the foregoing, it will be appreciated that the performance of any one material is highly dependent not only on its composition and microstructure, but also on the conditions to which it is subjected and how the effects of corrosion are measured. It is therefore particularly difficult to give a clear appreciation of the comparative performance of the huge range of materials now available. There are relatively few sources of directly comparative numerical data over a wide range of products.

If interested in the degradation of a specific ceramic in a specified environment, it is suggested that this reference (Morrell 2010) be consulted. In contrast, what is presented here is limited to, and focused on, some *fundamental principles* regarding degradation then show how they apply to certain materials and applications of technological importance where convincing—more or less—experimental data do exist. When degradation data are not available, this approach provides some guidance and predictability of how a specific ceramic material might be expected to behave in a defined corrosive environment based on available thermodynamic data and the application of kinetic principles successfully applied to other ceramics. Also, some common assumptions and/or misconceptions about how ceramics behave under certain conditions will be evaluated in light of potential mechanisms and available experimental data. Further justification for this approach, rationale, and the difficulty of developing and presenting a ceramic degradation "database" has been developed before (Munro 1997, 349):

> In the corrosion of ceramics, numerous distinct contributing processes occur in different regions of the specimen simultaneously. These different regions usually have different compositions, native microstructure, and environmentally intrusive agents. Both predictive modeling efforts and post-test diagnostic analyses, therefore, require chemical and physical property data that are related to the specific material substructures. There are significant problems in the practice of representing such information in general-purpose materials properties databases. The essential issues pertain to the form of the representation, the accessibility of the data, and the uniqueness of the value set.

In short, ceramics typically consist of more than one phase, each of which will probably respond to the corroding or degrading environment differently making it difficult to model how the overall material is degrading or what are the defining measurements that indicate the level of degradation. For example, an oxide phase used to enhance densification, present at grain boundaries in a nonoxide ceramic such as SiC, may leach much more rapidly in an acidic aqueous environment than the carbide itself. However, even though a small amount of material may have been corroded by the acid, the removal of the grain boundary oxide phase can leave behind crack-like regions greatly decreasing the fracture strength of the SiC. As a result, a small weight loss on leaching has little relevance to the more significant loss of strength of this specific material compared to another "SiC" processed in a different way or sintered with a different oxide composition. Therefore, the "modeling" aspects of the information necessary to understand the corrosion of ceramics are emphasized here without attempting to apply them to the behavior of the wide variety of ceramic materials that are available to the design engineer.

19.2 Background

19.2.1 Importance of Ambient Corrosion of Ceramics

The high-temperature corrosion behavior of ceramics has received considerably greater attention than ambient corrosion since many of the important applications for ceramics are at high temperatures. The fundamental mechanisms controlling the ambient degradation or corrosion of crystalline ceramics are not well understood except in a few isolated cases. Ceramics are comparatively inert so degradation by gaseous and aqueous corrosion has received little attention by ceramic engineers and scientists, and very few general principles have been developed. However, ceramics can corrode and do in aqueous environments, and this is where much of the experimental work has been focused. Recent experience with the environmental degradation of $YBa_2Cu_3O_{7-x}$ high-T_c superconductors (Fitch and Burdick 1989), the destabilization of yttria-stabilized zirconia by water (Lilley 1990), and the microstructural degradation during sintering caused by dissolution during powder processing (Anderson et al. 1988) emphasizes the recurring importance of ceramic corrosion (Clark and Zoitos 1992).

Ceramics have been considered for nuclear waste hosts for many years (Readey and Cooley 1977). From the time they were first considered as nuclear waste hosts, the aqueous corrosion of ceramics and glass as waste hosts became a paramount consideration. Glass has been the primary candidate as a potential waste host, and as a result, there have been many extensive studies performed on the corrosion behavior of silicate-based glasses. Much of this work has focused on corroding of specific silicate glass compositions in specific environments, and some progress has been made in generalizing the understanding of the fundamental mechanisms of leaching and degradation (Hench et al. 1986). Because of the alkali ion exchange that occurs between silicate glasses and solutions containing hydronium ions, the corrosion of glasses is complex and a poor choice as a model material to use for understanding wide-ranging fundamental principles of crystalline ceramics, the majority of which do not have singly charged and mobile alkali ions available to readily exchange with solution ions. Generally, in the case of aqueous solution corrosion, the question that needs to be answered is: "What are the thermodynamic and kinetic factors that control the rate of dissolution that can be used to predict the corrosion behavior of a given ceramic?"

For example, from a health hazard point of view, the biopersistence of ceramics is a critical issue. For particulates ingested into the lungs, a short residence time is desirable since a long residence time can lead to tissue damage and alteration. Because of the differences in dissolution or corrosion rates, the difference in persistence times of different types of asbestos contributes to their differences in toxicities (Mossman et al. 1990). As will be seen later, chrysotile asbestos, $Mg_6Si_4O_{10}(OH)_8$, might be expected to dissolve much more rapidly than amphibole types such as tremolite, $Ca_2Mg_5Si_8O_{22}(OH)_2$, due to the higher, and more readily soluble, alkaline earth-to-silica ratio in the former. As a result, chrysotile asbestos has a shorter residence time in the lungs and is less toxic (Mossman et al. 1990, Hume and Rimstidt 1992).

As another example, scrap chrome-magnesite refractory brick is a health hazard because of the discharge of chromium ions into the solution. In this case, dissolution or leaching is to be minimized. Chrome-magnesite refractories contain excess magnesium oxide (MgO) in addition to the magnesium chromite phase, $MgCr_2O_4$. Magnesia can dissolve relatively rapidly as will be seen later. Although no data are available, magnesium chromite probably dissolves much more easily than pure chrome oxide, Cr_2O_3, due to the presence of MgO in the compound. To minimize the biological hazard presented by scrap chrome-magnesite refractories, they could be reacted with excess aluminum oxide, Al_2O_3, or hematite, Fe_2O_3, to move the overall composition into the $MgCr_2O_4$-$(Al_2O_3$-$Cr_2O_3)_{ss}$ or $MgCr_2O_4$-$(Fe_2O_3$-$Cr_2O_3)_{ss}$ regions as shown in Figure 19.1 (Levin and McMurdie 1975, Figures 4570 and 4246). In doing so, the thermodynamic

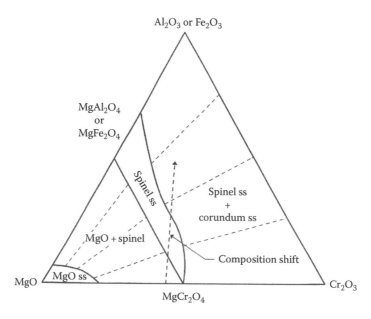

FIGURE 19.1 Combined schematic ternary diagram in the MgO–Cr_2O_3–$(Al_2O_3 + Fe_2O_3)$ system showing how the dissolution kinetics of $MgCr_2O_4$ could be decreased by adding Al_2O_3 or Fe_2O_3 to decrease the thermodynamic activity of MgO and the dissolution kinetics of $MgCr_2O_3$. (Based on data from Levin, E.M. and McMurdie, H.F., Eds., *Phase Diagrams for Ceramists*, Vol. III, The American Ceramic Society, Westerville, OH, 1975.)

activity of MgO is significantly reduced so the dissolution of $MgCr_2O_4$ is reduced in addition to putting most of the Cr_2O_3 into much more slowly dissolving corundum phases.

Finally, an area of current intense interest in ceramics is their use in fuel cells and as electrodes in high-energy density batteries. In these cases, particularly for battery electrodes, the rapid exchange of ions between the ceramic and electrolyte is of paramount importance in their selection as an electrode material: Li in LiV_2O_x, for example (McGraw et al. 1999). Such materials, by their very nature, are readily susceptible to ambient degradation and will behave more like a metal in terms of reactivity.

19.2.2 Limited Availability of Information and Application of General Principles

In this chapter, the possible fundamental mechanisms and factors affecting the kinetics of corrosion of ceramics under ambient conditions—near room temperature—are presented. The focus is primarily on dissolution kinetics as that is where the majority of the data and modeling exist. In addition, most of the available experimental data are for oxides, while information is just becoming available for nonoxides (Morrell 2010). However, most nonoxides oxidize in air, water vapor, or aqueous media so the dissolution behavior of the resultant oxide impacts the behavior of the parent nonoxide. Therefore, understanding the fundamental mechanisms of corrosion of oxides is important for understanding nonoxide degradation as well.

For both high-temperature and low-temperature corrosion of ceramics, there have been relatively few attempts at generalizing the behavior of crystalline ceramics, which is considerably different from the case of metals (Fontana 1986). In fact, most ceramic texts do not even mention corrosion (Barsoum 1997, Chiang et al. 1997), and if they do, the coverage is limited (Kingery et al. 1976, Richerson 2006).

Similar comments can be made about introductory materials science and engineering texts; they all give extensive coverage (usually a chapter or more) to oxidation and galvanic corrosion of metals while barely mentioning corrosion of ceramics (a paragraph), which typically refers to high-temperature corrosion in molten metals or slags (Callister 2000, Shackelford 2000, Askeland and Phulé 2003). Admittedly, most metals are thermodynamically unstable in an ambient air environment, and their degradation deserves attention because it is severe, obvious, costly, and pervasive, primarily because metals are used mainly for their mechanical properties, which are significantly degraded by corrosion. Furthermore, the main mechanism of ambient corrosion of metals is galvanic corrosion, and the development of its understanding and control has been helped considerably by the parallel development and general understanding of electrochemistry over the last 150 years (Bockris and Reddy 1970). Finally, a measure of the attention that corrosion of ceramics has received is that the recent review paper (Morrell 2010) covers 23 pages of a four-volume set comprising 3540 pages on corrosion: less than 1%.

In contrast, under ambient conditions, ceramics are considered to be oxidation and corrosion resistant. Certainly most oxides are relatively corrosion resistant, and it not surprising that the most corrosion-resistant metals—titanium and tantalum, for example—are those that form the most corrosion-resistant protective surface oxides. However, most oxides are not thermodynamically stable in gases with water vapor or in aqueous solutions, and both oxides and nonoxides are not stable in corrosive fluids containing HCl or HF. In near ambient aqueous solutions, most oxides are partially soluble and will dissolve or react to form hydroxides, which has led to extensive studies by geologists and geochemists on the stability and/ or weathering of oxides (and nonoxides such as sulfides), particularly silicates (Garrels and Christ 1990, Faure 1991, Pankow 1991, Stumm 1992, Stumm and Morgan 1996, Lasaga 1998).

19.2.3 Some Personal Experiences

19.2.3.1 Hydration Kinetics of CaO Powder

A personal first encounter with the environmental effects on the degradation of ceramics occurred during a quantitative analysis experiment carried out as an undergraduate. The experiment was to determine the amount of calcium in an unknown sample by calcination of a precipitate to CaO and then weighing—*with a manual chain balance*—the CaO produced. The experiment was carried out on a warm and humid Midwestern spring day. During the weighing process, the sample weight continually increased. This was less than ideal because if the results were not within a certain error band around the correct value, then the experiment had to be repeated. Since there simply wasn't sufficient time left in the semester to repeat the experiment, a weight was recorded that was a guess of what it might have been before the sample started gaining weight. It turned out to be close enough. Of course, the CaO was hydrating to $Ca(OH)_2$ with water in the atmosphere as (Roine 2002)

$$CaO(s) + H_2O(g) = Ca(OH)_2(s); \quad \Delta G^0(30°C) = -65.6 \text{ kJ/mol} \tag{19.1}$$

implying that the equilibrium partial pressure over CaO necessary to form $Ca(OH)_2$ needed to be only about 5×10^{-12} bar. However, at 30°C, on a humid day—say, 70% relative humidity—the water vapor pressure is calculated to be about 3×10^{-2} bar. For this and all following calculations of free energies, solubilities, and/or pressures of gases (unless noted otherwise), the software program *HSC Chemistry for Windows, Version 5.11* (Roine 2002) was used. This program has its own built-in database that is used for calculations and includes the data from most of the published thermodynamic databases. Clearly, *thermodynamics* suggests that the CaO could react with the ambient atmospheric water vapor to form $Ca(OH)_2$, and, since the calcined-sample weight was changing while being weighed, the *kinetics* of the reaction were quite rapid. This is not too surprising in retrospect in that the particle size of the calcined oxide was probably in the nanometer to micrometer range and presented a significant surface area where the reaction took place.

19.2.3.2 Hydration of Dense La$_2$O$_3$

Another personal encounter of the effects of the environment on the degradation of ceramics was during a research program investigating the microwave dielectric properties of the compounds in a variety of TiO_2–metal oxide binary systems in the search for materials having a high and temperature-independent dielectric constant at microwave frequencies. The literature suggested that, along with the end-members TiO_2 and La_2O_3 in the titanium oxide–lanthanum oxide binary system, there are three intermediate phases: $La_2O_3 \cdot TiO_2$, $La_2O_3 \cdot 2TiO_2$, and $La_2O_3 \cdot 9TiO_2$ stable at room temperature (Levin et al. 1969, Figure 2373, 103). In order to obtain optimum baseline properties, specimens with 100% of theoretical density were desired, and the individual compounds, including the end-members, were made by hot-pressing—at about 35 MPa ($\approx$357 bar) and 1200°C.

A cylinder of dense, pure La_2O_3—1.5 cm diameter by 3 cm tall—was produced and put on a desk late Friday afternoon. On Monday morning, after a hot and humid New England summer weekend, where the lanthanum oxide cylinder had been on Friday, there was now a pile of powder that proved to be $La(OH)_3$ by x-ray diffraction. The dense lanthanum oxide cylinder had completely reacted with the ambient water vapor even though its surface area was initially minimal. Certainly, the thermodynamics of the hydration reaction are favorable:

$$La_2O_3(s) + 3H_2O(g) = 2La(OH)_3; \quad \Delta G^0(30°C) = -164 \text{ kJ/mol} \tag{19.2}$$

but given the low surface area of the dense sample, it was surprising to find the kinetics of the reaction so fast and devastating. The microwave dielectric properties of pure La_2O_3 were not measured.

This is all the more surprising given the fact that MgO is thermodynamically similar to CaO and La_2O_3 powders in the presence of water vapor and reacts similarly as a powder;

$$MgO(s) + H_2O(g) = Mg(OH)_2(s); \quad \Delta G^0(30°C) = -35.0 \text{ kJ/mol} \tag{19.3}$$

FIGURE 19.2 Piece of large-grain MgO from the electrofusion process more than 50 years old but shows little surface degradation due to atmospheric corrosion/reaction. Numbered divisions = cm.

while dense samples are quite stable and can remain intact for long periods of time. MgO is used as a high-temperature refractory in several applications as are MgO–CaO refractories made from dolomite, $Mg_xCa_{1-x}CO_3$ (Chesters 1974). Figure 19.2 shows a piece of very large-grain-size MgO condensed from the vapor more than 50 years ago during the electrofusion process for making dense MgO particles for refractory bricks and shows very little surface degradation due to the formation of $Mg(OH)_2$ and/ or $MgCO_3$ by reaction with the atmosphere. Is the $Mg(OH)_2$ forming a thin protective layer in this case similar to the passive oxide layers on metals?

19.2.3.3 Hydration of Calcium Oxide Single Crystals

For several years, a few high-purity single crystals of calcium oxide (CaO) grown at Oak Ridge National Laboratory (Abraham et al. 1971) were kept in a desk drawer for future experiments. The samples were rapped loosely with paper. Periodically—every 6 months or so—the paper would have to be changed and the crystals wiped off because a layer of condensation formed on the surface, which presumably was a saturated $Ca(OH)_2$ solution that absorbed additional water from the atmosphere. Clearly, the CaO crystal was thermodynamically unstable in the ambient humid environment that typically occurs in summer. Nevertheless, dolomite, $Ca_xMg_{1-x}(CO_3)_2$, refractories that contain mixtures of MgO and CaO have been used for steelmaking for over 130 years. However, they do pose problems on furnace shutdown because of hydration of the CaO phase (Krause 1987, 31).

19.2.4 Implication of Ambient Degradation on Ceramic Processing

Most of today's high-technology ceramics are made by processing purified industrial inorganic chemicals in powder form. Because of reactions such as the hydration of CaO, a material such as calcium titanate ($CaTiO_3$), perovskite, would not be made by reaction of CaO and TiO_2 powders. In order to ensure the one-to-one Ca/Ti stoichiometry, the powders would need to be weighed in those proportions before reaction, which is a high-temperature calcination step (Jacob and Abraham 2009):

$$CaO(s) + TiO_2(s) \xrightarrow{\text{about } 1200^\circ C} CaTiO_3(s); \quad \Delta G^0(1200^\circ C) = -89.4 \text{ kJ/mol} \tag{19.4}$$

Because of the tendency of oxides such as CaO and MgO to hydrate—or react with the ever-increasing CO_2 concentration in the atmosphere to form carbonates,

$$MgO(s) + H_2O(g) = Mg(OH)_2(s); \quad \Delta G^0(30^\circ C) = -35.0 \text{ kJ/mol} \tag{19.5}$$

$$MgO(s) + CO_2(g) = MgCO_3(s); \quad \Delta G^0(30^\circ C) = -47.7 \text{ kJ/mol} \tag{19.6}$$

oxide powders will frequently partially transform to carbonates and/or hydroxides during storage prior to use. For the formation of the carbonate (Equation 19.4), the equilibrium CO_2 pressure is 6×10^{-9} bar ($\approx 6 \times 10^{-4}$ Pa), and the current CO_2 ambient in the atmosphere is about 4×10^{-4} bar (≈ 40 Pa) and rising. As a result, when making compounds such as $CaTiO_3$, $BaTiO_3$, $MgFe_2O_4$, and $Y_3Fe_5O_{12}$, the usual practice is to use carbonate powders—or another atmospherically stable compound of the reactive oxide—so that a typical calcining reaction might be

$$MgCO_3(s) + Fe_2O_3(s) \xrightarrow{1000^\circ C} MgFe_2O_4(s) + CO_2(g); \quad \Delta G^0(1000^\circ C) = -126 \text{ kJ/mol} \tag{19.7}$$

Otherwise, as in this example, the Mg/Fe ratio would not necessarily have the desired stoichiometric 1:2 ratio but would be either above or below stoichiometry because the MgO used was not pure MgO because some carbonate or hydroxide was present that would yield less MgO than desired. Both $MgFe_2O_4$ and $Y_3F_5O_{12}$ are magnetically soft oxides used for microwave and other applications where the magnetic properties are very critical and strongly dependent on the final microstructure of the ceramic. For example, Figure 19.3 shows an ideal single-phase microstructure of a modified $Y_3Fe_5O_{12}$ ceramic—containing four or five additional alloying oxides to achieve the desired magnetic properties

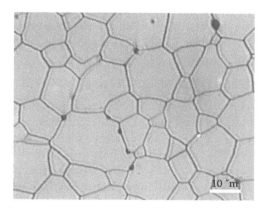

FIGURE 19.3 Ideal single-phase microstructure of a chemically modified $Y_3Fe_5O_{12}$ magnetic alloy having the ideal 3/5 Y/Fe ratio.

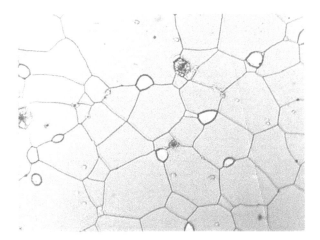

FIGURE 19.4 Undesirable two-phase microstructure of a chemically modified $Y_3Fe_5O_{12}$ magnetic alloy not having the ideal 3/5 Y/Fe ratio. The second phase particles—Fe_3O_4—are smaller and lighter in color.

for a microwave phase shifter element. The compositional variation in the ideal 3Y/5Fe ratio necessary to achieve a single-phase material is extremely limited—on the order of 0.1% (Paladino and Maguire 1970). If the ratio is above or below this, a second phase is present (Figure 19.4), which seriously degrades the desired magnetic properties of the material.

19.3 Corrosion Reactions in General

19.3.1 Oxidation of SiC

19.3.1.1 Passive Oxidation

The high-temperature oxidation of SiC is a good illustration of the kinds of degradation that ceramic materials can undergo: albeit this does not take place to any appreciable extent at low temperatures. Nevertheless, just as in the case of metals at low temperatures, a thin oxide layer of SiO_2 will exist at the surface of SiC at all temperatures since the free energy for the reaction

$$SiC(s) + 2O_2(g) = SiO_2(s) + CO_2(g); \quad \Delta G(0°C) = -1186 \text{ kJ/mol} \tag{19.8}$$

is very negative at all temperatures. This *passive oxidation*, as sketched in Figure 19.5, is limited by the diffusion of oxygen through the SiO_2 layer and is vanishingly small at ambient temperatures but becomes sufficiently rapid at elevated temperatures so that the CO_2 pressure at the SiC–SiO_2 interface may exceed 1 bar, causing bubbling and general degradation of the protective SiO_2 film (Mayer and Lau 1990, 257). Like all corrosion reactions, there are several series and parallel steps in the overall reaction kinetics. For this concrete example, the series steps are (1) diffusion of oxygen to the surface of the SiO_2, (2) diffusion of oxygen through the SiO_2 layer, and (3) the actual reaction at the SiC–SiO_2 interface of the SiC and oxygen. The fate of the CO_2 is not well known except that it apparently diffuses sufficiently rapidly out through the SiO_2 until the rate of oxidation becomes sufficiently fast that the CO_2 pressure reaches one bar at temperatures in the neighborhood of 1600°C. For any reactions in series, the *slowest* controls the rate of reaction, and, in this case at high temperatures, it is the diffusion of oxygen through the oxide layer that controls the rate. It should be noted that the diffusion through the SiO_2 layer is a parallel process in that either oxygen diffusion from the O_2–SiO_2 interface through the SiO_2 layer to the SiC–SiO_2 interface or silicon diffusion in the opposite direction can occur. These are parallel processes and the *faster* of these two controls the rate. In this case, the interstitial diffusion of the O_2 is much faster

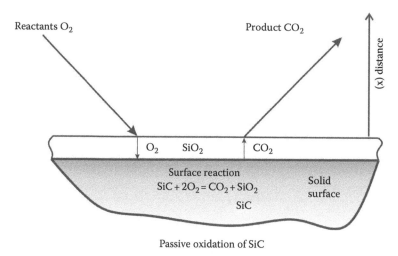

Passive oxidation of SiC

FIGURE 19.5 Schematic showing the passive oxidation—the formation of a protective SiO_2 layer—in oxygen.

than the "lattice" diffusion of the silicon in the SiO_2 so oxygen diffusion controls the overall reaction rate (Deal and Grove 1965, Mayer and Lau 1990).

19.3.1.2 Active Oxidation

At low oxygen pressures, *active oxidation* of SiC can occur by the reaction:

$$SiC(s) + O_2(g) = SiO(g) + CO(g); \quad \Delta G(0°C) = -189 \text{ kJ/mol} \tag{19.9}$$

as sketched in Figure 19.6. In this case, the oxidation continues at a constant linear rate since there is no passive oxide layer formed, only gaseous products. It is because of this reaction that SiC is not used in a vacuum or reducing gases as furnace fixtures or heating elements. In this case, there are again three steps in series that could control the reaction: (1) transport or diffusion of the reactant oxygen to the surface, (2) the actual reaction at the surface, and (3) transport or diffusion of the product species away from the surface. Again, the slowest of these three steps is rate controlling. Such active oxidation or *active corrosion* is more typical of the type that can cause degradation of ceramics at low temperatures, particularly in water vapor or aqueous solutions.

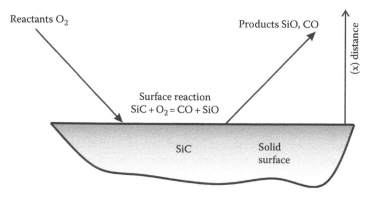

FIGURE 19.6 Schematic showing the active oxidation of SiC in oxygen.

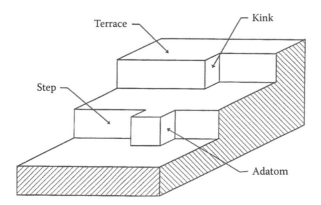

FIGURE 19.7 Schematic figure showing the important features on the surface of solid undergoing some solid–fluid reaction.

19.3.2 Surface Reactions

Figure 19.7 shows the typical schematic representation of a solid surface where the kink sites are thought to be the sites where the final reaction step takes place (Hudson 1995, 5). However, there are several possible atomistic steps that can be postulated to occur on the surface of a solid before the reaction can be complete. In reality, these are not known for most reactions, yet any one of these could be the rate-controlling step in the surface reaction since they are all series steps, and the slowest will control the rate. For the active oxidation of SiC of Equation 19.9, some possible steps that can be imagined include

- Adsorption of O_2 onto the solid surface
- Decomposition of O_2 into two O atoms on the surface
- Surface diffusion of the O atoms to the reaction/kink site
- Reaction of O with Si to form SiO
- Reaction of O with C to form CO
- Diffusion of CO and/or SiO from the reaction site
- Desorption of CO and/or SiO from the surface

This certainly is not an all-inclusive list nor is it known whether any of the listed steps actually occur, but there will usually be a single reaction step that controls the rate of surface reaction. From an energy standpoint, this is illustrated in Figure 19.8, where ΔG^0 is the free energy for the reaction of Equation 19.9 and ΔG^* is the activation energy for the surface reaction.

19.3.3 General Kinetic Considerations

Regardless whether the ambient corrosion of ceramics is taking place in a gaseous or aqueous environment, the same kinetic considerations can be applied. Furthermore, to completely understand the degradation of ceramics in hostile environments, quantitative measurements of degradation rates need to be performed to obtain values for the kinetic parameters so that material behavior can be confidently predicted. Unfortunately, there are relatively few instances where such studies have been performed on ceramics that have unambiguously defined the degradation mechanism let alone quantitatively measured kinetic parameters necessary for behavior prediction. Nevertheless, it is still instructive to consider the general principles involved in the series steps necessary for degradation to occur: (1) transport/diffusion of the reactants to the surface, (2) reaction at the solid surface, and (3) transport/diffusion of the products away.

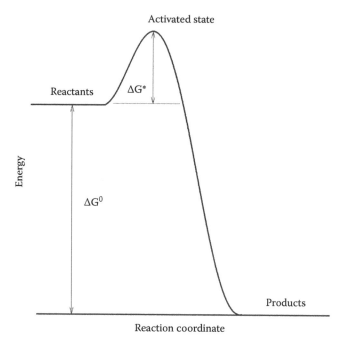

FIGURE 19.8 Free energy as a function of the reaction coordinate—extent of reaction—showing the activated state and the activation energy, ΔG^*, for the surface reaction.

19.3.3.1 Surface Reaction Infinitely Fast Compared to Transport/Diffusion

The assumption of an infinitely fast surface reaction compared to diffusion implies that the surface concentration of the reactants reaches an equilibrium value resulting in a concentration gradient and diffusion of the reactants to the solid surface. For the sake of simplicity, only diffusion of the reactant gases will be considered, but transport of the product gases gives a similar result. Figure 19.9 shows the

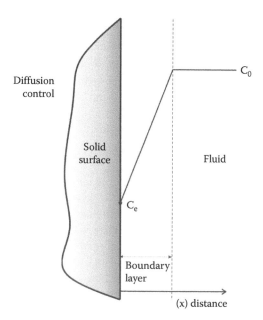

FIGURE 19.9 Model for diffusion-controlled kinetics.

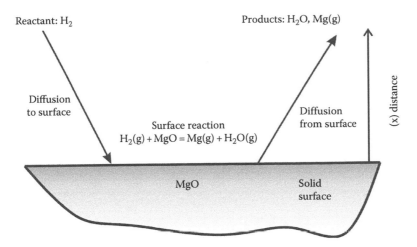

FIGURE 19.10 Schematic showing the active gas reaction of MgO in hydrogen.

concentration profile from the surface of the reacting solid to far out into the fluid phase. It is assumed that there is fluid flow parallel to the surface with a certain velocity that determines a flow boundary layer thickness, δ. The boundary layer approximates the distance over which the fluid velocity goes from zero at the surface to the stream velocity at δ. It is also generally assumed the diffusion in the fluid only occurs in the boundary layer and the concentration gradient through this layer is constant; that is, the concentration is a linear function of distance.

For the sake of concreteness, consider the corrosion of MgO in hydrogen by the following reaction shown schematically in Figure 19.10:

$$MgO(s) + H_2(g) = Mg(g) + H_2O(g) \tag{19.10}$$

Figure 19.11 shows all of the pressures of the various possible gas species in the MgO–H_2 system as a function of temperature and demonstrates that this is indeed the reaction since the Mg and H_2O product pressures are the highest at all temperatures. Therefore, the corrosion flux, J (mol/cm² s), of MgO is given by Fick's first law of diffusion and is

$$J(MgO) = J(H_2) = -D\frac{dC}{dx} \cong -\frac{D}{\delta}(C_0 - C_e) \tag{19.11}$$

where the concentrations (C) are for hydrogen in the gas phase and the fluxes of both the MgO and hydrogen are negative since they are both being consumed at the surface. Equation 19.11 can be written in terms of the hydrogen pressures and the rate of corrosion da/dt (cm/s) calculated:

$$J(MgO) = \frac{1}{\overline{V}(MgO)}\frac{da}{dt} = -\frac{D}{RT\delta}(p_0 - p_e)$$

$$\frac{da}{dt} = -\frac{\overline{V}(MgO)D}{RT\delta}(p_0 - p_e) \tag{19.12}$$

where $\overline{V}(MgO)$ is the molar volume of MgO (cc/mol) and D is essentially the self-diffusion coefficient of hydrogen in the gas phase—since the pressures of Mg and H_2O are small—and can be calculated

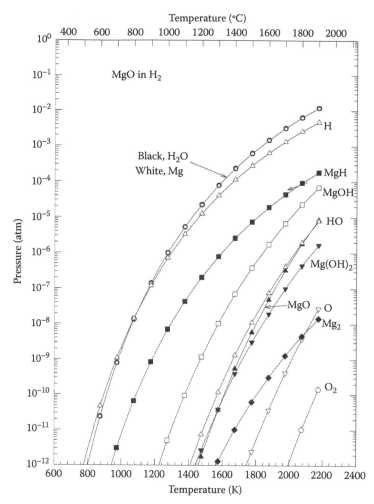

FIGURE 19.11 Vapor pressures of various gaseous species over MgO in 1 atm (1.0135 bar) H_2 as a function of temperature.

reasonably accurately from the kinetic theory of gases with the Chapman–Enskog equation (Geankoplis 1972, 2003, Cussler 1997, Bird et al. 2002):

$$D_{AB} = 1.86 \times 10^{-3} \left(\frac{1}{M_A} + \frac{1}{M_B} \right)^{1/2} \frac{T^{3/2}}{d_{AB}^2 \Omega_{AB} p} \; cm^2/s \qquad (19.13)$$

where

  D_{AB} is the gas diffusion coefficient for two gases, A and B, with molecular weights M_A and M_B (g/mol)
  T is the temperature (K)
  p is the pressure (atmosphere)
  $d_{AB} = (d_A + d_B)/2$ (in units of 10^{-8} m or Ångstrom) is the mean "collision diameter" of the interdiffusing gases
  Ω_{AB} is the collision integral that takes into consideration a real interatomic potential between the gas atoms rather than treating them simply as colliding spheres

Most of Equation 19.13 can easily be calculated from the kinetic theory of gases with the exception Ω_{AB} whose values are between about 0.4 and 2.6 and can be evaluated from data given in several references (Geankoplis 1972, 2003, Cussler 1997, Bird et al. 2002). Calculated from "d" and Ω values in the literature $D(H_2, 300 \text{ K}) = 1.46 \text{ cm}^2/\text{s}$ and at 1000°C, $D(H_2, 1273 \text{ K}) = 16.3 \text{ cm}^2/\text{s}$. Now $\bar{V}(MgO) = M/\rho = 40.31/3.58 = 11.26 \text{ cc/mol}$ where ρ = density (g/cc). Assume that the flow conditions are such that $\delta = 1$ cm, and the total pressure is one atmosphere ($\approx$1 bar); then, at 1000°C in pure hydrogen, the rate of corrosion of MgO can be calculated from Equation 19.12 to be

$$\frac{da}{dt} = -\frac{(11.26)(1 \times 10^{-6})}{1(8.314)(1273)} = 1.06 \times 10^{-9} \text{ cm/s}$$

Since there are about 3×10^7 s/year, the rate of corrosion of MgO in pure hydrogen at 1000°C is about 0.3 mm/year. This is probably small enough that MgO could be used with confidence in a hydrogen atmosphere at 1000°C without serious degradation. This calculation demonstrates one of useful features of identifying diffusion control as the rate-controlling step; that is, the actual corrosion rate can be calculated quite accurately for both liquids and gases since diffusion coefficients can be calculated or obtained from the literature, which is not the case for solids.

19.3.3.2 Diffusion Infinitely Fast Compared to the Surface Reaction Rate

In this case, the reactant concentration is constant with distance from the surface of the corroding material as shown in Figure 19.12. Generally, it is found—or assumed—that surface reactions on solids are first order with respect to the reactants, so, again with MgO reacting in hydrogen as the example, the rate of reaction is given by

$$J(MgO) = J(H_2) = -k(C_0 - C_e) \tag{19.14}$$

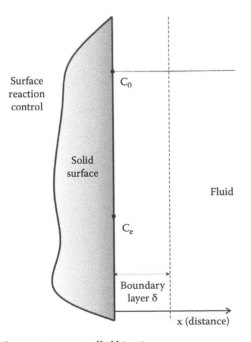

FIGURE 19.12 Model for surface reaction-controlled kinetics.

where "k" is the reaction rate constant and is of the form $k = k_0 e^{-(\Delta H^*/RT)} = k_0 e^{-(Q/RT)}$ where $\Delta H^* = Q$ is the activation enthalpy or the activation energy related to the ΔG^* of the activated state (Figure 19.8). The pre-exponential includes the number of atoms or molecules hitting the surface per unit area and time and an entropy term. What the rate constant essentially implies is that only a certain fraction of the atoms or molecules striking the surface have sufficient energy, Q, to react. In general, the values of k_0 and Q are not easily evaluated from first principles and must be determined by experiment. Nevertheless, the temperature dependence of the surface reaction is exponentially temperature dependent—much stronger than the $T^{3/2}$ dependence of gaseous diffusion and weak exponential temperature dependencies of liquid diffusion coefficients. As a result, surface reaction rates decrease rapidly as the temperature is decreased, much faster than the rates of diffusion. Therefore, it is expected that surface reactions control corrosion rates at low temperatures and diffusion at higher temperatures, which is generally the case.

19.3.3.3 Diffusion and Reaction Rates Comparable

This last point can be demonstrated by considering the situation where both diffusion and the surface reaction rates are comparable as sketched in Figure 19.13. In this case, the relevant equations are

$$J(MgO) = J(H_2) = -D\frac{dC}{dx} \cong -\frac{D}{\delta}(C_0 - C_S) \tag{19.15}$$

$$J(MgO) = J(H_2) = -k(C_S - C_e) \tag{}$$

where C_S is the *surface concentration* and is somewhere in between C_0 and C_e. Eliminating C_S in the earlier two equations gives

$$J(MgO) = J(H_2) = -h(C_0 - C_e) \tag{19.16}$$

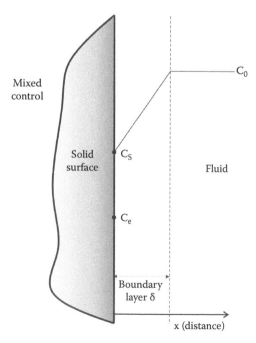

FIGURE 19.13 Model for mixed diffusion and reaction control.

where "h" is the mass transfer coefficient given by

$$\frac{1}{h} = \frac{1}{D/\delta} + \frac{1}{k} \tag{19.17}$$

so that when $k \gg D/\delta$, then $h \simeq D/\delta$ so Equation 19.16 becomes Equation 19.11—diffusion control—and when $k \ll D/\delta$, then $h \simeq k$ so Equation 19.16 becomes Equation 19.14—reaction control. That is, the slower step controls the overall rate of the reaction. The implication of this is shown schematically in Figure 19.14 where the weak temperature dependence of diffusion in the gas phase is plotted along with an exponentially temperature-dependent reaction rate as well as "h" from Equation 19.17 showing that the slower of the two is rate controlling when plotted as Equation 19.16.

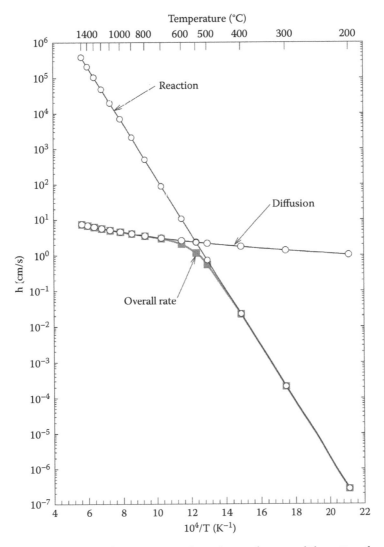

FIGURE 19.14 Schematic showing the temperature dependence of a gas–solid reaction showing the weakly temperature-dependent gas diffusion, the exponentially temperature-dependent surface reaction, and the overall reaction rate—gray boxes—demonstrating that the slower of series reactions is rate controlling.

19.4 Ceramic–Gas Reactions

19.4.1 Reactive Gas Corrosion: High Temperatures—Active Gas Corrosion

Active corrosion, as discussed earlier, usually means atmospheric degradation with the formation of nonpassivating corrosion products such as gaseous SiO and CO, which result from oxidizing SiC in water vapor:

$$SiC(s) + 2H_2O(g) = SiO(g) + CO(g) + 2H_2(g) \tag{19.18}$$

Kinetic data on this system (Kim and Readey 1989) and on the active corrosion of SiO_2 in hydrogen at temperatures above 1200°C (Readey 1991)

$$SiO_2(s) + H_2(g) = SiO(g) + H_2O(g) \tag{19.19}$$

show that, at these high temperatures, diffusion of the product gases away from the corroding surface controls the rate of corrosion. As a result, as discussed earlier, the kinetics of the reaction can be easily modeled, and quantitative comparisons can be made with experimental data to confirm diffusion control (Readey 1991). Thermodynamic data for the reaction and gas diffusion coefficients calculated from the kinetic theory of gases (Equation 19.13) give the necessary information to permit quantitative prediction of corrosion rates. Of course, if one of the many possible surface reactions controls the rate, then prediction of corrosion rates without experimental data is impossible. In any event, for any given material, at the highest temperatures, gas diffusion will control the corrosion rate, and predictions of corrosion behavior and corrosion rates can be made (Readey 1998a,b).

Water vapor, oxygen, and hydrogen are not the only corrosive gases of interest to ceramics. The corrosion of ceramics in halogen-containing gases such as Cl_2, HCl, F_2, and HF is of interest for several reasons, both scientific and technical. The number of ceramic compounds that can undergo active gas corrosion in halogens is much larger than those that react to form only gaseous species in either reducing or oxidizing environments. Therefore, the active gas corrosion of a wide variety of ceramics (oxides, carbides, nitrides, borides, etc.) can be studied to determine the systematics, similarities, and differences in their behavior.

In addition, gas corrosion of ceramics by halogens is technologically important and becoming increasingly so. Industrial chemicals such as TiO_2 pigments and fluorocarbons are processed in halogen environments at high temperatures, and furnace refractory corrosion is a problem. Ceramics are being used more frequently as supports and furnace fixturing for silicon wafers in integrated circuit manufacturing. Many of the processing steps are carried out in halogen-containing gases either at high temperatures or in plasmas. Not only is corrosion resistance important, but also the mechanism of degradation is critical in these applications. For example, preferential attack of the glass grain boundary phase in a high-alumina ceramic by a reactive atmosphere can lead to release of alumina particles, which pose a serious problem in subsequent wafer processing steps if these particles fall onto the wafer surface. The behavior of ceramics in incinerators is becoming of increasing importance. The existence of halogens in the combustion gases is assured by the presence of polyvinyl chloride and other halogen-containing polymers in the waste stream. Not only can ceramics corrode in the atmosphere and temperature conditions in an incinerator, but the corrosion of metal components in the same environment depends critically on the resistance of any passive oxide to the halogen gases that are present. Finally, any high-temperature system operating near saltwater ingests NaCl from the atmosphere and generates chlorine-containing gases that can cause ceramic and protective metal oxide degradation. Figure 19.15 shows the gaseous species over solid TiO_2 in equilibrium with a 50% HCl–50% Ar gas mixture. The main reaction is

$$TiO_2(s) + 4HCl(g) = TiCl_4(g) + 2H_2O(g) \tag{19.20}$$

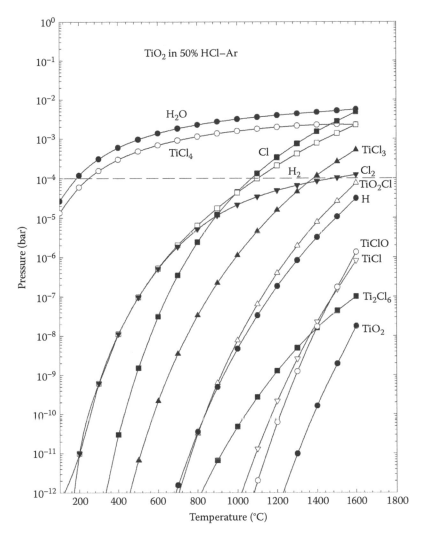

FIGURE 19.15 Gaseous species over TiO_2 in a 50% HCl–50% Ar atmosphere.

Note that active gas corrosion takes place at all temperatures since $TiCl_4$ is a liquid at room temperature and boils at 136.4°C (Weast 1979, B-137).

In another area of technical importance, the presence of reactive atmospheres, such as HCl, during sintering and densification of ceramic powders leads to enhanced vapor transport, little shrinkage, and significant grain growth (Readey 1990, 86–110), and this has been shown to be the case with TiO_2 in HCl as Equation 19.20 predicts (Readey and Readey 1987). Reactive vapor phase sintering can be used to fabricate materials with controlled porosity and pore size that can be used in filters, sensors, biomaterials, and the backbone for ceramic–metal composites (Ritland and Readey 1993, 896–907). Knowing whether vapor transport is controlled by gas diffusion or by a surface reaction is important in understanding the kinetics of vapor phase sintering and other applications of reactive vapor transport.

Therefore, based on previous experimental data (Kim and Readey 1989, Readey 1991) that demonstrate the ability to model gas-phase corrosion, thermodynamic calculations can be used to predict the high-temperature, gas diffusion-controlled corrosion of several different oxides and nonoxides (Readey 1998a,b).

19.4.2 Reactive Gases: Low Temperatures

In contrast to high-temperature gas corrosion, the behavior of different materials at ambient temperatures—less than about 100°C—is much more difficult to generalize, and there are precious few experimental results available. However, thermodynamics can, nevertheless, still be an important tool in determining whether active or passive corrosion should occur. Of course, if diffusion control is assumed, then for active corrosion it still would be possible to predict corrosion rates because the gas diffusion coefficient can be calculated (Equation 19.13). However, the likelihood of diffusion being the rate-controlling step is small given the exponential temperature dependencies of surface reactions as shown in Figure 19.14. For example, Figure 19.16 shows the gaseous species over SiC in HF as a function of temperature. In this case, at room temperature, the reaction

$$SiC(s) + 4HF(g) = SiF_4(g) + CH_4(g) \tag{19.21}$$

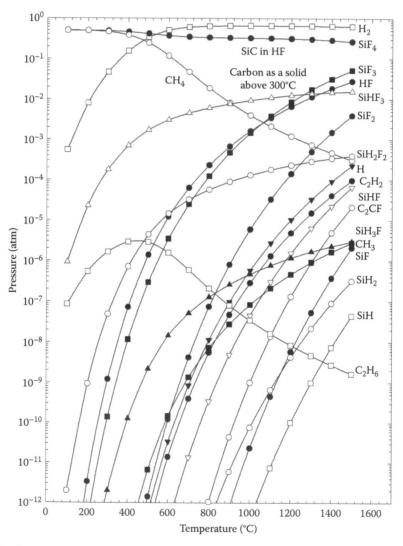

FIGURE 19.16 Gaseous species over SiC in 50% HF–50% Ar atmosphere.

essentially goes to completion, and above 300°C, carbon is stable and will be left behind requiring that the continued reaction will have to take place through the porous carbon. Nevertheless, these thermodynamic data suggest that SiC will undergo active gas corrosion to SiF_4 and CH_4—or carbon—at all temperatures. What is not known is whether or not the reaction at ambient temperatures is diffusion or reaction controlled. Literature data on the corrosion of SiC in HF atmospheres are not available.

In contrast, Figure 19.17 shows the gaseous species over MgO in HF and, in this case, solid MgF_2 (melting point T_{mp} = 1261°C) (Weast 1979, B-94) or liquid is formed at all temperatures and the vapor pressure of MgF_2 only becomes significant—p ∼ 10^{-4} bar—above about 1200°C. So in this case, at ambient temperatures—below a few hundred degrees Celsius—*passive corrosion* of MgO is occurring with the formation of an MgF_2 passive layer on the surface—similar to that shown in Figure 19.5. As a result, continued reaction to form the fluoride can only occur by diffusion through the MgF_2 layer if it forms a dense layer on the MgO surface.

A well-established principle in the active oxidation of metals is the Pilling–Bedworth ratio (West 1986, 183–191, Kofstad 1988, 234–246), which states that the volume of the oxide should exceed the volume

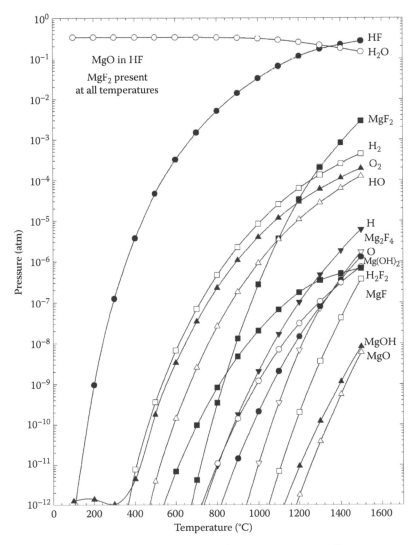

FIGURE 19.17 Gaseous species over MgO in HF showing that MgF_2 is present at all temperatures.

TABLE 19.2   Molar Volumes of MgO and Al_2O_3 and Possible Corrosion Products

Compound	Molecular Weight (g/mol)	Density (g/cc)	Molar Volume (cc/mol)	Melting Point (°C)
MgO	40.31	3.58	11.26	2852
$MgCl_2$	95.22	2.32	41.04	714
MgF_2	62.31	3.15	19.79	1261
$Mg(OH)_2$	58.33	2.36	24.72	$-H_2O$ 350
$MgCO_3$	84.32	2.96	28.51	$-CO_2$ 900
Al_2O_3	101.96	3.97	25.72	2050
$AlCl_3$	133.34	2.44	54.64	Sublimes 178
AlF_3	83.98	2.88	29.14	Sublimes 1291
$Al(OH)_3$	78.00	2.42	32.23	$-H_2O$ 300

Source: Weast, R.C., Ed., *CRC Handbook of Chemistry and Physics*, 60th edn., CRC Press, Boca Raton, FL, 1979.

of the metal oxidized so that the oxide forms a continuous layer on the metal. This leads to passive oxidation. However, as the volume of the oxide to the volume of the metal—Pilling–Bedworth ratio—becomes larger, stresses are built up in both the metal and oxide that can lead to cracking or spalling of the oxide layer leading to enhanced rates of oxidation. At high temperatures, both plastic deformation of the metal and the oxide layer may occur to limit stress buildup and cracking. However, for passive corrosion of ceramics near room temperature—such as the formation of MgF_2 on MgO—no such plastic deformation is going to occur. As a result, stress buildup in any layer will occur. Table 19.2 shows the molar volumes of MgO and Al_2O_3 with possible passive corrosion products and their calculated molar volumes (Weast 1979). For both MgO and Al_2O_3, the ratio of the volume of the corrosion product to that of the oxide is typically greater than 2.0—in the case of Al_2O_3, the volumes of the corrosion products must be multiplied by two to determine the ratio. These data suggest that any corrosion product layer would have to be very thin—on the order of a nanometer or less—to prevent cracking (West 1986). However, this might not be a major problem. Considering Equation 19.12 again, the rate of growth of a passive corrosion layer by diffusion through the layer of thickness "a" is given by

$$\frac{da}{dt} = -\frac{\bar{V}(MgF_2)D}{RTa}(p_0 - p_e) \tag{19.22}$$

This leads to the parabolic oxidation for passive oxide layers on metals, namely,

$$a^2 = \frac{\bar{V}(MgF_2)D}{RT}(p_0 - p_e)t \tag{19.23}$$

where "D" is the diffusion coefficient of the faster diffusing species through the product layer. Diffusion coefficients for MgF_2 were not found, but an idea of the thickness of such corrosion product layers can be obtained for the oxidation of silicon, which has been well modeled (Deal and Grove 1965) and studied (Mayer and Lau 1990) since the thickness of the SiO_2 layer on Si is critical for integrated circuit devices. This insulating SiO_2 layer is the main reason that silicon is the semiconductor material of choice even though its properties as a semiconductor are inferior to those of either germanium or gallium arsenide. For the reaction $Si(s) + O_2(g) = SiO_2(s)$ at 25°C, $\Delta G^0 = -856$ kJ/mol, which makes the equilibrium oxygen pressure, p_e, at the SiO_2–Si interface (Figure 19.5) (SiO_2–SiC interface in the figure) essentially zero. The molar volume of SiO_2 $\bar{V}(SiO_2) = M/\rho = 60.08/2.2 = 27.3$ cc/mol (Weast 1979), and (Mayer and Lau 1990)

$$D(O_2) = D_0 e^{-(Q/RT)} = 2.7 \times 10^{-4} e^{-(111.9kJ/mol/(RT))} = 6.44 \times 10^{-24} \text{ cm}^2\text{s}^{-1}$$

so in air, $p_0 = 0.21$ bar at 25°C, $a^2 = 1.52 \times 10^{-27}$ t cm^2. One year $\approx 3 \times 10^7$ s, so in a year, $a \approx 2 \times 10^{-10}$ cm, which is less than an atomic diameter, $\approx 3 \times 10^{-8}$ cm. This implies that at these length scales, the macroscopic diffusion laws really don't apply. As a result, the best that can be said is that there will be some absorption of the reactive species with some possible rearrangement within the first few atomic layers of the substrate. What this essentially implies, unless the diffusion through the passive layer is extremely fast, then any ceramic for which the thermodynamics indicates a passive solid layer will form, basically will form some sort of passive layer that will only be a few atoms thick, and the ceramic material is essentially not degraded by the environment, at least as far as extensive formation of a corrosion product phase is concerned.

19.5 Ceramic Corrosion in Aqueous Environments

19.5.1 General

Earlier, several examples of corrosion of ceramics in aqueous environments were mentioned. Of perhaps equal importance, most ceramics are processed as powders and most often in aqueous solutions of controlled pH to control the surface charge on the particles to keep them dispersed for optimum processing conditions (Funk and Dinger 1994, Reed 1995, Ring 1996, Lewis 2000). Here again, the possible fundamental mechanisms and factors affecting the kinetics of corrosion of ceramics are presented. Most of the available experimental data are for oxides, while information is just becoming available for nonoxides. However, most nonoxides oxidize in aqueous media to form at least a surface layer of oxide, as described earlier, so the dissolution behavior of the resultant oxide impacts the behavior of the parent nonoxide. Therefore, understanding the fundamental mechanisms of corrosion of oxides applies to nonoxide aqueous corrosion as well.

19.5.2 Aqueous Corrosion Thermodynamics

19.5.2.1 Types of Corrosion

Corrosion of oxide ceramics is essentially a dissolution process (Blesa et al. 1994). Two types of dissolution can be distinguished (Diggle 1973, 281–306). One is chemical dissolution involving no charge transfer or oxidation/reduction step such as

$$MgO_{(s)} + 2H^+ = Mg^{2+} + H_2O \qquad (19.24)$$

The other is electrochemical dissolution with electron transfer similar to that which occurs during the dissolution or aqueous corrosion of metals:

$$Fe_2O_{3(s)} + 3H^+ + 2e^- = 2Fe^{+2} + 3OH^- \qquad (19.25)$$

19.5.2.2 Chemical Dissolution

The thermodynamics of aqueous corrosion is concerned with the solubility of a ceramic. The solubility is determined by the stability of the constituent ions in solution relative to their stability in the solid or crystal. For purely ionic crystalline compounds, it is the relative values of the crystal lattice energy and the ion hydration energy in solution that determine solubility (Burgess 1988).

The solubility of sodium chloride in pure water at 25°C is about 0.63 mol/L for aq means aqueous solution

$$NaCl(s) = Na^+(aq) + Cl^-(aq) \qquad (19.26)$$

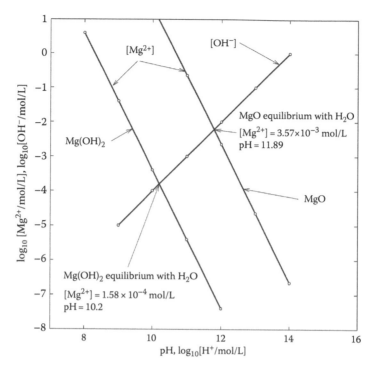

FIGURE 19.18 Calculated solubilities of MgO and Mg(OH)$_2$ as a function of pH at 25°C. The line labeled "[OH]" is the concentration of OH$^-$ at the given pH as determined by the ionization of water.

and, of course, is independent of the pH of the solution. On the other hand, the solubility of MgO depends strongly on the pH since the solubility reaction is

$$MgO(s) + H_2O(l) = Mg^{+2}(aq) + 2OH^-(aq) \tag{19.27}$$

Figure 19.18 shows the calculated solubilities of MgO and Mg(OH)$_2$ as a function of pH. Since the solubility of Mg(OH)$_2$ is less than that of MgO, MgO in aqueous solutions will form a layer of Mg(OH)$_2$ on its surface while it dissolves. Again, calling this a "layer" may be a matter of semantics since it really will be nothing more than some addition of water molecules to the first few surface layers as discussed earlier.

On the other hand, SiO$_2$ has a low solubility in acidic solutions (Pourbaix 1966, 458–463) because electrically neutral Si(OH)$_4$ is formed, which is not strongly solvated by water molecules. The Si(OH)$_4$ "molecule" is uncharged and nonpolar so there is virtually no hydration energy as it enters solution. Its solubility would not be expected to be much different than that of methane. At high pH, Si(OH)$_4$ behaves as a weak acid, it ionizes and hydrates, and the solubility of SiO$_2$ increases with pH. However, in solutions containing fluorine, the stable SiF$_6^{2-}$ ion is formed, and thermodynamics suggests that silica is quite soluble in acidic HF solutions:

$$SiO_2(s) + 6HF(aq) = 2H^+(aq) + SiF_6^{2-}(aq) + 2H_2O(l);$$

$$\Delta G^0(25°C) = -36.5 \text{ kJ/mol} \tag{19.28}$$

and this has been borne out by experiment (Liang and Readey 1987). So the composition of the solution can influence the nature of the aqueous ionic species and have significant effects on the solubility of a given ceramic material.

Most oxides are amphoteric (Pourbaix 1966, 168–176), acting as weak bases in acidic solutions and weak acids in basic solutions. For example, in acids, Al_2O_3 will dissolve as

$$Al_2O_{3(s)} + 6H^+ = 2Al^+ + 3H_2O \tag{19.29}$$

while in basic solutions,

$$Al_2O_{3(s)} + 2OH^- = 2AlO_2^- + H_2O \tag{19.30}$$

Because of this amphoteric dissolution behavior, there is frequently a minimum in oxide solubility as a function of pH as shown in Figure 19.19 (Pourbaix 1966, 174).

These examples of solubilities of both oxide and nonoxide ceramics can, in principle, be calculated from thermodynamic data (Latimer 1952, Bard et al. 1985), which are now included in the software database used for thermodynamic calculations. However, frequently, the species in solution might be considerably more complex than the simple dissolution equations given above (Browne and Driscoll 1992). For example, for Al_2O_3, equilibria are thought to exist between several polymeric species, particularly near the solubility minimum (Baes and Mesmer 1976):

$$Al^{+3} \approx \left[Al(OH)\right]^{+2} \approx \left[Al_2(OH)_2\right]^{+4} \approx \left[Al_3(OH)_4\right]^{+5} \approx$$

$$\left[Al_8(OH)_{20}\right]^{+4} \approx \left[Al_{13}O_4(OH)_{24}\right]^{+7} \tag{19.31}$$

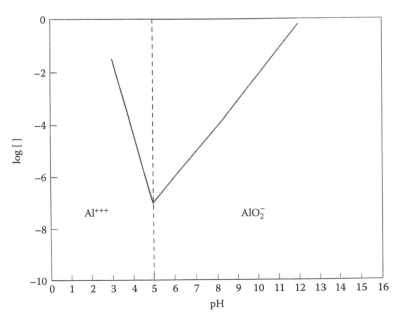

FIGURE 19.19 Amphoteric solubility behavior of Al_2O_3 as a function of pH. (From Pourbaix, M., *Atlas of Electrochemical Equilibria in Aqueous Solutions*, p. 174, Pergamon Press, New York, 1966; Readey, D.W., Aqueous corrosion of ceramics, in D.K. Peeler and J.C. Marra, Eds., *Environmental Issues and Waste Management Technologies in the Ceramic and Nuclear Industries, III, Ceramic Transactions*, Vol. 87, The American Ceramic Society, Westerville, OH, pp. 43–62, 1998. With permission of the American Ceramic Society.)

How pervasive and important such complex solution species are in understanding the basic thermodynamics and kinetics of ceramic degradation in aqueous environments is not well understood. Furthermore, thermodynamic data generally do not exist for all of the possible aqueous species so the solubilities that *can* be calculated depend on the available thermodynamic data.

19.5.2.3 Electrochemical Dissolution/Degradation

Corrosion of metals is an electrochemical phenomenon controlled by electron exchange, and the stable species in a given aqueous environment depends not only on the pH and the presence of certain ions but also on the presence of oxidation or reduction potentials supplied by either an applied voltage or reduction–oxidation—redox—couples either in solution or by the proximity of a dissimilar metal. This has led to extensive compilation of the so-called Eh–pH stability diagrams that plot the regions of stability of the various solid and solution species as a function of the solution pH and the applied voltage (Pourbaix 1966). Eh–pH diagrams can also be calculated with thermodynamic programs as shown in for magnesium in Figure 19.20. The only difference between the calculated diagram of Figure 19.20 and that in the literature (Pourbaix 1966, 141) is that the calculated diagram shows the presence of MgH_2 under strongly reducing conditions, while this species was not considered by the literature diagram that has this region assigned to magnesium metal. The boundary between Mg^{+2} and $Mg(OH)_2$ is where $[Mg^{+2}]$ (note: "[]" represents concentration, usually molar) becomes 1 molar. The dashed lines are the limits for the reduction and oxidation of water with the formation of hydrogen and oxygen, respectively. In general, for good ceramic insulators, such as MgO or Al_2O_3, the application of any applied voltage will typically be across the insulating ceramic since an aqueous solution will be a far better conductor than the ceramic. As a result, for insulating ceramics, the direct effects on dissolution by changing the stability of a given compound through applied voltages are not effective. However, if the ceramic is a

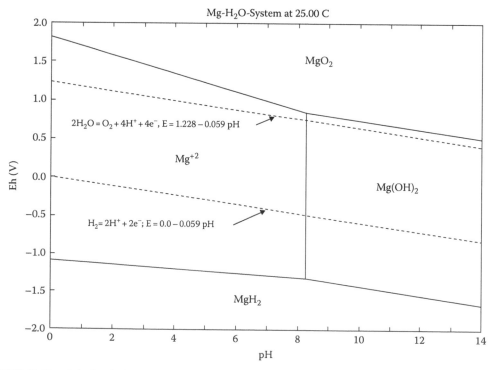

FIGURE 19.20 Calculated Eh–pH diagram for magnesium and H_2O showing the regions of stability of the various species.

semiconductor such as SiC or a metallic conductor like TiB_2, then the situation is quite different and the Eh can affect the kinetics of dissolution.

19.5.3 Aqueous Corrosion Kinetics

19.5.3.1 General

As discussed earlier for corrosion by gases, corrosion in liquids also requires three reaction steps in series: (1) diffusion of the reactant to the solid surface, (2) reaction at the surface, and (3) diffusion of the products away from the surface. In this case, because of the low temperatures involved, surface reactions are typically rate limiting. And here again, there are many possible steps in the surface reaction, any one of which could be rate controlling.

In aqueous solutions, the diffusion coefficient is approximated by the Stokes–Einstein equation (Robinson and Stokes 2002):

$$D = \frac{kT}{6\pi r\mu} \tag{19.32}$$

where
   r is the radius of the diffusing atom, ion, or molecule
   μ is the liquid viscosity (Pa-s)

Figure 19.21 is a plot of the diffusion coefficients for pure water based on tracer data—$H_2{}^{18}O$ and ^2H_2O—given in the literature (Robinson and Stokes 2002, 75). The activation energy, Q, in $D = D_0 e^{-(Q/RT)}$ is quite low and $Q \cong 19$ kJ/mol. There is no obvious reason for the difference between the two diffusion coefficients for the two different tracers. Also, the calculated size of the radius of the water molecule is smaller than expected from these data. In any event, for diffusion in water, diffusion coefficients are on the order of $D \sim 10^{-5}$–10^{-4} cm²/s and $Q \sim 19$ kJ/mol. Similarly, Figure 19.22 is a logarithmic plot of T/μ versus $1/T$, which is proportional to the aqueous diffusion coefficient (Equation 19.32). The viscosity as a function of temperature data was taken from the literature (Weast 1979, F-51). This plot gives an apparent activation energy—the solid regression line—for diffusion of $Q = 17.9$ kJ/mol, which is consistent with the tracer diffusion coefficients in Figure 19.21.

Typically, the activation energy for a surface reaction, Q_R, in the expression $k = k_0 e^{-(Q_R/RT)}$ for the reaction rate constant will be near 100 kJ/mol. Therefore, reproducing the combined effects of diffusion and surface reaction as done in Equations 19.16 and 19.17 over the potential range of 0°C–100°C where aqueous corrosion can take place results in the overall reaction plot of Figure 19.23. In this graph, the diffusion data are taken from Figure 19.21, and the reaction rate data were chosen with an activation energy near 100 kJ/mol and a pre-exponential, k_0, so that the two sets of data would cross somewhere between 0°C and 100°C. What is noteworthy is that over such a small temperature region, it can be quite difficult to experimentally adjust the conditions so that either diffusion or a surface reaction is clearly the dominant mechanism by simply measuring an activation energy. As a result, in analyzing aqueous corrosion data, different sets of data can lead to different conclusions about the rate-controlling mechanism.

19.5.3.2 Diffusion Control

Figure 19.24 shows the steps involved in dissolving a ceramic solid. Solution ions—ligands—must diffuse to the solid surface where they react with surface anion and cations, which then enter the solution surrounded by the solution ligands. The reaction products must then diffuse away from the interface into the solution. Figure 19.25 depicts the concentration profile away from the surface of a solid sphere—a sphere is chosen since many dissolution studies and the processing of ceramics involve small

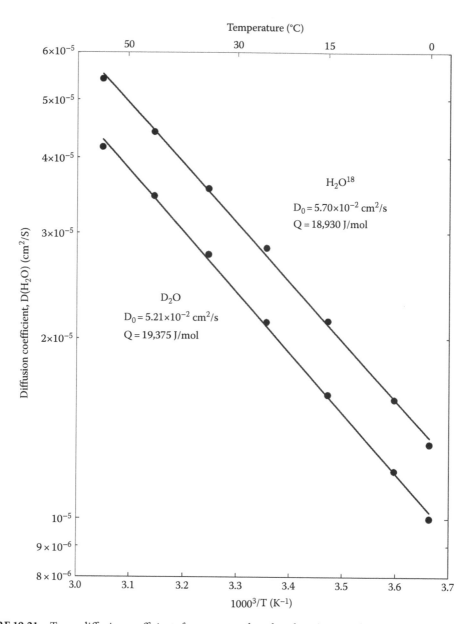

FIGURE 19.21 Tracer diffusion coefficients for pure water based on data. (From Robinson, R.A. and Stokes, R.H., *Electrolyte Solutions*, 2nd edn., Dover, Mineola, NY, 2002.)

"spherical" particles—assuming that solution diffusion of the products is rate controlling. For spherical particles, the steady-state diffusion flux, J, is given by

$$J = \frac{D}{a}(C_e - C_0) \tag{19.33}$$

where
   "a" is the particle radius
   C_e is the equilibrium concentration of the diffusing ion at the solid surface
   C_0 is the concentration in the solution far from the surface, that is, when $x \to \infty$

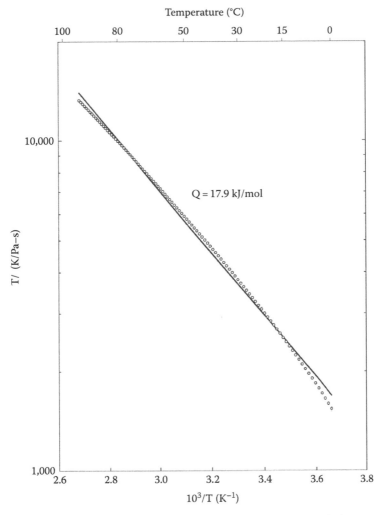

FIGURE 19.22 Temperature divided by the viscosity of water, μ, versus 1/T—which is proportional to the diffusion coefficients for ions in aqueous solutions—giving an apparent activation energy of 17.9 kJ/mol. (With data from Weast, R.C., Ed., *CRC Handbook of Chemistry and Physics*, 60th edn., CRC Press, Boca Raton, FL, 1979.)

This can be integrated to give the volume fraction, "f," of the sphere dissolved or reacted with time:

$$1-(1-f)^{2/3} = K_D t \quad \text{with} \quad K_D = 2\bar{V}D\frac{(C_e - C_0)}{a_0^2} \tag{19.34}$$

where
    t is the time
    $\bar{V}$ is the molar volume
    a_0 is the sphere radius at t = 0
    D is the diffusion coefficient in the fluid
    C_e is the equilibrium concentration in the fluid
    C_0 is the initial solution concentration

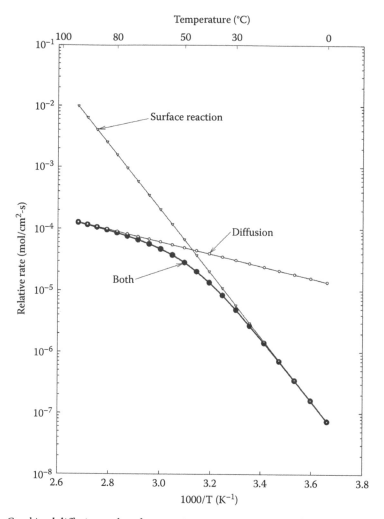

FIGURE 19.23 Combined diffusion and surface reaction rates in an aqueous solution.

If diffusion were rate controlling, the time to dissolve 1 μm Al_2O_3 ($\bar{V}$ = 25.72 cc/mol, D ~ 10^{-4} cm^2/s) particles in pure water of pH = 4 ([Al^{+3}] ~ 10^{-6} mol/L) would be about 30 min. If diffusion were rate controlling, then slip casting or any aqueous processing of alumina would be virtually impossible, which is certainly not the case. The actual dissolution rate of alumina, even in strong acids, is almost immeasurably slow—many orders of magnitude slower than predicted by diffusion—suggesting that some type of surface reaction is rate controlling. This is typically the case for many engineering ceramics.

19.5.3.3 Ambipolar Diffusion

Since the ions diffusing in the solution are charged, charge neutrality must be maintained. If the diffusion coefficients of the diffusing ions differ in magnitude, then an electric potential, Φ, will be generated that retards the motion of one of the ions and enhances the motion of the other so that they will diffuse together to maintain charge neutrality. Since the surface potential on ceramic particles plays an important role in both colloidal processing (Funk and Dinger 1994, Reed 1995, Ring 1996, Lewis 2000) and the dissolution kinetics of ceramics (Diggle 1973), these coupled diffusion coefficients and their attendant potential may be important in these processes. Such diffusion is frequently termed "ambipolar diffusion" (Chiang et al. 1997, 233). This phenomenon is far more important in ionic solids where the

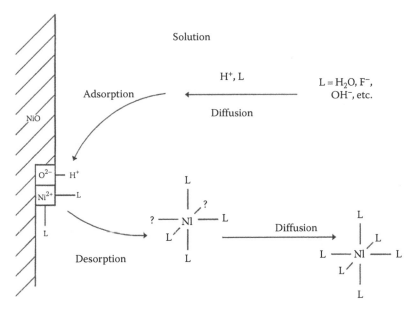

FIGURE 19.24 Schematic showing the three series steps involved in the dissolution of NiO with diffusion of solution ligands to the solid surface, a surface reaction of the solid ions exchanging solid ligands for solution ligands, and diffusion of the aqueous ion species away from the surface. (From Readey, D.W., Aqueous corrosion of ceramics, in D.K. Peeler and J.C. Marra, Eds., *Environmental Issues and Waste Management Technologies in the Ceramic and Nuclear Industries, III, Ceramic Transactions*, Vol. 87, The American Ceramic Society, Westerville, OH, p. 46, 1998a. With permission of the American Ceramic Society.)

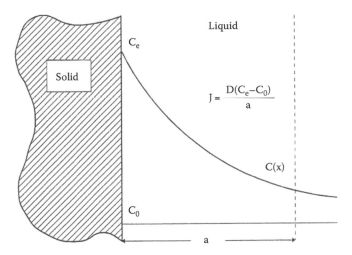

FIGURE 19.25 Schematic diffusion profile and flux away from a spherical particle of radius "a."

anion and cation diffusion coefficients may vary by many orders of magnitude. This is not the case for aqueous diffusion, and the effect may be small but, nevertheless, real and worth estimating since the effect will certainly be present if dissolution is controlled by diffusion.

From a practical standpoint, ambipolar diffusion will probably make little quantitative difference in dissolution rates since the diffusion coefficients of ions in an aqueous solution are all very similar—within an order of magnitude or less because their effective radii in solution are about the same (see Equation 19.32). For example, from the Stokes–Einstein relation (Equation 19.32), with the viscosity of water $\mu \approx 10^{-3}\,\text{Pa-s}$

at 298°C and the diffusion coefficient for Na^+ given as $D = 1.33 \times 10^{-5}$ cm²/s (Lasaga 1998), the radius of the sodium ion is calculated to be $r(Na^+) = 1.64 \times 10^{-8}$ cm compared to a literature value of $r(Na^+) = 0.98 \times 10^{-8}$ cm (Emsley 1998). A similar calculation for Cl^- with a literature value of $D = 2.03 \times 10^{-5}$ cm²/s (Lasaga 1998) gives $r(Cl^-) = 1.07 \times 10^{-8}$ cm compared to a literature value of $r(Cl^-) = 1.81 \times 10^{-8}$ cm (Emsley 1998). These differences may be due to the drag effects exerted by the solvating water molecules and/or the Debye–Huckel charge cloud of oppositely charged ions.

In ambipolar diffusion, an electrochemical potential, η_i, can be defined for a given diffusing species "i" that is a combination of the chemical potential, $\bar{G}_i$, and the electric potential, ϕ, that is built up due to the different rates of diffusion for the diffusing ions:

$$\eta_i = \bar{G}_i + z_i F\phi \qquad (19.35)$$

where
  z_i is the charge on the ion
  F is Faraday ($\approx 96{,}500$ coulombs)

Therefore, the flux of a given ionic species under the influence of this electrochemical potential is given by

$$J_i = -\frac{C_i D_i}{RT} \nabla \eta_i \qquad (19.36)$$

where
  C_i is the concentration of species "i"
  D_i is the diffusion coefficient of species "i"

This should be expressed as a vector equation, but it certainly is a safe assumption that diffusion will be isotropic in an aqueous solution.

For example, for the dissolution of NaCl, the appropriate one-dimensional equations for ambipolar diffusion become

$$J_{Na^+} = -\frac{C_{Na^+} D_{Na^+}}{RT} \left\{ \frac{d\bar{G}_{Na^+}}{dx} + F\frac{d\phi}{dx} \right\}$$

$$J_{Cl^-} = -\frac{C_{Cl^-} D_{Cl^-}}{RT} \left\{ \frac{d\bar{G}_{Cl^-}}{dx} - F\frac{d\phi}{dx} \right\} \qquad (19.37)$$

with the conditions that $J_{Na^+} - J_{Cl^-} = 0$ and $C_{Na^+} = C_{Cl^-}$ for electric neutrality. Solving these equations gives

$$F\frac{d\phi}{dx} = -\frac{RT}{C_{Na^+}} \frac{\left(D_{Na^+} - D_{Cl^-}\right)}{\left(D_{Na^+} + D_{Cl^-}\right)} \frac{dC_{Na^+}}{dx} \qquad (19.38)$$

which shows that a potential exists if the diffusion coefficients are not equal. Substitution of Equation 19.38 into 19.37 gives the desired result:

$$J_{Na^+} = -D_{NaCl} \frac{dC_{Na^+}}{dx}$$

where

$$D_{NaCl} = 2\frac{D_{Na^+}D_{Cl^-}}{D_{Na^+} + D_{Cl^-}} \tag{19.39}$$

Equation 19.39 shows that if the sodium and chlorine diffusion coefficients were the same, then the diffusion coefficient for NaCl (D_{NaCl}) is equal to the diffusion coefficient of both sodium and chlorine. If $D_{Cl^-} \gg D_{Na^+}$, then $D_{NaCl} \approx 2D_{Na^+}$, which says that the slower diffusing species controls the rate of diffusion, but it is faster because the faster moving chlorine ions are exerting a field pulling the sodium ions with them. Substituting the diffusion coefficients for Na^+ and Cl^- earlier, the calculated $D_{NaCl} = 1.68 \times 10^{-5}$ cm²/s is compared to the literature value of 1.61×10^{-5} cm²/s at infinite dilution (Robinson and Stokes 2002), which is probably closer than should be expected.

The potential difference between some point in the solution and the surface of a dissolving sphere of NaCl of radius "a" into pure water where $C_0 = 0$ can be calculated from Equation 19.38. For spherical coordinates, the steady-state solution for Fick's second law

$$\frac{\partial C}{\partial t} = \frac{D}{r^2}\frac{\partial}{\partial r}\left(r^2\frac{\partial C}{\partial r}\right) = 0$$

gives

$$C(r) = C_e\frac{a}{r} \tag{19.40}$$

For the values of the diffusion coefficients for sodium and chlorine given earlier, the potential difference between the surface of the sphere and five sphere radii into the solution is calculated to be 8.6 mV. This is a small potential difference compared to the potentials generated by adsorbed ions used to produce stable ceramic particle colloids for powder processing, which are on the order of several tens of millivolts and extend over much smaller distances, nanometers, compared to the size of the particles (Ring 1996).

Take MgO as an example, Figure 19.26 shows MgO dissolving in pure water with an initial pH = 7 or $[H^+]_0 = [OH^-]_0 = 10^{-7}$ mol/L at 25°C. In this case, there can be three diffusing species, Mg^{+2}, H^+, and OH^-, and for electric neutrality, $2J_{Mg^{2+}} + J_{H^+} - J_{OH^-} = 0$. However, the hydrogen ion and the hydroxyl ion fluxes are not independent since $[OH^-][H^+] = 10^{-14} = K_W$ at 298°C. This makes things more complicated, so for the sake of simplicity, ignore the hydrogen flux and consider only the diffusion of the magnesium and hydroxyl ions, then, with the same procedure as for NaCl,

$$J_{Mg^{+2}} = -D_{Mg(OH)_2}\frac{dC_{Mg^{2+}}}{dx}$$

where

$$D_{Mg(OH)_2} = \frac{3D_{Mg^{2+}}D_{OH^-}}{2D_{Mg^{2+}} + D_{OH^-}} \tag{19.41}$$

Here again, it is obvious that if the anion and cation diffusion coefficients are the same, then the overall diffusion coefficient is equal to either of them. Also, if one of the diffusion coefficients is much larger than the other, then the slower is rate controlling but the diffusion coefficient will be higher because of

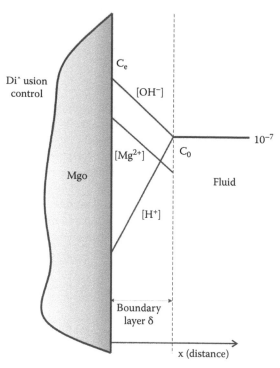

FIGURE 19.26 Dissolution of MgO into pure water showing the concentrations of the diffusing species, Mg^{+2}, OH^-, and H^+.

the field produced by the faster moving ion. A similar calculation could be done for the diffusion of the Mg^{2+} and H^+, and which of the two coupled diffusion coefficients were the larger, then it would be rate controlling since the fluxes of H^+ and OH^- are in parallel.

These results for ambipolar diffusion assume that the solutions are ideal, which is not the case for concentrated solutions. However, the theoretical effects of concentrated, nonideal solutions of electrolytes on ion migration are well established (Robinson and Stokes 2002) and could be used to calculate diffusion coefficients. Also, the effect on the concentration profile and the resultant corrosion rate caused by the motion of the solid interface during diffusion is not considered. This is a moving boundary problem and is more difficult to solve (Readey and Cooper 1966). The above calculations assume that diffusion in the liquid phase is sufficiently rapid so that a steady-state concentration profile essentially changes fast enough to keep pace with the motion of the dissolving interface. Both nonideality and moving boundary corrections are not large, and more involved diffusion calculations are not justified, particularly since the corrosion and degradation of most engineering ceramics are not controlled by diffusion but by surface reactions at low temperatures.

19.5.3.4 Surface Reaction Control

In most cases, a surface reaction controls dissolution with the rate determined by removal of ions from surface sites (Valverde and Wagner 1976). Figure 19.7 showed a typical crystalline surface with the kink sites being the most likely reaction site, and Figure 19.27 schematically depicts the removal of surface cations from surface kinks. As discussed earlier, if a surface reaction is rate controlling, then the activation energy for the reaction, Q_R, will usually be significantly higher than that for diffusion in the liquid solution as shown in Figure 19.23. So the various atomistic interpretations that lead to the activation energy, Q_R, for a surface reaction in $k = k_0 e^{-(Q_R/RT)}$ need to be explored.

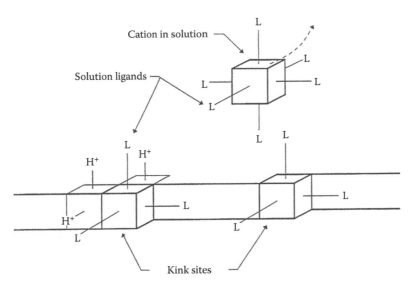

FIGURE 19.27 Schematic representation of the exchange of solution ligands for crystal ligands at surface kinks. (From Readey, D.W., Aqueous corrosion of ceramics, in D.K. Peeler and J.C. Marra, Eds., *Environmental Issues and Waste Management Technologies in the Ceramic and Nuclear Industries, III*, *Ceramic Transactions*, Vol. 87, The American Ceramic Society, Westerville, OH, p. 48. With permission of the American Ceramic Society.)

19.5.3.4.1 *Ligand Exchange*

The apparent activation energy for chemical dissolution is frequently the energy involved in forming an activated complex consisting of an ion changing its configuration between solution and crystal ligands at reactive sites (Figure 19.27). Homogeneous ligand exchange reactions in aqueous solutions have been extensively studied (Burgess 1988, 111–169), and two primary mechanisms are identified: the associative and dissociative mechanisms. In the dissociative ligand exchange reaction, an ion simply dissociates itself—breaks the bond—with one of its ligands, reducing its coordination number by one, before it recompletes its original coordination with the new ligand in the solution; for example, for the aqueous solution exchange of a water molecule for a chlorine ion by a nickel ion,

$$Ni(H_2O)_6{}^{+2} = Ni(H_2O)_5{}^{+2} + H_2O \qquad (19.42)$$

$$Ni(H_2O)_5{}^{+2} + Cl^- = Ni\left[Cl(H_2O)_5\right]^{+1}$$

The activation energy for this process is the energy necessary to remove one of the original water molecules from the coordination sphere of the Ni^{+2} ion. In the associative ligand exchange process, an extra ligand enters the coordination sphere of the cation increasing the coordination number by one to an unstable configuration. One of the ligands then leaves the coordination sphere; again, using the replacement of a water molecule with a Cl^- ion with the sixfold coordinated Ni^{+2} ion,

$$Ni(H_2O)_6{}^{+2} + Cl^- = Ni\left[Cl(H_2O)_6\right]^{+1}$$

$$Ni\left[Cl(H_2O)_6\right]^{+1} = Ni\left[Cl(H_2O)_5\right]^{+1} + H_2O \qquad (19.43)$$

In this case, the activation energy is the energy necessary to "squeeze" the extra ligand into the octahedral coordination sphere. Between these two extremes are indirect processes in which exchange takes

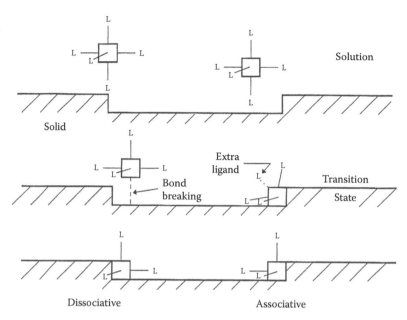

FIGURE 19.28   Schematic diagram showing the dissociative and associative ligand exchange processes involving solution ligands, L, and ions/atoms at kink sites on the solid surface. (From Readey, D.W., Aqueous corrosion of ceramics, in D.K. Peeler and J.C. Marra, Eds., *Environmental Issues and Waste Management Technologies in the Ceramic and Nuclear Industries, III, Ceramic Transactions,* Vol. 87, The American Ceramic Society, Westerville, OH, p. 49, 1998a. With permission of the American Ceramic Society.)

place between this inner coordination sphere and further into the solution and the distinction between the two processes becomes less clear. Nevertheless, similar principles and mechanistic models can be applied to the ligand exchange process at the solid–solution interface during the corrosion of ceramics. Figure 19.28 depicts these two ligand exchange processes taking place at kink sites on the surface of a solid. Unfortunately, these principles have only been applied sparingly to the interpretation of much of the data on rates of ceramic corrosion in aqueous solutions. It should be pointed out that ions that have rapid ligand exchange rates are considered "labile," while those that have slow exchange rates are "inert." This is to distinguish the kinetic behavior from the thermodynamics of a reaction in which a given species is either "stable" or "unstable" depending on whether the reaction is thermodynamically favorable or not (Huheey 1978, 489–510).

19.5.3.4.2 Ligand Exchange Rates

It is known from homogeneous solution chemistry that the rate of ligand exchange, in general, decreases as the cation charge increases and the cation radius decreases (Huheey 1978, 489–510). Ligand exchange kinetics would predict that the rates of dissolution of the following group of oxides would decrease in the sequence $Li_2O > MgO > Al_2O_3 > TiO_2 > Nb_2O_5$, for example. Thus, the presence of small, highly charged cations adsorbed onto the surface from impurities in the solution or present in solid solution in the solid could significantly alter the dissolution rate of dissolution of a given oxide. Deviations from stoichiometry or dopant additions that produce highly charged cations would also be expected to decrease the rate of dissolution. An example is the presence of Ni^{+3} in NiO produced by deviations from stoichiometry or by the addition of an acceptor dopant such as Li_2O (with Kröger–Vink notation [Kröger 1964, 194] for the defects):

$$2Ni_{Ni} + \tfrac{1}{2}O_2(g) \xrightleftharpoons{\text{NiO}} 2Ni^{\bullet}_{Ni} + V''_{Ni} + O_O \tag{19.44}$$

$$2Ni_{Ni} + \tfrac{1}{2}O_2(g) + Li_2O(s) \xrightleftharpoons{\;NiO\;} 2Ni^{\bullet}_{Ni} + 2Li'_{Ni} + 2O_O$$

where

Ni$_{Ni}$ represents a Ni^{+2} ion on a nickel site (and it is uncharged relative to the perfect crystal)

Ni$^{\bullet}_{Ni}$ is a Ni^{+3} ion on a nickel site (with the superscript dot indicating a single positive charge relative to the uncharged, perfect crystal)

V$''_{Ni}$ is a vacant nickel site (with the two apostrophes indicating a negative two charge relative to the perfect, uncharged crystal)

Li$'_{Ni}$ is a lithium ion Li^{+} on a nickel site (with the single apostrophe indicating a single negative charge relative to the pure, defect-free, uncharged crystal)

In both cases, the rate of dissolution should decrease because of the increase in the concentration of Ni^{+3} (Ni$^{\bullet}_{Ni}$ in Kröger–Vink notation) assuming that the ligand exchange rate of Ni^{+3} is less than that of Ni^{+2}. However, the ligand exchange rates of particularly Ni^{+3} are not well known (Blesa et al. 1994). Similarly, doping with a higher valence ion, such as aluminum, would be expected to reduce the rate of dissolution because of its slower rate of ligand exchange than Ni^{+2}. Therefore, for a ceramic such as NiO, which can become metal deficient, deviations from stoichiometry and aliovalent—different charge than the host—doping should always decrease the rate of dissolution.

For oxygen-deficient compounds, such as Fe$_2$O$_3$ or TiO$_2$, both deviations from stoichiometry, Fe$_2$O$_{3-x}$ and TiO$_{2-x}$ (some Fe^{+3} reduced to Fe^{+2} and Ti^{+4} to Ti^{+3}), and doping with lower valent cations than those of the host should have little effect on dissolution rates. Doping with a higher valent cation should decrease the rate.

19.5.3.4.3 Electronic Band Bending at the Surface

Other factors that change the *surface* concentration of highly charged cations should also increase or decrease the rate of dissolution. Many ceramics have relatively narrow band gaps—3 electron volts or less. Band bending at the surface of the solid induced by an applied voltage or the presence of a redox couple in solution can change the surface concentration of highly charged cations. For example, Figure 19.29 shows the band bending that can occur in either p-type NiO or n-type Fe$_2$O$_3$ in a solution

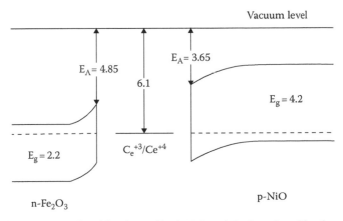

FIGURE 19.29 Surface electronic band bending of both NiO and Fe$_2$O$_3$ induced by the presence of Ce^{+3}/Ce^{+4} redox couple in solution. (From Readey, D.W., Aqueous corrosion of ceramics, in D.K. Peeler and J.C. Marra, Eds., *Environmental Issues and Waste Management Technologies in the Ceramic and Nuclear Industries, III, Ceramic Transactions,* Vol. 87, The American Ceramic Society, Westerville, OH, p. 52, 1998a. With permission of the American Ceramic Society.)

with a cerium redox couple at its standard potential. In such a solution, the concentration of the Ni^{+3} ions at the solid surface of p-type NiO would be reduced. This should increase the dissolution rate over that in a solution without the redox couple present. For Fe_2O_3, the opposite would occur, and the surface concentration of any Fe^{+2} ions caused by any nonstoichiometry would be reduced as would the rate of dissolution. Thus, the presence of redox couples in solution or the application of a potential can alter band banding and change the dissolution rate without any electron transfer between the solid and solution (Morrison 1980). The exact behavior of a given semiconducting ceramic in a specific solution will depend on the relative energies of the band edges, Fermi level, redox potential, applied voltage, etc. (Butler and Ginley 1980).

19.5.3.4.4 Surface Charges and Double Layers

The presence of the electric double layer at the surface of a ceramic will also affect the rate of chemical dissolution (Vermilyea 1966). Figure 19.30 schematically depicts this double layer at the surface, the presence of which is critical in the colloidal processing of ceramics (Funk and Dinger 1994, Reed 1995, Ring 1996, Lewis 2000). The conventional picture of this charge double layer is that there is some adsorbed charge on the solid surface (the positive charges in this figure) and then there are solvated "bound" counter ions of opposite (negative) charge that essentially move if the particle moves in the fluid leaving the *zeta potential*, ζ, at the surface of the shear plane followed by a diffuse double layer in the surrounding solution of counter ions of opposite charge (Adamson 1982, 626). Ignoring the

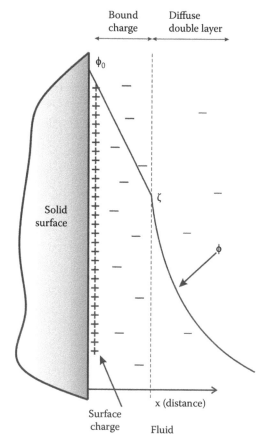

FIGURE 19.30 Schematic of a charge double layer present on the surface of a ceramic in an aqueous solution.

bound counter ions for the moment and assuming that all of the counter ions are in a diffuse double layer, then (Adamson 1982, 626)

$$\phi = \phi_0 e^{-\kappa x} \tag{19.45}$$

where ϕ_0 is the potential at the particle surface determined by the amount of adsorbed charge and κ^{-1} = the Debye length (m) given by (Hiemenz 1986, 691)

$$\kappa^{-1} = \sqrt{\frac{\varepsilon_0 \varepsilon_r RT}{1000 F^2 \sum_i z_i^2 C_i}} \tag{19.46}$$

where

ε_0 is the permittivity of free space (8.85×10^{-12} Farad/m)
ε_r is the relative dielectric constant of the liquid medium (for water $\varepsilon_r = 78$)
F is the Faraday ($\approx$96,500 coulombs)
z_i is the charge on ion species "i" in solution
C_i is the molar concentration of species "i" (mol/L)

For example, for a 0.01 molar concentration of NaCl, where $z^2(Na^+) = 1^2 = 1$ and $z^2(Cl^-) = (-1)^2 = 1$,

$$\kappa^{-1} = \sqrt{\frac{(8.85 \times 10^{-12})(78)(8.314)(298)}{1,000(96,500)^2 (1+1)(0.01)}}$$

$$\kappa^{-1} = 3.03 \times 10^{-9} \text{ m} \approx 3 \text{ nm}$$

which basically says that such double layers do not extend far from the solid surface into the aqueous solution.

The activation energy for dissolution, ΔG^*, will depend on the potential drop $\Delta\phi$ across the double layer. The double layer that forms on the surface of ceramics is thought to be due to adsorption of negative ions (F^-, Cl^-, OH^-, etc.) on cations and H^+ ions on the oxygen ions (Osseo-Asare 1984, 27–268), which behave as Lewis acid and base sites, respectively (Adamson 1982, 626). Specifically,

$$\Delta G^* = \Delta G_o^* + zbF\Delta\phi \tag{19.47}$$

where

z is the rate-controlling ion valence
F is Faraday
b is the symmetry coefficient
ΔG_o^* is the activation free energy at point of zero charge (charge at the surface is zero)

This predicts that the rate of dissolution will be affected by the amount of adsorbed surface charge. Typically, double-layer potentials are on the order of zero to a few 100 mV. A 100 mV potential drop corresponds to about a 10 kJ change in activation energy. This is to be compared with observed activation energies that are in the neighborhood of 100 kJ. This also corresponds to an adsorbed net charge of about 10^{12} cm^{-2} (Adamson 1982, 392), which can be compared to the total number of ions per unit surface of about 10^{15} cm^{-2}.

19.5.3.4.5 Electrochemical Dissolution

For a ceramic that is a good electronic conductor and that can undergo electrochemical dissolution by electron transfer, then an applied potential or the presence of a redox couple in solution can play a much more direct role. Application of a voltage or control of the solution Eh with a redox couple can change the stability of a compound (Diggle 1973, Blesa et al. 1994). This is the thermodynamic basis of leaching in many hydrometallurgical systems (Warren and Devuyst 1973, 229–263). For example, Fe_2O_3 can be reduced by Ti^{+3}/TiO^{+2} or V^{+3}/V^{+4} couples in acidic solutions (Warren and Devuyst 1973, 229–263, Bard et al. 1985):

$$Fe_2O_3(s) + 2Ti^{+3} + H^+ = 2Fe^{+2} + 2TiO^{+2} + OH^-; \quad \Delta G^0_{298} = -51\,kJ/mol \tag{19.48}$$

Presumably, Fe^{+3} is reduced to Fe^{+2} at the solid surface, and then dissolution of the lower charged Fe^{+2} cation proceeds rapidly (Frenier and Growcock 1984). For electrochemical dissolution, the charge transfer reaction, regardless of the detailed mechanism, is usually thought to be rate controlling. One postulated rate-controlling step in electrochemical dissolution of semiconducting compounds is minority charge carriers reaching the surface by emission over, or tunneling through, the surface Schottky barrier (Morrison 1977, 42). This is shown schematically in Figure 19.31.

In homogeneous solution chemistry, the charge transfer between ions surrounded by ligands depends greatly on the character of the ligands. Here again, two distinct mechanisms have been identified (Morrison 1977, 42). The first is the "inner sphere" mechanism in which electron charge transfer takes place during or after ligand exchange. The second mechanism is the "outer sphere" mechanism in which electron transfer takes place between the ligand shells surrounding the two cations without any ligand exchange. The latter process is facilitated if the ligands surrounding both ions have delocalized electrons.

Choosing a redox couple for leaching of ores in hydrometallurgy requires application of similar principles. To obtain the most rapid leaching by electrochemical dissolution, a redox couple not only must provide the proper Eh for the reaction but also should exchange electrons rapidly with the solid surface to produce a high current so that dissolution proceeds rapidly (Wadsworth and Miller 1979, 133–241).

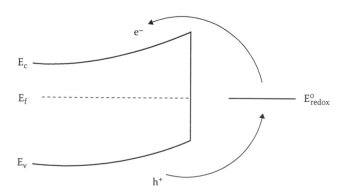

FIGURE 19.31 Band bending and electron and hole change between a ceramic and a solution redox couple. (From Readey, D.W., Aqueous corrosion of ceramics, in D.K. Peeler and J.C. Marra, Eds., *Environmental Issues and Waste Management Technologies in the Ceramic and Nuclear Industries, III, Ceramic Transactions*, Vol. 87, The American Ceramic Society, Westerville, OH, p. 52. With permission of the American Ceramic Society.)

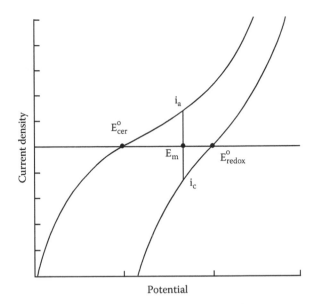

FIGURE 19.32 Current density curves for electronic exchange for both a ceramic undergoing electrochemical dissolution and a solution redox couple. The corrosion potential, E_m, occurs when the electron exchange currents, i_a and i_c, are equal. (From Readey, D.W., Aqueous corrosion of ceramics, in D.K. Peeler and J.C. Marra, Eds., *Environmental Issues and Waste Management Technologies in the Ceramic and Nuclear Industries, III, Ceramic Transactions*, Vol. 87, The American Ceramic Society, Westerville, OH, p. 55, 1998. With permission of the American Ceramic Society.)

This is reflected in a low value of the overpotential in the Butler–Volmer equation for electron charge transfer (Bockris and Reddy 1970):

$$i = i_0 \left\{ \exp\left[(1-b)\frac{F\eta}{RT} \right] - \exp\left[-b\frac{F\eta}{RT} \right] \right\} \qquad (19.49)$$

where
 i is the current density
 i_0 is the exchange current density
 η is the overpotential = (applied potential minus the equilibrium potential for the reaction) (volts)

This is illustrated in Figure 19.32 for a ceramic undergoing electrochemical dissolution in the presence of a redox couple. The same principles are applied to prevent corrosion of metals in which electrochemical dissolution essentially always occurs (West 1986, 183–191). In this case, however, it is desirable to have a low electron exchange rate or a high overpotential. Unfortunately, in many of the studies on electrochemical dissolution of ceramics, the differences between the *thermodynamics* of applied voltages and redox potentials have not been carefully separated from the *kinetics* of electron exchange and efforts to separate them have been difficult (Diggle 1973, 281–306, Blesa et al. 1994).

19.6 Degradation of Specific Systems

19.6.1 General

What follows is a discussion about individual ceramic materials and what is known—or assumed—about their degradation under ambient conditions: either in gases or aqueous solution at temperatures near room temperature. Here again, this will not be a critical review of all of the literature available as this has been done for many cases already as pointed out earlier (Diggle 1973, 281–306, Blachere and Pettit 1989, Fordham 1990, Tressler and McNallan 1990, Clark and Zoitos 1992, Gogotsi and Lavrenko 1992, Blesa et al. 1994, Nickel 1994, McCauley 2004, Salem and Fuller 2008, Morrell 2010). Rather, the focus here will be to present some of the most definitive results for selected important engineering ceramics and differences in either experimental results, or their interpretation, will be emphasized with the intent of demonstrating how the expected theoretical behavior is or is not actually observed for different materials. It should also be noted that relatively few of the studies on ceramic degradation have been carried out with the specific intent of elucidating the details of the various corrosion mechanisms discussed earlier. Furthermore, many of the studies that have been undertaken either have not carefully taken into consideration all of the factors involved in a reaction—particle size and degree of agglomeration, for example—or have made assumptions based on questionable data in the literature. As a result, generalizations about engineering ceramics in general, or even about a specific material, are frequently not possible. Examples where this is the case will be pointed out.

19.6.2 Sodium Chloride, NaCl

Sodium chloride and other alkali halides are often not considered as engineering ceramics. However, given the breadth of the definition of ceramics used here, they are indeed used as engineering materials: most commonly as optical elements in infrared and ultraviolet spectrometers. They quickly gained considerably more interest as an engineering ceramic when they became the materials of choice for the window for the high-energy, gas-dynamic CO_2 lasers being considered for the early embodiments of the Airborne Laser Laboratory (Figure 19.33). The CO_2 laser emits at 10.6 µm, and the most transparent materials at this wavelength are the alkali halides. As a result, a significant amount of effort was expended in attempting to develop alkali halides such as NaCl for the window—front dome in Figure 19.33—for the high-power CO_2 laser beam (Miles and Readey 1972, 507). Unfortunately, the corrosion of NaCl by water vapor in the atmosphere and drops in clouds eliminated it and other alkali halides from consideration,

FIGURE 19.33   Boeing NKC135 Airborne Laser Laboratory USAF. (U.S. Government photograph.)

and zinc selenide (ZnSe), a less than optimum optical material, but satisfactory, was chosen to be the window material, and it continues to be the window of choice for industrial CO_2 lasers.

The corrosion of sodium chloride appears to be diffusion controlled (Berner 1978). Experiments with natural convection on the dissolution kinetics of NaCl single crystals gave dissolution rates consistent with diffusion control (Wagner 1949). Other experiments with rotating disc samples of crystals dissolving in glycerin—with a dielectric constant about half that of water—also showed diffusion control (Cooper and Kingery 1962). The use of a rotating disc is common in corrosion experiments since it presents a uniform boundary layer thickness, δ (Figure 19.12), over the dissolving surface the thickness of which can be controlled by the rate of rotation of the sample: specifically,

$$\delta \simeq 4.0\left(\frac{\nu}{\omega}\right)^{1/2} \tag{19.50}$$

where
ω is the rotation rate (radians/s)
ν is the kinematic viscosity = μ/ρ = viscosity/density = Pa-s/kg/m³

Pure water at 25°C and a rotation rate of ω = 10 rad/s would give a boundary layer thickness of about 40 μm.

Figure 19.34 shows the concentration profile of a spherical NaCl particle dissolving by diffusion control into pure water. Taking $\bar{V}(NaCl)$ = 1/C(NaCl) = 27.03 cc/mol, D_{NaCl} = 1.68 × 10⁻⁵ cm²/s (Section 19.4.3.3), C_e = 6.1 × 10⁻³ mol/cc, from Equation 19.34, the time to dissolve roughly 1 mm particles, a_0 = 5 × 10⁻² cm, is about 12.3 s. Check how long it takes salt particles to dissolve next time you put salt in a beer to increase the head.

A numerical simulation of the dissolution of very small NaCl crystal (Ohtaki et al. 1988) showed that chlorine ions left from the corners—essentially kink sites—first and later-leaving chlorine ions left with

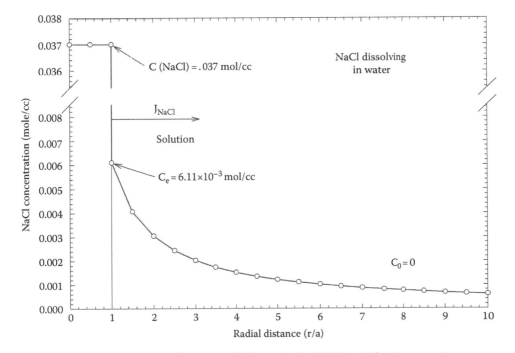

FIGURE 19.34 Concentration versus distance profile for a spherical NaCl particle into pure water.

smaller and smaller velocities since the crystal was developing a positive charge, consistent with the picture of dissolution of ionic compounds (Vermilyea 1966).

19.6.3 Magnesium Oxide, MgO

There is not an abundance of information on the corrosion of MgO in spite of the fact that it is an important refractory material and used in a wide variety of chemical applications where its interaction with the environment is critical (Shand 2006, 126). No information on the kinetics of formation of $MgCO_3$—or any other compound—by a gas–solid reaction appears to be readily available in the literature with the exception of some information of the hydration kinetics of MgO in water vapor with the specific intent of trying to determine the mechanism of hydration (Layden and Brindley 1963). The conclusion of this work was that the reaction was surface reaction control with the probable formation of a liquid layer on the particle surfaces. The measured activation energy for the reaction was 67.4 kJ/mol, which is much larger than the 18 kJ/mol for the activation energy for diffusion in water (Figures 19.21 and 19.22). This hydration from the vapor phase can pose problems for MgO-containing refractories such as magnesite-chrome refractories in that on reheating, the decomposition of any $Mg(OH)_2$ formed can lead to cracking and loss of strength of the refractory (Malarria and Tinivella 1997). How severe a problem this is no doubt depends strongly on the fired density and particle size of the MgO grain in the brick.

In contrast, there is a considerable body of literature on the dissolution of MgO in acidic solutions, most of which was developed in the 1980s and earlier (Vermilyea 1969, Macdonald and Owen 1971, Jones et al. 1984, Segall et al. 1978) and has been summarized (Blesa et al. 1994, Shand 2006) with the conclusion that at least in low pH acidic conditions, the dissolution appears to be diffusion controlled, while at higher pH, a surface reaction may control. One carefully done study that used the rotating disc technique described earlier for NaCl was able to separate the reaction and diffusion components and found an activation energy for surface reaction of 55 kJ/mol while that for diffusion was 15 kJ/mol (Macdonald and Owen 1971). The latter value is certainly consistent with diffusion in an aqueous solution. A more recent publication suggests that surface control dominates the dissolution behavior of MgO in acidic conditions (Suárez and Compton 1998). This uncertainty in the mechanism of dissolution of one of the simplest ceramic materials, depending on the details of the experimental conditions, emphasizes the inadequacy of our understanding the degradation behavior of engineering ceramics in general.

19.6.4 Silicon Dioxide, SiO_2

"There is a wealth of information on the dissolution behavior of silica and aluminosilicates, mainly in the geochemical literature. A comprehensive review of all the literature would be a formidable task" (Blesa et al. 1994). This is indeed the case since the behavior of silica-containing minerals in aqueous systems is of central interest to the geochemists concerned with their interaction with, and the alteration by, natural aqueous systems (Garrels and Christ 1990, Faure 1991, Pankow 1991, Stumm 1992, Stumm and Morgan 1996, Lasaga 1998). From an engineering material standpoint, there have been several studies on the corrosion of SiO_2 by HF-containing solutions since these solutions are used to etch silica in a number of industrial applications. The reaction of SiO_2 in HF is thought to be

$$SiO_2(s) + 6HF(aq) = 2H_2O(l) + 2H^+(aq) + SiF_6^{-2}(aq);$$

$$\Delta G^0(298\,K) = -36.5\,kJ/mol \tag{19.51}$$

Most of the literature on SiO_2 dissolution in HF has been reviewed (Liang and Readey 1987) and experimentally found that the rate of dissolution of SiO_2 was determined by a surface reaction with a relatively low activation energy of around 30 kJ/mol, not large but still significantly larger than would be expected by a diffusion-controlled dissolution. Furthermore, the dissolution rate was strongly dependent on crystallographic orientation varying by about three orders of magnitude from Z-cut quartz [(0001) face] to Y-cut quartz [($1\bar{1}00$) face], and the Z-cut face dissolves at about the same rate as amorphous silica (Liang and Readey 1980). This strong orientation effect on dissolution is not given a great deal of consideration in the literature (Diggle 1973, Blesa et al. 1994), but it suggests that the local structure surrounding a surface ligand-exchanging ion going into solution may play a very large—and largely unconsidered—role in the rate of dissolution.

19.6.5 Titanium Dioxide, TiO_2

Very few experiments on the dissolution of TiO_2 in aqueous solutions have been performed in spite of the fact that titanates have been suggested as possible nuclear waste host materials (Ringwood and Kesson 1979, 79) because TiO_2 is thought to thermodynamically stable relative to dissolution in aqueous systems (Pourbaix 1966, 213–222). Since TiO_2 should not readily dissolve, it was suggested that neither should the other major titanate compounds expected from the formation of "Synroc" nuclear waste compounds formed from the waste with TiO_2: zirconolite, $CaZrTi_2O_7$; hollandite, $BaAl_2Ti_6O_{17}$; and perovskite, $CaTiO_3$ (Ringwood and Kesson 1979). As will be seen later, how these other compounds will degrade in an aqueous system can be, and usually is, quite different than pure TiO_2; a phenomenon similar to that which occurs with the silicate minerals that is quite familiar to the geochemists. Furthermore, TiO_2 powders are produced in large quantities for white paint pigment where they are exposed to the ambient environment, and they are a critical component of dye-sensitized solar cells that offer the potential of low-cost production and reasonable efficiencies where they operate in aqueous media (Ginley et al. 2008). Therefore, the dissolution behavior of TiO_2 is of significant interest.

What the aqueous corrosion products in solution for the dissolution of TiO_2 in pure water is not known with any degree of certainly. Even if they were, because of the stability of TiO_2 in a pure aqueous system (Pourbaix 1966, 213–222), the dissolution would probably be much too slow to measure in reasonable laboratory times. As a result, the dissolution of TiO_2—and other ceramics—in aqueous HF solutions has been studied with the intent to clearly define an aqueous cation species (titanium in this case), increase the reaction rate, and try to determine the reaction mechanism that might be generalized to aqueous environments of different compositions. In addition, the stability of various oxides in HF solutions has merit from a strictly engineering material application standpoint where ceramics could be exposed to either gaseous or aqueous HF (Mikeska et al. 2000).

The dissolution kinetics of TiO_2 in HF–HCl solutions was studied since it was expected that the TiF_6^{2-} would be the solution species, which is reported to be quite stable (Cotton et al. 1999, 700). As a result, the thermodynamics should favor the reaction:

$$TiO_2(s) + 6HF(aq) = 4H_2O(l) + 2H^+ + TiF_6^{2-} \qquad (19.52)$$

From two databases (Bard et al. 1985, Roine 2002), $\Delta G^0(298) = 78.0$ kJ/mol for this reaction implying that TiF_6^{2-} is not a very stable ion for even concentrated HF solutions. The easiest way to explain this contradiction in the literature is that the problem may be simply the inadequacy in the free-energy values for the various constituents in Equation 19.52. The total free energies on either side of the equation are near 3×10^6 J/mol so two very large numbers are being subtracted and, given the uncertainty in the thermodynamic data, plus or minus a 100 kJ or more would not be unreasonable. Nevertheless, the reality is that TiO_2 does indeed dissolve in concentrated solutions of HF and H^+ near one molar (Bright and Readey 1987), and the experimental evidence is that TiO_2 is not stable in HF solutions.

The dissolution kinetics of TiO_2 in HF–HCl solutions suggest that the rate-controlling step is most likely an associative ligand rearrangement process between the fluorinated surface and adsorbed HF similar to that postulated for silica dissolution (Bright and Readey 1987). Figure 19.35 shows the dissolution of TiO_2 in HF–HCl solutions at 95°C. Further insight into the mechanism can be gained by comparing calculated rates with those obtained from experiment. The dissolution

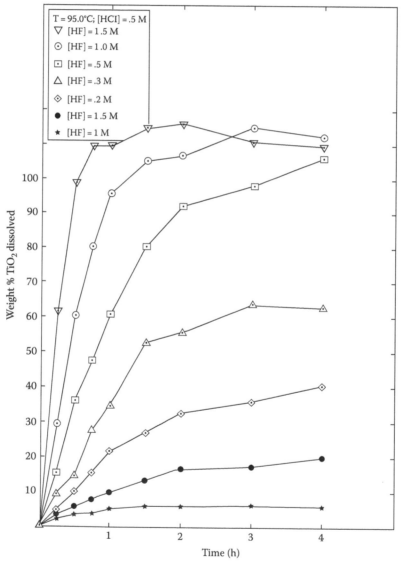

FIGURE 19.35 The effect of HF concentration on the dissolution kinetics of TiO_2 powder in 0.5 M HCl at 95°C. (From Bright, E. and Readey, D.W., *J. Am. Ceram. Soc.*, 70(12), 900, 1987. With permission of the American Ceramic Society.)

rate dependence on temperature and acid concentration can be compared to a model of associative surface ligand exchange:

$$J = k_0 e^{-(\Delta G^*/RT)}[HF] = [HF]\frac{kT}{h}fde^{\Delta S^*/R}e^{-(\Delta H^*/RT)}$$ (19.53)

where
  f is the fraction of reaction sites
  d is the intermolecular distance in the liquid
  ΔH^* is the enthalpy of formation of the activated complex
  ΔS^* is the entropy of formation of the activated complex
  $kT/h \sim 10^{13}$ s^{-1} is the frequency of collision of liquid molecules with the surface

The experimental data did indeed show that the dissolution rate was proportional [HF] as predicted from the model of the last ligand exchanged being the rate-controlling step (Figure 19.28). Also from the experimental rate of dissolution, the product $f \times \exp(\Delta S^*/R)$ is calculated to be about 3×10^{-6}. The major contribution to the entropy change is arguably that of the HF molecule giving up its translational and rotational degrees of freedom in the solution phase as it enters the activated surface complex ion. A rough estimate of the entropy change can be obtained from the heat of fusion of HF, about 20 J/mol/K (Diggle 1973). If this is roughly equal to $-\Delta S^*$, then the fraction of active sites is calculated to be about 3×10^{-5}. If the number of surface atoms is on the order of 10^{14} cm^{-2}, then the number of active sites is on the order of 10^9 cm^{-2}, a not unreasonable value for the concentration of kink sites. Even with very high specific area powders, this low surface concentration would make it very difficult to investigate directly the activated state with any kind of experimental probe. So the macroscopic dissolution experiments performed on TiO$_2$ (Bright and Readey 1987) and SiO$_2$ (Liang and Readey 1987) on well-characterized solids or powders can offer a great deal of insight into the details of the corrosion mechanism that might mitigate the need for some type of surface atomistic probe. The rate of dissolution was found to be independent on the concentration of HCl, and the activation energy for the dissolution process was about 75 kJ/mol, clearly indicating a surface reaction. Also, K$_2$TiF$_6$ was precipitated with KOH solutions showing that the TiF$_6^{2-}$ ion was indeed present in solution.

19.6.6 Aluminum Oxide, Al$_2$O$_3$

Aluminum oxide, alumina, is probably the most-studied engineering ceramic. Nevertheless, a review of the properties of alumina states, in a section called "Chemical Corrosion" (Morrell 1987, 9)

> In materials which contain significant amounts of secondary phase, i.e. in products which are than about 99.7% Al$_2$O$_3$, the resistance to attack is governed by the precise proportions of the various oxides that constitute this phase....Often the manufacturers can provide no comparative test data on which to judge performance, and it may be left to the purchaser to undertake tests to satisfy himself that particular product will give the desired life in particular conditions. It is not possible to give more than general guidelines in the Data Reviews because for the majority of products the tests generally have not been performed.

One difficulty in generating meaningful corrosion data on engineering aluminas is they frequently contain a "grain boundary" or more extensive silicate phase as shown in Figure 19.36. The light areas

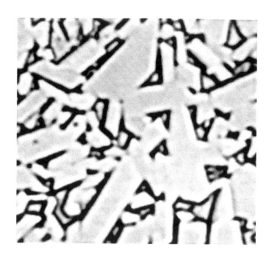

FIGURE 19.36 Microstructure of a typical high-alumina ceramic. The light phase consists of the alumina grains, and the darker phase is an aluminosilicate glass.

are grains of aluminum oxide, and the dark, intergranular regions are an aluminosilicate glass. Many different types of aluminas are available from different manufacturers, some with specified amounts of aluminum oxide such as "85% alumina." In this case, there is about 15% by weight of other compounds such as SiO_2 and CaO in the glass boundary phase. Typically, the volume fraction of the glass phase is about twice that of its weight percent because of the difference in density between the silicate glass and alumina. In any event, in testing the corrosion of aluminas in HF solutions, it was found that the glass phase preferentially was corroded or "leached" out the alumina (Mikeska and Bennison 1999, Mikeska et al. 2000). Typically, the silicate phase is a liquid at the firing temperatures of the alumina, and it forms a continuous glass phase on cooling. As a result, the leaching of the phase continues at essentially a linear rate being removed from deeper into the alumina ceramic with time.

The thermodynamics for the dissolution reactions of alpha and gamma alumina in HF solutions are quite similar:

$$\alpha\text{-}Al_2O_3(s) + 12HF(aq) = 3H_2O(l) + 6H^+(aq) + 2AlF_6^{3-}(aq)$$

$$\Delta G^0(298) = -149.3\,kJ/mol \tag{19.54}$$

and

$$\gamma\text{-}Al_2O_3(s) + 12HF(aq) = 3H_2O(l) + 6H^+(aq) + 2AlF_6^{3-}(aq)$$

$$\Delta G^0(298) = -167.8\,kJ/mol \tag{19.55}$$

where the (aq) suffix stands for an aqueous species. Clearly, both gamma and alpha alumina are essentially infinitely soluble in HF solutions. However, gamma alumina dissolves much more rapidly than alpha by about two orders of magnitude (Markhoff 1986). This suggests that crystal structure plays a role on dissolution kinetics, something that has not been considered much in the literature. Gamma alumina has a cubic spinel crystal structure that is formed from the decomposition of aluminum salts at low temperatures that transforms irreversibly to $\alpha\text{-}Al_2O_3$, which is rhombohedral—frequently called "hexagonal"—above about 1100°C. In the gamma structure, some of the Al^{+3} ions are in octahedral coordination, and some are in tetrahedral coordination. There is no valance difference between charges on the aluminum ion on the different sites: both contain Al^{+3}. The spinel gamma alumina formula could be written as

$$\left[Al_{8/9}V_{1/9}\right]_{tet}(Al_{16/9}V_{2/9})_{oct}O_4 \tag{19.56}$$

where
"tet" are the tetrahedral sites
"oct" are the octahedral sites in spinel (Evans 1964, 174)
"V" is the empty or vacant site

As a result, the γ-Al_2O_3 structure is considerably more open than that of corundum, α-Al_2O_3, with some of the Al^{+3} in tetrahedral coordination and empty cation sites. The dissolution rate of γ-Al_2O_3 was found to be linear with [HF] similar to TiO_2 and had an activation energy of about 60 kJ/mol. In addition, Na_3AlF_6 could be precipitated with NaOH again showing the Al^{+3} species in solution is AlF_6^{3-}.

The dependence on temperature and [HF] was not measured on alpha alumina since the dissolution rate was simply too slow. In addition to being about two orders of magnitude slower than that for gamma alumina, the dissolution rate of alpha alumina is about two orders of magnitude slower than that of TiO_2 and about five orders of magnitude slower than that of amorphous SiO_2 (Table 19.3; Markhoff 1986). The fact that alpha alumina dissolves much more slowly than either silica or titania is not consistent with the theory of ligand exchange rates since the Al^{+3} ion has a smaller charge than either Ti^{+4} and Si^{+4}, while they all have about the same ionic radius. However, there seem to be little or no data on the relative rates of ligand exchange between these three ions (Huheey 1978).

On the other hand, a correlation between the local "crystal" structures of these various materials can be made (Markhoff 1986). Both titania and silica have more open crystal structures than does alumina. The SiO_4^{4-} tetrahedra in silica share corners, the TiO_6^{8-} octahedra in rutile share edges, and the AlO_6^{9-} octahedra in alpha alumina form an interpenetrating lattice. Therefore, these dissolution rates correlate more with the structure surrounding the immediate coordinating ligands than they do with cation charge only. Gamma alumina has a defect spinel structure with some of the aluminum ions in tetrahedral coordination. These tetrahedra share only corners with the remaining octahedrally coordinated aluminum ions. As a result, gamma has a more open structure than alpha and would be expected to dissolve more rapidly. This proposed effect of local structure on dissolution rate is consistent with anisotropic dissolution of SiO_2 as well. However, in this case, it is less clear what the differences in local structure are on different crystal faces that would lead to a difference in dissolution rates.

TABLE 19.3 Dissolution Rates of Oxides in HF–HCl Solutions

Oxide	Rate (mol/cm²/s)
TiO_2, rutile	3.4×10^{-11}
SiO_2, amorphous	2.4×10^{-8}
Al_2O_3, gamma	2.8×10^{-11}
Al_2O_3, alpha	3.0×10^{-13}

Source: Markhoff, C.J., Dissolution of alumina in HF-HCl solutions, MS thesis, The Ohio State University, Columbus, OH, 1986.

19.6.7 Semiconducting Oxides and Charge Transfer: Fe_2O_3, NiO, and UO_2

19.6.7.1 Iron Oxide, Fe_2O_3

The transition element oxides have been popular materials for study, Fe_2O_3 (Valverde 1976, Frenier and Growcock 1984, Valverde 1988) and NiO (Lee et al. 1975, Lussiez et al. 1981, Pease et al. 1986), for example. Results on iron oxides (Valverde 1976, Frenier and Growcock 1984, Valverde 1988) show that the presence of either a strongly reducing redox couple or a strong cation-bonding ligand increases the rate of dissolution. These data suggest that Fe^{+3} is being reduced to Fe^{+2} ion on the solid surface and the rate of dissolution is increased because it is well documented that Fe^{+2} exchanges ligands many orders of magnitude faster than Fe^{+3} (Huheey 1978). Complexing agents also often increase the rate of dissolution. This can be partially explained in that some complexing agents, such as $C_2O_4^{2-}$, preferentially complex the oxidized state and change the potential of a redox couple (Frenier and Growcock 1984, Burgess 1988).

19.6.7.2 Nickel Oxide, NiO

The data for NiO are confusing and conflicting (Lee et al. 1975, Valverde 1976, Lussiez et al. 1981, Pease et al. 1986, Valverde 1988). Since Li_2O increases the Ni^{+3} concentration (Equation 19.44), Li_2O would be expected to lower the rate of dissolution, but just the opposite was observed (Lussiez et al. 1981). Application of an anodic potential to NiO increased the rate of dissolution (Lee et al. 1975). This was explained by the Vermilyea theory as a change in double-layer potential and not related to electrochemical dissolution. Redox couples in solution were also found to increase the rate of dissolution of NiO, which was explained by an electron transfer reaction reducing Ni^{+3} to Ni^{+2} on the solid surface (Valverde 1976, 1988). Thus, three different rate-controlling mechanisms have been proposed for the dissolution of NiO, one of the most-studied semiconducting oxide compounds. Yet, some of these mechanisms depend on the assumption and explanation that the ligand exchange rate of Ni^{+3} is faster than that of Ni^{+2} (Diggle 1973, Blesa et al. 1994) for which there is little or no evidence (Huheey 1978). These conclusions are based on two older literature references (Prasad and Tendulkar 1931, Bogatsky 1951) on the dissolution behavior of nickel oxide produced from the decomposition of salts at temperatures below 1000°C. (Note that the latter reference is given incorrectly in the literature [Diggle 1973].) Both studies find that the rate of dissolution depends on the temperature of decomposition of the nickel salt with higher temperatures leading to slower to almost no dissolution. One author attributes this to the formation of a solid solution of Ni_2O_3 in NiO with more Ni_2O_3 formed at low temperatures where the dissolution is greater (Bogatsky 1951). The other attributes this dissolution behavior to probable particle size differences: the higher the temperature of decomposition, the larger the particle size—smaller surface area—and therefore slower rate of dissolution (Prasad and Tendulkar 1931). Certainly, the latter explanation seems to be the better one. Phase equilibria data for the Ni–NiO phase boundary as a function of temperature and oxygen pressure exit but do not show data for other possible oxides of nickel, including Ni_2O_3 (McHale and Roth 1999, 47, Figure 9857). Yet another shows a Li_2O–NiO–Ni_2O_3 ternary system suggesting large amounts of solid solubility of Ni_2O_3 in NiO (Roth et al. 1981, 45, Figure 5071). Another reference states, "There is no good evidence for Ni_2O_3, but there are two proved crystalline forms of black NiO(OH)" (Cotton et al. 1999, 846). Sources of thermodynamic data are equally ambivalent about the thermodynamics of oxides in the Ni–O system. One gives data for Ni^{+2}(aq), NiO, $Ni(OH)_2$, $Ni(OH)_3$ (Latimer 1952), and NiO(OH), and another gives Ni^{+2}(aq), NiO, $Ni(OH)_2$, and $Ni(OH)_3$ (Bard et al. 1985), while a third gives data for Ni^{+2}, NiO, Ni_2O_3, Ni_3O_4, NiO_2, and $HNiO_2^-$ (Roine 2002). Certainly, the use of different databases would give different results for thermodynamics of nickel oxides. What does seem firmly established is NiO will become nonstoichiometric at high temperatures (Equation 19.44), leading to $Ni_{1-x}O$—essentially a solid solution of Ni_2O_3 in NiO—but "x" is typically small, $x \sim 10^{-4}$, at least at elevated temperatures 1000°C–1400°C (Smyth 2000, 247).

The principal reason that nickel oxidation chemistry is important is that nickel compounds have been used in electrochemical cells for a number of years and are currently being considered as cathodes in lithium ion batteries. For example, in the nickel–cadmium secondary—rechargeable—cell, the principal reaction is (Vincent and Scrosati 1997, 163)

$$Cd(s) + 2NiO(OH)(s) + 2H_2O(l) \underset{\text{charge}}{\overset{\overset{\text{discharge}}{\text{KOH(so ln)}}}{\rightleftarrows}} Cd(OH)_2(s) + 2Ni(OH)_2(s) \tag{19.57}$$

and that in the nickel metal hydride cell (Vincent and Scrosati 1997, 178),

$$MH(s) + NiO(OH)(s) \underset{\text{charge}}{\overset{\overset{\text{discharge}}{\text{KOH(so ln)}}}{\rightleftarrows}} M(s) + 2Ni(OH)_2(s) \tag{19.58}$$

Currently, $Li_{1-x}NiO_2$ is being considered both for the electrode material in molten carbonate fuel cells (Antolini 1993) and for an intercalation cathode in lithium ion cells (Thackeray 1995, Xie et al. 1995, Delmas and Croguennec 2002, Whittingham 2008). In the former case, it is because of its corrosion resistance in molten carbonate and its high electric conductivity of about 10 S/cm at room temperature. In the latter case, it is of interest as a cathode material because the lithium ion can easily enter and leave the lattice of the lithiated nickel oxide for cell reactions such as

$$xLi(s) + Li_{1-x}NiO_2(s) \underset{\text{charge}}{\overset{\text{discharge}}{\rightleftarrows}} LiNiO_2(s) \tag{19.59}$$

In the completely charged state, the nickel ions are all Ni^{+3}, while in the partially discharged state, some of the ions are Ni^{+4}. Apparently, Li_2O and NiO can form a solid solution all the way from NiO to $LiNO_2$. The reason that this material is an interesting cathode material is due to the crystal structure of the $LiNiO_2$, which is known and is essentially a modified NaCl—rock salt—structure with alternating (111) planes of Ni^{+3} and Li^+ ions in octahedral sites between close-packed (111) planes of O^{7-} (Wells 1984, 270). As a result, the Li^+ ions can easily move in and out of the oxygen sheets, and the electron reduces a Ni^{+4} to a Ni^{+3}. Note that if all of the Li were removed from $LiNiO_2$, the result would be NiO_2 with all of the nickel ions Ni^{+4} in octahedral coordination, between two close-packed oxygen planes giving the layered CdI_2 structure (Wells 1984, 258) with the nickel–oxygen sheets held together by van der Waals forces allowing easy intercalation of the Li^+ ions between the sheets. This is, of course, the reason that TiS_2 was chosen as one of the first cathodes for lithium ion cells; it has the same CdI_2 crystal structure (Whittingham 2008).

It is also worth noting that $NiO(OH)$ could be written as $HNiO_2$, and one would suspect that its structure would be similar to that of $LiNiO_2$ and would allow easy motion of hydrogen ions into and out of the crystal. $HNiO_2$ does indeed have a crystal structure very similar to that of $LiNiO_2$ (Wells 1984, 260).

Clearly, the dissolution kinetics of NiO and lithiated NiO demonstrate not only the importance of both the electric properties of the solid and the properties of the solution but also the details of the crystal structure on the rate of chemical dissolution. These aspects of ceramic corrosion clearly need to be more carefully examined. Two questions need answering. First, "Can chemical and electrochemical dissolution be clearly distinguished?" Second, "How do the electric properties of the solid, its crystal structure, the double layer, redox potential, and applied potential affect chemical dissolution (i.e., no surface charge transfer) in various systems?" It would seem that lithiated NiO would be an ideal candidate to study given the amount of uncertainty in its properties in spite of its potential importance for battery and fuel cell applications. One final comment, it has been found that carefully characterized powders of lithiated NiO do dissolve in 0.25 molar HCl more rapidly the higher the lithium content, but all have the same activation energy for dissolution—going from zero to about 19.2 mol% Li_2O—of about 76 kJ/mol (Yan and Readey 1994). Whether this is due to the greater lability of the Ni^{+3} ion or simply due to the fact that the Li^+ will readily enter the solution leaving surface sites where the surface nickel ions can be more easily removed by the solution ligands is not known.

19.6.7.3 Uranium Dioxide, UO$_2$

If careful consideration is given to the principles of electron exchange reactions, electrochemical dissolution controlled by surface charge transfer can be identified and quantified. For example, the study of dissolution of UO$_2$ has yielded one of the most definitive investigations (Nicol et al. 1975) on electrochemical dissolution of oxides. In this study, UO$_2$ was oxidized to the UO$_2^{+2}$ ion in solution. A one-to-one correspondence was found between the amount of UO$_2$ dissolved and the total electronic charge transported in the external circuit during anodic dissolution. Furthermore, the rate of dissolution could be correlated with both the redox potential of the solution and the *rate* of redox electron exchange with a solid UO$_2$ electrode. This experimental evidence clearly demonstrated that the rate-controlling step was electron transfer at the solid surface.

Few other studies on the dissolution of ceramics in the presence of applied potentials or redox couples provide such definitive evidence of electron transfer control of dissolution. As a result, much of the work on the effects of various redox couples on dissolution rates is difficult, if not impossible, to interpret because the ease with which the redox couple can exchange electrons with the solid was either not known or not considered. It is known from homogeneous electron exchange reactions that the coordinating ligands can have an important effect on the rate of electron exchange (Burgess 1988). Careful studies on the electrochemical dissolution with various ligands in solution have not been performed. Such a correlation might indeed be helpful in choosing the appropriate redox couple and the surrounding ligands to control and explain the mechanism of electrochemical dissolution of ceramics.

19.6.8 Aluminum Nitride, AlN

Little will be said about the degradation of nonoxides such as SiC and Si$_3$N$_4$ other than they have been studied and are fairly resistant to aqueous corrosion in various environments, particularly SiC (Morrell 2010). However, again with these ceramics, it is difficult to try to determine the mechanisms of corrosion given the variation in material preparation techniques and resulting microstructures as well as very different testing conditions.

One material that has been studied several times since it appears to be easily attacked by aqueous solutions but is an attractive material for many applications is aluminum nitride (AlN). AlN is of interest because thermal conductivities of AlN in excess of 200 w/m-K, roughly five times that of alumina and 60% that of copper, can be obtained. Thus, there is a strong interest in using AlN as an insulating packaging material to remove heat from semiconductor chips that have continually increasing device density and heat generation. The high thermal conductivity is achieved by removing Al$_2$O$_3$ impurities from solid solution in the AlN (which strongly decrease the thermal conductivity of AlN) during densification with the addition of Y$_2$O$_3$ to form Y$_3$Al$_5$O$_{12}$, yttrium aluminum garnet (YAG), as isolated pockets of a second phase at grain boundaries where it has little effect on the thermal conductivity. However, AlN is not very stable in aqueous environments; for example, the reaction of AlN with water vapor gives (just one of many possible reactions leading to hydrated alumina species, all of which lead to the same deleterious result)

$$AlN(s) + 3H_2O(g) = Al(OH)_3(s) + NH_3(g);$$

$$\Delta G^0(298 \text{ K}) = -182.4 \text{ kJ/mol} \tag{19.60}$$

On opening a bottle of commercial AlN powder, the ammonia odor is unmistakable, clearly indicating that a reaction such as this occurs. The reaction of AlN powder with water precludes colloidal processing of AlN ceramics, and undesirable nonaqueous systems must be used. However, even dense polycrystalline AlN is subject to degradation in aqueous media (Svedberg et al. 2000, Tamai et al. 2000). A careful study performed on powders found that the initial corrosion product was a porous layer of AlO(OH), which latter transformed into $Al(OH)_3$ with first-order kinetics. The data were fit to a model with a porous product layer with an unreacted core of AlN. It was deduced that a surface reaction at the AlN–H_2O interface controlled the reaction kinetics. The temperature dependence of the process was not investigated so no activation energy for the reaction was determined (Bowen et al. 1990). Oxidation of the surface of dense piece of AlN will form a protective Al_2O_3 layer but reduce the thermal conductance of the thermally insulating Al_2O_3 layer in series with the conducting AlN (Svedberg et al. 2000).

19.6.9 Multicomponent Systems

19.6.9.1 General

In this section, what is known—or believed—about the ambient degradation of a selected group of ceramic materials that have more than one cation in an oxide is discussed. Again, the major purpose will be to demonstrate different degradation phenomena and try to reach some conclusions about degradation behavior, particularly in aqueous environments. One of the problems that exists with crystalline materials is all too frequently, there is a tendency to apply the results from the aqueous degradation of silicate glasses to multicomponent crystalline ceramics. In some cases, the comparison is appropriate because similar degradation mechanisms are occurring, but this is frequently not the case. For example, Figure 19.37 shows what typically occurs for an alkali–silicate glass. The interstitial alkali ions in the glass are preferentially leached out and replaced with hydronium ions, H_3O^+, from the solution. If the corrosion is taking place in a confined volume, then the pH of the solution will increase, and the higher pH solution will begin to attack and dissolve the silicate framework structure precipitating new hydrous aluminosilicate phases—or others—on the leached glass surface (Clark and Zoitos 1992, McCauley 2004). This is "incongruent" dissolution and frequently occurs during the weathering of silicate minerals (Blesa et al. 1994). Similar things can occur in crystalline ceramics, particularly with monovalent cations present in the material that could easily be replaced by the hydronium ions in solution, for example, the lithium ions in the lithiated NiO discussed earlier. However, if hydronium ions cannot easily substitute for the cations in the crystalline materials—the cations may be divalent or trivalent—then preferential leaching is less likely to occur, and congruent dissolution will occur with the possible precipitation of one or more phases that exceed their solubility in the solution. Figure 19.38 schematically

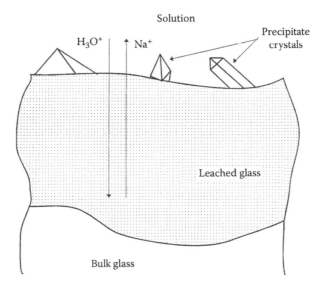

FIGURE 19.37 Schematic showing the corrosion of a sodium silicate glass with a Na-leached layer and surface recrystallization products.

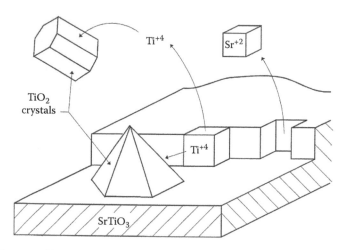

FIGURE 19.38 Schematic of "congruent" dissolution in which the entire material goes into solution and then one or more insoluble solids precipitates. SrTiO₃ is used as an example.

depicts congruent dissolution of $SrTiO_3$ with the precipitation of TiO_2 crystals since the concentration of titanium species in solution from the congruent dissolution exceeds the solubility of TiO_2.

19.6.9.2 Beta-Alumina, $Na_2O \cdot 11Al_2O_3$

Beta-alumina, $\beta\text{-}Al_2O_3$, is not a different crystallographic form of Al_2O_3 but $Na_2O \cdot 11Al_2O_3$ and has a layered crystal structure with the Al^{+3} ions in spinel-like layers with oxygen ions and with the sodium in layers in between these spinel "blocks." Only about one-third of the Na^+ sites are filled so the Na^+ can move quite freely in two dimensions in these layers giving rise to a high sodium ion electric conductivity at temperatures slightly above ambient. As a result, $\beta\text{-}Al_2O_3$ or $\beta''\text{-}Al_2O_3$ (more Na^+ and somewhat higher conductivity) is being used as the solid electrolyte in the sodium–sulfur battery (Moulson and Herbert 2003, 196):

$$2Na(l) + S(l) \xrightarrow[300°C]{\beta\text{-}Al_2O_3} Na_2S(l) \tag{19.61}$$

where here the reactants and products are all liquids and are separated by the solid electrolyte.

With the mobile Na^+ ion in this structure, it might be expected that it could easily be replaced by hydronium ions from solution and that is indeed the case. In fact, all of the sodium can be replaced in the structure (Breiter et al. 1977). However, hydronium replacement of the sodium lowers the sodium ionic conductivity and is therefore undesirable (Kaneda et al. 1979, Dunn 1981). As a result, materials such as beta-alumina must be processed in nonaqueous systems in order to produce dense ceramics for sodium–sulfur battery and other applications.

19.6.9.3 Strontium Titanate, $SrTiO_3$

An article summarizing the principles of aqueous corrosion of glass and ceramics states that $SrTiO_3$ is an example of incongruent dissolution in which the Sr^{+2} is preferentially leached from $SrTiO_3$ and gives a figure illustrating the sharp boundary between the residual surface TiO_2 layer and the $SrTiO_3$ core (White 1992). This would not be expected for if a single hydronium ion substitutes for the Sr^{+2} ion, then the Ti^{+4} would have to oxidize to the Ti^{+5} state, and there is no chemical evidence that titanium exists in the +5 state (Emsley 1988). Furthermore, experimental data clearly show that $SrTiO_3$ dissolves congruently in HCl–HF solutions with slow precipitation of TiO_2, the rate of which depending critically on the [HF].

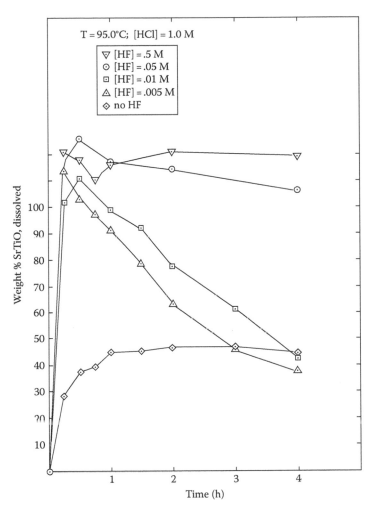

FIGURE 19.39 Dissolution of SrTiO$_3$ in HF–HCl solutions. (From Bright, E. and Readey, D.W., *J. Am. Ceram. Soc.*, 70(12), 900, 1987. With permission of the American Ceramic Society.)

SrTiO$_3$ exhibits congruent dissolution followed by TiO$_2$ precipitation as shown in Figure 19.39 (Bright 1982, Readey 1998b). The amount dissolved in this case was measured by the concentration of titanium in solution, and for low concentrations of HF, the titanium in solution decreases with because TiO$_2$ particles are precipitating from solution (Bright 1982). It should be noted that with the [HF] < 0.1 molar, there is insufficient fluorine in solution to have all of the titanium present as TiF$_6^{-2}$ with the size sample being dissolved. Therefore, the solution species must be some fluoro–chloro complex that is unstable, and TiO$_2$ precipitates with time. This is a significant result since it demonstrates that, under conditions of congruent dissolution, the amount of the "insoluble" species in solution, titania in this case, can far exceed its thermodynamic solubility. Also note that the SrTiO$_3$ dissolves much more rapidly—more than an order of magnitude faster—than TiO$_2$ under similar conditions. The data also show that both the dissolution and precipitation rates are strongly catalyzed by only a small concentration of fluoride ion suggesting that the fluoride ion plays a central role in the activated complex on the solid precipitate surface. These data, plus the temperature dependence, $\Delta H^* = 36.6$ kJ/mol with HF present and $\Delta H^* = 67.2$ kJ/mol in pure HCl, again suggest a ligand rearrangement rate-controlling process on the crystal surface. Finally, the results show that for a ternary compound or solid solution in which one of the constituent oxides dissolves much more rapidly than the other, whether there is preferential leaching,

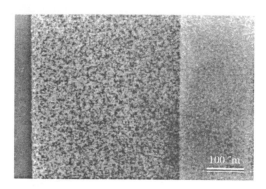

FIGURE 19.40   Dissolution—leaching—of SrTiO₃ from a two-phase mixture of TiO₂ and SrTiO₃ in 5 molar HCl at 117°C in 16 h. Left: residual TiO₂. Right: undissolved TiO₂–SrTiO₃ two-phase mixture. (From Yan, J.-K., Aqueous corrosion of multi-component Oxides. PhD thesis, Colorado School of Mines, St Golden, CO, 1997.)

recrystallization of the more insoluble constituent on the surface, or dissolution followed by precipitation will depend on the system and the experimental conditions.

Because of the differences in the dissolution rates of TiO_2 and $SrTiO_3$ in pure HCl, porous TiO_2 can be made by leaching the $SrTiO_3$ from a two-phase mixture of TiO_2 and $SrTiO_3$ (Figure 19.40; Yan 1997). Two main conclusions can be drawn from these results on $SrTiO_3$ dissolution. First, under the certain conditions, $SrTiO_3$ dissolves congruently with the titanium entering solution as some type of unstable aqueous species, which subsequently precipitates as TiO_2 particles homogeneously throughout the solution and not on the dissolving particle surface (Figure 19.38). Second, because one of the components in a multicomponent compound is insoluble in a given solution, TiO_2, in this case, does not preclude dissolution of the ternary compound under certain conditions of temperature and solution chemistry. However, these data do not preclude the possibility of leaching some SrO from $SrTiO_3$ leaving behind a probably porous TiO_2 layer on the surface of the $SrTiO_3$ (Nesbitt et al. 1981). But it is unlikely that the SrO can be replaced in the $SrTiO_3$ lattice with hydronium ions leaving the TiO_6^{-8} octahedra in a corner-shared framework as they do in the perovskite structure. It is far more likely that the framework will be destroyed with the dissolution of the SrO and the titanium octahedra will reform into a precipitate of some form of TiO_2 on the surface.

19.7 Stress-Corrosion Cracking and Strength Degradation in Humid Environments

19.7.1 Fracture of Ceramics

Most ceramics fail by brittle fracture with no ductile behavior. Fracture occurs when the stresses at an existing crack tip are high enough to break interatomic bonds and dissipate any energy, G, involved in propagating the crack. The fracture strength of ceramics is determined by preexisting flaws or cracks either in the bulk or at the surface of ceramics. As a result, the fracture strength of a ceramic is not a material property but depends on the number and size of preexisting flaws or cracks. The stress at the tip of a crack is dependent on the angular orientation, θ, around the tip of the crack and depends on the distance, r, from the tip of the crack in the form of (Lawn 1993, 25)

$$\sigma \propto \frac{K}{r^{1/2}} f(\theta) \tag{19.62}$$

where
   f(θ) is the function of the angle around crack tip
   "K" is the "stress intensity factor"

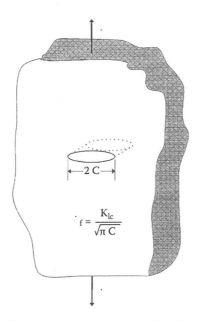

FIGURE 19.41 Ceramic sample with a thorough, internal crack of length 2C under a tensile stress, σ.

Of course, as r → 0, the stress at fracture cannot be defined, while K can be. When fracture occurs, the stress intensity factor becomes the "critical stress intensity factor," K_C. For a through elliptical crack of length 2C in a sheet (Figure 19.41), the ceramic fractures, $σ_f$, when the critical stress intensity, K_{Ic}, is reached (the subscript "I" indicates the type of stress on the crack and, in this case, indicates a normal stress as shown in Figure 19.41):

$$\sigma_f = \frac{K_{IC}}{\sqrt{\pi C}} \tag{19.63}$$

and it is K_{IC} that is the material parameter with units MPA-m$^{1/2}$, and it is called the "fracture toughness." The "C" refers to the fact that fracture will occur when the "critical stress intensity factor" is reached or that fracture will occur at a rapid rate: the so-called fast fracture.

Now, $K_{IC} = \sqrt{GE}$ where G is the energy for fracture, J/m², and E = elastic modulus, MPa. Figure 19.42 shows a double cantilever beam (DCB) specimen, a test sample geometry that is frequently used for

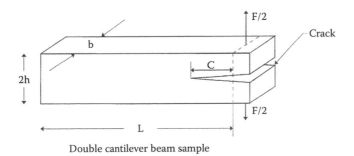

Double cantilever beam sample

FIGURE 19.42 DCB sample for determining the fracture toughness of ceramics. (From Davidge, R.W., *Mechanical Behavior of Ceramics*, Cambridge University Press, Cambridge, U.K., 1989.)

testing the fracture strength, or K_{IC}, of ceramic materials. For a DCB sample, the stress intensity, K_I, is given by (Davidge 1989, 42).

$$K_I = 3.45 \frac{FC}{bh^{3/2}} \left[1 + 0.7 \frac{h}{C} \right] \tag{19.64}$$

From the geometry of the sample and the applied force, F, at fracture, K_{IC}, can be measured.

19.7.2 Static Fatigue or Stress-Corrosion Cracking

However, it is found that the fracture strength, σ_f, decreases as the time that the load is applied increases as shown in Figure 19.43 (Kingery et al. 1976, 801). This phenomenon is known as "static fatigue" and has been recently reviewed particularly with respect to the mechanisms involved (Freiman et al. 2009). Static fatigue occurs because it has been found that cracks in ceramics grow slowly under an applied stress or stress intensity. In fact, it has been shown that the crack growth velocity, v, is given by

$$v = AK_I^n \tag{19.65}$$

where n is an integer and n = 30. In most systems, three different regions of crack propagation versus stress intensity have been found as shown in Figure 19.44 (Lawn 1993, 106). Region I is the "slow crack growth" region where Equation 19.65 applies and crack growth is controlled by bond breaking at the crack tip. Region II is where transport of chemical species that affect bond breaking controls the rate of crack growth. And, region III is fast fracture described by Equation 19.63 when K_I reaches K_{IC}.

Of significance for this discussion is that the slow crack growth rate is greatly influenced by water and water vapor and increases with the increasing availability of water (Figure 19.45; Wiederhorn 1967). A mechanism for the effect of water vapor and other polar solvents on crack growth has been proposed and verified (Michalske and Freiman 1983). Figure 19.46 schematically depicts this mechanism of bond breaking and crack growth. A molecule that has both an electron-donating region such as the p-orbital

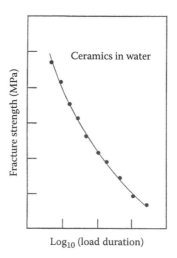

FIGURE 19.43 Decrease in fracture strength of a ceramic as a function of time under load in an aqueous environment. (From Kingery, W.D. et al., *Introduction to Ceramics*, 2nd edn., Wiley, New York, 1976.)

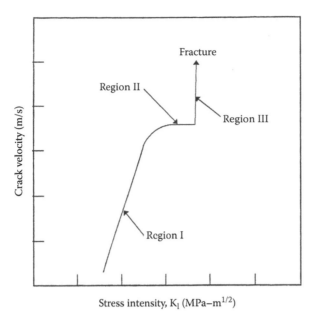

FIGURE 19.44 Environmentally affected crack velocity as a function of K_I typical for ceramic materials. (From Lawn, B., *Fracture of Brittle Solids*, 2nd edn., Cambridge University Press, Cambridge, U.K., 1993.)

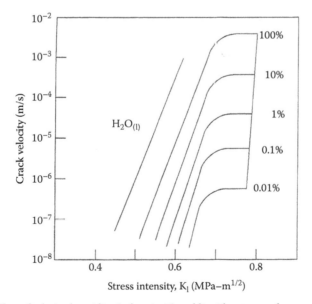

FIGURE 19.45 The effect of relative humidity (values in %) and liquid water on slow crack growth and the onset of fast fracture for silicate glass. (From Wiederhorn, S.M., *J. Am. Ceram. Soc.*, 50(8), 407, 1967.)

of the H_2O molecule and the ability to form a hydrogen bond is shown attacking an Si–O–Si bond leading to bond rupture and crack growth in a silicate (Michalske and Freiman 1983). The increase in slow crack growth velocity in the presence of water has been demonstrated for a number of oxides and nonoxides as well. Therefore, aqueous environments are not only a concern for corrosion and leaching of ceramics, but they can also accelerate mechanical degradation.

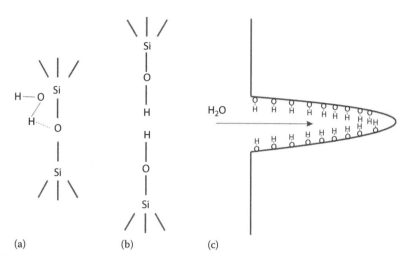

FIGURE 19.46 Proposed mechanism for the effect of water on slow crack growth in silicate glasses—and other ceramics. (a) Water molecules attack the Si–O–Si bonds. The dashed line represents a hydrogen bond. (b) Broken bonds now terminated by OH. (c) Crack tip with surface OHs and H_2O necessary to extend the crack now limited by its transport into the crack leading to region II in Figure 19.44. (From Michalske, T.A. and Freiman, S.W., *J. Am. Ceram. Soc.*, 66(4), 284, 1983.)

19.7.3 Degradation of Stabilized ZrO_2 by Water Vapor and Aqueous Environments

Pure zirconium oxide is of interest for a number of applications because of its chemical inertness and high melting point. However, pure ZrO_2 undergoes a couple of phase transitions at different temperatures:

$$\text{Monoclinic} \underset{}{\overset{\sim 1140°C}{\rightleftharpoons}} \text{Tetragonal} \underset{}{\overset{2370°C}{\rightleftharpoons}} \text{Cubic} \underset{}{\overset{2680°C}{\rightleftharpoons}} \text{Melt}$$

The reason for these transformations is that the high-temperature cubic form of ZrO_2 has the fluorite structure, but the Zr^{+4} ion is really too small to be in the necessary eightfold coordination by oxygen ions for the fluorite structure to be stable. As a result, on cooling, the zirconium ion tends to shift to one side or end of its coordination shell causing a distortion of the cubic structure into the tetragonal and monoclinic phases. The tetragonal to monoclinic transformation is accompanied by about a 3% volume expansion on cooling that causes cracking and disintegration of pure polycrystalline zirconia. Since most zirconia ceramics are sintered at temperatures below 2000°C, it is only the tetragonal to mono-clinic transition that is of interest. Over 100 years ago, it was discovered that the high-temperature cubic phase could be stabilized to lower temperatures with the addition of CaO, MgO, and Y_2O_3 (and others) additives in solid solution, hence the term "stabilized zirconia"; for example, "YSZ" means "yttria-stabilized zirconia." All of these additions create oxygen vacancies in the lattice:

$$Y_2O_3 \xrightarrow{\;ZrO_2\;} 2Y'_{Zr} + 3O_O + V_O^{\bullet\bullet} \tag{19.66}$$

both the presence of vacancies and yttrium ions randomly distributed over the lattice sites help to stabilize the cubic structure since they would have to move to different lattice sites to achieve either the tetragonal or monoclinic symmetry. This is a diffusion process so these phase transformation now are much slower than in the pure ZrO_2 and the cubic phase can be retained to room temperature. Furthermore, the presence of the oxygen ion vacancies makes stabilized zirconia a very good oxygen ion conductor at temperatures in excess of about 800°C. So stabilized zirconia has been used in the oxygen sensor in vehicles for the last 25 years saving considerable quantities of fuel by being the enabling material for controlling the combustion efficiency of the engine. This high oxygen conductivity makes it the material of choice for high-temperature fuel cells as well (Singhal and Kendall 2003).

If enough additive is present so that the ZrO_2 alloy is 100% cubic at room temperature, this is called "fully stabilized zirconia." With smaller amounts of additive, two-phase mixtures of cubic plus tetragonal are present at the typical firing temperatures, 1700°C. On cooling, the tetragonal phase may or may not transform to the monoclinic form. In any event, this material is "partially stabilized zirconia." If the particle size is small enough, then the tetragonal phase will not transform since the 3% volume expansion required is hindered by the surrounding matrix of cubic zirconia. The presence of the retained tetragonal phase can lead to a significant increase in the fracture toughness of the zirconia. If a crack forms and propagates, it releases the stress on the tetragonal phase in the cubic matrix allowing it to transform to monoclinic. The increase in volume on transformation partially closes the crack, decreasing K_I that leads to an increase in K_{IC}. This is called "transformational toughening," and stabilized zirconia can be a very strong material, and it is added to other ceramics such as Al_2O_3 making ceramic–ceramic composites with very high toughness and strength (Green et al. 1989).

Unfortunately, stabilized zirconia, particularly, 100% tetragonal phase zirconia, called "TZP" for "tetragonal zirconia polycrystals," is susceptible to severe humidity and aqueous degradation as shown in Figure 19.47 (Kube 1992). This behavior is again important for the colloidal processing of stabilized zirconia (Ho and Wei 1999) as well as the stability of stabilized zirconia components used for either their high strength or high ionic conductivity. There have been several mechanisms proposed for this degradation including dissolution of the stabilizing additive and bond breaking at the surface as sketched in Figure 19.46 that have been summarized by others (Gogotsi and Lavrenko 1992, Lepistö and Mäntylä 1992). The exact mechanism of the degradation is still uncertain, but the end result is transformation of the tetragonal phase to monoclinic with the formation of cracking and degradation as shown in Figure 19.47.

Several of the proposed mechanisms are similar to that proposed for the generation of proton conduction in various perovskite materials (Kreuer 1999). Take, for example, the case of barium cerate, $BaCeO_3$, with yttrium oxide, Y_2O_3, in solid solution (Coors and Readey 2002). Yttrium substitutes for cerium with the creation of oxygen vacancies:

$$2BaO(s) + Y_2O_3(s) \xrightarrow{\ BaCeO_3\ } 2Ba_{Ba} + 2Y'_{Ce} + V_O^{\bullet\bullet} + 5O_O \tag{19.67}$$

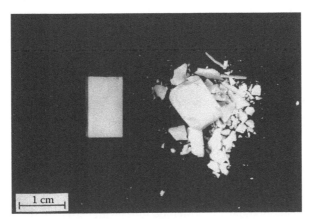

FIGURE 19.47 As-fabricated Y-TZP ceramic (left) and the same ceramic material degraded in 100°C 0.01 M HCl solution for 260 h. (From Kube, T.W., The degradation of Yttria-stabilized Zirconia in aqueous solutions. M.S. Thesis, Colorado School of Mines, St Golden, CO, 1992.)

Note that BaO must be added as well as Y_2O_3 to get the 1:1:3 ratio Ba:Ce:O sites, which is the usual case for the doping with aliovalent additives in ternary compounds. What makes this material a good proton conductor is the reaction between water and the oxygen vacancies:

$$H_2O(g) + V_O^{\bullet\bullet} + O_O \rightleftarrows 2(OH)_O^{\bullet} \tag{19.68}$$

where the hydrogen ion on the OH⁻ ion is now free to move from one oxygen to the next in the structure giving rise to high proton conductivity. Equation 19.68 essentially implies that some $Y(OH)_3$ is being produced in the lattice in the presence of water vapor.

It might be expected that a reaction similar to that in Equation 19.68 occurs in stabilized zirconia—surprising if it did not—and that the increase in molar volume of the $Y(OH)_3$ (the value of which or the density could not be found in the literature) over that of Y_2O_3 (e.g., Table 19.2) could lead to expansion and stresses causing transformation of the tetragonal to monoclinic phase, cracking, and degradation. It is well known that in these ionic conductors, there is a volume change with the concentration of charge carriers, protons in $BaCe_{1-x}Y_xO_{3-x/2}$ or oxygen ions in $Zr_{1-x}Y_xO_{2-x/2}$, that could lead to stresses and the tetragonal to monoclinic transformation.

19.8 Ceramic Degradation Measurement Standards

Given all of the forgoing discussion about the broad spectrum of ceramic materials, the differences in nominally the same material (e.g., SiC), the wide range of testing conditions, the separation of thermodynamics and kinetics, and the differences in potential applications, developing a database and applying standard tests are formidable tasks (Munro 1997, 352). The status of standards for corrosion testing of refractories (mainly) and advanced ceramics has recently been summarized (Rigaud 2011). Again, even with the focus narrowed to refractories (Rigaud 2011, 1117),

> … much effort has been expended in developing tests to measure the corrosion resistance of refractories under slagging conditions and of structural ceramics under hot gas corrosion and oxidation at high temperatures, especially for nonoxide materials. Numerous methods have been tried and some reasonable correlations have been obtained for very specific conditions, but very few methods have reached status of standard operating practices and none have yet been accepted for universal use.

Therefore, few general standards have been developed relating to the ambient degradation of ceramics. Nevertheless, several have been developed for specific materials under certain specific conditions. Some examples are (ASTM 2011)

ASTM C738-94(2011) Standard Test Method for Lead and Cadmium Extracted from Glazed Ceramic Surfaces

ASTM C724-91(2010) Standard Test Method for Acid Resistance of Ceramic Decorations on Architecture Type Glass

ASTM C225-85(2009) Standard Test Methods for Resistance of Glass Containers to Chemical Attack

ASTM C1220-10 Stand Test Method for Static Leaching of Monolithic Waste Forms for Disposal of Radioactive Waste

In addition, several standard tests have been developed to measure the mechanical properties of ceramics, and such tests certainly can be used to evaluate the degradation of mechanical properties after corrosion testing. These include (ASTM 2011)

ASTM C1161-02c(2008)e1 Standard Test Method for Flexural Strength of Advanced Ceramics at Ambient Temperature

ASTM C173-05(2010) Standard Test Method for Tensile Strength of Monolithic Advanced Ceramics at Ambient Temperatures

ASTM C1368-10 Standard Test Method for Determination of Slow Crack Growth Parameters of Advanced Ceramics by Constant Stress-Rate Strength Testing at Ambient Temperatures

ASTM C1421-10 Standard Test Methods for Determination of Fracture Toughness of Advanced Ceramics at Ambient Temperature

ASTM C1576-05(2010) Standard Test Method Determination of Slow Crack Growth Parameters of Advanced Ceramics by Constant Stress Flexural Testing (Stress Rupture) at Ambient Temperatures

In spite of all of the caveats given about the difficulty in developing general corrosion standards, there are two that do exist and are worth noting. The first of these (Morrell 2010, 2288) is ISO 17092:2005 Fine Ceramics (Advanced Ceramics, Advanced Technical Ceramics)—Determination of Corrosion Resistance of Monolithic Ceramics in Acid and Alkaline Solutions. In this standard, the test conditions are specified. The criteria used to evaluate corrosion in this standard are mass loss, dimension changes, and loss of mechanical strength. The second is NACE Standard TM0499-2009 Standard Test Method: Immersion Corrosion Testing of Ceramic Materials. In this case, the overall mass loss rate and the localized mass loss rate are used to evaluate corrosion susceptibility and possibly mechanical properties. This latter standard is very broad and suggests that the test conditions should reflect as closely as possible those that the ceramics will be subjected to during their use. In general, the use of these two standards has not been widely reported in the literature.

19.9 Conclusions

19.9.1 Material Selection

There are many possible degradation mechanisms for ceramics under ambient conditions particularly in the ubiquitous presence of water or water vapor. Unfortunately, the literature on degradation is scattered over several very different scientific disciplines, which makes summarizing data and determining rate-controlling degradation mechanisms extremely difficult. Furthermore, a given ceramic will have different behavior from different manufacturers depending on the different raw materials, additives, processing conditions, and final microstructures. In addition, there have been very few studies undertaken on a given material to demonstrate this variability, nor have there been many studies comparing different materials—whose compositions and microstructures may not have been well defined—under similar environmental conditions. Furthermore, some materials might be expected to be quite corrosion resistant such as tantalum oxide (Ta_2O_5) since tantalum metal is used in the chemical industry as a corrosion-resistant material largely because of the protection afforded by its native oxide. However, Ta_2O_5 is not a readily available commercial ceramic material. So it might be difficult to find a ceramic manufacturer that has either the experience or willingness to make these materials. As a result, it is difficult to recommend which ceramic to choose in a given environment, and methods to mitigate degradation are difficult to predict.

As an example of one of the few studies that have been performed, Morrell (2010, 2302) presents some unpublished data on the behavior of a variety of ceramic materials to different acids and bases at 140°C in terms of mass loss. In general, sintered Al_2O_3 of purity greater than 99.5% Al_2O_3 performed as well as any other material in most conditions. Some exceptions are hot-pressed SiC in acids, hot-pressed TiB_2 in bases, and hot-pressed B_4C, which performed better than alumina. However, all of these hot-pressed materials are very dense and presumably do not contain an easily corroded second phase. Nevertheless, because they are hot pressed rather than merely sintered, their costs are certain to be significantly higher than the sintered aluminas, and the range of shapes and sizes will be severely limited as well.

In another study focused specifically on ceramic behavior in high temperature (90°C) 20 M HF solutions (Mikeska et al. 2000), single crystal aluminum oxide, sapphire, was many orders of magnitude more corrosion resistant than any commercial aluminum oxides: even those with purities specified as being greater than 99% due to corrosion of a residual grain boundary phase. As mentioned earlier, thermodynamically, Al_2O_3 should be very soluble in such solutions, but, obviously, the kinetics of dissolution are very slow as was found before (Markhoff 1986). Here again, only high-density nonoxides such as SiC and B_4C were quite stable. Also performing well, compared to commercial aluminas was a hot-pressed MgO. As shown in Figure 19.17, in gaseous HF, MgF_2 is formed, and the same is true in HF solutions where the MgF_2 forms a "passive" coating on the MgO limiting the rate of further corrosion by diffusion through the solid MgF_2 layer, a very slow process. The results on sapphire and MgO illustrate the two important principles that have been the focus of this chapter. First, if the thermodynamics are unfavorable for corrosion, either by low product vapor pressures, solution concentrations, or the formation of a passive protective layer, then a ceramic for a given set of specific environmental conditions can be chosen. The importance of this is that such thermodynamic data are readily available today (Roine 2002). Conversely, if the thermodynamics suggest severe corrosion, such as Al_2O_3–sapphire in HF, it may not occur simply because the kinetics are much too slow, which is certainly the case for single crystal alumina. However, predicting reaction kinetics is not easy because of the many competing factors and usually must be determined by experiment.

One final study worth noting is the effect of small differences on composition on the corrosion resistance of ceramics. As mentioned earlier, high-purity polycrystalline alumina seems like a good first choice for a corrosion-resistant ceramic, assuming the absence of more easily corroded grain boundary phases. This research (Mikeska and Bennison 1999) showed that there is over a factor of 10^5 difference in the corrosion rate of some commercial polycrystalline aluminas and single crystal Al_2O_3 in 20 M HF at 90°C. Also, MgO was shown to greatly reduce the corrosion rate of commercial aluminas by removing the ubiquitous SiO_2-containing grain boundary phase postulating that the SiO_2 goes into solid solution in the Al_2O_3 with the MgO by the following point defect reaction:

$$SiO_2(s) + MgO(s) \xrightarrow{\quad Al_2O_3 \quad} Si^{\bullet}_{Al} + Mg'_{Al} + 3O_O \tag{19.69}$$

That is, the SiO_2 and MgO go into solid solution, and the charged silicon and magnesium defects just compensate each other allowing a much greater solubility of both compounds than either individually, which would create vacant lattice sites and is energetically unfavorable. As discussed earlier, any extra MgO at grain boundaries is probably not a problem because spinel, $MgAl_2O_4$, will form and may form a passive MgF_2 layer limiting dissolution.

19.9.2 General Principles and New Areas to Investigate

In all cases, however, as the earlier studies show, it is extremely important to separate thermodynamics from the kinetics of the corrosion process. If this can be done and the kinetic processes separated and analyzed, then plausible mechanisms of degradation and protection can be proposed. Another example of this is the case of multicomponent oxides such as $SrTiO_3$. Based on the data (Bright 1982, Yan 1997) (Figure 19.40), SrO is not preferentially leached from the perovskite structure leaving behind some charge-imbalanced framework of TiO_6^{-2} octahedra. Rather, the $SrTiO_3$ dissolves congruently in spite of the fact that TiO_2 should not dissolve. However, it does and the TiO_2 precipitates immediately at the surface or later from solution depending on the solution chemistry and temperature. The details of the reaction and its time dependence will depend on which of these is occurring and necessitate careful examination of the solution chemistry and the structure and location of the corrosion products.

The dissolution kinetics of $SrTiO_3$ also illustrates another important principle that even though one of the constituents, TiO_2 in this case, may dissolve much more slowly or not at all, this does not preclude the compound from dissolving much more rapidly. Although it was not discussed, it might be expected that Sr_2TiO_4 will dissolve even more rapidly: the higher the concentration of the more soluble and labile component, more rapid dissolution of a compound might be expected. As mentioned earlier, this apparently applies to the different types of asbestos and the suggestion of stabilizing waste $MgCr_2O_3$ refractories with Al_2O_3 or Fe_2O_3. A very useful set of experiments would be to evaluate the corrosion or dissolution kinetics for a series of compounds between two end-members: one labile and soluble and the other inert and stable. The La_2O_3–TiO_2 system comes to mind. This has not been done.

Finally, some of the data (Markhoff 1986) suggest that the surface exchange rate of ions seems to depend on the details of the structure of the compound—hence, the surface structure—which suggests that this is an area for more extensive investigation as well.

Acknowledgments

The American Ceramic Society has generously allowed several drawings previously published in their publications to be included in this article. Also, the works of several past graduate students are reflected in the discussions earlier: from the Ohio State University, Eric Bright, Da-Tung Liang, and Carole Markhoff and from the Colorado School of Mines, Jen-Kuo Yan and Todd Kube.

References

Abraham, M.M., C.T. Butler, and Y. Chen. 1971. Growth of high-purity and doped alkaline earth oxides: I. MgO and CaO. *Journal of Chemical Physics*, 35(8), 3752–3756.

Adamson, A.W. 1982. *Physical Chemistry of Surfaces*, 4th edn., Wiley, New York.

Anderson, D.A., J.H. Adair, D. Miller, J.V. Biggers, and T.R. Shrout. 1988. Surface chemistry effects on ceramic processing of $BaTiO_3$ powder, in G.L. Messing, E.R. Fuller Jr., and H. Hausner, Eds., *Ceramic Powder Science II: Ceramic Transactions*, Vol. I, The American Ceramic Society, Columbus, OH, pp. 485–492.

Antolini, E. 1993. A review study of the preparation of porous Lithium-Doped Nickel Oxide. *Journal of the European Ceramic Society*, 12, 139–145.

Askeland, D.R. and P.P. Phulé. 2003. *The Science and Engineering of Materials*, 4th edn., Brooks/Cole-Thomson Learning, Pacific Grove, CA.

ASTM Standards. 2011. http://www.astm.ort//Standards. Accessed August 15, 2011.

Baes, C.F. Jr. and R.E. Mesmer. 1976. *The Hydrolysis of Cations*, Wiley, New York.

Bard, A.J., R. Parsons, and J. Jordon. 1985. *Standard Potentials in Aqueous Solution*, Dekker, New York.

Barsoum, M. 1997. *Fundamentals of Ceramics*, McGraw-Hill, New York.

Berner, R.A. 1978. Rate control of mineral dissolution under earth surface conditions. *American Journal of Science*, 278(12), 1235–1252.

Bird, R., W. Stewart, and E. Lightfoot. 2002. *Transport Phenomena*, 2nd edn., Wiley, New York.

Blachere, J.R. and F.S. Pettit. 1989. *High Temperature Corrosion of Ceramics*, Noyes, Park Ridge, NJ.

Blesa, M.A., P.J. Morando, and A.E. Regazzoni. 1994. *Chemical Dissolution of Metal Oxides*, CRC Press, Boca Raton, FL.

Bockris, J.O.M. and A.K.N. Reddy. 1970. *Modern Electrochemistry, 1 and 2*, Plenum, New York.

Bogatsky, D.P. 1951. Diagram of the Ni-O_2 system and the physico-chemical nature of the solid phase in this system. *Journal of General Chemistry of the USSR (English Translation)*, 21, 1–8.

Bowen, P., J.G. Highfield, A. Mocellin, and T.A. Ring. 1990. Degradation of aluminum nitride powder in an aqueous environment. *Journal of the American Ceramic Society*, 78(3), 724–728.

Breiter, W.W., G.C. Farrington, W.L. Roth, and J.L. Duffy. 1977. Production of hydronium beta alumina from sodium beta alumina and characterization of the conversion products. *Materials Research Bulletin*, 12, 895–906.

Bright, B. 1982. Dissolution kinetics of TiO_2 and $SrTiO_3$. MS thesis. The Ohio State University, Columbus, OH.

Bright, E. and D.W. Readey. 1987. Dissolution kinetics of TiO_2 in HF-HCl Solutions. *Journal of the American Ceramic Society*, 70(12), 900–906.

Browne, B.A. and C.T. Driscoll. 1992. Soluble aluminum silicates: Stoichiometry, stability, and implications for environmental geochemistry. *Science*, 256(June), 667–670.

Burgess, J. 1988. *Ions in Solution: Basic Principles of Chemical Interactions*, Wiley, New York.

Butler, M.A. and D.S. Ginley. 1980. Review: Principles of photochemical solar energy conversion. *Journal of Materials Science*, 15, 1–19.

Callister, W.D. Jr. 2000. *Materials Science and Engineering, An Introduction*, 5th edn., Wiley, New York.

Chesters, J.H. 1974. *Refractories for Iron- and Steelmaking*, The Metals Society, London, U.K.

Chiang, Y.-M., D. Birnie, III, and W.D. Kingery. 1997. *Physical Ceramics*, Wiley, New York.

Clark, D.E. and B.K. Zoitos (Eds.) 1992. *Corrosion of Glass, Ceramics, and Ceramic Superconductors*, Noyes, Park Ridge, NJ.

Cooper, A.R. Jr. and W.D. Kingery. 1962. Kinetics of solution in high viscosity liquids: Sodium chloride-glycerol. *Journal of Physical Chemistry*, 66(4), 665–669.

Coors, W.G. and D.W. Readey. 2002. Proton conductivity measurements in Yttrium Barium Cerate by impedance spectroscopy. *Journal of the American Ceramic Society*, 85(11), 2637–2640.

Cotton, F.A., G. Wilkinson, C.A. Murillo, and M. Bochmann. 1999. *Advanced Inorganic Chemistry*, 6th edn., Wiley, New York.

Cussler, E.L. 1997. *Diffusion*, 2nd edn., Cambridge University Press, Cambridge, U.K.

Davidge, R.W. 1989. *Mechanical Behavior of Ceramics*, Cambridge University Press, Cambridge, U.K.

Deal, B.E. and A.S. Grove. 1965. General relationship for the thermal oxidation of Silicon. *Journal of Applied Physics*, 36(12), 3770–3778.

Delmas, C. and L. Croguennec. 2002. Layered $Li(Ni,M)O_2$ systems as the cathode material in Lithium-Ion batteries. *MRS Bulletin*, 27(8), 608–612.

Diggle, J.W. 1973. Dissolution of oxide phases, in J.W. Diggle, Eds., *Oxides and Oxide Films*, Vol. 2, Dekker, New York, pp. 281–306.

Dunn, B. 1981. Effect of air exposure on the resistivity of sodium beta and beta aluminas. *Journal of the American Ceramic Society*, 64(3), 125–128.

Emsley, J. 1998. *The Elements*, 3rd edn., Clarendon Press, Oxford, U.K.

Evans, R.C. 1964. *An Introduction to Crystal Chemistry*, Cambridge University Press, Cambridge, U.K., p. 174.

Faure, G. 1991. *Inorganic Geochemistry*, MacMillan, New York.

Fitch, L.D. and V.L. Burdick. 1989. Water corrosion of $YBa_2Cu_3O_{7-x}$ superconductors. *Journal of the American Ceramic Society*, 72(10), 2020–2023.

Fontana, M.G. 1986. *Corrosion Engineering*, McGraw-Hill, New York.

Fordham, R.J (Eds.) 1990. *High Temperature Corrosion of Technical Ceramics*, Elsevier, New York.

Fox, K., E. Hoffman, N. Manjooran, and G. Pickrell (Eds.) 2010. *Advances in Materials Science for Environmental and Nuclear Technology: Ceramic Transactions*, Vol. 222, Wiley, New York.

Freiman, S.W., S.M. Wiederhorn, and J.J. Mecholsky. Jr. 2009. Environmentally enhanced fracture of glass: A historical perspective. *Journal of the American Ceramic Society*, 92(7), 1371–1382.

Frenier, W.W. and F.B. Growcock. 1984. Mechanism of iron oxide dissolution—A review of recent Literature. *Corrosion*, 40(12), 663–668.

Funk, J.E. and D.R. Dinger. 1994. *Predictive Process Control of Crowded Particulate Suspensions Applied to Ceramic Manufacturing*, Kluwer Academic Publishers, Boston, MA.

Garrels, R.M. and C.L. Christ. 1990. *Solutions, Minerals, and Equilibria*, Jones and Bartlett, Boston, MA.

Geankoplis, C.J. 1972. *Mass Transport Phenomena*, Ohio State University Press, Columbus, OH.

Geankoplis, C.J. 2003. *Transport Processes and Separation Process Principles*, Prentice-Hall, Englewood Cliffs, NJ.

Ginley, D., M.A. Green, and R. Collins. 2008. Solar energy conversion toward Terawatt. *MRS Bulletin*, 33(4), 355–364.

Gogotsi, Y.G. and V.A. Lavrenko. 1992. *Corrosion of High-Performance Ceramics*, Springer-Verlag, Berlin, Germany.

Green, D.J., R.H.J. Hannink, and M.V. Swain. 1989. *Transformational Toughening of Ceramics*, CRC Press, Boca Raton, FL.

Hench, L.L., D.E. Clark, and A.B. Harker. 1986. Review: Nuclear waste hosts. *Journal of Materials Science*, 21, 1457–1478.

Hiemenz, P.C. 1986. *Principles of Colloid and Surface Chemistry*, 2nd edn., Dekker, New York.

Ho, F.-Y. and W.-C.J. Wei. 1999. Dissolution of Yttrium Ions and phase transformation of 3Y-TZP powder in aqueous solution. *Journal of the American Ceramic Society*, 82(6), 1614–1616.

Hudson, J.B. 1995. *Surface Science*, Wiley, New York.

Huheey, J.E. 1978. *Inorganic Chemistry*, 2nd edn., Harper and Row, New York.

Hume, L.A. and J.D. Rimstidt. 1992. The biodurability of chrysotile asbestos. *American Mineralogist*, 77, 1125–1128.

Jacob, K.T. and K.P. Abraham. 2009. Thermodynamic properties of calcium titanates: $CaTiO_3$, $Ca_4Ti_3O_{10}$, and $Ca_3Ti_2O_7$. *Journal of Chemical Thermodynamics*, 41(6), 816–820.

Jones, C.F. et al. 1984. Surface structure and dissolution rates of ionic oxides. *Journal of Materials Science Letters*, 3, 810–812.

Kaneda, T., J.B. Bates, J.C. Wang, and H. Engstrom. 1979. Effect of H_2O on the ionic conductivity of sodium beta-alumina. *Materials Research Bulletin*, 14, 1053–1056.

Kim, H.E. and D.W. Readey. 1989. Corrosion of SiC in hydrogen above 1400oC, in J.D. Cawley and C.E. Semler, Eds., *Silicon Carbide '87: Ceramic Transactions*, Vol. 2, The American Ceramic Society, Westerville, OH, pp. 301–312.

Kingery, W.D., H.K. Bowen, and D.R. Uhlmann. 1976. *Introduction to Ceramics*, 2nd edn., Wiley, New York.

Kofstad, P. 1988. *High Temperature Corrosion*, Elsevier, London, U.K.

Krause, C.A. 1987. *Refractories: The Hidden Industry*, The American Ceramic Society, Westerville, OH.

Kreuer, K.D. 1999. Aspects of the formation and mobility of protonic charge carriers and the stability of perovskite-type oxides. *Solid State Ionics*, 125, 285–302.

Kröger, F.A. 1964. *The Chemistry of Imperfect Crystals*, North-Holland, Amsterdam, the Netherlands.

Kube, T.W. 1992. The degradation of Yttria-stabilized Zirconia in aqueous solutions. M.S. Thesis, Colorado School of Mines, St Golden, CO.

Lasaga, A.C. 1998. *Kinetic Theory in the Earth Sciences*, Princeton University Press, Princeton, New York.

Latimer, W.M. 1952. *The Oxidation States of the Elements and their Potentials in Aqueous Solutions*, Prentice Hall, New York.

Lawn, B. 1993. *Fracture of Brittle Solids*, 2nd edn., Cambridge University Press, Cambridge, U.K.

Layden, G.K. and G.W. Brindley. 1963. Kinetics of vapor-phase hydration of magnesium oxide. *Journal of the American Ceramic Society*, 46(11), 518–522.

Lee, C.H., A. Riga, and E. Yeager. 1975. The dissolution kinetics of lithiated NiO in aqueous acid solutions, in A.R. Cooper and A.H. Heuer, Eds., *Mass Transport Phenomena in Ceramics, Materials Science Research*, Vol. 9, New York, Plenum, pp. 489–499.

Lepistö, T.T. and T.A. Mäntylä. 1992. Degradation of TZP ceramics in humid atmospheres, in D.E. Clark and B.K. Zoitos, Eds., *Corrosion of Glass, Ceramics and Ceramic Superconductors*, Noyes Publications, Park Ridge, NJ, pp. 49–513.

Levin, E.M. and H.F. McMurdie (Eds.) 1975. *Phase Diagrams for Ceramists*, Vol. III, The American Ceramic Society, Westerville, OH.

Levin, E.M., C.R. Robbins, and H.F. McMurdie (Eds.) 1969. *Phase Diagrams for Ceramist*, Vol. II, The American Ceramic Society, Westerville, OH.

Lewis, J.A. 2000. Colloidal processing of ceramics. *Journal of the American Ceramic Society*, 83(10), 2341–2359.

Liang, D.-T. and D.W. Readey. 1987. Dissolution kinetics of crystalline and amorphous silica in hydrofluoric-hydrochloric acid solutions. *Journal of the American Ceramic Society*, 70(8), 570–577.

Lilley, E. 1990. Review of low temperature degradation, in Y-TZPs. in R.E. Tressler and M. McNallan, Eds., *Corrosion and Corrosive Degradation of Ceramics: Ceramic transactions*, Vol. 10, The American Ceramic Society, Columbus, OH, pp. 387–410.

Lin, H.-T., J. Salem, and E.R. Fuller (Eds.) 2008. Corrosion, wear, fatigue, and reliability of ceramics. *Ceramic Engineering and Science Proceedings*, 29(3), 2009.

Lussiez, P., K. Osseo-Asare, and G. Simkovich. 1981. Effect of solid state impurities on the dissolution of nickel oxide. *Metall Trans B*, 12B(12), 651–657.

Macdonald, D.D. and D. Owen. 1971. The dissolution of magnesium oxide in dilute sulfuric acid. *Canadian Journal of Chemistry*, 49, 3375–3380.

Malarria, J.A. and R. Tinivella. 1997. Degradation of magnesite-chrome refractory brick by hydration. *Journal of the American Ceramic Society*, 80(9), 2262–2268.

Markhoff, C.J. 1986. Dissolution of alumina in HF-HCl solutions. M.S. Thesis. The Ohio State University, Columbus, OH.

Mayer, J.W. and S.S. Lau. 1990. *Electronic Materials Science: For Integrated Circuits in Si and GaAs*, MacMillan, New York.

McCauley, R.A. 2004. *Corrosion of Ceramic and Composite Materials*, 2nd edn., Marcel Dekker, New York.

McGraw, J., C.S. Bahn, P.A. Parilla, J.D. Perkins, D.W. Readey, and D.S. Ginley. 1999. Li ion diffusion measurements in V_2O_5 and $Li(Co_{1-x}Al_x)O_2$ thin-film battery cathodes. *Electrochmica Acta*, 45, 187–196.

McHale, A.E. and R.S. Roth (Eds.) 1999. *Phase Equilibria Diagrams*, Vol. XII, The American Ceramic Society, Westerville, OH.

Michalske, T.A. and S.W. Freiman. 1983. A molecular mechanism for stress corrosion in vitreous silica. *Journal of the American Ceramic Society*, 66(4), 284–288.

Mikeska, K.R. and S.J. Bennison. 1999. Corrosion of alumina in aqueous hydrofluoric acid. *Journal of the American Ceramic Society*, 82(12), 3561–3566.

Mikeska, K.R., S.J. Bennison, and S.L. Grise. 2000. Corrosion of ceramics in aqueous hydrofluoric acid. *Journal of the American Ceramic Society*, 83(5), 1160–1164.

Miles, P.A. and D.W. Readey. 1972. Polycrystalline halides as optical materials, in *Proceedings of the Conference on High Power Laser Window Materials (October 30–November 1, 1972)*, Report No. AFCRL-TR-73-0372, pp. 507–513.

Morrell, R. 1987. *Handbook of Properties of Technical and Engineering Ceramics. Part 2 Data Reviews. Section I: High Alumina Ceramics*, HMSO Books, London, U.K.

Morrell, R. 2010. Degradation of engineering ceramics, in B. Cottis, R. Lindsay, S. Lyon, D.J.D. Scantlebury, H. Stott, and M. Graham, Eds., *Shreir's Corrosion*, Elsevier, Amsterdam, the Netherlands, pp. 2282–2305.

Morrison, S.R. 1977. *The Chemical Physics of Surfaces*, Plenum, New York.

Morrison, S.R. 1980. *Electrochemistry at Semiconductor and Oxidized Metal Electrodes*, Plenum, New York.

Mossman, B.T., J. Bignon, M. Corn, A. Seaton, and J.B.L. Gee. 1990. Asbestos: Scientific developments and implications for public policy. *Science*, 247(19 January), 294–306.

Moulson, A.J. and J.M. Herbert. 2003. *Electroceramics*, 2nd edn., Wiley, New York.

Munro, R.G. 1997. Material specifications of advanced ceramics and other issues in the use of property databases with corrosion analysis models. *Journal of Testing and Evaluation*, 25(3), 349–353.

Nesbitt, H.W., G.M. Bancroft, W.S. Fyfe, S.N. Karkhanis, A. Nishijima, and S. Shin. 1981. Thermodynamic stability and kinetics of perovskite dissolution. *Nature*, 289(January), 358–362.

Nickel, K.G. 1994. *Corrosion of Advanced Ceramics*, Kluwer Academic Publishers, Norwell, MA.

Nicol, M.J., C.R.S. Needes, and N.P. Finkelstein. 1975. Electrochemical model for the leaching of Uranium Dioxide: 1-acid media, in A.R. Burkin, Eds., *Leaching and Reduction in Hydrometallurgy*, Institute of Mining and Metallurgy, London, U.K., pp. 1–11.

O'Bannon, L.S. 1984. *Dictionary of Ceramic Science and Engineering*, Plenum, New York.

Ohtaki, H. et al. 1988. Dissolution process of sodium chloride crystal in water. *Pure and Applied Chemistry*, 60(8), 1321–1324.

Osseo-Asare, K. 1984. Interfacial phenomena in leaching systems, in R.G. Bautista, Eds., *Hydrometallurgical Process Fundamentals*, Plenum, New York, pp. 27–268.

Paladino A.E. and E.A. Maguire. 1970. Microstructure development in Yttrium iron Garnet. *Journal of the American Ceramic Society*, 53(2), 98–102.

Pankow, J.F. 1991. *Aquatic Chemistry Concepts*, Lewis Publishers, Chelsea, MI.

Pease, W.R., R.L. Segall, R.S.C. Smart, and P.S. Turner. 1986. Comparative dissolution rates of defective nickel oxides from different sources. *Journal of the Chemical Society, Faraday Transactions 1*, 82(33), 759–766.

Perkins, W.P (Eds.) 1984. *Ceramic Glossary 1984*, The American Ceramic Society, Columbus, OH.

Pourbaix, M. 1966. *Atlas of Electrochemical Equilibria in Aqueous Solutions*, Pergamon Press, New York.

Prasad, M. and M.G. Tendulkar. 1931. CLXXXVIII—The preparation and properties of Nickelous Oxide. *The Journal of the Chemical Society*, 1403–1407.

Readey, D.W. 1990. Vapor phase sintering of ceramics, in C.A. Handwerker, J.E. Blendell, and W.A. Kaysser, Eds., *Sintering of Ceramics: Ceramic Transactions*, Vol. 7, The American Ceramic Society, Westerville, OH, pp. 86–110.

Readey, D.W. 1991. Gaseous corrosion of ceramics. *Ceramica Acta*, 3, 11–27.

Readey, D.W. 1998a. Aqueous corrosion of ceramics, in D.K. Peeler and J.C. Marra, Eds., *Environmental Issues and Waste Management Technologies in the Ceramic and Nuclear Industries, III, Ceramic Transactions*, Vol. 87, The American Ceramic Society, Westerville, OH, pp. 43–62.

Readey, D.W. 1998b. Modeling corrosion of ceramics in halogen-containing atmospheres, in D.K. Peeler and J.C. Marra, Eds., *Environmental Issues and Waste Management Technologies in the Ceramic and Nuclear Industries, III, Ceramic Transactions*, Vol. 87, The American Ceramic Society, Westerville, OH, pp. 23–32.

Readey, D.W. and C.R. Cooley (Eds.) 1977. *Ceramic and Glass Radioactive Waste Forms*, U. S. Energy Research and Development Administration, Report No. CONF-770102, Washington, DC.

Readey, D.W. and A.R. Cooper. Jr. 1966. Molecular diffusion with a moving boundary and spherical symmetry. *Chemical Engineering Science*, 21, 917–922.

Readey, M.J. and D.W. Readey. 1987. Sintering of TiO_2 in HCl atmospheres. *Journal of the American Ceramic Society*, 70(12), C358–C361.

Reed, J.S. 1995. *Principles of Ceramic Processing*, 2nd edn., Wiley, New York.

Richerson, D.W. 2006. *Modern Ceramic Engineering*, 3rd edn., CRC Press, Boca Raton, FL.

Rigaud, M. 2011. Corrosion testing refractories and ceramics, in R. Winston Revie, Eds., *Uhlig's Corrosion Handbook, Third Addition*, Wiley, New York, pp. 1117–1119.

Ring, T.A. 1996. *Fundamentals of Ceramic Powder Processing and Synthesis*, Academic Press, San Diego, CA.

Ringwood, A.E. and S.E. Kesson. 1979. Immobilization of high-level wastes in synroc titanate ceramic, in *Ceramics in Nuclear Waste Management*, U.S. Department of Energy, Report NO. CNF-790420, pp. 79–84.

Ritland, M.A. and D.W. Readey. 1993. Alumina-copper composites by vapor phase sintering. *Ceramic Engineering and Science Proceedings*, 14(9–10), 896–907.

Robinson, R.A. and R.H. Stokes. 2002. *Electrolyte Solutions*, 2nd edn., Dover, Mineola, New York.

Roine, A. 2002. *Outokumpu HSC Chemistry for Windows, Version 5.11*, Thermodynamic software program, Outokumpu Research Oy, P.O. Box 60, FIN-28102 PORI, Finland.

Roth, R.S., T. Negas, and L.P. Cook (Eds.) 1981. *Phase Diagrams for Ceramists*, Vol. IV, The American Ceramic Society, Columbus, OH.

Segall, R.L., R.S.C. Smart, and P.S. Turner. 1978. Ionic Oxides: Distinction between mechanisms and surface roughening effects in the dissolution of Magnesium Oxide. *Journal of the Chemical Society, Faraday Transactions*, 74, 2907–2912.

Shackelford, J.F. 2000. *Introduction to Materials Science for Engineers*, 5th edn., Prentice-Hall, Upper Saddle River, NJ.

Shand, M.A. 2006. *The Chemistry and Technology of Magnesia*, Wiley, New York.

Singhal, S.C. and K. Kendall. 2003. *High Temperature Solid Oxide Fuel Cells*, Elsevier, Oxford, U.K.

Smyth, D.M. 2000. *The Defect Chemistry of Metal Oxides*, Oxford University Press, Oxford, U.K., p. 247.

Stumm, W. 1992. *Chemistry of the Solid-Water Interface*, Wiley, New York.

Stumm, W. and J.J. Morgan. 1996. *Aquatic Chemistry*, 3rd edn., Wiley, New York.

Suárez, M.F. and R.G. Compton. 1998. Dissolution of Magnesium Oxide in aqueous acid: An atomic force microscopy study. *Journal of Physical Chemistry B*, 102, 7156–7162.

Svedberg, L.M., K.C. Arndt, and M.J. Cima. 2000. Corrosion of Aluminum Nitride (AlN) in aqueous cleaning solutions. *Journal of the American Ceramic Society*, 83(1), 41–46.

Tamai, H., T. Hamauzu, A. Ryumon, K. Aoki, K. Oda, and T. Yoshio. 2000. Hydrothermal corrosion and strength degradation of Aluminum Nitride ceramics. *Journal of the American Ceramic Society*, 83(12), 3216–3218.

Thackeray, M.M. 1995. Lithiated oxides for lithium-ion batteries, in S. Megahed, B.M. Barnett, and L. Xie, Eds., *Proceedings of the Electrochemical Society, Rechargeable Lithium-Ion Batteries*, Vol. 94–28, The Electrochemical Society, Pennington, NJ, pp. 233–244.

Tressler, R.E. and M. McNallan (Eds.) 1990. *Corrosive and Corrosive Degradation of Ceramics: Ceramic Transactions*, Vol. 10, The American Ceramic Society, Westerville, OH.

Valverde, N. 1976. Investigations on the rate of dissolution of Metal Oxides in acidic solutions with additions of redox couples and complexing agents. *Berichte der Bunsen Gesellschaft*, 80(4), 333–340.

Valverde, N. 1988. Factors determining the rate of dissolution of Metal Oxides in acidic aqueous solutions. *Berichte der Bunsengesellschaft für physikalische Chemie*, 92, 72–78.

Valverde, N. and C. Wagner. 1976. Considerations on the kinetics and the mechanism of the dissolution of Metal Oxides in acidic solution. *Berichte der Bunsen Gesellschaft*, 80, 330–333.

Vermilyea, D.A. 1966. The dissolution of ionic compounds. *Journal of the Electrochemical Society*, 113, 1067–1078.

Vermilyea, D.A. 1969. The dissolution of MgO and $Mg(OH)_2$ in aqueous solutions. *Journal of the Electrochemical Society*, 116(9), 1179–1183.

Vincent, C.A. and B. Scrosati. 1997. *Modern Batteries*, Arnold, London, U.K.

Wadsworth, M.E. and J.D. Miller. 1979. Hydrometallurgical processes, in H.Y. Sohn and M.E. Wadsworth, Eds., *Rate Processes in Extractive Metallurgy*, Plenum, New York, pp. 133–241.

Wagner, C. 1949. The dissolution rate of Sodium Chloride with diffusion and natural convection as rate-determining factors. *Journal of Chemical Physics*, 53(7), 1030–1033.

Warren, I.H. and E. Devuyst. 1973. Leaching of Metal Oxides, in D.J. Evans and R.S. Shoemaker, Eds., *International Symposium on Hydrometallurgy*, AIME, New York, pp. 229–263.

Weast, R.C (Ed.) 1979. *CRC Handbook of Chemistry and Physics*, 60th edn., CRC Press, Boca Raton, FL.

Wells, A.F. 1984. *Structural Inorganic Chemistry*, 5th edn., Clarendon Press, Oxford, U.K.

West, J.M. 1986. *Basic Corrosion and Oxidation*, 2nd edn., Ellis Horwood Ltd., Chichester, England.

White, W.B. 1992. Theory of corrosion of glass and ceramics, in D.E. Clark and B.Z. Zoitos, Eds., *Corrosion of Glass, Ceramics and Ceramic Superconductors*, Noyes Publications, Park Ridge, NJ, pp. 2–28.

Whittingham, M.S. 2008. Materials challenges facing electrical energy storage. *MRS Bulletin*, 33(4), 411–419.

Wiederhorn, S.M. 1967. Influence of water vapor on crack propagation in soda-lime glass. *Journal of the American Ceramic Society*, 50(8), 407–414.

Xie, L., W. Ebner, D. Fouchard, and S. Megahed. 1995. Electrochemical studies of $LiNiO_2$ for Lithium-Ion batteries, in S. Megahed, B.M. Barnett, and L. Xie, Eds., *Proceedings of the Electrochemical Society, Rechargeable Lithium-Ion Batteries*, Vol. 94–28, The Electrochemical Society, Pennington, NJ, pp. 263–276.

Yan, J.-K. 1997. Aqueous corrosion of multi-component oxides. PhD thesis, Colorado School of Mines, St Golden, CO.

Yan, J.-K. and D.W. Readey. 1994. Aqueous corrosion of Nickel Oxide-Lithium Oxide solid solutions, in D.F. Bickford, S. Bates, V. Jain, and G. Smith, Eds., *Environmental and Waste Management Issues in the Ceramic Industry, II: Ceramic Transactions*, Vol. 45, The American Ceramic Society, Westerville, OH, pp. 291–308.

20

Silicate Glasses

Larry L. Hench
University of Florida
and
University of Central Florida
and
Imperial College London

20.1 Introduction

Silicate-based glasses are one of the most important classes of industrial materials that include container (bottles), window, and automotive glasses and are also used for lighting, domestic ware, optics (lenses), and pharmaceutical ware applications. There is an enormous range of silicate glass compositions, ranging from ultrahigh purity silica (SiO_2) glass for fiber optics to borosilicate glasses for immobilization of high level radioactive wastes that contain more than 40 components. Understanding and generalizing the corrosion behavior and environmental resistance of such a large composition range of materials is a scientific and technical challenge. The approach taken in this chapter is that all silicate glasses can be described in terms of one of six surface conditions that may occur at any specific instant in its environmental history.

20.2 Structure of Glass Surfaces

The surface condition of a glass that exists at any specific time ($t = t_x$) depends upon many factors. The factors that influence environmental behavior can be divided into four categories, as summarized in Table 20.1. The factors are as follows: (1) chemical composition (the primary determinant in environmental resistance), (2) the environmental factors, (3) the physical factors, and (4) the specimen state of the glass. The interaction of all four classes of factors has a vital role in determining the rate of attack of a silicate glass surface.

TABLE 20.1   Factors Affecting Silicate Glass
Durability

1. Bulk glass composition
2. Environmental factors
   A. Generic
      a. Temperature
      b. Exposure time (continuous or cycled)
      c. External stresses upon specimen
      d. Radiation
   B. Atmospheric
      a. Relative humidity
   C. Condensed
      a. Solution pH
      b. Presence of inhibitors in the corrosion solution
      c. Solution composition
3. Physical factors
   A. Weathering versus aqueous corrosion
   B. Dynamic versus static corrosion
   C. Exposed surface area-to-solution volume ratio (SA/V)
   D. Corrosion behavior of bulk glass versus powdered glass
4. Specimen state
   A. Thermal history
      a. Degree of annealing for stress and density
      b. Phase separation (glass in glass)
      c. % Crystallization, if any
   B. Prior corrosion exposure history
   C. Surface features
      a. Surface roughness
      b. Surface composition
   D. Homogeneity of glass
      a. Defects, such as cords or stones
   E. Surface treatments

20.2.1 Five Types of Glass Surfaces

The five types of glass surfaces and six surface conditions resulting from glass–environment interactions are illustrated in Figure 20.1 (Hench and Clark 1978, Hench 1988).

The ordinate in Figure 20.1 represents the relative concentration of SiO_2 (or other oxides in Type III surfaces) in the glass. The abscissa corresponds to the depth into the glass surface. For glasses with very high silica content (fiber optics), the depth of surface change may be only a few nanometers or less. For other more reactive glasses, such as bioactive glasses or radioactive waste disposal glasses, the depth of surface reaction may be as large as tens to hundreds of micrometers.

When some compositional species are selectively dissolved from the glass surface, such as alkalis (Na, K), the relative SiO_2 concentration compared with the bulk glass will increase, producing a SiO_2-rich surface layer. This behavior is termed selective dissolution or incongruent dissolution. If all species in the glass are dissolved simultaneously (called congruent dissolution) the relative concentration of SiO_2 will remain the same as in the original glass. When combinations of selective dissolution, congruent dissolution, and precipitation from solution occur, then heterogeneous mixtures of the six surface conditions shown in Figure 20.1 are possible.

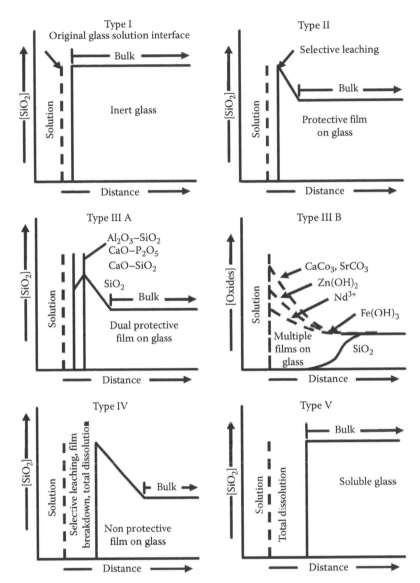

FIGURE 20.1 Five types of glass surfaces and six surface conditions resulting from glass–environment interactions.

The schematic illustration of the glass surface in Figure 20.1 shows that Type I glasses have undergone a surface reaction that is only a monolayer thick and no compositional profile is measurable. High purity vitreous silica exposed to neutral solutions is an example of a Type I surface. A Type II surface occurs when there is an exchange of alkali and alkaline earth ions with hydrogen and/or hydronium ions from an aqueous or humid environment (i.e., selective dissolution). However, there must be a sufficient concentration of network formers in the surface left behind to stabilize the silica-rich film. Type III surfaces are described later.

When there are not sufficient network formers (Si, Al, or B) in the glass, or if the environment is highly enriched in hydroxyls (OH⁻) or other species, such as F⁻, which can break Si–O–Si network bonds, the surface layer is unstable and a Type IV surface is produced. A glass that is undergoing total network attack (also referred to as congruent dissolution) is described as having a Type V surface. From the perspective of average surface composition, there is little distinction between

Type I and Type V surfaces. However, large quantities of ions are being lost from a Type V surface during corrosion and consequently extensive surface pitting can result due to localized heterogeneous attack. Additionally, large dimensional changes often accompany corrosion of glasses with Type V surfaces.

The transition from a Type I to Type V surface that results from static corrosion and a localized increase in solution pH is illustrated in Figure 20.2. The initial condition of the glass, pt (a), is one of very little surface reaction; only surface silanols (Si–OH) have formed on the outer layer of the silicate network. However, as the glass is exposed to water or a humid environment, selective leaching occurs resulting in a Type II surface, shown as pt (b). At longer times the interfacial pH of the glass and the solution or atmosphere becomes greater than pH > 9 and attack of the Si–O–Si network increases.

Consequently, the thickness of the ion-exchange layer on the surface reaches a maximum, pt (c), because at longer times (c → d) the surface SiO_2-rich layer is dissolved by network dissolution faster than it is formed by ion exchange. Eventually, a congruently dissolving Type V glass exists, shown in Figure 20.2 as pt (d).

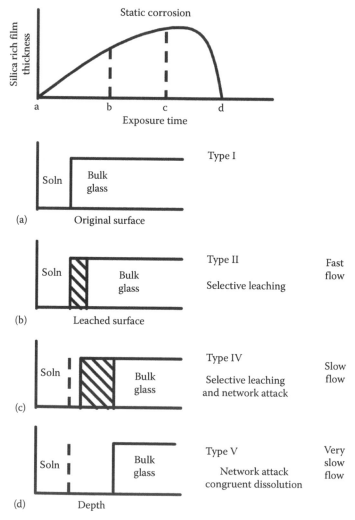

FIGURE 20.2   Time-dependent changes of glass surfaces with static leaching or flow: (a–d) correspond to the exposure time periods shown in the upper kinetics curve.

20.3 Degradation Principles

20.3.1 Kinetics of Silicate Glass Surface Reactions

The rate of attack of a silicate glass surface (kinetics) depends on all the factors summarized in Table 20.1. The general equation that describes the overall rate of glass surface reactions (R) has at least four terms:

$$R = -k_1 t^{0.5} - k_2 t^{1.0} + k_3 t^x + k_4 t^y + k_n t^z \qquad (20.1)$$

Stage 1 Stage 2 Stage 3 Stage 4

The first term in Equation 20.1, developed by Douglas and El-Shamy (1967), describes the rate of alkali extraction from the silicate glass. Thus, this term corresponds to what is often called a Stage 1 reaction (see Figure 20.3). A Type II glass surface shown in Figure 20.1 is primarily undergoing Stage 1 attack.

Stage 1: The initial (primary) stage of attack is a process that involves ion exchange between alkali ions from the glass and hydrogen ions (H⁺) (or hydronium (H₃O⁺) ions) from the solution, during which the remaining constituents of the glass are not altered. Reaction (1) in Figure 20.3 tends to dominate during Stage 1 attack. During Stage 1, the rate of alkali extraction from the glass is parabolic in character (Lyle 1943, Zagar and Schillmoeller 1960, Rana and Douglas 1961), as shown in Equation 20.1.

The second term in Equation 20.1 describes the rate of network dissolution that is associated with a Stage 2 reaction (Figure 20.3). A Type IV surface is experiencing a combination of Stage 1 and Stage 2 attacks. A Type V surface is dominated by a Stage 2 attack.

Stage 2: The second stage of attack is a process whereby breakdown of the silica network structure occurs and total glass dissolution is initiated. Reaction (2) in Figure 20.3 is the dominant reaction in Stage 2 processes. The rate of attack in Stage (2) is linear in time, as shown in Equation 20.2.

The chemical basis for understanding Stages 1 and 2 processes and the linear rate of attack is based upon studies by Wang and Tooley (1958a,b), Lyle (1943), Rana and Douglas (1961), Das (1969), Das and Douglas (1967), Bacon and Calcamuggio (1967), Douglas and Isard (1949), Tsuchihashi and Sekido (1959), and Weyl (1947) in addition to the classic paper of Douglas and El-Shamy (1967).

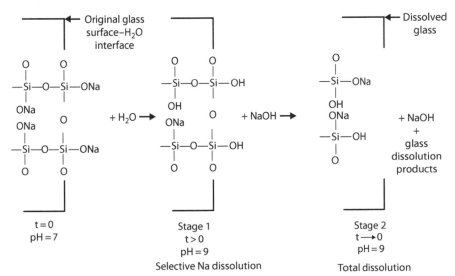

FIGURE 20.3 Mechanisms of glass corrosion for a soda–silica glass.

20.4 Degradation of Specific Glass Surfaces

20.4.1 Experimental Methods of Surface Compositional Analysis

The experimental evidence of formation of surface compositional profiles that arise as a consequence of Stage 1 and Stage 2 reactions was accumulated in the 1970s and early 1980s as new surface-sensitive analytical tools became available.

A number of studies of glasses of various silicate glass compositions produced depth compositional profiles characteristic of Type II, III, IV, and V surfaces. The techniques used are summarized in Figure 20.4. They include Auger electron spectroscopy (AES) (Chappell and Stoddart 1974, Rynd and Rastogi 1974, Pantano et al. 1975, Clark et al. 1979), secondary ion mass spectroscopy (SIMS) (Gossink et al. 1979), resonant nuclear reactions (RNR) (Lanford et al. 1979), electron spectroscopy for chemical analysis (ESCA) (Budd 1975, Hench et al. 1986), secondary ion photoemission spectroscopy (SIPS) (Bach and Bauke 1974, Baucke 1974), x-ray diffraction (XRD), and Fourier transform infrared reflection spectroscopy (FTIRRS) (Sanders et al. 1972, 1974, Clark et al. 1977). Each analytical method has a characteristic sampling depth, as illustrated in Figure 20.4. Thus, it is critical to select an analytical method commensurate with the expected range of surface attack. A method that is too limited in sampling depth, such as AES, is generally not suited to analyses of Type II or Type III glass surfaces, unless it is coupled with Argon ion milling to produce compositional profiles, as illustrated in Figure 20.5. A method such as FTIRRS is rapid and nondestructive but has a sampling depth that depends upon the refractive index of the glass (Darby et al. 2000).

Generally speaking, it is essential to establish a measurement protocol by trial and error of several analytical methods, and then select the one most suited to a particular type of composition and surface condition. Confirmation by a second method at the end of a study is often required to confirm the interpretation of results. The book "*Corrosion of Glass*" by Clark et al. (1979) provides a good example of the use of multiple analytical methods for investigating glass corrosion problems. Chapters 1, 8, 9, 10, 11, 13, 14, 18, 19, 20, 22 in the book "*Surface-Active Processes in Materials*" describe use of most of these methods (Darby et al. 2000, Lodding et al. 2000, Wicks 2000).

Film depths in the range of 0.01–1.0 μm are generally observed for Type II glasses. Type IV glasses typically exhibit films of 1.0–100 μm depth. Addition of network modifiers of high electric field strength to the glass, such as Ca^{+2} or Al^{3+}, changes Type IV surfaces to Type II, greatly decreases the thickness of the silica-rich films, and increases the film density and glass durability (Clark et al. 1977, 1979, Dilmore et al. 1978, 1979).

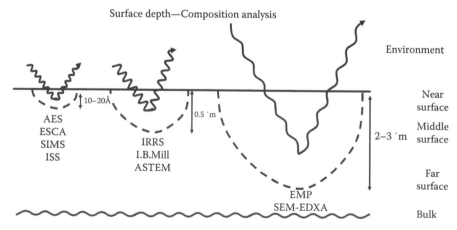

FIGURE 20.4 Instrumental methods for analyzing silicate glass surface compositions.

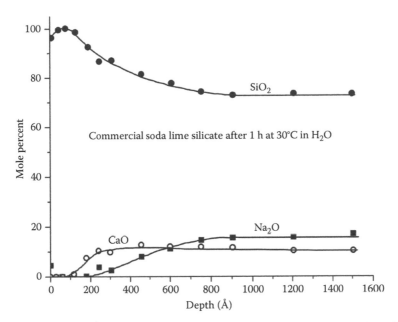

FIGURE 20.5 Surface compositional profile of a commercial soda lime–silica container glass after exposure to 37°C water for 1 h. Data obtained by AES analysis and Argon ion milling. Note the extended layer of alkali (Na) depletion and smaller depths of Ca ion depletion by ion exchange with Hydrogen ions. The glass has developed a Type II surface. (From Clark, D.E. et al., *Corrosion of Glass*, Books for Industry, New York, 1979.)

20.4.2 Type III Glass Surfaces

Soda-calcia-phospho-silicate glasses (Bioglass® composition 45S5) that formed a bond with living tissues were discovered in 1969. The seminal paper in 1971 indicated that the bonding was due to the formation of a special type of surface on the glasses when implanted in the body (Hench et al. 1971). The early work showed with transmission electron microscopy studies, IRRS and XRD, that the amorphous glass surface was converted into a silica gel layer (Hench and Paschall 1973, Wilson et al. 1981, Hench and Clark 1982, Hench 1991, Hench et al. 2004). The silica-rich film quickly became coated with a polycrystalline layer of hydroxyapatite (HA) that had the same XRD pattern as natural bone mineral (Hench et al. 1971, 2004). Figure 20.6 shows this sequence of surface reactions. The first five stages of reactions occur on the glass surface by the reactions described by Equation 20.1. The following seven stages of reaction shown in Figure 20.6 occur on the biological side of the interface.

Additional surface studies of bioactive glasses (Pantano et al. 1974, Ogino et al. 1980) and certain compositional ranges of $Li_2O–Al_2O_3–SiO_2$ glasses (Dilmore et al. 1978) showed that dual films developed on the glass. The secondary films were either hydrated $CaO–P_2O_5$ or $Al_2O_3–SiO_2$ layers. Chemical depth profiles using AES showed that the dual films formed on top of a silica-rich film, which resulted from rapid ion exchange of alkali for protons (or hydronium ions). Figure 20.7 shows an example of the surface compositional profile of the dual reaction layer characteristic of a Type III surface. Other work showed that such a dual film could be formed by addition of either phosphate (Ogino et al. 1980) or aluminum ions in solution (Dilmore et al. 1979) as well as through release of phosphate or aluminum ions from the glass.

Analysis of anion concentrations of the dual silica–apatite film formed on bioactive glasses showed exchange of CO_3^{2-} ions for OH^- ions similar to that found for bone (Hench et al. 2004). Incorporation of CaF_2 in the glass results in a fluorapatite film forming on the glass surface.

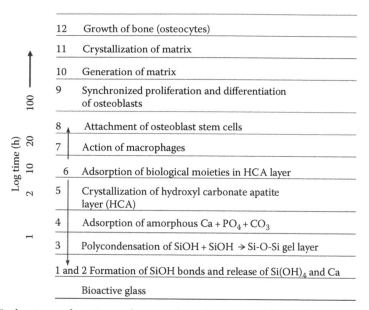

FIGURE 20.6 Twelve stages of reaction to form new bone (osteogenesis) bonded to a 45S5 bioactive glass surface. Stages 1–5 are controlled by the kinetics shown in Equation 20.1. (From Hench, L.L., *J. Am. Ceram. Soc.*, 74(7), 1487, 1991.)

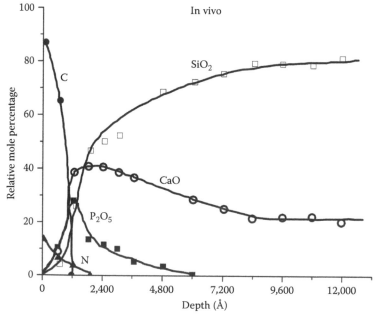

FIGURE 20.7 Example of stable bilayer surface films formed on a bioactive glass (45S5). The data were obtained by AES analysis of the surface after only 1 h exposure to a living rat bone. Ion exchange of the sodium in the glass has already progressed beyond the sampling depth of 1200 nm, obtained by Argon ion milling. The bulk composition of 42 mole% silica has been converted to a surface with a silica-rich gel layer containing 80 mole% silica. A calcium phosphate-rich HCA layer has formed on top of the silica layer to a thickness of nearly 800 nm (8000 Å). (From Ogino, M. et al., *J. Biomed. Mater. Res.*, 14(1), 55, 64, 1980.)

Studies also showed that the effectiveness of a number of alkaline corrosion inhibitors (Fujiu et al.), such as soluble calcium or beryllium salts, is due to the formation of dual films composed of insoluble alkaline earth silicate compounds (Oka et al. 1979, Oka and Tomozawa 1980).

This type of glass surface with a dual protective film is designated Type IIIA in Figure 20.1. The thickness of the secondary films can vary considerably, from as little as 0.01 μm for Al_2O_3–SiO_2-rich layers to as much as 30 μm for CaO–P_2O_5-rich layers on top of a >100 μm thick primary silica-rich layer.

The formation of a Type III surface is due to a combination of the re-polymerization of SiO_2 on the glass surface by the condensation of the silanols (Si–OH) formed from the Stage 1 reaction, for example,

$$Si-OH+OH-Si \rightarrow Si-O-Si+H_2O \qquad (20.2)$$

This Stage 3 reaction contributes to the enrichment of amorphous surface SiO_2 characteristic of Types II, III, and IV surface profiles (see Figure 20.1). It is described by the third term in Equation 20.1. This reaction is probably interface controlled with a time dependence of $+k_3t^{1.0}$, that is, x = 1.0, in Equation 20.1.

The fourth term in Equation 20.1, $+k_4t^y$ (Stage 4), describes the precipitation reactions, which result in multiple films characteristic of Type III glasses. When only one secondary film forms, the surface is Type IIIA.

20.4.3 Type IIIA Bioactive Glass Surfaces

One of the most important aspects of the CaO-P_2O_5-rich Type IIIA surface is that it is bioactive, that is, such compositions form a chemical bond to living bone (Hench et al. 1971, 2004, Hench and Paschall 1973, Wilson et al. 1981, Hench and Clark 1982, Hench 1991). When the bioactivity is sufficiently high, the material forms a bond to living soft tissues as well as bone (Wilson et al. 1981). The bonding is due to the formation of a hydroxycarbonate apatite (HCA) layer on top of the SiO_2-rich layer at the same rate as living cells are proliferating and differentiating. The kinetics of the Stage 4 precipitation reaction closely matches various biochemical reaction rates in repairing tissues. The biochemical reaction rates are controlled by a combination of genetics and local chemical and biomechanical factors. The reaction rates of the bioactive material in Stage 4 are controlled by composition of the glass and local chemical and biological environments. When the biochemical and physical chemical rates are equivalent, bonding occurs.

The HCA crystallites nucleate and bond to the interfacial metabolites, such as mucopolysaccharides, glycoproteins, and especially Type I collagen (Hench and Paschall 1973, Hench and Clark 1982, Hench et al. 2004). This mechanism of incorporation of organic biological constituents occurs very rapidly within the dynamic HCA surface and is the initial step in establishing bioactivity and bonding of tissues to the implant surface. Type IIIA compositions lie within the bioactive bonding region A in Figure 20.8 (Hench 1991).

Increasing the SiO_2 content of the glass toward 60 wt% reduces the rates of network dissolution, silica re-polymerization, and HCA precipitation and crystallization, that is, all the kinetics terms in Equation 20.1. Bioactivity is reduced proportionally; that is, as SiO_2 content of the glass increases, the values of k_1, k_2, k_3, and k_4 decrease and bioactivity decreases. At glass SiO_2 contents of 60 wt% and higher, the HCA layer does not form within 2–4 weeks and the glasses are not bioactive, region B in Figure 20.8. The glasses in region B have Type I and II surfaces. The biochemical reaction rates are much faster than Stage 4 kinetics for Type I and II surfaces. As a consequence, nonbonding scar tissue forms at the glass–tissue interface.

When there is insufficient SiO_2 or CaO, the glasses dissolve when implanted in tissues, region C in Figure 20.8. Such glasses exhibit Type IV and V surfaces. Stages 1 and 2 kinetics of Type IV and V glasses are much more rapid than the biochemical reaction rates and Stage 4 cannot occur. Therefore, the glasses in region C are unstable in living tissues.

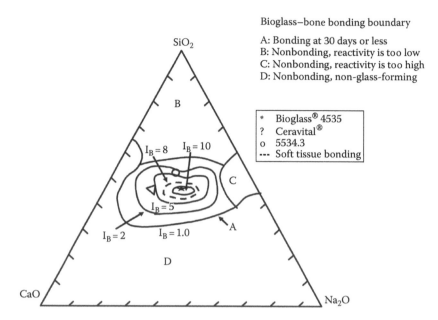

FIGURE 20.8   Ternary kinetics surface phase diagram showing index of bioactivity (I_B) as a function of composition. All compositions in region A contain 6 wt% P_2O_5. Bioglass® is a registered trademark held by the University of Florida, Gainesville, FL.

The thickness of the dual reaction layers (SiO_2 rich and CaO-P_2O_5 rich) also depends upon the initial glass composition. As the boundary between regions A and B in Figure 20.8 is approached, the interfacial thickness decreases, that is, the Type IIIA surface becomes a Type II surface at the boundary. The rate of bioactivity decreases similarly, where the index of bioactivity (I_B) is inversely proportional to the time required for more than 50% of the interface to be bonded to bone ($t_{0.5bb}$), for example, $I_B = \log_{10} 1/t_{0.5bb}$ (days^{-1}). The bioactivity contours shown in Figure 20.8 indicate the compositional dependence of I_B.

Thus, Type IIIA bioactive glasses have I_B values ranging from 12.5 to 1. All Type I, II, IV, or V glasses have an I_B value = 0. The I_B value of a given composition depends upon the relative rates of all four stages of reaction given in Equation 20.1. The diagram in Figure 20.8 is therefore a kinetics surface phase diagram, not a thermodynamic equilibrium phase diagram. The environmental behavior of all multicomponent silicate glass systems can be described with a similar kinetics surface phase diagram by using a similar approach based upon the relative reaction rates of Equation 20.1.

Recent studies show that it is the ionic dissolution products (soluble Si and Ca ions) released during Stage 1 and Stage 2 attacks of the bioactive glass that are critical for growth of new bone (osteogenesis) (Hench et al. 2002). The soluble ions are responsible for upregulating and activating seven families of genes that control the proliferation and differentiation of bone stem cells. The HCA layer on the glass surface provides a host for the stem cells and an anchor for the membranes of the growing cells that enhance formation of bone bridges between the bioactive glass particles (Hench et al. 2002, 2004).

20.4.4   Type IIIB Surfaces

The kinetics surface phase diagram approach developed for bioactive glasses was equally applicable in the development of stable glass systems for long-term storage of high level radioactive wastes. As illustrated in Figure 20.1, Type IIIB, multiple barrier surfaces constitute an important class of glass surface reactions that are characteristic of alkali borosilicate (ABS) nuclear waste glasses.

Nuclear waste glasses, also termed radwaste glasses, may contain as many as 40 differing constituents. A summary of the development and corrosion behavior of these glasses is given by George Wicks (2000). A number of ABS nuclear waste glasses that exhibit Type IIIB surface behavior have elemental leach rates as low as 0.02–0.2 g/m²/day with a time dependence of static leaching of $t^{0.5}$–$t^{0.2}$ or less after 28 days at 90°C.

Although a short (several days) period of alkali–hydrogen ion exchange may occur for Type IIIB glasses, the dominant, long-term mechanism controlling corrosion is a combination of matrix dissolution followed by incongruent dissolution and a series of solution/precipitation reactions (Hench 1988). The extent of matrix dissolution and onset of surface precipitation will depend on the time required for various species in the glass to reach saturation in solution. Saturation of a given species (i) is a function of the initial solution pH, amount of alkali in the glass and rate of alkali release, temperature, initial concentration of species (i) in the solution, exposed surface area (SA) to solution volume (V), for example, (SA/V) that influences solution concentration, or flow rate that also affects solution concentration. Until saturation of some species in solution is reached, the glass dissolves congruently at a rate proportional to $-k_2 t^{1.0}$, that is, Stage 2 in Equation 20.1 dominates.

When solution saturation of species (i) is reached for a Type IIIB glass, there is no longer any driving force for that species to leave the glass surface. Consequently, species (i) will accumulate at the glass–solution interface as the matrix dissolves, leaving the remaining concentration of species (i) behind in the glass. If the matrix dissolution releases alkali ions, as will be the case for most glasses, there will be a concomitant rise in pH proportional to the flow rate and SA/V of the system. An increase in pH can have several simultaneous effects on the glass, the solution, and the glass–solution interface. At the new pH, a second species (j) may reach solution saturation and subsequently be retained in the glass surface along with species (i). The extent of incongruent dissolution of the glass is thereby increased.

In addition, solution pH can have either one of three effects on species (i), previously in saturation: (1) It remains saturated but at a higher concentration; (2) it becomes supersaturated and precipitates either on the glass or other surface as a colloid; or (3) it becomes undersaturated and species (i) in the glass surface once again begins to be released by incongruent dissolution. The sequence of events that occurs is predictable based upon the solubility limits of each species at a given pH, as shown by Grambow (1982) and Wicks (2000). For example, Grambow's calculations showed that for ABS-based radwaste glasses, the $Fe(OH)_3$ solubility limit should be exceeded over a broad range of pH. Therefore, nuclear waste glasses containing Fe oxides should concentrate Fe within surface layers. Zinc, Nd, Sr, and Ca should be concentrated as well in nearly neutral or slightly alkaline solutions with Na and B depleted. Many studies confirm the prediction of solubility limited surface layers in ABS nuclear waste glasses (Hench et al. 1980, 1982, McVay and Buckwalter 1980, Grambow 1982, Lodding et al. 2000).

A consequence of the formation of the multiple barriers, for example, Type IIIB films, is slow overall leachability of many nuclear waste glasses over a pH range from 4.5 to 9.5. Wicks' data show that over the pH range expected for repository groundwaters, ABS glass leachability is lowest (Wicks 2000). Thus, we can conclude that the solubility limits that establish the equilibrium ionic concentrations for the groundwaters should also establish the multiple barrier films to protect nuclear waste glasses in contact with those groundwaters.

There is a compositional dependence of the leaching behavior of Type IIIB glasses that is similar to Type IIIA glasses. Figure 20.9 summarizes the regions of high leach rates (45 g/m²/day) vs very low leach rates (0.02 g/m²/day) for a large number of ABS nuclear waste glasses all tested under equivalent 90°C static leach conditions (Hench and Hench 1983). The multiple films form with most effectiveness in a narrow composition range where the SiO_2 content is approximately 45%–50% SiO_2. This result is similar to the findings of bioactive glass kinetics where the I_B values are highest at 50%–55% SiO_2. This is because both Type IIIA and IIIB glass surface behaviors are dominated by Stages 3 and 4 kinetics terms after a very rapid period of Stages 1 and 2 kinetics.

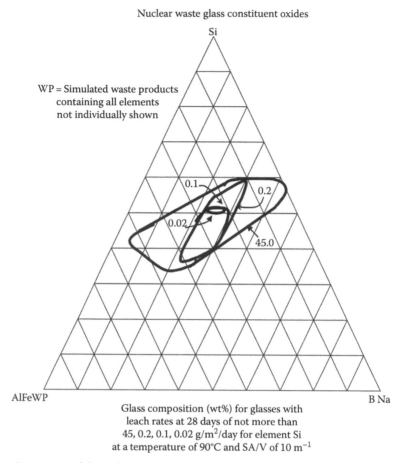

FIGURE 20.9 Compositional dependence of nuclear waste glass leaching. Note that the lowest levels of leaching occur in a narrow compositional range. (From Hench, A.A. and Hench, L.L., *Nucl. Chem. Waste Manage.*, 4(3), 231, 1983.)

20.4.5 Types of Glass Environment Interactions: Weathering

The prior discussion has generally been associated with exposure of the glass surface to an aqueous environment where the ion exchange and surface reactions can continue for an indefinite period of time. This condition is illustrated as cases (a) and (b) in Figure 20.10. When glass is exposed to an environment containing water vapor and reactive gases such as CO_2, the resulting surface alterations may be different from those observed when the glass is immersed in water (Clark et al. 1979). This type of exposure is called "weathering." Two alternative weathering conditions are shown schematically as cases (c) and (d) in Figure 20.10. The surface morphology and composition of a glass after weathering are often not homogeneous over the surface. Water droplets can form and result in aqueous corrosion with very high SA/V ratios. In contrast, other parts of the surface may be in contact with water vapor only. When part of the surface is not in contact with liquid, there is no means for the alkaline precipitates to be removed. The rate of weathering is controlled by the availability of water vapor and reactive gases capable of forming stable salts. Weathering is also strongly affected by cycles of increased and decreased relative humidity and washing of the reaction products from the surface (see Clark et al. 1979 and Hench and Clark 1978 for detailed discussions of weathering phenomena and the surface reaction layers that result).

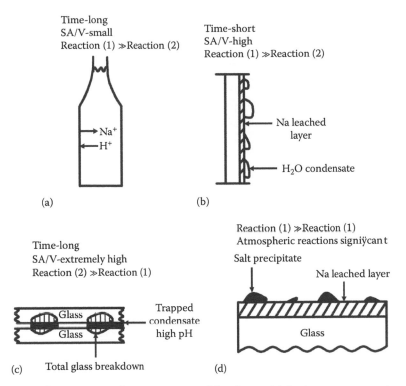

FIGURE 20.10 Typical exposure conditions encountered for glasses. (a) Static aqueous corrosion, (b) dynamic (flow) aqueous corrosion, (c) static weathering, (d) dynamic weathering. (From Clark, D.E. et al., *Corrosion of Glass*, Books for Industry, New York, 1979.)

20.5 Summary of Mechanisms of Glass Corrosion

To control silicate glass corrosion requires understanding of (1) formation of surface layers and (2) stability of surface layers.

The most important fact to recognize in understanding silicate glass corrosion is that many surface kinetics factors, for example, Stages 1–4, may occur simultaneously. Consequently, there is a high probability that heterogeneous attack may take place. The net effect is that at least ten differing mechanisms of glass corrosion can occur. Examples of many of these mechanisms follow as Figures 20.11 through 20.17.

20.5.1 The 10 Classical Mechanisms of Glass Corrosion

1. Ion exchange or selective leaching
   Involves the exchange of mobile species from the glass with protons or hydronium ions from the solution. Results in surface film formation.
2. Network dissolution (congruent and surface film dissolution)
   Involves breaking down of structural bonds in the glass or surface film. May occur uniformly or locally.
   Examples of both ion exchange and network dissolution are shown in Figure 20.11.
3. Pitting
   Localized network dissolution due to surface heterogeneities, stresses, or defects.
   Figure 20.12 is an example of glass pitting corrosion.

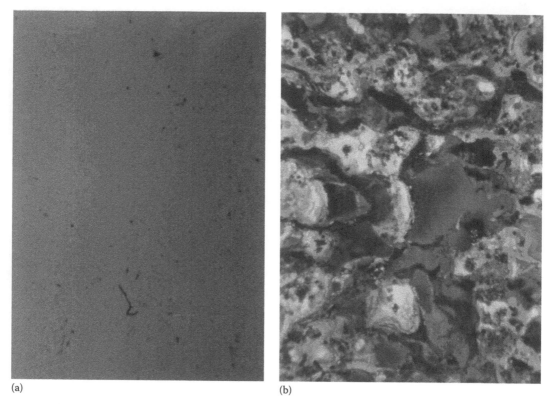

(a)   (b)

FIGURE 20.11 Examples of an ABS radwaste glass after exposure to two differing environments. Both samples were buried in a deep (>1 km) STRIPA granite mine shaft in Sweden for 1 year with 90°C heating. Both samples were identical compositions made from the same glass batch. (a) The sample on the left had a glass–glass interface, simulating a condition of cracks in a glass radwaste canister. (b) The sample on the right had a glass–bentonite clay interface, simulating use of an overpack in a waste storage site. There was very little surface reaction on the glass on the left (a). The Type IIIB surfaces were less than a micrometer deep, as revealed with SIMS profiling. In contrast, in (b) the ABS sample in contact with bentonite had very deep ion exchange and network dissolution due to the clay adsorption of the ions from the glass. The rate of depletion did not slow due to the continual removal of the ions, which prevented formation of a protective Type IIIB surface. (From Lodding, A.R. et al., Progress of corrosion and element kinetics of HLW glasses in Geological Burial; Evaluation by SIMS, in D.E. Clark, D.C. Folz, and J.H. Simmons, Eds., *Ceramic Transactions*, Vol. 101, The American Ceramic Society, Westerville, OH.)

4. Solution concentration

   Involves the concentration of the solution with respect to species from the glass. May result in a reduction of the corrosion rate.

5. Precipitation

   Involves the formation of insoluble compounds on the glass surface by reaction of the dissolved constituents from the glass with species already present in the solution. May be influenced by solution pH. Figure 20.13 is an example of surface precipitation of a crystalline phase on the corroded glass surface.

6. Stable film formation

   Involves the alteration of the glass surface composition by either diffusion processes or interfacial controlled reactions. Figure 20.14 is an example of formation of a stable layer on the glass to protect it from network dissolution.

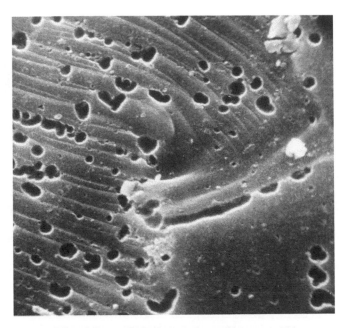

FIGURE 20.12 Example of pitting corrosion on a silicate glass surface.

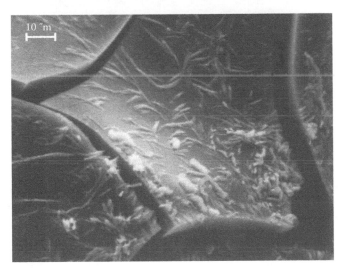

FIGURE 20.13 Example of precipitates forming on the surface of an ion exchanged layer on a silicate glass. Note the fracture of the silica-rich layer due to drying.

7. Surface layer exfoliation

Involves the flaking off of surface layers formed by ion exchange. Usually occurs after the glass has been removed from the solution due to dehydration and accompanying stresses. Figure 20.15 is an example.

8. Weathering

Involves the interaction of humidity and reactive gases from the atmosphere with the glass surface. Usually results in the accumulation of precipitates (both soluble and insoluble) on the glass. Figure 20.16 is an example.

FIGURE 20.14   Example of a stable surface film formed on a silicate glass. The composition is 70% SiO_2 and 30% Na_2O, which would normally be an unstable Type IV glass. However, this 4 cm long sample was buried in a geologic site in Ballidon, England, for 12 months. The site had a large concentration of soluble calcium salts, which formed a stable dual reaction layers on the surface converting the glass to a Type IIIA surface.

FIGURE 20.15   Example of a Type IV silicate glass that has a surface layer fractured and exfoliated due to drying after removal from the aqueous environment. Note the dimensions of the flakes are in the range of 200 μm wide and approximately 5 μm thick.

9. Stress corrosion

   Interaction of tensile stresses and chemical reactions leading to accelerated attack.

   The history and scientific basis of understanding these phenomena are reviewed by Freiman and Mecholsky (2012). Environmentally enhanced crack growth in silicate glasses is one of the major factors contributing to failure and limits the range of use of this important class of materials.

10. Erosion–corrosion

    This type of damage is due to a combination of mechanical abrasion and/or chemical reactions on the glass surface. Typical examples are high velocity wind or water carrying sand over glass surfaces. Figure 20.17 is an example.

Which of these 10 mechanisms are operative at any specific time depends upon the combination of factors listed in Table 20.1 and the reaction rate (Equation 20.1) that is dominant at that time. Various combinations of factors will yield specific types of corrosion mechanisms and a specific type of glass surface. However, changing glass composition by only a few percent can completely change the surface

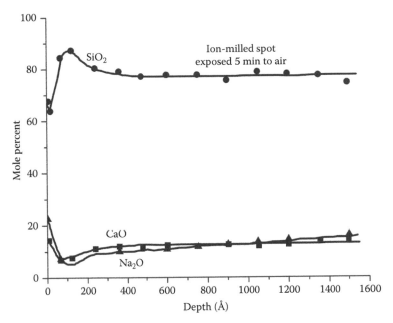

FIGURE 20.16 Example of rapid weathering of a commercial soda lime–silica commercial glass container. The surface has been cleaned by Argon ion beam etching and then exposed to ambient air for 5 min. The compositional profile shows that ion exchange (Stage 1 reaction) has already progressed to a depth of 100 nm (1000 Å). The Ca and Na ions are accumulated as a weathering layer on the surface on top of the silica-enriched layer created by the ion depletion. This is an example of a Type II glass surface, as depicted in Figure 20.1.

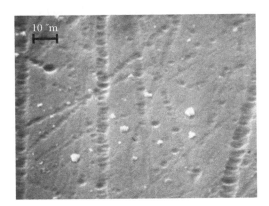

FIGURE 20.17 Example of selective corrosion attack of polishing surface scratches on a silicate glass. The combination of mechanical abrasion and chemical attack can be particularly severe and lead to pitting corrosion shown in Figure 20.12.

behavior, as shown in Figures 20.8 and 20.9. This offers the challenge, complexity, and at times the frustration of understanding and controlling glass corrosion.

20.6 Further Reading

A special symposium published in 2000 by the American Ceramic Society *"Surface-Active Processes in Materials"* edited by Clark, Folz, and Simmons (see Refs. [49–52], for example) provides an excellent source for additional reading that includes methods for protecting glass surfaces, discussion of numerous test methods, methods for molecular modeling of surface reactions with various atmospheres, and surface interactions of porous materials.

Acknowledgments

This chapter gives me the opportunity to acknowledge the creative research of many students that I have had the pleasure of supervising while working in the field of glass, glass–ceramic and gel–glass surface chemistry. They include Steve Freiman, Don Kinser, David Clark, David Sanders, Buddy Clark, Ed Ethridge, Morris Dilmore, Carlo Pantano, Fumio Ohuichi, Mike Ogino, Akio Fuwa, David Greenspan, Joy Barrett, Bill Lacefield, Walt McCracken, Ron Palmer, Larry Warren, Gerard Orcel, Bing-fu Zhu, Rounan Li, Shi-ho Wang, Martin Wilson, Wander Vasconcelas, Mark Ramer, Shuyan Liu, Steve Wallace, Jim Kunetz, Taipau Chia, Doug Parsell, Keith Lobel, Marivalda Periera, Jeff Thompson, Laura Elsberg, Rodrigo Orefice, Kevin Powers, Ian Thompson, Julian Jones, Liz Fielder, Russell Pryce, Maria Bellantone, Priya Saravanapavan, Jason Maroothynaden, Vicky Shirtliff, Dan Clupper, Ioan Notingher, and Chris Owen. Thanks to all of you for helping create this important aspect of materials science and engineering.

References

Bach, H. and F.G.K. Bauke. 1974. Investigations of reactions between glasses and gaseous phases by means of photon emission induced during ion beam etching. *Physics and Chemistry of Glasses*, 15, 123.

Bacon, F.R. and G.L. Calcamuggio. 1967. Effect of heat treatment in moist and dry atmospheres on chemical durability of soda-lime glass bottles. *American Ceramic Society Bulletin*, 46(9), 850–855.

Baucke, F.G.K. 1974. Investigation of surface layers, formed on glass electrode membranes in aqueous solutions, by means of an ion sputtering method. *Journal of Non-Crystalline Solids*, 14(1), 13–31.

Budd, S.M. 1975. In D.E. Dday, Ed., *Glass Surfaces*, North Holland Publishing Company, Amsterdam, the Netherlands.

Chappell, R.A. and C.T.H. Stoddart. 1974. An Auger electron spectroscopy study of float glass surfaces. *Physics and Chemistry of Glasses*, 15, 130.

Clark, D.E., E.C. Ethridge, M.F. Dilmore, and L.L. Hench. 1977. Quantitative analysis of corroded glass using infrared frequency shifts. *Glass Technology*, 18, 121.

Clark, D.E. Jr., C.G. Pantano, and L.L. Hench. 1979. *Corrosion of Glass*, Books for Industry, New York.

Darby, G., D.E. Clark, and J.H. Simmons. 2000. FT-IRRS analysis of surface reactions in alkali silicate glasses, in D.E. Clark, D.C. Folz, and J.H. Simmons, Eds., *Surface-Active Processes in Materials*, *Ceramic Transactions*, Vol. 101, The American Ceramic Society, Westerville, OH, pp. 111–122.

Das, C.R. 1969. Theoretical aspects of the corrosion of glass: Conclusion. *Glass Industry*, 50, 422–427 and 483–485.

Das, C.R. and R.W. Douglas. 1967. Studies on the reaction between water and glass. Part 3. *Physics and Chemistry of Glasses*, 8(5), 178–184.

Dilmore, M.F., D.E. Clark, and L.L. Hench. 1978. Aqueoous corrosion of Lithia-Alumina-Silicate glasses. *American Ceramic Society Bulletin*, 57(11), 1040–1048.

Dilmore, M.F., D.E. Clark, and L.L. Hench. 1979. Corrosion behavior of Lithia Disilicate glass in aqueous solutions of Aluminum compounds. *American Ceramic Society Bulletin*, 58(11), 1111–1114.

Douglas, R.W. and T.M.M. El-Shamy. 1967. Reactions of glasses with aqueous solutions. *Journal of the American Ceramic Society*, 50(1), 1–8.

Douglas, R.W. and J.D. Isard. 1949. The action of Sulphur Dioxide and of water on glass surfaces. *Journal of the Society of Glass Technology*, 33, 289.

Freiman, S.W. and J. Mecholsky. 2012. *Fracture of Brittle Materials*, John Wiley and Sons, NY.

Fujiu, T., M. Ogino, M. Kariya, and T. Schimura. Private Communication.

Gossink, R.G., H.A.M. de Grefte, and H.W. Werner. 1979. SIMS analysis of aqueous corrosion profiles in Soda-Lime-Silica glass. *Journal of the American Ceramic Society*, 62(1–2), 4–8.

Grambow, B. 1982. The role of metal ion solubility in leaching of nuclear waste glasses, in *5th International Symposium on the Scientific Basis for Radioactive Waste Management*, Berlin, Germany.

Hench, A.A. and L.L. Hench. 1983. Computer analysis of nuclear waste glass composition effects on leaching. *Nuclear and Chemical Waste Management*, 4(3), 231–238.

Hench, L.L. 1988. Corrosion of Silicate glasses: An overview. *MRS Online Proceedings Library*, 125, 189–200.

Hench, L.L. 1991. Bioceramics: From concept to clinic. *Journal of the American Ceramic Society*, 74(7), 1487–1510.

Hench, L.L. and D.E. Clark. 1978. Physical chemistry of glass surfaces. *Journal of Non-Crystalline Solids*, 28(1), 83–105.

Hench, L.L. and A.E. Clark. 1982. In D.F. Williams, Ed., *Biocompatibility of Orthopedic Implants*, CRC Press, Boca Raton, FL, pp. 100–122.

Hench, L.L., D.E. Clark, and A.B. Harker. 1986. Nuclear waste solids. *J. Mat. Sci.* 21, 1457–1478.

Hench, L.L., D.E. Clark, and E.L. Yen-Bower. 1980. Corrosion of glasses and glass-ceramics. *Nuclear and Chemical Waste Management*, 1(1), 59–75.

Hench, L.L., J.W. Hench, and D.C. Greenspan. 2004. Bioglass: A short history and Bibliography. *Journal of the Australian Ceramic Society*, 40(1), 1–42.

Hench, L.L. and H.A. Paschall. 1973. Direct chemical bond of bioactive glass-ceramic materials to bone and muscle. *Journal of Biomedical Materials Research*, 7(3), 25–42.

Hench, L.L., R.J. Splinter, W.C. Allen, and T.K. Greenlee. 1971. Bonding mechanisms at the interface of ceramic prosthetic materials. *Journal of Biomedical Materials Research*, 5(6), 117–141.

Hench, L.L., L. Werme, and A. Lodding. 1982. Burial effects on nuclear waste glass, in W. Lutze, Ed., *Scientific Basis for Radioactive Management V*, Elsevier Scientific Publication Company, New York, pp. 153–162.

Hench, L.L., I.D. Xynos, A.J. Edgar, L.D.K. Buttery, J.M. Polak, J.P. Zhong, X.Y. Liu, and J. Chang. 2002. Gene activating glasses. *Journal of Inorganic Materials*, 17(5), 897–909.

Lanford, W.A., K. Davis, P. Lamarche, T. Laursen, R. Groleau, and R.H. Doremus. 1979. Hydration of soda-lime glass. *Journal of Non-Crystalline Solids*, 33(2), 249–266.

Lodding, A.R., P. Van Iseghem, and L.O. Werme. 2000. Progress of corrosion and element kinetics of HLW glasses in Geological Burial; Evaluation by SIMS, in D.E. Clark, D.C. Folz, and J.H. Simmons, Eds., *Surface-Active Processes in Materials, Ceramic Transactions*, Vol. 101, The American Ceramic Society, Westerville, OH, pp. 123–140.

Lyle, A.K. 1943. Theoretical aspects of chemical attack of glasses by water. *Journal of the American Ceramic Society*, 26(6), 201–204.

McVay, G.L. and C.Q. Buckwalter. 1980. The nature of glass leaching. *Nuclear Technologies*, 51, 123.

Ogino, M., F. Ohuchi, and L.L. Hench. 1980. Compositional dependence of the formation of calcium phosphate films on bioglass. *Journal of Biomedical Materials Research*, 14(1), 55–64.

Oka, Y., K.S. Ricker, and M. Tomozawa. 1979. Calcium deposition on glass surface as an inhibitor to alkaline attack. *Journal of the American Ceramic Society*, 62(11–12), 631–632.

Oka, Y. and M. Tomozawa. 1980. Effect of alkaline earth ion as an inhibitor to alkaline attack on silica glass. *Journal of Non-Crystalline Solids*, 42(1–3), 535–543.

Pantano, C.G., A.E. Clark, and L.L. Hench. 1974. Multilayer corrosion films on bioglass surfaces. *Journal of the American Ceramic Society*, 57(9), 412–413.

Pantano, C.G. Jr., D.B. Dove, and G.Y. Onoda. Jr. 1975. Glass surface analysis by Auger electron spectroscopy. *Journal of Non-Crystalline Solids*, 19(0), 41–53.

Rana, M.A. and R.W. Douglas. 1961. The reaction between glass and water. Part 1: Experimental methods and observations. *Physics and Chemistry of Glasses*, 2(6), 179–195.

Rynd, J. and A.K. Rastogi. 1974. Auger electron spectroscopy—A new tool in the characterization of glass fiber surfaces. *American Ceramic Society Bulletin*, 53, 631–637.

Sanders, D.M., W.B. Person, and L.L. Hench. 1972. New methods for studying glass corrosion kinetics. *Applied Spectroscopy*, 26(5), 530–536.

Sanders, D.M., W.B. Person, and L.L. Hench. 1974. Quantitative analysis of glass structure with the use of infrared reflection spectra. *Applied Spectroscopy*, 28(3), 247–255.

Tsuchihashi, S. and E. Sekido. 1959. On the Dissolution of Na2O–CaO–SiO2 glass in acid and in water. *Bulletin of the Chemical Society of Japan*, 32(8), 868–872.

Wang, F.F.-Y. and F.V. Tooley. 1958a. Detection of reaction products between water and Soda-Lime-Silica glass. *Journal of the American Ceramic Society*, 41(11), 467–469.

Wang, F.F.-Y. and F.V. Tooley. 1958b. Influence of reaction products on reaction between water and Soda-Lime-Silica glass. *Journal of the American Ceramic Society*, 41(12), 521–524.

Weyl, W.A. 1947. Some practical aspects of the surface chemistry of glass: IV. *Glass Industry*, 28(8), 408–412, 428–432.

Wicks, G.G. 2000. Nuclear waste glasses-suitability, surface studies and stability, in D.E. Clark, D.C. Folz, and J.H. Simmons, Eds., *Surface-Active Processes in Materials, Ceramic Transactions*, Vol. 101, The American Ceramic Society, Westerville, OH, pp. 111–122.

Wilson, J., G.H. Pigott, F.J. Schoen, and L.L. Hench. 1981. Toxicology and biocompatibility of bioglasses. *Journal of Biomedical Materials Research*, 15(6), 805–817.

Zagar, L. and L. Schillmoeller. 1960. Physical-chemical processes in water extraction of glass surfaces. *Glastechnische Berichte*, 33(4), 409–416.

21

Cement and Concrete

Carolyn M. Hansson
University of Waterloo

Laura Mammoliti
Lafarge Canada Inc.

Tracy D. Marcotte
CVM Engineers

Shahzma J. Jaffer
*AECL–Chalk River
Laboratories*

21.1 Introduction

Although concrete, in various forms, has been used as a structural material for over 2000 years, today's Portland cement-based concrete is a far cry from the earlier versions. It is a highly sophisticated composite, with the possibility of tailoring the composition and microstructure to the specific application. The production of concrete exceeds that of any other engineered material: more than 400 billion cubic meters (or ~10 million tons) worldwide. Its popularity is due to a number of factors, including the availability of the raw materials (basically limestone, clay, water, sand, and stone), its formability into complex shapes at ambient temperatures, and its low cost compared with the alternatives.

The role of the Portland cement is to react with water to form a rigid product known as cement paste, which adheres to the particles of sand and stone together to form a hard solid mass of the appropriate shape. While the amount of cement paste needed is only that which is sufficient to coat all the aggregate particles (typically 10%–20% of the concrete by weight), it is the component that is most susceptible to environmental degradation of the concrete and, therefore, can undermine the durability of the structure.

21.2 Background

The American Concrete Institute (ACI) defines durability as "… the ability of a material to resist weathering action, chemical attack, abrasion, and other conditions of service" (ACI 2009). In the context of the extensive number of structures constructed worldwide, durability also encompasses serviceability in various environments (i.e., resisting excessive deflections or vibrations), and the performance of any

TABLE 21.1 List of Typical Sources of Concrete Degradation according to Service Environment

Environment	Degrading Component	Source of Degradation
Marine, Section 21.6.1	Concrete	Sulfate attack
		Biological action
		Freeze–thaw damage
		Salt scaling
		Abrasion
		Impact
	Reinforcement[a]	Carbonation
		Chlorides
Agriculture and Sewage, Section 21.6.2	Concrete	Sulfate attack
		Acid dissolution
		Abrasion
		Impact
	Reinforcement[a]	Corrosive contaminants
		Acid attack
Transportation Systems, Section 21.6.3	Concrete	Freeze–thaw damage
		Salt scaling
		De-icing chemicals other than chlorides (air transportation)
	Reinforcement[a]	Chlorides (de-icing salts and marine)
		Carbonation

[a] These topics are covered in detail in Chapter 22 and, therefore, are not considered here.

maintenance and repairs for a specified time. Table 21.1 outlines typical environmental degradation processes, according to the service environments considered in this chapter.

In spite of these complex and occasionally synergistic mechanisms (i.e., x accelerates the effect of y degradation rate), concrete structures are routinely designed, mixed, placed, and cured to resist these environmental conditions and, thereby, obtain a minimum duration of service, anywhere from a few years to decades. A holistic appraisal of "service life" depends upon three interrelated criteria (Somerville 1992):

- Technical service life—the time in service before some unacceptable state is reached
- Functional service life—the time in service during which the structure fulfills its designed purpose
- Economic service life—the time in service while expenditures on maintenance and repair are deemed less expensive than decommissioning and replacement construction

Ultimately, the safe and effective management of structures in service depends upon optimizing these criteria to provide the most durable and cost-effective structure over its lifetime. These types of analyses are termed "life cycle assessment" (LCA) as part of environmental management, and "life cycle cost analyses" (LCCA) have been generally described by authorities such as ASTM E 917 and the ISO 14000 family of standards (ASTM 2005; ISO 2010), but must be routinely tailored for the specific type of concrete structure and the service environment.

Since the 1980s, there has been increasing momentum in understanding the role of concrete in sustainability, where sustainability means "meeting the needs of the present without compromising the ability of future generations to meet their own needs" (Commission 1987). Overall, the economic, social, and environmental benefits of concrete seem obvious if durable structures are constructed and their maintenance is managed. Strategies and examples of the truly sustainable aspects of concrete and its limitations have been described by Schokker (2010), and considerable research in this area is under way.

21.3 Applications

When asked "What is concrete used for?" the most likely response would be the construction of infrastructure such as bridges, dams, and roadways. However, concrete is used for a much wider range of applications due to its versatility, durability, and "perceived" ease of use and placement.

Cast-in-place concrete is supplied by ready-mix concrete producers and transported to construction sites via concrete mixer trucks. Its ability to withstand environmental degradation is determined in part by its composition, described in the following text, but also by the quality of the placement, compaction, and curing by the construction crew and the climatic conditions during these processes. Precast concrete, on the other hand, is made in a factory under controlled conditions and is not, therefore, subject to the vagaries of the elements. Concrete from both suppliers can be tailored to meet the needs of the customer. This is achieved through adjustments in the concrete mix design; by choice of the type and amount of cement used; the replacement level of cement with supplementary cementitious materials (described in Section 21.4.3); the type, size, and proportion of coarse and fine aggregates; and the type of chemical admixtures incorporated into the mix, for example, to control workability and the rate of setting and hardening. One ready-mix concrete producer is capable of producing thousands of different concrete mixes. For example, compressive strengths ranging from as low as 0.4 MPa (megapascals) to in excess of 100 MPa are reproducibly achieved on a regular basis. However, many more properties can be modified and controlled, including the rate of strength development, resistance to specific environmental conditions, the placement method, and color and other aesthetic characteristics. Placement in both vertical and horizontal applications, and even under water, is possible.

As a result of the advancing science of concrete, it is now used for more and more nontraditional applications. In order to provide rapid repair of bridges, airfield runways, etc., with as little disruption to traffic as possible, concrete has been formulated such that its full strength is developed in as little as 4 h. The benefits of constructing residential homes predominantly of ready-mixed concrete are being recognized, expanding its use from traditional foundation walls and basement floors, to interior and exterior walls and floors on all levels and even utilizing concrete in kitchen countertops. Pervious concrete allows rainwater to percolate through the pavement, as shown in Figure 21.1, thereby reducing storm runoff and regenerating the aquifers in the ground below. There is even a concrete capable of self-cleaning and of removing pollutants from the air (Figure 21.2; Cassar et al. 2003; Hüsken et al. 2009).

FIGURE 21.1 Pervious concrete that allows water to penetrate to the soil below. (Courtesy of Ready Mix Concrete Association, Mississauga, Ontario, Canada.)

FIGURE 21.2 Photocatalytic cement in the concrete of the Dives in Misericordia Church, Rome. (Courtesy of Essroc Cement, Nazareth, PA.)

21.4 Structure

21.4.1 Concrete

Concrete is a composite material consisting of aggregates (stones and sand) bound together by hardened cement paste. Although the aggregates can comprise more than 80% of the concrete, they are natural materials and not generally conducive to engineering. Thus, they must simply be selected from available sources as the most appropriate for the application. For example, certain silica- or carbonate-containing aggregates can react adversely with alkalies from the cement or from the environment and, therefore, should be avoided if there is likely to be exposure to alkalies. Moreover, natural aggregates can be modified for certain purposes, for example, for low-density concrete, expanded shales, clays, or slate are produced as lightweight aggregate by firing the natural material. However, this topic is beyond the scope of this chapter.

21.4.2 Cement Paste

Portland cement is formed by heating the reactants, predominantly limestone and clay, to 1450°C–1550°C, at which temperatures the components react to form four major oxide phases. The oxides are generally referred by their short forms as follows: $C=CaO$, $S=SiO_2$, $A=Al_2O_3$, and $F=Fe_2O_3$, and the major Portland cement constituents are tricalcium silicate ($3CaO \cdot SiO_2$ or C_3S), dicalcium silicate ($2CaO \cdot SiO_2$ or C_2S), tricalcium aluminate ($3CaO \cdot Al_2O_3$ or C_3A), and tetracalcium aluminoferrite ($4CaO \cdot Al_2O_3 \cdot Fe_2O_3$ or C_4AF). Minor constituents include K_2O, Na_2O, and MgO. After calcining, the cement clinker is ground with gypsum (calcium sulfate $CaSO_4 \cdot 2H_2O$ or $C\bar{S}H_2$), which is added to control the setting time.

When mixed with water, these cement constituents react to form their hydrates. The C_3S is responsible for the hardening of the concrete and, together with the C_2S, produces calcium-silicate hydrate (C–S–H), shown in Figure 21.3 (Alizadeh 2010). C–S–H is responsible for the binding and strengthening of the cement paste and constitutes more than 50% by volume of the solid phases in the paste. The C–S–H can be considered as a colloidal solid with its particles bound together with water and increasing in polymerization over time (Jennings 2000; Thomas and Jennings 2006). A by-product of the hydration of C_3S and C_2S is calcium hydroxide (CH), which forms as large plate-shaped crystals (Figure 21.4; Walker et al. 2006).

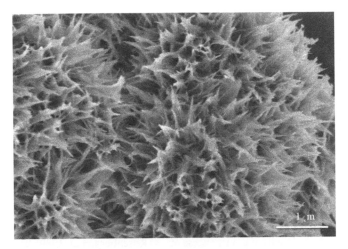

FIGURE 21.3 Scanning electron micrograph of calcium–silicate–hydrate (C–S–H), the main binding and strengthening phase of cement paste. (From Alizadeh, R., *Cement and Art* [Online]. Portland Cement Association, Skokie, IL, Available: http://www.cementlab.com/cement-art.htm, 2010.)

FIGURE 21.4 Calcium hydroxide plates and ettringite needles in cement paste. (From Walker, H.N. et al., Petrographic methods of examining hardened concrete: A petrographic manual, Virginia Transportation Research Council, Charlottesville, VA, 2006.)

Also shown in Figure 21.4 are needle-shaped crystals of ettringite produced by the reaction of C_3A and C_4AF with the gypsum. Ettringite (AFt) is calcium tri-sulfoaluminate hydrate in which iron substitutes for some of the aluminum. As the hardening progresses, the ettringite reacts with the remaining C_3A to form monosulfate (AFm, which is calcium mono-sulfoaluminate). As described in Section 21.6.2, this reaction can be reversed with the ingress of sulfates from the environment.

The amount of water required for the hydration is only ~23% by weight of cement (Neville 2006), but additional water is required to make a workable mix. The excess water is considered to be a separate phase in the hardened cement paste and plays a major role in both the structure and properties of the paste. The interlayer water in the C–S–H is believed to be held together and to hold the layers of the

TABLE 21.2   Classification of Pore Sizes according to Setzer

Name	Upper Radius	Type of Pore Water
Macro capillaries	2 mm	Bulk water
Meso capillaries	50 μm	Bulk water
Micro capillaries	2 μm	Bulk water
Meso pores	50 nm	Condensed water (physically absorbed)
Micro pores	2 nm	Structured surface water (the first monolayers of water chemisorbed in the surface of the pores)

Source: Setzer, M.J. (ed.), *Interaction of Water with Hardened Cement Paste,* The American Ceramic Society, Westerville, OH, 1990.

C–S–H together by hydrogen bonds. In the intervening spaces (the capillary pores) between the particles of the hydrated phases is a concentrated ionic solution containing the alkaline hydroxides (predominantly KOH and NaOH) with a pH in excess of 13 (Rosenberg et al. 1989). The solution is buffered to pH 12.6 by the less soluble $Ca(OH)_2$.

One classification of the different pores and the type of water they contain is given in Table 21.2 (Setzer 1990).

The consequence of the condensed and structured water is that it does not contribute to the ice formation at subzero temperatures, but remains as a liquid layer (Bager and Sellevold 1986). However, these pores are not regular cylindrical pores but are irregularly arranged often connected by narrow necks. The interconnected capillaries are responsible for the ingress of deleterious species from the environment and the leaching of components from the concrete.

The structure and composition of the cement paste phase is not uniform. Around each aggregate particle and along the surfaces of the formwork, there is an "interfacial zone" in which the capillary porosity and the concentration of CH and ettringite crystals are higher than in the bulk and the concentration of C–S–H is lower. Finally, the top surface of a concrete structure can play a major role in the environmental degradation. This is because, when the concrete is placed in the formwork, it is a slurry consisting of dense solid particles in a water matrix. After being placed and compacted, the solid particles have a tendency to settle, leaving a higher water/solids ratio at the top surface than in the bulk of the structure. This process is known as "bleeding" and the extra water as "bleed water." The greater the degree of bleeding, the weaker the resulting surface layers that are then more prone to abrasion and scaling as described in Section 21.6.1.

In summary, the "microstructure" of the cement paste component of concrete covers a large number of dimensions, from the colloidal C–S–H particles to the entrapped air voids, as illustrated schematically in Figure 21.5 (Mehta et al. 2006).

21.4.3 Supplementary Cementitious Materials

The cost of Portland cement, coupled with the environmental impact of the carbon dioxide released during its production and the need to find uses for by-products of other industrial processes, has led to the partial replacement of cement with supplementary cementitious materials (SCMs). The most common of these are ground granulated blast furnace slag (from integrated steel mills), fly ash (from coal-fired electrical power plants), silica fume (from the production of silicon and iron–silicon), and, in countries where rice is a major crop, rice husk ash. If not used in concrete, these materials would be regarded as waste materials. However, their incorporation in concrete actually has beneficial effects with regard to both the mechanical properties and durability.

As indicated earlier and shown schematically in Figure 21.6a, when Portland cement hydrates, water-filled pores are left in the interstices of the original clinker particles. If an SCM, such as fly ash, is present in the mix, it reacts with the dissolved CH in the pore solution formed from the cement to produce more

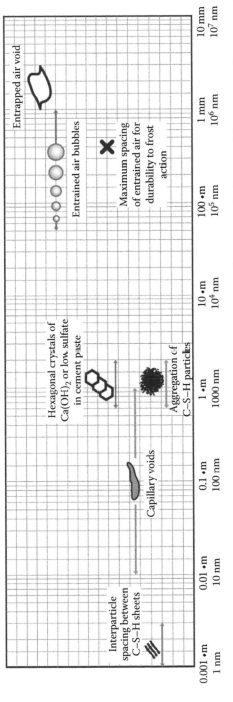

FIGURE 21.5 The range of dimensions of the features of cement paste microstructure. (From Mehta, P.K. et al., *Concrete: Microstructure, Properties and Materials,* McGraw Hill, New York, 2006.)

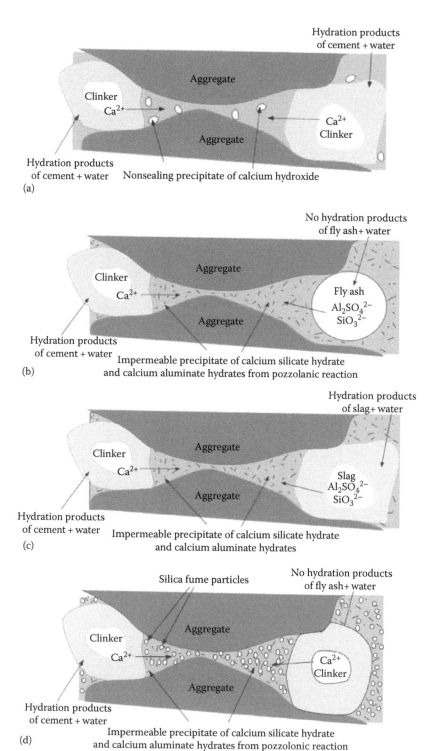

FIGURE 21.6 Hydration of (a) Portland cement–clinker, (b) Portland cement–fly ash, (c) Portland cement–blast furnace slag, and (d) Portland cement–silica fume mixtures, illustrating the reduction in porosity obtained by these supplementary cementing materials.

C–S–H and, by so doing, divides and reduces the porosity in the hardened cement paste, as illustrated in Figure 21.6b. Slag has an additional advantage in that it is itself hydraulic, reacting with water in the presence of alkalies, and producing both a shell of hydration products around the slag particle and also reacting in the interstices in a manner similar to that of fly ash (Figure 21.6c). Silica fume (also known as microsilica) has a much finer particle size (~0.1 μm) than those of either cement or the other SCMs (~50 μm). Consequently, it acts both as a filler and as an SCM (Figure 21.6d; Bache 1988; Neithalath et al. 2009).

21.5 General Properties

21.5.1 Physical Properties

As discussed earlier, concrete is composed of many individual materials some of which react together while others are usually inert. The hydration reaction involves volume changes, which, depending on the environment, the climatic conditions, and curing regime, can cause expansion or shrinkage (Aïtcin 1999). Chemical shrinkage is defined as "the reduction in absolute volume of solid and liquids in the paste resulting from cement hydration." An inadequate supply of water during and for some time after the setting and hardening period will result in "autogenous shrinkage: the macroscopic volume reduction of the paste, mortar or concrete" (Kosmatka et al. 2002) also known as self-desiccation (Aïtcin et al. 1997). An example of the volume changes is given in Figure 21.7. Because the shrinkage is restrained by the aggregates, this results in the development of microcracks throughout the cement paste and visible cracking such as that shown in Figure 21.8. Other mechanisms that cause volume changes leading to cracking are carbonation (Persson 1998) and thermal and moisture gradients. On the other hand, if the concrete is exposed to moisture for an extended period of time, concrete will expand (Aïtcin 1999; Igarashi et al. 2000). It is reasonable to assume that all concrete contains microcracks, which will propagate when subjected to tensile loads. Because it is impossible to anticipate the sizes, number, and location of these flaws in each location of each casting of concrete, the use of unreinforced concrete in situations where tensile strength is required is avoided. The tensile strength of concrete is only of the order of a few MPa. However, this limitation is usually overcome by the use of reinforcement either as unstressed bars, prestressed bars, posttensioned strand, or by fiber reinforcement.

Concrete is usually specified in terms of its 28 day compressive strength, which is dependent on a number of factors including the water/cementitious material (w/cm) ratio, SCMs, and aggregate characteristics. Nevertheless, one of the major limitations to its compressive strength is the weak interfacial

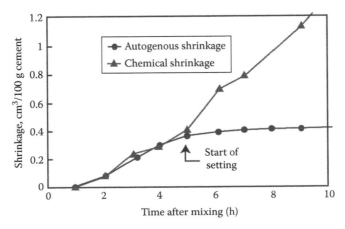

FIGURE 21.7 Relationship between autogenous shrinkage and chemical shrinkage of cement at early ages. (From Kosmatka, S.H. et al., *Design and Control of Concrete Mixtures*, Portland Cement Association, Ottawa, Ontario, Canada, 2002.)

FIGURE 21.8   Shrinkage cracking. (From Aïtcin, P.-C. et al., *Concr. Int.*, 19, 34, 1997.)

region between the paste and the aggregates. This means that the strength and toughness of concrete is, in fact, weaker than those of its individual components, as illustrated in Figure 21.9. The compressive strength of general-purpose concrete (e.g., concrete used in basements and sidewalks) usually varies from 15 to 35 MPa. On the other hand, high-strength concrete has a minimum compressive strength of 50 MPa. The compressive strength of concrete also influences its abrasion resistance. Higher-strength concrete is more resistant to abrasion than lower-strength concrete (Liu 1981).

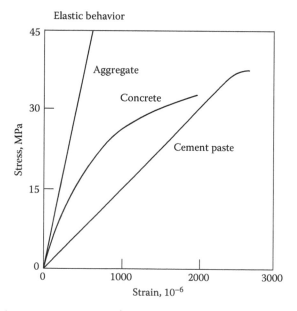

FIGURE 21.9   Compressive stress–strain curves for aggregates, hardened cement paste, and the composite, concrete, illustrating the limiting effect of the interfacial zone between aggregates and the cement paste. (From Mehta, P.K. et al., *Concrete: Microstructure, Properties and Materials*, McGraw Hill, New York, 2006.)

21.5.2 Chemical Properties

The response of concrete to chemical attack is discussed in detail in the following sections. In general, however, because of its high alkalinity, concrete can be generally considered to be susceptible to attack by acids, which react with the hydroxides in the cement paste as well as with any calcareous aggregates. Weak acids, such as acid rain, can etch the surface of concrete but do not affect the performance of the structure. A common strategy to counteract acid attack is to use a low to moderate cement content and include cementitious materials that are more acid resistant (Kosmatka et al. 2002).

21.5.3 Thermal Properties

When compared with metals, concrete is a poor conductor of heat. Thermal conductivity of concrete depends on its composition and can range from 1.4 to 3.6 J/m² S °C/m (Neville 2006). The amount of air in the concrete has an impact on its thermal conductivity because air is a poor conductor of heat. Therefore, for lightweight concrete, which contains more air than normal concrete, the thermal conductivity depends on the concrete's density. However, for normal concrete, thermal conductivity is not affected by density. A useful rule to follow, when considering the relative thermal conductivity of different types of concrete, is that the conductivity of air < conductivity of water < conductivity of cement paste. Therefore, the lower the w/cm ratio, the higher the thermal conductivity. On the other hand, for a given concrete, the higher the degree of saturation, the greater the thermal conductivity. For lightweight concrete, a 10% increase in moisture content increases its conductivity by up to 50% (Neville 2006). The use of crystalline aggregates also increases the thermal conductivity of concrete while temperature change in the ambient range has a slight influence.

Because concrete is a combination of aggregates and cement paste, its coefficient of thermal expansion (CTE) is a combination of the two. The CTE for concrete is typically $7.4–13.0 \times 10^{-6}$/°C. This value is very similar to that of steel ($11–12 \times 10^{-6}$/°C), making the thermal expansion of reinforced concrete a "nonissue."

21.5.4 Electrical Properties

Moist concrete acts as an ionic conductor with a resistivity that can be as low as 10 Ωm (i.e., in the range of semiconductors). On the other hand, oven-dried concrete is a good insulator with resistivity reaching 10^9 Ωm. If the w/cm ratio is constant, the conductivity of the concrete depends on the composition of the cement paste (particularly the alkali content). SCMs increase the resistivity of concrete because they decrease the amount of interconnected capillary porosity in the cement paste, which is essential for the movement of ions. Other substances that can be added to concrete to change its resistivity include acetylene carbon black (decreases resistivity) and finely divided bituminous material (increases resistivity) (Neville 2006).

21.6 Degradation of Concrete in Specific Environments

21.6.1 Marine and Coastal Environments

21.6.1.1 Introduction

Concrete is used extensively in marine and coastal environments, as it is readily adapted to a range of structures facilitating transportation, the petrochemical industry, and residential/commercial real estate. These structures can be relatively simple (breakwaters, bridge piers, wharfs) or very sophisticated (e.g., the offshore concrete oil platforms in the North Sea and off Newfoundland, the storm surge barriers protecting the Netherlands, or undersea tunnels like the Seikan Railway Tunnel in Japan, and the Channel Tunnel linking England and France).

TABLE 21.3   Major Ions within the Atlantic Ocean
Water, with a Corresponding pH in the Range of 7.8–8.3

Ion (S)	Concentration (g/L)
Cl^-	20
Na^+	11
SO_4^{2-}	2.9
Mg^{2+}	1.4
K^+, Ca^{2+}, Br^-, HCO_3^-	Each, 0.08

Source: Taylor, H.F.W., *Cement Chemistry,* Thomas Telford
Publishing, London, U.K., 1997.

Unlike freshwater, in which concrete can remain unscathed for centuries, seawater contains a range of ions and dissolved gases in sufficient concentrations to affect the durability of concrete structures over time by either direct contamination, or chemical interaction with the seawater altering the minerals of the concrete. These ions and gases include sodium ions (Na^+), magnesium ions (Mg^{2+}), chlorides (Cl^-), sulfates (SO_4^{2-}), bicarbonates (HCO_3^-), and dissolved gases: carbon dioxide (CO_2), hydrogen sulfide (H_2S), and oxygen (O_2). Their roles in the environmental attack of concrete are described in Section 21.6.1.2. Their relative concentrations vary by ocean over the globe, but as a frame of reference, the concentrations of the major ions within Atlantic Ocean water are listed in Table 21.3. Given the inherent variability of ocean waters, ASTM standardized a specification for simulated ocean water for use in laboratory research (ASTM 2008).

In addition to considerations of resistance to aggressive chemical species, concrete structures must also be designed to resist their environmental loads, including wave impacts and tide currents, which are too often strengthened by natural events like storms, hurricanes, and earthquakes. In addition, they also should be built to resist biological fouling from marine organisms, cycles of freezing and thawing temperatures within or above the tidal and splash zones, salt scaling, and erosion from floating debris. Some of these deterioration mechanisms are shared with other environments and are described in subsequent specific environment sections: sulfate attack (Section 21.6.2), freezing and thawing damage (Section 21.6.3.1), and salt scaling (Section 21.6.3.2). The remainder is briefly summarized in the subsequent paragraphs.

21.6.1.2 Chemical Attack by Seawater

21.6.1.2.1 *Magnesium, Sodium, and Bicarbonates*

Using Table 21.3 and ASTMD 1141 as references, seawater contains dissolved salts of Na_2SO_4 and $MgSO_4$, and the sulfate anions attack the solid phases of the cement paste in the manner described in Section 21.6.2. Overall, the magnesium salt is considered to be more aggressive than the sodium, because the magnesium cation also participates in the reactions with the cement paste: magnesium ions, in combination with bicarbonates, produce a surface skin of brucite ($Mg(OH)_2$) and aragonite ($CaCO_3$, a polymorph of the original calcite, also $CaCO_3$) (Taylor 1997). These reactions proceed from the surface of the concrete, inward, slowing with thickness and decreased permeability. Although this is an attack to the cement paste, the 30 μm thick skin can actually protect the concrete from additional magnesium and sulfate attack, chloride ingress, and carbonation from carbon dioxide and leaching of the cement components of the concrete, particularly $Ca(OH)_2$, if the skin remains intact (Taylor 1997). This beneficial effect is more likely to occur within the submerged portion of the concrete structure as erosion, impact, and salt scaling are more aggressive in the tidal and splash zones, shown schematically in Figure 21.10. On the other hand, continued reaction between the magnesium salts and the concrete can be highly detrimental. Because the $Mg(OH)_2$ has a much lower solubility than that of $Ca(OH)_2$, the reaction reduces the pH of the system significantly. Moreover, if the reserve of $Ca(OH)_2$ in the cement paste is consumed, the magnesium will then react with the C–S–H producing a magnesium-silicate hydrate, which is incompletely solid and has little strength.

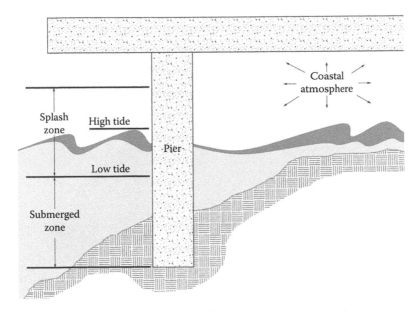

FIGURE 21.10 Schematic diagram of different zones affected by the marine atmosphere.

21.6.1.2.2 Chlorides and Carbon Dioxide

Although the major damage caused by these species is corrosion of the steel in reinforced concrete (Chapter 24), they can affect the integrity of the concrete independent of the steel. In a manner similar to that described earlier for the effect of magnesium salts, CO_2 can react with the $Ca(OH)_2$ to form $CaCO_3$, which has limited solubility and precipitates in the pores of the concrete. This can increase the strength of the concrete but also reduce the pH to a level at which steel is not passive (Chapter 22).

Sodium chloride is known to enhance the leaching of $Ca(OH)_2$ by ion exchange, and it has also been found that the sodium ions in NaCl can exchange with some of the calcium ions in C–S–H (Sugiyami 2008). The effects of chlorides on the damage produced by repeated freezing and thawing are described in Section 21.6.3, but it is relevant to note here that the concentrations of salts giving the most damage (Verbeck and Klieger 1957) are, coincidentally, those found in seawater.

21.6.1.2.3 Biological Fouling and Hydrogen Sulfide

Marine animals and plants are able to bind themselves to concrete if the surface is sufficiently rough or porous caused, for example, by mechanical action from debris in the tides, as shown in Figure 21.11. They can also cause degradation by leaching of minerals from the surfaces of the concrete (Mehta 1990). For example, it has been found that mussels use the calcium leached from the concrete in the building of their shells (Pérez et al. 2003). The buildup of these animals and plants depends upon temperature, light, pH, water currents, and oxygen (Mehta 1990), but since their specific gravity is not distinctly different from seawater, they do not usually provide significant mass (Gerwick 2007) that could add loads to a given structure. They do, however, interfere with regular inspections of affected elements, and some of the species secrete acids that dissolve the constituents of the cement paste (Mehta 1990).

Another form of acid attack arises from hydrogen sulfide (H_2S), which is naturally present in seawater from the underlying geology of the seafloor (Tuttle and Jannash 1974). Available H_2S is converted by *Thiobacillus* bacteria present in the ocean to sulfuric acid, which initiates an acid attack of the concrete (Section 21.6.2), the severity of which is governed by the concentrations of the H_2S and bacteria. This process is discussed in more detail in Section 21.6.2.

FIGURE 21.11   Example of biofouling of concrete pipeline. (Courtesy of Dalmaray Concrete Products Inc., Janesville, WI.)

21.6.1.3 Physical Attack by Seawater

21.6.1.3.1 Erosion

Although concrete is used extensively as a barrier to prevent erosion of the shoreline, it can itself be eroded, as illustrated by the undermining of the sea defenses protecting a major rail line shown in Figure 21.12 and the scoured bridge pier in Figure 21.13. Erosion can be caused by abrasion by solid particles, repeated impact by the seawater, or by cavitation (Preece 1979). Abrasion damage occurs where debris rolls and grinds across the surface of the concrete, typically by tidal action. Liquid impact produces the same result, but at the normal velocities experienced in oceans, it can take longer. Cavitation erosion arises from the collapse of tiny vapor cavities that are created in the low-pressure regions of the flow and collapse when they enter a higher-pressure region, emitting a shock wave into the water or into any adjacent solid. Proper design and materials selection minimizes the damage by these mechanisms (Woodson 2009).

FIGURE 21.12   Erosion of sea defenses built to protect a major rail line in North Wales. (Courtesy of Storm-Geomatics Ltd., London, U.K.)

FIGURE 21.13 Eroded bridge pier. (Courtesy of Levelton Engineering, Richmond, British Columbia, Canada.)

21.6.1.3.2 Impact

To enable a structure to meet its intended design life, its design and construction should account for loads (e.g., gravity, wind), forces (e.g., seismic, tidal currents), and the necessary material durability for in-service environmental conditions. Even in cold environments, structures can be designed to withstand iceberg collisions (e.g., the Confederation Bridge in eastern Canada). Nonetheless, unanticipated impacts can arise from collisions from sea vessels, equipment, or debris from other damaged structures during a natural event.

21.6.2 Agriculture and Sewage

Reinforced concrete is the material of choice for most agricultural and sewage systems. However, the environmental conditions to which the concrete in these structures are exposed are some of the most aggressive environments in which concrete is used. As an example, the liquid manure of various farm animals, such as cattle and pigs, contains compounds of ammonium (NH_4), phosphate (PO_4), chloride (Cl^-), and sulfate (SO_4^{2-}) as well as potassium, calcium, and magnesium oxides (K_2O, CaO, MgO), (Svennerstedt et al. 1999) typically in the range of 0.1–8.0 g/L. Many gases aggressive to concrete are also produced during the decomposition of the manure itself, such as methane (CH_4), carbon dioxide (CO_2), H_2S, and ammonia (NH_3) (Svennerstedt et al. 1999). In some conditions, the severity of the environment has required concrete replacement in less than 5 years (Assaad et al. 2002).

In agricultural applications such as animal housing structures and manure tanks, as well as in municipal sewage networks, the most significant mode of deterioration of the concrete is considered to be sulfate attack. As the waste, which contains dissolved sulfides, breaks down under anaerobic conditions, H_2S gas is produced by sulfate-reducing bacteria such as the *Desulfovibrio* species, which are naturally present within the manure and exist in the mud at the bottom of sewer pipes and in the slime on the sides (De Belie et al. 2000). The H_2S, which usually condenses on the walls of storage tanks or pipe walls, is then converted to sulfuric acid (H_2SO_4) by sulfur-oxidizing aerobic bacteria, mainly from the *Thiobacillus* species (De Belie et al. 2000; De Muynck et al. 2009).

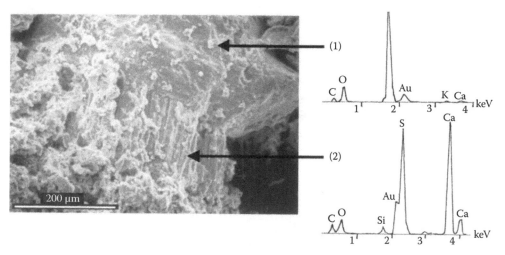

FIGURE 21.14 Scanning electron micrograph combined with EDS analysis of a degraded concrete sample: (1) quartz grain; (2) gypsum. (From Tulliani, J.-M. et al., *Cem. Concr. Res.*, 32, 843, 2002.)

As with concrete embedded in sulfate-containing soils, the concrete in agricultural and sewage systems can experience sulfate attack by a number of different mechanisms involving the sulfate anion (SO_4^{2-}). The sulfate can react with calcium hydroxide to form gypsum ($CaSO_4 \cdot 2H_2O$):

$$Ca(OH)_2 + SO_4^{2-} + 2H_2O \rightarrow CaSO_4 \cdot 2H_2O + 2(OH)^-$$

This leads to expansive pressures being exerted on the concrete as illustrated by the opening up of the interface between the cement paste and an aggregate particle in Figure 21.14 (Tulliani et al. 2002).

Due to the higher sulfate concentration produced by aerobic bacteria and the previously noted gypsum formation, coupled with the typical high humidity and temperature conditions generated in agricultural and sewage facilities, the sulfate can also cause monosulfate to convert back to ettringite, again leading to expansive pressure on the concrete, which, in turn, leads to cracking:

$$3CaO \cdot Al_2O_3 \cdot CaSO_4 \cdot 12H_2O + 2Ca^{2+} + 2SO_4^{2-} + 20H_2O \rightarrow 3CaO \cdot Al_2O_3 \cdot 3CaSO_4 \cdot 32H_2O$$

It is also possible that any C_3A or C_4AF that could not react during initial hydration of the cement due to a lack of gypsum will now react in the presence of the additional sulfate. Again, as the concrete is already in a hardened state, the hydration of the C_3A/C_4AF can lead to expansive pressure and cracking of the concrete.

In addition to deterioration of the concrete from the sulfate attack, the integrity of the reinforcing steel that is typically incorporated within these structures is also at risk. With the cracking of the concrete that results from the formation of gypsum and ettringite, the concrete becomes more permeable, thereby allowing species that can compromise the protective passive oxide layer on the reinforcement that forms in the alkaline environment of concrete. As noted earlier, highly acidic gases are present above the slurry in manure-holding tanks, which will cause corrosion if allowed to reach the level of the reinforcement. The formation of corrosion products will cause further deterioration of the concrete, as detailed in Chapter 22.

In some agricultural applications, lactic ($C_3H_6O_3$) and acetic (CH_3COOH) acids are also present. Many of the feed and water mixtures contain these acids, which are known to react with calcium hydroxide in the concrete to produce soluble calcium salts (De Belie et al. 1997). The depletion of calcium hydroxide from the pore solution can cause the primary phase of concrete (C–S–H) to become unstable as equilibrium between the solid phases and the pore solution is maintained. Cleaning of the concrete surface combined with wear and tear from the animals themselves can remove unstable material, causing further deterioration of the concrete.

The formation of thaumasite ($CaSiO_3 \cdot CaCO_3 \cdot CaSO_4 \cdot 15H_2O$) is also possible at lower temperatures (15°C) in the presence of sulfate, carbonate, and water and has been observed in concrete exposed to sewage (Tulliani et al. 2002). Thaumasite is formed at the expense of C–S–H, which again leads to deterioration of the basic building block of the concrete (Neville 2006).

21.6.3 Transportation Systems

The major cause of degradation of concrete in transportation structures, such as bridges and barrier walls and in parking garages, is the corrosion of reinforcing steel caused by de-icing salts or carbonation. As already mentioned, this is the subject of Chapter 22 and will not be discussed further in this chapter. The other major environmental degradation mechanisms facing concrete in transportation systems are (1) internal cracking due to repeated freezing and thawing, (2) salt scaling of the surface due to the combined effects of freezing and de-icing agents, and (3) chemo-mechanical degradation by de-icing agents.

21.6.3.1 Freezing and Thawing Damage

As described earlier, a major component of the cement paste phase of concrete is the pore solution, which is subject to repeated freezing and thawing in the winter months in many parts of the world. When water freezes, it expands by 9 volume%, and in concrete, this can cause cracking of the cement paste and a reduction in stiffness and strength of the concrete, resulting eventually in the destruction of the structure, as shown in Figure 21.15. De-icing agents are found to exacerbate this problem (Pigeon et al. 1996). The freezing does not occur spontaneously throughout the concrete as the temperature of the concrete reaches 0°C, but occurs over a range of temperatures because of the difficulties imposed on ice nucleation by the fineness of the pores and the depression in freezing temperature due to the solutes (Bager and Sellevold 1986, Marchand et al. 1995). Theories to account for the cracking range from the buildup of hydraulic pressure as the freezing pushes the remaining water into smaller spaces (Powers 1949) to an osmotic pressure theory (Powers and Helmuth 1953, Helmuth 1962) and, more recently, to

FIGURE 21.15 Damage due to repeated freezing and thawing. (Courtesy of Portland Cement Association, Ottawa, Ontario, Canada.)

the pressure imposed by crystallization (Scherer 1999). The osmotic pressure theory takes into account the fact that the ice formed from the pore solution is essentially pure water, and, therefore, the remaining unfrozen liquid becomes progressively more concentrated as the solutes are pushed ahead of the ice. This creates a concentration differential between the solution adjacent to the ice and the original solution in smaller pores where ice nucleation has not occurred. Water will then flow into the former pores adding to the increase in pressure. The greater the concentration of solutes in the pore solution, the greater will be this effect, which explains why de-icing chemicals have a detrimental effect on the frost resistance of concrete. Although these theories appear sound, they cannot account for all the field and laboratory observations of frost damage. Thus, the currently accepted theory (Scherer 1999) is based on the thermodynamics of crystallization in confined spaces, the observations that a thin layer of liquid (~1 nm) exists between the ice and the pore walls and the pressures that the crystals impose on the surrounding cement paste.

21.6.3.2 Salt Scaling

Salt scaling is defined as "superficial damage caused by freezing a saline solution on the surface of a concrete body. The damage is progressive and consists of the removal of small chips or flakes of material" (Valenza and Scherer 2007b) as exemplified in Figure 21.16. Any mechanistic explanation of this phenomenon must account for the following observations: (1) the scaling occurs only when there is a pool of the solution on the surface during freezing (Genin et al. 1997); (2) the maximum amount of damage occurs in ~3% solutions of the different de-icing agents, as illustrated in Figure 21.17 (Verbeck and Klieger 1957); (3) freezing the concrete with the salt solution is more destructive than applying the salt solution to an already frozen concrete (Rösli and Harnik 1980); (4) the freezing rate has less influence than the time at the low temperature; and (5) there is little damage if the minimum temperature is higher than −10°C.

One theory is that scaling is due to thermal shock induced by endothermic melting of ice in the surface by the de-icing agents, requiring heat to be drawn from the subsurface layers and, thereby, setting up a thermal gradient between the warmer surface and frozen bulk. Another is that precipitation of salts in the surface layers causes the spalling. A third theory is that, when ice forms in the surface and concentrates the salts in the remaining solution, the salts draw moisture from the surrounding concrete and from the environment and so the surface layers become more saturated than the bulk of the concrete. This is supported by the observation that scaling increases with increasing degree of saturation (Fagerlund 1977). However, none of these theories can account for the first two observations provided earlier. Valenza and Scherer (2006, 2007a) have, instead, likened salt scaling to the process of producing dimpled glass. They propose that, when the frozen de-icing solution cools, it contracts, cracks, and develops a tensile stress in the ice layer in the underlying concrete. This ice layer then forms a network of fractures, and the cracks propagate into the concrete to a degree dependent on the strength of the concrete surface layer. The propagation in the concrete depends on the microstructure, and as indicated in Figure 21.18, cracks have been

FIGURE 21.16   Small flakes of cement are repeatedly removed during salt scaling, leaving the aggregate in relief.

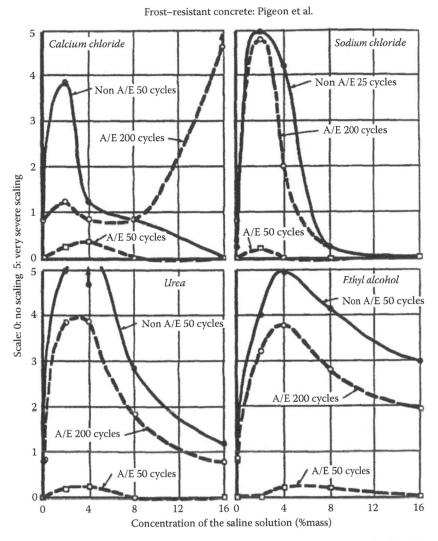

FIGURE 21.17 The maximum amount of salt scaling occurs at de-icer concentrations in the 3%–4% range. (From Verbeck, G.J. and Klieger, P., *Highw. Res. Board Bull.*, 150, 1, 1957.)

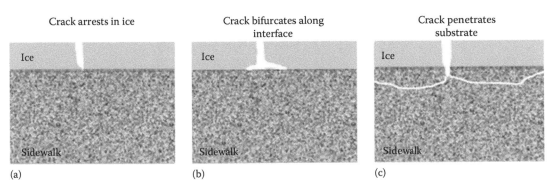

FIGURE 21.18 When an ice layer cracks, the crack may (a) arrest in the ice, (b) bifurcate along the ice/concrete interface, or (c) penetrate the cement surface. (From Valenza, J.J.I. and Scherer, G.W., *Mater. Struct.*, 40, 259, 2007a.)

shown to bifurcate and run parallel to the surface resulting in chips of concrete spalling off. This mechanism obviously accounts for the first observation, given earlier. The pessimum value of approximately 3% de-icing agent is attributed to the fact that ice formed in more concentrated solutions does not gain sufficient strength in the temperature range of scaling, whereas ice in more dilute solutions is stronger and does not crack.

21.6.3.3 Chemo-Mechanical Reaction with De-Icing Agents

The traditional de-icing agent is rock salt (i.e., NaCl with ionic impurities of Ca^+, K^+, Mg^{2+}, NO_3^-, and SO_4^{2-} totaling less than 1 wt.%). While the extensive use of rock salt is having major detrimental effects on groundwater systems and is the major cause of corrosion of reinforcing steel, it has not been found to have any significant chemical or mechanical deleterious effects on concrete. In recent years, however, several alternative de-icing agents have come on the market. The more common of these are $CaCl_2$ and $MgCl_2$, often mixed with organic materials such as "de-sugared beet juice" and "hop residue from beer breweries." These latter materials are added to reduce the freezing temperature.

There is no doubt that the de-icing and anti-icing (i.e., ice-prevention) capabilities of $CaCl_2$ and $MgCl_2$ are superior to those of NaCl, in large part because their eutectic temperatures with water are −30.3°C and −33.0°C, respectively, compared with −21.1°C for $NaCl–H_2O$, thereby allowing them to be effective at much lower ambient temperatures. Moreover, they are often used as concentrated brines to wet the rock salt and help adhere the anti-icer to the pavement and prevent being blow away by the traffic. Unfortunately, $CaCl_2$ has been shown to be very aggressive, causing a significantly higher scaling rate of concrete in freeze/thaw tests and in salt scaling tests than does NaCl (Wang et al. 2006, Juilo-Betancourt 2009, Shi et al. 2009). Furthermore, even in the absence of the freezing and thawing, both $CaCl_2$ and $MgCl_2$ cause deterioration of the concrete due to the formation of expansive oxychlorides (Neville and Aitcin 1998, Sutter et al. 2006, Poursaee et al. 2010). This deterioration to the concrete itself is exemplified by the cracked prisms illustrated in Figure 21.19 after only 6 week exposure to concentrated salt solutions (Poursaee et al. 2010). This effect has been attributed to the reaction of $CaCl_2$ with the $Ca(OH)_2$ in the cement paste forming a very expansive hydrated calcium-oxychloride, $3CaO \cdot CaCl_2 \cdot 12H_2O$.

FIGURE 21.19   Cracking of concrete after 6 week exposure to concentrated $CaCl_2$ solution.

In the case of $MgCl_2$, it is well known that the salt reacts with $Ca(OH)_2$ in the cement paste to form $Mg(OH)_2$ (brucite), which precipitates in the pores at the surface of the concrete. If the reaction stopped there, it would be beneficial because, by blocking the pores, the ingress of any deleterious species would be slowed. However, it does continue to react within the concrete, and if the $Ca(OH)_2$ becomes depleted, the magnesium will begin to replace the calcium in the C–S–H eventually forming M–S–H, which has little strength. Before this happens, however, the by-product of the brucite formation is $CaCl_2$, which continues to react with the $Ca(OH)_2$, as described earlier, resulting in cracking and destruction of the concrete.

Chloride salts cannot be used on airfields because of their corrosive effects on aircraft. Potassium acetate, sodium acetate, or sodium formate are, therefore, used as airfield de-icing and anti-icing agents. However, premature deterioration of the concrete has been observed and is attributed to an alkali-aggregate reaction between these chemicals and the aggregates in the concrete (Rangaraju et al. 2007).

21.7 Prevention and Protection

21.7.1 New Construction/Fresh Concrete

Durability of new structures requires the concerted effort of three parties: the designer, the contractor, and the owner. Protection begins with a designer considering (1) appropriate design parameters and materials selection for the specific service environment, (2) the structural performance requirements, and (3) the desired service life (defined by the owner). Building codes mandate minimum standards for safety and serviceability, but not necessarily durability or sustainable construction, which underscores the importance of a designer developing a range of options each with its corresponding LCA and LCCA data (Section 21.2) for consideration by the owner. Nonetheless, even with the most rigorous and thoughtful specifications for design and detailing, the quality of the materials and workmanship is managed by the contractor and is financially controlled by the owner. Together, these three entities must work together to build a durable structure.

Many durability problems in service arise from improper or low-quality materials selection, constructability issues (i.e., mismatching tolerances), and design, which is adequate for structural performance but not for durability. Nonetheless, the single most important durability consideration for any designer is that all concrete is inherently cracked (Section 21.5.1), and these cracks, at best, permit the easy ingress of deleterious species from the environment deeper into the concrete and, at worst, undermine the structural performance of the concrete. The number, size, and locations of the cracks, coupled with the aggressiveness of the environment, govern the long-term performance of a given structure. ACI 201.2 (ACI 2008) outlines in greater detail the considerations and design requirements necessary to mitigate the risks of cracking, and degradation arising from freezing and thawing damage, and chemical and abrasive environments, affecting plain and reinforced concretes. For reinforced concrete structures, the correct detailing of the reinforcement is one of the tools available to the designer that can reduce the extent of cracking, thereby managing the risk of environmental degradation.

21.7.1.1 Mixture Design and Proportions

The major goals of mixture design in providing durability are to minimize both the interconnected porosity and the cracks, which allow ingress of deleterious materials and allow leaching of constituents of the concrete. In general, concrete cast with a lower w/cm ratio performs better than concrete with a higher ratio, as described in Section 21.4. To reduce the water content and, therefore, the permeability of the concrete while retaining the workability of the fresh mix, most modern concrete designs contain surface-active agents known as low-, mid-, or high-range water-reducing agents.

Additions of SCMs also reduce the porosity as described in Section 21.4.3. Clearly, macro- and micro-structural cracks will undermine even the highest quality materials and workmanship, and research is ongoing, with the goal of better understanding the role of cracks in long-term durability.

In recent years, shrinkage-reducing admixtures (SRAs) have been developed to minimize the microcracking associated with cement hydration in low w/cm concretes (Mora-Ruacho et al. 2009). The SRAs are organic solutions, for example, neopentyl glycol or glycol-based polypropyl-ene (Collepardi et al. 2005). They function by reducing surface tension of the pore solution and, in turn, reduce the capillary forces causing the shrinkage. This slows the drying of the concrete sur-face (Berke et al. 2003), thereby maintaining a high internal relative humidity and minimizing the autogenous shrinkage associated with these concretes. An example of the effectiveness of the SRAs is shown in Figure 21.20 (Berke et al. 2003).

As described in Section 21.6.3, one of the major causes of cracking of concrete in cold climates is repeated freezing and thawing. The accepted method of minimizing such damage is to introduce a fine dispersion of minute air voids into the cement paste phase by the addition to the concrete mix of an air-entraining agent (AEA). This allows space for the expanding pore solution or the salt crystals to form without causing tensile stresses in the concrete. The amount of AEA is critical: too much and the concrete strength is jeopardized; too little and the concrete cannot resist the freezing and thawing. The "just right" is that amount needed to produce a dispersion of voids in the 0.01–1 mm size range and with an average distance between them of no more than ~200 μm, illustrated in Figure 21.21 (Kosmatka et al. 2002). The dosage required to produce this, however, varies with the concrete mix.

Chemical corrosion inhibitors are increasingly being incorporated into concrete that is exposed to chloride-based de-icing salts and seawater. These are described in more detail in Chapter 22.

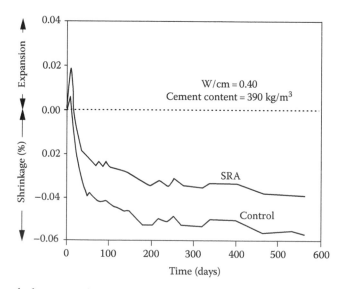

FIGURE 21.20 Length change as a function of time after casting for concretes with and without a shrinkage-reducing admixture. (From Berke, N.S. et al., Improving concrete performance with shrinkage reducing admix-tures, in V.M. Malhotra, Ed., *7th CANMET/ACI International Conference on Superplasticisers and other Chemical Admixtures in Concrete*, American Concrete Institute, Berlin, Germany, pp. 37–50, 2003.)

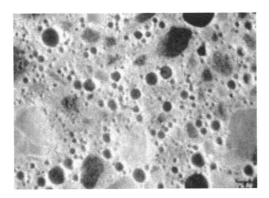

FIGURE 21.21 Air-entrained concrete. (From Kosmatka, S.H. et al., *Design and Control of Concrete Mixtures*, Portland Cement Association, Ottawa, Ontario, Canada, 2002.)

21.7.2 Structures in Service/Hardened Concrete

It is inevitable that a concrete structure will degrade over time from its initial construction, and the rate of degradation is an important consideration when optimizing the costs of maintaining a structure for its service life. Typically, it is far less expensive to maintain and repair a structure or transportation element, than building new, making concrete structures inherently sustainable. This strategy maximizes the use of existing materials and infrastructure, reduces construction waste, and maintains a material that has demonstrated resiliency against climate change (Schokker 2010). Thus, protection strategies can be more aptly described as mitigation and preservation strategies. Overall, the safe and effective management of a concrete structure in a given environment begins with a maintenance plan. This should include, but not be limited to

- Routine maintenance: the scheduled replacement of components at the end of their service lives, typically ancillary items like joint sealants, and drainage materials, that protect the concrete structural components
- Periodic inspection and assessment: a review and evaluation of a structure by a competent person, ideally an appropriate licensed design professional, such as an architect or engineer, experienced in forensic concrete investigations of the type under consideration
- Repairs and rehabilitation: remedial work that is over and above the normal routine maintenance, typically to the concrete materials and structural system themselves

When repairs and rehabilitation are needed at some frequency, the range of options available depends upon the degradation mechanism and the extent of the deterioration, and requires consideration of the materials selected for the repairs and whether intensive electrochemical protection strategies are merited. For further information, a comprehensive guide summarizing the insight and research from a number of worldwide practitioners is published by the ACI and International Concrete Repair Institute (ACI and ICRI 2008). As a counterpoint to the scientific research, a practical, illustrated guide summarizing concrete repairs was developed by Emmons (1994).

21.7.2.1 Materials Selection

For maintenance and repair of concrete structures, materials must be selected to be compatible with both the substrate and the environment to which they are exposed. Not only must consideration be given to the durability and maintenance requirements, they must not negatively impact the concrete substrate. For example, replacement of localized concrete sections undergoing chloride-induced corrosion with new concrete can exacerbate corrosion of the surrounding remaining reinforced concrete unless countermeasures are implemented. In addition, cracking represents another problem with repaired sections, as the new patches can shrink, leaving easy pathways for deleterious species to enter the concrete cover. Thus, materials selection is typically even more important for existing structures than new, and crack repair becomes critical as a structure ages.

21.7.2.2 Surface Treatments and Coatings

As a management strategy is implemented, it often becomes evident that some concrete elements are deteriorating at a faster rate than the remainder of the structure, or there is a desire to extend the service life of repairs for as long as possible. Surface treatments and coatings provide additional protection by limiting the exposure of the treated element to the environment. Typically, surface treatments are silanes or siloxanes and penetrate the surface of the concrete and enter the pore structure (De Vries and Polder 1997), while coatings remain on the surface and can be elastomeric paints, or even waterproofing systems based upon epoxy and polyurethane technologies (Almusallam et al. 2003). Surface treatments and coatings are not often applied to new structures, as they themselves become an additional maintenance item every 5–10 years. Thus, there is an optimal time in the service life of a concrete element when the treatment or coating provides an economic benefit by extension of the service life.

21.7.2.3 Localized Replacements and Overlays

The need for localized repairs is dictated by the volume of deteriorated material and the degree of contamination of the concrete cover with deleterious species. The long-term durability of localized repairs is governed by the materials selection (Section 21.7.2.1) and whether any additional protection options are installed. When the level of deterioration of the entire outer surface of the concrete becomes extensive or the concrete cover is insufficiently thick to provide protection, and the structural system can tolerate the additional mass of material, then overlays become a viable option for repairing structures, concrete decks, and pavements. Overlays can be new conventional concrete mixes, polymer-modified mixes, or fiber-reinforced concretes, and either bonded or unbounded to their substrate.

21.8 Specific Standard Tests

The following list of concrete testing standards in Table 21.4 is not, by any means, a complete list of available standards, but can be considered a starting point, introducing readily available technical standards, either international or regional. Overall, any standard for fresh concrete can be used in either the laboratory or the field, while tests for hardened concrete are often location specific. In general, though, tests suitable for in situ fieldwork can be used in the laboratory, but not vice versa unless noted otherwise. In addition, laboratory results might not have the same degree of testing variability that field concrete typically provides (i.e., the laboratory typically has a greater opportunity for quality control that field concrete lacks).

- AASHTO—American Association of State Highway and Transportation Officials (www.transportation.org)
- ASTM—American Society for Testing and Materials (www.astm.org)
- CSA—Canadian Standards Association (www.cement.ca)
- EN—European Committee for Standardization (CEN) (www.cen.eu)
- ISO—International Standards Organization (www.iso.org)

TABLE 21.4 Representative, Readily Available Concrete Standards from the Following Organizations

	Application		
		Hardened Concrete	
Standard Test	Fresh Concrete	Laboratory	Field
Abrasion resistance	n/a	*ASTM C944 Standard Test Method for Abrasion Resistance of Concrete or Mortar Surfaces by the Rotating-Cutter Method*	ASTM C418 Standard Test Method for Abrasion Resistance of Concrete by Sandblasting
Air content and void analysis	AASHTO TP-75: Air-Void Characteristics of Freshly Mixed Concrete by Buoyancy Change *ASTM C138 Standard Test Method for Density (Unit Weight), Yield, and Air Content (Gravimetric) of Concrete* *ASTM C231 Standard Test Method for Air Content of Freshly Mixed Concrete by the Pressure Method* *ASTM C1688 Standard Test Method for Density and Void Content of Freshly Mixed Pervious Concrete*	ASTM C457, Standard Test Method for Microscopical Determination of Parameters of Air-Void System in Hardened Concrete *ASTM C642-06 Standard Test Method for Density, Absorption, and Voids in Hardened Concrete* EN-480, Admixtures for concrete, mortar and grout. Test methods. Determination of air void characteristics in hardened concrete	n/a
Alkali aggregate reactivity	n/a	ASTM C856 Practice for Petrographic Examination of Hardened concrete *ASTM C1293 Standard Test Method for Determination of Length Change of Concrete Due to Alkali-Silica Reaction*	n/a
Carbonation depth	n/a	ASTM C856 Practice for Petrographic Examination of Hardened concrete EN 14630 Products and systems for the protection and repair of concrete structures. Test methods. Determination of carbonation depth in hardened concrete by the phenolphthalein method	n/a
Cement content	n/a	ASTM C1084 Test Method for Portland-Cement Content of Hardened Hydraulic-Cement Concrete	n/a
Chloride content	n/a	ASTM *C1152* Test Method for Acid-Soluble Chloride in Mortar and Concrete CSA A23.2-4B *Sampling and Determination of Water-Soluble Chloride Ion Content in Hardened Grout or Concrete*	n/a

(continued)

TABLE 21.4 (continued)	Representative, Readily Available Concrete Standards from the Following Organizations

	Application		
		Hardened Concrete	
Standard Test	Fresh Concrete	Laboratory	Field
Chloride penetration	n/a	AASHTO TP-64 Predicting Chloride Penetration of Hydraulic Cement Concrete by the Rapid Migration Procedure	n/a
		AASHTO T259 Method of Test for Resistance of Concrete to Chloride Ion Penetration	
		ASTM *C1202* Test Method for Electrical Indication of Concrete's Ability to Resist Chloride Ion Penetration	
		ASTM C1543 Standard Test Method for Determining the Penetration of Chloride Ion into Concrete by Ponding	
		ASTM C1556-04 Standard Test Method for Determining the Apparent Chloride Diffusion Coefficient of Cementitious Mixtures by Bulk Diffusion	
Compressive strength	n/a	ASTM C39 Test Method for Compressive Strength of Cylindrical Concrete Specimens	ASTM C805: Standard Test Method for Rebound Number of hardened Concrete
		ASTM C873 Standard Test Method for Compressive Strength of Concrete Cylinders Cast in Place in Cylindrical Molds	
		CSA A23.2-10C *Accelerating the Cure of Concrete Cylinders and Determining Their Compressive Strength* ISO 1920-6:2004 Testing of concrete—Part 6: Sampling, Preparing and Testing of Concrete Cores	EN 12504-2: Testing Concrete in Structures. Part 2: Non-Destructive Testing—Determination of Rebound Number
Defects (via ultrasonic pulse velocity)	n/a	ASTM C597 Standard Test Method for Pulse Velocity Through Concrete	Same as laboratory
Density (unit weight)	*ASTM C138 Standard Test Method for Density (Unit Weight), Yield, and Air Content (Gravimetric) of Concrete*	EN 12390-7: Testing Hardened Concrete—Part 7: Density of Hardened Concrete	ASTM C1040 Standard Test Methods for In-Place Density of Unhardened and Hardened Concrete, Including Roller Compacted Concrete, By Nuclear Methods
	ASTM C1688 Standard Test Method for Density and Void Content of Freshly Mixed Pervious Concrete		

TABLE 21.4 (continued) Representative, Readily Available Concrete Standards from the Following Organizations

	Application		
		Hardened Concrete	
Standard Test	Fresh Concrete	Laboratory	Field
Flexural strength	n/a	ASTM C293-08 Standard Test Method for Flexural Strength of Concrete (Using Simple Beam With Center-Point Loading)	n/a
Freeze–thaw resistance	n/a	ASTM C666 Standard Test Method for Resistance of Concrete to Rapid Freezing and Thawing	n/a
		EN 480 Admixture for Concrete—Test Methods—Determination of Air Void Characteristics in Hardened Concrete	
		EN 13687-1 Products and Systems for the Protection and Repair of Concrete Structures—Test Methods—Determination of Thermal Compatibility—Part 1: Freeze–Thaw Cycling with De-Icing Salt Immersion	
Humidity, relative	n/a	ASTM F2170 Standard Test Method for Determining Relative Humidity in Concrete Floor Slabs Using in situ Probes	Same as laboratory
Length change (expansion, shrinkage)	n/a	ASTM C157 Standard Test Method for Length Change of Hardened Hydraulic-Cement Mortar and Concrete	n/a
		ASTM C1581 Standard Test Method for Determining Age at Cracking and Induced Tensile Stress Characteristics of Mortar and Concrete under Restrained Shrinkage	
Modulus of rupture (flexural strength)	n/a	ASTM C78 Standard Test Method for Flexural Strength of Concrete (Using Simple Beam with Third-Point Loading)	n/a
		ASTM C293–08 Standard Test Method for Flexural Strength of Concrete (Using Simple Beam With Center-Point Loading)	
Modulus of elasticity (E modulus)	n/a	ASTM C469 Standard Test Method for Static Modulus of Elasticity and Poisson's Ratio of Concrete in Compression	n/a
Salt scaling	n/a	ASTM C672 Test Method for Scaling Resistance of Concrete Surfaces Exposed to Deicing Chemicals	n/a

(continued)

TABLE 21.4 (continued)   Representative, Readily Available Concrete Standards from the Following Organizations

| Standard Test | Fresh Concrete | Hardened Concrete | |
		Laboratory	Field
Slump	AASHTO TP-73: Slump Flow of Self-Consolidating Concrete (SCC)	n/a	n/a
	ASTM C143 Standard Test Method for Slump of Hydraulic-Cement Concrete		
	ASTM C1362 Standard Test Method for Flow of Freshly Mixed Hydraulic Cement Concrete		
	ASTM C1611 Standard Test Method for Slump Flow of Self-Consolidating Concrete		
	CSA A23.2-5C *Slump of Concrete*		
Temperature	*ASTM C1064 Standard Test Method for Temperature of Freshly Mixed Hydraulic-Cement Concrete*	*ASTM F2170 Standard Test Method for Determining Relative Humidity in Concrete Floor Slabs Using in situ Probes.* Note: some relative humidity sensors have integral temperature sensors, too	Same as laboratory
Tensile strength	n/a	ASTM C496 Standard Test Method for Splitting Tensile Strength of Cylindrical Concrete Specimens	ASTM C803 Standard Test Method for Penetration Resistance of Hardened Concrete
		CSA A23.2-13C *Splitting Tensile Strength of Cylindrical Concrete Specimens*	
		EN 12390-6: Testing Hardened Concrete—Part 6: Tensile Splitting Strength of Test Specimens	ASTM C1583 Standard Test Method for Tensile Strength of Concrete Surfaces and the Bond Strength or Tensile Strength of Concrete Repair and Overlay Materials by Direct Tension (Pull-off Method)
			EN 1542 Products and Systems for the Protection and Repair of Concrete Structures—Test Methods—Measurement of Bond Strength by Pull-Off

Source: AASHTO, American Association of State Highway and Transportation Officials (www.transportation.org); ASTM, American Society for Testing and Materials (www.astm.org); CSA, Canadian Standards Association (www.cement.ca); EN, European Committee for Standardization (CEN) (www.cen.eu); ISO, International Standards Organization (www.iso.org).

21.9 Summary

Concrete differs in many ways from other engineering materials, exhibiting not only several unique beneficial characteristics but also some detrimental ones. The advantages of this material are (1) it can be made from local components in almost all parts of the world; (2) it can be cast into complex shapes at ambient temperatures; (3) it does not require any post-casting treatment other than being kept moist for a few days; and (4) the cost of its production is significantly lower than any alternative material, such as steel. On the other hand, the micro- and macrostructures of concrete, consisting of natural stone bonded together with a more or less porous, highly alkaline, multiphase cement paste, can lead to a variability of properties unless the components, mixture proportions, and construction procedures are carefully controlled. These same characteristics also lead to the susceptibility of concrete to environmental degradation. The alkalinity means that the paste component is vulnerable to dissolution in acidic media. The solution-filled porosity allows ingress of deleterious species and leaching of soluble components of the cement paste. In those parts of the world that experience repeated freezing and thawing, the expansion of the pore solution on freezing can lead to cracking of the concrete and, where de-icing salts are used, scaling of the surface. Nevertheless, as mentioned earlier, appropriate mixture design and proportioning, careful placement, compaction, and sufficient curing can minimize the risk of premature deterioration. Together with the use of chemical admixtures to counteract some of these weaknesses, the concrete industry is able to produce a material tailored to the local environment and able to withstand the forces of both man and nature for its specified service life.

References

ACI. 2008. 201.2 *Guide to Durable Concrete*. American Concrete Institute, Farmington Hills, MI.

ACI. 2009. *ACI Concrete Terminology*, American Concrete Institute Committee 116, Farmington Hills, MI.

ACI and ICRI. 2008. *Concrete Repair Guide*. 3rd edn. American Concrete Institute and the International Concrete Repair Institute, Farmington Hills, MI.

Aïtcin, P.-C. 1999. Does concrete shrink or does it swell? *Concrete International*, 77–80.

Aïtcin, P.-C., A.M. Neville, and P. Acker. 1997. An integrated view of shrinkage deformation. *Concrete International*, 19, 34–41.

Alizadeh, R. 2010. *Cement and Art* [Online]. Portland Cement Association, Skokie, IL. Available: http://www.cementlab.com/cement-art.htm (accessed December 1, 2010).

Almusallam, A.A., F.M. Khan, S.U. Dulaljan, and O.S.B. AL-Almoudi. 2003. Effectiveness of surface coatings in improving concrete durability. *Cement and Concrete Composites*, 25, 473–481.

Assaad, V.F., J.C. Jofriet, S.C. Negi, G.L. Hayward, and L.J. Evans. 2002. *Corrosion of Concrete in Farm Buildings. 2nd Materials Specialty Conference*, Canadian Society for Civil Engineering, Montreal, Quebec, Canada.

ASTM. 2005. E917: *Standard Practice for Measuring Life-Cycle Costs of Buildings and Building Systems*. American Society for Testing and Materials, Philadelphia PA

ASTM. 2008. D1141: *Standard Practice for Preparation of Substitute Ocean Water*. American Society for Testing and Materials, Philadelphia PA.

Bache, H.H. 1988. The new strong cements: Their use in structures. *Physics in Technology*, 19, 43–50.

Bager, D.H. and E.J. Sellevold. 1986. Ice formation in hardened cement paste—Part I: Room temperature cured paste with variable moisture contents. *Cement and Concrete Research*, 16, 709–720.

Berke, N.S., L. Li, M.C. Hicks, and J. Bal. 2003. Improving concrete performance with shrinkage reducing admixtures, in V.M. Malhotra, Ed., *7th CANMET/ACI International Conference on Superplasticisers and Other Chemical Admixtures in Concrete*, American Concrete Institute, Berlin, Germany, pp. 37–50.

Cassar, L., C. Pepe, G. Tognon, G.L. Guerrini, and R. Amadelli. 2003. White cement for architectural concrete possessing photocatalytic properties, in *11th International Congress on the Chemistry of Cement*, Durban, South Africa.

Collepardi, M., A. Borsoi, S. Collepardi, J.J.O. Olagot, and R. Troli. 2005. Effects of shrinkage reducing admixture in shrinkage compensating concrete under non-wet curing conditions. *Cement and Concrete Composites,* 27, 704–708.

Bruntland Commission. 1987. *World Commission on Environment and Development, Our Common Future.* Oxford, U.K.

De Belie, N., M. Debruyckere, D. Van Nieuwenburg, and B. De Blaere. 1997. Attack of concrete floors in pig houses by feed acids: Influence of fly ash addition and cement-bound surface layers. *Journal of Agricultural Engineering Research,* 68, 101–108.

De Belie, N., J.J. Lenehan, C.R. Braam, B. Svennerstedt, M. Richardson, and B. Snock. 2000. Durability of building materials and components in the agricultural environment. Part III: Concrete structure. *Journal of Agricultural Engineering Research,* 76, 3–16.

De Muynck, W., N. De Belie, and W. Verstraete. 2009. Effectiveness of admixtures, surface treatments and antimicrobial compounds against biogenic sulfuric acid corrosion of concrete. *Cement and Concrete Composites,* 31, 163–170.

De Vries, J. and R.B. Polder. 1997. Hydrophobic treatment of concrete. *Construction and Building Materials,* 11, 259–265.

Emmons, P.H. 1994. *Concrete Repair and Maintenance,* R.S. Means Co., Kingston, MA.

Fagerlund, G. 1977. The international cooperative test of the critical degree of saturation method of assessing the freeze/thaw resistance of concrete. *Materials and Construction,* 10, 230–251.

Genin, J.M.R., A. Refait, and A. Rahaarinaivo. 1997. The intermediate green rust corrosion products formed on rebars in concrete in the presence of carbonation or chloride ingress, in R.M. Latinision and N.S. Burke, Eds., *Understanding Corrosion Mechanisms of Steel in Concrete,* MIT, Boston, MA.

Gerwick, B.C. 2007. *Construction of Marine and Offshore Structures,* CRC Press, Boca Raton, FL.

Helmuth, R.A. 1962. Discussion of paper: "Frost Action in Concrete" by P. Nerenst, in *4th International Congress on the Chemistry of Cement,* National Bureau of Standards, Boulder, CO.

Hüsken, G., M. Hunger, and H.J.H. Brouwers. 2009. Experimental study of photocatalytic concrete products for air purification. *Building and Environment,* 44, 2463–2474.

Igarashi, S., R.H. Kubo, and M. Kawamura. 2000. Long term volume changes and microcracks formation in high strength mortars. *Cement and Concrete Research,* 30, 943–951.

ISO. 2010. *ISO 14000 Environmental Management* [Online]. Available: http://www.14000.org (accessed July 1, 2010).

Jennings, H.M. 2000. A model for the microstructure of calcium silicate hydrate in cement paste. *Cement and Concrete Research,* 30, 101–116.

Julio-Betancourt, G.A. 2009. Effect of de-icer and anti-icer chemicals on the durability, microstructure and properties of cement-based materials. PhD thesis, University of Toronto, Toronto, Ontario, Canada.

Kosmatka, S.H., B. Kerkhoff, W.C. Panarese, N.F. Macleod, and R.J. Mcgrath. 2002. *Design and Control of Concrete Mixtures,* Portland Cement Association, Ottawa, Ontario, Canada.

Liu, T.C. 1981. Abrasion resistance of concrete. *Journal of the American Concrete Institute,* 78(5), 341–350.

Marchand, J., R. Pleau, and R. Gagné. 1995. Deterioration of concrete due to freezing and thawing, in J.S.A.S. Mindess, Ed., *Materials Science of Concrete,* The American Ceramic Society, Westerville, OH.

Mehta, P.K. 1990. *Concrete in the Marine Environment,* Taylor & Francis Group, New York.

Mehta, P.K., P.K. Mehta, and P.J.M. Monteiro. 2006. *Concrete: Microstructure, Properties and Materials,* McGraw Hill, New York.

Mora-Ruacho, J., R. Gettu, and A. Aguado. 2009. Influence of shrinkage-reducing admixtures on the reduction of plastic shrinkage cracking in concrete. *Cement and Concrete Research,* 39, 141–146.

Neithalath, N., J. Persun, and A. Hossain. 2009. Hydration of high-performance cementitious systems containing vitreous calcium aluminosilicate and silica fume. *Cement and Concrete Research,* 39, 473–481.

Neville, A.M. 2006. *Properties of Concrete,* Longman Scientific & Technical, New York.

Neville, A.M. and P.-C. Aïtcin. 1998. High performance concrete—An overview. *Materials and Structures,* 31, 111–117.

Pérez, M., M. García, L. Traversa, and M. Stupak. 2003. Concrete deterioration by golden mussels, in R.B. Silva, Ed., *PRO 34 Microbial Impact on Building Materials,* RILEM, Lisbon, Portugal.

Persson, B. 1998. Experimental studies on shrinkage of high performance concrete. *Cement and Concrete Research,* 28, 1023–1036.

Pigeon, M., J. Marchand, and R. Pleau. 1996. Frost resistant concrete. *Construction and Building Materials,* 10, 339–348.

Poursaee, A., A. Laurent, and C.M. Hansson. 2010. Corrosion of steel bars in OPC mortar exposed to NaCl, MgCl$_2$ and CaCl$_2$: Macro- and micro-cell corrosion perspective. *Cement and Concrete Research,* 40, 426–430.

Powers, T.C. 1949. The air requirement of frost resistance concrete. *Proceedings of the Highway Research Board,* 29, 184–211.

Powers, T.C. and R.A. Helmuth. 1953. Theory of volume changes in hardened cement paste during freezing. *Proceedings of the Highway Research Board,* 32, 285–297.

Preece, C.M. (ed.). 1979. *Erosion,* Academic Press, New York.

Rangaraju, R.R., K.R. Sompura, and J. Olek. 2007. Investigation into potential of alkali-acetate-based deicers to cause alkali-silica reaction on concrete. *Journal of the Transportation Research Board,* 1979/2006, 69–78.

RMCAO. *Pervious Concrete: When It Rains, It Drains* [Online], Ready Mix Concrete Association of Ontario, Mississauga, Ontario, Canada. Available: http://www.rmcao.org/db2file.asp?fileid=1799 (accessed March 31, 2009).

Rosenberg, A., C.M. Hansson, and C. Andrade. 1989. Mechanisms of corrosion of steel in concrete, in J. Skalny, Ed., *The Materials Science of Concrete,* The American Ceramic Society, Westerville, OH.

Rösli, A. and A.B. Harnik. 1980. Improving the durability of concrete to freezing and deicing salts, in P.J. Sereda and G.G. Litvan, Eds., *Durability of Building Materials and Components,* ASTM, Philadelphia, PA, pp. 464–473.

Scherer, G.W. 1999. Crystallization in pores. *Cement and Concrete Research,* 29, 1347–1358.

Schokker, A.J. 2010. *The Sustainable Concrete Guide—Strategies and Examples,* U.S. Green Concrete Council.

Setzer, M.J. (ed.). 1990. *Interaction of Water with Hardened Cement Paste,* The American Ceramic Society, Westerville, OH.

Shi, X., L. Fay, C. Galloway, K. Volkening, M.M. Peterson, T. Pan, A. Creighton, C. Lawlor, S. Mumma, Y. Liu, and T.A. Nguyen. 2009. Evaluation of alternative anti-icing and deicing compounds using sodium chloride and magnesium chloride as baseline deicers—Phase I. Colorado Department of Transportation, Report CDOT-2009-1.

Somerville, G. 1992. Some final reflections on design life, in G. Somerville, Ed., *The Design Life of Structures,* Blackie & Son Ltd., Glasgow, Scotland.

Sugiyami, D. 2008. Chemical alteration of calcium-silicate hydrate (C-S-H) in sodium chloride solution. *Cement and Concrete Research,* 38, 1270–1275.

Sutter, L., K. Peterson, S. Touton, T. Van Dam, and D. Johnston. 2006. Petrographic evidence of calcium oxychloride formation in mortars exposed to magnesium chloride solution. *Cement and Concrete Research,* 36, 1533–1541.

Svennerstedt, B., N. De Belie, C.R. Braam, J.J. Lenahan, M. Richardson, and B. Snock. 1999. *Durability of Building Materials and Components in Agricultural Environment.* Swedish University of Agricultural Sciences, Department of Agricultural Biosystems and Technology.

Taylor, H.F.W. 1997. *Cement Chemistry,* Thomas Telford Publishing, London, U.K.

Thomas, J.J. and H.M. Jennings. 2006. A colloidal interpretation of chemical aging of the C-S-H gel and its effects on the properties of cement paste. *Cement and Concrete Research,* 36, 30–38.

Tulliani, J.-M., L. Montanaro, A. Negro, and M. Collepardi. 2002. Sulfate attack of concrete building foundations induced by sewage waters. *Cement and Concrete Research*, 32, 843–849.

Tuttle, J.H. and H.W. Jannash. 1974. Occurrence and types of thiobacillus-like bacteria in the sea. *Limnology and Oceanography*, 17, 534–543.

Valenza, J.J.I. and G.W. Scherer. 2006. Mechanism for salt scaling. *Journal of the American Ceramic Society*, 89, 1161–1179.

Valenza, J.J.I. and G.W. Scherer. 2007a. Mechanism for salt scaling of a cementitious surface. *Materials and Structures*, 40, 259–268.

Valenza, J.J.I. and G.W. Scherer. 2007b. A review of salt scaling: II. Mechanisms. *Cement and Concrete Research*, 37, 1022–1034.

Verbeck, G.J. and P. Klieger. 1957. Studies of salt scaling of concrete. *Highway Research Board Bulletin*, 150, 1–17.

Walker, H.N., S. Lane, and P.E. Stutzman. 2006. *Petrographic methods of examining hardened concrete: A petrographic manual*. Virginia Transportation Research Council, Charlottesville, VA.

Wang, K., D.E. Nelson, and W.A. Nixon. 2006. Damaging effects of deicing chemicals on concrete materials. *Cement and Concrete Composites*, 28, 173–188.

Woodson, R.D. 2009. *Repair, Rehabilitation and Protection*, Butterworth-Heinemann.

Further Readings

Povindar, K., P. Mehta, K. Mehta, and Paulo J.M. Monteiro. 2006. *Concrete, Microstructure, Properties and Materials*, McGraw Hill, New York, ISBN 007146289.

Skalny, J., J. Marchand and I. Odler, eds. 2002. *Sulphate Attack on Concrete*, Spon Press, New York, ISBN 0-419-24550-2.

Ben C. Gerwick. 2007. *Construction of Marine and Offshore Structures*, 3rd Edn, CRC Press, Boca Raton, FL, 813pp.

Odd E. Gjørv. 2009. *Durability Design of Concrete Structures in Severe Environments*, Taylor & Francis Group, New York, 220pp.

P. Kumar Mehta. 1990. *Concrete in the Marine Environment*, Taylor & Francis Group, London, U.K., 214pp.

22

Cement and Concrete with Metals

Neal S. Berke
*Tourney Consulting
Group LLC*

Richard G. Sibbick
W. R. Grace & Co

22.1 Introduction

Concrete is the most widely used building material with over 8 billion tons being produced annually. Most of the concrete is made locally, and it can be cast and finished to be both decorative and functional.

Cement acts as the "glue" in concrete and is predominantly used in concrete and mortars. Much of the degradation of concrete is related to the cement deterioration and the transport properties of the cementitious paste. The emphasis in this chapter will be concrete as that has the biggest impact, but in discussing concrete, the role of cement will be included, and most of what applies to concrete degradation would occur in mortar or cement grouts without the aggregates of similar composition exposed to the same environment.

In general, the durability of concrete is excellent; however, this causes it to be often used in aggressive environments. These environments can cause early distress to the concrete, to embedded reinforcement, usually carbon steel, or both. In the case of corrosion of embedded steel, the expansive corrosion products can result in cracking of the concrete leading to spalling and eventual loss of mechanical properties if not addressed.

This chapter will provide a brief overview on concrete properties and primarily focus on the more common degradation mechanisms. The reference section includes several books on concrete that can provide a more in-depth background.

22.2 Background

Modern-day concretes are made using four basic ingredients with chemical admixtures and supplementary cementitious materials (SCMs). The four basic ingredients are (1) cement, (2) sand or fine aggregates, (3) stones (coarse aggregates), and (4) water. Cement typically makes up about 12%–20% of the concrete mass and water is approximately half of that. The hydrated cement provides the concrete with a high internal pH, as the concrete will not completely dry out under natural exposures. Note that cement and water alone is referred to as paste, and cement, water, and fine aggregate is a mortar.

Chemical admixtures are used to provide additional enhanced properties. These include a reduction in the water needed for a given fluidity and delaying or accelerating the hardening of the concrete. Additional admixtures are used to enhance durability such as air entraining agents for better freezing and thawing resistance, more water reduction for lower permeability, corrosion inhibitors to protect embedded metals from corrosion, dampproofing admixtures to reduce ingress of water and chlorides, shrinkage reducing admixtures (SRAs) for better dimensional stability, and admixtures to combat alkali–silica reactions (ASR) that can cause expansion when some aggregates react with the high-alkaline environment. Inorganic silicates are added to reduce permeability, improve hardness, or increase acceleration. The references will provide more details on these materials and others not mentioned. The following sections will go into more detail as to the effects of admixtures on durability.

SCMs are typically pozzolans (materials that are silica or alumina–silica fine particles, e.g., fly ash, silica fume, metakaolin) that are not reactive in water, but will react with the calcium hydroxide (CH) released as cement hydrates. Ground granulated blast furnace slag (GGBFS) is mildly cementitious but will react faster with cement. SCMs are typically used as a cement replacement to reduce the amount of cement needed for a given strength requirement or to reduce permeability as they react with CH to fill in void space.

Much of the degradation problems of concrete are due to making a concrete that has a relatively high permeability to aggressive elements that attack either the concrete or reinforcing steel. This often occurs because the strength requirements can be met without reducing permeability. The other cause of failures is due to using aggregates that can react at high pH, freezing and thawing damage (ice has a higher volume than water and creates high tensile forces), or impurities or additives in the mixing materials.

22.3 Applications

22.3.1 Uses

Concrete is used in numerous construction applications. It can be unreinforced or reinforced internally with reinforcing bars (predominantly carbon steel), wire meshes, or fibers. It can contain high-strength steels that are stretched and then released after the concrete hardens to provide a compressive stress or can be posttensioned after hardening to provide the compressive stress. Concrete can be cast-in-place at the job site, or be precast and shipped to the site.

Unreinforced concrete typically ends up in applications without high tensile or flexural requirements. The applications include, but are not limited to, residential foundations and sometimes basement walls, sidewalks, pavements, patios, and blocks. Many applications include the addition of colors or surface treatments for architectural aesthetics.

Reinforced concretes are used in areas where there is a need to have a more ductile behavior and crack opening resistance. These applications include bridge substructure and superstructure components, high loading pavements, parking structures, marine structures, septic tanks and grease separators, pipe, barrier walls, and many other applications.

22.3.2 Challenges

The main challenges facing concrete are related to sustainability and durability. On the sustainability side, the cement used has a relatively high carbon footprint due to the production process requiring significant energy for the cement kilns and the use of calcium carbonate that represents about 0.7–0.9 tons of CO_2/ton of cement. The industry is addressing this by using waste materials when possible for the energy, more efficient kilns and processes, and the use of the SCMs as mentioned earlier.

Durability is needed to have a truly sustainable material in that longer life cycles use less material over time. Challenges here are the need to occasionally use nonideal aggregates that can react in the high-pH environment as will be discussed later, as well as degradation caused by chemicals in the environment, freezing and thawing, and corrosion of embedded metals. These are discussed in the following sections.

22.4 General Properties

A very brief overview is given here as there are several good books that go into the properties in great detail.

22.4.1 Properties in the Plastic State

The properties of concrete before hardening can play a major role in the hardened concrete properties needed in its use. The workability of the concrete determines how well it can be consolidated and is a function of the water-to-cementitious ratio (w/cm) and the amount of water reducing additives added. A high degree of workability can be achieved with the use of a high-range water reducer or superplasticizer in lieu of added water. Other admixtures and SCMs can affect the viscosity and workability. As noted earlier, accelerators and retarders can be used to adjust the setting (time to harden) times, but depending on what is used can have beneficial or detrimental effects on durability.

22.4.2 Mechanical Properties

The mechanical properties of concrete are highly dependent on time, and increase rapidly within the first 14–90 days, and then tend to level off. In addition to time, the curing temperature and temperature affect strength development that can be predicted using Arrhenius equations with an activation energy or energies for the cementitious components. Low temperatures or lack of enough water can cause a delay or stop strength development.

In general, the flexural and tensile strengths of concrete are about an order of magnitude lower than the compressive strength. Formulas are given in ACI 363. Thus, concrete is a brittle material. To overcome this, reinforcement is added to provide load-bearing capacity after cracking and to improve toughness. Concrete compressive strengths typically range from 20 to 50 MPa, but high-performance concretes can have compressive strengths over 100 MPa.

The elastic modulus of concrete ranges from 14 to over 50 GPa. Unreinforced, concrete will fail at a strain of 0.06%, resulting in large cracks if sufficient reinforcement is not added. Cracks greater than 0.1 mm at the surface will significantly increase the migration of chlorides into concretes; whereas, there will be low chloride permeability in the noncracked state (Rapoport et al. 2002).

22.4.3 Physical Properties

Concrete is a composite material of cementitious components with aggregates and water. The typical density is about 2200–2300 kg/m³; however, lightweight versions can be as low as 1500 kg/m³ and some heavier versions are used in nuclear applications by replacing siliceous or limestone aggregates with heavier aggregates containing iron.

Concrete at a high w/cm and/or poor curing can have a connected pore system leading to rapid ingress of water and detrimental ions such as chlorides and sulfates. At lower w/cm and longer curing times, the pore system becomes discontinuous and migration shifts to a diffusion process. The effects on degradation will be shown in more detail later.

The internal pH of concrete is about 22.5–13.3 depending on the types of cementitious materials and mixture proportions. This environment is protective to steel in the absence of chlorides for most exposure conditions.

The temperature coefficient of expansion for concrete is similar to that of steel, which makes the combination effective in outdoor environments with temperature changes.

Concrete can shrink and lose water mass at RH below 90%, with increasing shrinkage as RH drops. If restrained, this can cause cracking, and shrinkage strain can induce stresses exceeding the tensile strength of the concrete. Means of addressing this will be shown later. If the w/cm is less than 0.4, this shrinkage can occur due to the reaction of water with the cementitious material.

Concrete will creep under load at a rate that drops logarithmically with time. The creep can help to offset shrinkage effects. In large spans, columns, walls, etc., creep and shrinkage must be accounted for to have the right final dimensions and to prevent cracking.

22.4.4 Thermal Properties

Concrete has a thermal conductivity of about 1–2 w/m°K. The heat capacity of concrete is approximately 0.8–0.9 kJ/kg – °K. As noted previously, the thermal coefficient of expansion ranges between 2.2 and 3.9 × 10^{-6}/°C. It maintains its strength up to about 200°C but could have half the strength by 500°C–600°C. Due to the thermal mass of concrete, and relatively low thermal conductivity, if properly designed, concrete can meet building fire ratings without additional protection.

22.5 Structure

22.5.1 Molecular Scale

On the molecular scale, the important features are in the hydration of the cement and SCMs. To simplify the writing of cement chemical formulas, one uses the nomenclature where C is calcium oxide, S is silicate, A is aluminum oxide, F is iron oxide, and H is water. Portland cement is predominantly oxides of $3CaO \cdot SiO_2$ (C_3S), $2CaO \cdot SiO_2$ (C_2S), $3CaO \cdot Al_2O_3$ (C_3A), and $4CaO \cdot Al_2O_3 \cdot Fe_2O_3$ (C_4AF). Gypsum is typically added in the grinding to control reaction rates. The SCMs are essentially the same chemistry, with much less to no calcium. Some cements are based on MgO but are not considered for this chapter.

Typical compositions for Portland cements are found in American Society for Testing and Materials (ASTM) C 150 and ASTM C 595. SCMs are covered in ASTM C 618, ASTM C 989, and ASTM C 1240. Specifications in other countries have similar chemistries.

When cement reacts with water, one forms a calcium silicate hydrate (C–S–H). This is the main glue that binds the aggregates to make mortar or concrete. The molecular structures of the C–S–H and other phases such as ettringite ($3CaO \cdot Al_2O_3 \cdot 3CaSO_4 \cdot 32H_2O$) that are formed are quite complex. Several references (Taylor 1997) (Bishnoi and Scrivener 2009, Juilliand et al. 2010) discuss this in detail and some newer studies are adding new light as to the actual structure of the C–S–H phases (Pellenq et al. 2009). Figure 22.1 shows a micrograph of early-age concrete containing cement, fly ash, and silica fume giving some idea as to reaction rates and morphology. Figure 22.2 shows ettringite.

What is important for degradation mechanisms is the reactivity of the numerous phases and the porosity of the paste and concrete overall. The hydration process results in gel pores that are less than 10 nm in diameter and are internal to the C–S–H structure. These play little role in transport.

FIGURE 22.1 Field example of severe sulfate attack.

FIGURE 22.2 Cube showing damage caused by the formation of ettringite.

22.5.2 Microstructure

The capillary pores are between 100 nm and 10 μm in diameter and are responsible for the bulk of transport through the concrete. These are influenced by the w/cm and degree of curing (Powers et al. 1959). In particular, the importance is whether the capillary structure or pore structure is continuous or discontinuous. Table 22.1 shows the typical wet curing times for a given water-to-cement ratio (w/c) to achieve a discontinuous pore structure in which bulk migration of ions and water is governed by diffusion versus sorption and permeability. This is of upmost importance for chemical attack and corrosion of embedded metals as well as waterproofing applications.

The aggregates present are of macroscopic dimensions (>1 mm). In a good concrete, they are evenly distributed with a good packing system due to selecting gradations that are given in the standards,

TABLE 22.1 Time to Discontinuous
Concrete Capillary Pore Structure as a
Function of Curing and w/c Based
on Work

w/c	Time in Days
0.4	3
0.45	7
0.50	14
0.55	28
0.60	180
0.70	365
>0.70	∞

Source: Powers, T.C. et al., *J. PCA Res.
Dev. Lab.*, 1(2), 38, 1959.

for example, ASTM C 33. Of interest on the microstructure side is the paste–aggregate interfaces. This interface can be more porous than that of normal paste/mortar due to the presence of CH or air bubbles. If this happens, there is an increase in the overall permeability or diffusion coefficients for ions in the concrete (Scali et al. 1987).

Concrete will often have entrapped air and this can result in voids over 500 μm in diameter. A well air-entrained concrete is produced with admixtures that produce air voids with a mean diameter of 200 μm and a spacing between the bubbles of less than 200 μm. As will be seen later, this significantly improves resistance to freezing and thawing.

22.6 Degradation of Concrete

Concrete can degrade due to chemical attack, freezing and thawing, or corrosion of embedded metals. This section will address the more common failure principles.

22.6.1 Chemical Attack

The two main forms of chemical attack are caused by sulfates in the environment or reaction of alkali in the concrete or environment with susceptible aggregates in the concrete known as ASR. As concrete is often used in severe exposures, it can be subjected to other chemicals in factories, docks, farms (fertilizers), and other applications where corrosive materials are handled. This section will concentrate on sulfates in the environment and ASR. Note that concrete will dissolve in acids and in highly concentrated phosphate solutions as well as in some cases in concentrated sugars (Plum and Hammersley 1984, Berke 1989). Given the wide range of potential chemical exposures, one needs to test individually for many of them.

22.6.1.1 Sulfate Attack

There are three major kinds of sulfate attack. The first type results in reactions that form ettringite, which is an expansive reaction causing cracking of the concrete and eventual failure. The second type involves the formation of thaumasite, which involves carbonate as well as sulfate. The third form occurs in high w/c concretes and is the precipitation of multi-hydrates of sodium sulfate.

22.6.1.1.1 Ettringite

The formation of ettringite due to sulfates entering the concrete from the environment or from a delayed ettringite formation (DEF) is the most common form of sulfate attack. Sulfates react with the calcium aluminate phases present in the unhydrated and hydrated cement and CH to produce calcium sulfoaluminate (ettringite) with the chemical composition as noted in Section 22.5.1. This is an expansive reaction that is not a problem when it occurs in newly placed concrete, but creates high stresses in older hardened concretes where there is no room for it to expand. Figure 22.1 shows how a bridge in the Middle East was ultimately destroyed from these reactions:

$$CH + SO_4^{2-} + H_2O \rightarrow CaSO_4 \cdot 2H_2O + 2OH^- \tag{22.1}$$

Equation 22.1 is mildly expansive. However, the products of this reaction will react with the calcium aluminate phase and water to produce the reaction forming the highly expansive ettringite as shown in the following equation:

$$C - A - H + CaSO_4 \cdot 2H_2O + 2H_2O \rightarrow C3A \cdot 3CaSO_4 \cdot 32H_2O \tag{22.2}$$

22.6.1.1.2 Thaumasite

Thaumasite is another sulfate reaction product that primarily occurs in colder conditions (0°C–15°C). The reaction for thaumasite is

$$C - S - H + CaSO_4 \cdot 2H_2O + 2H_2O + CaSO_4 \rightarrow CaSiO_3 \cdot CaSO_4 \cdot CaCO_3 \cdot 15H_2O \tag{22.3}$$

Unlike the ettringite reaction, this production of thaumasite does not result in an expansive reaction product but in a disintegrating reaction due to the consumption of the C–S–H, which is the binding part of the cement and attack and of limestone aggregates if present (Collepardi 2006). Figure 22.3 illustrates the effect of thaumasite formation.

Table 22.2 shows optical characteristics of ettringite and thaumasite that petrographers use to confirm the cause of degradation.

22.6.1.1.3 Effect of Elevated Temperatures

If concrete is heated to above 70°C in the production process or in field exposures, then the following reaction can take place as ettringite is converted to calcium aluminate hydrates (C–A–H) and gypsum:

$$C3A \cdot 3CaSO_4 \cdot 32H_2O \rightarrow C - A - H + CaSO_4 \cdot 2H_2O + 2H_2O \tag{22.4}$$

FIGURE 22.3 Thaumasite form of sulfate attack.

TABLE 22.2 Optical Properties of Thaumasite and Ettringite

Property	Thaumasite	Ettringite
Composition	$CaSiO_3 \cdot CaSO_4 \cdot CaCO_3 \cdot 15H_2O$	$3CaO \cdot Al_2O_3 \cdot 3CaSO_4 \cdot 32H_2O$
Crystal system	Hexagonal	Hexagonal
Crystal habit	Acicular	Acicular
Birefringence	0.036	0.006

The process reverses as in Equation 22.2 at 20°C causing an expansive reaction similar to sulfate coming in from the environment. This is known as DEF.

22.6.1.1.4 Precipitation of Sulfates

In high w/cm concretes, sodium sulfate from the environment can readily migrate into concrete. Expansive reactions occur, for example, due to the precipitation of multi-hydrates of sodium sulfate, which can cause cracking to occur (McMahon et al. 1992).

22.6.1.2 Alkali–Silica Reaction

ASR occurs in hardened concrete and requires aggregates that are susceptible to attack. It occurs in two phases: (1) formation of a gel from the reaction of silica in reactive aggregates with highly alkaline pore solution and (2) the gel imbibes moisture, expanding and, in the process, exerting internal pressure, which can in severe cases lead to cracking (Figure 22.4).

22.6.1.2.1 Mechanism of the Alkali–Silica Reaction

The ASR is a deleterious expansive chemical reaction that develops in hardened concretes. The reaction usually occurs between hydroxyl ions in the pore solution of concrete and certain forms of silica, which occur in some aggregates (Diamond 1975). ASR is not a reaction between the sodium and potassium ions and reactive silica. The ASR process can be divided into two separate phases: firstly the reaction forming the gel and secondly the expansion process.

22.6.1.2.2 Formation of ASR Gel

The hydration of normal ordinary Portland cement (OPC) results in the formation of a pore solution, which is found in the cement paste and contains calcium, potassium, and sodium hydroxides. Most of the CH produced during hydration is present as a sparingly soluble crystalline hydroxide (portlandite), but alkali metal hydroxides derived from the readily soluble sodium and potassium sulfates in cement

FIGURE 22.4 Field example of ASR.

phase minerals remain in the pore solution. When water is added to the cement, these alkali sulfates take part in complex reactions with the C_3A, C_4AF, and $Ca(OH)_2$ phases present in the cement. Ettringite is precipitated into the paste, and the aqueous pore solution is further enriched in Na^+, K^+, and OH^- ions. By this time, the pore solution consists almost entirely of sodium and potassium hydroxides (Lawrence 1966, Longuet et al. 1973), which can produce a pore solution pH value well in excess of 13. It is only in concretes with these high hydroxyl concentrations that significant attack of reactive silica particles can occur (Hobbs 1988). A high-alkali-content cement can have a hydroxyl concentration in the pore solution 10 times that of a low-alkali cement; therefore, it is usually with the high-alkali-content cements that ASR develops.

This highly alkaline pore solution then "attacks" the reactive silica. The surface of any silica particle displays a weak acid characteristic, which increases with increasing surface area or crystalline disorder. Common crystalline vein quartz is strongly ordered into bonded silicon–oxygen tetrahedra ($3[SiO]_2$) and is apparently stable in all alkaline conditions. At the other extreme of potential reactivity is opal, which is made up of a random network of tetrahedra with spaces in between the groups of molecules, which in normal environmental conditions are filled by water. In the highly alkaline conditions found in a concrete, these spaces are quickly filled by the alkaline pore solutions. At the reactive silica/pore solution interface, an acid/alkali reaction occurs and a hydrous silicate is formed. Hydroxyl ions are imbibed into the silica particle, and some of the silicon-to-oxygen linkages are attacked weakening the bonds locally. Sodium and potassium cations then diffuse in to maintain the electrical neutrality and attract more water to form a gelatinous alkali-metal-ion hydrous silicate. Once the Si–O–Si linkages are disrupted, more water and further hydroxyl ions are able to enter the system, being imbibed and forming more gel, which again results in further pressure. While the hydroxyl ion concentration remains high and the source of silica is still available, the expansive gel-forming processes are self-perpetuating. In saturated concrete, the gel product imbibes water and alkalis, at a rate that is dependent on the hydroxyl concentration (Hobbs 1988). The loss of hydroxyl ion into the ASR gel results in a progressive loss of this material from the pore solution until a point is reached at which no further gel formation is possible due to the pore solution alkali level now being depleted below the alkali threshold for significant ASR with that particular reactive aggregate mix. The chemical reaction can be shown simply by the equation shown in the following:

$$SiO_2 + 2NaOH \quad \text{or} \quad KOH + H_2O \rightarrow \left(\frac{Na}{K}\right)2 \cdot SiO_2 \cdot {}^*H_2O \tag{22.5}$$

* is the totally variable number of water molecules. (ASR gel is an "alkali-metal-ion hydrous silicate" [Hobbs 1988]).

Note that all reactive/expansive ASR gels appear to also contain a proportion of calcium (Sibbick 1993).

22.6.1.2.3 Mechanism of Expansion

The process by which the gel absorbs the water and expands has been debated since ASR research began back in the 1940s. Two main theories existed to explain the mechanism of expansion caused by ASR. In the first, the stresses that induce the cracking in the concrete are attributed to the growth of the gel caused by absorption of pore fluids (Vivian 1950), while the second, which is termed the osmotic cell theory, involves the development of internal stresses due to hydraulic pressure developed across a semipermeable membrane (Hansen 1944).

In the absorption theory, the expansion induced is dependent on the total gel volume, localized concentration, the rate of gel formation, and physical properties such as viscosity. If gel growth is slow, then the internal stresses may be dissipated by migration of the gel into the concrete. If the gel formation occurs relatively rapidly, the internal stresses can build up within the aggregate particle to a high enough level to cause cracking and expansion in the concrete. If the gel produced has a low viscosity, then it will dissipate before exerting any internal pressure.

In summary, at the present time, the mechanism of water uptake by the gel is considered to be by direct sorption or imbibition from the pore solution or other external water source (Pike et al. 1955, Gillot 1975, Dent Glasser 1979). The reaction between these two components forms a hygroscopic alkali–silica gel. This alkali–silica gel that forms and initially is located in or around the reactive aggregate particle absorbs/imbibes any free water and swells. This rapid swelling of the gel exerts an internal pressure within the aggregate particle, which may, in the more reactive cases, cause the aggregate and eventually the surrounding cement paste to crack.

22.6.1.2.4 Alkali–Silica Reaction Cracking

ASR is a long-term phenomenon, and it may take 4–5 years for a network of cracks to develop on the surface of affected structures. However, structures monitored in Denmark, the United States, and South Africa have shown microcracking within only 1–3.5 years (Hobbs 1988). In an unreinforced structure, these cracks will take on the form described as map cracking. This cracking appears to be a random divergent form often developing at single points from which three separate cracks will separate. In constrained or reinforced concretes, this pattern of map cracking is usually modified with larger cracks running parallel to the reinforcement bars. In the case of a concrete pavement (Sibbick and West 1992) constrained in all directions but upward, the major cracks develop in the concrete parallel to the top surface (direction of least compression).

Other deleterious activities that affect concretes in similar environmental situations can also produce expansive crack patterns. These features can be caused by frost attack, sulfate attack, severe plastic shrinkage, and shrinking aggregates. All these various deleterious activities can usually be separated by petrographical analysis. In extreme cases, the ASR gel is also exuded onto the outer surfaces of the concrete and is often associated with damp "sweaty" patches. As confirmed by petrographical analysis, the texture and color of the ASR gel is variable, dependent on other factors such as aggregate type, iron staining, and the pickup of fine rock debris by the gel.

It has been stated by some workers (Hobbs 1990) that only microcracking observed petrographically to exhibit the typical divergent cracking seen in chert-rich aggregate can be regarded as being representative of concrete affected primarily by ASR. The less well-documented microcracking observed in greywackes, siltstones, and silicified limestones, however, is just as typical for those specific aggregate types (Pettifer 1993).

22.6.1.2.5 Alkali–Silica Gel

The chemical composition of the alkali–silica gel has been studied first appearing in 1942 (Stanton 1940). The gel compositions are observed to vary greatly (Poole 1992a). Most present-day analyses are carried out using an electron microprobe (EMP), which requires the sample to be completely desiccated prior to analysis. It is worth pointing out first that elements with atomic number less than 11 such as carbon are more difficult to quantify depending on the detector type, so reliable analyses of gel carbonation ($CaCO_3$) as opposed to $Ca(OH)_2$ taken from the cement paste may be questioned, without backing up these results with localized alkalinity testing. Problems of excessive carbonation of the prepared samples can also lead to poor chemical analysis results.

Alkali–silica gel is composed of variable proportions of sodium, potassium, silicon, calcium, magnesium, oxygen, and water (H_2O), which is lost from the analyses. The role of these chemical constituents of the gel, particularly the calcium in determination of the overall expansion, is unclear. Experiments showed that the amount of calcium in the gel varied as it passed from aggregate to cement paste (Knudsen and Thaulow 1975). The gel in the aggregate was high in alkalis, while that found in the paste was higher in calcium. Verbeck and Gramlich (1955) suggested that if CH was present, then the ASR gel formed was calcium rich and the fluidity was reduced. Diamond et al. (1981) working with synthetic gels also confirmed that calcium-rich gel has reduced fluidity, but could find no relationship between gel composition and swelling. Chatterji et al. (1986a, 1986b) have also suggested that CH is a requirement of ASR, a view endorsed by Diamond (1989).

It is clear that when an alkali–silica gel first forms in the reactive aggregate particle, it will be very viscous (pressures of up to 11 MPa have been recorded), but as it absorbs more water from the pore solutions, it will swell and become less viscous becoming fluid enough to flow along the microcracks formed during the initial ASR-induced expansion and flowing into air voids and micropores in the cement paste. Much research has been undertaken to establish a composition to expansive behavior relationship (Moore 1978, Diamond et al. 1981). It seems clear now, however, that the behavior of the gel within a concrete undergoing ASR is most complex and not as dependent on chemical composition as was originally thought (Poole 1992b). The petrographical inspection of ASR reactive concretes often reveals a series of different-looking gels, as generations or pulses, carbonated gel material, and even gel that has apparently recrystallized into an alkali–silica mineral. The causes of these differences are still not fully understood.

Laing et al. (1992), carrying out gel composition analysis on synthetic and real structural concrete, observed a compositional progression and possible phase boundary between high-alkali, low-calcium gel in the aggregate and low-alkali, high-calcium gel in the cement paste. Both gels appeared petrographically identical. This work was carried out on Thames Valley gravel mixes. No work has as yet been carried out on the gel composition of different ASR reactive aggregates comparing the results to the established theories and degree of observed expansion taking factors such as initial alkali content and mix design as constant.

22.6.2 Freezing and Thawing

At subfreezing temperatures, the water in the capillary pores can freeze. The formation of ice from water results in approximately an 8% increase in volume. The stresses associated with the volume increase can crack the concrete, which has a low tensile strength. Figure 22.5 shows the macro effects of freezing damage.

Scaling damage, due to the use of deicing salts, is another degradation mode in freezing environments. The damage can be to the level of exposing aggregate surfaces, to the loss of aggregates, and to the crumbling of the surface in the most severe case. This can look similar to the outside of the failed beam in Figure 22.5 without salt.

22.6.3 Corrosion Mechanisms of Steel

Steel reinforced concrete is extensively used because of its durability and cost advantages over other structural materials. However, due to the aggressive use of deicing salts, or marine salt exposures, often

FIGURE 22.5 Freezing and thawing damage evident in top left beam without air entrainment.

coupled with high ambient temperatures, premature corrosion of embedded reinforcement occurs. These adverse conditions are well documented (Wilkins and Lawrence 1983, Weyers and Cady 1987, Somayaji et al. 1990, Bamforth and Price 1993). Carbonation of the concrete can also lead to corrosion of embedded steel; however, it is not specifically addressed in this chapter since the methods to reduce chloride-induced attack will prevent carbonation corrosion.

In this section, the emphasis will be on carbon steel, and other reinforcement will be addressed in the section on protection.

Steel in concrete is normally passive due to the protective oxide that forms in high-pH environments. A brief explanation of the protection and breakdown process is given later. More information can be found in the references in the Further Reading Section.

The normally occurring oxides on steel are either ferrous (Fe^{+2}) or ferric (Fe^{+3}) in nature. Both are chemically stable in concrete in the absence of carbonation or chloride. However, the ferric oxide is the most stable, especially in the presence of chloride. Over time, the ferrous oxide is converted to the more stable hydrated ferric oxide that is chemically referred to as γ-FeOOH. This process is never totally completed and is measured as a continuing very small passive corrosion rate, as will be discussed later. The development of the ferric oxide film proceeds according to the following anodic reactions:

$$2Fe \rightarrow 2Fe^{2+} + 4e^{-} \tag{22.6}$$

$$2Fe^{2+} + 4OH^{-} \rightarrow 2Fe(OH)_2 \tag{22.7}$$

$$2Fe(OH)_2 + \frac{1}{2}O_2 \rightarrow 2\gamma - FeOOH + H_2O \tag{22.8}$$

Neither oxide is protective at pH values below approximately 11. Thus, corrosion will occur if carbonation is present at the bars.

As noted previously, the ferric oxide is more resistant to chloride than the ferrous oxide. If chloride ions are present, they will induce corrosion when they come into contact with the reinforcing bar at a location where the ferrous oxide hasn't been converted. This results in pitting corrosion. This competes with the normal passivation process and will only proceed when the chloride content is sufficiently high compared to the oxygen and hydroxide content. This is the underlying theory as to why chloride-to-hydroxide ratios are related to the onset of pitting corrosion. In concrete, this corrosion threshold in mass of chloride to volume of concrete is ~0.9 kg/m³ (1.5 lb/yd³).

This passivation process can also be described in electrochemical terms. Figure 22.6 shows the anodic and cathodic reaction curves for passive steel in an alkaline environment such as concrete. Curve A in Figure 22.6 schematically shows the relationship between the anodic reaction rate and the potential of steel relative to a reference electrode in a concrete-type environment. The anodic reaction rate is represented as a current and describes the corrosion rate of the steel, with lower currents indicating lower corrosion rates. At very high negative potentials, the corrosion rate increases rapidly as the potential becomes less negative, then suddenly drops by several orders of magnitude. The potential at which this occurs is known as the primary passivation potential (E_{pp}) and is related to the development of the ferric oxide. Then, over a large range of potentials (typically from approximately −800 to +600 mV versus saturated calomel electrode, in the absence of chloride), the steel corrodes at a negligible rate and this is the passive region. At more positive potentials, the breakdown of water to produce oxygen occurs and severe corrosion can once again occur and passivity is lost. The potential at which this occurs is the transpassive potential (E_{tp}).

Curve A in Figure 22.6 describes the anodic behavior of steel, that is, it only looks at the oxidation reactions that are occurring. The actual corrosion potential will also depend upon the oxygen

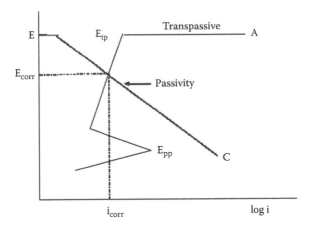

FIGURE 22.6   Steel in an alkaline environment.

content, which affects the cathodic reaction rate. For steel in concrete, the cathodic reaction is primarily the reduction of oxygen:

$$O_2 + 2H_2O + 4e^- \rightarrow 4OH^- \tag{22.9}$$

The cathodic behavior as a function of potential is schematically shown in Figure 22.6 as Curve C.

As in other chemical processes, the anodic reaction and the cathodic reaction have to be in balance. Thus, the point of intersection of the two curves portrays the only corrosion situation, which can exist for these conditions. This is shown in Figure 22.6 in which the anodic and cathodic curves have been superimposed and the rates have been shown as currents. Since this figure is representative of steel in non-carbonated chloride-free concrete, the intersection of the two reactions is in the passive zone, and only the low passive corrosion currents (I_{corr}) are reached. The potential at which the anodic and cathodic currents are of equal magnitude is the corrosion potential, E_{corr}. The I_{corr} is equal in magnitude to the anodic current and absolute value of the cathodic current. It is most useful to divide the I_{corr} by the area to obtain the corrosion current density (i_{corr}).

The anodic behavior of steel in concrete is changed in the presence of chloride as shown in Figure 22.7 with chloride causing pitting to initiate. When pitting is present, the i_{corr} in the pit is high and is only limited by the amount of cathode present. This is schematically shown in Figure 22.2, which shows that the anodic curve (A1) is almost horizontal at the pitting protection potential (E_p). At potentials negative to E_p, pits are not stable and pitting stops. Above E_p, the corrosion rate is orders of magnitude higher than the passive rate and severe local metal loss can occur.

Increasing the chloride content has the effect of facilitating the pitting process and results in a further lowering of the pitting E_p as shown with Curve A2 in Figure 22.7. For the same amount of available

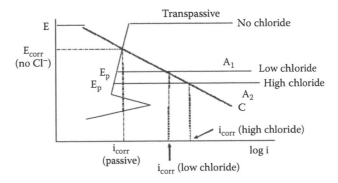

FIGURE 22.7   Effect of chloride on the corrosion of steel.

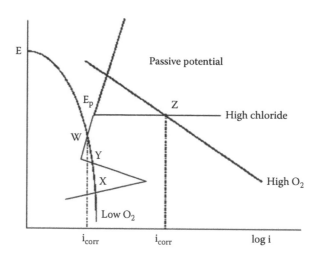

FIGURE 22.8 Role of oxygen on steel corrosion.

cathode (i.e., for identical curve C), the corrosion rate increases further as shown. This is where the rule of thumb about more negative corrosion potentials being associated with higher corrosion rates came from. However, even when chloride is present, and especially when it isn't, corrosion potential values versus a saturated calomel reference electrode below −280 mV are not necessarily indicative of corrosion as shown in Figure 22.8 discussed later.

The role of oxygen on the corrosion rate when pitting is present is illustrated in Figure 22.8. Variations in oxygen availability change the location of the cathodic reaction curve. Thus, even if sufficient chloride is present for pitting, corrosion rates can be low if the oxygen content or cathodic area is low. Points W and X are the possible conditions when oxygen levels are low (the intersection at point Y represents an unstable situation, which will convert to point W). If the steel has already been passivated due to higher oxygen conditions at some time point, W will be the potential. If passivation has not occurred, then point X will be the intersection point, and corrosion rates will be higher (active corrosion but at a low rate). In both cases, corrosion rates are negligible. This is commonly observed for concrete piles far below the water line in marine environments where oxygen levels are low.

At higher oxygen levels, pitting corrosion occurs since the oxygen reduction reaction intersects the anodic reaction above the E_p (Figure 22.8). This is illustrated as the intersection that occurs at point Z. Because current density is on a log scale, the i_{corr} is orders of magnitude higher than in the low-oxygen case.

22.7 Degradation Protection

As will be seen throughout this section, the first step in improving the performance of concrete against degradation is to reduce its permeability. This is accomplished by lowering the w/cm ratio with the use of superplasticizers and by the addition of SCMs as noted earlier.

22.7.1 Sulfate Attack

External sulfate attack can be mitigated by using a sulfate-resistant cement with a low C_3A content (i.e., C_3A should remain above 5% if chlorides are present) (Zhang et al. 2003). In addition, the lowering of the w/cm ratio to prevent the ingress of sulfate is highly affective (Monteiro and Kurtis 2008). The use of SCMs significantly reduces the amounts of available CH and further reduces permeability. The reduction in permeability by these methods will prevent ingress of sodium sulfate and expansive precipitation.

The previous measures will reduce the probability of thaumasite formation. However, for thaumasite prevention, moisture needs to be kept out, and this can be accomplished by adding waterproofing or membranes.

DEF can be prevented by keeping the maximum temperature of concrete below 70°C. If higher temperatures between 70°C and 85°C cannot be avoided, then the guidelines in ACI 201 should be followed.

22.7.2 Alkali–Silica Reaction

The first means of preventing ASR is to use aggregates that are nonreactive. There are several ASTM test methods that can be used to screen aggregates.

If one cannot get nonreactive aggregates, then reducing the alkalinity of the concrete can work for marginal aggregates. Methods that work are the addition of SCMs, in particular silica fume, Type F fly ash, metakaolin, and GGBFS. A similar approach is to use a low-alkali cement. The total alkali (Na_2O equivalent) should be less than 3 kg/m^3.

Another approach to reducing ASR is to use lithium-containing admixtures (Blackwell et al. 1997, Stokes et al. 2000, Adams and Stokes 2002, Balcom et al. 2005). The gel that forms is non-expansive. One of the more widely used additives is $LiNO_3$.

22.7.3 Freezing and Thawing

Degradation due to freezing and thawing is mitigated by adding air entrainment to concrete so that there is space for the ice to expand. Typically 4.5%–5% air is entrained into the plastic concrete. The goal is to have a spacing factor between air bubbles in the hardened concrete that is under 200 μm. Recommended minimum design strengths are over 27.5 MPa, with a w/cm ratio <0.45. Higher-strength concretes with strengths in excess of 35 MPa can often exhibit good behavior with spacing factors as high as 300 μm.

Lowering the w/cm ratio to under 0.25 will eliminate free water and thus provide freezing and thawing protection in many cases. This is not practical for most concretes unless ultrahigh strength or extremely low permeability is needed, as this is much more costly than adding air entrainment.

Other alternatives are solid air substitutes such as plastic spheres. However, these are expensive and require large storage space.

22.7.4 Corrosion

22.7.4.1 Reducing the Ingress of Chloride

22.7.4.1.1 Low-Permeability Concrete

Traditionally, the first steps in improving the durability of reinforced steel involved improving concrete quality. For example, ACI 318 and ACI 357 recommend w/cm ratios of 0.4 or under and minimum covers of 50 mm for nonmarine and 68 mm in marine exposures.

Additional means of reducing the ingress of chloride into concrete involves the addition of pozzolans or GGBFS. These materials are often added as cement substitutes and react with CH to reduce the coarse porosity of the concrete and to decrease the porosity at the paste–aggregate interfaces. Several conference proceedings exist documenting the positive benefits of these materials and one from the CANMET/ACI series is in the Further Reading Section.

Numerous references showed that even if concrete is produced to the most stringent of codes, chloride will ingress into the concrete and corrosion of the steel reinforcement will initiate (Browne 1982, Tuutti 1982, Pfeifer and Landgren 1987, Berke et al. 1992a,b, 1997). Berke et al. (1994) show that even for very low-permeability concretes, chloride will eventually ingress into the concrete and initiate corrosion. When chloride ingress is modeled as a function of w/cm and pozzolan contents for various geometries and environmental exposures, times to corrosion in excess of 40 years are difficult to achieve for low-permeability concretes (Berke and Li 2007). Thus, additional protection systems are necessary to meet extended design lives that are becoming increasingly specified.

22.7.4.1.2 Sealers and Membranes

Sealers and membranes work by either reducing the ingress of chloride into the concrete or by reducing the movement of moisture. They work by either producing a continuous film on the surface, creating a hydrophobic layer lining the pores and surfaces of the concrete, or by producing reaction products that block the concrete pores. Surface preparation and application techniques must be carefully followed to have good results. Several references discuss these protection systems in great detail (Pfeifer and Scali 1981, Keer 1992, Weyers et al. 1992, Bentur et al. 1997).

Products that provide a continuous film or coating on the surface include acrylics, butadiene copolymer, chlorinated rubber, epoxy resin, oleoresinous, polyester resin, polyethylene copolymer, polyurethane, and vinyl. Rubberized asphalt is a popular sheet membrane material. Membranes exposed to wear typically have protective surface coatings. Hydrophobic pore liners include the silanes, siloxanes, and silicones. The pore blockers include silicate, silicofluoride, and crystal growth materials in a cementitious slurry.

An extensive review of sealer and membrane field performance was conducted as part of the Strategic Highway Research Program (Weyers et al. 1992). It concluded that these systems had a useful effectiveness between 5 and 20 years. The importance of longer-term testing in evaluating these products was shown by Robinson (1987). His accelerated laboratory testing showed an early 100% reduction in chloride ingress, which did not hold up over time; however, most products did offer an improvement.

22.7.4.2 Corrosion Inhibitors

Corrosion inhibitors are chemical substances that reduce the corrosion of embedded metal without reducing the concentration of the corrosive agents. This definition paraphrased from ISO 8044-89 makes the distinction between a corrosion inhibitor and other additions to concrete that improve corrosion resistance by reducing chloride ingress into the concrete.

Corrosion inhibitors can influence the anodic, cathodic, or both reactions. Since the anodic and cathodic reactions must balance, a reduction in either one will result in a lowering of the corrosion rate. Figure 22.9 illustrates the effects of both types of inhibitors acting alone or in combination when the chloride concentration has not been changed. When no inhibitors are present, the anodic (A_1) and cathodic curves (C_1) intersect at point W. Severe pitting corrosion occurs. The addition of an anodic inhibitor (curve A_2) promotes the formation of γ-FeOOH (passive oxide), which raises the E_p, so that the anodic and cathodic curves now intersect at point X. The corresponding corrosion rate, i_{corr}, is reduced by several orders of magnitude and the steel is passive. Increasing quantities of anodic inhibitor will move curve A_2 to more positive E_p values.

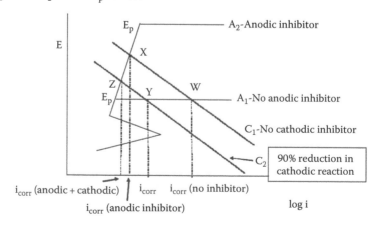

FIGURE 22.9   Comparison of anodic and cathodic inhibitors.

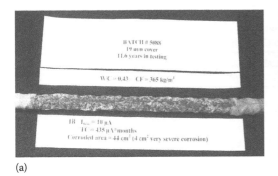

(a)

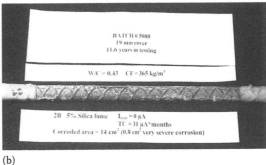

(b)

(c)

FIGURE 22.10 Improving corrosion performance in low-permeability concrete with combinations of silica fume and calcium nitrite: (a) Control: I_{mac} = 10 μA, corroded area = 44 cm², (b) +5% SF: I_{mac} = 0.43 μA, corroded area = 14 cm², (c) +5% SF + 10 L/m³CN: I_{mac} = 0 μA, no corrosion. (From Berke, N.S. et al., Holistic approach to durable high performance concrete, in Y. Yuan, S.P. Shah, and H. Lu, Eds., *Proceedings of the International Conference on ICACS 2003, Advances in Concrete and Structures*, RILEM Publications S.A.R.L., Paris, France, pp. 727–738, 2003.)

The addition of a cathodic inhibitor in the absence of an anodic inhibitor results in a new cathodic curve (C_2) as shown in Figure 22.10. The new intersection with the A_1 is at point Y. Though the corrosion rate is reduced, pitting corrosion still occurs, because the potential remains more positive than E_p. Therefore, a cathodic inhibitor would have to reduce cathodic reaction rates by several orders of magnitude to be effective by itself.

The case of combined anodic and cathodic inhibition is illustrated in Figure 22.10 as the intersection of the anodic (A_2) and cathodic (C_2) curves at point Z. The steel is passive as in the case of the anodic inhibitor alone (point X), but the passive corrosion rate is reduced further.

Commercially available inhibitors include calcium nitrite, sodium nitrite, and morpholine derivatives, amine and esters, dimethylethanolamine, amines, and phosphates. Several references discuss the performance of inhibitors in concrete (Berke 1991, 1993, Berke and Weil 1994, Nmai and Kraus 1994). With the exception of calcium nitrite, little to no long-term data beyond 10 years is available on the other products. Short-term data are available (Elsener 2001, Berke et al. 2003a, Bolzoni et al. 2005); however, caution is recommended in the evaluation of short-term results, which can be misleading (Berke et al. 1994).

The long-term performance benefits of calcium nitrite are well documented (Berke et al. 2003b, Berke and Hicks 2004, Aldykiewicz et al. 2005). Based upon these results, Table 22.3 was developed to indicate the level of chloride that a given addition of 30% calcium nitrite protects against.

Furthermore, as noted in these papers, the use of calcium nitrite or any other inhibitor is not a substitute for good-quality concrete, and guidelines for reducing chloride ingress must be followed. Figure 22.10 shows the benefits achieved when calcium nitrite was added to low-permeability concrete.

TABLE 22.3   Calcium Nitrite
(30% Solution) Dosage Rates Needed
to Protect to a Given Chloride Content

Calcium Nitrite (L/m^3)	Total Chloride Content (kg/m^3)
10	3.6
15	5.9
20	7.7
25	8.9
30	9.5

Source: Berke, N.S. and Hicks, M.C.,
Cement Concrete Comp., 26, 191, 2004.

Performance criteria for an amine and ester commercially available inhibitor are available (Johnson et al. 1996). This inhibitor at a dosage of 5 L/m^3 was stated to protect to 2.4 kg/m^3 of chloride. A reduction in the chloride diffusion coefficient of 22%–43%, depending on concrete quality, was determined using accelerated test methods.

22.7.4.3 Rebar Coatings

Rebar coatings that are in large-scale commercial use are epoxy and zinc. Epoxy acting as a barrier to chlorides and zinc (galvanized rebars) provides both a barrier and sacrificial anode.

Epoxy-coated steel is extensively used in the United States. ASTM Standards (D 3963, A 775, and A934) and a NACE Recommended Practice (RPO 395) are available to ensure good performance. These standards set requirements for the number of holidays (small non-visible defects), repair area allowed, handling, and coating performance. Numerous studies have been conducted that show improved performance versus black steel controls (Virmani et al. 1983, Kessler et al. 1986, Sagues and Zyad 1989, McKenzie 1993, Schiessl and Reuter 1993, Berke and Hicks 1995). However, several of these studies show that corrosion can occur over time and that the coating can be damaged during the placing and consolidation of the concrete. The damaged portions of the coating can be initiation sites for corrosion. The best performance is obtained when all of the steel is epoxy coated, which avoids large cathode-to-anode ratios.

Galvanized rebars are used more extensively outside of the United States. There is an applicable ASTM Standard (A 767). Several papers document performance in tests (Treadaway et al. 1980, Yeomans and Novak 1990, Yeomans 1993). In general, the coating thickness determines overall life, with thicker coatings having better performance.

22.7.4.4 Stainless Steel

Several studies exist on the performance of stainless steel alloys in concrete (Flint and Cox 1988, McDonald et al. 1995). Though performance is improved significantly, depending upon the alloy used, initial costs are quite high (Sorensen et al. 1990); therefore, its use is limited. Stainless clad reinforcing bars, which potentially reduce the cost of stainless by about a factor of two, have been tried. Some works (Fushuang et al. 2001) indicate that severe corrosion can occur at breaks in the cladding and that low-permeability, high-resistivity concrete is recommended. There are several new stainless bar materials being evaluated with improving performance as Pitting Resistance Equivalent Number (PREN) increases (Hartt et al. 2009).

22.7.4.5 Cathodic Protection

Cathodic protection if applied properly can prevent and stop corrosion of steel in concrete. It works by making the reinforcing steel a cathode by the use of an external anode. This anode can be either inert or sacrificial depending upon the exposure.

For cathodic protection to be effective, the reinforcement must be electrically continuous, the concrete conductivity between the anode and steel needs to be low, and alkali-reactive aggregates should not be used as the area next to the steel becomes more alkaline over time. More information is available in a joint document form NACE International and The Institute of Corrosion in the UK (CEA 54286). Broomfield (1992, 1993) and Bennett (1993) provide additional information.

Due to the expense relative to other corrosion protection systems, cathodic protection is most often used in repair. The reasons for this are quite evident once a life-cycle cost analysis is performed.

22.7.4.6 Additional Developments

SRAs significantly reduce drying shrinkage that often causes cracking in restrained concrete (Transportation Research Circular 1999). Figure 22.11 demonstrates the improvements in crack reduction on one side of a bridge that contained SRA, while the other side of the bridge with visible cracking did not contain SRA (Berke and Li 2007).

Another means to reduce the severity of concrete cracking involves the use of fiber reinforcement. Until recently, this was mostly accomplished with steel fibers, but now structural versions of synthetic fibers are available.

(a)

(b)   (c)

FIGURE 22.11 New London Turnpike bridge over I-95 at exit 7 in West Warwick, Rhode Island. (a) Bridge overall view, (b) view of deck from underneath, and (c) view of deck from top. Note in (b) and (c), deck on the left side contains SRA and is crack-free after 8 years of service, whereas deck on the right side does not contain SRA and developed evenly spaced transverse cracks within 4 weeks of placement. (From Berke, N.S. and Li, L., Holistic approach to sustainability of reinforced concrete exposed to chlorides, in S. Jacobsen, P. Jahren, and K.O. Kjellsen, Eds., *Proceedings of the International Conference on Sustainability in the Cement and Concrete Industry*, Lillehammer, Norway, pp. 452–466, 2007.)

22.8 Life-Cycle Analysis

Clearly, there are numerous methods to provide improved corrosion resistance to steel in concrete to improve long-term durability. A rational approach is needed to select the best method or combination of methods, that is, corrosion protection system based upon a total lifecycle cost of the structure. A couple of the numerous references are included (Berke et al. 1997, Thomas and Bentz 2000). In general, the models clearly show that several corrosion protection systems are highly cost-effective over the design life of the structure and that there is a good synergy with using low-permeability concretes with a corrosion protection system. As noted previously, cracking can be mitigated by using SRAs and fibers to maintain the low permeability of the structure. In addition, low-permeability concretes, as noted, provide enhanced resistance to sulfate and freezing and thawing degradation mechanisms.

22.9 Specific Standardized Tests

ASTM Book of Standards Volume 4.01 Cement. W. Conshohocken, PA: ASTM International.
ASTM Book of Standards Volume 4.02 Concrete. W. Conshohocken, PA: ASTM International.

22.10 Summary

Environmental degradation of concrete can come from several sources that either affect the concrete directly, cause corrosion of embedded metals, or both. Reducing the permeability of concrete by lowering the w/cm ratio and the use of SCMs is very effective in all cases. In more severe naturally occurring exposures, additional methods of protection are needed and are cost-effective over the life of the structure. If the proper protection measures are employed in construction, concrete and reinforced concrete will have excellent durability over the life of the structure.

Nomenclature and Units

Abbreviations

Ground granulated blast furnace slag	GGBFS
Water-to-cement ratio	w/c
Water-to-cementitious ratio	w/cm
Supplementary cementitious material	SCM
Protection potential	E_p
Primary passivation potential	E_{pp}
Transpassive potential	E_{tp}
Corrosion current density	i_{corr}

Cement Notation

CaO	C
SiO_2	S
Al_2O_3	A
Fe_2O_3	F
H_2O	H

Acknowledgments

The authors wish to thank their colleagues at Grace Construction Products for their support.

References

ACI 201.2R. *Guide to Durable Concrete*, American Concrete Institute, Farmington Hills, MI.

ACI 363R-10. *Report on High-Strength Concrete*, American Concrete Institute, Farmington Hills, MI.

Adams, N. and D.B. Stokes. 2002. Using advanced Lithium Technology to combat ASR in concrete. *Concrete International*, 24(8), 99–102.

Aldykiewicz, A.J. Jr., N.S. Berke, R.J. Hoopes et al. 2005. Long-term behavior of fly ash and Silica fume concretes in Laboratory and Field exposures to Chloride, *Corrosion 2005*, Paper No. 05253, NACE International, Houston, TX.

Balcom, B.J., T.W. Bremner, X. Feng et al. 2005. Studies on Lithium salts to mitigate ASR-induced expansion in new concrete: A critical review. *Cement and Concrete Research*, 35(9), 1789–1796.

Bamforth, P.B. and W.F. Price. 1993. Factors influencing chloride ingress into marine structures, Presented at *Economic and Durable Construction through Excellence*, Dundee, Scotland.

Bennett, J.E. 1993. Cathodic protection criteria under SHARP contract, *Corrosion'93*, Paper No. 323, NACE International, Houston, TX.

Berke, N.S. 1989. Resistance of microsilica concrete to steel corrosion, erosion and chemical attack, in V.M. Malhotra, Ed., *Proceedings of the Third International Conference, Fly Ash, Silica Fume, Slag, and Natural Pozzolans in Concrete*, SP-114, Trondheim, Norway, American Concrete Institute, Detroit, MI, pp. 861–886.

Berke, N.S. 1991. Corrosion inhibitors in concrete. *Concrete International*, 13(7), 24–27.

Berke, N.S., A.J. Aldykiewicz Jr., and L. Li. 2003a. What's new in corrosion inhibitors. *Structure*, July/August, 10–22.

Berke, N.S., M.P. Dallaire, and M.C. Hicks. 1992a. Plastic, mechanical, corrosion, and chemical resistance properties of silica fume (Microsilica) concretes, in V.M. Malhotra, Ed., *Proceedings of the Fourth International Conference on Fly Ash, Silica Fume, Slag, and Natural Pozzolans in Concrete*, SP 13, Istanbul, Turkey, American Concrete Institute, Detroit, MI, pp. 1125–1149.

Berke, N.S. and M.C. Hicks. 1995. Calcium nitrite corrosion inhibitor with and without epoxy-coated reinforcing bar for long-term durability in the Gulf, in G.L. Macmillan, Ed., *Concrete Durability in the Arabian Gulf*, Bahrain Society of Engineers, Manama, Bahrain, pp. 119–132.

Berke, N.S. and M.C. Hicks. 2004. Predicting long-term durability of steel reinforcement in concrete with calcium nitrite corrosion inhibitor. *Cement and Concrete Composites*, 26, 191–198.

Berke, N.S., M.C. Hicks, B.I. Abdelrazig et al. 1994a. A belt and braces approach to corrosion protection of steel in concrete, in *Proceedings of the International Conference on Corrosion and Corrosion Protection of Steel in Concrete*, University of Sheffield, Sheffield, U.K., pp. 893–904.

Berke, N.S., M.C. Hicks, and K.J. Folliard. 1997. Systems approach to corrosion durability, in V.M. Malhotra, Ed., *Proceedings of the Fourth CANMET/ACI International Conference on Durability of Concretes*, American Concrete Institute, Detroit, MI, pp. 1293–1316.

Berke, N.S., M.C. Hicks, R.J. Hoopes et al. 1994b. Use of laboratory techniques to evaluate long-term durability of steel reinforced concrete exposed to chloride ingress, in V.M. Malhotra, Ed., *Third Durability of Concrete International Conference*, SP 145, Nice, France, American Concrete Institute, Detroit, MI, pp. 299–329.

Berke, N.S., M.C. Hicks, L. Li et al. 2003b. Holistic approach to durable high performance concrete, in Y. Yuan, S.P. Shah, and H. Lu, Eds., *Proceedings of the International Conference on ICACS 2003, Advances in Concrete and Structures*, RILEM Publications S.A.R.L, Paris, France, pp. 727–738.

Berke, N.S., M.C., Hicks and P.G. Tourney. 1993. Evaluation of concrete corrosion inhibitors, in *Proceedings of the 12th International Corrosion Congress*, NACE International, Houston, TX, pp. 3271–3286.

Berke, N.S. and L. Li. 2007. Holistic approach to sustainability of reinforced concrete exposed to chlorides, in S. Jacobsen, P. Jahren, and K.O. Kjellsen, Eds., *Proceedings of the International Conference on Sustainability in the Cement and Concrete Industry*, Lillehammer, Norway, pp. 452–466.

Berke, N.S., M.J. Scali, J.C. Regan et al. 1992b. Long-term corrosion resistance of steel in silica fume and/ or fly ash concretes, in V.M. Malhotra, Ed., *Proceedings of the Second CANMET/ACI Conference on Durability of Concretes*, CANMET, Ottawa, Ontario, Canada, pp. 899–924.

Berke, N.S. and T.G. Weil. 1994. World wide review of corrosion inhibitors in concrete, in V.M. Malhotra, Ed., *Advances in Concrete Technology*, CANMET, Ottawa, Ontario, Canada, pp. 891–914.

Bishnoi, S. and K.L. Scrivener. 2009. Studying nucleation and growth kinetics of alite hydration using µIC. *Cement and Concrete Research*, 39, 849–860.

Blackwell, B.Q., M.D. Thomas and A. Sutherland. 1997. Use of lithium to control expansion due to ASR in concrete containing UK aggregates, in V.M. Malhotra, Ed., *Proceedings of the Fourth CANMET/ ACI International Conference on ACI SP-170, Durability of Concrete*, American Concrete Institute, Sydney, New South Wales, Australia, pp. 649–663.

Bolzoni, F., G. Fumagalli, M. Ormellese et al. 2005. Laboratory test results on corrosion inhibitors for reinforced concrete, *Corrosion 2005*, Paper No. 05265, NACE International, Houston, TX.

Broomfield, J.P. 1992. Results of a field survey of cathodic protection systems in North American bridges, *Corrosion'92*, Paper No. 203, NACE International, Houston, TX.

Broomfield, J.P. 1993. Five years research on corrosion of steel in concrete: A summary of the strategic highway research program structures research, *Corrosion '93*, Paper No. 318, NACE International, Houston, TX.

Browne, D. 1982. Design prediction of the life for reinforced concrete in marine and other chloride environments. *Durability of Building Materials*, 1: 113–125.

Chatterji, S., A.D. Jensen, N. Thaulow, and P. Christiansen. 1986a. Studies of alkali-silica reaction, Part 3. *Cement and Concrete Research*, 16, 246–254.

Chatterji, S., N. Thaulow, A.D. Jensen, and P. Christiansen. 1986b. Mechanisms of accelerating effects of NaCl and $Ca(OH)^2$ on alkali–silica reaction. In P. Grattan-Bellew, Ed., *Proceedings of the Seventh International Conference on Alkali-Aggregate Reaction*, Noyes Publications, Ottawa, Ontario, Canada, pp. 115–124.

Collepardi, M. 2006. *The New Concrete*, Grafiche Tintoretto, Castrette de Villorba TV, Italy.

Dent Glasser, L.S. 1979. Osmotic pressure and swelling of gels. *Cement and Concrete Research*, 9(4), 515–517.

Diamond, S. 1975. A Review of alkali-silica reaction and expansion mechanisms 1: Alkalis in cement and in concrete porte solutions. *Cement and Concrete Research*, 6(4), 329–346.

Diamond, S. 1989. ASR—Another look at mechanisms. In K. Okada, Ed., *Proceedings of the Eighth International Conference on Alkali Aggregate Reaction*, The Society of Materials Science, Kyoto, Japan, pp. 83–94.

Diamond, S., R.S. Barneyback, and L.J. Struble. 1981. in *Proceedings of the International Conference on Alkali Aggregate Reaction in Concrete*, Cape Town, National Building Research Institute, CSIR, Pretoria, South Africa, S252/22, p. 1.

Elsener, B. 2001. *Corrosion Inhibitors for Steel in Concrete*, European Federation of Corrosion Publications, No. 35, W.S. Manney and Son, Ltd, London, U.K.

Flint, G.N. and R.N. Cox. 1988. The resistance of stainless steel partly embedded in concrete to corrosion in seawater. *Magazine of Concrete Research*, 40(142), 13–27.

Fushuang, C., A.A. Sagues, and R.G. Powers. 2001. Corrosion behavior of stainless Clad Rebars, *Corrosion 2001*, Paper No. 1645, NACE International, Houston, TX.

Gillot, J.E. 1975. Alkali aggregate reactions in concrete. *Engineering Geology*, 9(4), 303–326.

Hansen, W.C. 1944. Studies relating to the mechanism by which alkali-aggregate reaction produces expansion in concrete. *Proceedings of the Journal of American Concrete Institute*, 40(3), 213–227.

Hartt, W.H., R.G. Powers, and R.J. Kessler. 2009. Performance of corrosion resistant reinforcements in concrete and application of results to service life projection, *Corrosion 2009*, Paper No. 09206, NACE International, Houston, TX.

Hobbs, D. 1990. British Cement Association, Crowthorne, UK. Personal communication.

Hobbs, D.W. 1988. *Alkali-Silica Reaction in Concrete*, Thomas Telford Ltd, London, U.K.

Johnson, D.A., M.A. Miltenberger, and S.L. Amey. 1996. Determining chloride diffusion coefficients for concrete using accelerated test methods, in *Proceedings of the Third CANMET/ACI International Conference on Concrete in Marine Environment*, St Andrews by-the-Sea, New Brunswick, Canada, August 4–9, 1996, Supplementary Papers, V.M. Malhotra, Ed., CANMET, Ottawa, Ontario, Canada, pp. 95–114.

Juilland, P., E. Galllucci, R. Flatt et al. 2010. Dissolution theory applied to the induction period in alite hydration. *Cement and Concrete Research*, 40, 831–844.

Keer, J.G. 1992. Surface treatments, in G. Mays, Ed., *Durability of Concrete Structures*, E & FN SPON, London, U.K., pp. 146–165.

Kessler, R.J. and R.G. Powers. 1986. Corrosion evaluation of substructure cracks in long key bridge, Corrosion Report 86-3, Florida Department of Transportation, Gainesville, FL.

Knudsen, T. and N. Thaulow. 1975. Quantitative micro-analysis of alkali silica gel in concrete. *Cement and Concrete Research*, 5, 443.

Laing, S.V., K.L. Scrivener, and P.L. Pratt. 1992. An investigation of alkali-silica reaction in seven-year old concretes using S.E.M. and E.D.S. In *Proceedings of the Ninth International Conference on 'Alkali Aggregate Reaction in Concrete'*, The Concrete Society, Slough, London, U.K., pp. 579–586.

Lawrence, C.D. 1966. Changes in composition of aqueous phase during hydration of cement pastes and suspensions, in *Proceedings of the Special Report 90, Symposium on Structure of Portland Cement Paste and Concrete*, Highway Research Board, Washington, DC, pp. 492–xxx.

Longuet, P., L. Burglen, and A. Zelwer. 1973. La Phase Liquide du Ciment Hydrate, Rev des Materiaux de Constructions et des Travaux Publics. *Ciment et Betons*, 676, 35–41.

McDonald, D.B., M.R. Sherman, D.W. Pfeifer et al. 1995. Stainless steel reinforcing as corrosion protection. *Concrete International*, 17(5), 65–70.

McKenzie, M. 1993. The effect of defects on the durability of epoxy-coated reinforcement, in *Highway Structures*, TRB Circular No 403, Transportation Research Board, National Research Council, Washington, DC, pp. 17–28.

McMahon, D.J., P. Sandberg, K. Folliard et al. 1992. Deterioration mechanisms of sodium sulfate, in J.D. Rodrigues, F. Henriques, and F.T. Jeremias, Eds., *Proceedings of the Seventh International Congress on Deterioration and Conservation of Stone*, Laboratorio Nacional de Engenharia Civil, Lisbon, Portugal, pp. 705–714.

Monteiro, P.J.M. and K.E. Kurtis. 2008. Experimental asymptotic analysis of expansion of concrete exposed to sulfate attack. *ACI Materials Journal*, 105(1), 62–71.

Moore, A.E. 1978. An attempt to predict the maximum forces that could be generated by alkali silica reaction, effects of alkalies in cement and concrete. In *Proceedings of the Fourth International Conference on the Effects of Alkalis in Cement and Concrete*, Purdue University, Purdue University CE-MT-1-78, West Lafayette, IN, pp. 363–365.

Nmai, C.K. and P.D. Kraus. 1994. Comparative evaluation of corrosion inhibiting chemical admixtures for reinforced concrete, in V.M. Malhotra, Ed., ACI SP 145, American Concrete Institute, Detroit, MI, pp. 245–262.

Pellenq, R.J-.M., A. Kushima, R. Shahsavari et al. 2009. A realistic molecular model of cement hydrates. *Proceedings of the National Academy of Sciences*, 106(38), 16102–16107.

Pettifer, K. 1993 Building Research Station, Watford, Personal Communication.

Pfeifer, D.W. and J.R. Landgren. 1987. Protective systems for new prestressed and substructure concrete, Report No. FHWA/ARP-86/193, Federal Highway Administration, Washington, DC.

Pfeifer, D.W. and M.J. Scali. 1981. Concrete sealers for protection of bridge structures, NCHRP Report 244, Transportation Research Board, National Research Council, Washington, DC.

Pike, R.G., D. Hubbard, and H. Insley. 1955. Mechanisms of alkali aggregate reaction. *Proceedings of the Journal of American Concrete Institute*, 52(1), 13–34.

Poole, A.B. 1992a. Introduction to the alkali-aggregate reaction in concrete. In R.N. Swamy, Ed., *Alkali-Aggregate Reaction in Concrete*, Publisher Blackie Ltd, Glasgow, Scotland, pp. 17–19.

Poole, A.B. 1992b. Alkali-silica reactivity mechanism of gel formation and expansion. In *Proceedings of the Ninth International Conference on 'Alkali Aggregate Reaction in concrete'*, The Concrete Society, Slough, London, U.K., pp. 782–789.

Plum, D.R. and G.P. Hammersley. 1984. Concrete attack in an industrial environment. *Concrete*, 18(5), 8–11.

Powers, T.C., L.E. Copeland, and H.M. Mann. 1959. Capillary continuity or discontinuity in cement pastes. *Journal of the PCA Research and Development Laboratories*, 1(2), 38–48.

Rapoport, J., C.-M. Aldea, S.P. Shah et al. 2002. Permeability of cracked fiber-reinforced concrete. *Journal of Materials in Civil Engineering*, 14(4), 355–358.

Robinson, H.L. 1987. An evaluation of silane treated concrete. *Journal of the Oil and Color Chemists Association*, 70, 163–172.

Sagues, A.A. and A.M. Zayed. 1989. Corrosion of epoxy-coated reinforcing steel—Phase 1, Report No. FL/DOT?SMO/89–419, University of South Florida, Tampa, FL.

Scali, M.J., D. Chin, and N.S. Berke. 1987. Effect of microsilica and fly ash upon the microstructure and permeability of concrete, in *Proceedings of the Ninth International Conference on Cement Microscopy*, International Cement Microscopy Association, Duncanville, TX, pp. 375–397.

Schiessl, P. and C. Reuter. 1993. Epoxy-coated rebars in Europe: Research projects, requirements and use, in *Highway Structures*, TRB Circular No 403, pp. 29–35, Transportation Research Board, National Research Council, Washington, DC.

Sibbick R.G. 1993. The susceptibility of various UK aggregates to alkali silica reaction. PhD thesis, Aston University in Birmingham, UK.

Sibbick, R.G. and G. West. 1989. Examination of concrete from the M40 Motorway, Research report 197, Transport and Road Research Laboratory, Crowthorne, U.K.

Sibbick, R.G. and G. West. 1992. Examination of concrete from the A6068, Padiham Bypass, Lancashire, Research report 304, Transport Research Laboratory, Crowthorne, U.K.

Somayaji, S., D. Keeling, and R. Heidersbach. 1990. Corrosion of reinforcing steel in concrete exposed to marine and freshwater environments, in D. Whiting, Ed., *Proceedings of the ACI SP 122: Paul Klinger Symposium on Performance of Concrete*, American Concrete Institute, Detroit, MI, pp. 139–172.

Sorensen, B., P.B. Jensen, and E. Maahn. 1990. The corrosion properties of stainless steel reinforcement, in C.L. Page, K.W.J. Treadaway, and P.B. Bamforth, Eds., *Corrosion of Reinforcement in Concrete*, Elsevier Applied Science, London, U.K., pp. 601–610.

Stanton, T.E. 1940. Expansion of concrete through reaction with cement and aggregate. *Proceedings of the American Society of Civil Engineers*, 66, 1781–1811.

Stokes, D.B., G.E. Foltz, and C.E. Manissero. 2000. US Patent 6022408.

Taylor, N.F.W. 1977. *Cement Chemistry*, 2nd edn., Thomas Telford, London, U.K.

Thomas, M.D.A. and E.C. Bentz. 2000. *Computer Program for Predicting Service Life and Life-Cycle Costs of Reinforced Concrete Exposed to Chlorides*, University of Toronto, Toronto, Ontario, Canada.

Transportation Research Circular No. 494. 1999. Durability of concrete, Transportation Research Board, National Research Council, Washington, DC.

Treadaway, K.W.J., B.L. Brown, and R.N. Cox. 1980. Durability of galvanized steel in concrete, in *ASTM STP 713*, American Society for Testing and Materials, Philadelphia, PA, pp. 102–131.

Tuutti, K. 1982. *Corrosion of Steel in Concrete*, Swedish Cement and Concrete Research Institute, Stockholm, Sweden.

Verbeck, G. and C. Gramlich. 1955. Osmotic studies and hypothesis concerning alkali-aggregate reaction, in *Proceedings of the American Society for Testing and Materials*, 55, 1110–1120.

Virmani, P., K.C. Clear, and T.J. Pasko. 1983. Time-to-corrosion of reinforcing steel in concrete slabs: Vol. 5—Calcium nitrite admixture or epoxy-coated reinforcing bars as corrosion protection systems, Report No. FHWA-RD-83/022, Federal Highway Administration, Washington, DC.

Vivian, H.E. 1950. Studies in cement-aggregate reaction, XV: The reaction product of Alkalis and Opal. *CSIRO Bulletin*, 256, 60–82.

Weyers, R.E., I.L. Al-Qadi, B.D. Prowell et al. 1992. Corrosion protection systems, Report to the Strategic Highway Research Program, SHRP Concrete C103, Transportation Research Board, National Research Council, Washington, DC.

Weyers, R.E. and P.D. Cady. 1987. Deterioration of concrete bridge decks from corrosion of reinforcing steel. *Concrete International*, 9(1), 15–20.

Wilkins, N.J.H. and P.F. Lawrence. 1983. The corrosion of steel reinforcement in concrete immersed in Sea Water, in E.P. Crane, Ed., *Corrosion of Reinforcement in Concrete Construction*, Society of Chemical Industry, London, U.K., pp. 119–142.

Yeomans, S.R. 1993. Corrosion testing of black, galvanized and epoxy-coated reinforcing steel in concrete, *Corrosion 93*, Paper No. 329, NACE International, Houston, TX.

Yeomans, S.R. and M.P. Novak. 1990. Further studies of the comparative properties and behavior of galvanized and epoxy-coated steel reinforcement, International Lead Zinc Research Organization Report, Project ZE-341.

Zhang, M. H., T.W. Bremner, and V.M. Malhotra. 2003. The effect of portland cement type on performance. *Concrete International*, 25(1), 87–94.

Further Readings

Bentur, A., S. Diamond, and N.S. Berke. 1997. *Steel Corrosion in Concrete: Fundamentals and Civil Engineering Practice*, E & FN Spon, London, U.K.

Berke, N.S. 2005. Concrete, in R. Baboian, S.W. Dean, H.P. Hack, E.L. Hibner, and J.R. Scully, Eds., *MNL 20, Manual on Corrosion Tests and Standards: Application and Interpretation*, 2nd edn., ASTM International, W. Conshohocken, PA, pp. 405–422.

Broomfield, J. 1997. *Corrosion of Steel in Concrete: Understanding Investigation and Repair*, Routledge mot E & FN Spon, London, U.K.

Collepardi, M. 2006. *The New Concrete*, Grafiche Tintoretto, Castrette de Villorba TV, Italy.

Lamond, J.F. and J.H. Pielert (eds.) 2006. *STP 169D Significance of Tests and Properties of Concrete and Concrete Making Materials*, ASTM International, W. Conshohocken, PA.

Mindess, S., J.F. Young, and D. Darwin. 2002. *Concrete*, 2nd edn., Lavoisier, Paris, France.

Taylor, N.F.W. 1997. *Cement Chemistry*, 2nd edn., Thomas Telford, London, U.K.

IV

Other Natural Materials

23

Wood

Jeffrey J. Morrell
Oregon State University

23.1 Introduction

From the time humans emerged on the plains in Africa, wood, in one form or another, has played an important role in everyday life. From the spears that brought down game to the wood fuel used to cook it and the poles that provided night time shelter, wood played a variety of roles in human development (Graham 1973). As human culture developed, wood continued to provide fuel, structural materials, and a variety of other products including food and medicinals. Over the past century, the role of wood has changed in many parts of the globe, but it remains one of our most important renewable structural materials. In comparison with other materials, wood is strong, light, and easily worked with simple tools. In terms of environmental impacts, wood harvest does necessitate cutting of trees; however, replanting and careful management creates an inexhaustible resource. In addition, wood sequesters or captures carbon from the air, thereby reducing the potential impacts of carbon dioxide releases and reducing the pace of global climate change. At the same time, wood is a natural material that is subject to physical and biological degradation over time and it is this susceptibility to degradation that is its weakest attribute.

23.2 Uses

Despite the risk of degradation, wood is used for houses, poles, marine piling, railway ties, composite panels, wood/plastic composites, and paper products. It is revered for its beauty and employed because of its utility. Wood has exceptional structural and thermal properties.

23.3 General Properties

23.3.1 Wood as a Material

Wood in the living tree serves two functions. First, it supports the foliage or canopy above, allowing it to reach above competing plants or trees to capture sunlight for photosynthesis. At the same time, the wood is a conduit for water and nutrients to move up and down the tree. If we look at a cross-sectional cut from a tree, we can see the bark, which protects the living tree from injury (Figure 23.1). Beneath the bark, we can see sapwood and heartwood (USDA 1996). Sapwood is the living part of the tree that conducts fluids. Inside that zone is the heartwood, which is the collection of nonliving cells that provides some support for the canopy above but no longer conducts fluids. If we look closer on our cross section, we see numerous cells with large openings called lumens. These hollow tubes or cells are both conducting elements and structural support for the tree.

There are two broad groups of trees, softwood and hardwoods. These terms are misnomers because many softwoods are quite hard, while many hardwoods are soft. Softwoods or gymnosperms comprise the oldest tree group, and we often call them conifers or evergreens because many retain their needles for long periods. The hardwoods or angiosperms are considered to be more advanced and have leaves in place of needles. In addition, the anatomy of these two groups varies.

Trees in temperate regions generally produce their conductive cells (vessels or tracheids) with larger openings or lumens early in their growing season and cells with smaller opening later as growth slows. The larger cells are called earlywood and the later cells are called latewood. Earlywood and latewood formed in a given season comprises one annual growth ring. Trees in tropical or subtropical climate sometimes lack visible rings or the rings may correspond to wet/dry cycles rather than a given year.

The wood of softwoods has relatively few cell types. The primary softwood cell is the longitudinal tracheid that provides structural support and conducts fluids between the roots and needles. Longitudinal tracheids are long, extending to several millimeters in length in some species. Ray tracheids are oriented from the bark to the pith of center of the living tree. They are primarily used to move fluids laterally. Ray parenchyma are similarly oriented but primarily serve to store starches, lipids, proteins, and other compounds that might be used to regenerate foliage or roots. In addition, some softwoods contain resin canals. These are essential holes in the wood structure surrounded by epithelial cells that produce resins. The resins are released when the tree is wounded and serve to limit entry of fungi and insects into

FIGURE 23.1 Cross section of a Douglas fir tree showing bark, sapwood, and hardwood with easily seen annual rings consisting of dark latewood and lighter earlywood.

the wood. These resins are often extracted from wood for production of turpentine and other products that were formerly called naval stores because of their use in protecting wood used in ships. All cells in the wood are connected with one another through openings termed pits. Pits are openings in the wood cell that allow fluids and some smaller particles to move between cells.

Hardwoods contain more cells types. The cells that transport liquids longitudinally are called vessels. These cells are connected with one another through sieve plates that allow liquids to be transported. Fibers are thick-walled cells that surround the vessels. They are also oriented longitudinally and provide structural support. Hardwoods also have ray parenchyma that store nutrients. As with softwoods, individual cells are connected with one another through pits.

If we look closer at an individual cell, we can see that there are layers. In softwoods, the middle lamella separates individual cells and is integrated with the primary cell wall layer (Figure 23.2). The secondary cell wall is composed of three layers, the S1, S2, and the S3. These layers differ in thickness, the percentages of polymers present, and the orientation of these polymers (Harada and Cote 1985). All cells are composed of three primary polymers, cellulose, hemicellulose, and lignin. Cellulose is composed of repeating units of β-D-glucose. Cellulose chains vary in length but usually contain approximately 2000 glucose units in wood. Cellulose chains tend to interact with one another to form microfibrils.

Microfibrils contain areas where the cellulose chains are highly ordered and crystalline in nature as well as other areas that are less organized and amorphous. The structure and orientation of the cellulose microfibril gives wood its exceptional longitudinal strength. It also makes wood more resistant to degradation than other organic materials. Hemicellulose is a heteropolymer consisting of glucose, mannose, galactose, arabinose, and a host of other sugars. Unlike cellulose, which is a homopolymer and lacks side chains, hemicellulose is highly branched and much more susceptible to degradation. Lignin is composed of phenyl propane units with a variety of linkages between individual units. The variety of linkages makes lignin one of the most recalcitrant natural polymers. Lignin compounds can last for decades in soil. Lignin, cellulose, and hemicellulose are integrated in a matrix that takes advantage of the properties of each polymer. Cellulose provides exceptional strength, lignin provides resistance to degradation, and hemicellulose acts as an encrusting medium that binds the mixture.

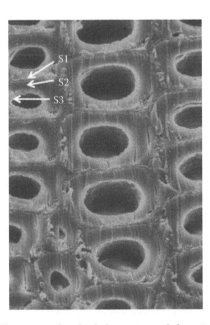

FIGURE 23.2 Wood cells typically consist of multiple layers termed the primary and middle lamella and three layers of secondary cell wall, termed S1, S2, and S3. The orientation of the cellulose microfibrils varies within these layers and gives wood its unique properties.

The arrangement of the cells makes wood exceptionally strong in tension and compression along the longitudinal plane but weaker in other directions. These variable properties make wood anisotropic with different properties in the tangential, radial, and longitudinal directions. In addition, wood tends to have differential properties depending on moisture content. As wood sorbs moisture, the water binds to the cellulose microfibrils, which expand. This expansion reduces the material properties. Many wood properties decrease as the moisture content increases from 0 to approximately 27%–30%. At this point, the hydroxyl groups on the cellulose are largely saturated and free water will begin to accumulate in the cell lumens. This point is termed the fiber saturation point (fsp). Wood properties are largely stable above the fsp, although the presence of free water in the wood can create other problems as will be discussed later in this chapter.

Wood cells can also contain other materials including minerals, sugars, starches, proteins, lipids, fatty acids, and resins that are typically found in the rays in the sapwood. These materials are termed extractives and have no structural function. As the sapwood ages, however, the cells in the sapwood begin to die and the stored materials can be converted to a variety of other compounds including phenolics and glycosides. The resulting wood is termed heartwood. Some materials in the heartwood can be exceptionally toxic to various agents of deterioration, rendering the wood resistant to degradation (Hillis 1962). Minerals, such as silica, may also increase wood hardness and improve durability against insects or marine borers. Not all woods have durability, and the degree of durability can vary widely within heartwood of a given tree or within trees in a given stand. Sapwood is generally considered to be nondurable, regardless of species. During heartwood formation, the pits connecting individual cells close and become encrusted with materials that sharply reduce fluid flow. This can have important implications for drying and impregnation.

23.4 Degradation Agents

While the structure and organization of cellulose, hemicelluloses, and lignin makes wood much more resistant to deterioration than other biological materials, a variety of physical and biological agents can affect wood properties. Anything that negatively affects wood properties is termed deterioration and this damage can be caused by both living and nonliving agents (Hale and Eaton 1992; Zabel and Morrell 1992). Often, deterioration is caused by a combination of agents, making it difficult to determine an exact cause.

23.4.1 Nonliving Agents of Degradation: A Variety of Nonliving Agents Include Chemicals, Sunlight, and Mechanical Damage (The Damage Can Be Either Visual or Structural)

The ultraviolet (UV) component of sunlight can be especially damaging to wood surfaces (Feist 1990; Evans et al. 1996) (Figure 23.3). The energy in the UV range is released into the wood surface and can cause the release of free radicals that then react with various wood components. Lignin is most sensitive to UV light and begins to degrade within hours of sun exposure. The weakened lignin opens up the other wood components to degradation. The weakened wood is then susceptible to either wind or water erosion. In general, dark woods will tend to lighten and light wood will tend to become darker with UV exposure. While UV damage is shallow, over time, erosion of the weaker wood and further damage to freshly exposed wood can lead to uneven wood surfaces. In some applications, however, UV-exposed wood is considered an attractive attribute and weathered barn siding can sell for a premium. In general, UV damage to wood is shallow and cosmetic and, by itself, only reduces wood dimensions a few mm over a century.

Wood can also be damaged by prolonged exposure to high temperature. This effect is often exacerbated when the wood is wet. In general, hemicellulose is most sensitive to heat followed by cellulose and finally lignin. Wood combusts in the presence of oxygen when heated over 451°F, but prolonged exposure to lower temperatures (>100°C) will produce darkening and strength loss.

Although wood is generally resistant to many chemicals and wood tanks are often used to store dilute acids or bases, a variety of chemical agents can affect wood properties. Strong acids attack cellulose and

FIGURE 23.3 UV light can damaged the wood surface; the damage is shallow as shown by the exposed wood on this western red cedar pole.

hemicelluloses and can sharply reduce wood properties. Strong bases attack lignin, reducing wood to a mass of cellulose fibers. Application of strong bases to wood forms the basis for the paper industry. Acid and base degradation are often associated with specific environments. For example, application of high-pH stucco to the exterior of a house can produce shallow degradation of the underlying wood surface. Chemical degradation is most common in industrial environments where wood is subjected to prolonged or repeated exposures.

Strong salts can also affect wood and this is often evident in wood used in marine environments or in wood used in salt storage facilities. Wood will absorb saltwater and, as this water evaporates, it leaves the salt behind. Repeated wetting with saltwater and subsequent drying will result in cells so filled with salt that they literally burst. The presence of such fibers on the wood surface can be disconcerting, but the damage is very shallow.

Wood can also be damaged mechanically through abrasion or repeated heavy loading. Wood stairs often show signs of abrasion on the treads as do wood decks in high traffic areas. Wood can also be damaged by repeated heavy loading. The best example is the railroad tie that is periodically subjected to millions of tons of loading from passing trains. Ties can last 40–70 years but eventually must be replaced often because of mechanical damage.

23.4.2 Biological Deterioration

23.4.2.1 Requirements for Degradation Agents

Wood can also be attacked by a variety of living agents. Some of these agents enzymatically degrade the wood polymers, while others tunnel or excavate the wood. Nearly all of these agents need to have four basic requirements acting concurrently to cause damage: adequate temperature, adequate oxygen, water, and a food source. Preventing deterioration requires limiting one of these four requirements.

Most wood-degrading organisms can survive at a wide range of temperatures (Zabel and Morrell 1992). For example, they may not be active at temperatures below freezing but can survive prolonged periods at low temperatures. As temperatures increase, organisms become more active but most organisms cease activity as temperatures reach 40°C–45°C. All wood-degrading organisms succumb to exposures at 67°C after 75 min. Survival at temperatures between 50°C and 67°C depends on the length of that exposure and stage of the organism. For example, some fungi produce temperature-resistant resting stages that require the prolonged heating at elevated temperatures to succumb, while an active insect will succumb at a much lower temperature. In general, temperature lethality occurs as enzymes and other proteins necessary for respiration are irreversibly denatured.

The majority of organisms that degrade wood are aerobic; however, most require far less oxygen than the 20% we find in air. Oxygen serves as the terminal electronic acceptor in the electron transport chain. It is generally not feasible to exclude oxygen to inhibit decay. The exception is wood that is continually submerged in or sprayed with freshwater. In these cases, the water will fill the wood cell lumens to the point where oxygen is limiting. Although anaerobic bacteria will begin to attack the sapwood, the rate of attack is generally very slow and the wood beneath remains sound. There are numerous examples of logs being excavated from rivers or buried in mud that are found to be sound after centuries of immersion. Similarly, untreated foundation pilings used to support buildings often perform well because the wood is driven below the water table and is thus protected from most degradation. Foundation piling can last centuries in this type of exposure. As noted, however, oxygen is generally not a limiting factor in decay of wood structures.

Water is essential for all living organisms. In wood, water serves as a swelling agent, it is a reactant in the degradation of cellulose and hemicelluloses and it is essential for cell respiration. In general, water becomes limiting in wood when the moisture content declines below 20% (wt/wt%), although there are exceptions to that rule. Wood decay generally begins when the wood is at the fiber saturation point, but most decay fungi have moisture optima that are closer to 40%–80% moisture content. Oxygen becomes limiting as wood moisture contents increase beyond 100%–140%, depending on the wood species, the environment, and the agent of deterioration. On the opposite end of the moisture scale, most wood-degrading organisms can survive in dry wood and some can survive for prolonged periods. A number of wood-degrading fungi produce long-term survival structures and are able to survive for a decade or more in dry wood. Moisture control is the most common method for limiting biological deterioration and most buildings are designed to limit wood moisture content to less than 20%.

A nutrient source is the final requirement and this is generally provided by the wood. Some organisms can only use the proteins and free sugars that are stored in the ray cells, others use hemicelluloses and cellulose, and still others can utilize all wood components. There are also wood degraders that do not use wood as a nutrient, but instead mine or tunnel through it to create galleries in which to live. These organisms then leave their nests to forage for food elsewhere. When it is not possible to limit temperature, oxygen, or water, we protect wood by modifying the nutrient source to make it either toxic or unusable and this approach forms the basis for wood preservation.

23.4.2.2 Bacteria

Bacteria are generally not significant wood degraders, but they can become important in some environments, especially those with limited oxygen and elevated moisture levels. Bacteria are single-celled prokaryotes that do not reproduce sexually. They are among the most common organisms on the planet and are present in virtually every environment. Some bacteria attack the pectin substances in the pits connecting cells, making the wood more permeable. If these bacteria-attacked logs are then stored in water, the resultant increased water absorption can lead to those logs sinking. Other bacteria are capable of degrading the secondary wood cell wall and their damage is characterized by either erosion of the cell wall or the production of elongated tunnels in the S2 cell wall layer. The former bacteria are termed tunneling bacteria and can be very common in submerged wood or wood in very moist forest soils. Bacterial damage is generally shallow and most prevalent in the sapwood, but tunneling bacteria can be preservative tolerant and may play a role in initial colonization of preservative-treated wood in soil contact.

23.4.2.3 Fungi

Fungi are eukaryotic, filamentous organisms and are saprotrophs on a variety of materials, including wood. The general fungal life cycle begins with a spore that germinates to produce fungal hyphae that

FIGURE 23.4   Molds produce pigments spores on the wood surface and typically grow on sapwood as shown on this Douglas fir pole.

grow through the substrate secreting enzymes that degrade the material. The degradation products then diffuse back to the fungus where they are absorbed and metabolized. There are literally hundreds of thousands of fungal species and many have adapted to utilize one or more wood components. Wood-inhabiting fungi can be classified on the basis of morphology, DNA, or the damage they cause to wood. For the purposes of this chapter, we will consider them on the basis of the damage they cause because that has more direct bearing on wood use.

Fungi that inhabit wood can be broadly classified as mold, stain fungi, or decay fungi. Decay fungi can be further divided into brown, white, or soft rot fungi.

Mold fungi are classified as those fungi that produce pigmented spores on the wood surface (Figure 23.4). These fungi generally do not attack any of the wood polymers, but instead live on the materials stored in the ray cells. The primary impact of molds on wood is visual, and those spores can generally be brushed away. The hyphae of mold fungi are clear or hyaline and grow through the sapwood through pits. Brushing away spores will not eliminate hyphae inside the wood. The damaged pits increase the permeability of the wood that can affect drying rates as well as the ability to evenly finish the surface with paints or coatings. The primary issue associated with mold, however, is public concern about potential health risks associated with some of these fungi. Some species of molds can produce mycotoxins, and there is concern that the presence of these fungi on wood in houses may cause health issues. There has been tremendous speculation about the risk of molds; however, most studies have concluded that molds are primarily allergens that can aggravate existing conditions such as asthma. Some people are exceptionally sensitive to molds, and there is no way to predict who among the general population will have issues with these fungi. The best approach with molds is to control moisture levels in the wood to reduce the risk of fungal attack. Where fungal attack has occurred, cleaning the surface with soapy water and then making sure the wood is allowed to dry below 20% moisture content can eliminate future risk.

Stain fungi are similar to molds in that they primarily attack stored compounds in the ray cells, but they can also begin to attack hemicelluloses and cellulose if allowed to continue to grow (Scheffer and Lindgren 1940). Stain fungi primarily attack the sapwood, but unlike molds, the hyphae of these fungi become covered with a brownish pigment called melanin. The brownish color inside the wood is seen as blue on the wood surface and damage caused by these fungi is commonly called blue stain. Stain fungi may also produce spores or masses of hyphae on the wood surface and are often found in the same locations as molds. The primary difference between mold and stain damage is that stain damage

(a) (b)

FIGURE 23.5 White (a) and brown (b) rot cause severe structural damage and are usually found on the interior of wood products.

is not easily removed, while mold damage can be brushed away. Stain fungi alter wood appearance and increase permeability.

Decay fungi have a more profound effect on wood properties because they depolymerize and modify lignin, cellulose, or hemicelluloses. These fungi are the most prominent cause of deterioration in wooden structures (Duncan and Lombard 1965). Brown rot fungi preferentially attack hemicelluloses and cellulose and leave behind a modified lignin matrix, causing up to 60%–70% mass loss (Figure 23.5). Brown rot fungi tend to attack softwoods, although they will attack all types of wood. Wood damaged by brown rot fungi tends to be darkened with numerous breaks across the grain. Brown rot fungi tend to be more important in structural applications because they degrade cellulose far faster than they utilize the breakdown products. Thus, at very low mass losses, the wood will experience substantial losses in material properties. This means that wood with decay that is not yet visible to the naked eye may have lost 60%–70% of its flexural properties.

White rot fungi utilize all three wood polymers and cause up to 95%–97% weight loss. White rot fungi leave the wood a bleached white color at the advanced stages of attack. While white rot fungi can attack all woods, they tend to be more aggressive on hardwoods. Some white rot fungi preferentially attack lignin and there have been efforts to use these fungi for biopulping for paper products. These fungi have also been explored for degradation of a variety of pesticides. White rot fungi tend to cause losses in structural properties that are proportional to the mass loss. As a result, decay is often visible when properties decline to the point where replacement might be advisable. White rot fungi tend to be more of a problem in tropical countries because of the prevalence of hardwood use in those areas.

Soft rot fungi tend to attack the cellulose and hemicelluloses but differ from both white and brown rot in their mode of attack. White and brown rot fungi tend to be found away from the surface and inside the wood where the moisture levels are more stable. Soft rot fungi tend to be found in more extreme environments, either on the wood surface where moisture levels fluctuate or in very wet environments where low oxygen levels limit white or brown rot attack. They also tend to be more damaging to the sapwood and on hardwoods. In addition, brown and white rot fungi cause their damage from the cell lumens outward, while soft rot fungi can either grow directly through the secondary cell wall or erode the cell wall surface from the lumen outward (Figure 23.6). Soft rot attack causes sharp drops in material properties and this damage becomes more important because it tends to occur on the surface. This becomes more important for wood used in applications where bending strength is important because most of the bending strength of the wood is found in the outer 25–50 mm. Soft rot fungi sharply reduce strength in that area, thereby reducing the effective circumference and magnifying their potential impacts.

FIGURE 23.6 Soft rot damage is usually characterized by softening of the wood surface as illustrated by this screwdriver penetrating into a pole.

23.4.2.4 Insects

A variety of insects have evolved to use wood as either a food source or as a habitat. Members of the orders Isoptera, Coleoptera, and Hymenoptera are among the most important agents of damage for wood in service. Depending on the insect, damage can occur in the living tree, as the tree is processed and dries or once it is placed in service. Understanding the biology of the insect is critical for developing strategies that render the wood unsuitable for insect utilization.

23.4.2.4.1 Isoptera

The termites or Isoptera are social insects that use wood as a food source. They are mostly confined to those equatorial and temperate zone areas between 50 N and S latitude. Termites have a highly structured caste system that includes the queen, workers, and soldiers. It is estimated that termites cause up to $9 billion dollars in damage to wood per year in the United States. This figure represents not only the wood loss but the cost of repairing the damage. There are three groups of wood-damaging termites. The Rhinotermitidae or subterranean termites live in large colonies (one to several million workers) in soil contact. They excavate tunnels through the soil searching for wood (Figure 23.7). They can also move upward over non-wood surfaces by constructing earthen tubes. This allows them to move over

FIGURE 23.7 Termite workers on severely attacked wood.

masonry foundations to attack wood in houses. There are a variety of methods for limiting subterranean termite attack including soil poisoning, physical soil barriers, and the use of preservative-treated wood. Dampwood termites, as the name implies, live in very wet wood and do not require soil contact to initiate a colony. The relative size of workers tends to be much larger than those for subterranean termites, but the colonies are much smaller, numbering only hundreds to a few thousand workers. Dampwood termite attack is usually prevented by controlling moisture. Drywood termites live in very dry wood, usually with less than 12% moisture content. Like most termites, they are rarely seen outside of the nest, but their damage can be detected by the presence of their frass outside the nest. Periodically, workers will tunnel to the surface and kick out frass that has built up inside the tunnels and these piles of frass are sometimes the only indicator of attack. Drywood termites tend to be more of a problem in drier climates.

23.4.2.4.2 Hymenoptera

The order Hymenoptera includes bees, wasps, and ants. Of the many species in this order, the most important from a wood damage perspective are carpenter bees and carpenter ants.

Carpenter bees attack dry wood in structures and do not use the wood as a food source. The adult female tunnels into the wood and places eggs along with food for the developing larvae. She then seals the hole and allows the eggs to develop. Once the egg proceeds through larval stages, it pupates and emerges through the original hole. Carpenter bees can reuse the same chambers, extending them each season. Attack can be limited by painting the wood. Preservatives have little effect because the insects do not consume the wood.

Carpenter ants are social insects that also do not use wood as a food source (Figure 23.8) (Hansen and Klotz 2005). The colony starts when male and female reproductives leave a mature colony and mate. The females then seek out a sheltered area where she lays eggs and uses her fat reserves to feed the developing larvae. These workers then feed and raise the next batch of eggs. It is only then that the workers begin to move outside the colony to feed on sugars, proteins, and other materials. Colonies can vary in size from a few hundred to upward of 50,000 workers. Carpenter ants are common in temperate forests and their tunneling plays an important role in wood recycling. They are also prevalent in houses in forested areas as well as timbers and utility poles. Controlling carpenter ants can be difficult because they do not consume the wood and because the entire colony can move to a new location when disturbed and that new location may be in the same structure. Carpenter ant treatments mostly consist of barrier sprays around structures to keep the workers from foraging inside.

FIGURE 23.8   Carpenter ant attack on a creosote-treated southern pine pole.

The Coleoptera or beetles contained the largest number of species in the order of insects. Beetles can attack living trees, freshly fallen trees, or even dried, finished products, depending on their moisture requirements. While a range of beetles can attack wood products, the most important are bark beetles (Scolytidae), ambrosia beetles, metallic wood borers (Buprestidae), long-horned borers (Cerambycidae), and powderpost beetles (Lyctidae, Bostrichidae, and Anobiidae).

Bark beetles attack live or freshly fallen trees, laying their eggs in galleries between the bark and wood. The larvae tunnel in this same area, cutting off the flow of nutrients to the living tree. Bark beetles do not cause appreciable wood damage, but they are associated with blue stain fungi that can discolor the wood, making the resulting lumber less valuable. Ambrosia beetles also attack freshly cut trees as well as wet lumber. The females tunnel into the wood for varying depths and deposit their eggs along with a bit of a fungus. The fungus grows into the wood and when the larvae hatch, they feed on the fungus. The fungus stains the wood, reducing the value of the finished product. Ambrosia beetles and bark beetles are both eliminated by drying the wood after which they do not continue to cause damage in the finished product.

Long-horned and metallic wood borers both attack freshly fallen trees and lay their eggs on or in the bark. The larvae tunnel inward and then mine between the bark and the wood for a period of 1 to several weeks. They then tunnel into the wood and have life cycles that last for 1–20 or more years, depending on the wood condition. These larvae can survive through the sawing process and then continue their life cycle in the structure. They will then pupate and emerge into the building. These beetles do not reinfest dry wood, so their damage is generally more of a nuisance than a structural issue unless there are too many beetle emergence holes on the same board. In that case, replacing or reinforcing the board may be necessary.

Powderpost beetles attack drier wood and are often a problem in houses, barns, and bridges. There are three broad groups of powderpost beetles with differing requirements for wood species and moisture content. These beetles prefer sapwood and do not penetrate materials with coatings such as paint or varnish on the surface. The adult beetles lay their eggs on the wood surface, and the larvae then tunnel inside. The initial entry hole can be very small and difficult to detect. The damage is often only discovered when the adult insect emerges from the wood through a new and larger hole (Figure 23.9). In some cases, the adults can mate and lay eggs inside the same piece of wood, and the damage can continue until the entire sample is reduced to a powdery mass surrounded by a thin veneer of sound wood. Once inside the wood, heat treatment (130°F for 30 min) or fumigation are the only effective ways to eliminate the infestation. Neither of these processes, however, can prevent reinfestation. The best way to keep the beetles out without painting is to brush on a 1%–2% solution of boric acid.

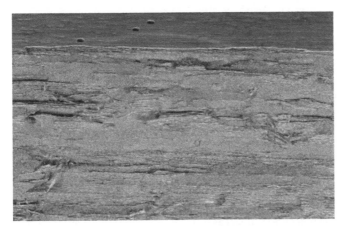

FIGURE 23.9 Powderpost beetles produce a powdery frass on the interior of the wood while leaving a thin veneer of solid wood on the surface.

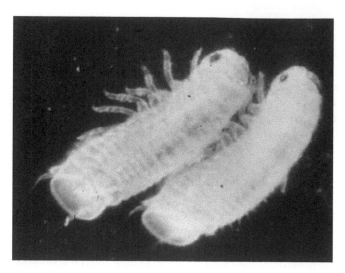

FIGURE 23.10 *Limnoria* or gribbles are free-swimming crustaceans that attack wood surfaces in salt or brackish water.

23.4.2.5 Marine Borers

Wood submerged in marine environments (i.e., salt or brackish water) is usually too wet for substantial fungal or insect attack, but it can be damaged by a number of marine wood borers. *Limnoria*, shipworms, and pholads are the three most important marine wood borers. In the age of wooden sailing vessels, marine borers were a major concern for seafarers and many ships that sank did so because of marine borer attack. Marine borers are less of a concern these days, but they still cause substantial damage to coastal infrastructure.

Limnoria or gribbles are small (3–6 mm long), mobile crustaceans that tunnel in wood (Figure 23.10). While the tunnels they produce are small, they weaken the wood surface to the point where wave action can wear away the surface. Over time, *Limnoria* attack can result in substantial wood loss and collapse of a piling.

Shipworms are mollusks that begin life as free-swimming larvae. Eventually, the larvae settle on a wood surface and burrow inward. As they burrow, they become wormlike with a pair of rasping shells at the head that tunnel into the wood and a pair of siphons on the opposite end that extend outward from wood surface to exchange oxygen and nutrients with the water. The shipworm withdraws into the hole at any sign of danger, leaving little sign of its presence. Inside, however, the shipworm removes large quantities of wood and can reach 25 mm in diameter and 0.3–1.5 m in length (Figure 23.11). Eventually, so much wood is removed that the structure collapses.

Pholads are also mollusks that begin life as free-swimming larvae. However, pholads do not use wood as a nutrient source and they remain within their shells near the wood surface. Like Limnoria, pholads weaken the wood surface, leading to gradual loss of cross section and eventual failure. Pholads tend to be more prevalent in subtropical to tropical waters.

Marine borer attack is generally prevented by the use of naturally durable or preservative-treated woods. Barriers such as polyurethanes can also protect wood but must remain intact in order to remain effective.

23.4.2.6 Birds and Mammals

Although they do not use wood as a food source, some birds and mammals can cause wood damage. Woodpeckers use wood poles and timbers as nesting sites. Since they have poor senses of taste and smell, these birds are not dissuaded by the presence of preservative treatments and can cause extensive damage to utility poles. The resulting cavities expose untreated wood to water, fungal spores, and insect attack.

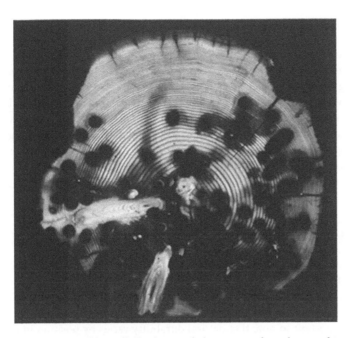

FIGURE 23.11   Shipworms are wormlike mollusks that attack the interior of wood exposed in salt or brackish water.

The result can be a small hole leading to extensive decay pockets (Figure 23.12). Many utilities wrap their poles in metal hardware cloth to discourage attack.

A variety of animals ranging from voles to horses can chew treated wood in search of salts. As with woodpeckers, these animals are not discouraged by treatment. Instead, farmers use untreated wood where chewing is likely to occur or they use oil-based treatments that are less susceptible to attack.

While bird or mammal attack can be costly in individual instances, it is generally not widespread.

FIGURE 23.12   Woodpecker galleries can allow water, fungi, and insects to enter untreated wood.

23.5 Wood Protection

The best approach to keeping wood sound is to limit one of those four previously listed essential requirements where the most common approach is to limit moisture. Engineers and architects avoid wood wetting by creating drainage planes that shift water away from a structure. Drainage planes can be steep sloping roofs with long overhangs that limit the risk of wetting of the siding. They include the use of caulking and coatings to limit moisture entry between joints, gutters to channel water away from the structure, and foundations to lift wood out of soil contact. It is important to note that wood, even wood that is integrated in other materials, remains susceptible to degradation. For example, the wood in a wood–plastic composite will decay under the proper moisture conditions (Mankowski and Morrell 2000). Where moisture control is not possible, the wood must be either naturally durable or be supplementally protected by chemicals.

23.5.1 Natural Durability

As mentioned earlier, the heartwood of some species contains a potential array of chemicals that render the wood resistant to various agents of deterioration (Hillis 1962; Scheffer and Cowling 1966; Taylor et al. 2002). The use of naturally durable woods is attractive from a public perception perspective, but it is important to remember that the compounds that provide this protection are still potent antimicrobials. It is also important to note that natural durability varies by position in the tree and between individual trees. The most durable heartwood is found in the base of the tree at the heartwood–sapwood interface. Durability declines inward toward the pith and upward in a given growth ring. There is also evidence that natural durability declines in second-growth trees of some species, notably teak and redwood. Naturally durable wood from temperate climate species tends to perform best in non-soil contact applications. Many tropical species also perform well in soil contact, but care should be taken to ensure that these materials originate from sustainably managed forests.

While naturally durable woods remain attractive for many applications, the supply does not meet the demand for durable wood products. Supplemental treatment, usually by pressure impregnation, provides an alternative method for prolonging the useful life of wood products used under adverse conditions.

23.5.2 Preservative Treatment

The goal of wood treatment is to deliver a sufficient quantity of protectant to a depth sufficient to protect the product over its useful life (Mac Lean 1952; Hunt and Garratt 1967). The amount of chemical deposited is usually expressed on either a weight per unit mass (pounds per cubic foot or kilograms per cubic meter) or on a mass/mass basis, depending on the country. The depth to which the chemical is delivered is generally expressed as either an absolute depth or a percentage of the sapwood.

Wood protectants can be either biocides that inhibit or are toxic to the various agents of decay or they may merely alter the wood to change moisture-holding capabilities or make the wood unrecognizable to those organisms. In most countries, biocides are regulated by their governments. For example, the U.S. Environmental Protection Agency (EPA) regulates pesticides under authority of the Federal Insecticide, Fungicide, and Rodenticide Act (FIFRA). The EPA reviews all pertinent health and safety data to make sure that the risks of use do not outweigh the benefits. The best example of this process was the 1970s review of all wood preservatives (USDA 1980), but the EPA continually reappraises all registered pesticides.

Preservatives can be formulated as either oil- or water-soluble systems, depending on the chemical characteristics of the biocide as well as the desires of the treater (Freeman et al. 2003; Schultz et al. 2007). Oil-based systems are preferred in some applications because they tend to make the wood more water repellent and this can reduce the effects of weathering. They are also used on utility poles because the oil makes the pole easier to climb. However, oil-based systems are more costly and they make it difficult to apply additional layers of paints or other coatings. This makes them less attractive for applications such as decks and other residential applications. Waterbornes are widely used for residential applications because the resulting treated wood can be painted or coated and lacks unsightly surface deposits.

Currently used oil-based preservatives include creosote, pentachlorophenol, and copper naphthenate. Creosote is the oldest wood preservative currently in use and was first patented in 1836 (Graham 1973). It was originally produced as a by-product of coking coal for steel. Gases released as the coal was heated were captured and condensed to produce various tars and creosote. Creosote is a mixture of polycyclic aromatic hydrocarbons and its composition will vary depending on source. For this reason, the preservative is largely defined by the percentages of its compounds with various boiling point ranges. Creosote is effective against fungi, insects, and most marine borers. Creosote remains widely used for railroad ties but is also used to treat marine piling and utility poles. It is classified as a restricted use pesticide in North America and can only be applied by those who are licensed by the appropriate state or federal agency. Use of wood treated with creosote is not restricted, although some regulatory bodies restrict its use over or in aquatic environments.

Pentachlorophenol (penta) was developed in the 1930s as the first synthetic organic wood preservative. It is widely effective against fungi and insects. Penta was an attractive alternative to creosote because it was cheap and available in steady supply, while creosote supplies tended to vary with the health of the steel industry. The primary negative issue with penta was the presence of the contaminant dioxin that was produced as a by-product of the chlorination process. A series of reviews by the U.S. EPA has led to penta being listed as a restricted use pesticide. Use of this chemical has been banned or severely restricted in many countries; however, continuing data development in the United States has shown that the chemical can be used safely when the manufacturing process was modified to reduce the dioxin content. The primary uses for penta in the United States include utility poles, bridge timbers, and foundation piling.

Copper naphthenate (CuNap) was developed at the end of the nineteenth century but saw little use as a wood preservative until the 1980s. This preservative is a complex between copper and naphthenic acid that is derived from petroleum. CuNap has a lower toxicity profile than penta, and numerous field trials have shown that it is just slightly less effective than penta in soil contact. The primary use for CuNap is treatment of utility poles although it can be used wherever penta is employed.

In addition to these heavy-duty wood preservatives, a variety of organic preservatives can be dissolved in either oil or water, depending on how they are formulated. These include tributyltin oxide (TBTO), 3-iodo-2-propynyl-butylcarbamate (IPBC), tebuconazole, propiconazole, and didecyldimethylammonium chloride (DDAC).

TBTO and IPBC are effective against most fungi and insects when used away from direct soil contact. Both are used in their oil-soluble forms for treatment of window frames. Tebuconazole and propiconazole are triazole compounds that have exceptionally low toxicity to nontarget organisms. These systems can be formulated in oil for application to window frames, while DDAC is used as a co-biocide with IPBC in some window treatments. The lower toxicity of these components makes them less effective in direct soil contact.

Water-based preservatives: The most common heavy-duty water-based preservatives contain one or more heavy metals. Copper is common to most of the systems that include chromated copper arsenate (CCA), ammoniacal copper zinc arsenate (ACZA), ammoniacal copper quaternary (ACQ), and copper azole (CA) (Freeman and McIntyre 2008). All of these systems use copper as the primary biocide but incorporate a supplemental biocide to protect against copper-tolerant organisms. CCA was developed in India in the 1930s and became the most widely used preservative in the world. The chromium in CCA reacts with the wood as well as the copper and arsenic to help immobilize the preservative (Dahlgren and Hartford 1972a,b,c; Dahlgren 1974). CCA is currently used to treat utility poles, marine piling, timbers, and a range of other products for nonresidential applications. It is listed as a restricted use pesticide.

ACQ and CA were developed in the late 1980s and early 1990s. Both use copper solubilized in either ammonia or ethanol amine coupled with either DDAC or tebuconazole as a co-biocide. The copper is immobilized as the ammonia or amine evolves from the wood. In addition to these solubilized formulations, there have been recent developments to create stable particulate systems of both ACQ and CA. These systems are created by grinding the solid material into fine particles that remain suspended in the treatment solution. One advantage of this approach is that the copper is less likely to leach; however, there is a fierce debate about the effects of using non-soluble copper on preservative performance. ACZA was developed in the 1930s specifically for the treatment of refractory or difficult to treat woods and uses ammonia to solubilize the copper, zinc, and arsenic. The ammonia also helps to swell the wood and remove debris from the pits, improving treatment. ACZA is primarily used in the Western United States for treatment of Douglas fir for utility poles, marine piling, and timbers.

In addition to heavy metal-based systems, boron and fluoride have long been used for wood protection. Both chemicals are water soluble and one of their primary advantages is their ability to diffuse into wood that is above the fsp. Conversely, however, both chemicals are susceptible to leaching if they are used in wetter environments. Fluoride is primarily used as a supplemental treatment that is applied to wood that is already in service; however, boron is also used as an initial treatment for wood that is exposed inside buildings where it is unlikely to be subjected to wetting. The primary goal in these cases is to provide protection against insects, particularly termites.

23.5.3 Non-Biocidal Treatments

Concerns about the safety hazards of pesticides have sparked increasing interest in other methods of wood protection. Most of these strategies involve altering the wood to change its moisture-holding characteristics or make it unrecognizable to fungal enzymes. The two most important approaches are acetylation and heat treatment. Acetylation involves treating the wood with agents that react with the hydroxyls present on cellulose and hemicelluloses, thereby making the wood less able to absorb water. Acetylation is currently used in Europe and has been explored in North America. The primary limit has been cost since effective acetylation requires substantial amounts of treatment to be effective (20%–30% weight gains). Heat treatment involves heating wood for long periods in a limited oxygen atmosphere. The primary effect of heating is to reduce the amount of hemicelluloses and free sugar in the wood and this in turn limits the ability of many fungi to colonize the wood. The heating can also reduce the material properties of the final product. Heat treating is used in parts of Europe, but it has not performed as well in North America.

23.5.4 Developing New Wood Protectants with Standardized Testing Protocols

Developing a new wood protectant requires extensive testing to ensure that the system will perform for prolonged periods of time. The U.S. EPA does not assess the efficacy of a preservative. This role is instead supported by the American Wood Protection Association (AWPA) (2009), a nongovernmental body whose members consist of chemical producers and users as well as those with a general interest in the subject. The AWPA has developed a series of protocols for standardization that lists the various tests for consideration by the technical committees. The protocol is voluntary, but it provides guidance for those contemplating standardizing a system. Among the standards that are especially important for this process is the E10 soil block test, which is the first step in assessing a new chemical under controlled laboratory conditions (Figure 23.13). Next, a proponent would treat small stakes with varying amounts of applied chemicals and then expose these in soil contact under AWPA Standard E7. At the same time, the proponent might treat and expose other samples to various aboveground exposures under Standards E15or E22 with the goal of developing data for specific applications such as decking or window frames (Figure 23.14). The proponent would also need to develop data on the ability of the system to migrate out

FIGURE 23.13 Example of a soil block test that is the one of the first standard methods used in North America to assess the efficacy of a new wood preservative.

FIGURE 23.14 Example of a l-joint test that is used to assess the efficacy of preservatives for aboveground wood protection.

from the wood, cause corrosion of metal fasteners, and/or to impact wood properties. Once a system has been proven to be effective, the proponent must develop methods for analyzing the active ingredients in the wood and must develop data showing that wood of various species can be treated to the target retention levels under commercial conditions.

The process of developing a wood protectant can take 3–5 years. Some proponents shorten this time by going directly to the building codes without AWPA Standardization through the International Code Council Evaluation Service (ICC-ES). ICC-ES reports can be accepted by building officials, but the level of scrutiny to which these systems have been subjected is weaker and the processes by which treatment quality is verified are not public.

23.5.5 Methods of Wood Treatment

By its nature, wood is a differentially permeable medium, and flow through this material is primarily driven by the smallest openings that are pits. Treatment tends to be easiest longitudinally and is often several orders of magnitude lower in the radial and tangential directions. Sapwood is more easily treated, reflecting a tendency for the pits in the heartwood to be either closed or occluded with extractives that limit fluid flow. The methods used to deliver chemical into wood depend on the wood species, the chemical involved, the deterioration hazard to which the wood will be exposed, and the length of time the product is expected to perform.

The two primary methods for wood treatment are pressure and non-pressure processes (Mac Lean 1952). Non-pressure processes included brushing, spraying, dipping, and soaking. These processes are primarily designed for surface protection and the depth of chemical penetration is generally shallow. As might be expected, the best treatment will occur along the grain, while treatment depth in the radial or tangential directions may extend to no more than a millimeter or two.

Dipping is used extensively in North America for treatment of wood used in windows and door frames. Since decay in windows tends to occur at end joints, the tendency for treatment to be better in this direction and the fact that the windows are painted or clad help the relatively shallow treatment perform. Dipping and spraying are also used for applying fungicides to freshly sawn lumber to prevent fungal stain and mold. In this case, the treatment provides a shallow, prophylactic barrier against attack by fungal spores that is only needed for 1–6 months when the wood has dried and is no longer at risk of fungal attack.

The vast majority of wood treated in North America is subjected to some combination of vacuum and/or pressure treatment designed to drive chemicals more deeply into the wood. These processes are usually performed in a vessel called a retort. Attached to the retort are pressure and vacuum pumps along with various storage tanks from which chemicals can be transferred (Figure 23.15).

There are three basic treatment processes: vacuum, full cell, and empty-cell processes (Figure 23.16). In the vacuum process, the wood is placed into the retort, the door is closed, and a vacuum is drawn over the wood. The length of time the vacuum is drawn depends on the wood species and the treatment. The chemical is then added into the retort, the vacuum is released, and the treatment solution is then drawn into the wood. This process may be repeated one additional time for a vac–vac process. This process produces acceptable treatment of dry sapwood and is used in Europe for treating wood for use in windows and doorframes.

The full cell process was developed in the 1830s by a British engineer, John Bethel (Graham 1973). Wood is placed into the retort, a vacuum is drawn over the wood, and then the treatment solution is added. The vacuum is released and then the pressure is increased to the desired level (usually 100–200 psi). The pressure is held until gauges show that the wood has absorbed a sufficient amount of chemical then the pressure is released. A series of vacuums may then be drawn to relieve any residual pressure and recover excess chemical. The full cell process is designed to produce the maximum amount of treatment solution uptake for a given depth of treatment. It is most often used for treatment of wood for use in marine environments where high chemical loadings are required or for water-based solutions where there is little or no solvent cost and solution strength can be adjusted to achieve a given target.

FIGURE 23.15 A retort or treating cylinder used to pressure impregnate wood with preservatives.

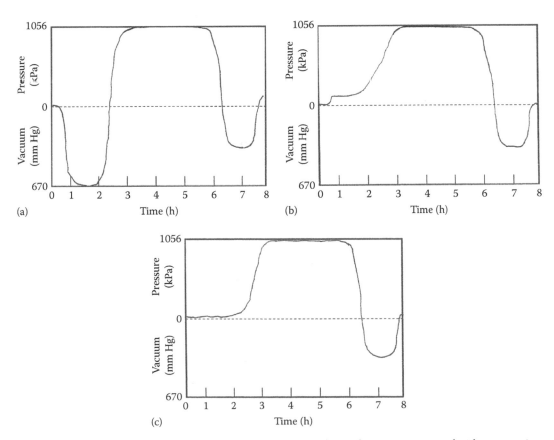

FIGURE 23.16 Examples of (a) full, (b) Lowry, and (c) Rueping cycles used to impregnate wood with preservatives.

The empty-cell processes were designed for use with more costly oil-based chemicals in terrestrial exposures. The Lowry and Rueping processes both eliminate the vacuum and use only pressure to drive chemical into the wood. In the Lowry process, treatment solution is added; the pressure is raised and held until the desired amount of chemical has been absorbed. In the Rueping process, pressure is raised to 30–50 psi before the treatment chemical is added and then the pressure is raised to the desired level. In both cases, air is trapped into the wood at the start of the process. This air is compressed as the pressure is raised. At the end of the process, this cylinder of compressed air expands outward, carrying with it excess chemical. Vacuums are then used to hasten this release and recover excess solution. The empty-cell processes produce much lower retentions or loadings than a comparable full cell process.

The AWPA provides the primary standard for pressure treatment in the United States. These standards are based upon the hazard to which the product will be exposed (called the Use Category) and the commodity to be treated. The AWPA Standards provide minimum target retentions and specify the depth to which the treatment must be delivered for a given species and decay risk. There are five Use Categories beginning with the lowest risk of wood inside a house protected from wetting (Use Category 1) to Use Category 5 where the wood is exposed in a tropical marine environment. The Canadian Standards differ slightly but follow a similar approach for hazard classes. Assessing the risk of deterioration can be complicated, but there are a number of guides for the process based upon prior performance data or climatic conditions (Scheffer 1971, AWPA 2009).

While the treatment processes are relatively well-defined, there have been efforts to develop new methods for impregnating wood-based materials. Two recent examples are vapor phase boron and supercritical fluid impregnation. Vapor phase boron was simultaneously developed in the United Kingdom and New Zealand (Murphy and Turner 1989) and was briefly used commercially for treatment of composite panels. Trimethyl borate was introduced, under vacuum, into a chamber containing the wood and then volatilized to penetrate into the panels where it reacted with water, leaving boric acid. The process was effective but too costly to be competitive. Supercritical carbon dioxide is an excellent, non-swelling solvent that has been shown to be able to carry a variety of biocides into wood-based materials (Sahle-Demessie et al. 1995). The process is used commercially in Denmark for treating window frames, but it remains too costly for general use. These processes not only illustrate the potential for developing new treatment methods but also demonstrate the difficulty of bringing them to widespread commercial use.

23.5.5.1 Treatment Quality

Treatment quality can be assessed under a variety of standards. For pressure-treated wood, the two primary standards used in North America are those of the AWPA (2009) and the Canadian Standards Association. The AWPA is one of the oldest standards-writing bodies in the world and it produces result-oriented standards for wood treatment. Results oriented means that within certain parameters, the treater can use any process they wish to use but they must achieve a desired outcome in terms of chemical penetration and retention. The Canadian Standards are similarly structured. Treatment quality under these standards is usually assessed by removing increment cores from selected wood samples in a given treatment charge, measuring preservative penetration, one for each core, and then cutting a specific assay zone from each core. These assay zone core segments are then combined from a given charge, ground to a dust, and quantitatively analyzed for the treatment chemical using methods that are listed in the standards. Lumber treatments under these standards are also overseen by the American Lumber Standards Committee (ALSC) that operates under the auspices of the U.S. Department of Commerce. The ALSC also oversees the lumber grade stamp program. The primary goal of the quality program is to protect the consumer.

23.5.5.2 Environmental Impacts of Treatment

While wood treatments have the positive environmental benefit of prolonging the useful life of an already renewable material, there is also the potential for negative impacts as a result of the chemicals employed. Most of this risk occurs at the treatment facilities where large quantities of preservatives are used and this is where most regulatory authorities have instituted the tightest controls. Once the treated wood leaves the plant, there is still a risk of environmental contamination. Virtually all wood protectants have some degree of water solubility and will migrate from the treated product into the surrounding environment. This water solubility is essential for the chemical to move into a target organism (fungus, insect, or marine borer) to exert an inhibitory effect (Chou et al. 1971). However, problems can arise when too much of the chemical migrates from the wood.

Migration of oil-based preservatives is a function of the degree of water solubility of the active ingredient. In most cases, the degree of migration into the soil is limited to 150–300 mm away from the wood. This reflects a combination of the relatively low levels of chemical migrating from the wood and the tendency for the soil microflora to degrade organic compounds. As a result, treated wood poses little risk in soil contact. The potential impact of migration from wood exposed to aquatic environments is a function of the amount of treated wood, the water conditions, the current, and the organisms of concern (Lebow et al. 2000). The greatest impact of treated wood in an aquatic environment occurs shortly after installation when any excess chemical on the wood surface can be more easily dislodged. Chemical losses then decline to a low steady state. A number of models have been developed to predict the risk of using treated wood in aquatic environments. In general, these models show that treated wood can pose risk when used in poorly circulating waters with minimal current but pose little risk in other applications.

Migration of metal-based preservatives is also a function of water solubility, but the difference is that most water-based systems are metallic and are, therefore, not degraded. Soil sampling around utility poles and fence posts, however, reveals that metal levels are elevated only for zones 150–300 mm away from the structure. In aquatic applications, metal losses are greatest shortly after installation and decline sharply with time.

23.5.5.3 Mitigation of Environmental Risks

In response to the obvious migration of preservatives from treated wood used in aquatic environments, treaters in some parts of the United States have developed a series of Best Management Practices (BMPs) designed to reduce the impact of using treated wood (WWPI 2006). BMPs were primarily developed for wood used in aquatic environments, but they also have application for wood used in other sensitive areas. BMPs vary with the treatment chemical but they include efforts to limit overtreatment, posttreatment processes to clean the wood surface of chemical deposits, and, in the case of water-based systems, efforts to immobilize the chemical near the wood surface to reduce the initial flush of chemical when the wood is immersed in water.

23.6 Summary

Wood is one of the world's most important renewable natural fibers, and it is used for fuel, housing, railroads, bridges, utility poles, and a host of other applications. Like all materials, wood is also susceptible to degradation by a variety of living and nonliving agents. Living agents require adequate moisture, oxygen, and temperature in order to utilize wood, and an important component of wood protection involves limiting access to one or more of these requirements. The most important requirement is to limit moisture and most buildings are designed to limit wetting. Where moisture control is not possible, the wood can be selected from among those species that contain chemicals that are toxic or inhibitory to degrading organisms or the wood can be supplementally protected. Wood protection can involve impregnation with pesticides, but it can also utilize compounds that alter wood structure to reduce the ability to absorb water or make the wood unrecognizable to fungal enzymes. These materials are typically delivered using combinations of vacuum and pressure to force chemicals into the wood. In most cases, the treatment is limited to the sapwood, but this limited penetration can be very effective as long as the protective barrier remains intact. Successful application of protective chemicals can increase wood service life 8- to 10-fold over untreated wood.

References

American Wood Protection Association (AWPA). 2009. *Annual Book of Standards*, AWPA, Birmingham, AL.

Chou, C.K., J.A. Chandler, and R.D. Preston. 1971. Uptake of metal toxicants by fungal hyphae colonising CCA-impregnated wood. *Wood Science and Technology*, 7, 206–211.

Evans, P.D., P.D. Thay, and K.J. Schmalzl. 1996. Degradation of wood surfaces during natural weathering. Effects on lignin and cellulose and on the adhesion of latex primers. *Wood Science and Technology*, 30, 411–422.

Feist, W.C. 1990. Outdoor wood weathering and protection, in R.M. Rowell and R.J. Barbour, Eds., *Archaeological Wood: Properties, Chemistry, and Preservation*, Advances in Chemistry Series 225, American Chemical Society, Washington, DC, pp. 263–298.

Freeman, M.H. and C.R. McIntyre. 2008. A comprehensive review of copper based wood preservatives. *Forest Products Journal*, 58(11), 6–27.

Freeman, M.H., T.F. Shupe, R.P. Vlosky, and H.M. Barnes. 2003. The future of the wood preservation industry. *Forest Products Journal*, 53(10), 8–15.

Dahlgren, S.E. 1974. Kinetics and mechanisms of fixation of Cu-Cr-As wood preservatives. IV. Conversion reactions during storage. *Holzforschung*, 28(2), 58–61.

Dahlgren, S.E. and W.H. Hartford. 1972a. Kinetics and mechanisms of fixation of Cu-Cr-As wood preservatives. III. Fixation of Tanalith C and comparison of different preservatives. *Holzforschung*, 26(4), 142–149.

Dahlgren, S.E. and W.H. Hartford. 1972b. Kinetics and mechanisms of fixation of Cu-Cr-As wood preservatives. II. Fixation of Boliden K33. *Holzforschung*, 26(3), 105–113.

Dahlgren, S.E. and W.H. Hartford. 1972c. Kinetics and mechanisms of fixation of Cu-Cr-As wood preservatives. I. pH behaviour and general aspects of fixation. *Holzforschung*, 26(2), 62–69.

Duncan, C.G. and F.F. Lombard. 1965. Fungi associated with principal decays in wood products in the United States. U.S. Forest Service Research Paper WO-4, USDA, Washington, DC.

Graham, R.D. 1973. History of wood preservation, in D.D. Nicholas, Ed., *Wood Deterioration and Its Prevention by Preservative Treatments*, Syracuse University Press, Syracuse, NY, pp. 1–25.

Hale, M.D.C. and R.A. Eaton. 1992. *Wood: Decay, Pests and Protection*, Chapman & Hall, London, U.K.

Hansen, L.D. and J.H. Klotz. 2005. *Carpenter Ants of the United States and Canada*, Comstock Publishing Associates, Cornell University Press, Ithaca, NY.

Harada, H. and W.A. Cote, Jr. 1985. Structure of wood, in T. Higuchi, Ed., *Biosynthesis and Biodegradation of Wood Components*, Academic Press, San Diego, CA, pp. 1–42.

Hillis, W.E. 1962. *Wood Extractives*, Academic Press, New York.

Hunt, G.M. and G.A. Garratt. 1967. *Wood Preservation*, McGraw-Hill, New York, 457p.

Lebow, S.T., P.K. Lebow, and D.O. Foster. 2000. Environmental impact of preservative-treated wood in a wetland boardwalk. Part I. Leaching and environmental accumulation. U.S. Department of Agriculture Forest Products Laboratory Research Paper FPL-RP-582. Madison, WI, 72p.

Mac Lean, J.D. 1952. Preservative treatment of wood by pressure methods, in *USDA Agricultural Handbook 40*, Washington, DC, 160p.

Mankowski, M. and J.J. Morrell. 2000. Patterns of fungal attack in wood-plastic composites following exposure in a soil block test. *Wood and Fiber Science*, 32, 340–345.

Murphy, R.J. and P. Turner. 1989. A vapor phase preservative treatment of manufactured wood-based board materials. *Wood Science and Technology*, 23(3), 273–279.

Sahle-Demessie, E., K.L. Levien, and J.J. Morrell. 1995. Impregnation of wood with biocides using supercritical fluid carriers, in K.W. Hutchenson and N.R. Foster, Eds., *Innovations in Supercritical Fluids: Science and Technology*, American Chemical Society, Washington, DC, pp. 415–428.

Scheffer, T.C. 1971. A climate index for estimating potential for decay in wood structures above ground. *Forest Products Journal*, 21(10), 25–31.

Scheffer, T.C. and E.B. Cowling. 1966. Natural resistance of wood to microbial deterioration. *Annual Review of Phytopathology*, 4, 147–170.

Scheffer, T.C. and R.M. Lindgren. 1940. Stains of sapwood and sapwood products and their control. U.S. Department of Agriculture, Technical Bulletin 714, Washington, DC.

Schultz, T.P., D.D. Nicholas, and A.F. Preston. 2007. A brief review of the past, present and future of wood preservation. *Pest Management Science*, 63(8), 784–788.

Taylor, A.M., B.L. Gartner, and J.J. Morrell. 2002. Heartwood formation and natural durability—A review. *Wood and Fiber Science*, 34(4), 587–611.

U.S. Department of Agriculture (USDA). 1980. The biologic and economic assessment of pentachlorophenol, inorganic arsenicals, and creosote, Volume I. Wood preservatives. USDA Technical Bulletin 1658-1. USDA, Washington, DC, 435p.

U.S. Department of Agriculture (USDA). 1996. Wood handbook: Wood as an engineering material. USDA Forest Service Forest Products Laboratory General Technical Report FPL-GTR-113. Madison, WI, 463p.

Western Wood Preservers' Institute. 2006. *Best Management Practices: For the Use of Treated Wood in Aquatic and Other Sensitive Environments*, WWPI, Vancouver, WA.

Zabel, R.A. and J.J. Morrell. 1992. *Wood Microbiology: Decay and Its Prevention*, Academic Press, San Diego, CA.

24

Asphalt

Didier Lesueur
Lhoist Recherche et Développement

Jack Youtcheff
United States Department of Transportation-Federal Highway Administration

24.1 Introduction

In a world concerned with environmentally friendly construction technologies and sustainability, asphalt durability is taking on an added significance. Worldwide asphalt consumption was on the order of 108 Mt in 2010 (United States 33 Mt), which accounted for 2.4% of the total oil demand. Asphalt's use as a paving material accounts for roughly 80% of this and is the focus of this chapter.

Asphalt mixtures have greatly evolved from their beginning in the late nineteenth century to modern times (Lay 1992). They are now becoming very technical materials obtained by careful selection of raw materials among which are asphalt binder and aggregate, and then mixed in dedicated plants whose operations are computer-controlled in order to optimize product quality and minimize energy consumption, waste, and emissions (Lee and Mahboub 2006). Today, roughly 90% of the roads in the United States are paved with asphalt.

Due to the rising cost of asphalt binders, there is a keen interest in the concept of sustainable asphalt mixtures. However, the environmental aspects are generally covered with a somewhat narrow view, consisting essentially of (1) minimizing wastes through the recycling of reclaimed asphalt pavements (RAPs) in new production (Dunn 2001); (2) supplementing the asphalt through the use of other by-products such as reclaimed asphalt shingles (RAS), sulfur, and recycled motor oil; and (3) using lower manufacturing temperatures in the manufacturing of the so-called warm or even semi-warm mixtures, thus decreasing CO_2 emissions and energy consumption (Button et al. 2007, D'Angelo et al. 2008). Clearly, the basic principle of waste management, that is, preventing and reducing waste generation, is not fully taken into account by the industry. This is all the more surprising as the European Waste Directive (Directive 2008/98/EC) or the waste hierarchy rule known as the 4 R's (Reduce, Reuse, Recycle, and Recover); all converge on the primary objective: make your mixtures more durable!

Therefore, this chapter tries to fill the gap in our knowledge about asphalt binders and mixtures by addressing the main factors affecting their durability, namely, their ability to maintain satisfactory performance and structure in long-term service. We review the current understanding of relevant parameters and provide some advice on how to maximize durability.

24.2 Background

Asphalt's usage dates back 180,000 years ago to the El Kowm Basin of Syria, where it was applied to stick flint implements as handles for various tools; this application persisted up until Neolithic time (Connan 1999). In this earliest record and in many of the later uses, the adhesive and waterproofing properties of local sources of natural asphalt were generally recognized. Examples such as the waterproofing of Noah's arch, of the Babel tower, or of the cradle of Moses are given in the Bible (Abraham 1960). Medical uses were also reported, with asphalt acting as a remedy for various illnesses (trachoma, leprosy, gout, eczema, asthma, etc.) as well as a disinfectant or as an insecticide (Abraham 1960). Another well-studied historical application was for the embalming of mummies by the Egyptians (Abraham 1960).

The first mention of the use of asphalt in road construction dates back to Nabopolassar, King of Babylon (625-604 BC); an asphalt-containing mortar cemented both the foundation made of three or more courses of burned bricks and the stone slabs that were placed on top (Abraham 1960). However, asphalt essentially disappeared from pavement applications until the early nineteenth century, when a rediscovered European source of natural asphalt led to the development of the modern applications for this material (Lay 1992). The use of natural asphalt in road construction started to decline in the 1910s with the advent of vacuum distillation, which made possible the manufacture of asphalt from crude oil. Nowadays, paving-grade asphalt is almost exclusively obtained as the vacuum residue of petroleum distillation (Lesueur 2009).

The word bitumen is mostly used in Europe as a synonym for the terms asphalt binder and asphalt cement used in the United States. In this chapter, we will adhere to the American usage and refer to it as asphalt binder.

The asphalt binder is blended with mineral aggregates to form asphalt mixtures (also called bituminous mixes, asphalt concrete, or bituminous concrete). The binder serves two purposes: one is to glue the aggregate together, and the other is to protect the aggregate from environmental distresses. Hot mix asphalt (HMA), the most prevalent paving product, is generally fabricated by heating typically 5 wt.% asphalt binder up to around 160°C in order to decrease its viscosity and then blending it with 95 wt.% of hot aggregate (Monismith 2006). The specifications for paving asphalt generally follow this application. Other techniques to manufacture asphalt mixes using asphalt emulsions or foamed asphalt also exist; these amount to less than 5% of the total asphalt mix production, but their percentages are increasing.

To produce HMA that resists climate and increasing traffic demands, specifications on paving-grade asphalts are becoming more stringent (Read and Whiteoak 2003). The properties that are needed to obtain a suitable asphalt binder are largely rheological. First, the asphalt has to be fluid enough at high temperature (around 160°C) to be pumped and workable to enable a homogeneous coating of the aggregates upon mixing. Second, it has to be stiff enough at the highest pavement temperature to resist rutting (around 60°C, depending on local climate). Third, it must be pliable enough and sufficiently strong at the lowest pavement temperature (down to around –20°C, depending on local climate) to resist cracking. All these properties operate at cross-purposes, and it is therefore difficult to obtain an asphalt binder that performs well under all possible climates. As a consequence, different paving grades exist, the softer being generally suitable for cold climates and the harder, for hotter regions. In order to widen the temperature range of asphalt, additives such as polymers and/or acids are finding increased usage (PIARC 1999, Lesueur 2009, McNally 2011).

24.3 Pavement Deterioration

24.3.1 Deterioration due to Material Chemical Aging

The chemical aging process can be separated into two stages: one during the manufacture of HMA, and the second during the service life of the pavement. The consequence of both processes is the same: there is a global hardening of the asphalt binder (Wright 1965, Bell 1989), which in turn increases its cracking propensity (Isacsson and Zeng 2003) and embrittlement of the asphalt mixture. The risk of cracking increases, whether attributed to thermal factors or traffic-induced fatigue. In some instances, oxidation products are formed that may enhance the moisture sensitivity of the asphalt mix and induce stripping or rutting (as discussed later).

24.3.2 Deterioration due to Climate

24.3.2.1 Moisture Damage and Frost

Moisture-induced damage and the effect of freeze–thaw cycles are common phenomena with asphalt mixtures. They generally materialize through the progressive loss of aggregate as illustrated in Figure 24.1. The asphalt–aggregate bonds weaken in the presence of water to the point that the cohesive and adhesive bonds are no longer strong enough to hold the aggregate in place. This results in aggregate stripping or raveling when it is limited to the surface. This is induced by environmental factors and worsened by traffic. Flushing is another type of water damage that similarly leads to the loss of aggregate; this mechanism occurs in the bulk of the material as a consequence of the traffic-induced internal water pressures. If left untreated, such damage can deteriorate into potholes. Frost and freeze thaw cycles tend to enhance these detrimental effects, thus a tough winter can directly generate potholes.

According to a U.S. survey taken in the early 1990s (Hicks 1991), water-induced damage on average is evident in untreated HMA between 3 and 4 years post construction; sometimes this can be manifested in the very first year.

Moisture damage arises when water forms a thermodynamically favored film at the asphalt–aggregate interface thereby displacing the asphalt binder from the aggregate surface. In addition, water scour can

FIGURE 24.1 Aggregate stripping as a consequence of moisture induced-damage (for scale, the pavement marking is 4 in. wide). (From Miller, J.S. and Bellinger, W.Y., *Distress Identification Manual for the Long-Term Pavement Performance Program*, FHWA-RD-03-031, FHWA, Lanthum, MD, 2003.)

subject the pavement to both adhesive- and cohesive-induced damage (Kringos et al. 2008a,b). Ice formation is also a contributor as the associated volume change can damage the material (Mauduit et al. 2010) and enhance the further uptake of water.

While some asphalt oxidation products may contribute to the adhesiveness of the binder to the aggregate (Curtis et al. 1989a,b), other functional groups may be converted to hydrophilic and surface-active compounds and attract moisture such as sulfones (functionalities found in soaps, etc.). The presence of such groups can adversely affect the adhesive properties and freeze–thaw tolerance of the binder.

The presence of surface-active species in the aggregate, typically clay contaminants, has been linked to moisture damage (Aschenbrener et al. 1995). Selective adsorption of the viscosity-building components of the asphalt binder (asphaltenes) by clays can lead to the softening of the binder and result in rutting of the pavement.

Thus, moisture damage has multiple effects and multiple impacts on pavement durability from cracking to rutting. Consequently, attributing and assessing damage to this distress mode are difficult.

24.3.2.2 Thermal Cracking

Thermal cracking is largely observed in cold climatic zones. In these regions, the low temperatures experienced by the asphalt make it essentially perform in its glassy state where it is brittle. As a consequence, the thermal shrinkage occurring during cooling leads to the development of stresses that can overcome the materials strength, hence generating transverse cracks in the pavement.

Thermal cracking is not limited to cold regions. Large day–night amplitudes can also generate cracking patterns with the crack propagating from the top of the layer to the bottom. This has been observed in Southern France (GNB 1997) and elsewhere (Ferne 2006).

24.3.3 Deterioration due to Traffic

24.3.3.1 Rutting

Rutting has been observed in asphalt pavements since the very beginning of their use, but became increasingly significant following World War II as traffic volumes and loads started to increase rapidly (Sousa et al. 1991). Rutting occurs when the traffic load over the asphalt mixture exceeds its viscoplastic limit, hence generating permanent plastic deformations evidenced by depressions in the wheel path as illustrated in Figure 24.2. The rutting attributed to viscoplastic permanent deformation is inversely proportional to the traffic speed and asphalt binder stiffness. As a result, rutting is more predominant under low speed, high

FIGURE 24.2 Rutting in an asphalt mixture (for scale, the pavement marking is 4 in. wide). (From Miller, J.S. and Bellinger, W.Y., *Distress Identification Manual for the Long-Term Pavement Performance Program*, FHWA-RD-03-031, FHWA, Lanthum, MD, 2003.)

loads, and at high temperatures (Ould-Henia et al. 2004). (Typical pavement temperatures in Europe and North America range from 40°C to 60°C.) However, rutting remains a complex phenomenon, because the asphalt mixes deform in a viscoelastoplastic way under these conditions (Gibson 2006).

Rutting in asphalt mixes is associated with several factors such as higher than optimum asphalt content, high natural sand content, round aggregate shape (e.g., uncrushed gravel), or high binder deformability (Sousa et al. 1991). Consequently, factors favoring the stiffening of the mixtures should also increase the rutting resistance. Rutting on average accumulates for untreated mixes over the 5 years post construction, but sometimes premature rutting may occur during the very first year (Hicks 1991).

24.3.3.2 Fatigue Cracking

Fatigue cracking of asphalt pavements is a more recently studied phenomenon, though it had been recognized back in the 1950s as a possible failure mode for asphalt mixtures (Duriez and Arrambide 1954) and later demonstrated in the highly visible American Association of State Highway Officials (AASHO) trials from 1957 to 1961 (AASHO 1962).

Fatigue cracking occurs when the repeated traffic loads progressively damage the asphalt mixtures generating cracks that propagate from the bottom to the top of the layer as shown in Figure 24.3. As a consequence, fatigue cracking is more pronounced in thin thickness layers or where poor adhesion exists between the successive layers; both of these conditions promote high flexural stresses at the bottom of the asphalt layers (LCPC-SETRA 1997). Significantly aged pavements will experience top-down cracking whereby the top layer is significantly embrittled and the stresses overcome the strain tolerance of the pavement layer. The cracks initiating in this brittle surface can then propagate to softer layers below.

24.3.3.3 Loss of Skid Resistance

The top layer of the asphalt pavement, called the wearing course, has specific requirements that go beyond the mechanical properties assuring that the pavement resists traffic-induced damage.

In particular, skid resistance is a big issue, as it ensures that the tires of the riding vehicles maintain good contact with the pavement surface, hence contributing to the safety of the road users. Skid resistance is controlled largely by the aggregate sizing and properties and by the type of mixture and, to a lesser extent, binder used. More precisely, microtexture, that is, the roughness of the mixture at the micrometer level (Brosseaud et al. 2005, Masad et al. 2009), governs this property and is essentially related to the aggregate texture and hardness. As a result, the wear of the aggregate due to traffic decreases the skid resistance as shown in Figure 24.4 (Stasse 2000) and therefore makes it essential to use wear-resistant aggregates in the surface or wearing courses.

FIGURE 24.3 Fatigue cracking (alligator cracking) in an asphalt mixture (these interconnected cracks can be from 1 to 6 in. in length). (From Miller, J.S. and Bellinger, W.Y., *Distress Identification Manual for the Long-Term Pavement Performance Program*, FHWA-RD-03-031, FHWA, Lanthum, MD, 2003.)

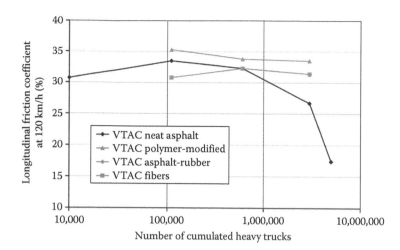

FIGURE 24.4 Longitudinal skid resistance at 120 km/h versus cumulative traffic (number of heavy trucks) for different kinds of 0/10 very thin asphalt concrete: with neat asphalt binder, with polymer-modified asphalt, with asphalt-rubber, with fibers. (Data from the Carat database adapted from Stasse, G., Caractérisation d'adhérence de revêtements de chaussées routières, Report Chaussées CR 25, LCPC, Paris, France, 2000.)

24.3.3.4 Other Distress Modes

Bleeding occurs when higher than optimum asphalt binder exceeds the mix void capacity in the pavement and moves upward under traffic or with thermal expansion. This is a nonreversible cumulative process, which is a safety-related issue as it creates a slippery surface.

Block cracking results from the shrinkage and hardening of aged asphalt binders. This consists of a series of large interconnected rectangular cracks on the asphalt pavement surface, which can range in size from 1 cm to over 10 cm in length.

Reflective cracking occurs when cracks from the base or underlying layer propagate to the top layer; this can be observed in asphalt mixtures overlaying rigid (cement concrete) pavements. Another significant distress mode results from the rutting or deformation of the subbase, especially when this consists of granular materials or soils. This deformation may not be limited to the wheel paths and results in a general uneven pavement.

Other distress modes can also be found that are somewhat related to traffic. In such cases, these failure modes are observed on the top asphalt layer but are really due to structural problems underneath. A thorough cure for these distresses would necessitate a major treatment of the responsible layers. However, the higher processing costs of such operations make repeated milling and replacing the modus operandi, which does not eliminate the primary cause of such disorders.

24.3.4 Observed Field Durability

Given all the earlier factors, asphalt pavements gradually deteriorate and lose their strain tolerance or develop load associated cracks and finally need to be preserved via partial milling and overlaying or ultimately full-depth replacement at the end of their service life. In general, the wearing course is the most distressed pavement layer, and the tendency is to design pavement structures that can withstand 20 years or more of cumulative traffic where only the wearing course needs to be repaired.

At present, the durability of European wearing course materials has been studied, and service lives were found to range from 8 to 25 years, depending on mixture type. The results of this field survey are presented in Figure 24.5 (EAPA 2007). Care is needed in interpreting these data, as the definition of service life was not clearly stated in this study. Some countries reported the expected service life, others provided the real service life (i.e., when distresses necessitated rehabilitation of the pavement

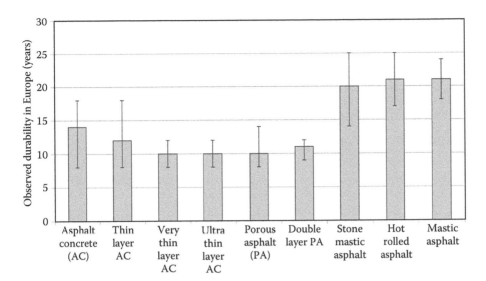

FIGURE 24.5 Durability of asphalt layers on major European roads as a function of the type of wearing course. The error bar gives the 15% lowest and 85% highest observed durability, respectively. (Adapted from EAPA, *Long-Life Asphalt Pavements—Technical Version*, EAPA, Brussels, Belgium, 2007.)

to maintain the riding quality), and some defined this as the usual time between maintenance operations, regardless of the level of pavement deterioration. In addition, climate and traffic vary from one country to another. Nevertheless, this shows that wearing courses are expected to last 8–25 years, and shorter durations are then considered premature failures. The most frequently observed distress modes in Europe are detailed in Figure 24.6, where it appears that rutting is the main issue, followed by the loss of skid resistance and top-down cracking (FEHRL-ELLPAG 2004).

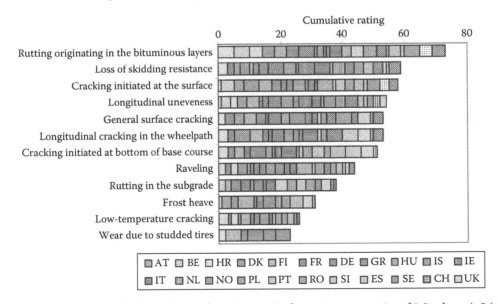

FIGURE 24.6 Most observed degradation modes in Europe. Each country gave a rating of 0 (irrelevant)–5 (major issue) to quantify the importance of this damage on its road network. AT, Austria; BE, Belgium; HR, Croatia; DK, Denmark; FI, Finland; FR, France; DE, Germany; GR, Greece; HU, Hungary; IS, Iceland; IE, Ireland; IT, Italy; NL, Netherlands; NO, Norway; PL, Poland; PT, Portugal; RO, Rumania; SI, Slovenia; ES, Spain; SE, Sweden; CH, Switzerland; UK, United Kingdom. (From ELLPAG, ELLPAG Phase 1 report, 2004, Reproduced with permission of FEHRL.)

Improvements needed to restore the pavement condition will depend on type of damage that has accumulated. In the case of fatigued or moisture damaged materials, milling and replacing of the surface lift is generally the best option.

24.4 Laboratory Evaluation of Asphalt Mixture Durability

24.4.1 Deterioration due to Material Chemical Aging

24.4.1.1 Asphalt Binder Conditioning

The resistance of the asphalt binder to the short-term (asphalt mix production) and long-term (in service life) chemical aging has been evaluated in the laboratory through the conditioning of binders at elevated temperatures and or pressures for extended periods of time.

The short-term aging of asphalt binder is simulated in the laboratory by conditioning a thin asphalt film at a high temperature, for a short duration; the most prevalent specification is the Rolling Thin Film Oven Test (RTFOT—ASTM D2872—EN 12607), which exposes the binder in 1.25 mm thick moving films to air at 163°C for 85 min. This simulated "average processing condition" typically results in a nominal doubling of the viscosity after cooling; although the extent of hardening is asphalt dependent, 1.5- to 4-fold increases in viscosity at 60°C have been reported (Bell 1989, Anderson et al. 1994). Note that part of the hardening originates from the loss of volatile compounds, and asphalt specifications generally require that mass loss after short-term aging remain below 0.5%. Current European and U.S. specifications indirectly eliminate asphalts whose aging upon mixing is deemed too rapid.

The long-term aging occurs over the service life of the pavement, which can extend to several decades. It is largely dependent on the diffusion of oxygen and the pavement temperature, the latter being dependent on the local climate. The exposed layer will obviously be subjected to a greater level of oxidation; however, the mix formulation and the achievement of the desired level of compaction (pavement density) also come into play, as do asphalt film thickness and mix porosity. All these factors influence asphalt aging and make it quite complicated to accurately describe in situ aging. One widely used conditioning procedure, the Pressure Aging Vessel Test (PAV—ASTM D6521—EN 14769), was shown to match the increase in viscosity of approximately 4–8 year-old asphalt binders in surface sources taken from two diverse locations, namely, Wyoming and Florida (Anderson et al. 1994). However, the PAV does not accurately simulate the in situ aging of the binder in the pavement.

24.4.1.2 Asphalt Mixture Conditioning

Plant-produced mix samples taken from paver can be compacted into lab specimens using a gyratory or linear kneading compactor or can employ some other mode of compaction. Lab-produced mixes, which mimic this level of aging, involve conditioning the loose mix for 2 h at 135°C prior to compaction when designing mixes; 4 h conditioning is used to produce specimen for performance-related testing. Long-term aging of asphalt mixtures is simulated by aging the compacted cores for 5 days at 85°C (AASHTO R 30-02). While this conditioning is to reflect 7–10 years of field aging, the post conditioning appearance of the compacted samples do not visually simulate the field-aged samples obtained from aged pavements.

24.4.2 Deterioration due to Climate

24.4.2.1 Moisture Damage and Frost

Many test methods have been developed to evaluate the moisture resistance of asphalt mixtures. Table 24.1 lists those most frequently referenced in the literature. Their predictive power is still debated, and there is no clear consensus on which method is best suited for predicting moisture damage in the field. The Lottman test (AASHTO T-283—Table 24.1) evaluates the loss of indirect tensile

TABLE 24.1 Most Used Testing Methods in order to Evaluate the Resistance to Moisture Damage or Frost of Asphalt Mixtures

Test Method	Standard	Type of Specimen	Testing Method	Conditioning	Test Result
Hamburg Wheel Tracking	EN 12697-22—Method B Under Water	260 mm × 300 mm rectangular slabs with final thickness	Wheel tracking device under water	Testing under water at 50°C	Rut depth in mm
Indirect Tensile Strength Ratio (ITSR)	EN 12697-12—Method A	100 mm diameter (or 150 or 160 for large aggregate size) cylindrical specimen of the asphalt mixture to be tested	Indirect tensile strength (ITS) at 25°C and 50 mm/min	Specimens in vacuum (7 kPa) for 30 min 70 h in water at 40°C 2 h at 25°C	ITS ratio in % (after conditioning/no conditioning)
Duriez	EN 12697-12—Method B	100 mm diameter (or 80 or 120 or 150 or 160) cylindrical specimen of the asphalt mixture to be tested	Compressive strength at 18°C and 55 mm/min	Specimens in vacuum (47 kPa) for 120 min; 7 days in water at 18°C	ITS ratio in % (after conditioning/no conditioning)
Cantabro	EN 12697-17	101.6 mm diameter × 63.5 mm cylindrical specimen of the asphalt mixture to be tested (generally a porous asphalt)	Mass loss after 300 revolutions in the Los Angeles test (without steel balls)	Generally the same as ASTM D1075	Mass loss ratio (after conditioning/no conditioning)
Saturation Aging Tensile Stiffness (SATS)	U.K. Specification for Highway Works—Clause 953	100 mm diameter × 60 mm cylindrical specimen of the asphalt mixture to be tested, cored from a slab with 8% voids	ITS modulus measured at 20°C using the Nottingham Asphalt Tester	Specimens in vacuum (55 kPa) for 30 min 65 h at 85°C and 2.1 MPa in water saturated vessel 24 h at 30°C and 2.1 MPa	ITS ratio in % (after conditioning/no conditioning)
Lottman	AASHTO T283 Tex 531-C	101.6 mm diameter × 63.5 mm cylindrical specimen of the asphalt mixture to be tested compacted to 6.5%–7.5% voids	ITS at 25°C and 50.8 mm/min	70%–80% pore saturation 16 h at 17.8°C 24 h in water at 60°C 2 h in water at 25°C	ITS ratio (after conditioning/no conditioning)
Repeated Lottman	—	101.6 mm diameter × 63.5 mm cylindrical specimen of the asphalt mixture to be tested compacted to 6.5%–7.5% voids	ITS at 25°C and 50.8 mm/min	Lottman conditioning but with consecutive freeze–thaw cycles (generally from 1 to 20)	ITS ratio vs. number of freeze–thaw cycle

(*continued*)

TABLE 24.1 (continued)   Most Used Testing Methods in order to Evaluate the Resistance
to Moisture Damage or Frost of Asphalt Mixtures

Test Method	Standard	Type of Specimen	Testing Method	Conditioning	Test Result
Texas Freeze–Thaw Pedestal		41.3 mm diameter × 19 mm cylindrical briquette of 0.4/0.8 sand coated with asphalt at optimum + 2% compacted by static pressure of 27.58 kN for 20 min	Visual (crack)	Briquette immerged in distilled water 15 h at 12°C 45 min in water at 24°C 9 h at 49°C then repeat	Number of freeze–thaw cycles to failure
Retained Tensile Strength (Tensile Splitting Ratio/ Indirect Tensile Strength/ Root–Tunnicliff Test)	ASTM D4867	101.6 mm diameter × 63.5 mm cylindrical specimen of the asphalt mixture to be tested compacted by any means (static/ Marshall, etc.) to 6%–8% air voids	ITS at 25°C and 50.8 mm/min	55%–80% pore saturation 24 h in water at 60°C 1 h at 25°C	ITS ratio (after conditioning/ no conditioning)
Immersion/ Compression	ASTM D1075 AASHTO T165	101.6 mm diameter × 101.6 mm cylindrical specimen of the asphalt mixture to be tested compacted by static compaction on both sides (3000 psi during 2 min)	Compressive strength at 25°C and 5 mm/min	4 days in water at 48.9°C or 1 day in water at 60°C	Compressive strength ratio (after conditioning/ no conditioning)
Retained Marshall		101.6 mm diameter × 76.2 mm cylindrical specimen of the asphalt mixture to be tested compacted by impact compaction (50 or 75 blows)	Marshall stability at 60°C and 50.8 mm/min	24 h in water at 60°C	Stability ratio (after conditioning/ no conditioning)
Texas Boil Test	ASTM D3625 Tex 530-C	(300 g + asphalt content) of asphalt mixture to be tested or (100 g + asphalt) of 4.8/9.8 aggregate	Visual (aggregate surface covered in asphalt)	Asphalt mixture in boiling water for 10 min	% of retained asphalt after boiling

strength following one freeze–thaw cycle; this was found to be one of the more effective tests in use in the United States as shown in Figure 24.7 (Hicks 1991).

The Hamburg Wheel Tracking test (AASHTO 324, EN 12697-22) was not addressed in this survey and has since gained significant popularity. This employs a loaded steel wheel that reciprocates over samples submerged in a heated water bath. The magnitude of the deformation as a function of loading passes is captured. Typical experimental curves are shown in Figure 24.8.

24.4.2.2 Thermal Cracking

Thermal cracking is generally studied in the lab by measuring stresses and specimen temperature in a restrained specimen subjected to a cooling cycle (10°C/h is typical). The dimensions of the prismatic specimens are maintained constant such that thermal stresses build up as the specimen shrinks. The temperature at which the specimen breaks is then recorded.

Indirect tension testing has been used to determine creep compliance and strength of HMA in tension, which allows a computation of the thermal shrinkage, stress buildup, and theoretical failure temperature, phenomena that occur physically in the thermal stress restrained specimen test (TSRST).

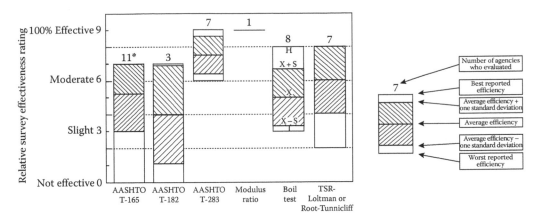

FIGURE 24.7 Comparison of the effectiveness of several test methods in order to predict moisture damage as evaluated by state agencies' experience. *Number of responses rating effectiveness. H = high; L = low; X = mean; S = standard deviation. (From Hicks, R.G., Moisture damage in asphalt concrete, *NCHRP Synthesis of Highway Practice* 175, Transportation Research Board, Washington, DC, 1991.)

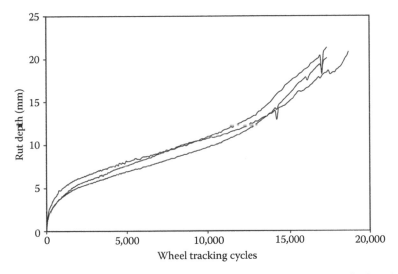

FIGURE 24.8 Hamburg Wheel Tracker Rut depth is plotted on the vertical axis versus wheel tracking cycles plotted on the horizontal axis. Three replicate curves from a single mixture are plotted together where the typical trend is a rapid increase in rut depth (compaction), which then grows nearly linearly and then transitions into a tertiary phase where rut depth growth accelerates with loading cycles. (From Gibson, N. et al. 2012. Full-scale accelerated performance testing for Superpave and structural validation. Final report transportation pooled fund study TPF-5(019) and SPR-2(174). Accelerated pavement testing of crumb rubber modified asphalt pavements, FHWA, Lanthum, MD.)

Roque et al. (1999) have applied this to determine crack growth rate parameters for HMAs in intermediate temperature range in conducting their fatigue cracking analysis.

24.4.3 Deterioration due to Traffic

24.4.3.1 Rutting

Several test methods are available to evaluate the rutting resistance of asphalt mixtures. Most are traffic simulators, such as wheel trackers; others are mechanical tests quantifying the permanent deformation accumulated by the material under repeated loads at high temperature (generally in the 40°C–60°C

FIGURE 24.9   Photograph of pneumatic wheel in the French Pavement Rutting Tester and rutted test specimen (slab dimensions are 20 in. [500 mm] long, 7 in. [180 mm] wide, and 4 or 2 in. [100 or 50 mm] thick). (From Gibson, N. et al. 2012. Full-scale accelerated performance testing for Superpave and structural validation. Final report transportation pooled fund study TPF-5(019) and SPR-2(174). Accelerated pavement testing of crumb rubber modified asphalt pavements, FHWA, Lanthum, MD.)

range). The European standard EN 12697-22 incorporates several test setups in testing asphalt mixtures for rutting resistance. The Asphalt Pavement Analyzer, the Hamburg Rut Tester, and the French Pavement Rutting Tester are examples of traffic simulators. The latter is shown in Figure 24.9. Mechanical tests in use impose creep or dynamic compression and measure irrecoverable deformation and stiffness. An example of the former is the asphalt mixture performance tester (AMPT); the data from this are incorporated into pavement designs.

24.4.3.2 Fatigue Cracking

Fatigue life is generally studied in the lab by submitting a specimen to repeated loads of constant intensity. The load can be either stress or strain controlled. In stress-controlled experiments, failure is easily detected as the breaking point of the specimen. In strain-controlled experiments, failure is conventionally defined as the point where the specimen modulus is decreased by 50%. The number of cycles to failure is measured as a function of loading intensity. Representative fatigue curves of both modes are shown in Figure 24.10. European standard EN 12797-24 describes various test methods in order to evaluate the fatigue resistance of asphalt mixtures. Apart from the inherent low repeatability and reproducibility of the fatigue tests, the different setups used in the industry yield quite different conclusions (Di Benedetto et al. 2003), making it difficult to assess the fatigue life in sound theoretical terms.

Currently, models based on viscoelastic continuum damage theories are being developed and applied. These theories take into account the development of small microcracks and their coalescence and growth into macrocracks (Lee and Kim 1998a,b). Cyclic axial fatigue characterization tests can be conducted on field- and laboratory-produced specimens to generate a damage characteristic curve that can be used to describe the damage and cracking response at any temperature and under any generalized loading (i.e., stress control and strain control) (Kutay et al. 2008).

Healing is also another key phenomenon that affects the fatigue life of asphalt mixtures. This phenomenon has been observed in the early studies on asphalt mixture fatigue (Saunier 1968), and as a

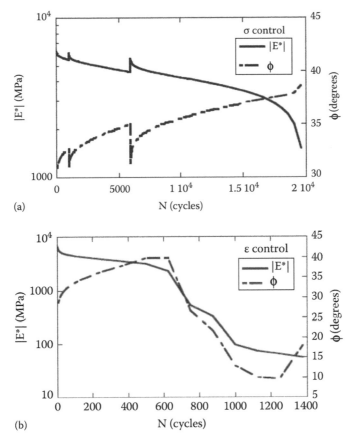

FIGURE 24.10 Dynamic modulus |E*| is plotted on the left vertical axis, and phase angle is plotted on the right vertical axis versus number of fatigue cycles on the horizontal axis. (a) Typical trends observed during a stress control fatigue test are illustrated with a gradual reduction of modulus and gradual increase in phase angle as loading cycles. (b) Typical trends observed during a strain control fatigue test is illustrated with a gradual reduction of modulus and gradual increase and then decrease in phase angle as loading cycles accumulate. (From Gibson, N. et al. 2012. Full-scale accelerated performance testing for Superpave and structural validation. Final report transportation pooled fund study TPF-5(019) and SPR-2(174). Accelerated pavement testing of crumb rubber modified asphalt pavements, FHWA, Lanthum, MD.)

result, the occurrence of rest periods between loading sequences has been observed to increase the fatigue life of asphalt mixtures (Kim et al. 2003). The rest periods allow for the microcracks to heal and therefore decrease fatigue damage. At the moment, there is no standardized test method where healing effects can be taken into account.

24.4.4 Practical Use of the Lab Tests: Specifications

Specifications on asphalt mixtures are generally based on three complementary elements. First, specifications on the binder are largely purchase based that allow for an adequate choice with respect to foreseen climate and traffic conditions. Second, specifications on the aggregate exist to guarantee that the most important component in terms of weight will fulfill minimum requirements. Finally, specifications on the asphalt mixture constituents make sure that the proportions and qualities of the raw materials are fit for their intended use. For example, hot-mix asphalt mix design specifications are used to ensure that an optimum amount of asphalt binder is used in a mix along with an appropriate gradation that ensures adequate volumetric properties.

24.4.4.1 Binder Specifications

European paving-grade asphalts are still defined using empirical tests such as penetration and ring-and-ball softening temperature, whereas the United States has adopted specifications that control asphalt rheology, which were developed during the Strategic Highway Research Program (SHRP) in the early 1990s (AASHTO T315). The current U.S. specifications for bituminous binders are collectively known as the performance grade (PG) specification.

The PG system relies essentially on two test methods: one characterizing the binder in the high-temperature range, and the other one for the low-temperature range. The binder is said to be PG H–L, where H is the limiting high temperature and –L is the limiting low temperature. For example, a binder with PG 58–28 grade means that its limiting high field service temperature is 58°C while its limiting low temperature is –28°C.

The limiting high field service temperature corresponds to the temperature at which the inverse viscous compliance (G*/sin δ) of the binder is equal to 1 kPa when measured at a frequency of 10 rad/s (AASHTO T315). The test is typically performed on the unaged binder. Grades are defined by 6°C steps starting from 46°C up to 82°C. The idea behind this criterion is that too fluid a binder generates a risk of rutting for the pavement. Thus, the limiting high temperature can be thought of as the temperature above which the risk of rut formation becomes significant for a hot mix made with the corresponding binder (Asphalt Institute 1994).

The limiting low field service temperature is based on the more restrictive of two criteria. One refers to the m-value, which controls 98% of the U.S. binders. The other corresponds to the temperature at which the flexural creep modulus of the binder is less than or equal to 300 MPa when measured at a loading time of 60 s (AASHTO T315). Grades are defined by 6°C steps starting from –10°C down to –46°C. The limiting temperature stated in the paving grade is in fact 10°C lower than that of the creep test. In other words, if the binder has a creep modulus of 300 MPa or lower at –12°C and 60 s, the limiting low temperature will be –22°C. The idea behind this criterion is that too rigid a binder generates a risk of cracking for the pavement. Thus, the limiting low temperature can be thought of as the temperature below which the risk of cracking becomes significant for a hot mix made with the corresponding binder (Asphalt Institute 1994). Since the risk increases with binder aging, the test is typically performed on binders subjected to the short-term aging (RTFOT—ASTM D2872, EN 12607) followed by the long-term conditioning (PAV—ASTM D6521, EN 14769).

A refinement to the PG specification is the Multiple Stress Creep Recovery (MSCR) test procedure (AASHTO TP70), which more accurately predicts the performance of polymer-modified binders as it relates to rutting. The test captures the nonrecoverable creep compliance (J_{NR}) and percentage of recovery during each loading cycle (D'Angelo et al. 2007).

24.4.4.2 Aggregate Specifications

The choice of proper aggregate is critical to the performance of the HMA. The aggregate accounts for roughly 95 wt.% of the mixture. The specifications classify aggregates in terms of particle size and shape, mechanical properties, and surface activity.

The geometrical criteria are of the utmost importance. Aggregates are separated into fractions depending on their particle size by sieving. In Europe, aggregate fractions are generally designated by a d/D value, where d is the maximum sieve size (in mm) that retains all of the aggregate, and D is the smallest sieve size through which all of the aggregate passes. For example, a 6/12 aggregate passes through a 12 mm sieve but is retained on a 6 mm one. The fraction finer than 75 µm (63 µm in Europe) is called the filler. The fraction passing the 2 mm sieve is referred to as sand and generally includes most of the filler. Coarser fractions are called gravel. Asphalt mixtures generally do not use aggregate larger than 30 mm, and more often no larger than 20 mm, because of the high risk of segregation.

Aggregate shape also plays a critical role. Ideally, cubic aggregate would be desirable in order to maximize aggregate–aggregate interlocking, which in turn maximizes the rutting resistance. In the case of rounded materials such as river gravels, the level of crushing (number of fracture faces) is often stipulated. Flat or elongated particles are not very favorable, as they tend to break into smaller pieces during compaction and trafficking. Consequently, specifications are in place to restrict the amount of such particles.

Finally, good mechanical properties are very important; these include strength and resistance to polishing. The closer the asphalt layer is to the surface, the more severe the specifications. Mechanical resistance is generally evaluated using the Los Angeles abrasion test (ASTM C131 and C535—EN 1097-2). Abrasion-resistant materials, with Los Angeles values below 20% are generally preferred for surface courses. In base or subbase courses, higher Los Angeles values, up to 40% in some countries, are accepted. Also the wear resistance of aggregate is of primary importance for wearing courses because of its strong effect on skid resistance (Brosseaud et al. 2005, Masad et al. 2009). This is generally assessed through the Polishing Stone Value (PSV) test (EN 1097-8). The PSV reflects the microtexture and is a measure of the resistance of an aggregate to polishing. In order to maintain a good skid resistance, the most demanding specifications require a PSV above 55.

The specifications based on LA or PSV generally prohibit the use of soft aggregates, such as limestone, in the wearing courses, especially for high-traffic roads, though they have been blended with a hard aggregate to achieve a good texture due to their differential wearing (Abdallah et al. 2008).

An approach that has been developed to quantify three-dimensional shape, angularity, and texture of coarse aggregate particles as well as the angularity of fine aggregate particles is the aggregate imaging system. This has been used to identify accelerated aggregate polishing and changes in microtexture (Masad 2005) with more objectivity than conventional techniques like fine aggregate angularity (AASHTO T304).

Finally, specifications on cleanliness are generally invoked. They limit the amount of very fine materials (generally clays) in the aggregate, where finer particulates have a detrimental effect on the asphalt binder and its adhesion to the aggregate. As a matter of fact, clay materials have the tendency to form layers around the aggregate such that the asphalt binder sticks to this layer instead of the aggregate. It is similar in concept to the use of wheat flour as a demolding agent in the kitchen where the flour does not stick to the walls of the pan, so the cake can be easily removed. The same happens with asphalt binder when it adheres to the clay or fines instead of directly to the aggregate particles. Sand equivalent or methylene blue values are generally used to quantify the amount of surface-active clays.

24.4.4.3 Asphalt Mixture Specifications

The specifications on asphalt mixtures help the formulator choose the right components and quantity. Most specifications worldwide give the particle size distribution of the aggregate for a given mixture formula, for example, an asphalt concrete for a wearing course (Monismith 2006). The Bailey Method and gyratory locking point are used to optimize the aggregate packing and structure; the locking point is defined as the point where mix density is not affected by consecutive gyrations (Mohammad and Shamsi 2007, Li and Gibson 2011). Interestingly, the French devised a local method where no particle size distribution is given, only maximum aggregate size (Delorme et al. 2007). This is because the full curve is validated indirectly through its compaction behavior, for which tight specifications are present.

Once preliminary binder contents and grades, and an aggregate gradation are selected, asphalt testing is performed on mixtures to ensure their constructability and durability (Asphalt Institute 2001, Monismith 2006, Delorme et al. 2007). Depending on the local requirements, specifications are generally based on moisture resistance, density and compaction behavior, and mechanical properties. Testing the latter may entail evaluating one or more of the following properties: strength, modulus, fatigue resistance, rutting resistance.

Other specific properties may be considered in a specification, for example, the permeability of porous asphalt may be evaluated.

While specifications are striving to become more performance based (Asphalt Institute 2001, Monismith 2006, Delorme et al. 2007), we are far from being able to anticipate the service life of an asphalt mixture from a laboratory study. Therefore, a lot of additional research is needed to assess the field relevance of most test methods.

24.5 General Properties of Asphalt Mixtures

24.5.1 Composition

HMA is generally fabricated by first heating the asphalt binder to around 160°C in order to decrease its viscosity and then blending it with heated aggregate. The specifications on asphalt binders for pavements are mostly based on this application. Other techniques to manufacture asphalt mixes using asphalt binder emulsions, warm and cold mix technologies, are also available, but they amount to less than 5% of the total asphalt mix production.

24.5.1.1 Asphalt Binder: Definitions

As the name suggests, asphalt mixtures are usually obtained by blending aggregate and asphalt binder. Asphalt cement, generally called bitumen in Europe, is defined as a *virtually involatile, adhesive and waterproofing material derived from crude petroleum, or present in natural asphalt, which is completely or nearly completely soluble in toluene, and is very viscous or nearly solid at ambient temperatures* in the current European specifications (EN 12597). A solubility value in trichloroethylene above 99% is required in the paving specifications (EN 12591).

Historically, natural sources of asphalt were initially used. In Europe, the deposits of Seyssel in France and Val de Travers in Switzerland were at first the principal asphalt sources for road applications (although not exactly as asphalt mixtures). The reference materials for paving applications in the United States at the beginning of the twentieth century were mostly the Trinidad Lake asphalt and, to a lesser extent, the Bermudez Pitch Lake asphalt from Venezuela, both of which constitute some of the largest asphalt deposits in the world.

Although they have very similar uses and properties, asphalt binder is not to be confused with *coal tar*, which is the residue of the pyrolysis of coal. Unfortunately, the name coal tar remains widely used in everyday language as a general term for a black paving material. Health issues associated with the use of coal tar have largely eliminated its use. These issues arise from the higher content of carcinogenic polynuclear aromatic hydrocarbons found in coal tars, whereas they are only present in trace amounts in asphalt binders (Burstyna et al. 2000).

Currently, asphalt binder is essentially obtained by the distillation of crude oil (Corbett 1969, Lesueur 2009). Not every crude oil source yields sufficient amounts of asphalt binder. Typically, the heavier the crude oil, the higher is its yield of asphalt binder (Corbett 1969, Read and Whiteoak 2003). Native asphalt binders such as Trinidad Lake asphalt and Gilsonite are still in use in the paving industry but accounts for small and very specific markets, generally as an additive to straight-run asphalt binder.

24.5.1.2 Asphalt Cement: Physicochemical Properties

Asphalt cement densities at room temperature typically range between 1.01 and 1.04 g/cm³, depending on the crude source and paving grade (Read and Whiteoak 2003). As a rule of thumb, the harder it is, the denser the asphalt binder.

Asphalt binder exhibits a glass transition temperature around −20°C, although it varies over a very wide range from +5°C down to −40°C depending essentially on the crude origin and to a lesser extent on the processing history. The transition range typically spans 30°C–45°C where −20°C corresponds to the typical midpoint value (Lesueur 2009). From a thermodynamic standpoint, asphalt binder is a very viscous liquid at room temperature.

The complexity of asphalt binder chemistry lies in the fact that many different species are present. Typical functional groups found in asphalt binder are shown in Figure 24.11. As an overall descriptor, the chemical nature of the crude oil is generally described as paraffinic, naphthenic, or aromatic if a majority of saturate, cyclic, or aromatic structures, respectively, are present. This classification of the petroleum is sometimes applied to the corresponding asphalt binder. For example, Venezuelan asphalt binders are generally known as naphthenic asphalt binders.

The elemental composition of an asphalt binder depends primarily on its crude source (Mortazavi and Mouthrop 1993). The data in Table 24.2 illustrate this fact. These come from the extensive research

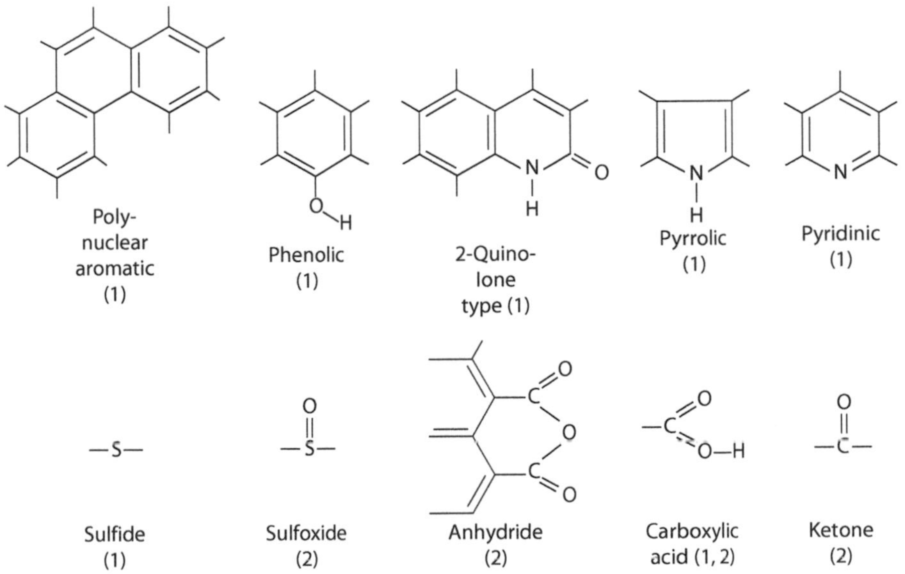

FIGURE 24.11 Functional groups present in asphalt binder. (1) Naturally occurring and (2) formed on oxidative aging. (From Branthaver, J.F. et al., Binder characterization and evaluation—Vol. 2 chemistry, SHRP report A-368, National Research Council, Washington, DC, 1994.)

TABLE 24.2 Elemental Analysis for the Core SHRP Asphalt Binders

		AAA-1	AAB-1	AAC-1	AAD-1	AAF-1	AAG-1	AAK-1	AAM-1
Origin		Canada	USA	Canada	USA	USA	USA	Venezuela	USA
C	wt.%	83.9	82.3	86.5	81.6	84.5	85.6	83.7	86.8
H	wt.%	10.0	10.6	11.3	10.8	10.4	10.5	10.2	11.2
H + C	wt.%	93.9	92.9	97.8	92.4	94.9	96.1	93.9	98.0
H/C	molar	1.43	1.55	1.57	1.59	1.48	1.47	1.46	1.55
O	wt.%	0.6	0.8	0.9	0.9	1.1	1.1	0.8	0.5
N	wt.%	0.5	0.5	0.7	0.8	0.6	1.1	0.7	0.6
S	wt.%	5.5	4.7	1.9	6.9	3.4	1.3	6.4	1.2
V	ppm	174	220	146	310	87	37	1480	58
Ni	ppm	86	56	63	145	35	95	142	36
Mn	g/mol	790	840	870	700	840	710	860	1300

Source: Data from Mortazavi, M. and Moulthrop, J.S., The SHRP materials reference library, SHRP report A-646, National Research Council, Washington, DC, 1993.

effort on asphalt binder chemistry, structure, and properties undertaken as part of the SHRP. The coding (AAA-1, etc.) refers to the various crude sources and grades for the asphalts used within the SHRP (Mortazavi and Mouthrop 1993).

As shown in Table 24.2, asphalt binder mainly consists of carbon (typically 80–88 wt.%) and hydrogen atoms (8–12 wt.%). This gives a hydrocarbon content generally above 90 wt.% with the hydrogen-to-carbon molar ratio H/C around 1.5. This H/C ratio is therefore intermediate between that of aromatic structures (benzene has H/C = 1) and that of saturate alkanes (H/C ~ 2) (Branthaver et al. 1994).

In addition, heteroatoms such as sulfur (0–9 wt.%), nitrogen (0–2 wt.%), and oxygen (0–2 wt.%) are generally present. Traces of metals are also found, usually as metallic porphyrins, the most ubiquitous being vanadium, up to 2000 parts per million (ppm), and nickel (up to 200 ppm) (Mortazavi and Moulthrop 1993, Lesueur 2009).

Sulfur is generally the most abundant heteroatom. In the neat asphalt, this appears in the form of sulfides, thiols, and thiophenes; upon oxidation, many of these groups are converted to sulfoxides and occasionally sulfones. Oxygen is typically present in the form of ketones, phenols, and, to a lesser extent, carboxylic acids. Nitrogen exists typically in pyrrolic and pyridinic structures and also forms amphoteric species such as 2-quinolones (Branthaver et al. 1994), which have both an acid and base functionality on the same molecule.

Given the concentration of polar atoms, functional groups generally do not amount to more than a few multiples of 0.1 mol/L for straight-run asphalt binders (Branthaver et al. 1994); however, their concentration can increase significantly upon aging.

The number-average molecular weight of asphalt binder falls typically in the range 600–1500 g/mol range (Table 24.2—Traxler 1961, Lesueur 2009). The values provided in Table 24.2 were measured by vapor pressure osmometry in toluene and pyridine at 60°C (ASTM D2503).

Average molecular structures for asphalt binder have been proposed (Jennings et al. 1993). The molecular weight of the average structure equals the average molecular weight of the studied asphalt binder, and the atoms are distributed to account for NMR spectral data obtained for the corresponding asphalt binder (Jennings et al. 1992). Two examples are provided in Figure 24.12. Note that asphalt

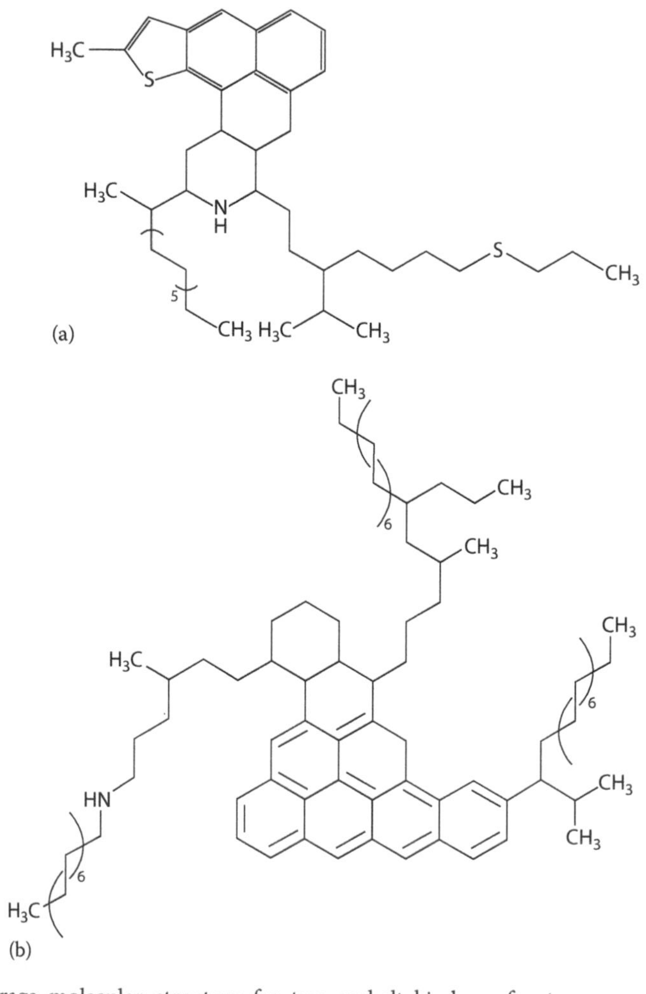

FIGURE 24.12 Average molecular structure for two asphalt binders of extreme compositions: California Coastal (AAD-1) (a) and West Texas Intermediate (AAM–1) (b). (After Jennings, P.W. et al., Binder characterization and evaluation by nuclear magnetic resonance spectroscopy, SHRP report A-335, National Research Council, Washington, DC, 1993.)

binder molecules are not macromolecules in the polymeric sense. As a consequence, care must be taken when trying to compare the properties of polymers to that of asphalt binder, especially when it comes to the modeling of the viscoelastic properties based on this averaged molecular approach.

Given these molecular weights and the proportion of polar atoms, it is clear that only—one to three polar atoms are present on average in each asphalt binder molecule, as illustrated in Figure 24.12.

However, approaching asphalt binder chemistry on a global basis is inadequate when one tries to understand the properties of asphalt binder. Thus, the molecules are generally operationally separated into different chemical families, depending on their solubility in given aromatic or nonpolar solvents.

Using the separation procedure such as ASTM D-4124, asphalt is separated into asphaltenes and maltenes based on the solubility in n-heptane. Maltenes are further separated into resins (polar aromatics), aromatics, and saturates. Subsequently, the composition of asphalt binder is usually given in terms of the relative quantity of its so-called SARA fractions for saturates, aromatics, resins, and asphaltenes. Specific SARA values for a diverse suite of binders are provided in Figure 24.13, and typical ranges for their chemical properties are provided in Table 24.3.

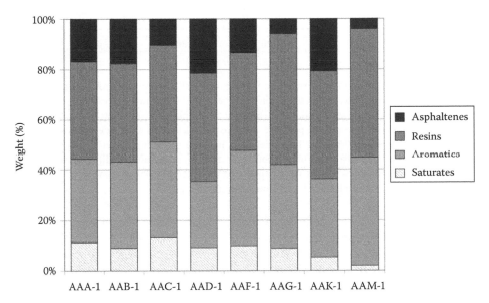

FIGURE 24.13 Separation into SARA fractions for the core SHRP asphalt binders (core sample identifying acronyms same as in Table 24.2). (Data from Mortazavi, M. and Moulthrop, J.S., The SHRP materials reference library, SHRP report A-646, National Research Council, Washington, DC, 1993.)

TABLE 24.3 Chemical Properties of Asphalt Binder and the SARA Fractions: Typical H/C, Elemental Analysis, Number Average Molar Mass and Solvent Used in ASTM D-4124

	H/C	C (%)	H (%)	O (%)	N (%)	S (%)	M_n (g/mol)	Solvent in ASTM D4124
Asphalt cement	1.5	80–88	8–12	0–2	0–2	0–9	600–1500	—
Saturates	1.9	78–84	12–14	<0.1	<0.1	<0.1	470–880	n-Heptane
Aromatics	1.5	80–86	9–13	0.2	0.4	0–4	570–980	Toluene and toluene/methanol 50/50
Resins	1.4	67–88	9–12	0.3–2	0.2–1	0.4–5	780–1400	Trichloroethylene
Asphaltenes	1.1	78–88	7–9	0.3–5	0.6–4	0.3–11	800–3500	n-Heptane insoluble

Source: Data from Lesueur, D., *Adv. Colloid Interface Sci.*, 145, 42, 2009.

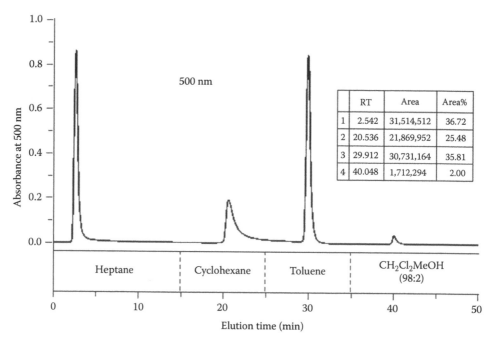

FIGURE 24.14 Asphaltene determinator separation profile for SHRP core asphalt AAK-1 (Boscan) using 500 nm absorbance detectors. The maltenes are eluted with the heptanes, and the asphaltenes are fractionated using solvents of increasing polarity. The cyclohexane peak is a measure of the highly alkyl substituted pericondensed aromatic components of the asphaltenes; the toluene peak is a measure of the less alkyl substituted pericondensed aromatic molecules and the methylene chloride: methanol peak is a measure of the highly pericondensed aromatic molecules. (From Western Research Institute, Fundamental properties of asphalts and modified asphalts, III. Quarterly technical progress report January 1–March 31, 2010 prepared for Federal Highway Administration under contract no. DTFH61-07-D-00005, Western Research Institute, Laramie, WY, 2010.)

The polar asphaltenes are essentially the viscosity builders, whereas the low-polarity components comprising the maltenes appear to control the low-temperature performance.

More recently, researchers have been fractionating the asphaltenes using high-performance column chromatography (WRI 2010). A profile from such a separation is provided in Figure 24.14. Recent work using Fourier transform ion cyclotron resonance mass spectrometry indicates that the upper molecular weight range is below 2000 g/mol (WRI 2010).

Some researchers concluded that asphalt is a simple homogeneous fluid, calling this model the dispersed polar fluid (Christensen and Anderson 1991). Others then claimed that asphaltenes in asphalt binder form a molecular solution based on their solubility parameters (Redelius 2006). All these models have been shown to be incomplete, and the most accepted view now is that asphaltenes form micelles of radius 2–8 nm dispersed in the maltenes (Lesueur 2009).

24.5.1.3 Manufacturing Process

HMA is manufactured in dedicated plants, sometimes continuous, sometimes batch (Lee and Mahboub 2006). In all cases, the manufacturing steps include aggregate drying and then asphalt/aggregate mixing. In order to ensure a good wetting of the aggregate surface by the asphalt binder, a binder viscosity around 200 mPa s is sought. This is generally obtained by having the asphalt binder at temperatures close to 160°C. The aggregate exits the dryer at a similar temperature whereupon it may be stored in heated and insulated silos; therefore, the HMA is manufactured at temperatures close to 160°C.

The HMA is then transported hot by truck to the jobsite. Compaction is performed immediately after laying; the compaction temperature is generally 30°C lower than manufacturing temperature.

Note that additives or innovative processes are now used in order to manufacture the asphalt mix at lower temperatures. They are called warm mixes when the manufacturing temperature is lowered by 30°C–50°C and sometimes half-warm mixes when the temperature is below 100°C. In all cases, the objective is to maintain the properties of HMA while using lower manufacturing temperatures (Button et al. 2007).

24.5.2 Properties

24.5.2.1 Mechanical Properties

Pavements are essentially designed in order to withstand traffic-induced damage. This is done by first calculating the stresses and strains induced by traffic and then computing the stress distribution at any point of the different pavement layers (LCPC-SETRA 1997). From this exercise, it is possible to optimize the thickness and properties of each of the layers such that the corresponding materials can withstand the anticipated stresses for the design period, usually between 10 and 40 years.

As a consequence, the material parameters that are used for pavement design are mostly based on mechanical properties. More precisely, modulus and fatigue resistance are the key parameters. Construction materials are also widely characterized by strength measurements. This is also the case for asphaltic materials, and considerable data are available in the literature on the subject (Taylor 2000).

Modulus is a fundamental mechanical property of materials (Timoshenko 1976). It is the ratio between the stress applied to the material and the resulting deformation (or the opposite if the material is tested under a deformation-controlled mode). Young's or tension modulus is measured using tensional forces (compression thus is a negative tension). It is a Coulomb or shear modulus when measured using shear forces (though pure shear is never really generated).

The mechanical properties of an asphalt mixture are known to be temperature sensitive and loading time dependent; this is a consequence of their viscoelastic behavior (Francken 1998). So the modulus is temperature and time (or frequency) dependent, and it is generally expressed in terms of a complex number, the complex modulus (Figure 24.15).

From a mix formulation standpoint, modulus has been found to peak at optimum asphalt binder content; it also increases with the modulus of the binder and decreases as the air void content increases (Francken 1998). Modulus is of critical importance in the design of pavement layers, because it governs the stress distribution inside each pavement layer. For given load and thickness, higher moduli translate to the presence of lower stresses in each respective layer.

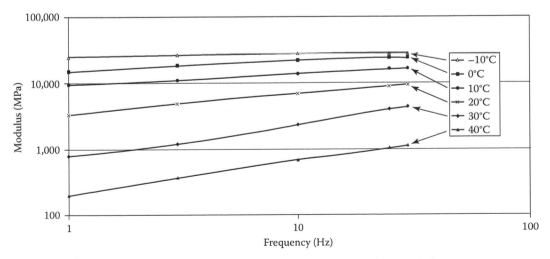

FIGURE 24.15 Complex tensile modulus versus temperature and frequency for an asphalt mixture.

As the modulus is an intrinsic property, that means it should be essentially independent of the specific testing setup, however, small differences are generally observed whenever that modulus is measured in compression, flexion, tension, or indirect tension testing modes. Testing geometry, that is, specimen shape and dimensions, as well as the signal type, that is, that controls the application of deformation or force, or if cyclic, its amplitude and its sinusoidal frequency, also affects somewhat the magnitude of the specific data measured. Therefore, it is necessary to be aware of the measurement conditions and parameters when evaluating those asphalt mixture moduli values. A European standard exists in order to limit the differences (EN 12697-26).

Strength is another important mechanical property of materials (Timoshenko 1976). Here it is specified as the maximum applied stress that causes that material to break apart or fail. Strength for asphalt mixtures is usually measured using either compressive or indirect tensile modes of deformation; these are generally conducted at controlled temperatures close to room temperature.

In general, modulus and strength are somewhat related when measured under the same temperature and loading conditions, although one is an intrinsic property (modulus) and the other strongly depends on specimen shape and dimensions and is therefore not intrinsic (Timoshenko 1976). However, it is a lot easier to measure strength than modulus, hence its reporting predominance in materials engineering. Stiffer asphalt mixes or soils tend to have smaller rutting. In other words, there can be trends between modulus and strength with varying degrees of reliability. For example, a popular relationship used in practice for unbound pavement materials is the correlation between the California bearing ratio (AASHTO T193) and the resilient modulus (AASHTO T 307). However, this relationship is generally used for analyses where the criticality is small and does not warrant materials testing.

Because of this almost constant ratio between strength and modulus, mix variables affect the strength in the same way as the modulus. Therefore, strength is known to peak at an optimum asphalt binder content, to increase with the modulus of the binder and to decrease as the air void content increases (Francken 1998).

As previously detailed, fatigue cracking occurs when the repeated traffic loads progressively damage the asphalt mixtures, generating cracks propagating from the bottom of each layer to its top. As a consequence, fatigue cracking is favored by smaller layer thicknesses or poor adhesion between the successive adjacent layers; both factors promote high tensile strains at the bottom surfaces of each asphalt layer.

From the mix formulation standpoint, fatigue resistance is known to be enhanced by higher asphalt binder contents or the use of high-performance binders (Tangella et al. 1990, Francken 1998). Depending on the method of measuring fatigue, a soft binder can increase the fatigue life (strain controlled) or decrease it (stress controlled).

Fatigue cracking is the main failure mode considered in the design of pavement structures. More precisely, the bituminous layers are designed to be thick enough to ensure that fatigue cracking does not appear until the end of design life, which can vary from 10 to 40 years in Europe (FEHRL-ELLPAG 2004). The design period has even been extended to 50 years for the so-called perpetual pavements (Newcomb et al. 2001).

Interestingly though, as the asphalt binder is fundamentally a very viscous liquid at room temperature, it can heal small cracks. Factors affecting healing rate are those favoring asphalt autodiffusion; higher pavement temperatures and lower viscosity binders will tend to heal these cracks more rapidly (Bhasin et al. 2011). However, healing is not taken into account in current pavement design.

The resistance of mixes to thermal cracking is largely binder dependent. Soft binders increase the cracking resistance; hence they are preferred in Nordic-like climates. As with fatigue cracking, this resistance can also be enhanced by formulating mixes with higher binder contents or high-performance binders (Marasteanu et al. 2004).

24.5.2.2 Physical Properties

Apart from their mechanical properties, some other physical properties are of great importance for asphalt mixture durability.

As mentioned earlier, the wearing course of a pavement must ensure a good skid resistance performance. This is evaluated using appropriate methods; the simplest one uses a pendulum (Masad et al.

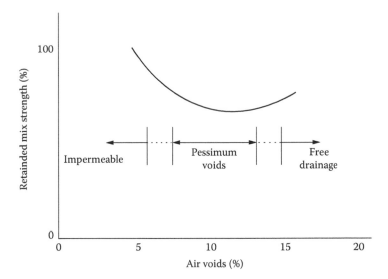

FIGURE 24.16 Concept of pessimum voids stemming from relationship between the strength of mixtures and the air void content. (From Terrel, R.L. and Al-Swailmi, S., Water sensitivity of asphalt–aggregate mixes: Test selection, SHRP report A-403, National Research Council, Washington, DC, 1994.)

2009). More advanced methods use a braking wheel with controlled slip ratio and a controlled surface water layer thickness in order to record the transversal or longitudinal friction coefficient at varying speeds (Brosseaud et al. 2005). Good correlation between skid resistance and braking distance has been found (Wallman and Åström 2001).

Second, reduced water permeability is also quite important. In fact, the wearing course generally provides waterproofing for the rest of the pavement. Exceptions are open-graded friction courses (OGFC; porous asphalt) that are designed to improve the safety of wet surfaces by reducing splash and spray; here the waterproofing function is done below that porous layer. Permeability is largely influenced by the aggregate size. Most asphalt mixes have a nominal maximum aggregate size between 9.5 and 12.5 mm. For asphalt mixtures with less than 4% air voids, the mixtures are generally believed to be impervious to water. When the air voids reach ~6% to 8%, permeability starts to increase significantly from 10^{-5} up to 10^{-4} m/s for continuous-graded mixtures (Huang et al. 1999). The term pessimum voids was coined for this most commonly targeted range of air voids for HMA (Terrel and Al-Swailmi 1994); this is schematically shown in Figure 24.16. In contrast, for fine aggregate mixtures such as −4.75 mm mixtures, these are considered essentially nonporous even with 12% air voids.

Newly constructed pavements of porous asphalt must have permeability above 10^{-4}–$4\cdot10^{-3}$ m/s to be acceptable (Delorme et al. 2007). Interestingly, permeability of continuous-graded mixtures tends to increase with service time; this is attributed to microcracking as well as moisture damage. However, under certain conditions, the permeability of porous asphalt may decrease with service time, especially under low traffic speed conditions, because of dirt and debris accumulation (Gal 1992).

24.6 Distress Mechanisms

24.6.1 Asphalt Chemical Aging

As noted in Section 24.5.1.3, asphalt binder is a complex mixture of mainly hydrocarbon molecules. Some of these molecules may irreversibly evolve through chemical aging. The latter process is generally thought to be a combination of oxidation and polymerization reactions and, to a lesser extent, the evaporation of lighter components (Traxler 1961, Petersen 2009).

This aging can be separated into two conditions. First, during the manufacture of HMA, there is a rapid chemical aging with some loss of volatiles when the hot aggregate is coated with a thin film of asphalt. Second, there is an in situ aging during the service life of the pavement whereby high temperatures and UV exposure accentuate chemical aging (Petersen 2009).

The short-term aging results in the asphaltene content of the asphalt binder typically increasing by 1–4 wt.% (Lesueur 2009) and a doubling of the asphalt binder viscosity. Most of the sulfides are converted to sulfoxides, and there is a jump in the carbonyl content.

During the service life of the pavement that can extend to several decades, the asphalt composition changes (e.g., increase in asphaltenes) and the increases in carbonyl content slow down though remain linear with time. It depends of course on the position of the asphalt inside the pavement, the top layers being more exposed than the base course. Consequently, pavement characteristics and environmental factors make it quite complicated to accurately describe in situ aging.

24.6.1.1 Oxidative Hardening

In chemical terms, aging leads first to a decrease in aromatic content and subsequent increase in resin content, together with a higher asphaltene content. Therefore, it is generally accepted that the aromatics are converted into resins, and the resins in turn generate asphaltenes. The saturates remain essentially unchanged, as could be assessed from their low chemical reactivity. All these changes result in a slightly higher but almost unchanged glass transition temperature. Of note and consistent with changes in the distributions of these fractions is that the width of the glass transition range increases with aging (WRI 2011).

The rate of asphaltene formation was found to be essentially linear with time in RTFOT laboratory experiments at 163°C with 6–7 wt.% asphaltenes formed in 340 min. This linear increase was also observed for in situ aging of asphalt recovered from real pavement sites, with increases between 2 and 10 wt.% in a 90 month period in southern France (Lamontagne et al. 2001).

Asphaltenes produced upon aging may be somewhat different than the initial ones. Increased molecular weights of the asphaltenes upon air blowing were reported, suggesting the presence of polymerization reactions (Lesueur 2009).

Aging results first in the formation of sulfoxides, followed by carbonyls although at a somewhat slower rate. Some of the carbonyls formed may end up as anhydrides and carboxylic acids (Petersen 2009). The sulfoxides are rapidly formed but are thermally unstable. Therefore, they reach a steady-state level that depends not only on the initial sulfur content of the asphalt, but also on the oxygen diffusion into the asphalt. Ketones and carboxylic acids are more stable and do not reach an asymptotic value in laboratory aging experiments. Anhydrides may form once a significant amount of ketones have been generated. They are thought to derive from the oxidation of benzylic carbons at the 1,8 bridgehead position of a naphthalene ring. The in situ aging seems to yield a steady-state level not only of sulfoxides but also of carboxylic acids, as observed after 2 years of service life in Southern France. As a result, the amount of functional groups in asphalt after aging may increase by more than 1 mol/L.

In the pavement, the aging rate will depend both on mixture parameters such as porosity and environment and on depth and local climate (temperature, UV exposure, etc.). Asphalt aging rates are slower with lower air void contents and at colder climates (Petersen 2009). In one set of field monitoring trials, the aging extent as quantified by viscosity ratio (aging index) was almost doubled when air void contents rose from 3%–5% to 7%–9% (Petersen 2009). Note that this last value is generally believed to be at the threshold where interconnected voids are formed, in agreement with the data on permeability. Site location can have a much bigger effect with aging indices increasing from 5 to 10 when going from milder locations (Lake Tahoe with yearly average temperature of 42°F) to hotter U.S. desert conditions (Indio with yearly average temperature of 73°F) (Petersen 2009).

For low-air-void dense mixtures, aging is seen to predominantly occur in the top 1 cm (half inch) of the layer (Petersen 2009). The intensity of aging in this top section is a lot more pronounced than what is generally observed when thick cores are taken from the pavement and asphalt from the topmost section mixes with that from the bottom during the recovery procedure. This effect explains much of the top--down cracking phenomenon in hot regions (see Section 5.3.2.2 on thermal cracking).

In addition, ultraviolet light is known to increase the oxidation process by activating photooxidation reactions. Photooxidation is believed to generate polymerization reactions, not only among the asphaltene molecules, but also for the less polar fractions. Photooxidation is strongly radiation intensity dependent and is almost temperature independent, whereas thermal oxidation is highly temperature dependent.

Finally, and as a consequence of its black color, the mean temperature of newly placed asphalt pavements are generally about 10°C higher than the mean air temperature (Han et al. 2011). Therefore, oxygen diffusion, UV exposure, and thermal effects all combine to increase the aging intensity of the upper part of the asphalt layer.

24.6.1.2 Physical and Steric Hardening

When monitoring the viscosity or any mechanical property of asphalt versus time at negative or room temperature, a slow heat-reversible hardening is indeed observed and generally referred to as physical hardening for low temperatures (below 0°C) and steric hardening for higher temperatures. At low temperature, depending on the crude source and paving grade, the stiffness modulus can double after 3 days at −15°C, while some asphalts do not experience any significant hardening (Anderson et al. 1994). At room temperature, the viscosity increase follows a power law as a function of time with a slope between 0.017 and 0.183, corresponding to a viscosity increase of between 10% and 200% after 2000 h at 25°C, where the magnitude depends on bitumen type (Lesueur 2009).

Physical hardening was first reported by SHRP researchers (Anderson et al. 1994), but steric hardening was observed by Traxler and Schweyer as early as 1936 when they showed that it is not a consequence of chemical aging as described earlier, and especially due to its reversible nature (Traxler and Schweyer 1936).

Physical hardening was initially thought to be largely a consequence of free volume contraction (Anderson et al. 1994), but this interpretation is not consistent with the observation that it even occurs at temperatures above the glass transition temperature and that its extent is highly asphalt dependent, whereas all asphalts exhibit quite similar features in terms of mechanical behavior in the vicinity of the glass transition temperature. A more acceptable explanation was proposed relating it to the crystallization of the waxy components of the asphalt (Claudy et al. 1992). Still, the relationship between wax content and physical hardening is not straightforward (Edwards et al. 2006).

Similarly, the proposed interpretation for steric hardening generally relies on the sol–gel effect related to the buildup of a stronger asphaltene network with time (Traxler and Schweyer 1936); however, this effect might indeed be consistent with crystallization kinetics showing that both steric and physical hardening are in fact due to the slow kinetics of wax crystallization (Lesueur 2009).

24.6.1.3 Exudative Hardening

Exudative hardening is one possible mechanism for the accelerated hardening of the asphalt binder. This occurs when mixes are made with porous aggregate. In such cases, some of the lighter compounds slowly diffuse into the aggregate, leaving a somewhat stiffer binder with reduced film thickness on the aggregate. Both factors contribute to increasing the cracking probability of the mixture.

Lime treatment of porous aggregates can diminish this effect by allowing calcium carbonate to precipitate within the aggregate porosity. Moreover, porous aggregates are generally harder to fully dry at the asphalt plant, thus the presence of lime, especially quicklime, can reduce the influence of water that would otherwise be trapped inside the aggregate, causing a "soup" phenomenon that prevents a proper laying process.

24.6.2 Microb/Vegetal Attack

Asphalt binder is generally considered as being a quite stable material with respect to biodegradability. In fact, asphalt itself is the result of the slow degradation under high pressure and temperature of organic compounds (Durand 2003). Still, it is not chemically inert as previously noted in the discussion on chemical aging. As a hydrocarbon, it is not surprising to find that biological agents can have an effect on it. This effect is very limited from a practical standpoint, and to our knowledge, no road failures due to asphalt biodegradability have been recorded so far. Asphalt-containing materials from the archeological ages confirm that the probability of biodegradation under normal outside conditions is quite low (Connan 1999).

Even though biodegradation is probably not so relevant for asphalt durability in pavements, it has been studied carefully. It was found that aerobic degradation occurs 100 times faster than anaerobic degradation under optimum laboratory conditions, where measured degradation rates were of the order of 27–55 g of asphalt/m²/year (Wolf and Bachofen 1991). *Pseudomonas aeruginosa* and species of *streptomyces* and *alcaligenes* were identified as the main microorganisms responsible for the aerobic degradation. Still, only the light components of asphalt are biodegradable, and a large part, consisting mostly of resins and asphaltenes, is resistant to biodegradation (Potter and Duval 2001). In more representative situations, but still highly favorable for microbial growth (burial under active compost), only the first millimeter of an asphalt lining was found to degrade after 3 years in a canal (Read and Whiteoaks 2003).

Fungi, molds, and yeasts have been observed to occasionally develop on asphalt roofs under hot and humid climates (Read and Whiteoaks 2003). However, their growth remains largely unreported in asphalt pavements.

Plants can grow in contact with bituminous materials (Read and Whiteoaks 2003). Under normal pavement conditions, the low permeability of asphalt mixtures makes it unlikely that plants will develop inside the paved area. However, for porous asphalt mixes and highly damaged pavements, with numerous cracks and potholes, water and dirt can accumulate and make a suitable environment for plant growth.

24.7 Enhancing Longevity Performance of Asphalt

24.7.1 Improved Mix Design and Construction

24.7.1.1 Mix Design

Mix design consists of defining the following:

- The aggregate type and relative quantity
- Asphalt binder type and relative quantity
- Additive(s) type and content, when applicable
- Manufacturing temperature
- Final relative amount of air voids of the in-place mixture (density)

Adequate selection of these parameters helps to ensure that the final properties of the mixtures are the ones sought. Here, specifications come into play, and depending on the local practice and on the final use of the material (base course, wearing course, etc., for high/low traffic road), specific specification metrics are generally chosen. These are based on lab-scale tests to measure

- Mechanical properties (modulus and/or rutting resistance and/or fatigue and/or strength)
- Moisture damage and/or frost resistance
- Density

Other important properties such as permeability, noise emission, skid resistance, etc., are generally not measured at the lab scale and are therefore left out of the formulation process. Still, an appropriate choice of adequate mix design and components ensures that the desired properties will be obtained in the field, due in part to a thorough experience linking mixture type to in situ properties.

In all cases, the first way to increase durability is to ensure that the specifications take into account minimization of the relevant failure modes for the in-service conditions and use situations at stake (e.g., environmental factors, traffic load). Then, provided that the specifications are sound (this means that the testing methods are representative of the field behavior, which is not always the case), ensure that tight quality controls are in place to assure that the manufactured and placed product meet these specifications.

Specific asphalt plant controls include means to validate the quality of the raw materials, ensure proper and reproducible operational conditions (raw materials content, manufacturing temperature, etc.), and validate asphalt mixture composition.

In terms of mix design, it is generally accepted that the higher the asphalt binder content, the higher the durability. However, for contents above the optimal level, excessive amounts can lead to conditions inducing rutting and thus must be avoided. But for a given set of performance metrics, including rutting, increasing the asphalt binder content generally increases the durability. In fact, this helps explain the very high durability performance shown by mixtures such as split mastic asphalts (SMAs).

24.7.1.2 Construction

On the jobsite, additional controls further assure that the mixture complies with the requirements of the formulation study. It generally suffices to base them on control of

- Asphalt mixture composition
- Processing temperatures (especially during compaction)
- Thicknesses of the layers
- In-place density

If all these steps are performed correctly, together with adequate material selection and pavement design, then premature failures should be avoidable.

Thickness control is very important in pavement design. As a matter of fact, too thin a layer will lead to premature fatigue failure, whereas too thick a layer is also not desirable, as very thick layers can be challenging to compact.

Density is critical, because the desired mixture properties are obtained only when the design density is attained. A lower density than expected generally results in a lower moisture resistance, lower modulus, and lower fatigue resistance, whereas an over-compacted mixture is generally more sensitive to rutting.

Note that pavement design generally implies that the overlaying layers of asphaltic materials are glued together, so that stresses propagate through all the layers. This is obtained through the tack coat, usually made of a bituminous emulsion with application of 300–500 g/m² of residual asphalt binder. However, if the tack coat is missing or has been poorly applied, the top layers will experience much higher stresses at their bottom surface, generating a premature fatigue failure in the very first years following construction. New solutions such as the use of harder grades of asphalt in the tack coat or the spreading of a protecting layer of diluted milk of lime prevent the tack coat from being tracked by traffic off the jobsite.

Finally, it is well known in the industry that the transverse joints formed between the ends of adjacent laid asphalt layers form weak spots that can easily degrade into potholes. One simple way to limit their occurrence is to place extension devices along the full width of the pavement. Note that this needs a constant asphalt mix supply in order to avoid unevenness issues. Then, while joints still exist, the application of a tack coat emulsion on the cold surface can be used to strengthen the bond. For longitudinal joints, upon laying a new layer, an excess 3–4 cm of asphalt mixture on the old existing surface is recommended. This excess material is then compacted from the cold surface, with the roller mostly on the old surface and only a few centimeters on the new one, leaving a nicely closed joint. A last recommendation regarding transverse joints is to avoid superimposing joints in subsequent layers. A minimum separation of 20 cm between joints guarantees that the possible degradation of one joint will not contaminate other nearby joints.

24.7.1.3 Warm Mix Asphalts

As previously mentioned, warm mix asphalts (WMA) are increasingly used today. Because of their very recent development, with few sections more than 6 years old, the impact of these new technologies on durability remains to be observed, as there are both positive and negative aspects that make it difficult to predict whether or not WMA durability will be similar to that of HMA.

First, the current WMA technologies facilitate the compaction of the mix during laydown. Second, the decreased manufacturing temperature means that the asphalt binder experiences less aging during fabrication. With aging proceeding in an autocatalytic way, WMA would be less sensitive to aging. Consequently, both factors favor higher WMA durability.

On the negative side, decreasing the manufacturing temperature tends to decrease the moisture resistance (Traxler 1961, Aschenbrener 1995). This is particularly the case when using porous aggregates, which may not be completely dried. Additives such as surfactants or hydrated lime can be used to counteract this effect, but only time will tell whether or not the problem is definitely solved through their use.

24.7.1.4 Recycling

The asphalt mixtures that are milled or removed during resurfacing, rehabilitation, and reconstruction operations are readily recyclable. With increasing binder costs and aggregate availability issues, the inclusion of this RAP in new asphalt mixtures is becoming the norm. Currently, most State highway agencies in the United States allow the use of greater than 30% RAP, though the actual utilization of RAP at this level is considerably less (Copeland 2011).

Utilization of RAP sometimes includes a crushing step in order to limit the size of the particle agglomerations. In most cases, addition of 10–20 wt.% RAP has little effect on the overall mixture formula and only slightly changes the final properties (Dunn 2001). This can be performed with limited equipment modifications at the plant.

High rates of RAP can also be used, even attainments close to 100% recycling, but this requires special equipment and very tight control of the RAP source (Dunn 2001); applications for these high RAP mixes have generally been limited to lower-volume roads, lower pavement lifts, or shoulders.

The durability of high RAP mixes is largely determined by the ability to control the aggregate gradation and address a number of binder-related issues. First, the RAP binder is considerably stiffer than the original binder. The compatibility or the ability of the new and old binders to blend during manufacturing of the mix may affect the cohesive strength of the binder. This is likely to be more of an issue for WMA and RAP, which are produced at lower temperatures. Second, determining the optimum binder content of these mixes is not straightforward. A portion of the RAP binder is considered in this determination. The amount of RAP binder performing as a black rock (not blending) but merely acting as a precoating of the aggregate is a subject of investigation. Ascribing too large an amount will result in binder lean, less durable pavements, whereas an excessive amount of binder will increase the propensity for rutting.

24.7.2 Additives

Increasing demands on the pavement and implementation of the Superpave system (discussed in Section 24.4.4.1) have increased the use of unconventional binder additives, air blowing, blending, and chemical modification. PG operating range spreads (limiting high temperature minus limiting low temperature) greater than 95°C are not achievable using most crudes (Youtcheff and Jones 1994) and require some form of asphalt binder modification.

24.7.2.1 Polymer

Polymer additives are now commonly used to modify asphalt binders (PIARC 1999, McNally 2011). The idea of mixing asphalt binder and natural rubber dates back to 1843 (Thomson 1964). There the original objective was more to find a substitute for rubber than to modify asphalt binder. At the turn of the twentieth century, field trials with rubber-modified asphalt binders were performed and continued for several decades until polymer-modified asphalt binders gained significant commercial interest in the late 1970s (Thomson 1964).

The typical amount used today ranges between 3 and 6 wt.%. Most of the currently used polymers include elastomers, plastomers, and reactive polymers.

- Elastomers include natural or synthetic rubbers such as styrene–butadiene (SB) copolymers (random SBR, diblock SB, or triblock SBS) and others (e.g., styrene–isoprene)
- Plastomers include ethylene–vinylacetate random copolymers or related molecules (e.g., ethylene–methacrylate, ethylene–butylacrylate), polyolefins such as polyethylenes and polypropylenes
- Random terpolymers comprising ethylene, glycidyl methacrylate (GMA), and an ester group (usually methyl, ethyl, or butylacrylate) represent reactive polymers. These are generally referred to as reactive ethylene terpolymers because of the chemical reactions that are thought to occur between functional groups on the asphaltenes and the polymer (Polacco et al. 2004).

Many grades of each of these polymers are available (PIARC 1999, McNally 2011), but it is generally sufficient for the asphalt binder industry to characterize a polymer by its monomer composition and molar mass or, in the case of reactive polymers, the number of reactive groups per repeat unit.

Different processes are in place to produce modified bituminous mixes (PIARC 1999). Polymers may be added either directly to the asphalt binder prior to the mixing with the aggregate (wet process) or to the mix at the same time asphalt binder is blended with the aggregates (dry process). These dry process mixes have different properties from those obtained via the wet process. While some details might apply to the dry process, the following discussion focuses only on the wet process product.

Adding polymer to asphalt binder greatly expands its PG temperature range, namely, the low temperature can be established and the high temperature greatly increased. More specifically, the applicable rule of thumb states that for every 1% of added polymer, there is a 2°C gain in the high-temperature PG (PIARC 1999). On the low-temperature end, the rule becomes more that 1% of added polymer produces a 1°C rise in low-temperature PG for SB-type modifiers. Since the PG classes are based on 6°C steps, the typical 3% polymer content modification generally allows a one-step high-temperature-class increase leaving the low-temperature-class step sometimes unchanged.

The use of crumb rubber originating from the recycling of scrap tires is another environmentally friendly polymer modification technology with growing interest due to the large amount of old tires to dispose of (Caltrans 2003) and their low cost relative to conventional polymer modification and even the binder itself (as a low-cost binder extender).

The gains in high-temperature PG with the level of loading are not as marked as noted earlier for conventional polymers. This assumes that the rubber has been well dispersed within the asphalt binder, and with the caveat that given the coarse particle size (at least a few 100 μm sometimes up to 1 mm) of crumb rubber, rheological testing with the tools currently used for asphalt binder evaluation might not be adequate (Kim et al. 2001).

24.7.2.2 Acids

Polyphosphoric acid (PPA) $[H_{n+2}P_nO_{3n+1}]$ is a polymer of orthophosphoric acid $[H_3PO_4]$. This has been used for over 30 years to stiffen asphalt binders (Baumgardner 2010). Recently, PPA has been used to supplement polymer modification to improve the handling and performance. While PPA is hygroscopic, the addition of small amounts (<0.75%) has not been demonstrated to adversely affect the moisture resistance of mixes (Arnold et al. 2009).

24.7.2.3 Hydrated Lime

Hydrated lime has been incorporated in asphalt mixtures since their very beginning. At the end of the nineteenth century, the National Vulcanite Company paved roads in Washington, DC, and Buffalo, NY, with a proprietary asphalt mixture called Vulcanite containing hydrated lime (Lesueur 2010). Other proprietary asphalt mixtures of the time using hydrated lime as a filler included Warrenite and Amiesite (Lesueur 2010). A few decades later, hydrated lime was still not widely used and was merely listed as a possible filler component in asphalt mixtures in the United States (Asphalt Institute 1947). In the early 1950s, in France, Duriez and Arrambide recommended the use of hydrated lime as a way to improve asphalt binder–aggregate adhesion (Duriez and Arrambide 1954). However, hydrated lime did not experience a renewed interest until the 1970s in the United States, partly as a consequence of a general decrease in asphalt binder quality due to the petroleum crisis of 1973 when moisture damage and frost became some of the more pressing pavement failure modes of the time (Hicks 1991).

The various additives to asphalt mixtures available to limit moisture damage were tested both in the laboratory and in the field, and hydrated lime was observed to be the most effective additive (Hicks 1991). As a consequence, hydrated lime is now specified in many states, and it is estimated that 10% of the asphalt mixtures produced in the United States now contain hydrated lime (Hicks and Scholz 2003).

Given its extensive use in the past 30 years in the United States, hydrated lime has been seen to be more than an additive to mitigate moisture damage (Little and Epps 2001, Lesueur 2010). Hydrated lime is known to reduce chemical aging of the asphalt binder. Furthermore, it generally stiffens the mechanical properties of the asphalt mixture, which has a positive impact on the rutting resistance of the mixtures. In parallel, the resistance to cracking is also mentioned to be improved.

The National Lime Association survey of 2003 gave some precise numbers on the changes in asphalt mixture durability associated with the use of hydrated lime (Hicks and Scholz 2003). The survey was performed by sending a questionnaire to all the agencies that are experienced in the use of hydrated lime. The full results are given in Table 24.4.

From these data, it can be seen that the life expectancy for all types of roads is increased by 2–10 years when hydrated lime is added. Given that the life expectancy of untreated roads ranges from 5 to 20 years, the relative improvement goes from 20% up to 50% higher durability.

The European experience is not yet as developed as in the United States, but the beneficial effect of hydrated lime on asphalt mixture durability has also been largely reported. As an example, the Sanef motorway company, managing 1740 km of highways in Northern France, currently specifies hydrated lime in the wearing courses of its network (Raynaud 2009). Sanef observed that hydrated lime-modified asphalt mixtures have a 20%–25% higher durability (Raynaud 2009). Similar observations led the Netherlands to specify hydrated lime in porous asphalt, a type of mix that now covers 70% of the highways in the country. As a result, hydrated lime is being increasingly used in asphalt mixtures in most European countries, in particular Austria, France, the Netherlands, the United Kingdom, and Switzerland (Lesueur 2010).

24.7.2.4 Others (Liquid Antistrips, Fibers, Epoxy Asphalts)

Other additives are also used in asphalt mixtures to improve their durability; these include liquid antistrips, fibers, and epoxy asphalts. Adhesion promoters, such as polyamines and polyphosphates often referred to as antistrips, are used when the moisture resistance of the mixture is not sufficient. Although each formula has a distinct dosage, a typical content is 0.5 wt.% based on asphalt binder. Adhesion promoters are generally surface-active agents. The polar head is typically amine based, but other chemistries are available. A key issue is the thermal stability of the molecule in the asphalt, and care must be taken not to store the treated asphalt binder for a day or more at elevated temperatures.

Fibers are used in some mixtures, especially in the German SMAs, to prevent draindown of the asphalt binder. The fiber enables the use of thicker binder films (i.e., higher binder content mixes) with no risk of binder drainage. Fibers are also used as a reinforcement aid to improve the fatigue performance of the mix (Button and Hunter 1984, McDaniel and Shah 2003). In an accelerated pavement testing experiment (Gibson et al. 2012), the fiber-modified section was found to be considerably more durable than that of the control.

The use of epoxy asphalt has been limited to bridge decks and small critical surfaces (i.e., dangerous curves, and intersections) due to its higher cost. This is the subject of an ongoing study of Long-Life Surfaces for Busy Roads by the OECD (2008), where its application is as a thin asphalt mix overlay and for use in an OGFC.

TABLE 24.4 Life Expectancy (Years) of Hydrated Lime Treated and Untreated Mixes in the United States

Agency	Lime Treated			Non-lime Treated		
	10%	Average	90%	10%	Average	90%
Interstate roads						
Arizona	13	15	17	10	12	14
California	8	10	12	6	8	10
Colorado	8	10	12	6	8	10
Georgia	7	10	15		N/A	
Mississippi	7	10	15		N/A	
Nevada[a]	7	8	9	3	4	7
Oregon	10	15	20	8	12	15
South Carolina	10	12	15		N/A	
Texas	8	12	15	7	10	12
Utah	15	20	25	7	10	15
State roads and U.S. highways						
Arizona	18	20	22	15	17	20
California	8	10	12	6	8	10
Colorado[a]	8	10	12		8	
FHWA	15	20	25		N/A	
Georgia	8	10	14		N/A	
Mississippi	12	15	17		N/A	
Nevada	10	12	14	6	8	10
Oregon	15	17	20	8	12	15
South Carolina	8	10	12		N/A	
Texas	10	12	15	8	10	12
Utah	15	20	25	7	10	15
Low-volume roads						
Arizona	20	25	30	15	20	25
California		N/A			N/A	
Colorado	10	12	15	8	10	12
FHWA	15	20	25		N/A	
Georgia	8	10	15	8	10	15
Mississippi	12	15	17		N/A	
Nevada	18	20	22	12	15	18
Oregon	15	20	25	7	10	15
South Carolina	10	15	20		N/A	
Texas	8	12	15	7	10	15
Utah	7	10	15	3	5	7

Source: Hicks, R.G. and Scholz, T.V., *Life Cycle Costs for Lime in Hot Mix Asphalt*, 3 volumes, National Lime Association, Arlington, VA, 2003.

24.8 ASTM/AASHTO/EN Method Specifications and Tests

24.8.1 Aggregate

- Aggregate Durability Index, AASHTO 210-10.
- Resistance to Degradation of Small-Size Coarse Aggregate by Abrasion and Impact in the Los Angeles Machine, AASHTO 96-02; ASTM C131 and C535, EN 1097-2.

- Resistance of Coarse Aggregate to Degradation by Abrasion in the Micro-Deval Apparatus, AASHTO T327-09; EN 1097-1.
- Polishing Stone Value, EN 1097-8.
- The Qualitative Detection of Harmful Clays of the Smectite Group in Aggregates Using Methylene Blue, AASHTO T 330-07.
- Sizes of Aggregate for Road and Bridge Construction, AASHTO M43-05.
- Uncompacted Void Content of Fine Aggregate, AASHTO T 304.

24.8.2 Asphalt Binder

- Performance-Graded Asphalt Binder, AASHTO M 320, EN 12591.
- Performance-Graded Asphalt Binder Using Multiple Stress Creep Recovery (MSCR) Test, AASHTO MP 19-10.
- Determining the Rheological Properties of Asphalt Binder Using a Dynamic Shear Rheometer (DSR), AASHTO T315.
- Determining the Flexural Creep Stiffness of Asphalt Binder Using the Bending Beam Rheometer (BBR), AASHTO T313.
- Determining the Rheological Properties of Asphalt Binder in Direct Tension (DT), AASHTO T314.
- Multiple Stress Creep Recovery (MSCR) Test of Asphalt Binder Using a Dynamic Shear Rheometer, AASHTO TP 70.
- Determining the Cracking Temperature of Asphalt Binder Using the Asphalt Binder Cracking Device, AASHTO TP92-11.

24.8.2.1 Aging

- Effect of Heat and Air on a Moving Film of Asphalt Binder (Rolling Thin-Film Oven Test), AASHTO T240-09; ASTM D2872; EN 12607.
- Accelerated Aging of Asphalt Binder Using a Pressurized Aging Vessel (PAV), AASHTO R28; ASTM D 6521; EN 14769.

24.8.2.2 Moisture

- Boiling Water Test, ASTM D-1075.

24.8.2.3 Chemistry

- Terminology, EN 12597.
- Corbett Separation, D4124.
- Solubility of Bituminous Materials, AASHTO T44-03.
- Test Method for Molecular Weight of Hydrocarbons by Thermoelectric Measurement of Vapor Phase, ASTM D2503.
- SARA Separation, ASTM D4124.

24.8.3 Asphalt Mix

- California Bearing Ratio, AASHTO T 193.
- Resilient Modulus Test, AASHTO T 307.
- Determining Dynamic Modulus of Hot mix Asphalt (HMA), AASHTO T-342-11; EN 12697-26.
- Determining the Creep Compliance and Strength of Hot Mix Asphalt (HMA) Using the Indirect Tensile Test Device, AASHTO 322-07.
- Compressive Strength of Hot Mix Asphalt, AASHTO T167-10.
- Dynamic modulus test, ASTM D3497.

24.8.3.1 Aging

- Mixture Conditioning of Hot Mix Asphalt (HMA), AASHTO R30.
- Cantabro Test, EN 12697-17.

24.8.3.2 Moisture

- Resistance of Compacter Hot Mix Asphalt (HMA) to Moisture-Induced Damage (Indirect Tensile Strength Ratio with one Freeze–Thaw Cycle—Lottman 1978), AASHTO T283.
- Hamburg Wheel-Track Testing of Compacted Hot Mix Asphalt, AASHTO 324; EN 12697-22 Method B Under Water.
- Freeze–Thaw Pedestal Test (Plancher et al. 1980).
- Resistance of Compacted Hot Mix Asphalt (HMA) to Moisture-Induced Damage, AASHTO 283–07.
- Retained Tensile Strength or Indirect Tensile Strength Ratio, ASTM D4867; EN 12697-12-Method A.
- Immersion/Compression Test; AASHTO T165; ASTM D1075; EN12697-12 Method B (Duriez).

24.8.3.3 Rutting

- Determining the Dynamic Modulus and Flow Number for Hot Mix Asphalt (HMA) Using the Asphalt Mixture Performance Tester (AMPT), AASHTO TP79-10.
- Developing the Dynamic Modulus Master Curves for Hot Mix Asphalt (HMA) Using the Asphalt Mixture Performance Tester (AMPT), AASHTO PP61-10.
- Determining the Creep Compliance and Strength of Hot Mix Asphalt (HMA) Using the Indirect Tensile Test Device, AASHTO T322; EN 12697-22.
- Rutting Resistance of Asphalt Mixtures, EN 12697-22.

24.8.3.4 Thermal Cracking/Fatigue

- Determining the Fatigue Life of Compacted Hot mix Asphalt (HMA) Subjected to Repeated Flexural Bending, AASHTO T321-07.
- Resilient Modulus by Indirect Tension (IDT) Subjected to Repeated Flexural Bending, AASHTO TP31.
- Indirect Tensile Strength/Resilient Modulus Test, ASTM D4123.
- Thermal Stress Restrained Specimen Tensile Strength (TSRST), AASHTO TP10.
- Determining the Fatigue Life of Compacted Hot Mix Asphalt (HMA), AASHTO T321.
- Fatigue Resistance of Asphalt Mixtures, EN 12697-24.

24.9 Summary

In a world concerned with environmentally friendly construction technologies and sustainability, asphalt durability takes on an added significance. Asphalt mixtures are now very technical materials obtained by careful selection of raw materials among which are asphalt binder and aggregate and then mixed in dedicated computer-controlled plants.

Due to the rising cost of asphalt binders, there is a keen interest in the concept of sustainable asphalt mixtures. However, the environmental aspects are mostly covered by (1) minimizing wastes through the recycling of RAP in new production; (2) supplementing the asphalt through the use of other by-products such as RAS, sulfur, and recycled motor oil; and (3) using lower manufacturing temperatures in the manufacturing of the so-called warm or even semi-warm mixtures, thus decreasing CO_2 emissions and energy consumption. Hence, the basic principle of waste management, that is, preventing and reducing waste generation, is not fully taken into account by the industry, and more durable asphalt pavements should be a clear objective.

Therefore, this chapter addresses the main factors affecting the durability of asphalt pavements, namely, their ability to maintain satisfactory performance and structure in long-term service under increasingly demanding conditions. Considering that asphalt pavement deterioration is attributed to combined effects of chemical aging, climate, and traffic, some guidelines emerge in order to enhance their durability. First, a careful selection of the ingredients, especially the asphalt binder, ensures a proper function (climate and loading) and resistance to chemical aging. Tests exist that simulate long-term aging, and they are therefore highly recommended to be part of any binder specification.

Then, the resistance to climate, in particular moisture damage, must be evaluated using appropriate test methods on the asphalt mixture and setting sound specifications. Finally, the resistance to traffic is first obtained by an appropriate pavement design and then by the validation of the mechanical properties of the mixture. Still, the construction stage is critical in guaranteeing good pavement durability, because too thin a layer, badly executed joints or lack of tack coat are critical factors that can lead to premature failure.

From a mixture formulation standpoint, several solutions are available to the designers in order to improve any of the mentioned key properties. In particular, hydrated lime and polymers have been the most widely used additives to enhance the durability of asphalt mixtures.

In all cases, a tight control of the raw material properties (asphalt binder, aggregate, recycled materials, and additives), of the manufacturing process (quantities and qualities of materials, and temperature), and of the laying operations (thickness and density) is necessary in order to obtain long-lasting roads.

Improvements in the durability of asphalt are arising from multiple fronts. Improved testing and classification of asphalt binders are enhancing the screening and identification of poor performing binders for a given climatic region or traffic loading. Additives and polymer modification are increasingly being used to target potential distresses, and together with improvements in asphalt compaction, this is leading to more durable binders and pavements. Ultimate goals are to be able to predetermine the asphalt pavement's lifetime and utilize pavement preservation strategies to minimize or offset embrittlement of the pavement.

Thus from environmental as well as economic perspective, consideration of asphalt durability is paramount to the sustainability of asphalt and ensuring that asphalt remains a cost-effective paving material.

Nomenclature

Asphalt binder: an asphalt-based cement that is produced from petroleum residue either with or without the addition of nonparticulate organic modifiers.
Asphalt modifier: any material of suitable manufacture that is used in virgin or recycled condition and that is dissolved, dispersed, or reacted in asphalt binder to enhance its performance.

Acknowledgment

The authors wish to acknowledge Drs. Raj Dongre and Nelson Gibson for their critical review of the manuscript.

References (ASTM, AASHTO, and EN Specs and Tests Listed Separately in Section 24.8)

AASHO (American Association of State Highway Officials). 1962. *AASHO Road Test Report 5*, Highway Research Board special report 61E, Publication 954, National Research Council, Washington, DC.

Abdallah, I., E. Mahmoud, S. Nazarian, and E. Masad. 2008. Quantifying the role of coarse aggregate strength on resistance to load in HMA for blended aggregates, Research report 0-5268-3, Center for Transportation Infrastructure Systems, The University of Texas El Paso, El Paso, TX.

Abraham, H. 1960. *Asphalt and Allied Substances*, 6th Edn., Van Nostrand, New York.

Anderson, D.A., D.W. Christensen, H.U. Bahia et al. 1994. *Binder Characterization and Evaluation, Volume 3: Physical Characterization*, SHRP report A-369, National Research Council, Washington, DC.

Arnold, T., S. Needham, and J. Youtcheff. 2009. The use of phosphoric acid as a modifier for hot mix asphalt, Transportation Research Board, annual meeting CD-ROM.

Aschenbrener, T., R. McGennis, and R. Terrel. 1995. Comparison of several moisture susceptibility tests in pavements of known field performance. *Journal of the Association of Asphalt Paving Technologists*, 66, 163–208.

Asphalt Institute. 1947. *The Asphalt Handbook*, Asphalt Institute, Lexington, KY.

Asphalt Institute. 1994. *Superpave: Performance Graded Asphalt Binder Specification and Testing*/SP-1, Asphalt Institute, Lexington, KY.

Asphalt Institute. 2001. *Superpave Mix Design*, Asphalt Institute, Lexington, KY.

Baumgardner, G.L. 2010. Why and how of polyphosphoric acid modification—An industry perspective. *Journal of the Association of Asphalt Paving Technologists*, 79, 663–678.

Bell, C. 1989. Summary report on aging of asphalt–aggregate systems, SHRP report A-305, National Research Council, Washington, DC.

Bhasin, A., R. Bommavaram, M.L. Greenfield et al. 2011. Use of molecular dynamics to investigate self-healing mechanisms in asphalt binders. *Journal Materials Civil Engineering*, 23, 485–492.

Branthaver, J.F., J.C. Petersen, R.E. Robertson et al. 1994. Binder characterization and evaluation—Vol. 2 Chemistry, SHRP report A-368, National Research Council, Washington, DC.

Brosseaud, Y., V. Le Turdu and Y. Delanne. 2005. Adhérence des revêtements de chaussées routières. *Bulletin Laboratoires Ponts Chaussées*, 255, 71–90.

Burstyna, I., H. Kromhouta, and P. Boffettab. 2000. Literature review of levels and determinants of exposure to potential carcinogens and other agents in the road construction industry. *Journal American Industrial Hygiene Association*, 61, 715–726.

Button, J., C. Estakhri, and A. Wimsatt. 2007. A synthesis of warm-mix asphalt, TTI report SWUTC/07/0-5597-1, Texas Transportation Institute, College Station, TX.

Button, J. and T. Hunter. 1984. Synthetic fibers in asphalt paving mixtures. FHWA report TX-85-73-319-1F, Arlington, VA.

Caltrans (California Department of Transportation). 2003. *Asphalt Rubber Usage Guide*, Caltrans, Sacramento, CA.

Christensen, D.W. and D.A. Anderson. 1991. Rheological evidence concerning the molecular architecture of asphalt cements. *Proceedings Chemistry of Bitumen: Rome*, 2, 568–595.

Claudy, P., J.-M. Létoffé, G.N. King et al. 1992. Characterization of asphalt cements by thermomicroscopy and differential scanning calorimetry: Correlation to classic physical properties. *Fuel Science Technology International*, 10, 735–765.

Connan, J. 1999. Use and trade of bitumen in antiquity and prehistory: Molecular archaeology reveals secrets of past civilizations. *Philosophical Transactions of the Royal Society B*, 354, 33–50.

Copeland, A. 2011. Reclaimed asphalt pavement in asphalt mixtures: State of the practice, Report FHWA-HRT-11-021, Federal Highway Administration, Arlington, VA.

Corbett, L.W. 1969. Composition of asphalt based on generic fractionation using solvent deasphalteneing, elution-adsorption chromatography and densiometric characterization. *Analytical Chemistry*, 41, 576–579.

Curtis, C.W., Y.W. Jeon, and D.J. Clapp. 1989a. Adsorption of asphalt functionalities and oxidized asphalts on aggregate surfaces. *Fuel Science and Technology International*, 7, 1225–1228.

Curtis, C.W., Y.W. Jeon, D.J. Clapp et al. 1989b. Adsorption behavior of asphalt functionalities, AC-20, and oxidized asphalts on aggregate surfaces. *Transportation Research Record*, 1228, 112–127.

D'Angelo, J., E. Harm, J. Bartoszek et al. 2008. Warm-mix asphalt: European practice, Report FHWA-PL-08-007, Federal Highway Administration, Arlington, VA.

D'Angelo, J., R. Kluttz, R.N. Dongre et al. 2007. Revision of the Superpave high temperature binder specification: The multiple stress creep recovery test. *Journal of the Association of Asphalt Paving Technologists*, 76, 123–162.

Delorme, J.-L., C. de La Roche, and L. Wendling, Eds. 2007. *LCPC Bituminous Mixtures Design Guide*, LCPC, Paris, France.

Di Benedetto, H., C. de La Roche, H. Baaj et al. 2003. Fatigue of bituminous mixtures. *Materials and Structures*, 37, 202–216.

Dunn, L. 2001. *A Basic Asphalt Recycling Manual*, Asphalt Recycling and Reclaiming Association, Annapolis, MA.

Durand, B. 2003. A history of organic geochemistry. *Oil & Gas Science and Technology—Review IFP*, 58, 203–231.

Duriez, M. and J. Arrambide. 1954. *Liants Hydrocarbonés*, Dunod, Paris, France.

Edwards, Y., Y. Tasdemir, and U. Isacsson. 2006. Influence of commercial waxes and polyphosphoric acid on bitumen and asphalt concrete performance at low and medium temperatures. *Materials and Structures*, 39, 725–737.

EAPA (European Asphalt Pavement Association). 2007. *Long-Life Asphalt Pavements—Technical Version*, EAPA, Brussels, Belgium.

FEHRL-ELLPAG (Forum of European National Highway Research Laboratories–European Long-life Pavement Group). 2004. *Making Best Use of Long-Life Pavements in Europe—Phase 1: A Guide to the Use of Long-Life Fully-Flexible Pavements*, FEHRL, Brussels, Belgium.

Ferne, B. 2006. Long-life pavements—A European study by ELLPAG. *International Journal of Pavement Engineering*, 7, 91–100.

Francken, L., Ed. 1998. *Bituminous Binders and Mixes*, RILEM report 17, E & FN Spon, London, U.K.

Gal, J.-F. 1992. Evolution de la perméabilité des enrobés drainants. *Revue Générale des Routes et Aérodromes*, 702, 118–119.

Gibson, N.H. 2006. A viscoelastoplastic continuum damage model for the compressive behavior of asphalt concrete, Ph.D. dissertation, University of Maryland, College Park, MD.

Gibson, N., X. Qi, A. Shenoy et al. 2012. Full-scale accelerated performance testing for Superpave and structural validation, Final report transportation pooled fund study TPF-5(019) and SPR-2(174), Accelerated pavement testing of crumb rubber modified asphalt pavements, FHWA, Lanthum, MD.

GNB (Groupe National Bitume). 1997. Susceptibilité au vieillissement des bitumes—Expérimentation A08, Research report CR19, LCPC, Paris, France.

Han, R., X. Jin, and C.J. Glover. 2011. Modeling pavement temperature for use in binder oxidation models and pavement performance prediction. *Journal Material Civil Engineering*, 23, 351–359.

Hicks, R.G. 1991. Moisture damage in asphalt concrete, in *NCHRP Synthesis of Highway Practice 175*, Transportation Research Board, Washington, DC.

Hicks, R.G. and T.V. Scholz. 2003. *Life Cycle Costs for Lime in Hot Mix Asphalt*, 3 volumes, National Lime Association, Arlington, VA.

Huang, B., L.N. Mohammad, A. Raghavendra et al. 1999. Fundamentals of permeability in asphalt mixtures. *Journal of the Association of Asphalt Paving Technologists*, 68, 479–493.

Isacsson, U. and H. Zeng. 2003. Cracking of asphalt at low temperature as related to bitumen rheology. *Journal Materials Science*, 33, 2165–2170.

Jennings, P.W., M.A. Desando, M.F. Raub et al. 1992. NMR spectroscopy in the characterization of eight selected asphalts. *Fuel Science Technology International*, 10, 887–907.

Jennings, P.W., J.A. Pribanic, M.A. Desando et al. 1993. Binder characterization and evaluation by nuclear magnetic resonance spectroscopy, SHRP report A-335, National Research Council, Washington, DC.

Kim, Y., D.N. Little, and R.L. Lytton. 2003. Fatigue and healing characterization of asphalt mixtures. *Journal Materials Civil Engineering*, 15, 75–83.

Kim, S., S.-W. Loh, H. Zhai et al. 2001. Advanced characterization of crumb-rubber modified asphalts, using protocols developed for complex binders. *Transportation Research Record*, 1767, 15–24.

Kringos, N., A. Scarpas, A. Copeland et al. 2008a. Modelling of combined physical-mechanical moisture-induced damage in asphaltic mixes. Part 2: Moisture susceptibility parameters. *International Journal Pavement Engineering*, 9, 129–151.

Kringos, N., A. Scarpas, C. Kasbergen et al. 2008b. Modelling of combined physical-mechanical moisture-induced damage in asphaltic mixes. Part 1: Governing processes and formulations. *International Journal Pavement Engineering*, 9, 115–128.

Kutay, M.E., N.H. Gibson, and J. Youtcheff. 2008. Conventional and viscoelastic continuum damage (VECD) based fatigue analysis of polymer modified asphalt pavements. *Journal of the Association of Asphalt Paving Technologists*, 77, 395–434.

Lamontagne, J., P. Dumas, V. Mouillet et al. 2001. Comparison by Fourier transform infrared (FTIR) spectroscopy of different aging techniques: Application to road bitumens. *Fuel*, 80, 483–488.

Lay, M.G. 1992. *Ways of the World*, Rutgers University Press, New Brunswick, NJ.

LCPC-SETRA (Laboratoire Central des Ponts et Chaussées—Service des Etudes Techniques des Routes et Autoroutes). 1997. *French Design Manual for Pavement Structures*, LCPC, Paris, France.

Lee, H.J. and Y.R. Kim. 1998a. Viscoelastic constitutive model for asphalt concrete under cyclic loading. *Journal of Engineering Mechanics, ASCE*, 124, 32–40.

Lee, H.J. and Y.R. Kim. 1998b. Viscoelastic continuum damage model for asphalt concrete with healing. *Journal of Engineering Mechanics, ASCE*, 124, 1224–1232.

Lee, W. and K. Mahboub, Eds. 2006. *Asphalt Mix Design and Construction: Past, Present, and Future*, American Society of Civil Engineers, Reston, VA.

Lesueur, D. 2009. The colloidal structure of bitumen: Consequences on the rheology and on the mechanisms of bitumen modification. *Advances in Colloid Interface Science*, 145, 42–82.

Lesueur, D. 2010. *Hydrated Lime: A Proven Additive for Durable Asphalt Pavements—Critical Literature Review*, European Lime Association, Brussels, Belgium.

Li, X. and N. Gibson. 2011. Mechanistic characterization of aggregate packing to assess gyration levels during HMA mix design, Asphalt Paving Technology, *Journal of the Association of Asphalt Paving Technologists*, 80, 33–64.

Little, D.N. and J.A. Epps. 2001. *The Benefits of Hydrated Lime in Hot Mix Asphalt*, National Lime Association, Arlington, VA.

Lottman, R.P. 1978. Predicting moisture-induced damage to asphaltic concrete, NCHRP report 192, Transportation Research Board, Washington, DC.

Marasteanu, M.O., X. Li, T.R. Clyne et al. 2004. Low temperature cracking of asphalt concrete pavements, Report MN/RC 2004-23, MN Department of Transportation, St. Paul, MN.

Masad, E.A. 2005. Aggregate Imaging System (AIMS): Basics and applications, FHWA/TX-05/5-1707-01-1, Texas Transportation Institute, College Station, TX.

Masad, E., A. Rezaei, A. Chowdhury et al. 2009. Predicting asphalt mixture skid resistance based on aggregate characteristics, TTI report 0-5627-1, Texas Transportation Institute, College Station, TX.

Mauduit, C., F. Hammoum, J.-M. Piau et al. 2010. Quantifying expansion effects induced by freeze–thaw cycles in partially water saturated bituminous mix: Laboratory experiments. *Road Materials Pavement Design*, 11, 443–457.

McDaniel, R. and A. Shah. 2003. Asphalt additives to control rutting and cracking, FHWA report IN-JTRP-2002-29, FHWA, Lanthum, MD.

McNally, T. Ed. 2011. *Polymer Modified Bitumen: Properties and Characterization*, Woodhead Publishing, Cambridge, U.K.

Miller, J.S. and W.Y. Bellinger. 2003. Distress identification manual for the long-term pavement performance program. FHWA-RD-03-031, FHWA, Lanthum, MD.

Mohammad, L.N. and K.A. Shamsi. 2007. A look at the bailey method and locking point concept in superpave mixture design. Transportation Research Circular E-C124, Practical approaches to hot-mix asphalt mix design and quality control testing, Transportation Research Board, Washington, DC.

Monismith, C.L. 2006. State of the art: Bituminous materials mix design, in W. Lee and K. Mahboub, Eds., *Asphalt Mix Design and Construction: Past, Present, and Future*, American Society of Civil Engineers, Reston, VA.

Mortazavi, M. and J.S. Moulthrop. 1993. The SHRP materials reference library, SHRP report A-646, National Research Council, Washington, DC.

Newcomb, D.E., M. Buncher, and I.J. Huddleston. 2001. Concepts of perpetual pavements. *Transportation Research Circular*, 503, 4–11.

OECD (Organisation for Economic Co-Operation and Development). 2008. *Long-Life Surfaces for Busy Roads*, OECD/ITF, Paris, France.

Ould-Henia, M., M. Rodriguez, and A.-G. Dumont. 2004. Élaboration d'une méthode prédictive de l'orniérage des revêtements bitumineux, Ecole Polytechnique Fédérale de Lausanne, Lausanne, Switzerland.

Petersen, J.C. 2009. A review of the fundamentals of asphalt oxidation—Chemical, physicochemical, physical property and durability relationships, Report EC140, Transportation Research Board, Washington, DC.

PIARC (World Road Association). 1999. Use of modified bituminous binders, special bitumens and bitumens with additives, in *Road Pavements, Routes/Roads 303*, also available from LCPC, PIARC, Paris, France.

Plancher, H., G. Miyake, R.L. Venable, and J.C. Petersen. 1980. A simple laboratory test to indicate the susceptibility of asphalt–aggregate mixtures to moisture damage from repeated freeze-thaw cycling. *Proceedings of the Canadian Technical Asphalt Association*, 25, 246–262.

Polacco, G., J. Stastna, D. Biondi et al. 2004. Rheology of asphalts modified with glycidylmethacrylate functionalized polymers. *Journal of Colloid Interface Science*, 280, 366–373.

Potter, T.L. and B. Duval. 2001. Cerro Negro bitumen degradation by a consortium of marine benthic microorganisms, *Environmental Science Technology*, 35, 76–83.

Raynaud, C. 2009. L'ajout de chaux hydratée dans les enrobés bitumineux. *BTP Matériaux*, 22, 42–43.

Read, J. and D. Whiteoak. 2003. *The Shell Bitumen Handbook*, 5th Edn., Thomas Telford Publishing, London, New York.

Redelius, P. 2006. The structure of asphaltenes in bitumen. *Road Materials Pavement Design*, Special issue EATA, 143–162.

Roque, R., Z. Zhang, and B. Sankar.2009. Determination of crack growth rate parameters of asphalt mixtures using the Superpave IDT. *Journal of the Association of Asphalt Paving Technologists*, 68, 404–433.

Saunier, J. 1968. Autoréparation des enrobes bitumineux. *Revue Générale des Routes et Aerodromes*, 435, 69–72.

Sousa, J.B., J. Craus, and C.L. Monismith. 1991. Summary report on permanent deformation in asphalt concrete, SHRP-A/IR-91-104 report, National Research Council, Washington, DC.

Stasse, G. 2000. Caractérisation d'adhérence de revêtements de chaussées routières, Report Chaussées CR 25, LCPC, Paris, France.

Tangella, R., J. Craus, J.A. Deacon et al. 1990. Summary report on fatigue response of asphalt mixtures, SHRP report A-312, National Research Council, Washington, DC.

Taylor, G.D. 2000. *Materials in Construction: An Introduction*, 3rd Edn., Longman Publisher, Chicago, IL.

Terrel, R.L. and S. Al-Swailmi. 1994. Water sensitivity of asphalt–aggregate mixes: Test selection, SHRP report A-403, National Research Council, Washington, DC.

Thomson, D.C. 1964. Rubber modifiers, in A.J. Hoiberg, Ed., *Bituminous Materials: Asphalts, Tars and Pitches*, Vol. 1, InterScience Publishers, New York, pp. 375–414.

Timoshenko, S. 1976. *Strength of Materials*, 3rd Edn., 2 volumes, Krieger Publishing Company, Malabar, FL.

Traxler, R.N. 1961. *Asphalt: Its Composition, Properties and Uses*, Reinhold Publishing, New York.

Traxler, R.N. and H.E. Schweyer. 1936. Increase in viscosity of asphalts with time. *Proceedings of the American Society for Testing Materials*, 36, 544–551.

Wallman, C.-G. and H. Åström. 2001. Friction measurement methods and the correlation between road friction and traffic safety: A literature review. VTI report 911A, VTI, Stockholm, Sweden.

Waste Framework Directive (Directive 2008/98/EC) of the European Parliament and of the Council of 19 November 2008 on waste.

Western Research Institute (WRI). 2010. Fundamental properties of asphalts and modified asphalts, III. Quarterly technical progress report January 1–March 31, 2010, Prepared for Federal Highway Administration under contract no. DTFH61-07-D-00005, Western Research Institute, Laramie, WY (www.westernresearch.org).

Wolf, M. and R. Bachofen. 1991. Microbial degradation of bitumen matrix used in nuclear waste repositories. *Naturwissenschaften*, 78, 414–417.

Wright, J.R. 1965. Weathering: Theoretical and practical aspects of asphalt durability, in A.J. Hoiberg, Ed., *Bituminous Materials: Asphalts, Tars and Pitches*, Vol. 2, Part 1, InterScience Publishers, New York, pp. 249–306.

Youtcheff, J.S. and D.R. Jones. 1994. Guidelines for asphalt refiners and suppliers, SHRP report A-686, National Research Council, Washington, DC.

Further Readings

American Association of State Highway and Transportation Officials. 2011. Standard specifications for transportation materials and methods of sampling and testing: Part 2A Tests. AASHTO, Washington, DC.

Annual Book of ASTM Standards. 2012. American Society Testing Materials, Philadelphia, PA.

Asphalt Institute. 1979. Manual Series 2 (MS-2), Asphalt Institute, College Park, MD.

SHRP National Research Council. 1994. Fatigue response of asphalt–aggregate mixes, SHRP A-404, Washington, DC, 309pp.

Index

Milton Keynes UK
Ingram Content Group UK Ltd.
UKHW051901071024
449327UK00025B/2040